T0245201

Advances in Intelligent Systems and Computing

Volume 418

Series editor

Janusz Kacprzyk, Polish Academy of Sciences, Warsaw, Poland
e-mail: kacprzyk@ibspan.waw.pl

About this Series

The series "Advances in Intelligent Systems and Computing" contains publications on theory, applications, and design methods of Intelligent Systems and Intelligent Computing. Virtually all disciplines such as engineering, natural sciences, computer and information science, ICT, economics, business, e-commerce, environment, healthcare, life science are covered. The list of topics spans all the areas of modern intelligent systems and computing.

The publications within "Advances in Intelligent Systems and Computing" are primarily textbooks and proceedings of important conferences, symposia and congresses. They cover significant recent developments in the field, both of a foundational and applicable character. An important characteristic feature of the series is the short publication time and world-wide distribution. This permits a rapid and broad dissemination of research results.

More information about this series at http://www.springer.com/series/11156

Luís Paulo Reis · António Paulo Moreira
Pedro U. Lima · Luis Montano
Victor Muñoz-Martinez
Editors

Robot 2015: Second Iberian Robotics Conference

Advances in Robotics, Volume 2

Editors
Luís Paulo Reis
University of Minho, School of Engineering
Information Systems Department
Guimarães
Portugal

António Paulo Moreira
University of Porto, Faculty of Engineering
INESC-TEC
Porto
Portugal

Pedro U. Lima
University of Lisbon, Instituto Superior Técnico
Institute for Systems and Robotics
Lisboa
Portugal

Luis Montano
University of Zaragoza, School of Engineering and Architecture
Computer and Systems Engineering Department
Zaragoza
Spain

Victor Muñoz-Martinez
University of Malaga, Superior Technical School of Industrial Engineers
Automatic Control and Systems Engineering Department
Málaga
Spain

ISSN 2194-5357 ISSN 2194-5365 (electronic)
Advances in Intelligent Systems and Computing
ISBN 978-3-319-27148-4 ISBN 978-3-319-27149-1 (eBook)
DOI 10.1007/978-3-319-27149-1

Library of Congress Control Number: 2015955886

Springer Cham Heidelberg New York Dordrecht London

Printed on acid-free paper

Springer International Publishing AG Switzerland is part of Springer Science+Business Media
(www.springer.com)

Preface

This book contains a selection of papers accepted for presentation and discussion at ROBOT 2015: Second Iberian Robotics Conference, held in Lisbon, Portugal, November 19th–21st, 2015. ROBOT 2015 is part of a series of conferences that are a joint organization of SPR – "Sociedade Portuguesa de Robótica/Portuguese Society for Robotics", SEIDROB – Sociedad Española para la Investigación y Desarrollo de la Robótica/Spanish Society for Research and Development in Robotics and CEA-GTRob – Grupo Temático de Robótica/Robotics Thematic Group. The conference organization had also the collaboration of several universities and research institutes, including: University of Minho, University of Porto, University of Lisbon, Polytechnic Institute of Porto, University of Aveiro, University of Zaragoza, University of Malaga, LIACC, INESC-TEC and LARSyS.

Robot 2015 builds upon several successful events, including three biennal workshops (Zaragoza- 2007, Barcelona – 2009 and Sevilla – 2011) and the first Iberian Robotics Conference held in 2013 at Madrid. The conference is focussed on the Robotics scientific and technological activities in the Iberian Peninsula, although open to research and delegates from other countries.

Robot 2015 featured three plenary talks by:

- Manuela Veloso, Herbert A. Simon University Professor at Carnegie Mellon University, USA, on "Symbiotic Autonomous Mobile Service Robots";
- Bill Smart, director of the Personal Robotics Group at Oregon State University, USA on "How the Law Will Think About Robots (and Why You Should Care)"; and
- Jon Agirre Ibarbia, co-ordinator of R&D projects at TECNALIA Research & Innovation, Spain, on "Applications in Flexible Manufacturing with Humans and Robots".

Robot 2015 featured 19 special sessions, plus a main/general robotics track. The special sessions were about: Agricultural Robotics and Field Automation; Autonomous Driving and Driver Assistance Systems; Communication Aware Robotics; Environmental Robotics; Social Robotics: Intelligent and Adaptable AAL

Systems; Future Industrial Robotics Systems; Legged Locomotion Robots; Rehabilitation and Assistive Robotics; Robotic Applications in Art and Architecture; Surgical Robotics; Urban Robotics; Visual Perception for Autonomous Robots; Machine Learning in Robotics; Simulation and Competitions in Robotics; Educational Robotics; Visual Maps in Robotics; Control and Planning in Aerial Robotics, the XVI edition of the Workshop on Physical Agents and a Special Session on Technological Transfer and Innovation.

In total, after a careful review process with at least three independent reviews for each paper, but in some cases 4 or 5 reviews, a total of 118 high quality papers were selected for publication, with a total number of authors over 400, from 21 countries, including: Brazil, China, Costa Rica, Croatia, Czech Republic, Ecuador, France, Germany, Italy, India, Iran, The Netherlands, Poland, Portugal, Serbia, Singapore, Spain, Switzerland, United Kingdom, USA and Viet Nam.

ROBOT 2015 was co-located with the RoCKIn Competition 2015, which took place in the Parque das Nações, Lisboa, between 19 and 23 November, nearby the conference venue. RoCKIn is a Coordination Action funded by the European Commission FP7, and its main goal is to foster robotics research, education and dissemination through robot competitions. Thirteen teams from seven countries, including two teams from Mexico, were qualified and competed in RoCKIn@Home and RoCKIn@Work Challenges. Participants from both events had the opportunity to join in social events and to visit both venues, taking advantage of an extraordinary opportunity to follow presentations and actual robot systems showing recent results in this exciting field.

We would like to thank all Special Sessions' organizers for their hard work on promoting their special session, inviting the Program Committee, organizing the Special Session review process and helping to promote the ROBOT 2015 Conference. This acknowledgment goes especially to Vitor Santos, Angel Sappa, Miguel Oliveira, Danilo Tardioli, Alejandro Mosteo, Luis Riazuelo, João Valente, Antonio Barrientos, Luís Santos, Jorge Dias, Raul Morais Santos, Filipe Santos, Germano Veiga, José Lima, Guillermo Heredia, Anibal Ollero, Manuel Silva, Cristina Santos, Manuel Armada, Vicente Matellán, Miguel Ángel Cazorla, Rodrigo Ventura, Nicolas Garcia-Aracil, Alicia Casals, Elena García, José Pedro Sousa, Marta Malé-Alemany, Paulo Gonçalves, Jose Maria Sabater, Jorge Martins, Pedro Torres, Tamás Haidegger, Alberto Sanfeliu, Juan Andrade, João Sequeira, Anais Garrell, Andry Maykol Pinto, Aníbal Matos, Nuno Cruz, Brígida Mónica Faria, Luis Merino, Nuno Lau, Artur Pereira, Bernardo Cunha, Armando Sousa, Fernando Ribeiro, Eduardo Gallego and Oscar Reinoso Garcia.

We would also like to take this opportunity to thank the rest of the organization members (Carlos Cardeira, Brígida Mónica Faria, Manuel Fernando Silva, Daniel Castro Silva and Pedro Fonseca) for their hard and fine work on the local arrangements, publicity, publication and financial issues. We also express our gratitude to the members of all the Program Committees and additional reviewers, as they were crucial for ensuring the high scientific quality of the event and to all the authors and delegates whose research work and participation made this event a success. Last, but not the least, we acknowledge and thank our editor, Springer, that was in charge of these proceedings, and in particular to Dr. Thomas Ditzinger.

November 2015

Luís Paulo Reis
António Paulo Moreira
Pedro U. Lima
Luis Montano
Victor Muñoz-Martinez

Organization

General/Program Chairs

Luís Paulo Reis	University of Minho, Portugal
António Paulo Moreira	University of Porto - FEUP/INESCTEC, Portugal
Pedro U. Lima	University of Lisbon – IST, Portugal
Luis Montano	University of Zaragoza, Spain
Victor F. Muñoz	University of Malaga, Spain

Organizing Committee

Luís Paulo Reis	University of Minho, Portugal (General/Program Chair)
António Paulo Moreira	University of Porto - FEUP/INESCTEC, Portugal (General/Program Chair)
Pedro U. Lima	University of Lisbon – IST, Portugal (General/Program Chair)
Luis Montano	University of Zaragoza, Spain (General/Program Chair)
Victor F. Muñoz	University of Malaga, Spain (General/Program Chair)
Carlos Cardeira	University of Lisbon - IST, Portugal (Local Arrangements Chair)
Brígida Mónica Faria	Polytechnic Institute of Porto, Portugal (Publicity Chair)
Manuel Fernando Silva	Polytechnic Institute of Porto, Portugal (Publications Chair)
Daniel Castro Silva	University of Porto - FEUP, Portugal (Publications Chair)
Pedro Fonseca	University of Aveiro, Portugal (Financial/Registration Chair)

General Robotics Session Organizing Committee

Luís Paulo Reis	University of Minho, Portugal
António Paulo Moreira	University of Porto - FEUP/INESCTEC, Portugal
Pedro U. Lima	University of Lisbon – IST, Portugal
Luis Montano	University of Zaragoza, Spain
Victor F. Muñoz	University of Malaga, Spain

General Robotics Session Program Committee

Alberto Sanfeliu	University Politécnica de Cataluña, Spain
Alexandre Bernardino	Universiy of Lisbon, Portugal
Alfonso García-Cerezo	University Málaga, Spain
Alícia Casals	University Politécnica de Cataluña, Spain
Americo Azevedo	FEUP/INESCTEC, Portugal
Aníbal Matos	University of Porto, Portugal
Anibal Ollero	CATEC-Universidad Sevilla, Spain
Antonio Barrientos	CAR CSIC-UPM, Spain
Antonio Fernando Ribeiro	University of Minho, Portugal
António José Neves	University of Aveiro, Portugal
Antonio R Jiménez	CAR-CSIC, Spain
António Pedro Aguiar	University of Porto – FEUP, Portugal
António Valente	University of Tras dos Montes e Alto Douro, Portugal
Armando Jorge Sousa	University of Porto, Portugal
Artur Pereira	University of Aveiro, Portugal
Brígida Mónica Faria	I.P. Porto, Portugal
Bruno Guerreiro	Universiy of Lisbon, Portugal
Carlos Cardeira	Universiy of Lisbon, Portugal
Carlos Cerrada	UNED, Spain
Carlos Rizzo	University of Zaragoza, Spain
Carlos Sagüés	University Zaragoza, Spain
Cristina Santos	University of Minho, Portugal
Eduardo Zalama	University of Valladolid, Spain
Estela Bicho	University of Minho, Portugal
Eugenio Aguirre	University Granada, Spain
Fernando Caballero	University Sevilla, Spain
Fernando Torres	University Alicante, Spain
Filipe Santos	FEUP/INESCTEC, Portugal
Filomena Soares	University of Minho, Portugal
Francisco Melo	University of Lisbon, Portugal
Hugo Costelha	I.P. Leiria, Portugal

Javier Pérez Turiel	CARTIF, Valladolid, Spain
João Calado	I.P. Lisboa, Portugal
Jon Aguirre	Tecnalia, Spain
Jorge Dias	University of Coimbra, Portugal
Jorge Lobo	University of Coimbra, Portugal
José A. Castellanos	University Zaragoza, Spain
José Luís Azevedo	University of Aveiro, Portugal
José Luis Magalhães Lima	I.P. Bragança, Portugal
José L. Villarroel	University of Zaragoza, Spain
José M. Cañas	University Rey Juan Carlos, Spain
José Nuno Pereira	University of Lisbon, Portugal
José Santos Victor	University of Lisbon, Portugal
Josep Amat	University of Politécnica de Cataluña, Spain
Lino Marques	University of Coimbra, Portugal
Luis Almeida	University of Porto, FEUP, Portugal
Luis Basañez	University Politécnica de Cataluña, Spain
Luis Merino	University Pablo Olavide, Sevilla, Spain
Luis Moreno	University Carlos III de Madrid, Spain
Luis Seabra Lopes	University of Aveiro, Portugal
Manuel Armada	CAR CSIC-UPM, Spain
Manuel Bernardo Cunha	University of Aveiro, Portugal
Manuel Fernando Silva	I.P. Porto, Portugal
Manuel Ferre	CAR CSIC-UPM, Spain
Marcelo Petry	Univ. Federal Santa Catarina, Brazil
Maria Isabel Ribeiro	Universiy of Lisbon, Portugal
Miguel A. Cazorla	University de Alicante, Spain
Nicolás García-Aracil	University Miguel Hernández, Spain
Nuno Lau	University of Aveiro, Portugal
Oscar Reinoso	University Miguel Hernández, Spain
Pascual Campoy	CAR CSIC-UPM, Spain
Paulo Costa	University of Porto, Portugal
Paulo Gonçalves	I.P. Castelo Branco, Portugal
Paulo Jorge Oliveira	University of Lisbon, Portugal
Pedro Costa	University of Porto, Portugal
Pedro Fonseca	University of Aveiro, Portugal
Pedro J. Sanz	UJI, Castellón, Spain
Pere Ridao	University of Girona, Spain
Raul Morais	University of Trás dos Montes e Alto Douro, Portugal
Rafael Sanz	University of Vigo, Spain
Rodrigo Ventura	Universiy of Lisbon, Portugal
Rui Rocha	University of Coimbra, Portugal
Urbano Nunes	University of Coimbra, Portugal
Vicente Feliú	University Castilla la Mancha, Spain
Vicente Matellán	University León, Spain
Vitor Santos	University of Aveiro, Portugal

Agricultural Robotics and Field Automation Session Organizing Committee

Raul Morais Santos	UTAD University, Portugal
Filipe Santos	INESC-TEC, Portugal

Agricultural Robotics and Field Automation Session Program Committee

Angela Ribeiro	CAR-CSIC, Spain
Antonio Valente	UTAD University, Portugal
Armando Sousa	University of Porto - FEUP, Portugal
Carrick Detweiler	University of Nebraska, United States
Dimitrios S. Paraforos	University of Hohenheim, Germany
Eduardo Solteiro Pires	UTAD University, Portugal
Filipe Neves dos Santos	INESC TEC, Portugal
Joris Ijsselmuiden	Wageningen UR, Netherlands
Joaquín Ferruz-Melero	University of Seville, Spain
Raul Morais Dos Santos	INESC TEC/CROB/UTAD, Portugal
Manuel Silva	Inst. Superior de Engenharia do Porto, Portugal
Nieves Pavon	University of Huelva, Spain
Paulo Costa	U. of Porto, FEUP, Portugal
Paulo Moura Oliveira	UTAD University, Portugal
Robert Fitch	Australian Centre for Field Robotics, The University of Sydney, Australia
Tiago Nascimento	Federal University of Paraíba - UFPB, Brazil
Timo Oksanen	Aalto University,Finland

Autonomous Driving and Driver Assistance Systems Session Organizing Committee

Vitor Santos	Universidade de Aveiro, Portugal
Angel Sappa	CVC, Barcelona, Spain
Miguel Oliveira	INESC-TEC, Porto, Portugal

Autonomous Driving and Driver Assistance Systems Session Program Committee

Angelos Amanatiadis	Democritus University of Thrace, Greece
Antonio Valente	UTAD, Portugal
Antonio M. López	CVC and UAB, Barcelona, Spain
Arturo De La Escalera	Universidad Carlos III de Madrid, Spain
Bernardo Cunha	Universidade de Aveiro, Portugal
Carlos Cardeira	IST, Lisboa, Portugal
Cristina Peixoto Santos	Universidade do Minho, Portugal
David Vázquez Bermúdez	CVC, Barcelona, Spain
Fadi Dornaika	University of the Basque Country UPV/EHU & IKERBASQUE, Spain
Frederic Lerasle	LAAS-CNRS, France
Jorge Almeida	DEM, Universidade de Aveiro, Portugal
José Azevedo	Universidade de Aveiro, Portugal
José Álvarez	RSCS, ANU College, Australia
Jose A. Castellanos	University Zaragoza, Spain
Luis Almeida	Universidade do Porto, Portugal
Miguel Angel Sotelo	University of Alcala, Spain
Paulo Dias	IEETA - Universidade de Aveiro, Portugal
Procópio Stein	INRIA, France
Rafael Sanz	Universidad de Vigo, Spain
Ricardo Pascoal	Universidade de Aveiro, Portugal
Ricardo Toledo	CVC and UAB, Barcelona, Spain
Susana Sargento	IT, University of Aveiro, Portugal
Urbano Nunes	University of Coimbra, Portugal

Control and Planning in Aerial Robotics Organizing Committee

Guillermo Heredia	Universidad de Sevilla, Spain
Anibal Ollero	Universidad de Sevilla, Spain

Control and Planning in Aerial Robotics Program Committee

Abdelkrim Nemra	Ecole Militaire Polytechnique, Algiers
Alessandro Rucco	University of Porto, Portugal
Begoña Arrue	Universidad de Sevilla, Spain
Bruno, Guerreiro	Instituto Superior Técnico, Univ. Lisboa, Portugal
Elena Lopez Guillen	Universidad de Alcala, Spain

Eugenio Aguirre	University of Granada, Spain
Fernando Caballero	University of Seville, Spain
Luis Merino	Pablo de Olavide University, Spain
Mario Garzon	Universidad Politécnica de Madrid, Spain
Rita Cunha	LARSyS, Instituto Superior Técnico, Univ. Lisboa, Portugal

Communication Aware Robotics Organizing Committee

Danilo Tardioli	Centro Universitario de la Defensa de Zaragoza, Spain
Alejandro Mosteo	Centro Universitario de la Defensa de Zaragoza, Spain
Luis Riazuelo	University of Zaragoza, Spain

Communication Aware Robotics Program Committee

Carlos Rizzo	University of Zaragoza, Spain
Domenico Sicignano	University of Zaragoza, Spain
Eduardo Montijano	Centro Universitario de la Defensa de Zaragoza, Spain
Enrico Natalizio	Universitè de Technologie de Compiègne, France
Jesus Aisa	University of Zaragoza, Spain
Jorge Ortin García	Centro Universitario de la Defensa de Zaragoza, Spain
Luis Merino	Pablo de Olavide University, Spain
Lujia Wang	The Chinese University of Hong Kong, Hong Kong, China
María T. Lázaro	Sapienza University of Rome, Italy
María-Teresa Lorente	University of Zaragoza, Spain
Pablo Urcola	University of Zaragoza, Spain

Educational Robotics Session Organizing Committee

A. Fernando Ribeiro	University of Minho, Portugal
Armando Sousa	University of Porto - FEUP, Portugal
Eduardo Gallego	Complubot, Spain

Educational Robotics Session Program Committee

A. Fernando Ribeiro	University of Minho, Portugal
Armando Sousa	University of Porto - FEUP, Portugal
Eduardo Gallego	Complubot, Spain
Gil Lopes	University of Minho, Portugal
José Goncalves	ESTiG – I.P. Bragança, Portugal
Paulo Costa	University of Porto - FEUP, Portugal
Paulo Trigueiros	I.P. Porto, Portugal

Environmental Robotics Special Session Organizing Committee

João Valente	Universidad Carlos III de Madrid, Spain
Antonio Barrientos	Universidad Politécnica de Madrid, Spain

Environmental Robotics Special Session Program Committee

Achim J. Lilienthal	Örebro University, Sweden
Angela Ribeiro	Centre for Automation and Robotics (CAR) UPM-CSIC, Spain
Carol Martinez	Pontifical Xavierian University, Colombia
David Gomez	Nat. Res. Inst. Science and Technology for Environment and Agriculture, France
Gonzalo Pajares	Complutense University of Madrid, Spain
Jaime Del Cerro	Polytechnic University of Madrid, Spain
Marc Carreras	University of Girona, Spain
Mario Andrei Garzón	Polytechnic University of Madrid, Spain
Mohamed Abderrahim	Carlos III University of Madrid, Spain
Pablo Gonzalez de Santos	Spanish National Research Council (CSIC), Spain
Paloma de la Puente	Vienna University of Technology, Austria
William Coral	Polytechnic University of Madrid, Spain

Future Industrial Robotics Systems Organizing Committee

Germano Veiga	INESC TEC - Robotics and Intelligent Systems, Portugal
José Lima	INESC TEC - Robotics and Intelligent Systems, Portugal

Future Industrial Robotics Systems Program Committee

Andry Pinto	Universidade do Porto, INESC-TEC, Portugal
António Paulo Moreira	Universidade do Porto, INESC-TEC, Portugal
Fabrizio Caccavale	UNIBAS, Italy
Joerg Roewekaemper	University of Freiburg, Germany
José Barbosa	I.P. Bragança, Portugal
José Miguel Almeida	I.P. Porto, Portugal
Klas, Nilsson	Lund University, Sweden
Luis Rocha	Universidade do Porto, Portugal
Manuel Fernando Silva	I.P. Porto, Portugal
Nuno Mendes	Universidade de Coimbra, Portugal
Pedro Neto	Universidade de Coimbra, Portugal
Simon Bogh	AalBorg University, Denmark
Ulrike Thomas	TU Chemnitz, Germany

Legged Locomotion Robots Session Organizing Committee

Manuel Silva	ISEP/IPP - School of Engineering, Polytechnic Institute of Porto and INESC TEC, Portugal
Cristina Santos	Industrial Electronics Department and ALGORITMI Center, University of Minho, Portugal
Manuel Armada	Centre for Automation and Robotics - CAR (CSIC-UPM), Spain

Legged Locomotion Robots Session Program Committee

Carla M.A. Pinto	Instituto Superior de Engenharia do Porto, Portugal
Filipe Silva	University of Aveiro, Portugal
Filomena Soares	University of Minho, Portugal
Gurvinder S. Virk	University of Gävle, Sweden
José Machado	University of Minho, Portugal
Lino Costa	University of Minho, Portugal
Lino Marques	University of Coimbra, Portugal
Paulo Menezes	University of Coimbra, Portugal
Pedro Figueiredo Santana	Escola de Tecnologias e Arquitetura, Portugal
Rui P. Rocha	University of Coimbra, Portugal
Yiannis Gatsoulis	University of Leeds, United Kingdom

Machine Learning in Robotics Session Organizing Committee

Brígida Mónica Faria	Polytechnic Institute of Porto (ESTSP-IPP), Portugal
Luis Merino	Pablo de Olavide University (UPO), Spain

Machine Learning in Robotics Session Program Committee

Ana Lopes	Institute for Systems and Robotics, Portugal
Armando Sousa	University of Porto, Portugal
Daniel Castro Silva	Univesity of Porto - FEUP, Portugal
Fernando Caballero Benítez	University of Seville, Spain
João Fabro	UTFPR - Federal University of Technology Parana, Brazil
João Messias	Institute for Systems and Robotics, Portugal
Marcelo Petry	Federal University of Santa Catarina, Brazil
Noé Pérez-Higueras	Pablo de Olavide University (UPO), Spain
Nuno Lau	University of Aveiro, Portugal
Pedro Henriques Abreu	University of Coimbra, Portugal
Rafael Ramón-Vigo	Pablo de Olavide University (UPO), Spain

Rehabilitation and Assistive Robotics Organizing Committee

Alicia Casals	Institute for Bioengineering of Catalonia, Spain
Nicolás García-Arazil	Universidad Miguel Hernandez, Spain
Elena García	Centre for Automation and Robotics (CSIC-UPM), Spain

Rehabilitation and Assistive Robotics Program Committee

Aikaterini D. Koutsou	CSIC, Spain
Alicia Casals	Institute for Bioengineering of Catalonia, Spain
Arturo Bertomeu-Motos	UMH, Spain
Diana Ruiz Bueno	University of Saragoza, Spain
Elena García	Centre for Automation and Robotics (CSIC-UPM), Spain
Eloy Urendes Jimenez	CSIC, Spain
Iñaki Diaz	CEIC, Spain
Javier P. Turiel	University of Valladolid, Spain

Jose Maria Sabater-Navarro	UMH, Spain
Luis Daniel Lledó Pérez	UMH, Spain
Nicolás García-Arazil	Universidad Miguel Hernandez, Spain

Robotic Applications in Art and Architecture Session Organizing Committee

Manuel Silva	ISEP/IPP - School of Engineering, Polytechnic Institute of Porto and INESC TEC, Portugal
José Pedro Sousa	Digital Fabrication Lab (DFL/CEAU), Fac. Arquitetura, Universidade do Porto, Portugal
Marta Malé-Alemany	Architect, Researcher and Curator + Polytechnic University of Catalonia, Spain

Robotic Applications in Art and Architecture Session Program Committee

Alexandra Paio	ISCTE - Instituto Universitário de Lisboa, Portugal
André Dias	Polytechnic Institute of Porto and INESC TEC, Portugal
Andrew Wit	Ball State University, USA
António Mendes Lopes	Faculty of Engineering, University of Porto, Portugal
Filipe Coutinho Quaresma	ECATI - Universidade Lusófona, Portugal
Germano Veiga	INESC TEC, Portugal
Gonçalo Castro Henriques	Universidade Federal do Rio de Janeiro, Brazil
Leonel Moura	Universidade de Lisboa, Portugal
Mauro Costa	Universidade de Coimbra, Portugal
Paulo Fonseca de Campos	Faculdade de Arquitetura e Urbanismo, Universidade de São Paulo, Brazil
Sancho Oliveira	ISCTE - Instituto Universitário de Lisboa, Portugal
Wassim Jabi	Welsh School of Architecture, Cardiff University, United Kingdom

Simulation and Competitions Session Organizing Committee

Artur Pereira	Universidade de Aveiro/IEETA, Portugal
Nuno Lau	Universidade de Aveiro/IEETA, Portugal
Bernardo Cunha	Universidade de Aveiro/IEETA, Portugal

Simulation and Competitions Session Program Committee

Antonio Morales	Universitat Jaume I, Spain
Armando Sousa	Universidade do Porto, Portugal
Eurico Pedrosa	Universidade de Aveiro, Portugal
Jorge Ferreira	University of Aveiro, Portugal
José Luís Azevedo	Universidade de Aveiro, Portugal
Luis Moreno	Universidad Carlos III de Madrid, Spain
Nicolas Jouandeau	University Paris8, France
Paulo Goncalves	Polytechnic Institute of Castelo Branco, Portugal
Paulo Trigueiros	Instituto Politécnico do Porto, Portugal
Pedro Fonseca	Universidade de Aveiro, Portugal
Rosaldo Rossetti	Universidade do Porto, Portugal

Social Robotics: Intelligent and Adaptable AAL Systems Organizing Committee

Luís Santos	Institute of Systems and Robotics – University of Coimbra, Portugal
Jorge Dias	Khalifa University Robotics Institute and Institute of Systems and Institute of Systems and Robotics, University of Coimbra, Portugal

Social Robotics: Intelligent and Adaptable AAL Systems Program Committee

Filippo Cavallo	The BioRobotics Institute, Italy
Friederike Eyssel	Bielefeld University, Germany
João Sequeira	Instituto Superior Técnico de Lisboa, Portugal
Jorge Lobo	ISR - University of Coimbra, Portugal

Surgical Robotics Organizing Committee

Alicia Casals	Universidad Politecnica de Catalunha, Spain
Paulo Gonçalves	Instituto Politécnico de Castelo Branco, Portugal
Nicolas Garcia	Universidad Miguel Hernandez de Elche, Spain
Jose Maria Sabater	Universidad Miguel Hernandez de Elche, Spain
Jorge Martins	Instituto Superior Técnico, Univ. Lisboa, Portugal

Surgical Robotics Program Committee

Alicia Casals	Universidad Politecnica de Catalunha, Spain
Jorge Martins	Instituto Superior Técnico, Univ. Lisboa, Portugal
Jose Maria Sabater	Universidad Miguel Hernandez de Elche, Spain
Nicolas Garcia	Universidad Miguel Hernandez de Elche, Spain
Paulo Gonçalves	Instituto Politécnico de Castelo Branco, Portugal
Pedro Torres	Instituto Politécnico de Castelo Branco, Portugal
Tamás Haidegger	ABC Center for Intelligent Robotics, Óbuda University

Urban Robotics Organizing Committee

Alberto Sanfeliu	IRI (CSIC-UPC), Spain
Juan Andrade	IRI (CSIC-UPC), Spain
Joao Sequeira	ISR - Univ. Lisboa, Portugal
Anais Garrell	IRI (CSIC-UPC), Spain

Urban Robotics Program Committee

Alberto Sanfeliu	IRI (CSIC-UPC), Spain
Anais Garrell	IRI (CSIC-UPC), Spain
Joao Sequeira	ISR - Univ. Lisboa, Portugal
Juan Andrade	IRI (CSIC-UPC), Spain

Visual Maps in Robotics Session Organizing Committee

Oscar Reinoso Garcia	Miguel Hernandez University, Spain

Visual Maps in Robotics Session Program Committee

Arturo Gil Aparicio	UMH, Spain
Carlos Sagües	University of Zaragoza, Spain
Fernando Torres Medina	University of Alicante, Spain
Javier González Jiménez	University of Malaga, Spain
Jose Mª Sebastian Zuñiga	UPM, Spain

Jose María Martínez Montiel	University of Zaragoza, Spain
Luis Payá Castelló	UMH, Spain
Pablo Gil	University of Alicante, Spain

Visual Perception for Autonomous Robots Session Organizing Committee

Andry Maykol Pinto	INESC-TEC and Faculty of Engineering of the University of Porto, Portugal
Pedro Neto	University of Coimbra, Portugal
Luis Rocha	INESC-TEC, Portugal

Visual Perception for Autonomous Robots Session Program Committee

Aníbal Matos	INESC-TEC and Faculty of Engineering of the University of Porto, Portugal
António Neves	University of Aveiro, Portugal
Armando Pinho	University of Aveiro, Portugal
Bernado Cunha	University of Aveiro, Portugal
Brígida Mónica Faria	Polytechnic Institute of Porto – ESTSP/IPP, Portugal
Hélder Oliveira	INESC-TEC and Faculty of Engineering of the University of Porto, Portugal
Marcelo Petry	Federal University of Santa Catarina and INESC P&D Brasil, Brasil
Nuno Cruz	INESC-TEC and Faculty of Engineering of the University of Porto, Portugal

WAF – XVI Workshop on Physical Agents Organizing Committee

Vicente Matellán	Universidad de León, Spain
Miguel Ángel Cazorla	Universidad de Alicante, Spain
Rodrigo Ventura	Instituto Superior Técnico, Universidade de Lisboa, Portugal

WAF – XVI Workshop on Physical Agents Program Committee

Domenec Puig	Universitat Rovira i Virgili, Spain
Eugenio Aguirre	University of Granada, Spain
Francisco Javier Rodríguez Lera	Universidad de León, Spain
Ismael García-Varea	Univ. de Castilla-La Mancha, Spain
J. Francisco Blanes Noguera	Universidad Politécnica de Valencia, Spain
Joaquin Lopez	University of Vigo, Spain
Jose Manuel Lopez	Basque Country University, Spain
José María Armingol	Universidad Carlos III de Madrid, Spain
Josemaria Cañas Plaza	Universidad Rey Juan Carlos, Spain
Lluís Ribas-Xirgo	Universitat Autònoma de Barcelona, Spain
Pablo Bustos	Universidad de Extremadura, Spain
Rafael Muñoz-Salinas	University of Cordoba, Spain
Roberto Iglesias Rodriguez	University of Santiago de Compostela, Spain

Additional Reviewers

Abdelkrim Nemra
Albert Palomer Vila
Alberto Vale
Ali Marjovi
Alireza Asvadi
Ammar Assoum
Ana Maria Maqueda
André Farinha
André Mateus
Arpad Takacs
Begoña Arrúe
Carlos Martinho
Clauirton Siebra
Cristiano Premebida
Daniel Silvestre
David Fornas
Diana Beltran
Diego Faria
Dolores Blanco
Eduard Bergés
Enric Cervera
Fernando Martín
Friederike Eyssel
Giovanni Saponaro
Guillem Alenya
Guillem Vallicrosa
Iñaki Maurtua
Isabel García-Morales
Iván Villaverde
Jan Veneman
João Quintas
Johannes Kropf
José Antonio Cobano
José Barbosa
José María Martínez Montiel
Juan Carlos García Sánchez
Lorenzo Jamone
Manel Frigola
Miguel Garcia-Silvente
Narcís Palomeras
Pablo Lanillos
Pedro Casau
Pedro Lourenço
Raul Marin
Santiago Garrido
Sedat Dogru
Sten Hanke
Vijaykumar Rajasekaran
Xavier Giralt

Contents

Part I
Environmental Robotics

A UGV Approach to Measure the Ground Properties of Greenhouses

Alberto Ruiz-Larrea, Juan Jesús Roldán, Mario Garzón, Jaime del Cerro and Antonio Barrientos

Abstract Greenhouse farming is based on the control of the environment of the crops and the supply of water and nutrients to the plants. These activities require the monitoring of the environmental variables at both global and local scale. This paper presents a ground robot platform for measuring the ground properties of the greenhouses. For this purpose, infrared temperature and soil moisture sensors are equipped into an unmanned ground vehicle (UGV). In addition, the navigation strategy is explained including the path planning and following approaches. Finally, all the systems are validated in a field experiment and maps of temperature and humidity are performed.

Keywords Environmental monitoring · Agriculture · Greenhouse · Robotics · UGV · Sensory system · Navigation system

1 Introduction

The agriculture in greenhouses is an appropriate field to implement innovative technologies. In fact, there are many proposals of autonomous systems for the production monitoring, crop irrigation and nutrition, or ventilation and heating in greenhouses (e.g. [14]). Nevertheless, the application of these technologies in the greenhouses is usually restricted to those with experimental purpose or large production. In the common facilities, the implementation of these technologies may be difficult due to their cost and their complexity.

A. Ruiz-Larrea · J.J. Roldán(✉) · M. Garzón · J. del Cerro · A. Barrientos
Centre for Automation and Robotics (UPM-CSIC), Madrid, Spain
e-mail: alberto.ruiz-larrea.guillen@alumnos.upm.es,
{jj.roldan,ma.garzon,j.cerro,antonio.barrientos}@upm.es
http://www.car.upm-csic.es/

L.P. Reis et al. (eds.), *Robot 2015: Second Iberian Robotics Conference*,
Advances in Intelligent Systems and Computing 418,
DOI: 10.1007/978-3-319-27149-1_1

The monitoring of the environmental variables is fundamental in greenhouses. This information is useful to perform the climate control, which can be at global or local level, according to the resolution of the information and the features of the greenhouse. In addition, this information is interesting for studying the production and traceability of products. A complete log with spatial and temporal information of the environmental variables allows to know the conditions of each area of the greenhouse and to determine the optimal ones for plant growth and maturation.

The aim of this paper is to use an Unmanned Ground Vehicle (UGV) as a platform for measuring the properties of the soil of a greenhouse. This robot allows moving and placing the sensors in the desired point of the ground and performing a path in the greenhouse to collect the measures over time. The challenge of the navigation in a closed and occupied environment is considered in the proposal of this work.

Section 2 describes the state of art of robotics in greenhouse farming. Section 3 studies the variables that influence the greenhouse climate or the plant growth and describes the selection of sensors and robot and their integration. Section 4 addresses the navigation strategy in the greenhouse environment, which encompasses the path planning and following. Section 5 describes the experiments performed for validating the platform and their results. Finally, section 6 summarizes the conclusions of the work and the proposals for future ones.

2 State of Art

The application of technology in greenhouse farming and environmental monitoring is relatively common. The wired or wireless sensor networks have been used both in outdoor and indoor agriculture.

The wired sensor networks have been used in the automation of productive systems for years. In the context of indoor agriculture, they were used for monitoring variables as temperature, humidity, etc. The main disadvantage of this technology is the need of wires that increases its costs and make difficult its maintenance.

Wireless Sensor Networks (WSNs) overcome the problems of previous sensor networks [21]. A WSN is a system consisting of a set of motes, which are autonomous and intelligent nodes, that are able to interact and cooperate among them [1]. These motes have small sizes and low power consumptions. The main advantages of WSNs are their modular character, their ability to manage the information and their absence of wires. On the other hand, their main disadvantage is that each node requires - in addition to the sensors - a controller, a battery and a communication module. Therefore, the cost of these systems may grow strongly with the size of the greenhouses.

There are many cases of application of WSNs in agriculture and food industry [18]. There are also some proposals for the implementation of these systems in indoor farming such as [2], [16], [15] and [22]. In these works, WSNs with motes that measure temperature, humidity or luminosity are deployed in greenhouses.

On the other hand, the application of robots in the context of greenhouse farming has evolved in last years [6]. These robots have different configurations (both industrial manipulators and mobile robots [4] [12] [19]) and diverse applications (planting and harvesting [4] [19], supply of water and nutrients [19], application of fertilizers and pesticides [12] [19]...).

In previous works [17], an Unmanned Aerial Vehicle (UAV) was proposed for monitoring variables, such as air temperature and humidity, luminosity and carbon dioxide concentration. This solution provides to the sensor platform with capabilities for moving and accessing to any point of the three dimensional space. Nevertheless, the measurement of the ground properties is more suitable with a ground platform than an aerial one, particularly if the soil has to be prepared in order to take the measures.

This work continues this research line and proposes a UGV as a platform for measuring ground properties, building maps of them and monitoring production. In the following sections, the sensory and navigation systems are described with detail, and the experiments and results are presented.

3 Sensory System

The design of sensory system takes into account the environmental variables that must be monitored. Different climate models for greenhouses and growth models for plants were studied in order to determine these variables.

The climate model proposed in [7] considers the heating power, ventilation flow, wind speed, solar radiation, air temperature, air humidity and carbon dioxide concentration. On the other hand, the model of [11] takes into account outdoor temperature, solar radiation, heating power and ventilation flow for performing the climate control. Finally, another model [20] includes as variables the transpiration of plants, the cover condensation and the vapor flow in the ventilation.

Other publications [10] and [5] show the relation between the maturation of fruits and the ethylene concentration in the air. Ethylene behaves as a hormone that controls the processes of growth and maturation in the plants.

The influence of water in the growth of plants is fundamental. It transports nutrients and other substances required for their development. In addition, it takes part in the chemical processes that control the metabolism of the plants. As pointed in [8], a water deficit implies a reduction of the size of roots, tails and leaves and, therefore, a reduction in the fruit production and quality.

Meanwhile, the ground temperature is a factor that limits the germination of seeds, the growth of roots and the production of organic material. Therefore, it is a fundamental variable for the development of plants and their processes of nutrition and transpiration.

After this study, the variables of ground temperature and moisture were selected for the sensory system. It should be noted that air variables, such as temperature, humidity or carbon dioxide concentration, were measured in previous works [17]. In addition, the measurement of the nutrient concentration in the soil is not considered, due to the sensors have a high response time (around 30 minutes) and, therefore, the monitoring of these values is not possible. Other variables as ethylene emissions or chlorophyll fluorescence are posed for future works, due to the complexity of the sensors and their integration.

3.1 Sensors

The sensors selected in this work are described below:

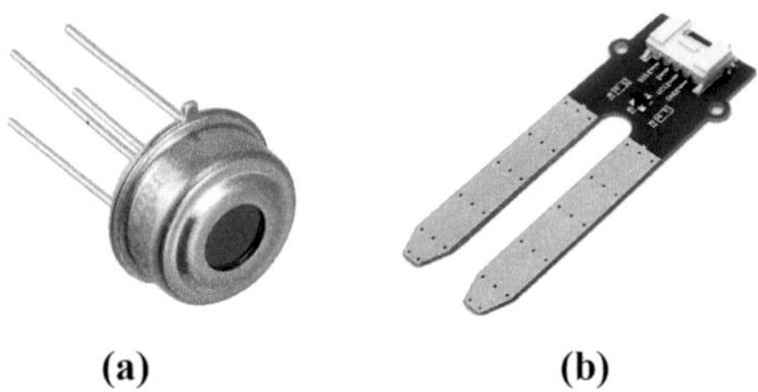

(a) **(b)**

Fig. 1 Selected sensors: (a) MLX9614 for temperature. (b) SEN92355P for humidity.

Temperature Sensor: The MLX90614 sensor (figure 1 a) is selected for measuring the ground temperature. This sensor is an infrared thermometer that can measure the temperature of surfaces without contact. It provides both the temperature of the object and the temperature of the environment. Its range of temperatures is from -40 to 125 °C for environment and -70 to 380 °C for objects. Its accuracy is around 0.5 °C over the range of temperatures and 0.1 °C in the central values. Its angle of measure is around 90 °and it is the main limitation.

This sensor has been selected according to the following reasons:

- The sensor is able to measure the ground temperature without contact, which reduces the complexity of system and the risk of damage caused by impacts or scratches.
- The temperature range and accuracy are appropriate for the application, due to they allow to measure the potential values of temperature in the greenhouse with high precision.
- The cost is low in comparison to other non-contact temperature sensors available in market.

Humidity Sensor: The SEN92355P sensor (figure 1 b) is chosen to measure the soil humidity. This sensor has two exposed electrodes that can measure the resistivity of the ground and then infer the moisture of it. It is able to work in dry and humid soils or directly in water. However, it should be introduced completely in the terrain in order to obtain correct measures and avoid potential noises and biases.

This sensor has been selected according to these criteria:

– The sensor can be integrated easily in some controllers such as *Arduino* or *Raspberry Pi*, which reduce the complexity and the cost of the sensory system.
– The size and performance of the sensor are adequate for the application: i.e. the range of humidity in the greenhouse is covered by the sensor.
– The cost is low in comparison to other alternatives of market.

3.2 Robot

The UGV used as sensor platform is a *Robotnik Summit XL* (figure 2). This robot has a size of 722x610x392 mm and a weight of 45 kg. It has motors in the four wheels, so it is able to rotate in place. The robot reaches a speed of 3 meters per second. It is controlled by an embedded computer with Linux and Robot Operating System (ROS). The autonomy of its batteries is about 180 minutes.

The UGV is equipped with an Inertial Measurement Unit (IMU) and a Global Navigation Satellite System (GNSS) receiver for performing navigation. It also has a linear laser scanner, which has range of 30 meters and angle of 270 degrees, for recognizing the perimeter and avoiding the obstacles. Finally, it has a pan-tilt-zoom camera that is useful for manual control or autonomous navigation.

The robot has a load capacity of 20 kg, so it is able to transport the sensors and the components required to their work. The following section describes the integration of sensors into the robot at hardware and software levels.

(a) (b) (c)

Fig. 2 (a) Robot coming through the plants, (b) Robot with sensory system, (c) Integration of sensors.

3.3 Integration

The integration of the sensors in the robot has required the following actions:

- The sensors are connected to a controller that reads the signals, computes the values and stores the information. In this work an *Arduino UNO* has been chosen for its compatibility with the sensors (the temperature sensor is connected by I2C and the moisture sensor is connected directly) and its ease of use.
- The controller is connected to the embedded PC of the robot, in order to synchronize the sensor and path following algorithms. Thus, when the path following algorithm arrives to a measure point, it passes the control to the sensor algorithm. On the other hand, when the sensor algorithm obtains the measures, it returns the control to the path following algorithm. In this work, the sensor controller is connected to the robot computer via USB and it acts as a ROS node in the robot architecture.
- An autonomous tool is developed in order to drill the ground and place the moisture sensor for obtaining measures. This tool is shown in figure 2 and is based on a threaded bar, which converts the motor rotation to a vertical movement that drills the soil and place the sensor. The bar is moved by a stepper motor, which is connected through a driver to the *Arduino UNO*. An ultrasonic sensor controls the depth of the hole, while a contact sensor detects the collecting of the tool. A structure of bars and plates ensures the stability of the mechanism.
- The sensor algorithm is executed in the *Arduino UNO*. It takes the control when the robot arrives to a measure point. At first, it controls the deployment of the measurement tool and the placement of moisture sensor. Later, it takes ten measures of temperature and humidity and computes the average values. Finally, it sends the values of temperature and humidity and returns the control to the path following algorithm.

4 Navigation System

A greenhouse is a closed place that commonly has high occupancy. It has regular properties such as the crop layout, which is usually composed of crop lines (as seen in figure 3), but it also has irregular elements such as the plants, which are planted in regular places but grow irregularly.

Figure 3 shows the common exploitation in Almería (Andalucia, Spain), where there is the largest agglomeration of greenhouses in the world. The greenhouses of Almería have an average surface of 6,200 m^2 [9], which is clearly less than the fields of outdoor agriculture. These greenhouses often present a front side with one or more doors that can be used for input and output of machinery, a set of main corridors with a width of around 2 meters, and a series of side corridors with a width of around 1 meter.

Fig. 3 Greenhouse views: (a) Outdoor, (b) Main corridor, (c) Side corridor

4.1 Path Planning

A back and forth strategy [3] is selected for path planning. This strategy allows the robot to cover all the surface of greenhouse and pass next to all the plants for obtaining the measures (as shown in figure 4). Back and forth motion is described in the algorithm 1. The algorithm has the following inputs: l, the length of the crop lanes, w, the width of the sets of two crop lanes, L, the length of greenhouse, and W, the width of greenhouse.

Algorithm 1. Path planning.

```
function PATH PLANNING(l, w, L, W)
  for i = 0 to L/l do
    if i mod 2 = 0 then
      for j = 0 to W/w do
        List ← (i, j)
      end for
    else
      for j = 0 to W/w do
        List ← (i, W/w − j)
      end for
    end if
  end for
  return List
end function
```

The path planning algorithm has been implemented in a ROS node, which receives the parameters of the greenhouse and their corridors and returns the trajectory as a list of points. The path must pass through all the points, with the shortest possible longitude and without changing over time. In fact, the monitoring of the ground properties requires measuring at the same points over time.

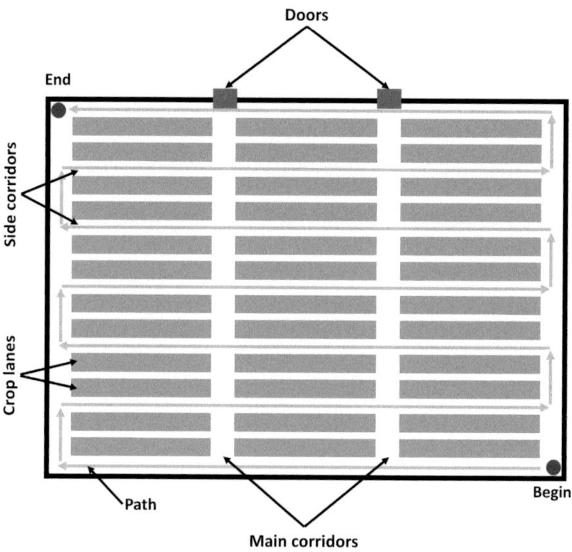

Fig. 4 Robot path on generic greenhouse layout.

4.2 *Path Following*

There are some possible differences between theoretical and real paths. On the one hand, the robot can find obstacles in its path that it should avoid. On the other hand, it should stop at the measure nodes, in order to obtain the ground temperature and humidity. Figure 5 shows the ROS architecture that performs the perception, guidance, navigation and control functions.

The navigation is performed by using the navigation stack of ROS [13]. It receives the position and orientation through the integration of IMU, GPS and odometry data in a Extended Kalman Filter (EKF). This method enhances the accuracy to the order of a few decimeters. It also receives the location of the obstacles around the robot detected by the laser scanner. This information allows compensating the position errors. Finally, it receives the coordinates of the next goal from the path planning and following module. The navigation stack computes the trajectory that reaches the goal and converts it to speed commands for the controller of the robot.

The path following node sends the goals to the navigation stack one by one. This node controls the stop and start in the measure points, which allows the deployment and work of sensors. In addition, it is able to cancel the goals when the robot cannot reach them: e.g. when the goal is unaccessible or when the time to reach it is high.

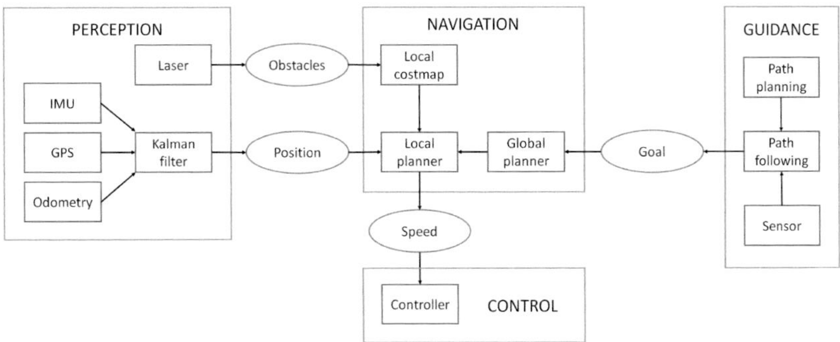

Fig. 5 ROS architecture for robot guidance, navigation and control.

5 Experiments and Results

Some experiments were performed in a test field with height of 12 meters and width of 12 meters, where the distance among the measure points was about 4 meters in both directions. These experiments were developed in the summer of 2015 in Arganda del Rey (Community of Madrid, Spain), which is not a period of production due to the high temperatures and low humidities. The robot covered the path in around 12 minutes (15 movements between points of about 4 seconds and 16 stops of about 40 seconds, in order to place the moisture sensor into the soil to obtain the measures). Therefore, the temperatures and humidities should not vary significantly during the path.

During this coverage path, measurements of ground temperature and humidity were collected by the sensory system. The maps are shown in figure 6 and show the functionalities of the developed system. This figure shows the maps of environmental variables. As shown, the ground temperature was between 40 and 60 °C (when the air temperature was around 35 °C), while the relative humidity was low, due to the season and the absence of irrigation.

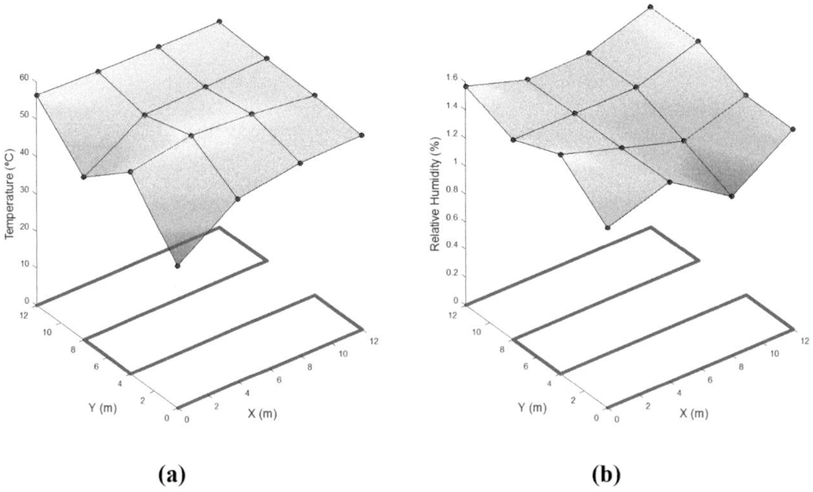

(a) (b)

Fig. 6 (a) Map of ground temperature, (b) Map of soil humidity.

6 Conclusions

Following the line of the previous works, this paper addresses the application of robotics in greenhouse farming. Specifically, it proposes the use of a UGV as a platform for measuring the ground properties of greenhouses.

A study of the literature about both greenhouse climate models and crop growth models is performed in order to select the ground variables that should be measured: the ground temperature and the soil humidity. The MLX90614 temperature sensor and SEN92355P moisture sensor are selected and integrated through an Arduino UNO controller and a mechanism to pick and place them.

A back and forth strategy is proposed for the movement in the greenhouse. The path planning and following algorithms are implemented in Robot Operating System (ROS) and connected to the sensor control algorithms. These algorithms are adapted to the navigation in a closed and occupied place such as a greenhouse.

Finally, the systems are validated in a test field, where maps of ground temperature and humidity are performed. The application of this system in a productive greenhouse and the integration with other systems (e.g. climate control or irrigation) are proposed for future works.

Acknowledgments The research leading to these results has received funding from the RoboCity 2030-III-CM project (Robótica aplicada a la mejora de la calidad de vida de los ciudadanos. fase III; S2013/MIT-2748), funded by Programas de Actividades I+D en la Comunidad de Madrid and cofunded by Structural Funds of the EU, and from the DPI2014-56985-R project (Protección robotizada de infraestructuras críticas) funded by the Ministerio de Economía y Competitividad of Gobierno de España.

References

1. Akyildiz, I.F., Su, W., Sankarasubramaniam, Y., Cayirci, E.: A survey on sensor networks. IEEE Communications Magazine **40**(8), 102–114 (2002)
2. Antonio, P., Grimaccia, F., Mussetta, M.: Architecture and methods for innovative heterogeneous wireless sensor network applications. Remote Sensing **4**(5), 1146–1161 (2012)
3. Choset, H., Pignon, P.: Coverage path planning: the boustrophedon cellular decomposition. In: Field and Service Robotics, pp. 203–209. Springer (1998)
4. Correll, N., Arechiga, N., Bolger, A., Bollini, M., Charrow, B., Clayton, A., Dominguez, F., Donahue, K., Dyar, S., Johnson, L., et al.: Building a distributed robot garden. In: IEEE/RSJ International Conference on Intelligent Robots and Systems, IROS 2009, pp. 1509–1516. IEEE (2009)
5. Ecker, J.R.: The ethylene signal transduction pathway in plants. Science **268**(5211), 667 (1995)
6. García, M.A., Gutiérrez, S., López, H.C., Rivera, S., Ruiz, A.C.: Estado del arte de la tecnología de robots aplicada a invernaderos. Avances en Investigación Agropecuaria **11**(3), 53–61 (2007)
7. van Henten, E.J.: Greenhouse climate management: an optimal control approach. Landbouwuniversiteit te Wageningen (1994)
8. Kirnak, H., Kaya, C., Tas, I., Higgs, D.: The influence of water deficit on vegetative growth, physiology, fruit yield and quality in eggplants. Bulg. J. Plant Physiol. **27**(3–4), 34–46 (2001)
9. Langreo, A.: La agricultura mediterránea en el siglo xxi. Méditerraneo Económico **2**, 101–123 (2002)
10. Lieberman, M., Baker, J.E., Sloger, M.: Influence of plant hormones on ethylene production in apple, tomato, and avocado slices during maturation and senescence. Plant Physiology **60**(2), 214–217 (1977)
11. Linker, R., Seginer, I.: Greenhouse temperature modeling: a comparison between sigmoid neural networks and hybrid models. Mathematics and Computers in Simulation **65**(1), 19–29 (2004)
12. Mandow, A., Gomez-de Gabriel, J.M., Martinez, J.L., Munoz, V.F., Ollero, A., García-Cerezo, A.: The autonomous mobile robot aurora for greenhouse operation. IEEE Robotics & Automation Magazine **3**(4), 18–28 (1996)
13. Marder-Eppstein, E., Berger, E., Foote, T., Gerkey, B., Konolige, K.: The office marathon: Robust navigation in an indoor office environment (2010)
14. Martínez, M., Blasco, X., Herrero, J.M., Ramos, C., Sanchis, J.: Monitorización y control de procesos. una visión teórico-práctica aplicada a invernaderos. RIAII **2**(4), 5–24 (2005)
15. Park, D.H., Kang, B.J., Cho, K.R., Shin, C.S., Cho, S.E., Park, J.W., Yang, W.M.: A study on greenhouse automatic control system based on wireless sensor network. Wireless Personal Communications **56**(1), 117–130 (2011)
16. Pawlowski, A., Guzman, J.L., Rodríguez, F., Berenguel, M., Sánchez, J., Dormido, S.: Simulation of greenhouse climate monitoring and control with wireless sensor network and event-based control. Sensors **9**(1), 232–252 (2009)
17. Roldán, J.J., Joossen, G., Sanz, D., del Cerro, J., Barrientos, A.: Mini-uav based sensory system for measuring environmental variables in greenhouses. Sensors **15**(2), 3334–3350 (2015)
18. Ruiz-Garcia, L., Lunadei, L., Barreiro, P., Robla, I.: A review of wireless sensor technologies and applications in agriculture and food industry: state of the art and current trends. Sensors **9**(6), 4728–4750 (2009)
19. Sánchez-Hermosilla, J., González, R., Rodríguez, F., Donaire, J.G.: Mechatronic description of a laser autoguided vehicle for greenhouse operations. Sensors **13**(1), 769–784 (2013)
20. Stanghellini, C., de Jong, T.: A model of humidity and its applications in a greenhouse. Agricultural and Forest Meteorology **76**(2), 129–148 (1995)
21. Valdiviezo, D.V.: Diseño de una red de sensores inalámbrica para agricultura de precisión. PhD thesis (2009)
22. Zhang, Q., Yang, X., Zhou, Y., Wang, L., Guo, X.: A wireless solution for greenhouse monitoring and control system based on zigbee technology. Journal of Zhejiang University Science A **8**(10), 1584–1587 (2007)

An Aerial-Ground Robotic Team for Systematic Soil and Biota Sampling in Estuarine Mudflats

Pedro Deusdado, Eduardo Pinto, Magno Guedes, Francisco Marques, Paulo Rodrigues, André Lourenço, Ricardo Mendonça, André Silva, Pedro Santana, José Corisco, Marta Almeida, Luís Portugal, Raquel Caldeira, José Barata and Luis Flores

Abstract This paper presents an aerial-ground field robotic team, designed to collect and transport soil and biota samples in estuarine mudflats. The robotic system has been devised so that its sampling and storage capabilities are suited for radionuclides and heavy metals environmental monitoring. Automating these time-consuming and physically demanding tasks is expected to positively impact both their scope and frequency. The success of an environmental monitoring study heavily depends on the statistical significance and accuracy of the sampling procedures, which most often require frequent human intervention. The bird's-eye view provided by the aerial vehicle aims at supporting remote mission specification and execution monitoring. This paper also proposes a preliminary experimental protocol tailored to exploit the capabilities offered by the robotic system. Preliminary field trials in real estuarine mudflats show the ability of the robotic system to successfully extract and transport soil samples for offline analysis.

P. Deusdado · M. Guedes · A. Silva · R. Caldeira · L. Flores
INTROSYS SA, Setúbal, Portugal
e-mail: pedro.deusdado@introsys.pt

E. Pinto · F. Marques(✉) · P. Rodrigues · A. Lourenço · R. Mendonça · J. Barata
CTS-UNINOVA, Universidade Nova de Lisboa (UNL), Lisbon, Portugal
e-mail: fam@uninova.pt

P. Santana
ISCTE - Instituto Universitário de Lisboa (ISCTE-IUL), Lisbon, Portugal

P. Santana
Instituto de Telecomunicações (IT), Lisbon, Portugal

J. Corisco · M. Almeida
Centro de Ciências e Tecnologias Nucleares (C2TN),
Instituto Superior Técnico, Universidade de Lisboa, Lisbon, Portugal

L. Portugal
Agência Portuguesa do Ambiente (APA), Lisbon, Portugal

L.P. Reis et al. (eds.), *Robot 2015: Second Iberian Robotics Conference*,
Advances in Intelligent Systems and Computing 418,
DOI: 10.1007/978-3-319-27149-1_2

Keywords Multi-robot system · Field robots · UGV · UAV · Environmental monitoring · Radiological monitoring · Estuarine mudflats

1 Introduction

Primary deposition of contaminants, like radionuclides and heavy metals, on underwater surface sediments can occur through physical settling of particulate matter or through direct chemical sorption from the water. The process is ruled by physical conditions, such as water turbulence, contact time, sediment surface topography, and by the chemical nature and physical form of contaminants in the water column. All of these variables tend to be spatially complex, leading to heterogeneous distribution patterns of contaminants. In time, the hydrodynamics-induced redistribution of particles will intensify the spatial heterogeneity created by a primary deposition.

The presence of radionuclides and heavy metals in the mudflats of estuarine bays may be an issue of increased public concern and environmental relevance, suggesting the need for an extensive survey of the intertidal mudflat. Continuous dredging of the navigation channel and fishery activities, such as intensive clam harvesting, promote the re-suspension of both surface and anoxic bottom sediments, which might be a cause for remobilisation of adsorbed toxicants [1].

Traditionally, surveys in estuarine mudflats are performed by experts who handle manual sampling tools and carry the samples from the site to the lab for an offline analysis. Walking in the mudflat, handling the sampling tools, transporting the samples, and ensuring those are properly tagged and geo-referenced, are just a few of the

Fig. 1 The robotic system. (a) The unmanned ground vehicle turning around its geometric centre. The robotic arm is moving a sample container to the drilling tool. (b) The UAV taking off for an aerial survey in the Samouco region.

several physically demanding, costly, and time-consuming challenges humans must face in these sampling campaigns. To mitigate these difficulties, this paper presents a robotic system (see Fig. 1) developed to support sampling operations in mudflats and, as a result, facilitate thorough spatio-temporal sampling campaigns therein.

Robotic radiological monitoring is most often considered from an emergency-response perspective [2]. Conversely, the robotic team herein presented targets routine radiologic monitoring campaigns. The robotic system is composed of a wheeled Unmanned Ground Vehicle (UGV), capable of performing soil sample acquisition and delivery, and a multirotor Unmanned Aerial Vehicle (UAV), providing to the UGV and human experts an aerial perspective of the operation site. Via custom drilling tools, the robotic team is able to extract up to 9 cylindrical soil samples, each with a section of 6 cm and a depth of 45 cm. A solution based on a robotic arm manipulating smaller sampling containers (e.g., [3]) would not be able to reach such sample depth and volume, hampering offline analysis in non-emergency scenarios. Self-burying robots (e.g., [4, 5]) are a promising alternative to drilling tools, yet, it is still unclear how these robots can extract sample volumes for sufficiently significant routine radiological monitoring.

The ability to sample near-the-surface seaweeds is also key, as they supply organic food to a variety of dependent food webs and act as nursery ground for animal species [6]. Sampling clams is also relevant as these can influence human health directly. Bearing this in mind, the herein proposed robotic team is equipped with a second extraction tool adapted to sample seaweeds and clams.

The robotic team is a sampling tool, which, as other sampling tools, must be properly encompassed by experimental protocols. Consequently, this paper adapts well established environmental monitoring experimental protocols to exploit the robotic system's capabilities and to demonstrate the robotic system as an environmental survey robotic tool, contributing to the monitoring of the actual distribution patterns of radionuclides and heavy metals in an estuarine bay.

This paper is organised as follows. Section 2 presents the use-case that motivates the experimental protocol described in Section 3. Then, the robotic system, developed bearing into account the experimental protocol, is presented in Section 4. Finally, some conclusions and future work avenues are drawn in Section 5.

2 A Motivating Example: The Tagus River's Estuary

The specification, development, and validation of the robotic system relies on Tagus river's estuarine bay as the main case-study (see Fig. 2(a)). The selection of this case-study stems from two main observations. First, as it will be shown below, studying Tagus' estuary is a remarkably relevant problem from an environmental monitoring standpoint. Second, Tagus' estuary is representative of a vast set of estuaries influenced by major littoral cities, such as Lisbon.

The muddy sediments of the Tagus river's estuary have been exposed to decades of contaminants' deposition from local industries. The runoff and wind spreading

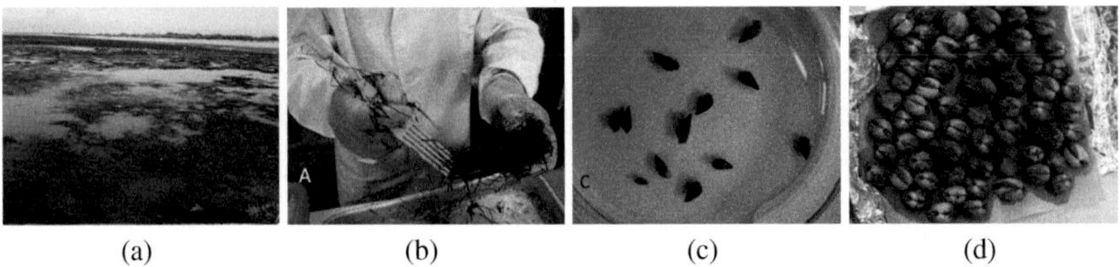

(a) (b) (c) (d)

Fig. 2 Biota at Tagus river's estuary mudflats. (a) Mudflat covered with seagrass. (b) Working on a sample of seagrass *Z. noltii*. (c) small gastropods removed from the seagrass' stems. (d) A clam (*R. philippinarum*) sample.

of particulate materials coming from the phosphogypsum stockpile of a disabled phosphate plant, near the city of Barreiro, have been a source for localised enhanced concentrations of natural radioisotopes of the uranium family. Phosphogypsum is an industrial waste primarily resulting from phosphate rock reacting with sulphur acid to produce phosphoric acid. Uranium rich phosphate rock from the north of Africa was used in the plant, so the waste product has enhanced levels of radionuclides form the uranium decay series, usually classified as Normally Occurring Radioactive Material (NORM). The radiological impact of the Barreiro phosphate industry due to uranium 238 descendants, lead 210, and polonium 210, in the bottom sediments and in the water column particulate matter has been described in [7].

Other industries set on both sides of the estuarine bay contributed to the dispersion of toxic metals like mercury, cadmium or arsenic, and a variety of other different contaminants including Polycyclic Aromatic Hydrocarbons (PAH) and organometallic compounds that have been previously reported suggesting a deterioration of water and sediment quality in some critical areas of the Tagus estuary [7]. The construction of the Vasco da Gama Bridge (roughly 25 km upstream the estuary mouth), from September 1994 to December 1998, caused additional disturbance, promoting the remobilisation of anoxic contaminated sediments. This fact led to the temporary solubility of toxicant metals followed by re-adsorption to the particulate phase [8].

The decade 2000-2010 was a time for some observable changes in both the presence of human activity and the physiognomy of the estuarine mudflat extending through the shoreline from Barreiro to Alcochete. The introduction of the invasive Asian clam *Ruditapes philippinarum* (Fig. 2(d)) and its massive population expansion triggered an intense activity of clam harvesting for human consumption without any control of toxicants. At the same time, a notorious and progressive green coverage of several areas of the mudflat could be witnessed. The sea grass *Zostera noltii* (see Fig. 2(a) and Fig. 2(b)) was then identified in the course of an exploratory sampling initiative and appeared to be the residence substrate for small gastropods (see Fig. 2(c)). There is also evidence about the presence of squids.

In sum, there are several factors potentiating the presence of radionuclides and heavy metals in the estuarine bay, which, in turn, contaminate the food-chain that ultimately impacts human health. The robotic team herein presented is expected to foster accurate spatio-temporal soil and biota sampling so that environment researchers can study these phenomena in detail.

3 The Experimental Protocol

The sampling procedure will follow the principle of transect sampling generally described by the International Commission on Radiation Units and Measurements, for the purpose of estimating spatial distribution patterns of radionuclides in large areas with closely spaced sampling locations [9]. All actions performed by the robotic system will be the result of the fine tuned interaction of both aerial and terrestrial devices, the latter being capable of acting accordingly to the information transmitted by the former, and also in the context of human-robot interaction to the direct instructions of an operator.

3.1 Chain of Robot-Assisted Tasks

During a sampling process in the estuarine mudflat, the operational tasks will be performed according to the following sequence (refer to Fig. 3 for a graphical representation of each of the following steps):

1. The aerial robot takes off and performs a scan on the operational area defined by the user. The result is a high resolution geo-referenced mosaic built from a set of mutually registered aerial images. On top of this aerial mosaic, the system (assisted by the human operator) maps potential dead-ends, safe paths, and sampling points.
2. At the control centre, the high resolution mosaic is segmented, either manually or automatically, so as to obtain the main features of the estuary to be sampled. Water ponds and channels, sea grass coverage, salt marsh vertical vegetation, sand banks, and all sorts of physical obstacles are examples of such features. Based on the segmented aerial mosaic, the user prepares the mission by specifying a set of transects to be sampled by the robot.
3. With the information collected in 2, the ground robot traverses the transects and periodically samples the terrain. While moving, the robot avoids unexpected obstacles and sends current telemetry data to the control station. From there, the user supervises the robots' operation and sends corrective commands whenever required. At each sampling point, the ground robot extracts the terrain samples while the UAV provides images augmenting the operator perception about the current mission.
4. When the robot has either its reservoir filled-out, or visited all the pre-defined sampling locations, it returns to the base. The user operator may help the robot with this procedure.
5. Back on the base, the robot unloads the sample containers into isothermal boxes with cooling pads. This way, the samples are kept frozen until they are brought to the lab for post-processing.
6. The robot executes a process of self-washing to clean its sensors and tools.

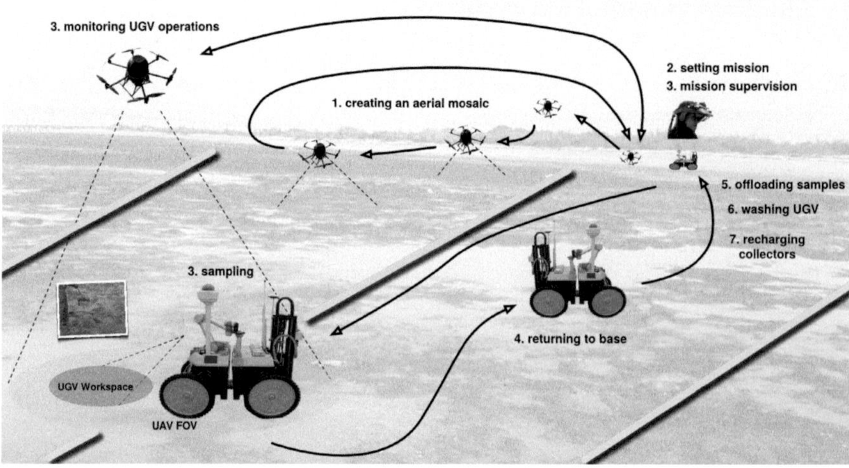

Fig. 3 Diagram representing a typical mission work-flow, whose sequence is represented by the illustrated enumeration. Tasks with the same enumeration run in parallel. Robot's motion is represented by the black curved arrows. The yellow thick lines depict the transects to be executed by the robot.

7. Finally, the robot is recharged with empty collectors for a new mission. At this point, the user may be called upon to execute some maintenance procedures, such as recharging batteries or re-inflate tyres.

3.2 Sampling Procedures

A proper sampling campaign must ensure that samples are both spatially distributed, and their volume is sufficient for the subsequent offline analysis to provide significant results. In non-emergency radiological situations the density of contaminants per sample is low. As a consequence, large sample volumes are required if contaminants are to be detected therein. The following paragraphs specify the volumes for a typical radionucleids and heavy metals characterisation.

Sediment cores are extracted both in bare and sea grass covered mudflat, to support a posteriori partition analysis of metal and radionuclides in sediments and sea grass. For each sampling location, 500 m long transversal transects are defined perpendicularly to the shore line. The ground vehicle stops every 100 m along the transect. Each stopping position defines the centre of a circular sampling area with 6 m diameter, from which 9 sediment cores (45 cm depth and 6 cm diameter) are randomly extracted by the robot.

Sea grasses and bivalves are collected with the grid-like dredgers present in the ground vehicle. Sea grass sampling is done by dredging along 100 m transects defined on the sea grass bed, desirably accumulating no less than 2 kg, including attached sediments and debris. Bivalves resident in the upper layers of the bottom sediments are collected similarly. The path extension has necessarily to be larger than for sea grass sampling and the procedure must be executed in several trials until an amount of 4 kg is reached (including attached sediments and debris).

3.3 *Sample Processing*

At the lab, the cores are unfrozen and sectioned in depth layers (0 cm – 5 cm; 5 cm – 15 cm; 15 cm – 25 cm; 25 cm – 35 cm; 35 cm – 45 cm) and all sections of a specified depth range are mixed into a composite sample. Composite samples from specified depth layers are oven dried at a temperature of 60 °C. The fine grain size fraction composed of silts and clays are separated from sand particles in a mechanical sieving system (silt, clay $< 64\,\mu$m, sand $< 200\,\mu$m). Samples are kept dried in tagged plastic containers for further radiological and trace metal analysis.

Fresh seagrasses are washed and rinsed in water for the removal of attached sediments, debris (shell fragments, stones, etc.) and small invertebrates. Clean subsamples are separated for fresh/dry weight measurements. All samples are oven dried at 60 °C, homogenised in a knife mill to a small particle size and kept dried in plastic containers.

Fresh bivalves are washed and rinsed in water and then kept frozen in tagged plastic containers. Individual size classes are determined by biometric analysis of shells' samples. Samples are processed by unfreezing the individuals and separating edible parts from shells. Unfrozen edible parts are weighed fresh and freeze dried. The dried edible parts are homogenised in a knife mill and shells are crushed and homogenised in a cutting mill. Dried samples of edible parts and crushed shells are kept in tagged plastic containers for further analysis.

4 The Robotic System

This section describes the two unmanned vehicles that compose the proposed robotic team. The robots have been devised in order to meet the requirements imposed by the experimental protocol presented in Section 3.

4.1 *Mission Control*

For improved interoperability, the whole robotic system runs on top of the Robot Operating System (ROS) [10]. ROS provides a publish-subscribe inter-process messaging service on top of a master-slave communications framework. To avoid a single point of failure, a multi-master configuration is used by including the *rosbridge* extension [11]. Additionally, *rosbridge* provides a JSON API to ROS functionality, which enables interoperability between robots and control centre over web-based communication channels.

The control centre is based on a Getac V200 with an Intel Core i7-620LM 2.0 GHz and 4 GB of RAM. The laptop docks onto the control centre allowing it to charge and establish a secure communication path to the robots. Two joysticks allow the smooth and precise control of the robot and pan-tilt-zoom camera. A 1200 cd/m^2 sunlight readable display allows the user to supervise in real-time the robots' progress.

4.2 The Unmanned Ground Vehicle (UGV)

The UGV is a 1100 × 1525 × 1368 mm (width x length x height) four-wheeled based on the general-purpose INTROBOT robotic platform [12]. Its front and rear wheels are decoupled through a passive longitudinal joint so as to comply with uneven and muddy terrain (see Fig. 1(a)). The robot's chassis is made of aluminium alloy to reduce weight and increase durability and thermal conductivity. To prevent corrosion caused by salt water, the robot is covered with an hydrophobic coating. Several parts, such as the robot's top covers, are made of composite materials in order to achieve the lowest weight possible while still offering the required robustness. With payload, the overall weight of the robot amounts to 240 kg.

The UGV is equipped with individual 250 W steer and 550 W drive motors, providing quasi-omnidirectional locomotion capabilities. The robot is able to move in Ackerman and double Ackerman configurations, to rotate around its own centre (see Fig. 1(a)), and to move linearly in a wide range of directions without rotation. This multi-modal locomotion becomes of special importance whenever the robot needs to make fine adjustments to its pose to, for instance, align its drilling tool to a specific sampling spot. The wheels have a diameter of 0.3 m and a nominal section width of 0.1 m. The wheels are partially deflated to two thirds of their recommended pressure so as to increase the tyres footprint and, consequently, reduce the chances of slippage in sand and mud. Although deflating the tyres is expected to largely solve the slippage problem in estuarine environments, we are currently assessing whether tyres with larger nominal width would not provide a more robust and energetically efficient solution.

Interchangeable sampling tools can be appended to the robot's rear. Currently, two tools are available, one for drilling the terrain for soil samples and another for dredging seaweeds and clam (see Fig. 4). An actuator in the drilling tool revolves a cylindrical hollow metallic tube with internal section 45 mm and length 500 mm, which is simultaneously pushed downwards by a linear actuator. The whole drilling process takes roughly 60 sec. To cope with the contingency of drilling in surfaces with buried hard elements (e.g., rocks), the linear actuator's current and position is monitored continuously throughout the process. A similar linear actuator is also used in the dredging tool. In this case, the linear actuator pushes a 1.4 cm^3 dredger towards the ground, which is then filled with seaweeds and clam by moving the robot forwards.

An Universal Robotics UR5 compliant robotic arm with 6 degrees of freedom is responsible for autonomously moving the hollow metallic tubes and dredger between their storage sockets and their corresponding sampling tools. The compliance

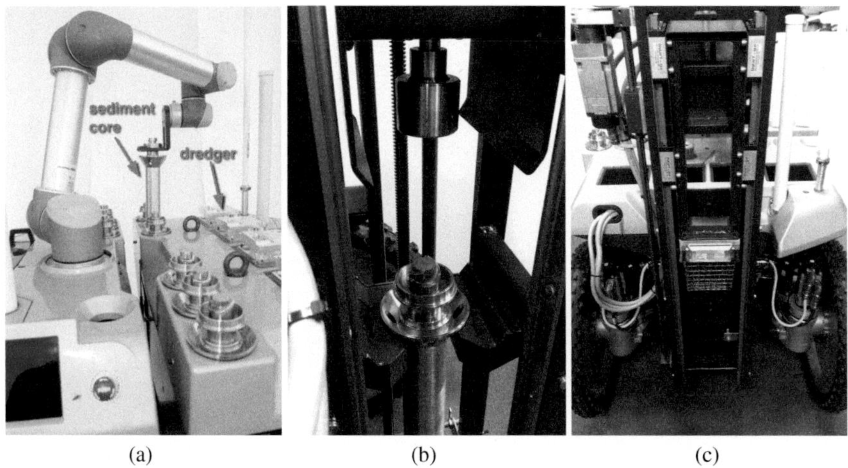

Fig. 4 (a) The robotic arm grasping a hollow tube to insert it into the drilling tool. (b) An hollow tube being attached to the drilling tool. (c) A dredger being pushed downwards by the dredging tool.

characteristics of the robotic arm are crucial to ensure the safety of people working in the robot's vicinity. The ground vehicle is able to store 9 hollow metallic tubes and 4 dredgers. Once the available containers are filled, the soil samples in the hollow metallic tubes must be removed as a pack, that is, without mixing soil from different depths. To facilitate this task, each hollow metallic tube encompasses two half-hollow PVC tubes (separated longitudinally), which easily slide out of the metallic tube with the help of an operator. Fig. 5 depicts a sampling sequence, from setting the hollow tube in the drilling tool to the point when the expert removes the sample's inner PVC container.

The robustness of the robotic platform and its payload was validated with a standard finite element analysis in SolidWorks. Quantitative results are not herein provided due to space limitations. Energy is supplied by eight lithium ion cells with a total capacity of 100 Ah, allowing more than 4 hours of operation. Robot's inter-process communications are ensured by a gigabit Ethernet, whereas wireless communications are available via Ubiquiti airMAX 2.4/5.0 GHz dual-band and a GSM uplink. This setup enables communications up to 1 km in the mudflat, under line of sight.

A SICK LMS111 2D laser scanner mounted on a Robotis Dynamixel tilting unit and a pair of DragonFly cameras for stereo-vision deliver 3D point clouds, which are integrated onto a probabilistic octree [13]. Localisation is estimated with an Extended Kalman Filter, fed by a PhidgetSpatial Inertial Measurement Unit (IMU), from Phidgets, and a GPS-RTK Proflex 800, from Ashtec SAS. Motion planning is carried out using conventional motion and path planning techniques [14] operating on a cost map [15].

Fig. 5 Sequence of a sampling run. (a) The robot's 6-DOF arm grasps one hollow tube to be used as a soil sediments container. (b) The container is then placed on the drilling tool. (c) The tool begins drilling to obtain a core sample. (d) The half-hollow PVC inner tube being removed from the outer hollow metallic tube.

4.3 The Unmanned Aerial Vehicle (UAV)

The aerial team member, the UAV, was designed to withstand the difficult operational environment of estuarine environments, where robustness and reliability are key. Taking this into account, a 6-rotor configuration with vertical take-off and landing capabilities was chosen (see Fig. 1(b)). This configuration combines a good thrust-to-weight ratio and redundancy to enable safety landings in the event of a malfunctioning motor.

The UAV's control system is supported by two computational units, one dedicated to low-level motion control and, the other, to high-level mission execution functions. Communication between controllers is assured by the MAVLink Micro Air Vehicle Communication Protocol. The low-level control unit is a VRBrain from Virtual Robotix, interfaced with an on-board IMU based on the MPU6500, a MS5611 barometer, and a GPS device from Ublox. These sensors are Kalman-filtered for pose estimation. The high-level functions are ROS-enabled and run on the top of a Xubuntu's 14.04 lightweight Linux distribution. The computational unit supporting high-level functions is an Odroid-XU from Hardkernel equipped with an Exynos octa-core CPU.

The low-level stabilisation and basic navigation software is assured by a modified version of the open-source Arducopter platform, whereas high-level navigation and interaction features are handled by dedicated ROS nodes. To take aerial images, which support the creation of aerial mosaics for a proper mission planning, the UAV uses

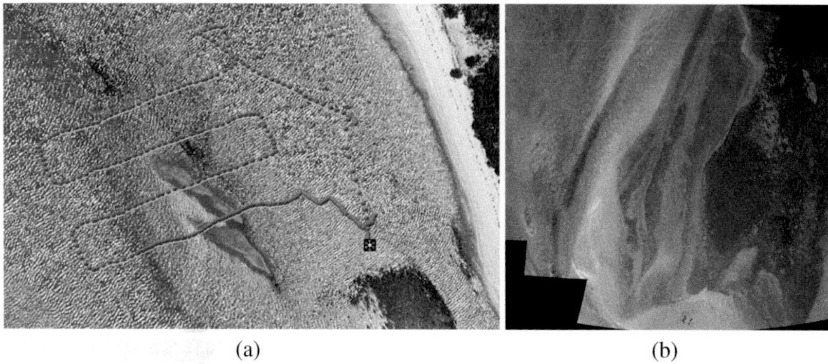

(a) (b)

Fig. 6 An aerial mosaic, built by the UAV, of an operations site nearby Samouco, in the south margin of Tagus' estuary bay. (a) Satellite imagery of the site with UAV's executed path overlaid. (b) The resulting aerial panorama.

a SJ4000 camera with diagonal 170° field of view mounted on an active gimbal. Communications with the UGV and the control centre are ensured by a 2.4 GHz Ubiquiti Picostation.

The main function of the UAV is to autonomously execute a line-sweep pattern so as to build an aerial mosaic of a designated operational area. As can be seen in Fig. 6, the mosaic's appearance differs greatly from the satellite imagery of the same site, which clearly shows the value of relying on up-to-date imagery taken by the UAV. The UAV takes a set of aerial images, which are then registered to each other to build up an aerial mosaic using the Hugin Panorama open-source software [16]. Then, this aerial mosaic is used to specify the sampling mission by the operator in the control centre. Currently, the user simply specifies the transects by visually analysing the mosaic. In the future, we expect to include automated terrain classification algorithms to aid the operator in this task. Once this initial process is completed, the UAV lands and waits for additional user-requested monitoring of the UGV's sampling activity.

5 Conclusions

A ground-aerial robotic team for soil and biota sample collection and retrieval in estuarine mudflats, was presented. To frame the robotic system in radionucleids and heavy metals environmental monitoring campaigns, a preliminary experimental protocol tailored to exploit the capabilities offered by the robotic system was also proposed. A set of preliminary field trials in Tagus' estuarine bay showed the ability of the robotic system prototype to navigate, extract samples, and retrieve them for subsequent offline analysis. We are currently preparing a full environmental monitoring campaign to validate the proposed experimental protocol and further assess the robustness and accuracy of the robotic system when facing the burdens of long field operations. As future work, the robotic system will be extended so as to include water surface unmanned vehicles for joint water/land sampling. The experimental protocol will also be adapted to account for the extended robotic team.

Acknowledgements This work was co-funded by ROBOSAMPLER project (LISBOA-01-0202-FEDER-024961). The authors wish to thank the fruitful comments provided by the anonymous reviewers.

References

1. Eggleton, J., Thomas, K.V.: A review of factors affecting the release and bioavailability of contaminants during sediment disturbance events. Environment International **30**(7), 973–980 (2004)
2. Murphy, R.R., Peschel, J., Arnett, C., Martin, D.: Projected needs for robot-assisted chemical, biological, radiological, or nuclear (cbrn) incidents. In: Proc. of the IEEE International Symposium on Safety, Security, and Rescue Robotics (SSRR), pp. 1–4. IEEE (2012)
3. Guzman, R., Navarro, R., Ferre, J., Moreno, M.: Rescuer: Development of a modular chemical, biological, radiological, and nuclear robot for intervention, sampling, and situation awareness. Journal of Field Robotics (2015). doi:10.1002/rob.21588
4. Winter, A.G., Deits, R.L., Dorsch, D.S., Hosoi, A.E., Slocum, A.H.: Teaching roboclam to dig: the design, testing, and genetic algorithm optimization of a biomimetic robot. In: Proc. of the IEEE/RSJ International Conference on Intelligent Robots and Systems (IROS), pp. 4231–4235. IEEE (2010)
5. Darukhanavala, C., Lycas, A., Mittal, A., Suresh, A.: Design of a bimodal self-burying robot. In: Proc. of the IEEE International Conference on Robotics and Automation (ICRA), pp. 5600–5605. IEEE (2013)
6. Larkum, A., Orth, R.J., Duarte, C. (eds.): Seagrasses: Biology, Ecology and Conservation. Springer (2006)
7. Carvalho, F.P., Oliveira, J.M., Silva, L., Malta, M.: Radioactivity of anthropogenic origin in the tejo estuary and need for improved waste management and environmental monitoring. International Journal of Environmental Studies **70**(6), 952–963 (2013)
8. Caetano, M., Madureira, M.J., Vale, C.: Metal remobilisation during resuspension of anoxic contaminated sediment: short-term laboratory study. Water, Air, and Soil Pollution **143**(1–4), 23–40 (2003)
9. ICRU: Sampling to estimate spatial pattern. Journal of the ICRU **6**(1), 49–64 (2006)
10. Quigley, M., Gerkey, B., Conley, K., Faust, J., Foote, T., Leibs, J., Berger, E., Wheeler, R., Ng, A.: Ros: an open-source robot operating system. In: Proc. of the ICRA Open-Source Software Workshop (2009)
11. Mace, J.: Rosbridge. http://wiki.ros.org/rosbridge_suite (2015) (accessed: September 07, 2015)
12. Marques, F., Santana, P., Guedes, M., Pinto, E., Lourenço, A., Barata, J.: Online self-reconfigurable robot navigation in heterogeneous environments. In: Proc. of the IEEE International Symposium on Industrial Electronics (ISIE), pp. 1–6. IEEE (2013)
13. Wurm, K.M., Hornung, A., Bennewitz, M., Stachniss, C., Burgard, W.: Octomap: a probabilistic, flexible, and compact 3d map representation for robotic systems. In: Proc. of the ICRA 2010 Workshop on Best Practice in 3D Perception and Modeling for Mobile Manipulation, vol. 2 (2010)
14. Gerkey, B.P., Konolige, K.: Planning and control in unstructured terrain. In: Proc. of the IEEE ICRA Workshop on Path Planning on Costmaps (2008)
15. Lourenço, A., Marques, F., Santana, P., Barata, J.: A volumetric representation for obstacle detection in vegetated terrain. In: Proc. of the IEEE Intl. Conf. on Robotics and Biomimetics (ROBIO). IEEE Press (2014)
16. D'Angelo, P.: Hugin-panorama photo stitcher. http://hugin.sourceforge.net (accessed: September 07, 2015)

Autonomous Seabed Inspection for Environmental Monitoring

Juan David Hernández, Klemen Istenic, Nuno Gracias, Rafael García, Pere Ridao and Marc Carreras

Abstract We present an approach for navigating in unknown environments, while gathering information for inspecting underwater structures using an autonomous underwater vehicle (AUV). To accomplish this, we first use our framework for mapping and planning collision-free paths online, which endows an AUV with the capability to autonomously acquire optical data in close proximity. With that information, we then propose a reconstruction framework to create a 3-dimensional (3D) geo-referenced photo-mosaic of the inspected area. These 3D mosaics are also of particular interest to other fields of study in marine sciences, since they can serve as base maps for environmental monitoring, thus allowing change detection of biological communities and their environment in the temporal scale. Finally, we evaluate our frameworks, independently, using the SPARUS-II, a torpedo-shaped AUV, conducting missions in real-world scenarios. We also assess our approach in a virtual environment that emulates a natural underwater milieu that requires the aforementioned capabilities.

Keywords Path planning · Mapping · Photo-mosaics · Online computation constraints · Monitoring · Underwater environments · AUV

J.D. Hernández(✉) · K. Istenic · N. Gracias · R. García · P. Ridao · M. Carreras
Underwater Vision and Robotics Research Center (CIRS), Computer Vision and Robotics Institute (VICOROB), University of Girona, C\Pic de Peguera, 13 (La Creueta), 17003 Girona, Spain
e-mail: {juandhv,ngracias}@eia.udg.edu,
{klemen.istenic,rafael.garcia,pere.ridao,marc.carreras}@udg.edu
http://cirs.udg.edu/

J.D. Hernández—This work was supported by the MORPH and ROBOCADEMY EU FP7-Projects under the Grant agreements FP7-ICT-2011-7-288704 and FP7-PEOPLE-2013-ITN-608096, respectively, and partially supported by the Colombian Government through its Predoctoral Grant Program offered by Colciencias.

L.P. Reis et al. (eds.), *Robot 2015: Second Iberian Robotics Conference*, Advances in Intelligent Systems and Computing 418,
DOI: 10.1007/978-3-319-27149-1_3

1 Introduction

During the last years, there has been a growing interest on monitoring different environmental variables that permits not only to have a better understanding of our planet, but also to estimate the human impact on it. This, consequently, has propelled research in various sectors of the scientific community, but especially in the development of robotic systems, which are capable of conducting autonomously data gathering tasks in different environmental media, such as air, soil and water [1]. In this latter, for instance, most of the underwater robotics applications include an autonomous underwater vehicle (AUV) that follows a sequence of pre-calculated waypoints in order to collect data. These underwater surveys are normally conducted in a previously explored area so that the vehicle navigates at a constant and safe altitude from the seafloor. In a typical application, the vehicle uses its onboard sensors, such as multibeam and imaging sonars, to gather information that is used to build bathymetric maps (elevation maps of the seabed). These maps can be used for safe surface and sub-surface navigation, thus enabling a variety of applications, such as underwater archeology [2].

More recent applications, on the other hand, require the AUV to navigate in close proximity to underwater structures and the seafloor. Such applications are especially dedicated to imaging and inspecting different kinds of structures such as underwater boulders [3] or confined natural structures (*e.g.,* underwater caves). In some of these cases, preliminary information about the structure to be inspected, such as its location and shape, permits determining in advance a region of interest, so that a coverage path is pre-calculated, while information obtained during the inspection is used to correct online the path in order to, for instance, adapt to the real structure's shape [3]. There is, nonetheless, a group of applications in which no previous information is available or cannot be obtained autonomously. In such cases, preliminary works have been focused on gathering data to characterize such environments.

On that basis, the purpose of this paper is twofold: first, we build up on our previous work [4] a framework to endow an AUV with the capability to navigate autonomously in unknown environments while exploring the ocean floor; secondly, to build a framework that allows the construction of 3-dimensional (3D) geo-referenced photo-mosaics of the seafloor. Therefore, the use of both frameworks represents

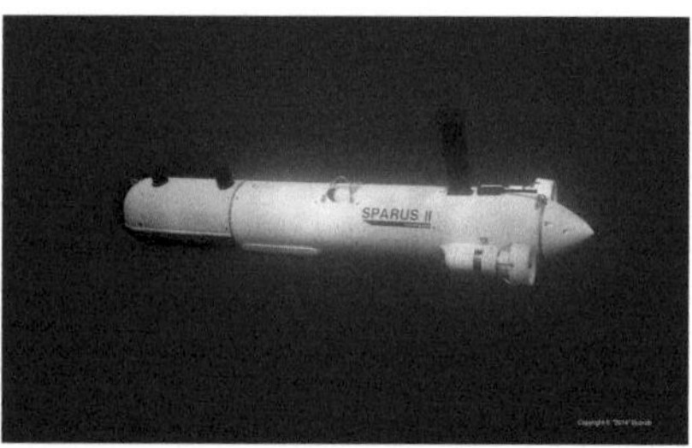

Fig. 1 SPARUS-II, a torpedo-shaped AUV.

the main contribution of this work, which endows an AUV with the capability to navigate autonomously an underwater milieu, while gathering optical information. Such information is used for building video-based 3D mosaics, which will serve as base maps for environmental monitoring of interest areas, allowing change detection of biological communities and their environment in the temporal scale, and enabling a new way to visualize the evolution of wide areas in that temporal scale. The planning and reconstruction frameworks have been tested with both synthetic and real-world scenarios using the SPARUS-II AUV (see Fig. 1).

2 Path Planning Framework

This section reviews our path planning framework for solving start-to-goal queries online for an AUV in an unknown environment (see Fig. 2) [4]. Additionally, we explain how to incorporate a criterion to maintain a desired distance to guarantee visibility constraints while conducting a mission in close proximity [5].

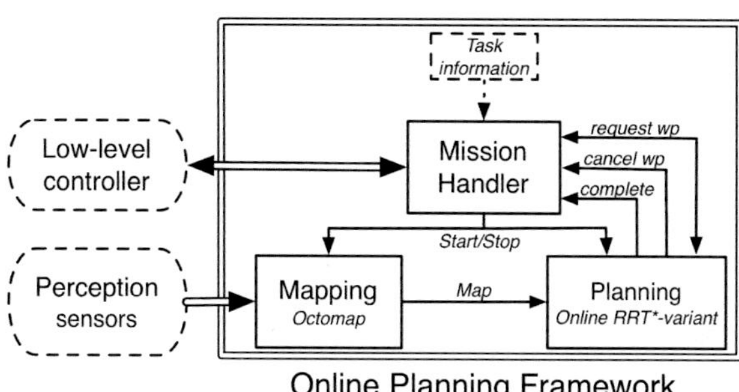

Fig. 2 Framework for online path planning with its three main functional modules and corresponding inputs and outputs [4].

2.1 *Mission Handler*

The *mission handler* controls the general flow of the proposed path planning framework. To ensure that the vehicle is prepared for solving and conducting a task, this module communicates with other vehicle's functional modules and verifies that navigation data is being received, perception sensors are providing valid data and vehicle's controllers are not conducting any low-level safety maneuver. When the mission has started, a bidirectional communication with the *planning* module is established. The *mission handler* requests waypoints, when required, and receives and adapts them for the vehicle's low-level controller. Additionally, the *planning* module can notify the *mission handler* to cancel ongoing waypoint requests.

2.2 *Mapping*

The *mapping* module incrementally builds a representation of the environment using information received from different perception sensors, such as multibeam or mechanically scanned profiling sonars, echosounders, etc. Such sensors provide range information about nearby obstacles and, combined with the vehicle navigation (position and orientation), defines the free and occupied space with respect an inertial coordinate frame. To process this information, we use an octree-based representation, named Octomap [6], which is a framework for modeling volumetric information with three main characteristics.

2.3 *Planning*

The *planning* module receives a query to be solved, which is specified as a start and goal vehicle configuration, as well as additional planning parameters, such as the available computing time, minimum distance to the goal, and boundaries of the workspace. This module contains a modified version of the RRT* [7], which, as any other RRT-variant, has as a main characteristic the rapid and efficient exploration of the configuration space (C-Space), but also includes the asymptotic optimality property. Finally, our modified version of the RRT* incorporates concepts of *anytime* algorithms and *lazy collision evaluation* that, together with its incremental nature, make it suitable for online (re)planning applications (see [4] for further details).

2.4 *Costmap for Surveying at a Desired Distance*

In addition to common challenges in images processing, visibility plays a critical role when gathering optical information in underwater environments. Having this in mind, we propose to define a costmap over the C-Space that attempts to meet the distance constraint required to guarantee visibility with respect to the inspected structure. This approach is based on our previous work, where the main objective was to maintain a desired altitude from the seafloor [5]. The cost associated to each configuration q, $0 \leq Cost \leq 100$, where 0 and 100 are the minimum and maximum cost respectively, is presented in Eq. 1. The Cost depends on the distance (d) and includes adjustable parameters. One of those parameters is the expected distance d_e to the inspected structure. Another is an admissible range of distance Δd_a, which permits to define an interval of distance in which the associate cost has its minimal value (clearly observed in Fig. 3). Finally, this cost function defines the optimization objective for the modified RRT* used in the planning module previously explained.

$$Cost(d) = \begin{cases} \left(1 - \frac{d}{d_e}\right) 100, & d < d_e - \frac{\Delta d_a}{2} \\ 0, & d_e - \frac{\Delta d_a}{2} \leq d \leq d_e + \frac{\Delta d_a}{2} \\ \left(\frac{d}{d_e} - 1\right) 100, & d > d_e + \frac{\Delta d_a}{2} \end{cases} \quad (1)$$

Nonetheless, it is important to remark that defining high-cost values to certain zones does not restrict or limit them as possible paths, but it will attempt to avoid them as much as possible. In other words, this approach is different than defining restricted areas, where the vehicle would not be allowed to move through. An example of these situations is when the only possible path coincides with the restricted zone (highest cost), in which our approach will permit that path as a valid solution.

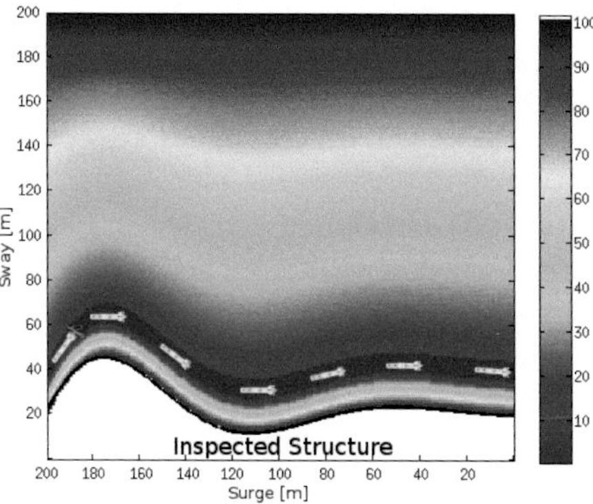

Fig. 3 Costmap projected in vehicle's X(surge)-Y(sway) plane. Dark blue indicates the zone that meets visibility constraints.

3 3-Dimensional (3D) Reconstruction Framework

While conducting an autonomous underwater mission, optical cameras continuously acquire images in order to gather information of the observed environment. Once the mission is completed, a reconstruction pipeline processes the images through a series of steps (see Fig. 4), which results is a textured 3D triangle mesh representation of the observed scene. While this is a preferable representation for visualization purposes, a smaller subset of successive steps can be conducted, if the subsequent usage of the results requires solely dense or even sparse point cloud representation, enabling the reduction in the computational cost. Different pipeline processing steps are described in the following subsections.

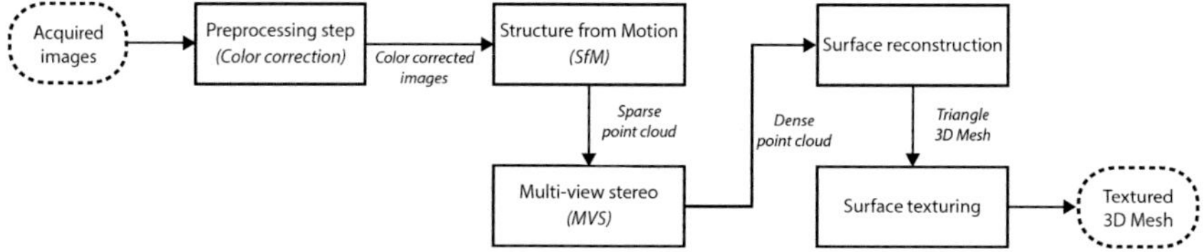

Fig. 4 3D reconstruction framework pipeline.

3.1 Color Correction

The acquisition of the optical images in underwater environments significantly differs from the conventional imagery, due to the specific properties of the medium. Light attenuation and scattering are the two main phenomena that strongly affect the image quality and consequently the acquisition task [8]. The progressive attenuation of the light combined with the various degrees of absorption at different wavelengths results in characteristic predominance of the blue and green color in farther objects (see Fig. 6a). As this phenomenon affects not only the ability to reconstruct a visually accurate representation of the scene, but also the quality and quantity of scale-invariant feature transform (SIFT) matchings [9] used in the subsequent steps, an automated preprocessing step is performed to ensure the color consistency. By equalizing the color contrast through contrast stretching of RGB (red, green, blue) channels, and subsequent saturation and intensity stretching of HSV (hue, saturation, value) the increment of true colors is achieved, as proposed by Iqbal *et al.* [10]. An example of a preprocessed image is shown in Fig. 6b.

3.2 Structure from Motion

Using the set of previously preprocessed images, the 3D information of the observed scene is obtained by the process known as Structure from Motion (SfM). The simultaneous estimate of the structure, simplified to 3D points, and the poses of the cameras, from which the images were acquired, is inferred entirely from the texture features extracted and matched across the image set.

In the proposed framework, an incremental implementation of SfM - VisualSfM [11, 12] is used, due to its efficient and reliable performance. While the algorithm enables the recovery of both extrinsic (*i.e.,* the position and orientation of the camera at the moment of the acquisition) and intrinsic (*i.e.,* focal length and radial distortions of the lens) cameras parameters, we use predetermined values for the intrinsics obtained through standard calibration procedure [13]. This significantly reduces the complexity of the problem and subsequently minimizes the possibility of wrong convergence in the process of optimization of the rest of the parameters. Additionally, all images are undistorted prior to their usage, as radial distortion introduces an additional ambiguity into SfM [14]. As a final result, an estimate of the camera poses and 3D structure in a form of sparse 3D point cloud are obtained.

3.3 Multi-view Stereo

To obtain a globally consistent dense representation of the scene, a window-based voting approach introduced by Goesele *et al.* [15] is used. Initially, the individual depth maps for each of the views are reconstructed using estimation of camera poses. This is followed by a merging step, which fuses the independent depth maps using a hierarchical signed distance fields approach of Fuhrmann *et al.* [16]. While the overlapping of certain views produces an excessive redundancy in storage and computation, the avoidance of averaging the geometry at different resolutions of the maps enables the estimation of highly detailed geometry in a form of dense 3D point cloud. Additionally, as only a small subset of neighbouring images are required for the computation of each depth map, the process is easily scalable to longer sequences and larger scenes.

3.4 Surface Reconstruction

Using a dense set of unorganized 3D points, the surface reconstruction process estimates the underlying surface in a form of a triangle mesh. As described by Kazhdan *et al.* in [17], the Poisson surface reconstruction, used in our approach, forms a unique implicit surface representation through the reconstruction of an indicator function. This is achieved by blending local contributions of implicit locally approximated functions at sets of points. As the approach assumes the normals to be known, these are computed using the connectivity information in the process of fusion of depth maps in the previous MVS step. Given the fact that Poisson method returns a closed surface, relevant information is subsequently extracted by eliminating the triangles with edges longer than certain threshold (triangles tend to increase size in non-sampled parts) [18].

3.5 Texturing

A vitally important step in providing the scientists with the realistic representation of the observed scene, is enhancing previously obtained 3D triangle mesh with a texture. Using the approach of Waechter *et al.* [19] realistic texture is computed by ensuring the photo consistency together with the minimization of the discontinuities and the visibility of seams between patches. Together with the previously described steps this enables the reconstruction of texturized 3D mesh representation of the observed scene.

4 Results

To validate our proposed approach, we evaluate both the path planning [4] and 3D reconstruction frameworks, independently, in a real-world scenario. Additionally, we propose and demonstrate our approach suitability in a virtual scenario that emulates a natural underwater environment, which requires navigating autonomously with no previous surroundings information.

4.1 Online Mapping and Path Planning in Unknown Environments

To evaluate our path planning framework, we used the SPARUS-II AUV (see Fig. 1), a torpedo-shaped vehicle with hovering capabilities, rated for depths up to $200m$. The robot has three thrusters (two horizontal and one vertical) and can be actuated in surge, heave and yaw degrees of freedom (DOF). The vehicle is equipped with a navigation sensor suite including a pressure sensor, a doppler velocity log (DVL), an inertial measurement unit (IMU) and a GPS to receive fixes while at surface. To perceive the environment, a set of five echosounders are located within the vehicle payload (front) area. Four of them are in the horizontal plane, three are separated by 45°, with the central one looking forward and parallel to the vehicle's direction of motion, while the fourth one is perpendicular to the central one. In order to assess the effectiveness of our approach, we used the external and open area of the harbour of Sant Feliu de Guíxols in Catalonia (Spain) as test scenario (see Fig. 5a), which is composed of a series of concrete blocks of $14.5m$ long and $12m$ width, separated by a four-meter gap with an average depth of $7m$.

In this scenario, the SPARUS-II AUV had to move amidst the concrete blocks without any previous knowledge of their location. All queries have been defined to conduct

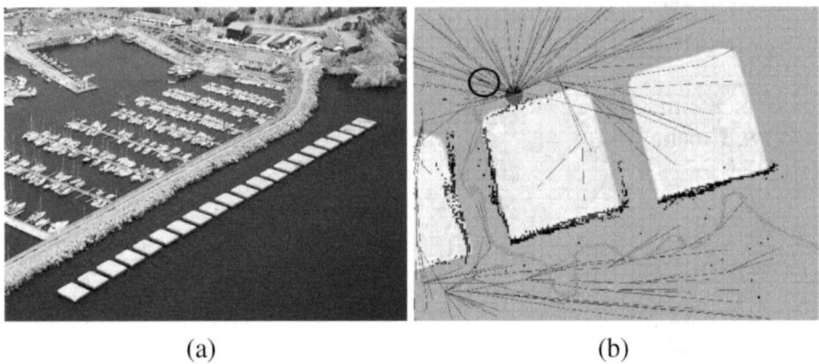

(a) (b)

Fig. 5 (a) Experiments scenario: a breakwater structure composed of concrete blocks in the harbor of Sant Feliu de Guíxols in Catalonia, Spain. (b) the vehicle has moved through the four-meter gap between the second and third concrete block and approaches to the last waypoint.

missions with a constant depth, since most of perception sensors (echosounders) are located to cover the horizontal plane, thus the motion is restricted to a 2-dimensional (2D) task.

4.2 3D Reconstruction

To evaluate the reconstructed 3D model, a real-world dataset of images obtained using the SPARUS-II AUV during a demonstration in Breaking the Surface (BtS) 2014 (Biograd Na Moru, Croatia)[1] has been used. In the demo, the SPARUS-II conducted an autonomous mission navigating at constant depth while gathering information with onboard sensors, including a pre-calibrated optical camera oriented downwards. Images were obtained at a frequency of $4Hz$ and recovered from the AUV after the mission for offline processing. This section presents the results of a 3D reconstruction using such images.

To illustrate the possibility of such approach, the results of processing an approximately a minute long sequence containing 213 images acquired while observing a geometrically diverse scene due to the presence of rocks and underwater vegetation are presented. An example of an acquired image is shown in Fig. 6a. The predominance of blue and green tones is evident, together with the low contrast of the colors. Using color correction step, as presented in Section 6a, the true colors are enhanced as can be observed in Fig. 6b.

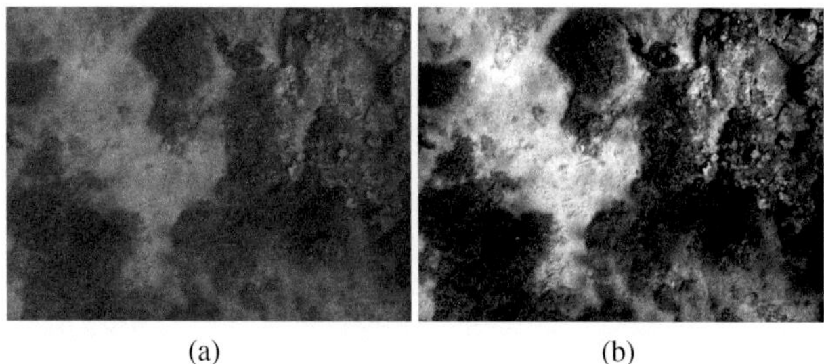

(a) (b)

Fig. 6 Example of an image acquired with a pre-calibrated optical camera oriented downwards (Biograd Na Moru, Croatia). (a) original image. (b) image after color correction.

Using the pipeline described in Section 3, the sequence of images was processed obtaining a textured 3D triangle mesh. While the data can be visualized in a standard top-down mosaic like perspective, as shown in Fig. 7a, the obtained reconstruction also permits observations from arbitrary user-defined poses, as shown with two examples in Fig. 7b.

---

[1] https://bts.fer.hr/

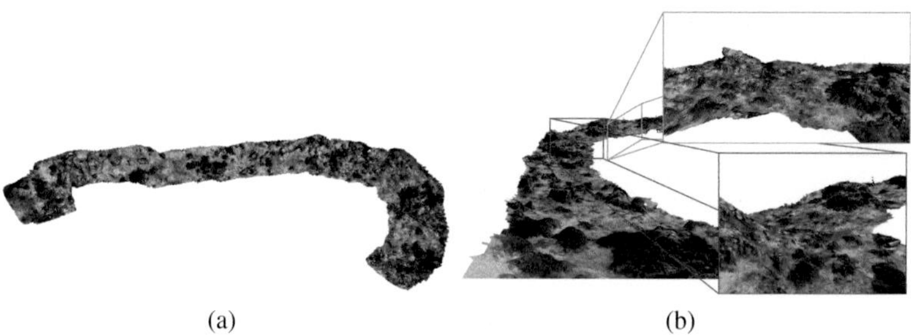

(a) (b)

Fig. 7 (a) Top-down view of the reconstructed scene. (b) Arbitrary user-defined views of reconstructed scene

As aforementioned, the reconstruction process can consist of a smaller subset of successive steps, enabling the user to reduce the computational time. The intermediate results (sparse/dense point cloud, triangle mesh and textured triangle mesh) are presented in Figs. 8a - 8d.

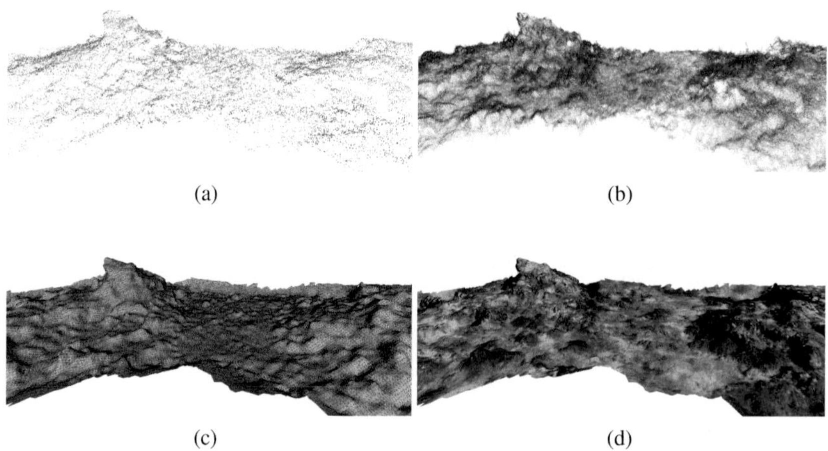

(a) (b)

(c) (d)

Fig. 8 Intermediate reconstruction results. (a) Sparse point cloud. (b) Dense point cloud. (c) Triangle mesh. (d) Textured triangle mesh.

It should be mentioned that given the fact that the reconstruction is performed solely based on the information from a set of acquired images, the area of interest should be imagined multiple times to ensure sufficient information and observation of areas which might be occluded in certain views.

4.3 *Navigating Autonomously through a Natural-like Environment*

SPARUS-II, as well as the other AUVs developed at the underwater vision and robotics research center (CIRS)[2], is controlled through the component oriented layer-based architecture for autonomy (COLA2) [20], a control architecture that is completely integrated with the robot operating system (ROS). Besides operating aboard real robots, COLA2 can interact with the underwater simulator (UWSim) [21], which can import 3D environment models and simulate the vehicle's sensors and dynamics with high fidelity. We used UWSim with a 3D model that emulates an underwater canyon, in which we tested our path planning framework while gathering optical images. Results validated the AUV capability to create a representation (map) of a complex and unknown environment that was used, simultaneously, to incrementally plan collision-free paths (see Fig. 9).

Fig. 9 SPARUS-II AUV in UWSim [21] conducting an autonomous mission. (a) virtual scenario that resembles an underwater canyon. (b) SPARUS-II navigates through the canyon. (c) intermediate states and resulting path of the RRT* solving the start-to-goal query. (d) SPARUS-II successfully completes the task.

[2] http://cirs.udg.edu/

5 Conclusions and Further Work

In this paper, we presented a new approach for inspecting autonomously underwater structures in close proximity to the seabed using AUVs. To do so, we proposed to use our framework for mapping and planning collision-free paths simultaneously and online, thus permitting an AUV to autonomously inspect an area of interest by acquiring optical images. Using that data, we then proposed a reconstruction framework to create 3D geo-referenced photo-mosaics, which can be used as base maps for environmental monitoring.

We validated, independently, both the path planning and 3D reconstruction frameworks in real-world scenarios. We also assessed our approach in a virtual scenario that emulates an underwater canyon, in which a simulated SPARUS-II AUV was able to navigate without previous knowledge of the surroundings while acquiring optical images. Finally, we will focus our immediate efforts to replicate this latter experiment, but now conducting the mission in an real-world underwater corridor-like (canyon) environment. Additionally, longer sequences will be used in the reconstruction process, to obtain the reconstruction of larger areas. This will permit us to validate our capability to reconstruct complex 3D environments.

References

1. Dunbabin, M., Marques, L.: Robots for Environmental Monitoring: Significant Advancements and Applications. IEEE Robot. Autom. Mag. **19**, 24–39 (2012)
2. Bingham, B., Foley, B., Singh, H., et al.: Robotic tools for deep water archaeology: Surveying an ancient shipwreck with an autonomous underwater vehicle. J. Field Robot. **27**, 702–717 (2010)
3. Galceran, E., Campos, R., Palomeras, N., et al.: Coverage Path Planning with Real-time Replanning and Surface Reconstruction for Inspection of Three-dimensional Underwater Structures using Autonomous Underwater Vehicles. J. Field Robot. (2014)
4. Hernández, J.D., Vidal, E., Vallicrosa, G., Galceran, E., et al.: Online path planning for autonomous underwater vehicles in unknown environments. In: IEEE Int. Conf. Robot. (ICRA), Seattle, pp. 1152–1157 (2015)
5. Hernández, J.D., Vallicrosa, G., Vidal, E., et al.: On-line 3D path planning for close-proximity surveying with AUVs. In: IFAC NGCUV, Girona (2015)
6. Hornung, A., Wurm, K.M., Bennewitz, M., et al.: OctoMap: an efficient probabilistic 3D mapping framework based on octrees. Auton. Robot. **34**, 189–206 (2013)
7. Karaman, S., Frazzoli, E.: Sampling-based Algorithms for Optimal Motion Planning. Int. J. Robot. Res. **30**, 846–894 (2011)
8. Garcia, R., Nicosevici, T., Cufí, X.: On the way to solve lighting problems in underwater imaging. In: MTS/IEEE OCEANS, vol. 2, pp. 1018–1024 (2002)
9. Pramunendar, R.A., Shidik, G.F., Supriyanto, C., et al.: Auto level color correction for underwater image matching optimization. Jurnal Informatika (2014)
10. Iqbal, K., Abdul Salam, R., Osman, M., et al.: Underwater Image Enhancement Using An Integrated Colour Model. IAENG Int. J. Comp. Sci. **32**, 239–244 (2007)
11. Wu, C.: Towards linear-time incremental structure from motion. In: 2013 International Conference on 3D Vision-3DV, pp. 127–134 (2013)

12. Wu, C., Agarwal, S., Curless, B., et al.: Multicore bundle adjustment. In: IEEE Conf. Comp. Vis. Patt. Recog. (CVPR), pp. 3057–3064 (2011)
13. Bouguet, J.-Y.: Camera Calibration Toolbox for Matlab
14. Wu, C.: Critical configurations for radial distortion self-calibration. In: IEEE Conf. Comp. Vis. Patt. Recog. (CVPR), pp. 25–32 (2014)
15. Goesele, M., Snavely, N., Curless, B., et al.: Multi-view stereo for community photo collections. In: IEEE Int. Conf. Computer. Vision. (ICCV) (2007)
16. Fuhrmann, A., Goesele, M.: Floating scale surface reconstruction. ACM Transactions on Graphics (TOG) **33**, 46 (2014)
17. Kazhdan, M., Hoppe, H.: Screened poisson surface reconstruction. ACM Transactions on Graphics (TOG) **32**, 29 (2013)
18. Campos, R., Garcia, R., Nicosevici, T.: Surface reconstruction methods for the recovery of 3D models from underwater interest areas. In: IEEE OCEANS (2011)
19. Waechter, M., Moehrle, N., Goesele, M.: Let there be color! large-scale texturing of 3D reconstructions. In: Computer Vision–ECCV, pp. 836–850 (2014)
20. Palomeras, N., El-Fakdi, A., Carreras, M., et al.: COLA2: A Control Architecture for AUVs. IEEE Journal of Oceanic Engineering **37**, 695–716 (2012)
21. Prats, M., Perez, J., Fernandez, J.J., et al.: An open source tool for simulation and supervision of underwater intervention missions. In: IEEE/RSJ Int. Conf. Intel. Robot. Syst. (IROS), pp. 2577–2582 (2012)

Integrating Autonomous Aerial Scouting with Autonomous Ground Actuation to Reduce Chemical Pollution on Crop Soil

Jesús Conesa-Muñoz, João Valente, Jaime del Cerro, Antonio Barrientos and Ángela Ribeiro

Abstract Many environmental problems cover large areas, often in rough terrain constrained by natural obstacles, which makes intervention difficult. New technologies, such as unmanned aerial units, may help to address this issue. Due to their suitability to access and easily cover large areas, unmanned aerial units may be used to inspect the terrain and make a first assessment of the affected areas; however, these platforms do not currently have the capability to implement intervention.

This paper proposes integrating autonomous aerial inspection with ground intervention to address environmental problems. Aerial units may be used to easily obtain relevant data about the environment, and ground units may use this information to perform the intervention more efficiently.

Furthermore, an overall system to manage these combined missions, composed of aerial inspections and ground interventions performed by autonomous robots, is proposed and implemented.

The approach was tested on an agricultural scenario, in which the weeds in a crop had to be killed by spraying herbicide on them. The scenario was addressed using a real mixed fleet composed of drones and tractors. The drones were used to

---

J. Conesa-Muñoz · Á. Ribeiro(✉)
Centre for Automation and Robotics, CSIC-UPM, Arganda del Rey, 28500 Madrid, Spain
e-mail: {jesus.conesa,angela.ribeiro}@csic.es

J. Valente · J. del Cerro · A. Barrientos
Centre for Automation and Robotics, UPM-CSIC,
José Gutierrez Abascal 2, 28006 Madrid, Spain
e-mail: {joao.valente,j.cerro,antonio.barrientos}@upm.es, jvalente@ing.uc3m.es

J. Valente
Carlos III University, Av. Universidad 30, 28911 Leganés, Spain

L.P. Reis et al. (eds.), *Robot 2015: Second Iberian Robotics Conference*,
Advances in Intelligent Systems and Computing 418,
DOI: 10.1007/978-3-319-27149-1_4

inspect the field and to detect weeds and to provide the tractors the exact coordinates to only spray the weeds. This aerial and ground mission collaboration may save a large amount of herbicide and hence significantly reduce the environmental pollution and the treatment cost, considering the results of several research works that conclude that actual extensive crops are affected by less than a 40% of weed in the worst cases.

Keywords Collaborative inspection and intervention mission · Aerial and ground fleet · Autonomous fleet · Site-specific weed treatment · Precision agriculture

1 Introduction

Many environmental problems require surveillance or scouting stages previous to the intervention phase that alleviates or solves the problem. Many cases require coverage of large areas, often in rough terrain constrained by natural obstacles, which makes continuous inspections difficult. New technologies, such as unmanned aerial units, may help in this issue due to their suitability to access and easily cover large surfaces. Thus, environmental actuation can be split into two stages: aerial inspection with drones and ground intervention with typically more powerful platforms. The aerial inspection may provide a quick and easy assessment of the affected areas to be used for ground intervention to implement the work more efficiently. The proper integration of aerial and ground units would make the use of the current autonomous robots more efficient for treating environmental disasters, such as oil spills [1], forest fires [2] or earthquakes [3]. Such integration could be applied even in agriculture, where some agricultural tasks, such as weed treatment, might be accomplished by ground units only in the affected zones by following a weed distribution map obtained from the information provided by the aerial units. This site-specific weed management has clear environmental benefices, mainly in extensive crops where research work reported weed infestations around the 40% in the worst cases [4,5]. In other words, more than a 60% of herbicide could be potentially saved with the proper technology.

In many contexts, inspection and actuation would be greatly enhanced if performed by autonomous robots and, in particular, for large areas, with fleets of autonomous robots. Moreover, the entire work to be accomplished by the fleet would be more efficient if the autonomy of the whole system was complete, i.e., the fleet of aerial and ground autonomous robots works together without human intervention, which would only be in charge of supervising the work of the fleet. In the following sections, a system designed and developed to accurately treat weeds in field crops with herbicides is described.

In the agricultural context, herbicide application is an important economic and environmental issue. Herbicides are chemical products used to control unwanted plants (weeds) interfering with crops. EU countries used approximately 135,000 tonnes of herbicides in 2007 [6]. These products make a significant contribution to maintaining food production; according to [7], each euro (€) invested in herbicides (and pesticides in general) returns 4 euros in crops saved. Considering that the total sales of herbicides in Europe is currently approximately 3,390 million € per year [8], we can estimate that, in Europe, pesticides may provide over 13,500 million € per year in saved crops. However, such assessments do not consider the indirect, but substantial, environmental and economic costs associated with herbicide use. For example, it has been estimated that only 5% of herbicides reach the target weeds [9], whereas the bulk of each application (over 95%) is left to impact the surrounding environment. The economic value of pesticide environmental impact has been estimated to total approximately 8,000 million $ per year in the USA [7], and approximately 50% of pesticide usage consists of herbicide treatments.

To mitigate the abusive use of herbicides and the consequent chemical pollution on crop soils, precision agriculture was developed as a more environmentally careful way to manage fields. Precision agriculture is the application of technologies and principles to manage the spatial and temporal variability associated with all aspects of agricultural production for the purpose of improving crop performance and environmental quality [10]. In this context, aerial inspection missions may be used to easily acquire the variability in fields (that is, the distributions of the crop, weeds, insects, humidity, and soil fertility), and farmers may use these data to work selectively on the fields (also known as site-specific treatments), significantly decreasing the use of agrochemical products (herbicides, insecticides and fertilizers), which are highly dangerous for the environment. There are several studies devoted to crop inspection by analyzing and processing aerial images, for example, to detect weeds [11], and there are works devoted to developing tools [12] and site-specific treatments based on previously acquired knowledge [13,14]. Nevertheless, only the RHEA project [15], in which this work is framed, has linked the two steps to completely automate the site-specific herbicide treatments. To achieve this goal, this paper uses an autonomous and heterogeneous fleet to implement the entire process autonomously and accurately. The inspection step is accomplished by an aerial team composed of 2 drones and a treatment step using a ground team composed of 3 medium autonomous tractors.

The use of a collaborative heterogeneous fleet for selective treatments is a novel approach that presents several advantages. The benefits of this solution over the conventional large vehicles equipped with many different actuators and sensors arise from different facts summarized in Table 1.

In the following sections, the architecture of the overall system (Mission Manager) designed and developed to integrate aerial scouting missions with ground treatment missions is explained. The employed robot platforms used to implement site-specific weed treatments are described. Finally, the results section explains how the overall system, fleet and implemented Mission Manager, performed an accurate selective treatment in a real crop in an autonomous way.

Table 1 Advantages of using a fleet of small/medium sized robots over one large agricultural vehicle

	Traditional big machine	**A fleet of small/medium robots**
Safety in autonomous operation mode	Becomes a safety problem in case of failure	Small/medium sized robots can interact with humans in a safer way
Fault impact on mission completion	A failure will stop the entire mission until the machine is repaired	Robot teams allow for mission re-planning in case of failure of one vehicle
Impact on the field	High damage by soil compaction	Lower compaction (lighter vehicles) and more precision movements (farming at plant level)
Personnel	An operator for each vehicle	An operator can supervise several vehicles

2 Mission Manager Architecture

In general, even if the robotic platforms used are autonomous, software is required to manage the entire process, that is, an overall system to generate the directions for the units to follow to accomplish their missions, to send them to the platforms,

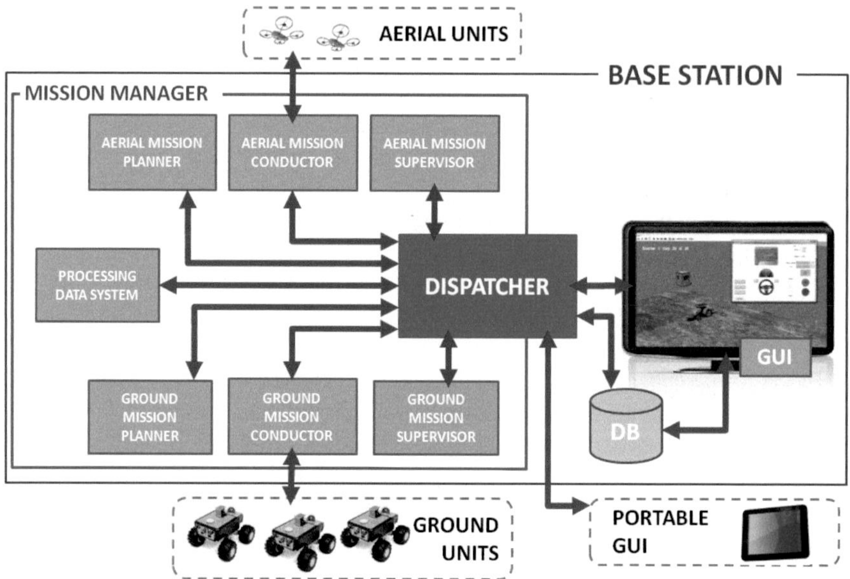

Fig. 1 Architecture of the Mission Manager and its connections with external elements/systems

to coordinate the fleet (the vehicles may interfere with each other), to supervise the fleet while working, to report failures to the operator in charge of the fleet, and to process the data acquired by the inspection missions.

Fig. 1 shows the generic architecture that the proposed system, called hereinafter Mission Manager, should have.

Thus, the generic Mission Manager is composed of the following modules:

- An aerial and a ground mission planner: To generate the plans that the units have to follow to complete the missions. They cannot be unified in a single planner because of the inherent differences of the aerial and ground units, as well as the different characteristics of their missions (surveillance/scouting and intervention/treatment).
- An aerial and a ground conductor: To automate the mission at the fleet level. Although the units, considered separately, may be autonomous, the fleet has to be coordinated, for instance, to launch/pause/resume/stop the mission for all members simultaneously. Additionally, these conductors are in charge of decoding the calculated plans (and transmitting them) to the exact commands supported by the units.
- An aerial and a ground mission supervisor: To monitor and corroborate that the missions are executed according to the generated plans. Because the units work in an uncontrolled environment, subject to unpredictable conditions (wind, light, terrain roughness, animals that may suddenly appear, etc.), there may be differences between the planned mission and the execution, for example, small deviations in the trajectories and speed due to wind or the terrain. Once deviations are detected, the supervisors report to the operator by issuing alarms that may be displayed on a GUI (graphical user interface).
- A processing data system: To receive and analyze the raw data acquired by the scouting mission to extract knowledge to be used in the intervention mission. For instance, in an agricultural mission, this module may consist of a mapping system to process the images taken by the aerial units and to detect and obtain the exact coordinates of the weed patches within the field.
- A dispatcher: To manage the workflow required to complete the entire process. To do this, the dispatcher encapsulates the connections to all the modules included into the Mission Manager and redirects the process to the appropriate modules when required. Moreover, it gathers and processes and redirects the queries (plans, executions, pauses, resumes, and aborts) from the external systems (GUI) if the operator wants to actively control the workflow.
  This component is particularly important because it allows the connection of new modules to the Mission Manager in order to support new functionalities.

In addition to the Mission Manager internal modules, there are some external systems that may interact with it.

- GUI (Graphical User Interface): Allows the operator to access the Mission Manager. The GUI also displays all the information generated by the Mission

Manager (plans, execution states, alarms, etc.) and guides the operator through the different workflow steps.

- Portable GUI: The Mission Manager is intended to be run on a computer hosted in a base station (a cabin with some antennas and a router to create a Wi-Fi network to access the units) next to the affected area. Thus, for those situations in which a breakdown forces the operator to move to the units, it is useful to have a portable GUI to control a particular unit of the fleet outside the cabin.
- Database: Allows register data about the mission, such as plans or the acquired data, to interrupt and resume the process, or even to process offline when the units are not working (for example, the case of processing images or any other big data acquired during the inspection).

3 Fleet Robots

In this section, the available fleet of robots used in the former results section is described. The fleet used is the fleet of the European project RHEA [15].

3.1 Aerial Fleet

The aerial fleet was composed of two six-rotor drones (AR200 model), developed by the AirRobot company [16]. Each one was able to carry a sensor-payload up to 1.5 kg with a fly autonomy of around approximately 40 minutes. Six-rotor units were used to provide certain safety redundancy in case of failure in one motor.

The drones were equipped with two cameras, visible and near infrared spectrum (two Sigma DP2 Merril models, one of them modified to record NIR images), mounted on a gimbal system (see Fig. 2) to reduce vibrations and to allow the cameras to point down when the drones perform steady flights.

The drones accept plans mainly composed of a list of ordered way-points where the drone has to take a picture, and the drones then autonomously fly to the way points.

Drones are able to provide telemetry information during the flight, including information required for supervision, such as position estimation and battery level. After finishing the mission, the drones return to their home points.

Fig. 2 AR200 drone in flight with a detail of the camera mounting

3.2 Ground Fleet

The ground fleet was composed of three medium tractors (see Fig. 3), based on a restructured New Holland Boomer 3050 (50 hp, 1270 kg) [17], in which the cabin was reduced to mount some of the computer equipment required for the perception, actuation, location, communication and safety systems.

Several sensor systems, such as an RTK-GPS receiver, an RGB camera and a LiDAR, allow autonomous and safe navigation.

The RTK-GPS receiver, a Trimble BX982 model, is a multi-channel, multi-frequency OEM GNSS receiver that enables OEM and system integrators to rapidly integrate centimeter-level positioning. The receiver supports two antennas connected in such way that the independent observations from both antennas are passed to the processor, where multi-constellation RTK baselines are computed and compared with the positions provided by both antennas. Because the real physical distance and their positions on the vehicle are known, it is possible to calculate the vehicle's heading with high accuracy. Therefore, a single connection to the tractor receiver (via RS232, USB, Ethernet or CAN) delivers both centimeter-accuracy positions and a heading that is accurate to less than a tenth of a degree (2 m baseline). In this manner, both the position and heading of the vehicles are provided with high precision at a maximum frequency of 20 Hz.

The camera onboard each tractor is an SVS4050CFLGEA model from SVS-VISTEK (Seefeld, Germany) with a CCD Kodak KAI 04050M/C sensor and a GR Bayer color filter, which provides high-resolution images (2,336 by 1,752 pixels with a 5.5 by 5.5 μm pixel size) to accurately determine in real time the locations of the weeds, obstacles and crop lines. The camera was placed inside a housing unit with a fan controlled by a thermostat for cooling purposes, which allows it to work even when it is raining or when the temperature is above 50 °C. The description of how the camera detects weed and crop rows (appropriate strategy for wide-row crops, such as maize) is out of the scope of this paper. Actually, the considered scenario only takes into account the weed detection by remote sensing, since it is the proper example to illustrate the integration of the whole elements of the fleet, in other words the scouting mission with the intervention mission.

The LiDAR sensor, an LMS 111 (SICK AG, Waldkirch, Germany), was installed in the middle of the vehicle's front with a push–broom configuration (4° inclination) and was used to detect obstacles along the vehicle trajectory with a ground clearance of 70 cm.

To perform the treatment, the tractor was equipped with a selective sprayer bar developed by Agrosap [12]. This tool is a 6-m spray boom with 12 nozzles, which can be independently activated, and 2 tanks, one to store water (200 L) and the other, smaller tank to store the herbicide. The sprayer is equipped with a direct injection system that mixes the agrochemical product and water just when a single or several nozzles are opened, which reduces herbicide waste.

Fig. 3 Ground unit

Finally, the tractors are equipped with an on-board computer, a CompactRIO model 9082 from National Instruments (Austin, TX, USA), which runs the internal control system that manages the sensors and actuators and allows remote control of the unit. Similarly to the aerial units, this internal controller allows autonomous execution of some remote commands: move, pause, resume, and stop, including performing a treatment plan. The plan is mainly composed of a list of the way-points the vehicles have to cover and also contains the states for the spraying bar (that is, the nozzles that must be opened and closed) for each point and other mission parameters, such as the speeds. More details about the ground units and their capabilities, such as navigation and control techniques utilized, can be found in [18].

4 Results

To test the complete set of steps implemented in the Mission Manager, a winter cereal field was prepared containing weed patches. The idea was to autonomously and sequentially execute all the steps required to perform a site-specific herbicide treatment via the Mission Manager running on a computer placed in a cabin situated next to the field and using the aerial and ground units presented in Section 3. The field was located in the experimental CSIC farm "La Poveda" [40°18′51.102″N, 3°29′03.379″W] in Arganda del Rey. The field was 2,400 m^2 and was treated using a pre-emergency herbicide, except for nine 3 m x 3 m square areas (see Fig. 4), where some weeds (*Sinapis arvensis*) were seeded.

Fig. 4 Winter cereal field prepared to contain nine weed patches

The field contour (yellow rectangle on Fig. 5) was acquired using a GPS and was stored in the database. Via the GUI, an aerial scouting plan was requested. The aerial planner automatically built a safety border (green contour on Fig. 5) expanding some margins of the field contour (this border cannot be exceeded by the drones) and calculated (from the contour of the field, the flight attitude, the resolution and the size of the images provided by the cameras) the way-points where the drones need to take images to sample the entire field. This information is used later to create the weed distribution map. The obtained routes for each drone are the red and blue lines represented in Fig. 5a.

Once planned, the aerial mission conductor requested the launch of the scouting mission, and the operator in charge of supervision approved the start of the mission. The plans were automatically loaded into the units, and the supervisor pilot was asked (via GUI) to approve take-off (required by the Spanish drone regulation) until the initial attitude specified in the plan was reached. Then, the drones executed the inspection following the trajectories shown in Fig. 5b. The aerial supervisor (module that is part of the Mission Manager) monitored the mission, and non-failures were detected during the execution of the aerial mission.

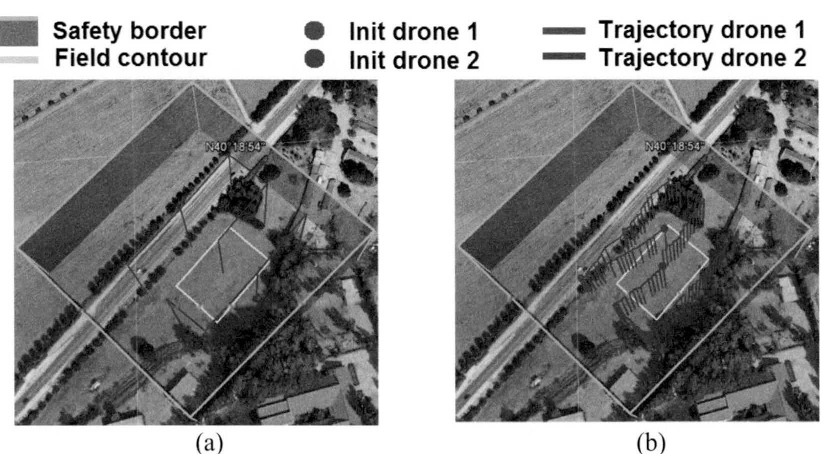

Fig. 5 Inspection aerial mission: a) planned trajectories and b) real trajectories

Once the drone's plans were completed, the Mission Manager requested a landing maneuver from the supervisor pilot and, once on the ground, the cameras' cards with the images were manually removed and inserted into the computer running the Mission Manager. The processing data system was invoked by the Dispatcher (module that is part of the Mission Manager). The processing data, in this case, consists of a weed detection system composed by a mosaicking system [19] and mapping module [11] developed by the IRSTEA and IAS-CSIC groups, respectively. The system outputs a weed distribution map, which is used by the ground Planner to develop the plan for treatment. Unfortunately, the weeds did not grow as expected and did not have the shape of the expected patches (see Fig. 6a). Consequently, the obtained distribution map, although it contained the real patch shapes, did not have the expected squares, making it difficult to determine whether the herbicide was sprayed on the appropriate areas. For this reason, the expected map was built artificially (Fig. 7) and was used to generate the treatment plan. The trajectories were optimized to reduce fuel, so the planner decided to use only one tractor.

The expected patches were covered with paper (Fig. 6b). A total of five paper strips were used in each patch, arranged in parallel and spaced 1 meter (three strips inside and two outside of the patch), for measuring the on/off time lag and therefore the percentage of the target area sprayed and not sprayed by water mixed with colorant. Then the treatment mission was executed. The ground supervisor (module that is part of the Mission Manager) monitored the mission, and non-failures were detected during the execution of the treatment mission; in fact the real trajectories were nearly the same (deviations of less than 7 cm) as the planned trajectories (Fig. 7). Moreover, the sprayed surface accurately matched the weeds, in fact the results showed that the spraying operation successfully sprayed more than 97% of the target area (i.e., weed patches) without any spraying in non-target areas (i.e., weed-free areas).

Only six of the nine patches were covered because the right part of the field was reserved for intermediate tests. The entire test can be played on [20] as part of an RHEA project demo.

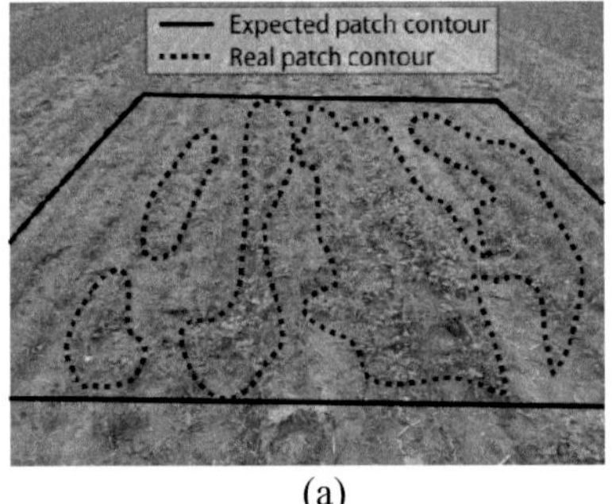

(a)

(b)

Fig. 6 a) Real patch vs. expected patch y b) paper strips along the field

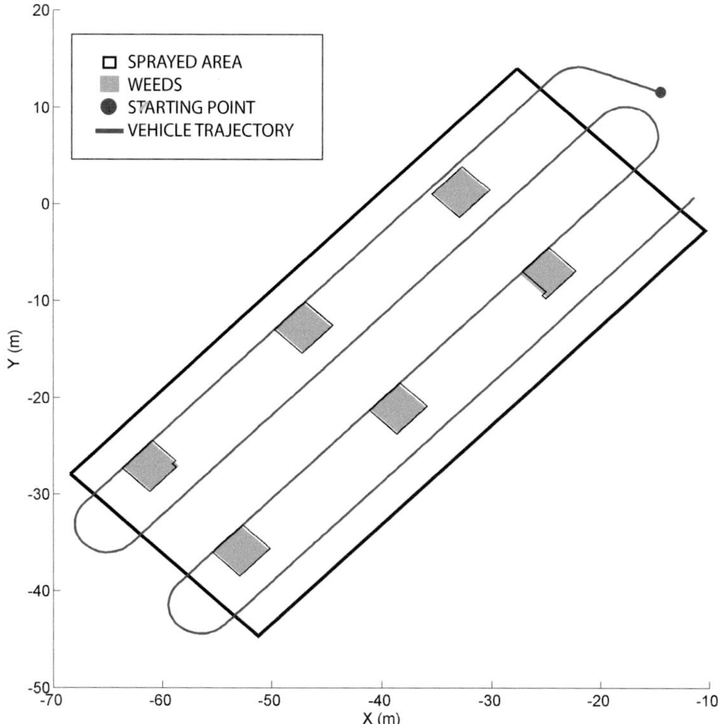

Fig. 7 Ground mission trajectory and sprayed surface

5 Conclusions and Future Work

An approach to properly combine autonomous aerial inspection and autonomous ground intervention missions to address environmental problems was proposed. The approach involves a Mission Manager that allows a single operator to supervise the entire process and manage the workflow required to autonomously complete a mission.

The proposed system was tested by performing a real site-specific weed treatment, in which the scouting mission was used to acquire the data to detect the weed patches positions that allowed the intervention treating only the infested areas and, consequently, reducing the cost of the treatment and the chemical pollution.

All the steps needed to achieve the site-specific weed treatment as well as the management of the workflow required to complete the entire process were entirely automated. Human intervention was only required to launch/land the aerial units (due to the current Spanish regulations) and to input the aerial images into the base station computer, since the camera characteristics did not allow the real-time output of high quality images directly to the computer during the acquisition step.

In future, work image transmission and human intervention will be automated, and the new approach will be tested for other types of applications.

Acknowledgements This project was funded in part by the 7[Th] Framework Programme of the European Union under Grant Agreement No. 245986. The authors recognise the following RHEA beneficiaries: CSIC (Spain), CogVis (Austria), FTW (Austria), Cyberbotics (Switzerland), University of Pisa (Italy), University Complutense of Madrid (Spain), Tropical (Greece), AGROSAP (Spain), Polytechnic University of Madrid (Spain), AirRobot (Germany), University of Florence (Italy), IRSTEA (France), CNH (Belgium), Bluebotics (Switzerland) and CM (Italy).

References

1. Gaudin, S.: MIT builds swimming, oil-eating robots. Computerworld, August 26, 2010. http://www.computerworld.com/article/2514966/emerging-technology/mit-builds-swimming–oil-eating-robots.html
2. Casbeer, D.W., Beard, R.W., McLain, T.W., Li, S.M., Mehra, R.K.: Forest fire monitoring with multiple small UAVs. In: American Control Conference 2005 (2005)
3. Nathan, M., Shaojie, S., Kartik, M., Yash, M., Vijay, K., Keiji, N., Yoshito, O., Seiga, K., Kazuki, O., Kazuya, Y., Kazunori, O., Eijiro, T., Satoshi, T.: Collaborative mapping of an earthquake-damaged building via ground and aerial robots. Journal of Field Robotics **29**(5), 832–841 (2012)
4. Marshall, E.J.P.: Field-scale estimates of grass weed populations in arable land. Weed Research **28**(3), 191–198 (1988)
5. Johnson, G.A., Mortensen, D.A., Martin, A.R.: A simulation of herbicide use based on weed spatial distribution. Weed Research **35**(3), 197–205 (1995)
6. ECPA. Report Annual Review 2007 (2008). http://www.ecpa.be/files/ecpa/documentslive/22/18192_ECPA%202008%20Annual%20report.pdf
7. Pimentel, D., Acquay, H., Biltonen, M., Rice, P., Silva, M., Nelson, J., Lipner, V., Giordano, S., Horowitz, A., D'Amore, M.: Environmental and economic costs of pesticide use. BioScience **42**, 750–758 (1992)
8. ECPA. Report Annual Review 2011 (2012). http://www.ecpa.eu/files/attachments/AnnualReport_web.pdf
9. Miller G.T.: Sustaining the Earth, 6th edn. Thompson Learning, Inc. Pacific Grove, California. Chapter 9, pp. 211–216 (2004)
10. Pierce, F.J., Nowak, P.: Aspects of Precision Agriculture. Advances in Agronomy **67**, 1–85 (1999)
11. Torres-Sánchez, J., López-Granados, F., De Castro, A.I., Peña-Barragán, J.M.: Configuration and specifications of an unmanned aerial vehicle (UAV) for early site specific weed management. PLoS ONE **8**(3) (2013)
12. Carballido, J., Perez-Ruiz, M., Gliever, C., Agüera, J.: Design, development and lab evaluation of a weed control sprayer to be used in robotic systems. In: First International Conference on Robotic and Associated High-Technologies and Equipment for Agriculture, Pisa, vol. 1, pp. 23–29 (2012)
13. Gerhards, R.: Managing weeds with respect to their spatial and temporal heterogeneity. In: 2nd Conference on Precision Crop Protection (2007)
14. Ruiz, D., Escribano, C., Fernandez-Quintanilla, C.: Assessing the opportunity for site-specific management of Avena sterilis in winter barley fields in Spain. Weed Research **46**, 379–387 (2006)
15. RHEA project Website. http://www.rhea-project.eu

16. AirRobot Website. http://www.airrobot.de
17. Boomer 3050 Tractor Website. http://agriculture1.newholland.com/nar/en-us/equipment/products/tractors-telehandlers/boomer-3000-series/models
18. Emmi, L., Gonzalez-de-Soto, M., Pajares, G., Gonzalez-de-Santos, P.: Integrating Sensory/Actuation Systems in Agricultural Vehicles. Sensors **14**, 4014–4049 (2014)
19. Rabatel, G., Labbé, S.: A fully automatized processing chain for high-resolution multispectral image acquisition of crop parcels by UAV. Precision agriculture **15**(1), 135–141 (2015)
20. RHEA demo. http://www.rhea-project.eu/img/Videos/DEMO_short_version_01.mp4

Part II
Future Industrial Robotics Systems

Force-Sensorless Friction and Gravity Compensation for Robots

Santiago Morante, Juan G. Victores, Santiago Martínez and Carlos Balaguer

Abstract In this paper we present two controllers for robots that combine terms for the compensation of gravity forces, and the forces of friction of motors and gearboxes. The Low-Friction Zero-Gravity controller allows a guidance of the robot without effort, allowing small friction forces to reduce the free robot motion. It can serve to aid users providing kinesthetic demonstrations while programming by demonstration. In the present, kinesthetic demonstrations are usually aided by pure gravity compensators, and users must deal with friction. A Zero-Friction Zero-Gravity controller results in free movements, as if the robot were moving without friction or gravity influence. Ideally, only inertia drives the movements when zeroing the forces of friction and gravity. Coriolis and centrifugal forces are depreciated. The developed controllers have been tuned and tested for 1 DoF of a full-sized humanoid robot arm.

Keywords Robotics · Control · Dynamics · Friction · Force · Humanoid

1 Introduction

Most robots are hard and heavy. They are built with metallic mechanical links and electric/hydraulic motors attached to heavy gearboxes that introduce high frictions. This fact makes it very difficult to physically interact with the robot. With the advent of paradigms such as Programming by Demonstration (PbD) [1], where physical movements are used to program the robot, there has been an increasing necessity to improve the existing physical interaction mechanisms.

We have developed two different types of controllers for robots which combine gravity compensation and motor friction compensation. Our motivation is to study

S. Morante(✉) · J.G. Victores · S. Martínez · C. Balaguer
Robotics Lab Research Group Within the Department of Systems Engineering and Automation, Universidad Carlos III de Madrid (UC3M), Madrid, Spain
e-mail: smorante@ing.uc3m.es

L.P. Reis et al. (eds.), *Robot 2015: Second Iberian Robotics Conference*,
Advances in Intelligent Systems and Computing 418,
DOI: 10.1007/978-3-319-27149-1_5

new forms of physical human-robot interaction. In kinesthetic teaching, a popular choice in PbD, the robot's motors are set to a passive mode where each limb can be driven by the human demonstrator [2]. Some authors suggest that kinesthetic demonstrations are more intuitive for naive users, but that this fact changes when facing with high degree of freedom (DoF) robots [3]. They present an alternative, called keyframe demonstration, where key positions of the task are recorded, while the intermediate movements are interpolated. For instance, Baxter robot uses this technique of recording frames to be programmed, aided by gravity compensation [4]. While gravity compensation is useful for providing kinesthesic demonstrations, one of our controllers, called Low-Friction Zero-Gravity controller (LFZG) adds an additional friction compensation term for aiding keyframe demonstration. We aim to create even simpler interactions with robots, as this controller makes the robot move in the direction indicated by small forces applied, eventually stopping. Additionally, our approach does not require torque or force sensors to be implemented. The second developed controller, formally Zero-Friction Zero-Gravity controller (ZFZG), makes the robot move similarly as if it were floating in space. As the forces that make the robot reduce its motion (mainly gravity and friction) are compensated, the final output is the free movement of the robot, driven by inertia.

2 State of the Art

The main fields of study of this work are related with friction and gravity compensation. Only selected works will be mentioned, as the literature in friction compensation in robots is extensive. A review can be found in [5].

On one side, friction is described as the resistance of motion of two contacting sliding surfaces [5]. To measure friction accurately is extremely difficult. Exact models of friction do not exist, and instead approximations obtained through experiments are used (Coulomb, viscous friction, Stribeck, Dahl, LuGre, Leuven, etc.). No specific model has proven better than others [6]. Canudas et al. [7] focused on modeling non-linear effects of friction in DC motor drives. They combine a linear model for viscous friction with a parameter estimation algorithm, which recalculates linear model parameters in a feedback loop to reduce the error in velocity commands. Some methods for friction identification in robotics consider elements in isolation, or do not consider mechanical limitations [8][9]. A low-velocity approach allows obtaining friction models depreciating inertia in [10]. As modeling motor frictions involves non-linearities (Stribeck effect, hysteresis, pre-sliding displacement, etc.), some authors [11] have delegated this problem to learning algorithms such as Neural Networks. Gearboxes also have high frictions, and additionally increase motor frictions from the link's point of view (due to the reduction factor). The most popular gearboxes in humanoid robotic platforms are Harmonic Drives, because of their compactness and reduction factor. Authors [12] have tried to model Harmonic Drives' frictions, finding similar problems of non-linearities as those of the motor case.

Regarding humanoid robots, in [13] they identify friction parameters on an iCub robot, aided by 6-axis force/torque sensors.

On the other side, gravity compensation is computed using the dynamic model of the robot. By analyzing the kinematic configuration and the masses of links and motors, it is possible to calculate the influence of gravity in each motor, and compute the torque value necessary to compensate it. In [14] they compensate gravity by projecting gravity forces on each joint of a robot arm. First, they translate all joint coordinates to the base frame. Then, they project on each joint, the torque generates by gravity forces on the rest of links and motors. This method is a simple and methodical procedure to compensate gravity in rigid links. In classical literature, the inclusion of a gravity compensation term in robot manipulation control schemes was used for improving a PD position control [15]. Including gravity compensation performed as well as a full feedforward controller with full inertial terms. However, the possibility of simulating free movements was not studied.

3 Theoretical Foundations

The design of our controllers is related with the fundamental laws of dynamics for serial rigid multibody systems. We consider the Euler-Lagrange equations of motion of multibody rigid links in the robot joint space as:

$$B(q)\ddot{q} + C(q, \dot{q})\dot{q} + F_v\dot{q} + F_s \operatorname{sgn}(\dot{q}) + g(q) = u + \tau_{ext} \tag{1}$$

Where B is the inertia matrix, C represents the centrifugal and Coriolis forces, $F_v\dot{q}$ is the viscous friction torque, $F_s \operatorname{sgn}(\dot{q})$ is the Coulomb friction torque, F_v is the matrix of viscous friction coefficients, $g(q)$ is the gravity term, u is the actuation torque, and finally τ_{ext} is the torque originated by external forces. An ideal friction and gravity compensator could be expressed as a u with the following form:

$$u = F_v\dot{q} + F_s \operatorname{sgn}(\dot{q}) + g(q) \tag{2}$$

Due to the low speeds applied in robotics, the Coriolis and centrifugal forces C are negligible. Substituting (2) in (1):

$$B(q)\ddot{q} = \tau_{ext} \tag{3}$$

Which means that the mechanism would offer a resistance to external forces (e.g. pushing or pulling) equivalent only to its inertia. When applied to a robotic system, this controller would make the whole mechanism behave as if it were in free movement. Our controllers combine friction and gravity compensation terms to provide new forms of physical interaction with robots. Let us formally describe the equations governing the controllers. Let $g(q)$ be the term of gravity compensation, with q as the actual joint configuration. Let $\tau_f(q, \dot{q})$ be the term of friction compensation, where

$\dot{q}$ is the joint angular velocity. Then, a generic friction and gravity compensation controller can be expressed as:

$$u = g(q) + \tau_f(q, \dot{q}) \tag{4}$$

A block scheme of this generic friction and gravity compensation controller can be seen on Fig. 1. Let us now describe how the gravity and friction compensation terms can be determined.

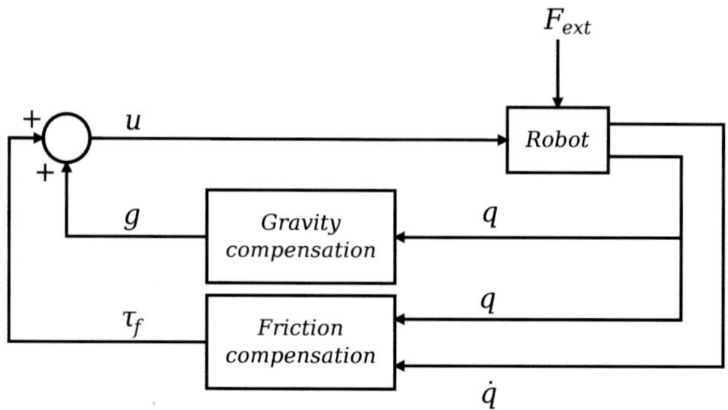

Fig. 1 Block scheme of friction and gravity compensation. In this type of control, there is no external reference, the robotic system is moved by the external perturbations caused by the user.

3.1 Gravity Compensation

The potential energy of a robot, assuming rigid links and punctual masses, can be defined as [16]:

$$U = \sum_{i=1}^{n} (U_{l_i} + U_{m_i}) \tag{5}$$

Where U_{l_i} is the sum of potential energy contributions of each link, U_{m_i} is the contribution of each motor, and i is an index for each link or motor. The first term U_{l_i} is defined as:

$$U_{l_i} = -m_{l_i} g_0^T p_{l_i} \tag{6}$$

Where m_{l_i} is the mass of the center of masses of link i, g_0 is the gravity vector expressed in base frame (e.g. $g_0 = [0 \quad 0 \quad -9.81]^T$), and p_{l_i} is the set of coordinates of the center of masses of link i expressed in the base frame. Similarly, the motor contributions U_{m_i} are defined as:

$$U_{m_i} = -m_{m_i} g_0^T p_{m_i} \tag{7}$$

Substituting (6) and (7) in (5), U becomes:

$$U = -\sum_{i=1}^{n}(m_{l_i} g_0^T p_{l_i} + m_{m_i} g_0^T p_{m_i}) \tag{8}$$

Where p_{l_i} and p_{m_i} depend on the joint configuration q. The torque $g(q)$ exerted by gravity can be computed as [17]:

$$g(q) = \frac{\partial U}{\partial q} \tag{9}$$

And is thus the torque required for gravity compensation. In the real world, determining the influence of each element in the potential energy equation is a non-trivial issue. For instance, the distinction between motor and link mass contribution is blurry, as the mass contribution between motors includes the parts of the motors located between the axes of rotation. This is the reason why we will use a simplified dynamic model of U. In this simplified model, the terms of link and motor contributions are mixed, and their masses are concentrated in the intermediate point between each pair of axes of rotation. This dynamic model is commonly used in humanoid robot research, and is usually called 'mass concentrated model'.

3.2 *Friction Compensation*

The static friction forces, $F_v\dot{q} + F_s\,\mathrm{sgn}(\dot{q})$, from (1) can be compacted into a joint friction term, $\tau_{fj}(\dot{q})$. It can be computed with a model-based identification procedure inspired by [18]. Among the available friction models, they have assumed the one including Coulomb friction (initial opposing torque) and viscous friction (friction dependent on velocity). Their aim is to model the friction of an electric motor. The motion of an electric motor can be described as:

$$\tau_m(t) - \tau_{fm}(\dot{\theta}) = J\ddot{\theta} \tag{10}$$

Where τ_m is the motor torque, $\tau_{fm}(\dot{\theta})$ is the motor friction torque, $\ddot{\theta}$ is the motor angular acceleration and J is the inertia of the motor. If the angular velocity $\dot{\theta}$ is stabilized, then $\ddot{\theta} = 0$, so the torque of the motor is used exclusively to compensate the friction:

$$\tau_m(t) = \tau_{fm}(\dot{\theta}) \tag{11}$$

Measuring the different velocities where the motor stabilizes for several torques applied, the stabilized velocities for these different torques can be plotted. The friction model selected by [18] becomes a piecewise linear model:

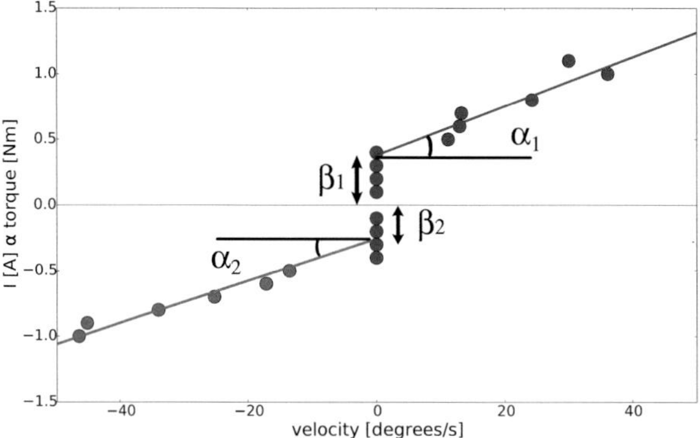

Fig. 2 This friction model includes Coulomb friction and viscous friction. The variables α and β represent the parameters of the linear model assumed.

$$\tau_{fm}(\dot{q}) = \begin{cases} \alpha_1\dot{\theta} + \beta_1 : \dot{\theta} > 0 \\ \alpha_2\dot{\theta} - \beta_2 : \dot{\theta} < 0 \end{cases} \tag{12}$$

Where model parameters α and β are obtained by linear regression on the plot (Fig. 2).

In the original procedure, they measure the motor velocity $\dot{\theta}$ in isolation. In our proposed identification procedure, we measure the velocity $\dot{q}$ with the motor within the robot, including the gearboxes and the mechanical structure. Modeling each part independently (motor, gearbox, structure, construction) would result in intractable combinations of models to be evaluated and coordinated, specially for many DoF. Our assumptions are the following:

– We use joint velocity $\dot{q}$ instead of motor velocity $\dot{\theta}$. The joint friction model selected becomes the following piecewise linear model:

$$\tau_{fj}(\dot{q}) = \begin{cases} \alpha_1\dot{q} + \beta_1 : \dot{q} > 0 \\ \alpha_2\dot{q} - \beta_2 : \dot{q} < 0 \end{cases} \tag{13}$$

Where $\tau_{fj}(\dot{q})$ is the torque necessary to compensate the friction generated in function of the joint angular velocity $\dot{q}$.
– We assume that the motors have a symmetrical behavior, they oppose to movement in both directions with the same strength. Therefore, $\alpha_1 = \alpha_2$ and $\beta_1 = \beta_2$.
– When applying constant torques, we limit the time given for velocity stabilization due to the mechanical constraints of the robot joints.
– As we have to deal with gravity forces, which may influence friction, we add an additional term when opposing gravity.

Robot joints have mechanical constraints, so there is a limit in the time the joint velocity can be recorded. This time may not be enough for the velocity to stabilize. In these cases, the velocity achieved before reaching the joint limit must be used instead of the stabilized velocity. This causes a steeper slope of the posterior linear regression. The final parameters of the linear regression should be further adjusted in these cases.

As stated in our final assumption, we add a term in addition to $\tau_{fj}(\dot{q})$. We assume that an additional mechanical friction is generated in the motor axle and gearbox due to gravity. This is the reason why we have added a term τ_{fg} dependent on the joint position and the velocity:

$$\tau_{fg} = f(q, \dot{q}) \tag{14}$$

The term $\tau_{fg}(q, \dot{q})$ is purely experimental, as it depends on the mechanical design and construction of the robot. In our model, we only add this term when the gravity opposes the direction movement of the arm. To see whether the gravity is in favor or against this movement, the variation of the potential energy U can be used. When $\Delta U > 0$, the movement is against gravity. The final friction compensator can be expressed as:

$$\tau_f(q, \dot{q}) = \begin{cases} \tau_{fj}(\dot{q}) + \tau_{fg}(q, \dot{q}) : & \Delta U > 0 \\ \tau_{fj}(\dot{q}) & : \Delta U < 0 \end{cases} \tag{15}$$

3.3 *Friction and Gravity Compensation Controllers*

Different applications may require different behaviors of the robot. Hence, two controllers have been derived from the generic friction and gravity compensation controller (4).

Low-Friction Zero-Gravity Controller (LFZG). This controller can improve the physical interaction with robots. In this controller, a new parameter ξ has been incorporated. This parameter attenuates the influence of the friction compensation on the system. Introducing ξ in the controller, it becomes:

$$u = g(q) + \xi \tau_f(q, \dot{q}) \tag{16}$$

By setting $0 < \xi < 1$, this controller allows the robot to move easily, without effort, but eventually stopping due to the low friction. This controller can be useful in paradigms such as keyframe demonstration and PbD, where there is a direct physical contact with the robot. For instance, when aiming to record a task using keyframe demonstration, different robot configurations must be recorded. In many cases, a demonstrator may have to use both hands to move a single robot joint, due to its individual friction. Therefore, in robots with many DoF, it can be difficult to physically move the robot between the different desired configurations. Using our controller, one has to simply push the robot in the desired direction, and stop it when desired. The attenuated friction serves as an aid for stopping at the desired target keyframes.

Zero-Friction Zero-Gravity Controller (ZFZG). This controller, when applied to all joints, ideally makes the robot move as if only the external dynamic forces and inertia would modify the motion. To achieve this behavior, we can use the generic friction and gravity compensation controller (4):

$$u = g(q) + \tau_f(q, \dot{q}) \tag{17}$$

A robotic platform using this control could be employed to test how devices would behave in complete absence of friction and gravity.

4 Experiments

The experiments have been performed using the arm of the humanoid robot Teo [19]. A single 1 DoF robot joint was tested, in order to avoid the high-dimensionality and coupling effects of many DoF (similarly to [20]).The humanoid robot joint used was the robot's left shoulder, which is moved by a Maxon brushless EC flat motor. It has a torque constant of 0.0706 Nm/A. The motor driver has an internal current loop with a PI regulator, with constant $K_p = 0$ and $K_i = 0.1651$. The gearbox is a Harmonic Drive CSD-25 with a reduction factor of 160. Joint position is measured using an optical relative encoder attached to the motor. Velocity is obtained by numerical differentiation of the position signal.

The robot arm weight m is 4.446 kg (including hand and electronics), and it has a length L of 0.82 m. The control algorithms were implemented in C language. The gravity compensation term of the control was computed as the torque caused by the arm modeled as a punctual mass at its center of gravity. Considering h as the height of the center of gravity with respect to its lowest position, being a single joint, this term is trivial to be calculated. Assuming q_1 as the angle between the arm and the trunk, the potential energy of a mass situated at $L/2$ from the shoulder is:

$$U = mg_0 h = mg_0(L/2)(1 - cos(q_1)) \tag{18}$$

Then, the gravity torque term is:

$$g(q_1) = \frac{\partial U}{\partial q_1} = mg_0(L/2)\sin(q_1) \tag{19}$$

The friction compensation term was determined by the procedure indicated in previous sections. When high torques are applied to the motor, leading to high velocities, the motor is not able to stabilize its velocity before reaching the mechanical limit. This results in a steeper slope on the posterior regression. This effect can be seen in the shortest curves in Fig. 3. Fourteen different constant motor torques were tested, including both movements against and in favor of gravity, ranging between −0.0706 Nm and 0.0706 Nm (from -1 A to 1 A). In our case, positive velocities go against gravity, and negative velocities are in favor of gravity. A summary of the process of friction identification for the joint can be seen first on Fig. 3, where the stability velocities are measured. Fig. 4 depicts the performed linear regressions.

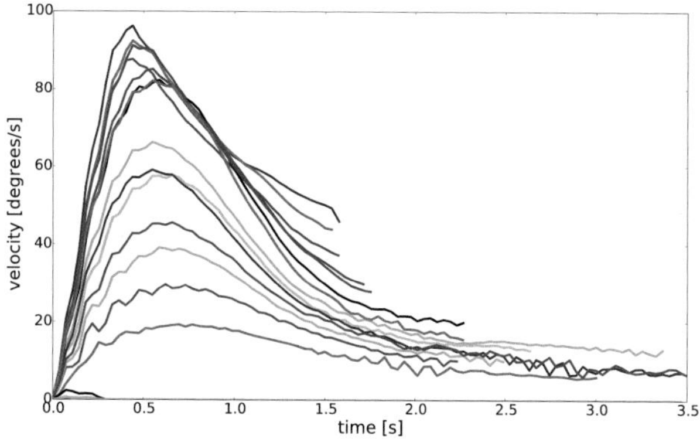

Fig. 3 Velocity vs. time curves, with constant torque applied in each curve. There is a proportional linear dependence between current in the motor and torque applied.

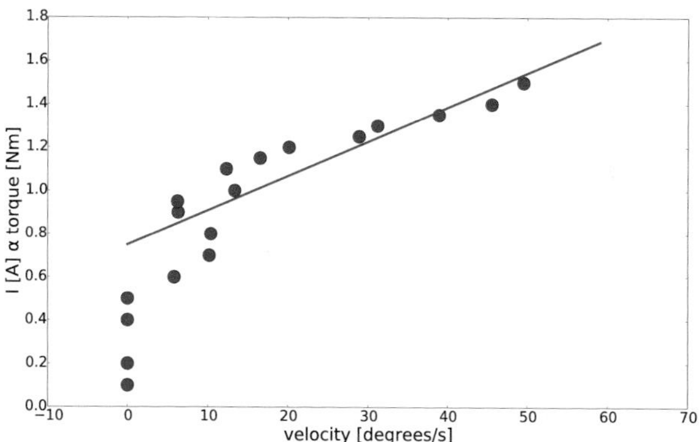

Fig. 4 Stabilization velocity vs. the torque applied. The Coulomb friction (blue) and the viscous friction (red) can be seen. The same procedure is applied for negative velocities.

The linear regressions obtained without manual tunning resulted in:

$$\tau_f(\dot{q}_1) = \begin{cases} 0.009\,\dot{q}_1 + 0.490: & \Delta U > 0 \\ 0.005\,\dot{q}_1 - 0.586: & \Delta U < 0 \end{cases} \tag{20}$$

A posterior manual adjustment of these linear regression parameters resulted in:

$$\tau_f(\dot{q_1}) = \begin{cases} 0.006\,\dot{q_1} + 0.4: & \Delta U > 0 \\ 0.001\,\dot{q_1} - 0.7: & \Delta U < 0 \end{cases} \tag{21}$$

To adjust to a symmetric joint friction model, either of the equations may be selected to fix the parameters of the model. Here, we have used the parameters of $\Delta U < 0$ case of (21), resulting in the joint friction term τ_{fj} of the controller:

$$\tau_{fj}(\dot{q_1}) = \begin{cases} 0.001\,\dot{q_1} + 0.7: & \Delta U > 0 \\ 0.001\,\dot{q_1} - 0.7: & \Delta U < 0 \end{cases} \tag{22}$$

All other frictions will be considered part of the gravity friction term τ_{fg}. The position-dependent parameter of τ_{fg} was experimentally adjusted to $0.0025\,q_1$. The final expression of τ_{fg} is computed as (21) minus (22) plus the position-dependent parameter τ_{fg}.

$$\tau_{fg}(q_1, \dot{q_1}) = 0.005\,\dot{q_1} + 0.0025\,q_1 - 0.3 \tag{23}$$

The compensators were evaluated activating the ZFZG controller. A well designed Zero-Friction Zero-Gravity controller would maintain constant, or tightly bounded, velocities in absence of external perturbations (beyond the one initiating the movement). To test whether these conditions are applicable to our system, several interactions with the arm were performed. A single push was given to the arm, letting it move freely while recording its velocities. This experiment was repeated while pushing the robot arm with different forces. Several velocity profiles for different pushes can be seen on Fig. 5.

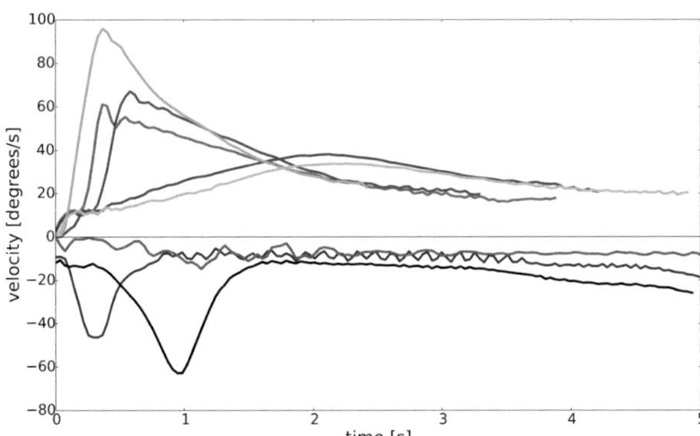

Fig. 5 Velocity profiles for several initial 'pushes' with the Zero-Friction Zero-Gravity controller (ZFZG).

Curves reaching higher peak values in the figure represent larger forces exerted by the human. The peak of each curve roughly represents the instant where the robot arm is let free. An example of one of these interactions using the ZFZG controller can be seen in Fig. 6.

Fig. 6 Sequence of the movement of the robot arm using a Zero-Friction Zero-Gravity controller. When the user pushes the arm, it moves freely in the direction of the applied force.

Results from Fig. 5 showed that the combination of friction compensation and gravity compensation was successful at maintaining bounded velocities in absence of external perturbations for velocities below 30-35 degrees/s. Results also show that the friction model is not adequate for velocities above 30-35 degrees/s. For instance, the light blue and purple curves do not maintain their values after their peaks. This could be explained because of the unmodeled non-linearities of the friction function at these velocities. A video of the implementation was shown in Humanoids 2014 conference [21], and can be seen online[1].

Our linear friction model may look simplistic, as each part in the mechanism includes its own non-linearities. However, it accounts for physical components (such as gearboxes and joint limits) that are found in robotic systems. Ultimately, the controllers are able to manifest the friction and gravity compensation expected behavior for the tested robotic joint. Our efforts are directed to implement these control algorithms in our full humanoid robot.

5 Conclusions

In this paper, the authors have presented a new set of controllers for robots which aim at compensation of static friction and gravity. The experiments show, in general, an acceptable performance of the ZFZG controller tested. More accurate friction models and identification procedures could lead to improved controller behaviors under high joint velocity conditions, and would also aid in maintaining the stability of low velocities. We also consider using different dynamic robot models (pendulum-like models, or even the complete dynamic model).

With the potential increase in complexity of the complete humanoid robot model, we consider using machine learning algorithms which lighten the efforts necessary to obtain a reliable friction and gravity compensation. They could also capture the non-linearities present in the system.

Acknowledgements This work was supported by RoboCity2030-III-CM project (S2013/MIT-2748), funded by Programas de Actividades I+D in Comunidad de Madrid and EU.

[1] http://dai.ly/x2vjrfs

References

1. Calinon, S., D'halluin, F., Sauser, E.L., Caldwell, D.G., Billard, A.G.: Learning and reproduction of gestures by imitation. IEEE Robotics & Automation Magazine **17**(2), 44–54 (2010)
2. Billard, A.G., Calinon, S., Guenter, F.: Discriminative and adaptive imitation in uni-manual and bi-manual tasks. Robotics and Autonomous Systems **54**(5), 370–384 (2006)
3. Akgun, B., Cakmak, M., Yoo, J.W., Thomaz, A.L.: Trajectories and keyframes for kinesthetic teaching: a human-robot interaction perspective. In: Proceedings of the Seventh Annual ACM/IEEE International Conference on Human-Robot Interaction, pp. 391–398. ACM (2012)
4. Guizzo, E., Ackerman, E.: The rise of the robot worker. IEEE Spectrum **49**(10), 34–41 (2012)
5. Olsson, H., Åström, K.J., de Wit, C.C., Gäfvert, M., Lischinsky, P.: Friction models and friction compensation. European Journal of Control **4**(3), 176–195 (1998)
6. Bona, B., Indri, M.: Friction compensation in robotics: an overview. In: 44th IEEE Conference on Decision and Control, 2005 and 2005 European Control Conference, CDC-ECC 2005, pp. 4360–4367. IEEE (2005)
7. Canudas, C., Astrom, K., Braun, K.: Adaptive friction compensation in dc-motor drives. IEEE Journal of Robotics and Automation **3**(6), 681–685 (1987)
8. Kostic, D., de Jager, B., Steinbuch, M., Hensen, R.: Modeling and identification for high-performance robot control: an rrr-robotic arm case study. IEEE Transactions on Control Systems Technology **12**(6), 904–919 (2004)
9. Papadopoulos, E.G., Chasparis, G.C.: Analysis and model-based control of servomechanisms with friction. Journal of Dynamic Systems, Measurement, and Control **126**(4), 911–915 (2004)
10. Kermani, M.R., Wong, M., Patel, R.V., Moallem, M., Ostojic, M.: Friction compensation in low and high-reversal-velocity manipulators. In: Proceedings on IEEE International Conference on Robotics and Automation, ICRA 2004, vol. 5, pp. 4320–4325. IEEE (2004)
11. Na, J., Chen, Q., Ren, X., Guo, Y.: Adaptive prescribed performance motion control of servo mechanisms with friction compensation. IEEE Transactions on Industrial Electronics **61**(1), 486–494 (2014)
12. Gomes, S.C.P., Santos da Rosa, V.: A new approach to compensate friction in robotic actuators. In: Proceedings IEEE International Conference on Robotics and Automation, ICRA 2003, vol. 1, pp. 622–627. IEEE (2003)
13. Traversaro, S., Del Prete, A., Muradore, R., Natale, L., Nori, F.: Inertial parameter identification including friction and motor dynamics. In: IEEE-RAS International Conference on Humanoid Robots (Humanoid 2013), Atlanta, USA (2013)
14. Luo, R.C., Yi, C.Y., Perng, Y.W.: Gravity compensation and compliance based force control for auxiliarily easiness in manipulating robot arm. In: 2011 8th Asian Control Conference (ASCC), pp. 1193–1198. IEEE (2011)
15. An, C.H., Atkeson, C.G., Hollerbach, J.M.: Model-based control of a robot manipulator, vol. 214. MIT press Cambridge, MA (1988)
16. Sciavicco, L., Villani, L.: Robotics: modelling, planning and control. Springer (2009)
17. De Luca, A., Panzieri, S.: Learning gravity compensation in robots: Rigid arms, elastic joints, flexible links. International Journal of Adaptive Control and Signal Processing **7**(5), 417–433 (1993)
18. Virgala, I., Kelemen, M.: Experimental friction identification of a dc motor. International Journal of Mechanics and Applications **3**(1), 26–30 (2013)
19. Martínez, S., Monje, C.A., Jardón, A., Pierro, P., Balaguer, C., Muñoz, D.: Teo: Full-size humanoid robot design powered by a fuel cell system. Cybernetics and Systems **43**(3), 163–180 (2012)
20. Mallon, N., van de Wouw, N., Putra, D., Nijmeijer, H.: Friction compensation in a controlled one-link robot using a reduced-order observer. IEEE Transactions on Control Systems Technology **14**(2), 374–383 (2006)
21. Morante, S., Victores, J.G., Martinez, S., Balaguer, C.: Sensorless friction and gravity compensation. In: 2014 14th IEEE-RAS International Conference on Humanoid Robots (Humanoids), pp. 265–265. IEEE (2014)

Commanding the Object Orientation Using Dexterous Manipulation

Andrés Montaño and Raúl Suárez

Abstract This paper presents an approach to change the orientation of a grasped object using dexterous manipulation teleoperated in a very simple way with the commands introduced by an operator using a keyboard. The novelty of the approach lays on a shared control scheme, where the robotic hand uses the tactile and kinematic information to manipulate an unknown object, while the operator decides the direction of rotation of the object without caring about the relation between his commands and the actual hand movements. Experiments were conducted to evaluate the proposed approach with different objects, varying the initial grasp configuration and sequence of actions commanded by the operator.

Keywords Dexterous manipulation · Teleoperation · Tactile sensors · Grasping

1 Introduction

Teleoperation of robots is a challenging subject in applications in which an operator takes decisions and the robots perform actions following the commands of the operator. Some application fields where the teleoperation is relevant are: handling hazardous material, telesurgery, underwater vehicles, space robots, mobile robots, among others [1]. Object dexterous manipulation is a problem which is involved in this fields. A detailed discussion of the general problems related to teleoperation as well as a description of typical applications was presented in [2].

A. Montaño(✉) · R. Suárez
Institute of Industrial and Control Engineering,
Universitat Politècnica de Catalunya, Barcelona, Spain
e-mail: {andres.felipe.montano,raul.suarez}@upc.edu

R. Suárez—Work partially supported by the Spanish Government through the projects DPI2011-22471, DPI2013-40882-P and DPI2014-57757-R.

L.P. Reis et al. (eds.), *Robot 2015: Second Iberian Robotics Conference*,
Advances in Intelligent Systems and Computing 418,
DOI: 10.1007/978-3-319-27149-1_6

Autonomy of the robotic system in teleoperation has been addressed following different approaches. In some of them, the operator has the control of the movements and actions of the robot (a fully teleoperated system), but the control can also be shared between the operator and the robot [3]. Multiple examples of fully teleoperated systems have been introduced, several of them teleoperating robot arms and hands, which are commanded using input interfaces as trackers [4], wiimotes [5] for the arm, and gloves [4, 6], multi touch interfaces [7], or video based systems [5] for the hand; these approaches are oriented to solve the mapping between the human pose and movements, and those of the robot.

The manipulation of unknown objects has been addressed using different strategies. A virtual object frame is used to change the pose of the object varying the triangular fingertip configuration of a three-fingered hand. A control law to manipulate the object was introduced in [8], however the lack of sensorial feedback limits the accuracy of the method. A composite position-force control scheme was presented in [9]. The relative position of the object with respect to the hand is changed following an input trajectory; the control scheme is evaluated in simulations introducing noise on the sensor measurements to simulate a real environment, however other grasp aspects, as the initial grasp configuration or the stability of the grasp, are not addressed. In a previous work [10], the shape of an unknown object was recognized using tactile information obtained during the object manipulation.

In the case of dexterous telemanipulation, the terminal element (gripper or robotic hand) can also be fully controlled by the operator or the control can be shared with the robot. In the first case, a mapping between the terminal element and the hand of the operator is required. Three mapping methods can be distinguished in the literature [11]: joint-to-joint mapping, which is applied to anthropomorphic hands [12]; pose mapping, which tries to find robot hand poses correlated with human hand poses[13]; and point-to-point mapping, which is the most used common approach, where the fingertip positions of the human hand are mapped to the fingertip positions of robot hand [14]. However, in dexterous telemanipulation a natural approach is to share control of the object manipulation while giving the human operator direct access to remote tactile and force information at the slave fingertips [15]. In this work we use the shared control scheme, the operator commands the robotic hand to manipulate an object (performs the grasping and rotation), while the robotic system uses the tactile and kinematics information to control the forces and movements in order to avoid object falls. It must be remarked that the geometric model of the object is unknown and that the rotation limits are given by the friction constraints, which are evaluated during the manipulation to avoid object falls and by the kinematic constraints of the fingers during the manipulation.

The paper is arranged as following. After this introduction, Section 2 introduces the approach overview. Section 3 presents the dexterous manipulation details and the motion strategy to avoid object falls. Experimental results are described in Section 4. Finally, Section 5 presents the summary and future work.

2 Approach Overview

Teleoperated dexterous manipulation is the main problem addressed in this work, the teleoperation is performed by an operator using an input interface like a keyboard, while the dexterous manipulation is executed by the hand following inputs from the operator. The hand is fixed on a base over a table, but it can be assembled in a robotic arm as well, this is not relevant for the proposed approach. The teleoperation of the robotic arm is out of the scope of this work. The manipulation task is focused on changing the object orientation (rotating the object) when this is grasped by the hand, thus the operator can command the hand to close, open, turn clockwise or counter clockwise when the object is grasped. The commands are sent using four keys of a keyboard. In this work we consider that the absolute position of the object in the space can be controlled by the arm, and therefore only the orientation will be controlled by the fingers of the hand.

The Schunk Dexterous Hand (SDH2) is the robotic hand used for the experimentation. This is a three finger hand (gripper) with seven active degrees of freedom (*dof*). The SDH2 has tactile sensors attached to the surface of the proximal links and the distal links, thus the tactile sensor system has six sensor pads. Two fingers of the hand are coupled and can be rotated on the base to work opposite to each other in the same plane (see Figure 1). Using the coupled fingers it is possible to perform a prismatic precision grasp [16], which is comparable with a human grasp using the thumb and index fingers. Thus, the two coupled fingers of the SDH2 are rotated on their bases to work opposite to each other on the same plane, as it is shown in Fig. 1c.

The manipulation algorithm is described by the state machine shown in Fig. 2, we distinguish the following states with their associated actions:

S_{init} : The hand is in the initial configuration, ready to perform a grasp.

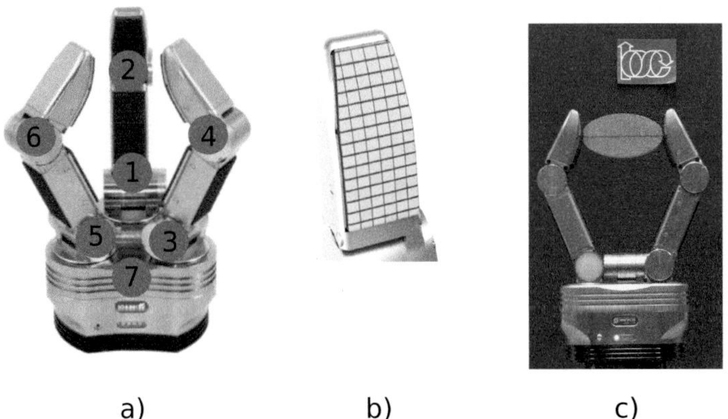

Fig. 1 a) Schunk Dexterous Hand (SDH2) with the joints labels. b) detail of the sensor pad showing the texels on the fingertip. c) Grasp configuration with two fingers working in a opposite way.

S_{close} : The fingers are closed until reaching a desired grasp force.
S_{open} : The fingers are opened to release the grasped object.
S_{grasp} : The object is grasped and the hand is waiting for a command.
S_{turnC} : The next configuration for a clockwise rotation of the graped object is computed.
S_{turnCC} : The next configuration for a counterclockwise rotation of the graped object is computed.
S_{move} : The hand executes the next configuration rotating the object.

The state transitions are determined by:

Keyboard signals: These are four signals generated by the operator using a standard keyboard, each signal is simply generated by pressing a predetermined key. The signals command the four available actions during the teleoperation: close (K_c), open (K_o), rotate clockwise (K_{tc}) and rotate counterclockwise (K_{tcc}).

Force signal: It is a binary signal that is activated when $F_k > F_d$, where F_k is the grasp force in the k-th iteration and F_d is a desired maximum grasping force. This condition is reached when the hand has been closed and the object is in contact with the sensor pads.

Friction signals: It is a binary signal (G_c) that is activated when the friction constraints allows the grasp to firmly hold the object (these constraints are detailed in Section 3). The binary complement of G_c is represented as $\overline{G_c}$.

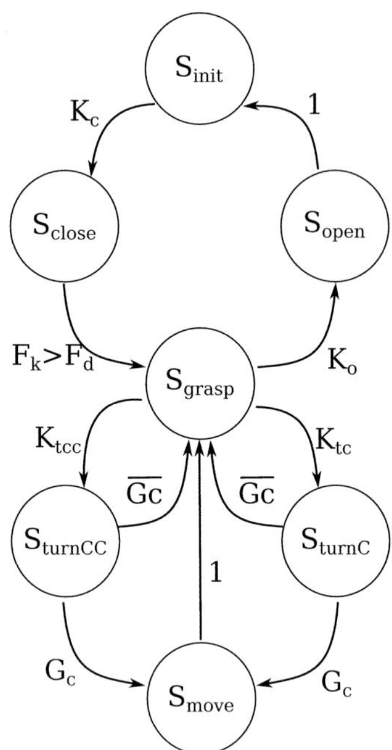

Fig. 2 State machine for the teleoperation approach.

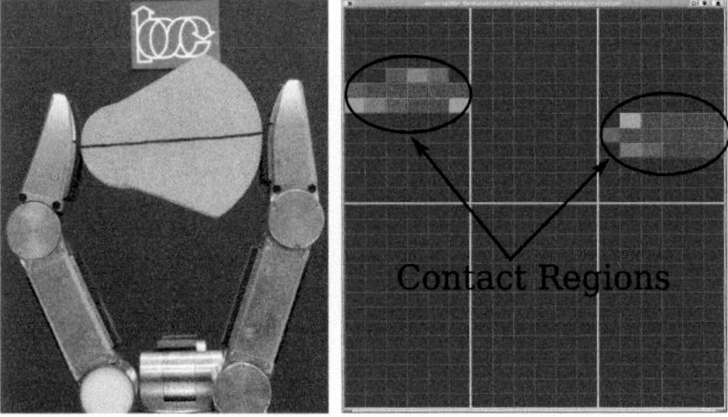

Fig. 3 Detected contact region on sensor pads when an object is grasped.

The state machine for the teleoperation starts in the state S_{init}, where the hand is waiting for the command K_c, to be introduced by the operator in order to close the fingers. When the command K_c is introduced, the system evolves to the state S_{close}. The system remains in the state S_{close} until the measured force on the sensor pads is greater than a desired grasp force, and the object has been actually grasped. Once the grasp is done and the proper force is detected, the system evolves to state S_{grasp}. In the state S_{grasp} the system is waiting for the commands K_{tc}, K_{tcc} or K_o, in order to rotate clockwise, counterclockwise or to open the fingers to release the object, respectively. The finger movements to rotate the object are computed in the states S_{turnC} and S_{turnCC} depending on the direction of rotation indicated by the operator. In these states an autonomous dexterous manipulation algorithm is applied, which is introduced in Section 3. If the friction constraints are satisfied then the system evolves to the state S_{move}, where the movements of the fingers are performed and then the system comes back to S_{grasp}, or else, there is not a possible movement and the system comes back to S_{grasp} without producing any movement.

3 Dexterous Manipulation

Consider a global reference system located at the base of the finger f_1, the initial grasp produces two contact points, namely P_1 on finger f_1 and P_2 on finger f_2. Usually the contact with the object produces a contact region, we consider the barycenter of this region as the contact point and the average force over all the region as the contact force, as proposed in [17] (See Fig. 3). The absolute positions of P_1 and P_2 are computed using the sensor contact information and the hand kinematics.

Dexterous manipulation is based on a reactive control scheme, in which the current information of the contacts and the kinematic information of the hand are used as inputs. The reactive control action is applied to update the distance d_k between the contact points P_1 and P_2. The controlled distance d_{k+1} is a function of the measured force F_k, d_{k+1} is computed as,

$$d_{k+1} = d_k + \Delta d \tag{1}$$

with Δd being a function of the force measured by the tactile sensors according to the follow relationship,

$$\Delta d = \begin{cases} 0 & \text{if } F_{\min} < F_k < F_{\max} \\ +\lambda & \text{if } F_k \leq F_{\min} \\ -\lambda & \text{if } F_k \geq F_{\max} \end{cases}$$

where the constant values $F_{\min}$, $F_{\max}$ and λ are empirically determined based on the sensors response, and k denotes a manipulation iteration.

The distance d_k between contact points is given by,

$$d_k = \sqrt{(P_{x1_k} - P_{x2_k})^2 + (P_{z1_k} - P_{z2_k})^2} \tag{2}$$

The grasping force F_k is computed as the average of the two contact forces F_{1_k} and F_{2_k} measured, respectively, by the sensors of each fingertip, so that potential errors are minimized,

$$F_k = \frac{F_{1_k} + F_{2_k}}{2} \tag{3}$$

In order to compute the expected position of the contact points, we consider the hypothesis that the fingers are moved over a circular path whose diameter is given by the controlled distance between contact points, d_{k+1}, as shown in Figure 4. Both fingers are moved to $P_{1_{k+1}}$ and $P_{2_{k+1}}$ at same time. Let's call ϕ the object orientation. The orientation resulting from the initial grasp is considered as the reference orientation (i.e. $\phi = 0$). The expected contact points $P_{1_{k+1}}$ and $P_{2_{k+1}}$, are computed as,

$$P_{x1_{k+1}} = C_{x_k} - (d_{k+1}/2)\cos(\phi + \Delta\phi) \tag{4}$$

$$P_{z1_{k+1}} = C_{z_k} - (d_{k+1}/2)\sin(\phi + \Delta\phi) \tag{5}$$

$$P_{x2_{k+1}} = C_{x_k} + (d_{k+1}/2)\cos(\phi + \Delta\phi) \tag{6}$$

$$P_{z2_{k+1}} = C_{z_k} + (d_{k+1}/2)\sin(\phi + \Delta\phi) \tag{7}$$

where $\Delta\phi$ is chosen positive to turn the object clockwise or negative to turn the object counterclockwise. $\Delta\phi$ is chosen small enough to assure small movements of the object on each manipulation step. The point C_k is the center of the circular path followed by the fingers, and it is given by,

$$C_{x_k} = \frac{P_{x2_k} - P_{x1_k}}{2} + P_{x1_k} \tag{8}$$

$$C_{z_k} = \frac{P_{z2_k} - P_{z1_k}}{2} + P_{z1_k} \tag{9}$$

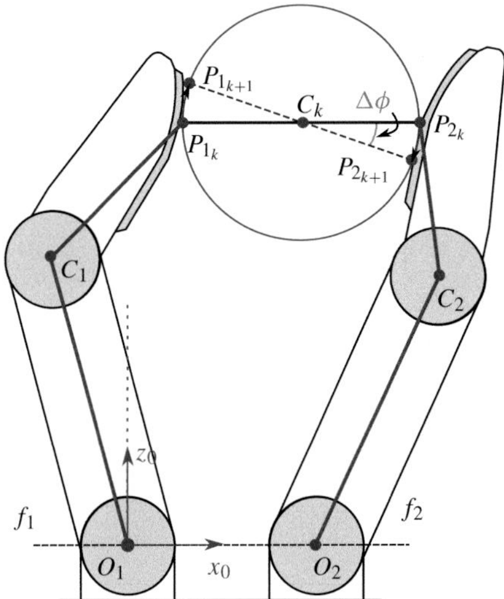

Fig. 4 Two-finger model used for the control of the object orientation.

In order to avoid sliding, each force applied on the object must be located within the friction cone centered at the direction normal to the sensor surface at the contact point. When this condition is satisfied the binary signal G_c is activated allowing a transition to the state S_{move} in the manipulation state machine where motions are executed. When G_c is not activated (i.e. $\overline{G_c}$ is activated) the state changes to S_{grasp} and the system waits for new commands. A planar grasp with two frictional contact points is force-closure when the segment connecting the contact points lies inside the friction cone at both contact points, as shown in Figure 5. The friction cone is given by $\alpha = \arctan \mu$, with μ being the friction coefficient, following the Coulomb friction model. Any applied force that belongs to the friction cone will not produce slippage, therefore the angle β_i $i = 1, 2$, between the normal direction at each contact point and the segment between the two contact points must satisfy $\beta_i < \alpha$. Then, the above condition can be expressed as,

$$\pi/2 - \alpha < \omega_i < \pi/2 + \alpha \tag{10}$$

where ω_i, $i = 1, 2$ is computed for both contact points as,

$$\omega_1 = \arccos\left(\frac{-|C_1P_2|^2 + r_1^2 + |P_1P_2|^2}{2r_1|P_1P_2|}\right) - \theta_3 - \pi/2 + \varphi_1 \tag{11}$$

$$\omega_2 = \arccos\left(\frac{-|C_2P_1|^2 + r_2^2 + |P_1P_2|^2}{2r_2|P_1P_2|}\right) - \theta_5 - \pi/2 + \varphi_2 \tag{12}$$

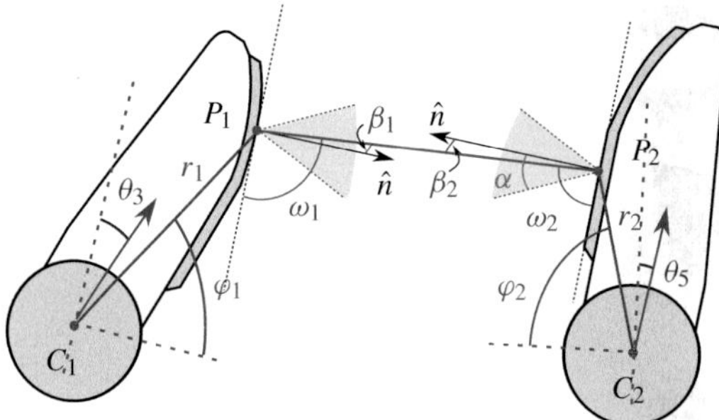

Fig. 5 Detail of the fingertips and angles considered to compute the friction constraints.

Fig. 6 Set of manipulated objects.

The points C_1 and C_2, the distances r_1 and r_2, and the angles φ_1 and φ_2 are computed using the kinematics of the hand and the information of the contact points. A complete description of the hand kinematics for the SDH2 can be found in [18].

The limits for the change of the object orientation depend on the friction constraints and the kinematic constraints of the fingers, which in turn depend on the shape of the object being manipulated.

4 Experimental Results

The described approach has been fully implemented using C++ for unknown object telemanipulation with the SDH2 hand. Figure 6 shows the set of objects used in the experimentation. Each object is held between the two coupled fingers of the SDH2, then the operator introduces the command, clicking the proper key in the keyboard, to close the fingers until the measured force reaches a desired value $F^d = 2\,\text{N}$. Note that the initial contact points are unknown, i.e. the initial grasp configuration changes at each execution of the experiment. After this, the operator can manipulate the object, rotating it clockwise or counterclockwise by means of a simple teleoperation.

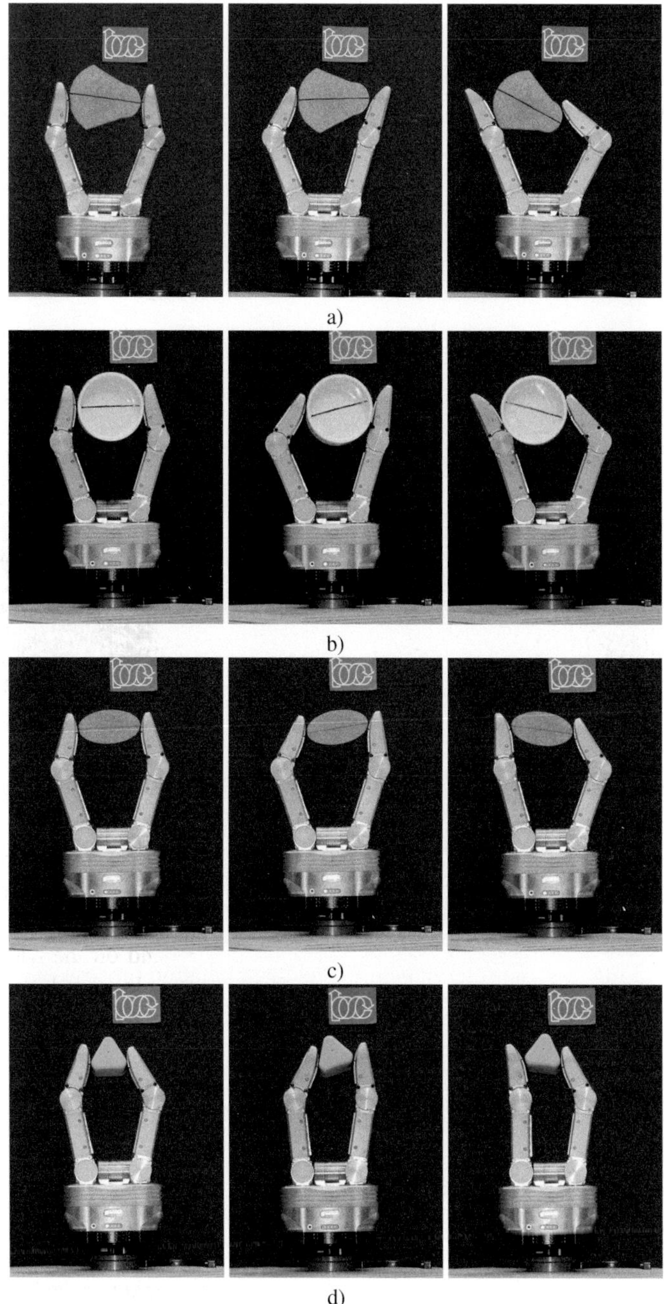

Fig. 7 Snapshots of the telemanipulated objects.

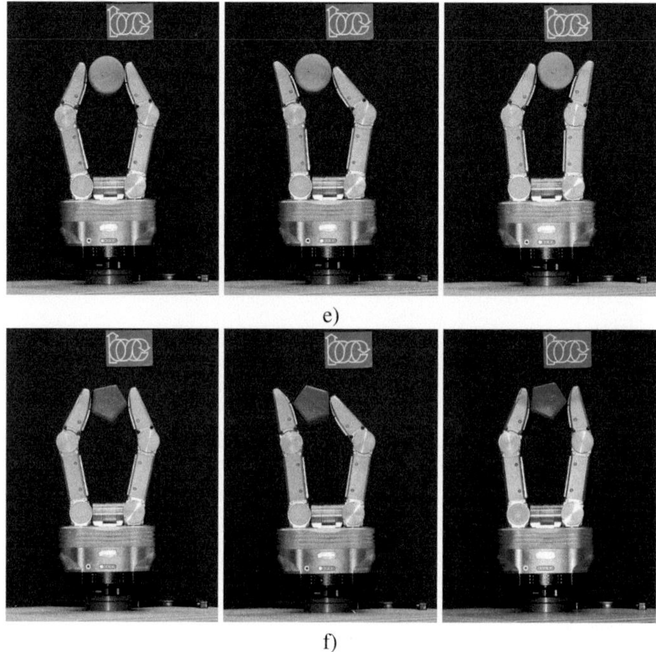

Fig. 8 Snapshots of the telemanipulated objects.

The material of the sensor pads is rubber and the material of the objects is wood, cardboard or plastic, thus we consider a worst case friction coefficient $\mu = 0.4$, which is lower than the friction coefficient between rubber and wood $\mu = 0.7$, and rubber and cardboard $\mu = 0.5$ [19], the friction coefficient between rubber and plastic is greater than the aforementioned. The constant λ to adjust the distance d_k is set to 1 mm.

Fig. 7 and Fig. 8 show snapshots of the telemanipulation process, the left snaps show the initial configuration of each object, the center snaps show the configuration when the limit of rotation is reached in counterclockwise direction and the right snaps show the configuration when the same limit is reached in clockwise direction.

5 Summary and Future Work

This work has proposed a telemanipulation method to rotate unknown objects based on tactile information and force feedback autonomously managed by the system. The telemanipulation allows to change the orientation of the object rotating it in a plane according to the commands introduced by an operator using a keyboard. The experimental results showed that the approach is effective to manipulate different objects.

An extension of the implemented work is to consider the use of three fingers in the manipulation process, which would allow to consider other motion strategies addressing the 3-D rotational problem.

References

1. Hokayem, P.F., Spong, M.W.: Bilateral teleoperation: An historical survey. Automatica **42**(12), 2035–2057 (2006)
2. Basañez, L., Suárez, R.: Teleoperation. In: Nof, S. (ed.) Springer Handbook of Automation, pp. 449–468. Springer-Verlag (2009)
3. Sheridan, T.B.: Telerobotics, Automation, and Human Supervisory Control. MIT Press, Cambridge (1992)
4. Rosell, J., Suárez, R., Pérez, A.: Safe teleoperation of a dual hand-arm robotic system. In: Armada, M.A., Sanfeliu, A., Ferre, M. (eds.) ROBOT2013: First Iberian Robotics Conference. AISC, vol. 253, pp. 615–630. Springer International Publishing (2014)
5. Ciobanu, V., Popescu, N., Petrescu, A., Noeske, M.: Robot telemanipulation system. In: Int. Conf. on System Theory, Control and Computing (ICSTCC), pp. 681–686 (2013)
6. Kuklinski, K., Fischer, K., Marhenke, I., Kirstein, F., aus der Wieschen, M., Solvason, D., Kruger, N., Savarimuthu, T.: Teleoperation for learning by demonstration: data glove versus object manipulation for intuitive robot control. In: Int. Cong. on Ultra Modern Telecommunications and Control Systems and Workshops (ICUMT), pp. 346–351 (2014)
7. Toh, Y.P., Huang, S., Lin, J., Bajzek, M., Zeglin, G., Pollard, N.: Dexterous telemanipulation with a multi-touch interface. In: IEEE-RAS Int. Conf. on Humanoid Robots (Humanoids), pp. 270–277 (2012)
8. Tahara, K., Arimoto, S., Yoshida, M.: Dynamic object manipulation using a virtual frame by a triple soft-fingered robotic hand. In: Proc. IEEE Int. Conf. on Robotics and Automation, pp. 4322–4327 (2010)
9. Li, Q., Haschke, R., Ritter, H., Bolder, B.: Towards manipulation of unknown objects. In: IFAC Symposium on Robot Control, IFAC (2012)
10. Montaño, A., Suárez, R.: Object shape reconstruction based on the object manipulation. In: Proc. IEEE Int. Conf. Advanced Robotics, pp. 1–6. November 2013
11. Colasanto, L., Suarez, R., Rosell, J.: Hybrid mapping for the assistance of teleoperated grasping tasks. IEEE Transactions on Systems, Man, and Cybernetics: Systems **43**(2), 390–401 (2013)
12. Kyriakopoulos, K., Van Riper, J., Zink, A., Stephanou, H.: Kinematic analysis and position/force control of the anthrobot dextrous hand. IEEE Transactions on Systems, Man, and Cybernetics, Part B: Cybernetics **27**(1), 95–104 (1997)
13. Kjellstrom, H., Romero, J., Kragic, D.: Visual recognition of grasps for human-to-robot mapping. In: Proc. IEEE/RSJ Int. Conf. on Intelligent Robots and Systems, pp. 3192–3199, September 2008
14. Peer, A., Einenkel, S., Buss, M.: Multi-fingered telemanipulation - mapping of a human hand to a three finger gripper. In: The 17th IEEE International Symposium on Robot and Human Interactive Communication, RO-MAN 2008, pp. 465–470 (2008)
15. Griffin, W.B., Provancher, W.R., Cutkosky, M.R.: Feedback strategies for telemanipulation with shared control of object handling forces. Presence: Teleoperators and Virtual Environments **14**(6), 720–731 (2005)
16. MacKenzie, C., Iberall, T.: The Grasping Hand. Advances in Psychology. Elsevier Science (1994)
17. Worn, H., Haase, T.: Force approximation and tactile sensor prediction for reactive grasping. In: World Automation Congress, pp. 1–6 (2012)
18. Montaño, A., Suárez, R.: Getting comfortable hand configurations while manipulating an object. In: Proc. IEEE Conf. Emerging Technology and Factory Automation, pp. 1–8 (2014)
19. Kutz, M.: Mechanical Engineers' Handbook. A Wiley-Interscience publication, Wiley (1998)

Validation of a Time Based Routing Algorithm Using a Realistic Automatic Warehouse Scenario

Joana Santos, Pedro Costa, Luís Rocha, Kelen Vivaldini, A. Paulo Moreira and Germano Veiga

Abstract Traffic Control is one of the fundamental problems in the management of an Automated Guided Vehicle (AGV) system. Its main objectives are to assure efficient conflict free routes and to avoid/solve system deadlocks. In this sense, and as an extension of our previous work, this paper focus on exploring the capabilities of the Time Enhanced A* (TEA*) to dynamically control a fleet of AGVs, responsible for the execution of a predetermined set of tasks, considering an automatic warehouse case scenario. During the trial execution the proposed algorithm, besides having shown high capability on preventing/dealing with the occurrence of deadlocks, it also has exhibited high efficiency in the generation of free collision trajectories. Moreover, it was also selected an alternative from the state-of-art, in order to validate the TEA* results and compare it.

Keywords Multi-robot coordination · Routing · AGV

1 Introduction

The increase of the logistics technology leads to the need to constantly adapt the production flow (production of small or customized series). Fully automated systems

---

J. Santos(✉) · P. Costa · L. Rocha · A.P. Moreira · G. Veiga
INESC-TEC, Porto, Portugal
e-mail: {joana.r.santos,germano.veiga}@inesctec.pt

P. Costa · A.P. Moreira
Department of Electrical Engineering and Computers,
Faculty of Engineering of the University of Porto, Porto, Portugal
e-mail: {pedrogc,amoreira}@fe.up.pt

K. Vivaldini
Mechatronics Group - Mobile Robotics Lab, São Carlos, Brazil
e-mail: kteixeira@sc.usp.br

L.P. Reis et al. (eds.), *Robot 2015: Second Iberian Robotics Conference*,
Advances in Intelligent Systems and Computing 418,
DOI: 10.1007/978-3-319-27149-1_7

(Automated Guided Vehicles - AGVs) were considered, instead manual or mechanical (forklifts, conveyors, ...) solutions.

The wide range of AGVs applications makes them suitable for industrial environments as a specific task or as part of a global automated system. These vehicles could be used in different stages of an industry, such as production lines, end-of-line automation chain, warehouse and distribution. Thus the flexibility of these systems allows them to execute different tasks in different industry types: manufacturing, food and beverage, warehouse and distribution.

Path planning for automated guided vehicles in industrial environments can be quite challenging and the current state of industrial practice is most of the times based on predetermined routes and fine parameter tuning for a specific scenario. Therefore, robust path planning algorithms can reduce the level of engineering efforts needed and increase the efficiency during runtime.

Industrial environments have some characteristics that must be taken into consideration, such as the safety of operators and other autonomous vehicles; compliance with production schedules' and task sequence. For these reasons a routing algorithm to ensure collision free routes must be considered.

In this sense, the work developed in this paper addresses this industry's need, and emerges as an extension of our previous work,[3]. The TEA* (Time Enhanced A*) algorithm was developed which consists of an incremental algorithm that builds the path of each vehicle taking into account the movements of other AGVs. This feature allows the algorithm to produce conflict free routes and at the same time deal with deadlock situations.

Some changes on the map representation were performed to this algorithm (our publication[3]), such as:

- a graph implementation. This will allow not only to reduce the processing times but also facilitates its integration on a real industrial environment. Using large scenarios this implementation requires a smaller memory representation than a grid map.
- Another extension from the original TEA* is concerned with the livelocks resolution. Similar with the deadlocks, there are the livelocks that lead the system to an undefined state.

Several simulation experiments were performed considering 6 AGVs and different scheduling algorithms. These results were compared with a state-of-art alternative[4]. The paper is organized as follows. Section 2 presents a short review of the state-of-art in the field. Section 3 describes the proposed algorithm and the minor changes performed to the original proposed TEA* algorithm. Section 4 focus on presenting the behavior of the algorithm with deadlocks and livelocks situations. Section 5, describes the industrial case scenario used on the simulation experiments. A state-of-art alternative was also considered to show the potentiality of the developed algorithm. Section 6 presents the results achieved. Finally, Section 7 discusses about the contributions of the presented work and future work.

2 Literature Review

The vehicle routing problem consists on computing trajectories that minimize the total distance traveled by AGVs taking into consideration constraints such as the carrying capacity of each vehicle and the actual plant layout where vehicles can circulate. In real applications, other restrictions, particularly associated with the task scheduling system can complicate the model.

Typically, the goals of scheduling are related to the task processing times, use of resources (minimize the number of AGVs involved or minimize the total travel time of all vehicles, maximize the system throughput, etc) under certain constraints, such as deadlines, priorities and others. For scheduling purposes and considering the state-of-art, the main approaches are based on integer/linear/mixed programming [13] (meta) heuristics [14] and dispatching rules [12]. Scheduling System ([1]) typically aims to minimize an objective function that includes several system features. In [14] is proposed a formal analysis which defines simple heuristics at the 'coalition' level, like 'MinProcTime' and 'MinStepSum'. The Routing algorithm aims to optimize the paths performed by each AGV, avoiding the occurrence of collisions and deadlocks during the navigation. Some approaches are based in a Time Windows like in [11]. In these approaches a route is constructed considering that one point can only be visited one time for only one robot at a given time interval. This problem is similar to the Vehicle Routing Problem with Time Windows constraints (VRPTW) in which a time window $[s, t]$ is given so that each vehicle can perform the transportation tasks [6]. Mhring et al. [8] proposes an algorithm for the problem of AGVs routing without conflicts at the time of route calculation. These routing algorithms are on-line approaches with a better performance than static routing methodologies as demonstrated in [8]. The most used methodologies uses a static approach to calculates the routes offline in a static graph. In scenarios with many changes in the system or a high traffic volume, a dynamic approach is considered. Vivaldini et al. [4] proposed an algorithm based on the shortest path (enhanced Dijkstra Algorithm) and time windows to solve the bottlenecks and generate optimized conflicted free routes. Considering more general problems other approaches have been developed. Maza S. and Castagna P. [7] propose a predictive method for the routing problem without conflict.

Deadlocks are a difficult problem that any routing algorithm should consider. In [9] a control zone approach was used to solve the deadlock problem. The paths are separated in several disjoint zones inside the AGVs control zone. The most used rule is that only one vehicle can occupy a particular zone, in each time. Another interesting approach consists in a Time Windows used in [10]. Similar with the deadlock situation there is the livelock This differs from the other, because the processes involved in the livelock constantly change with regard to one another, none progressing. In this case, both the processes change its state in order to solve the situation, but fails. In the deadlock situation both the processes are stopped waiting one from another.

3 TEA* Description

During the routes' execution, several changes on the task information could occur, like obstacles and delays in the production or transportation systems. Account with this, on-line routing algorithms guarantee a better performance because recalculate its paths according with the new map's information. In the TEA* methodology, the paths are constantly recalculated in a loop, making it an on-line method. Collisions are avoided once TEA* updates the map information in each iteration. This way, unpredictable events are considered in the input map allowing to avoid the main challenges of any multi-AGV approaches like collisions and deadlocks.

The proposed algorithm is based on the known A* heuristic search. This method is largely used in many study areas of the Artificial Intelligence, concerning path finding problems.

A third dimension, the time, was added on the traditional A*. This component is an important contribution to a better prediction of the vehicles' movements during the execution time. Each node on the map have three dimensions: the Cartesian coordinates (x, y) and a representation of the discrete time as can be seen in the Figure 1. The time is represented with temporal layers, $k = [0, T_{Max}]$ (T_{max} denotes the maximum number of layers). Each temporal graph is a set of free and occupied/obstacles vertices.

3.1 Methodology

The route for each AGV is calculated during the temporal layers. The position of each robot in each temporal layer is known and shared with all vehicles. This grants the identification of future possible collisions, allowing them to be avoided at the beginning of the paths' calculation.

Consider a graph G with a set of vertices $V = \{v_0, v_1, ..., v_{NUM\_VERTICES}\}$ and edges $E = \{e_0, e_1, ...e_{NUM\_EDGES}\}$ (links between the vertices), with a representation of the time [0, T_{max}] (as can be seen in Figure 1). Each AGV can only starts and stops in vertices and a vertex can only be occupied by one vehicle at a temporal layer.

Considering a **single AGV approach**, similarly to the A* Algorithm described in[2], during the path calculation the next analyzed neighboring cell is dependent of a cost function, given by the sum of two terms: the distance to the initial vertex and the distance until the final point. In our approach the heuristic is given by the euclidean distance.

Due to the fact that the time is being considered, some modifications to the traditional A* were implemented. These modifications are important features to understand how TEA* calculates free-collision routes for a fleet of AGVs.

– *Definition 1: The neighbor vertices of a vertex j in the temporal layer k belongs to the next temporal layer given by* $k + 1$ (Figure 1)

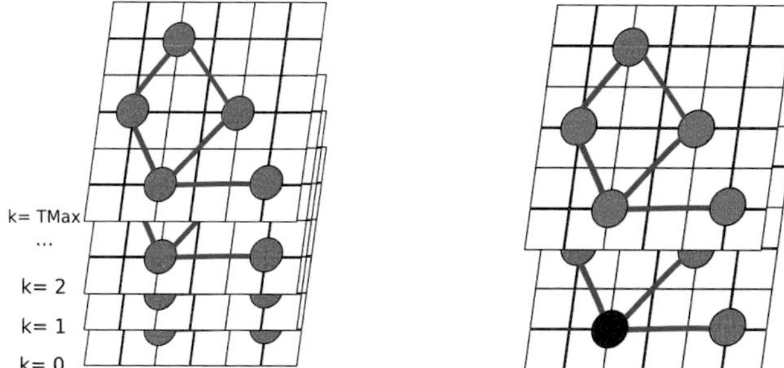

Fig. 1 The input map and the analyzed neighbors cells focusing the cell with the same position of the AGV

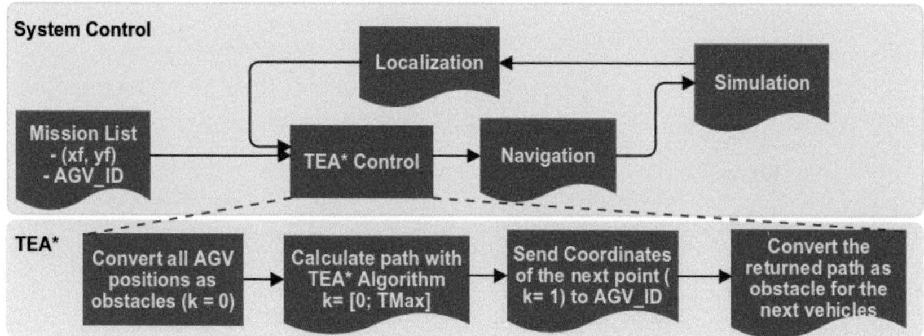

Fig. 2 Control Diagram for each algorithm iteration considering a Multi-AGV approach.

The total number of temporal layers depends of the required iterations to achieve the final point of the mission and the map complexity. Note that the larger the map is, more time layers are required. Also higher the system entropy, more temporal layers are required to achieve the mission goal.

– *Definition 2:The neighbor vertices of vertex j (v^j_{adj}) include the vertex containing the AGV current position, and all adjacent vertices in the next time component.* The set of neighbor vertices includes not only the adjacent vertices but also the vertex corresponding to the position in analysis. This property allows a vehicle maintain its position between consecutive time instants if any neighbor vertex is free, as in Figure 1. In this case, is not considered a zero value to the euclidean distance, it is assigned a constant heuristic value corresponding to the stopped movement.

In Figure 2 it is represented the system control diagram for each algorithm iteration for a **multi-AGV approach**. The first step is to convert the AGVs current positions as obstacles. This allows a vehicle to consider the other vehicles' positions as occupied vertices. In order to avoid deadlocks, this vertices are placed as obstacles only in $k = \{0, 1\}$, as explained in a previous publication [3]. The list of missions is analyzed and the path for each vehicle is calculated by the TEA* algorithm. Before moving

```
O ← o^0_vi
loop
  let λ^k_j = min_O{o^k_j.cost}
  if λ^k_j == vf then
    return
  end if
  for v^j_adj adjacent vertices of j do
    if λ^{k+1}_{v^j_adj} == 0 then
      {Only the non-visited vertices have heuristic zero}
      if heu^{k+1}_{v^j_adj} == 0 then
        CalculateHeuristic(heu^{k+1}_{v^j_adj});
        p_{j,k} = (v^j_adj, k + 1)
        CalculateCost(o^{k+1}_{v^j_adj})
        O ← o^{k+1}_{v^j_adj}
      else if dis^{k+1}_{v^j_adj} > dis^k_v + weight(j, v^j_adj) then
        UpdateCost(o^{k+1}_{v^j_adj});
        O ← o^{k+1}_{v^j_adj}
      end if
    end if
  end for
end loop
{Through the parent vertices to find the calculated path; }
```

Fig. 3 TEA* loop running in each AGV.

to the next mission, the complete path is converted as obstacle for the following missions and respective AGVs. This way, the coordination between the vehicles is ensured without collisions.

3.2 The Algorithm

The description of the TEA* for a single AGV is presented in Figure 3.

Parameters:

- λ^k_j : Value of vertex j in the time layer k (Free - 0 or Occupied/Obstacle - 1).
- $r^k_{i,j}$: AGV i occupies the vertex j in the k time layer.
- $O = \{o^k_j, ..\}$: Open list contains the vertex j in k instant. Each item contains the respective cost value, $o^k_j.cost$.
- vi : initial vertex.
- vf : final vertex.

- v_{adj}^{j} : adjacent vertex of j vertex.
- $p_{j,k} = (l, \tau)$: The vertex l in the time layer τ is the parent vertex of the j vertex in the instant k.
- heu_j^k : Heuristic Value for vertex j in k temporal layer.
- dis_j^k : Distance Value for vertex j in k temporal layer.
- $weight(j1, j2)$: Weight value of the edge $(j1, j2)$.

4 Deadlocks and Livelocks

A deadlock occurs when a set of shared resources are simultaneously needed by two or more entities (robots). All of those entities' execution will be frozen indefinitely. Similar with the deadlock situation is the livelock. In the livelock case, the robots involved execute actions in order to reverse the situation without any progress because both repeatedly move in the same way.

The formulation of TEA* avoids the occurrence of a deadlock however if some unpredictable situation cause it, the algorithm is capable to recover. Some examples with deadlocks were exposed in [3].

In this article will be presented an improvement of the TEA* in order to solve livelocks situations. This case, was not considered in the previous work.

Consider the Figure 4, left, with the initial and final positions given by the numbers and the stars respectively.

Fig. 4 Example of a Livelock Situation

Considering that the TEA* methodology executes missions sequentially for each robot, AGV0 only considers on its path calculation the starting positions from other AGVs. However, the following/remaining vehicles (AGV1, AGV2. ...) know the routes from all previous vehicles, for example: AGV1 knows the path of AGV0 and AGV2 considers both AGV0 and AGV1's paths.

Looking to the example situation of the Figure 4, left, AGV2 (blue) firstly arrives to its final position than AGV1 (green). A livelock occurs when AGV2 is in the final position and AGV1 wants to go to its destination point. When AGV2 finishes the time

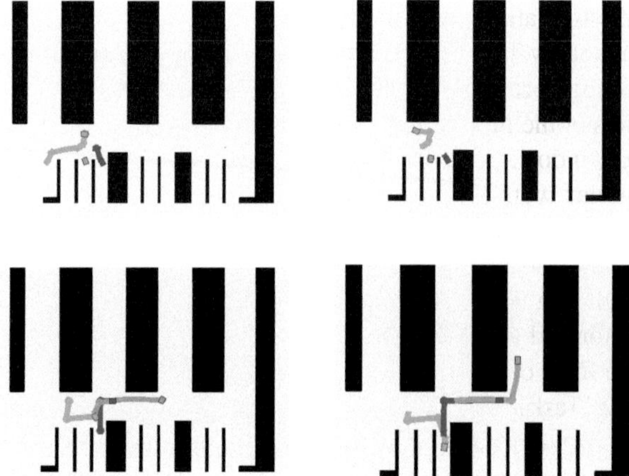

Fig. 5 Results Livelock Sequence: (1) top left, (2) top right, (3) bottom left, (4) right left.

to load/unload, and wants to proceed on its mission's list, AGV1 blocks its passage in the beginning of the corridor(Figure 4, right).

Remember that current AGV's positions are placed as obstacles only for $k = \{0, 1\}$. If these positions were placed as obstacles vertices for all temporal layers (from $k = 0$ until $k = T_{Max}$), AGV1 would not find any possible free path and the two vehicles would stand stopped forever.

This way, besides AGV1 knows AGV2 position, it cannot recalculates a new path because it does not know the desired AGV2 course. Although AGV2 knows the AGV1 path, he does not have any free route to escape.

A solution is to change the order in which the vehicles are analyzed. In this case, if AGV2 is the first vehicle to be analyzed, it knows that the adjacent vertices for $k = \{0, 1\}$ are occupied by AGV1, but could find a possible route in the following temporal layers $k = 2, \ldots$. Then AGV1 knows the path of AGV2, and deviates/calculates another free collision path. In TEA* implementation, the livelock resolution consists in verifying if a vehicle stand in the same position during a pre-defined time, and according with that, the order which the vehicles are analysed is changed.

Figure 5 represents a sequence of frames that shows the resolution of a livelock with three AGVs. It demonstrates that AGV1 (green) gives passage to the AGV3 (pink) and AGV2 (blue).

5 Experiments

In order to validate the capabilities of the TEA* Algorithm it was decided to compare it with a state-of-art alternative: the enhanced Dijkstra Algorithm accordingly with [4]. The main goal of a Multi-AGV system is the efficient and safety execution of a set of tasks. The simulation experiments were validated considering the distances of the paths calculated by TEA* for each vehicle.

The methodology proposed by [4] comprises two phases: in the first stage it calculates the routes of the forklifts considering the costs and then a Dijkstra algorithm based in a time-windows approach is executed to minimize the number of curves.

The experiments were performed considering a typical industrial scenario. The warehouse layout used for the simulation has 9 shelves, 3 docks and 6 depots in a total of 349 vertices and 330 edges. As previously mentioned each vehicle only starts and stops its path in a vertex, and a vertex can only be occupied by one vehicle at a time. The performed tests include 2 tasks that consists in the release of an array of 3x3 pallets in front of each dock, resulting in a total of 27 sub-tasks. These tasks are mentioned as 'Task1' and 'Task2' in Tables 1 and 2. The average speed for each AGV is $1.5m/s$. It was also considered the time that an AGV take to load/unload a pallet as 10 seconds.

Table 1 Values of the distance, in meters, for the TEA* Algorithm. SJF and TS represents the Shortest-Job-First and Tabu Search Methods, respectively.

		Task 1		Task 2	
		SJF	TS	SJF	TS
1 AGV	AGV1	1052	1077	994	1039
		1052	**1077**	**994**	**1039**
2 AGV	AGV1	461	533	453	515
	AGV2	563	510	566	493
		1024	**1043**	**1019**	**1008**
3 AGV	AGV1	301	351	297	302
	AGV2	394	303	357	351
	AGV3	325	364	349	327
		1020	**1018**	**1003**	**980**
4 AGV	AGV1	273	251	268	217
	AGV2	243	270	254	246
	AGV3	249	245	243	276
	AGV4	229	223	229	227
		994	**989**	**994**	**966**
5 AGV	AGV1	153	218	237	238
	AGV2	261	211	207	166
	AGV3	152	159	145	201
	AGV4	216	195	153	191
	AGV5	209	191	226	172
		991	**974**	**968**	**968**
6 AGV	AGV1	181	187	130	170
	AGV2	131	196	142	133
	AGV3	191	168	171	185
	AGV4	115	120	156	136
	AGV5	161	136	166	141
	AGV6	195	159	204	170
		974	**966**	**969**	**935**

The list of missions was generated using the following scheduling methods: Shortest-Job-First and Tabu Search accordingly with [5].

6 Results

Table 1 present the results for TEA* Algorithm. This table includes the traveled distance (in meters) by each vehicle for each task and each scheduling method. In article [5] are presented the results for the Enhanced Dijkstra Algorithm.

The evaluation of the results for the first task shows that the best result for the Enhanced Dijkstra Algorithm was achieved using the Tabu Search Method and 3 AGVs at $1045m$. The same case for TEA* corresponds to $1018m$. However the best result of TEA* for the task 1 was achieved for 5 AGVs at $974m$ that corresponds

Table 2 TEA* Comparison with the Enhanced Dijkstra Algorithm. The values represents the percentage of the difference between the values of Table 1 and the values presented in [4]

		Task 1		Task 2	
		SJF	TS	SJF	TS
1 AGV	AGV1	-0.43%	0.29%	1.82%	1.51%
		-0.43%	**0.29%**	**1.82%**	**1.51%**
2 AGV	AGV1	-1.72%	0.16%	0.22%	1.78%
	AGV2	-2.42%	-3.11%	1.07%	0.41%
		-2.10%	**-1.46%**	**0.69%**	**1.10%**
3 AGV	AGV1	-3.00%	4.78%	-1.33%	-2.27%
	AGV2	-4.61%	-13.54%	-0.28%	0.29%
	AGV3	-2.37%	1.13%	0.29%	-1.21%
		-3.43%	**-2.62%**	**-0.40%**	**-1.01%**
4 AGV	AGV1	-5.48%	-7.27%	-1.47%	-21.66%
	AGV2	-4.71%	6.46%	-2.31%	-0.81%
	AGV3	-5.11%	-16.15%	-8.65%	-0.72%
	AGV4	-5.80%	-8.88%	-4.58%	-2.16%
		-5.28%	**-6.80%**	**-4.24%**	**-6.67%**
5 AGV	AGV1	-10.90%	-5.93%	0%	2.15%
	AGV2	-2.39%	-4.89%	-5.48%	-5.68%
	AGV3	-10.38%	-15.69%	-7.64%	-8.22%
	AGV4	-9.96%	-3.34%	-7.83%	-5.91%
	AGV5	-3.29%	-8.79%	-8.50	-9.95%
		-6.92%	**-7.52%**	**-5.65%**	**-5.28%**
6 AGV	AGV1	-13.47%	-2.60%	-10.96%	-9.57%
	AGV2	-8.01%	0.28%	-12.35%	-6.99%
	AGV3	-11.09%	-13.51%	-8.56%	-5.61%
	AGV4	-14.63%	-27.38%	-1.27%	-11.69%
	AGV5	-7.76%	-11.85%	-11,23%	-7.24%
	AGV6	-3.66%	8.10%	-5.56%	-6.08%
		-9.66%	**-7.87%**	**-8.24%**	**-7.79%**

in the Dijkstra based Algorithm to 1054m. The analysis for the second task shows that the Enhanced Dijkstra Algorithm achieve the best solution for 1 AGV using the Shortest-Job-First at 976m. However the best result of TEA* for the task 2 was obtained using the Tabu Search and 6 AGVs at 935m.

Notes that is not straightforward to say that the increase of the number of AGVs, reduces the distance traveled by each vehicle. The system may become congested and could not compensate a higher number of AGVs, forcing them to calculate longer paths.

Generally, the TEA* Algorithm gives lower distances than Enhanced Dijkstra, except for 1 vehicle where the results are very similar or higher in some cases.

Table 2 compares the two routing algorithms presenting the difference, in percentage, of the values obtained with TEA* and Enhanced Dijkstra. The main conclusion is that with the increase of the number of AGVs, TEA* gives better results, optimizing the routes for each vehicle. This is due to the fact how TEA* grants the coordination for a multi-AGV system. The fact that a vehicle consider the paths of the other vehicles as obstacles and a representation of the time, gives it the possibility to wait some instants for the traveling of the previous AGV, instead of calculates a longer deviation. The major difference is obtained for the first task using the Shortest-Job-First and 6 AGVs where the difference between algorithms is -9.66%.

7 Conclusion and Future Work

In this article was presented a new routing algorithm suitable for Multi-AGV Systems. The increased use of AGVs in warehouses environments brings to the need in to select algorithms that fulfill its requirements considering takt time, and an optimized execution of a set of tasks. The fact that the time to be considered in TEA* allows it to prevent collisions and solve deadlock situations. These are the major challenges of any coordination system, and are essential conditions in an industrial scenario. TEA* seems to be a safety solution providing at the same time optimized routes.

The major contributions of this work are: (i) Validation of TEA* comparing it with a state-of-art alternative in a realistic layout of a warehouse environment; (ii) Analysis of the TEA* algorithm behavior with the increase number of vehicles (until 6 AGVs) avoiding collisions and deadlocks.

As part of the ongoing work a larger industrial scenario will be used, testing the robustness of the TEA* using a higher number of AGVs.

Acknowledgements The authors would like to thanks to the STAMINA Project. This project has received funding from the European Unions Seventh Framework Programme for research, technological development and demonstration under grant agreement no 610917. The authors also thanks to Project NORTE-07-0124-FEDER-000057 and NORTE-07-0124-FEDER-000060 financed by the North Portugal Regional Operational Programme (ON.2 O Novo Norte), under the National Strategic Reference Framework (NSRF), through the European Regional Development Fund (ERDF), and by national funds, through the Portuguese funding agency, Fundação para a Ciência e a Tecnologia (FCT).

This work is also financed by the ERDF European Regional Development Fund through the COMPETE Programme (operational programme for competitiveness) and by National Funds through the FCT Fundação para a Ciência e a Tecnologia (Portuguese Foundation for Science and Technology) within project FCOMP-01-0124-FEDER-037281.

References

1. Pinto, A.M., Rocha, L.F., Moreira, A.P., Costa, P.G.: Shop floor scheduling in a mobile robotic environment. In: Portuguese Conference on Artifcial Inteligence (EPIA) (2011)
2. Costa, P., Moreira, A., Costa, P.G.: Real-time path planning using a modified A* algorithm. In: Robotica 2009–9th Conference on Mobile Robots and Competitions (2009)
3. Santos, J., Costa, P., Rocha, L.F., Moreira, A.P., Veiga, G.: Time enhanced A*: towards to the development of a new approach for multi-robot coordination. In: IEEE International Conference on Industrial Technology (ICIT) (2015)
4. Vivaldini, K.C.T., Galdames, J.P.M., Pasqual, T.B., Sobral, R.M., Arajo, R.C., Becker, M., Caurin, G.A.P.: Automatic routing system for intelligent warehouses. In: IEEE International Conference on Robotics and Automation, vol. 1, pp. 1–6 (2010)
5. Vivaldini, K., Rocha, L.F., Martarelli, N.J., Becker, M., Moreira, A.P.: Integrated tasks assignment and routing for the estimation of the optimal number of AGVs (2015)
6. Larsen, A.: The dynamic Vehicle Routing Problem. PhD, Lyngby: Institute of Mathematical Modeling; Technical University of Denmark: Bookbinder Hans Meyer (2000)
7. Maza, S., Castagna, P.: Conflict-free AGV routing in bi-directional network. In: IEEE Int. Conf. On Emerging Tech. And Factory Automation, New York, pp. 761–764 (2001)
8. Möhring, R.H., et al.: Conflict-free real-time AGV routing. In: Hein, F., Dic, H., Peter, K. (eds.) Operations Research Proc. 2004, Berlin, pp. 18–24. Springer, Heidellberg (2004)
9. Yeh, M.-S., Yeh, W.-C.: Deadlock prediction and avoidance for zone-control AGVs. International Journal of Production Reserach **36**(10), 2819–2889 (1998)
10. Smolic-Rocak, N., Bogdan, S., Kovacic, Z., Petrovic, T.: Time Windows Based Dynamic Routing in Multi-AGV Systems. IEEE Transactions on Automation Science and Engineering **7**(1), 151–155 (2010)
11. Brysy, O., Gendreau, M.: Vehicle Routing Problem with Time Windows, Part I: Route Construction and Local Search Algorithms. Transportation Science **39**(1), 104–118 (2005)
12. Naso, D., Turchiano, B.: Multicriteria meta-heuristics for AGV dispatching control based on computational intelligence Systems Man and Cybernetics. IEEE Transactions on Part B: Cybernetics **35**(2), 208–226 (2005)
13. Udhayakumar, P., Kumanan, S.: Task scheduling og AGV in FMS using non-traditional optimization techniques. Int. J. simul. model. **9**, 28–39 (2010)
14. Cheng, Y.-L., Sen, H.-C., Natarajan, K., Teo, C.-P., Tan, K.-C.: Dispatching Automated Guided Vehicles in a Container Terminal Supply Chain Optimization. Applied Optimization **98**, 355–389 (2005)

Online Robot Teleoperation Using Human Hand Gestures: A Case Study for Assembly Operation

Nuno Mendes, Pedro Neto, Mohammad Safeea and António Paulo Moreira

Abstract A solution for intuitive robot command and fast robot programming is presented to assemble pins in car doors. Static and dynamic gestures are used to instruct an industrial robot in the execution of the assembly task. An artificial neural network (ANN) was used in the recognition of twelve static gestures and a hidden Markov model (HMM) architecture was used in the recognition of ten dynamic gestures. Results of these two architectures are compared with results displayed by a third architecture based on support vector machine (SVM). Results show recognition rates of 96 % and 94 % for static and dynamic gestures when the ANN and HMM architectures are used, respectively. The SVM architecture presents better results achieving recognition rates of 97 % and 96 % for static and dynamic gestures, respectively.

Keywords Gesture spotting · Robot programming · Robotic assembly · Industrial robot

1 Introduction

Robot programming is a time consuming and monotonous task. In order to speed up this task several approaches have been proposed, the majority of them work similarly to a traditional joystick and make use of accelerometers, electromyography and vision systems, besides that CAD-based systems have been extensible

---

N. Mendes(✉) · A.P. Moreira
Centre for Robotics and Intelligent Systems,
Institute for Systems and Computer Engineering Technology and Science, Porto, Portugal
e-mail: nuno.m.mendes@inesctec.pt, amoreira@fe.up.pt

P. Neto · M. Safeea
Centre for Mechanical Engineering of the University of Coimbra,
University of Coimbra, Coimbra, Portugal
e-mail: pedro.neto@dem.uc.pt, safeea@student.dem.uc.pt

L.P. Reis et al. (eds.), *Robot 2015: Second Iberian Robotics Conference*,
Advances in Intelligent Systems and Computing 418,
DOI: 10.1007/978-3-319-27149-1_8

explored. The big problem in these systems is the fact that they are limited to perform robot movements, or its user has to carry uncomfortable devices, and in some cases they are not intuitive. In order to make robots more user friendly and expand its use in industry, the way that a user interacts with a robot needs to be intuitive and no complex technical knowledge must be required.

The purpose of this study is the development of an intuitive robot programming system based on performed gestures. In order to achieve a feasible solution able to identify gestures performed by users, a vision system is used and gestures performed by a user are recognized by three different methods. The effectiveness of the proposed system is assessed through practical experiments identifying the recognition method that presents the best performance. Additionally, the system is tested in an industrial task of pin assembly in car doors.

2 State of the Art

Vision technology has been used in the recognition of human hands, face and body behaviors [1,2,3]. A major advantage of this technology is the fact that a user does not have to carry any device during the interaction process making the process more natural. In the last developments neither a mark is needed to be held by the user. A necessary requirement is the human presence in the vision sensor's field of view. Vision systems have difficulty producing robust information when facing cluttered environments. Some vision-based systems are view dependent, require a uniform background and illumination, and a single person (full-body or part of the body) in the camera field of view. In addition, occlusion of some reference points can occur frequently which need to be coped with.

There have been an increasing interest for gesture recognition using vision-based interfaces, for hand, arm and full-body gesture recognition [4]. Vision-based solutions have been used for real-time gesture spotting applied to the robotics field. A recent study presents a motion tracking system combining vision, inertial and magnetic sensing for spatial robot programming using gestures [5]. The main limitation of this approach is related to complex path execution. An American sign language (SL) word recognition system, that uses as interaction devices both a data glove and a motion tracker system, is presented in [6]. Inertial sensors have also been explored for different gesture-based applications [7], [8].

A number of machine learning techniques have been proposed to deal with gesture recognition, the most explored techniques rely on artificial neural network (ANN), hidden Markov models (HMM) and support vector machine (SVM). Mitra and Acharya provide a complete overview of techniques for gesture pattern recognition [9]. ANN-based problem solving techniques have been demonstrated to be a reliable tool in gesture recognition, presenting good learning and generalization capabilities. ANNs have been applied in a wide range of situations such as the recognition of continuous hand postures from gray-level video images, gesture recognition having acceleration data as input [10], full-body motion recognition for robot teleoperation and SL recognition [6]. The capacity of recurrent neural networks (RNNs) for modeling temporal sequence learning has also been demonstrated [11]. Nevertheless, RNNs are still difficult to train. HMMs are stochastic methods known for their application in temporal pattern recognition, including gesture spotting [12]. A survey on human-computer interaction using gesture recognition and vision-based systems as interaction technology was presented in [13]. It is concluded that research efforts are required to reliably recognize gestures in continuous and in relatively large libraries of gestures.

In order to recognize dynamic gestures different approaches have been employed, for example using discrete HMM to recognize online dynamic human hand gestures [14] and using HMM for full body gesture recognition [15]. Recognitions rates (RR) above 84 % were reported for a collection of seven dynamic gestures. A real-time system based on HMM was proposed by Kurakin et al. [16] for dynamic hand gesture recognition. This system takes into account variations in speed and human behavior as well as in hand orientations. A recognition rate of at least 76 % was achieved with this approach. Automatic recognition of facial emotion based on feed forward ANN and support vector regressors was presented by Zhang et al. [17]. An interesting study in the field reports a neural-based classifier for gesture recognition with an accuracy of over 99% for a library of only 6 gestures [18]. Other authors concluded that a hand contour-based neural network training is faster than complex moment-based neural network training but in the other hand the former proved to be less accurate (71%) than the latter (86%) [19].

3 Proposed Approach

A gesture recognition system is proposed in this study to command and program an industrial robot. This gesture recognition system relies on an infra-red stereo vision system that provides features about the human hands of a user to an intelligent recognition architecture. Three recognition architectures are approached and their results are compared among them. One of the recognition architectures based on ANN is used to recognize static gestures. On the other hand dynamic gestures are recognized in a second architecture based on HMM. Finally, a third architecture based on SVM is used to recognize the same static and dynamic gestures.

A Leap Motion Controller (LMC) provides a set of features x, also called a data frame, about human hands and wrists. In this study some of these features were chosen to be used in recognition of human hand gestures. The features used are presented in Table 1 for static gestures and in Table 2 for dynamic gestures. Each set of features is acquired at a rate of 40 Hz. The static gesture recognition is carried out at 40 Hz and the dynamic gesture recognition is carried out at 3 Hz. The frame acquisition rate for recognition of dynamic gestures is lower because thirteen sets of features are required to perform the recognition of the dynamic gesture while just one set of features is required to perform the recognition of a static gesture. Different features are used for static and dynamic gestures, this is because:

- Number of gestures in the considered libraries;
- Number of differentiation variables required;
- Methods take a lot of processing time when the number of variables is higher;
- These gestures are better characterized/represented by different features.

Table 1 Features of static gestures

Data	Description
x_1, x_2, x_3	x, y and z components of the normal vector to the palm hand, respectively
x_4, x_5, x_6	x, y and z components of the palm hand direction vector, respectively
x_7, x_8, x_9	x, y and z components of the thumb finger direction vector, respectively
x_{10}, x_{11}, x_{12}	x, y and z components of the index finger direction vector, respectively
x_{13}, x_{14}, x_{15}	x, y and z components of the middle finger direction vector, respectively
x_{16}, x_{17}, x_{18}	x, y and z components of the ring finger direction vector, respectively
x_{19}, x_{20}, x_{21}	x, y and z components of the pinkie finger direction vector, respectively

Table 2 Features of dynamic gestures

Data	Description
x_1, x_2, x_3	x, y and z components of the normal vector to the palm hand, respectively
x_4	grab property
x_5	x component of the thumb finger direction vector

3.1 *Artificial Neural Network*

The ANN architecture used in this study consists of 21 input neurons, a hidden layer with 21 neurons and in the output layer is used 12 neurons. Table 3 presents a detailed parametrization of the ANN which is shown in Fig. 1. $\mathbf{W}$ represents a weight matrix, $\mathbf{b}$ is a weight vector, φ represents an activation function and y is the output of the ANN.

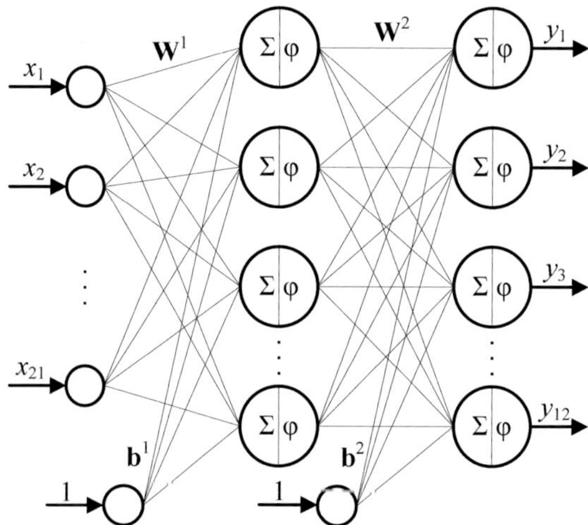

Fig. 1 ANN architecture used in the recognition of static gestures

Table 3 ANN parameters

Parameter	Value
Number of neurons in the input layer	21
Number of neurons in the hidden layer	21
Number of neurons in the output layer	12
Activation function in the hidden layer	asymmetric sigmoid function, σ = 1
Activation function in the output layer	asymmetric sigmoid function, σ = 1
Learning coefficient	0.25
Momentum	0.1
Number of training cycles	10 000
Updating rate (ms)	25

3.2 Hidden Markov Model

To recognize dynamic gestures the architecture based on HMMs shown in Fig. 2 is used. s represents the states of the HMM and a is the state transition probability. Table 4 presents the parametrization used in this architecture. A HMM based on left-right model is used for each dynamic gesture.

Table 4 HMMs parameters

Parameter	Value
Number of HMM	10
Number of states	20
Number of symbols	3
Stopping criterion - tolerance	0.01

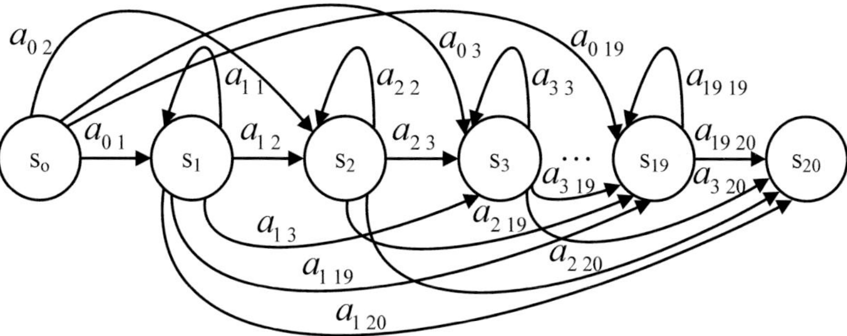

Fig. 2 Left-right model of a HMM

3.3 Support Vector Machine

An algorithm based on SVM was trained to recognize static and dynamic gestures. The number of SVMs used in this system is the same as the number of predefined gestures, i.e. 12 SVMs for static gestures and 10 SVMs for dynamic gestures. A comparative analysis among the outputs of the different SVMs is done.

In order to reduce the size of the data and at the same time preserve its information, a transformation of the coordinate system was implemented. The data were normalized and transposed from Euclidian coordinate system to spherical coordinate system resulting in unit vectors represented only by two angles.

SVM is inherently a binary classifier that provides always an output as validated. However, the validated output can be wrongly classified and thus leading to false validations. In order to overcome these false classifications, the one versus one method was used and extended to deal with multi-class classification problems.

n SVM algorithms are trained for n gestures, each algorithm is taught to distinguish one of the gestures apart from the rest. During the prediction the input feature vector feeds all of the n pre-trained SVMs and each one provides an outcome. The output of each SVM algorithm can assume one of the three following cases:

- A positive outcome from only one of the n pre-trained SVM algorithms. In this case the gesture associated with that algorithm is assumed as recognized;
- Two or more positive predictions from the n algorithms. In this case the result is assumed as uncertain considering the gesture as unrecognized;
- Negative predictions from all of the n algorithms, then the gesture is considered as unrecognized.

4 Experiments

In order to test and assess the feasibility of the different proposed architectures, they are used in a practical experiment that consists in performing each gesture an hundred times and estimate its recognition rate. Figures 3 and 4 illustrate the static and dynamic gestures, respectively, used in this experiment.

Additionally, each gesture was associated to a robot command or function as can be seen in Figs. 3 and 4. The system was used to generate of robot code for an industrial task of pin assembly in car doors. In a first stage a data file is generated on-line with all information required to program an industrial robot, such as position, orientation and speed. After that, in a second stage, the previous data is pros-processed and the robot code is generated.

In order to develop the gesture recognition system, execute it and perform the tests, a personal computer with the characteristics shown in Table 5 was used as well as an industrial robot ABB IRB140 equipped with IRC5 controller, a Leap Motion Controller and mechanical tools for the assembly task.

Table 5 Personal computer characteristics used in this study

Processor	Intel® Core™ i7-4700HQ CPU @ 2.40 GHz (8CPUs)
Memory	8GB
Operating System	Microsoft Windows 8.1 64 bits

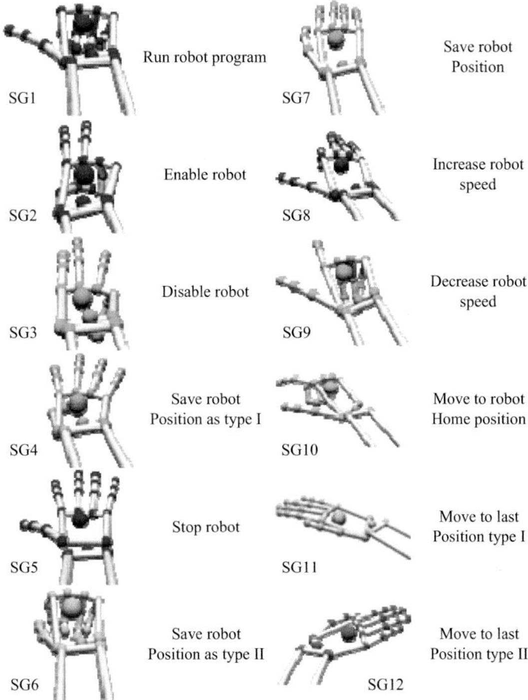

Fig. 3 Static gestures

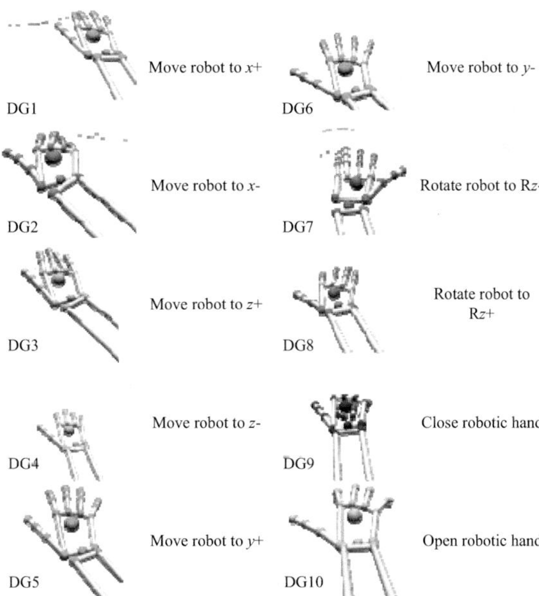

Fig. 4 Dynamic gestures

5 Results and Discussion

The results of the experiment are presented in Tables 6, 7, 8 and 9. In general the SVM architecture provides better RR than the other architectures (ANN and HMM). For the library of static gestures the ANN architecture provided a RR of 96 % while the SVM architecture provided a RR of 97 %. These results are clearly better than the study carried out by Badi et al. [19] who achieved a RR of 86% for a library of just six gestures. The full recognition was not achieved due to inability of the system in differentiate dissimilarities of the gestures. This imposes to the user a more accurate manner in performing the gesture. Another reason for wrong recognition is because the fact that the user performs some involuntary movements which provoke some dissimilarity in gestures. Finally, the occurrence of occlusion between the user hands or among his fingers leads the LMC to make wrong judgments about the scene and outputting wrong features. This effect is clearly visible in the recognition of the SG2 and SG3 which are wrongly recognized between them. Other problem that frequently happens is the recognition of a gesture in the transition between two different gestures. The SG7 is frequently recognized between SG11 and SG12 and vice versa. In fact, the SG7 is recognized because it is really performed during a fraction of milliseconds. In order to cope with this situation, a condition was introduced. This condition consists of validation a static gesture just after it to be continuously recognized during a period of time. In resume, the SVM architecture are able to better differentiate the 12 static gestures proposed in this test case.

In relation to the library of dynamic gestures a RR rate of 94 % was achieved with the HMM architecture while the SVM architecture provided a RR of 96 %. These are good results comparing to the study carried out by Bertsch and Hafner [15], which achieved a RR of 84 % for a library of just seven gestures, and Kurakin et al. [16] that achieved a RR of just 76 %. The hand shake introduce some wrong judgements about a DG this effect is higher with increasing distance between human hand and the center of the LMC. Some gestures trend to be recognized even when there is no intention in perform them, an example of this kind of gesture is DG4. This occurrence is because there is a general tendency to perform a hand moving downward (DG4) when perform any gestures.

It is clear that the SVM architectures provide better outcomes for the gesture recognition system. However, the adopted SVM architecture takes longer processing time than any of the other two architectures leading to slower responses. If the gesture recognition system was not limited by aquisition of data frame, robust static gesture recognitions could be obtained for the SVM architecture at a rate of 10 Hz while the ANN architecture could run at a rate of 80 Hz. Although the SVM architecture provides a slower system, it is effective and quick enough to be used as robot interaction technology.

Table 6 Confusion matrix for SG – ANN architecture

	Recognized gestores											
Performed gestures	#1	#2	#3	#4	#5	#6	#7	#8	#9	#10	#11	#12
#1	90	5	5	0	0	0	0	0	0	0	0	0
#2	0	80	20	0	0	0	0	0	0	0	0	0
#3	0	5	95	0	0	0	0	0	0	0	0	0
#4	0	0	0	100	0	0	0	0	0	0	0	0
#5	0	0	0	0	100	0	0	0	0	0	0	0
#6	0	0	0	0	0	95	5	0	0	0	0	0
#7	0	0	0	0	0	0	100	0	0	0	0	0
#8	0	0	0	0	0	0	0	100	0	0	0	0
#9	0	0	0	0	0	0	0	0	100	0	0	0
#10	5	0	0	0	0	0	0	0	0	95	0	0
#11	0	0	0	0	0	0	2	0	0	0	98	0
#12	0	0	0	0	0	0	0	0	0	0	0	100

Table 7 Confusion matrix for DG – HMM architecture

	Recognized gestures									
Performed gestures	#1	#2	#3	#4	#5	#6	#7	#8	#9	#10
#1	83	0	0	17	0	0	0	0	0	0
#2	0	93	0	5	2	0	0	0	0	0
#3	0	0	95	5	0	0	0	0	0	0
#4	0	0	0	100	0	0	0	0	0	0
#5	0	0	0	2	98	0	0	0	0	0
#6	0	0	0	9	2	89	0	0	0	0
#7	0	0	1	1	0	0	98	0	0	0
#8	0	0	0	4	0	0	0	96	0	0
#9	0	0	0	5	3	1	0	0	90	1
#10	0	0	0	0	0	0	0	0	0	100

Table 8 Confusion matrix for SG – SVM architecture

	Recognized gestores											
Performed gestures	#1	#2	#3	#4	#5	#6	#7	#8	#9	#10	#11	#12
#1	100	0	0	0	0	0	0	0	0	0	0	0
#2	0	93	7	0	0	0	0	0	0	0	0	0
#3	0	7	93	0	0	0	0	0	0	0	0	0
#4	0	0	0	100	0	0	0	0	0	0	0	0
#5	6	0	0	0	94	0	0	0	0	0	0	0
#6	7	0	0	0	0	93	0	0	0	0	0	0
#7	0	0	0	0	0	0	100	0	0	0	0	0
#8	0	0	0	0	7	0	0	93	0	0	0	0
#9	0	0	0	0	0	0	0	0	100	0	0	0
#10	0	7	0	0	0	0	0	0	0	93	0	0
#11	0	0	0	0	0	0	0	0	0	0	100	0
#12	0	0	0	0	0	0	0	0	0	0	0	100

In order to obtain a feasible system with high recognition rates, the same static gesture has to be recognized by one of the recognition architectures for ten times consecutively before the gesture is considered as validated. On the other hand, a dynamic gesture is just validated after it is recognized by the system twice. Thirteen data frames, photos of the gesture, are required to proceed dynamic gesture recognition.

The test bed performed with the presented system to command and program an industrial robot in the execution of an assembly task was carried out successfully. The introduction of conditioning in the recognition of the gestures, i.e. a SG and a DG is just considered as recognized after the system have already identify it five and two times consecutively, respectively, leads to achieve a RR of 100 %. Fig. 5 illustrated the execution of the robotic assembly tasks being the industrial robot commanded by the proposed gesture recognition system.

Instructing the robot with gestures results in a dramatic decrease in programming time, compared to traditional robot programming methods (using a teach pedant). Recurring to the proposed gesture recognition system to program the industrial robot in the execution of the proposed pin assembly task allowed the user to save about 20 % of the time that would be required if the traditional programing system was used. In addition, this system is intuitive and reduced setup time is required.

To the best of our knowledge, approaches similar to ours have never been successfully deployed in real industrial or other scenarios. Thus, we see the main value of our system, as witnessed by our experiment, to operate robustly and solve real industrial problems.

Table 9 Confusion matrix for DG – SVM architecture

		Recognized gestures									
		#1	#2	#3	#4	#5	#6	#7	#8	#9	#10
Performed gestures	#1	91	0	0	9	0	0	0	0	0	0
	#2	0	90	0	7	3	0	0	0	0	0
	#3	0	0	93	7	0	0	0	0	0	0
	#4	0	0	0	100	0	0	0	0	0	0
	#5	0	0	0	0	100	0	0	0	0	0
	#6	0	0	0	10	0	90	0	0	0	0
	#7	0	0	0	0	0	0	100	0	0	0
	#8	0	0	0	2	0	0	0	98	0	0
	#9	0	0	0	4	0	0	0	0	96	0
	#10	0	0	0	0	0	0	0	0	0	100

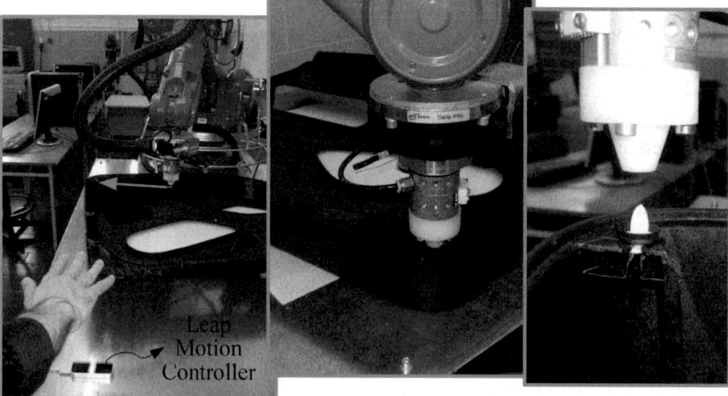

Fig. 5 Robot executing pin assembly task

6 Conclusions and Future Work

A gesture recognition system was proposed to command and program an industrial robot. Static and dynamic gesture features are provided to the system that has implemented three different recognition architectures, i.e. an ANN to static gestures, a HMM to dynamic gestures and a SVM architecture to static and dynamic gestures. All of the architectures provided high RR being the SVM the best one with 97 % and 96 % for a library of 12 static gestures and 10 dynamic gestures respectively. The similarity of the gestures and occlusion problems are preventing the RR improvement.

Finally, as any other system intended for industrial use, an industrial robot equipped with this gesture recognition system has been deployed and tested in an industrial pin assembly task, showing its robustness and effectiveness. We believe that our gesture recognition system constitutes a significant step towards achieving more friendly robots, and that such an approach can ultimately increase competitiveness of manufacturing companies.

Acknowledgements This research was supported by FEDER funds through the program COMPETE (Programa Operacional Factores de Competitividade), under the project NORTE-07-0124-FEDER-000060 (COOPERATION).

References

1. Ren, Z., Yuan, J., Meng, J., Zhang, Z.: Robust Part-Based Hand Gesture Recognition Using Kinect Sensor. IEEE Trans. Multimed. **15**, 1110–1120 (2013)
2. Huang, P.-C., Jeng, S.-K.: Human body pose recognition from a single-view depth camera. In: 2012 IEEE International Conference on Systems, Man, and Cybernetics (SMC), pp. 2144–2149. IEEE (2012)
3. Seal, A., Bhattacharjee, D., Nasipuri, M., Basu, D.K.: Thermal human face recognition based on GappyPCA. In: 2013 IEEE Second International Conference on Image Information Processing (ICIIP-2013), pp. 597–600. IEEE (2013)
4. Kirishima, T., Sato, K., Chihara, K.: Real-time gesture recognition by learning and selective control of visual interest points. IEEE Trans. Pattern Anal. Mach. Intell. **27**, 351–364 (2005)
5. Lambrecht, J., Kruger, J.: Spatial programming for industrial robots based on gestures and Augmented Reality. In: 2012 IEEE/RSJ International Conference on Intelligent Robots and Systems, pp. 466–472. IEEE (2012)
6. Oz, C., Leu, M.C.: Linguistic properties based on American Sign Language isolated word recognition with artificial neural networks using a sensory glove and motion tracker. Neurocomputing **70**, 2891–2901 (2007)
7. Neto, P., Pires, J.N., Moreira, A.P.: High-level programming and control for industrial robotics: using a hand-held accelerometer-based input device for gesture and posture recognition. Ind. Robot. An. Int. J. **37**, 137–147 (2010)
8. Neto, P., Pires, J.N., Moreira, A.P.: Accelerometer-based control of an industrial robotic arm. In: RO-MAN 2009 - The 18th IEEE International Symposium on Robot and Human Interactive Communication, pp. 1192–1197. IEEE (2009)

9. Mitra, S., Acharya, T.: Gesture Recognition: A Survey. IEEE Trans. Syst. Man Cybern. Part C Applications Rev. **37**, 311–324 (2007)
10. Yang, J., Bang, W., Choi, E., Cho, S., Oh, J., Cho, J., Kim, S., Ki, E., Kim, D.: A 3D hand-drawn gesture input device using fuzzy ARTMAP-based recognizer. J. Syst. Cybern. Informatics **4**, 1–7 (2006)
11. Yamashita, Y., Tani, J.: Emergence of functional hierarchy in a multiple timescale neural network model: a humanoid robot experiment. PLoS Comput. Biol. **4** (2008)
12. Peng, B., Qian, G.: Online gesture spotting from visual hull data. IEEE Trans. Pattern Anal. Mach. Intell. **33**, 1175–1188 (2011)
13. Badi, H.S., Hussein, S.: Hand posture and gesture recognition technology. Neural Comput. Appl. **25**, 871–878 (2014)
14. Wang, X., Xia, M., Cai, H., Gao, Y., Cattani, C.: Hidden-Markov-Models-Based Dynamic Hand Gesture Recognition. Math. Probl. Eng. (2012)
15. Bertsch, F.A., Hafner, V. V.: Real-time dynamic visual gesture recognition in human-robot interaction. In: 9th IEEE-RAS International Conference on Humanoid Robots, pp. 447–453. IEEE (2009)
16. Kurakin, A., Zhang, Z., Liu, Z.: A real time system for dynamic hand gesture recognition with a depth sensor. In: 20th European Signal Processing Conference (EUSIPCO 2012), pp. 1975–1979 (2012)
17. Zhang, Y., Zhang, L., Hossain, M.A.: Adaptive 3D facial action intensity estimation and emotion recognition. Expert Syst. Appl. **42**, 1446–1464 (2015)
18. El-Baz, A.H., Tolba, A.S.: An efficient algorithm for 3D hand gesture recognition using combined neural classifiers. Neural Comput. Appl. **22**, 1477–1484 (2012)
19. Badi, H., Hussein, S.H., Kareem, S.A.: Feature extraction and ML techniques for static gesture recognition. Neural Comput. Appl. **25**, 733–741 (2014)

Generic Algorithm for Peg-In-Hole Assembly Tasks for Pin Alignments with Impedance Controlled Robots

Michael Jokesch, Jozef Suchý, Alexander Winkler, André Fross and Ulrike Thomas

Abstract In this paper, a generic algorithm for peg-in-hole assembly tasks is suggested. It is applied in the project GINKO were the aim is to connect electric vehicles with charging stations automatically. This paper explains an algorithm applicable for peg-in-hole tasks by means of Cartesian impedance controlled robots. The plugging task is a specialized peg-in-hole task for which 7 pins have to be aligned simultaneously and the peg and the hole have asymmetric shapes. In addition significant forces are required for complete insertion. The initial position is inaccurately estimated by a vision system. Hence, there are translational and rotational uncertainties between the plug, carried by the robot and the socket, situated on the E-car. To compensate these errors three different steps of Cartesian impedance control are performed. To verify our approach we evaluated the algorithm from many different start positions.

Keywords Peg-In-Hole · Impedance control · Electric vehicles · Charging

1 Introduction

Many assembly tasks exist which are similar to peg-in-hole tasks. A specialized peg-in-hole tasks occurs when the robot takes a plug to connect it automatically to an E-car at a recharging station. In our project which is part of the GINKO project [3] we intent to build a force-fit and form-fit connection according to the norm DIN EN 62196-2. It is a big difference to recharging conventional fuels [1] [2] where the fuel hose diameter is much smaller than the tank opening. Considering this, we have high requirements on the accuracy of the positioning in our task (Fig. 1). The task can be

M. Jokesch(✉) · J. Suchý · A. Winkler · A. Fross · U. Thomas
Technische Universität Chemnitz, Reichenhainer Straße 70, 09126 Chemnitz, Germany
e-mail: {michael.jokesch,jozef.suchy,ulrike.thomas}@etit.tu-chemnitz.de,
winkler3@hs-mittweida.de, fross@coremountains.de
http://www.tu-chemnitz.de

This paper is supported by the Federal Ministry of Education and Research (Germany), under the funding code 03IPT505A. Responsibility for content lies with the authors.

L.P. Reis et al. (eds.), *Robot 2015: Second Iberian Robotics Conference*,
Advances in Intelligent Systems and Computing 418,
DOI: 10.1007/978-3-319-27149-1_9

classified as a complex peg-in-hole problem. The peg-in-hole task is well known from [5] [6] [7], representative for many more papers. Unfortunately, most of these solutions assume a cylindrical peg and a slightly enlarged cylindrical hole. Whitney [4] published an interesting approach with the idea of passive compliance that is also used in this work. In general, the insertion process is mostly divided into two phases. The search phase and the insertion phase.

During the search phase, the hole is detected and the peg center is moved toward the center of hole. The accuracy of this phase usually has to be within the clearance which is defined as the difference between the peg diameter and the hole diameter [5]. Approaches suggested to align the peg with the hole are using vision systems, force/torque information with blind-search-algorithms [6] or neural networks. An approach that uses a neural network was published by Newman [7]. The neural network takes a pre-recorded map of positions vs. torques as a basis.

Approaches with force/torque feedback, e.g. force/torque maps with particle filters [8], are very interesting when the vision system reaches its limits which could be caused by the illumination of the environment, the camera resolution, the field of view and/or the accuracy of hole detection. One disadvantage of the blind-search strategies can be the time duration. Depending on the size of the workspace and the uncertainties of prior positioning blind searching can be very time consuming. For smarter and more effective search, Newman et al. developed the strategy of position vs. torque map [7]. When the peg comes into contact with the surface the torques are measured and compared with the map. With this information it is calculated which direction has the highest probability for finding the hole. The big problem of this approach is its inflexibility. Already a small change in the workspace requires a complete recalibration of the map. Additionally, it is assumed that the peg has always the same orientation to the surface. In order to understand the peg-in-hole problems better and find new approaches the human behavior was analyzed several times, e.g. in [9], [10]. Although the peg-in-hole is an old problem in robotics [11] it is still an actual problem. This is why many groups still perform research into it [12], [13], [14].

The approaches presented above are not quite useful in solving our problem because the workspace changes steadily (different cars) and a simple blind-search is too time-consuming. Furthermore, we deal with valuable objects – cars, i.e. it has to be worked with high caution. Another important issue is the special shape of the charging plug (Fig. 1).

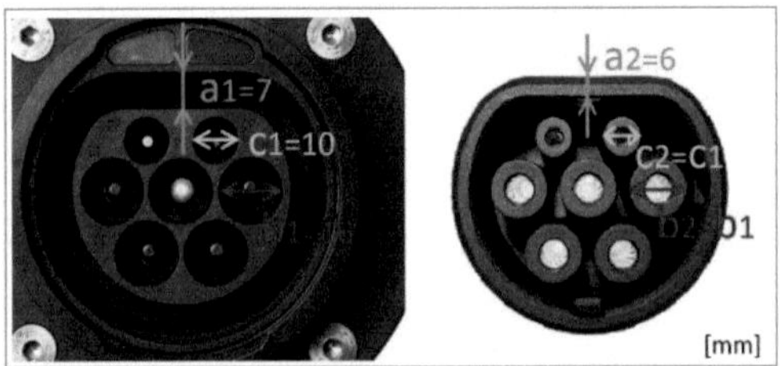

Fig. 1 Right image: EU standardized charging plug (blue) and plug socket (black) with 7 pins and a flattened side. The socket is situated at the car and the plug is moved by the robot. Left image: frontal view of the socket and the plug. b1/2 and c1/2 fit exactly (no clearance). Between a1 and a2 clearance is 1 mm which makes threading easier.

It has 7 different sized pins and a flattened side for protection against polarity reversal. During our research, we did not find work concerning the peg-in-hole task with such a high number of different pegs. The algorithm needs to compensate for translational and rotational displacements between the connector and the socket; otherwise the connection will not be possible. In [15] they use a kind of impedance control for peg-in-hole assembly of complex shaped workpieces. The impedance control method is used in our work in combination with a kind of blind-search algorithm. This gives rise to search strategy, which can be transferred to many other peg-in-hole problems with very small effort.

In the next section, the problem is discussed in more detail and the experimental set-up is presented. In Section 3 the proposed force/torque based control algorithm is explained and in Section 4 it is tested and evaluated. At the end the results are summarized and the conclusions are drawn.

2 Problem Formulation and Test Set-Up

The initial situation of this work is the prior coarse positioning by means of a vision system. The goal is to align plug's pose with socket pose and thus enable the robot to insert the charging plug into the plug socket. Differences between the poses are possible in every degree of freedom (DOF), see Fig. 2. These misalignments appear together, that means it is important that compensation works for every DOF simultaneously. There is no clearance between the pins and the holes. Only the external shape has a clearance value of 1 mm (shown in Fig. 1). After the alignment a force of 160 to 200 Newton in insertion direction is necessary for full connection. It results in a force closure connection. In summary, following steps need to be implemented:

- Apply vision for rough positioning (not presented in this paper)
- Using search strategies (align plug with socket)
- Establishing contact and insertion (generate required force in insertion direction)

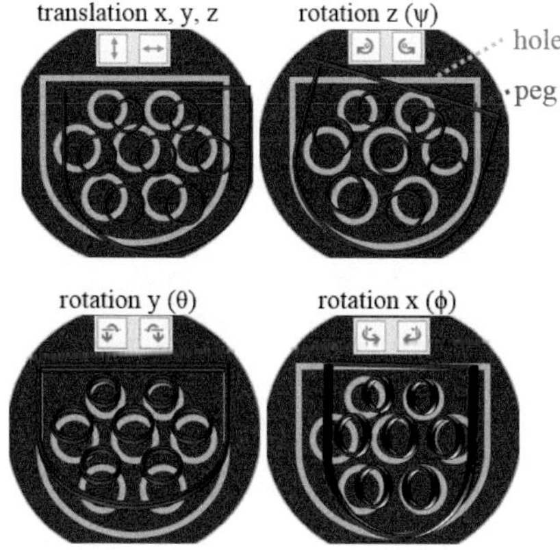

Fig. 2 Possible positioning errors due to the vision system

To analyze the algorithms we have built a test station in laboratory and also can work at a real electric car (Fig. 3). The position of the socket can easily be changed to simulate different initial situations. The Robot used for experiments is KUKA LWR iiwa 7 R800 with seven joints. Every joint is equipped with a torque sensor. Thus, it is possible to implement compliance control [16], which enables the proposed algorithm. The impedance control supports the insertion process and simultaneously protects the environment, e.g. car or human, from being damaged.

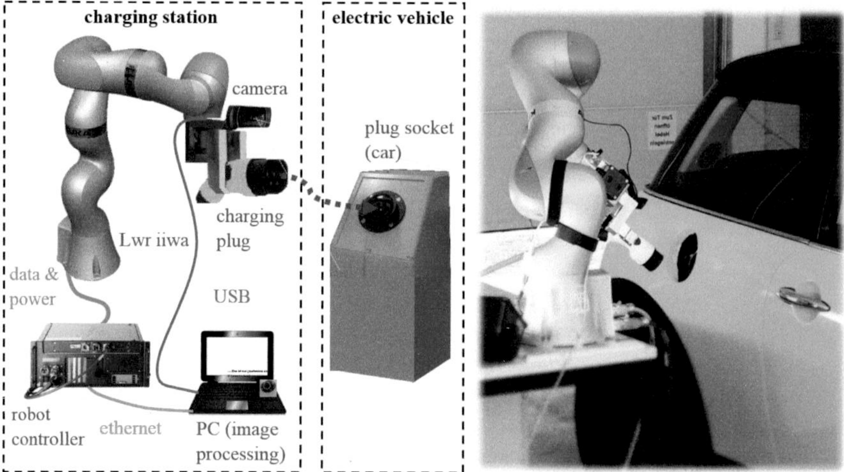

Fig. 3 Schematic test set-up in the laboratory (left) and at the real electric vehicle (right)

3 Insertion Algorithm

Before definition and parametrization of the search algorithm, first step is to roughly determine the error values of poses estimated by vision system to learn the size of the necessary search area. In Table 1 there are the average values, the standard deviation and the maximum values of the measured displacements for all translational and rotational DOFs. The values were determined during 25 repetitions of visual detection. These displacements have to be compensated for successful insertion. Humans perform similar tasks absolute intuitively and can even do it without any visual information. Humans feel the contact forces if the plug does not fit and adjust it to the socket without much problems. The idea is to use the compliant behavior of the LWR iiwa to imitate human behavior. For making robot's behavior compliant we use impedance control. Here, the impedance Z models the relationship between forces and positions. It is defined by means of the inertia matrix M, the damping matrix D and the stiffness matrix K. The controller input is the desired force $\vec{F}_d$ and the actual measured force $\vec{F}$. Concerning the inputs and the desired impedance, the corrective position $\vec{x}_c$ results. This position and the desired position $\vec{x}_d$ add up to the commanded target position $\vec{x}_t$ (Fig. 4). The robot's behavior is based on a 6-dimensional spring-damper system. The controller strains the spring until $\vec{F}_d$ is reached. If $\vec{F}_d$ is exceeded the robot yields to the forces.

Table 1 Displacement Values due to Vision System

DOF	Average of errors	Standard deviation	Maximum Error
x	0.61 mm	1.09 mm	4.10 mm
y	1.16 mm	1.77 mm	3.03 mm
z	2.36 mm	2.16 mm	7.35 mm
Ψ	3.01 °	0.64 °	5.10 °
Θ	2.81 °	0.88 °	4.51 °
ϕ	2.40 °	0.72 °	4.67 °

$$\vec{x}_t = \vec{x}_d + \vec{x}_c$$
$$\vec{x}_c = Z(s)^{-1} \cdot \left(\vec{F}_d - \vec{F}\right) = (Ms^2 + Ds + K)^{-1} \cdot \left(\vec{F}_d - \vec{F}\right) \tag{1}$$

Our algorithm (Algorithm 1) consists of three steps. Initially (line 5)) it is searched for the contact between the plug and the surface of the socket in z-direction. The second step (line 6) is a compensation of displacement between a pose of the plug and a pose of the socket, i.e. in this step we implement the main part of the search algorithm. In the third step (line 8) the required force is generated for complete insertion while small remaining displacements are compensated. Below these three steps will be explained in detail.

Algorithm 1. Insert the plug into the socket

```
   Input: Stiffness and damping values for compliance
          plus frequencies and force amplitudes for the
          search motion & values of break conditions
1    K[],D[];  //stiffness and damping values
2    sz_f;     //distance in z for successful searching
3    sz_in;    //required distance for complete insertion
   Output: was insertion process successful or not
4    success=false;
5    move towards the socket until impact force occurs;
6    compensate uncertainties by executing compliant
     search motion;
7    if sz>=sz_f then
8        generate required force for complete insertion;
9        if sz==sz_in then
10           success=true;
11       endif;
12   endif;
13   return success;
```

3.1 Find Contact with Socket Surface

After the visual system localized a rough pose, the plug should be 30 mm in front of the plug socket (z-direction). As Table 1 illustrates this value will contain errors. Searching for the first contact compensates this error. First impact is an older research problem and many approaches deal with it, e.g. [17], [18]. By means of impedance control, bouncing due to the impact is avoided. Making the robot's behavior very compliant in contact direction ($K_z = 50\, N/m$) leads to smooth contact until the force *Fz* is equal or greater than a defined contact force (break condition, *Fz*=3N). The robot already reacts to quite low forces and thereby avoids bouncing.

3.2 Compensate Pose Displacements

The control system of LWR iiwa allows Cartesian impedance control with additional forces/torques $\vec{F}_{add}$ [16]. In the following 'forces' includes also 'torques'. The additional forces lead to desired deviations from the proper path. This can be used as a search strategy. The additional forces influence $\vec{F}$ (compare eq. (2) and (1)):

$$\vec{x}_c = Z(s)^{-1} \cdot \left[\vec{F}_d - \left(\vec{F} + \vec{F}_{add}\right)\right] \tag{2}$$

It is also possible to add constants or changing forces in every degree of freedom. Thus, many different paths can be created on the tip of the charging plug (TCP). A number of parameters (stiffness K_i, frequency f_i, force amplitude A_i and the relative robot velocity v_{rel}) affect the path. This will be explained in the following. In our work, we use additional forces to generate a compliant blind-search strategy. For active compensation of the errors we add sinusoidal forces in x- and y- directions and a sinusoidal torque around *z*, i.e. ψ (eq. (3)). The errors in the other DOFs are compensated in a passive way [13], i.e. the robot behaves compliantly in these DOFs, but there are no additional forces and thus no active searching motions. That is why we talk about active and passive compensation. The following equations result according to [19].

$$\begin{bmatrix} F_{add_x} \\ F_{add_y} \\ F_{add_\psi} \end{bmatrix} = \begin{bmatrix} A_x \cdot sin\left(\frac{2\pi \cdot f_x}{v_{rel}} \cdot t\right) \\ A_y \cdot sin\left(\frac{2\pi \cdot f_y}{v_{rel}} \cdot t\right) \\ A_\psi \cdot sin\left(\frac{2\pi \cdot f_\psi}{v_{rel}} \cdot t\right) \end{bmatrix} \tag{3}$$

From the above forces, the search paths for the active coordinates arise:

$$\begin{bmatrix} x(t) \\ y(t) \\ \psi(t) \end{bmatrix} = \begin{bmatrix} \frac{1}{2}\left(F_{add_x}/K_x\right) \\ \frac{1}{2}\left(F_{add_y}/K_y\right) \\ \frac{1}{2}\left(F_{add_\psi}/K_\psi\right) \end{bmatrix} = \begin{bmatrix} \frac{1}{2}(A_x/K_x) \cdot sin\left(\frac{2\pi \cdot f_x}{v_{rel}} \cdot t\right) \\ \frac{1}{2}(A_y/K_y) \cdot sin\left(\frac{2\pi \cdot f_y}{v_{rel}} \cdot t\right) \\ \frac{1}{2}(A_\psi/K_\psi) \cdot sin\left(\frac{2\pi \cdot f_\psi}{v_{rel}} \cdot t\right) \end{bmatrix} \tag{4}$$

To get a phase shifted Lissajous formed path in the xy-plane the frequency ratio is defined as:

$$f_y/f_x = 0.4 \tag{5}$$

This ratio effects a periodic wave motion, which covers a complete rectangle (Fig. 6) [20]. For comparison, if we used a spiral motion the search area would only cover a circle inside the rectangle without any corners.

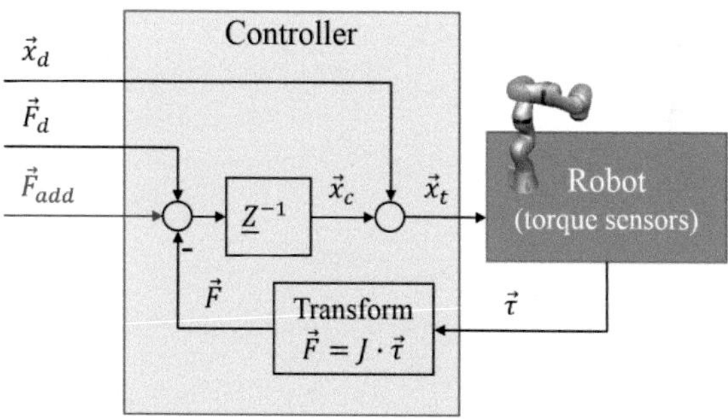

Fig. 4 Block Diagram of Impedance Control with Additional Force $\vec{F}_{add}$ ($\vec{\tau}$: measured joint torques, J: Jacobian)

To obtain a robust search strategy the size of the search area is greater than the maximum displacements listed in Table 1. We use the following values for the impedance parameters carry out the passive and the active compensation. The search path shown in Fig. 5 is generated because of the red marked values.

$$(K_x\ K_y\ K_z)^T = (300\ \ 300\ \ 1000)\ N/m \quad (K_\psi\ K_\theta\ K_\phi)^T = (10\ \ 10\ \ 10)\ Nm/rad$$
$$(A_x\ A_y\ A_z)^T = (6\ \ 6\ \ 0)\ N \quad (A_\psi\ A_\theta\ A_\phi)^T = (3.4\ \ 0\ \ 0)\ Nm$$
$$(f_x\ f_y\ f_z)^T = (5\ \ 2\ \ 0)\ Hz \quad (f_\psi\ f_\theta\ f_\phi)^T = (6\ \ 0\ \ 0)\ \text{Hz}$$

$$\Delta x = \Delta y = \frac{A_{x/y}}{K_{x/y}} = \frac{6}{300}\ m = 20\ mm \tag{6}$$

$$\Delta\psi = \frac{A_\psi}{K_\psi} = \frac{3.4}{10}\ rad \approx 19.5° \tag{7}$$

The stiffness Ki and the force amplitude Ai of x and y determine the maximum deviation of the search path (eq. (6)). It is similar to ψ (eq. (7)). The frequency of sine wave in fx -direction is chosen because of experimental tests. Thereby fy results from eq. (5) and $f\psi$ is chosen higher than fx in order to have a phase shift between the Lissajous figure in xy-plane and the sinus wave in ψ.

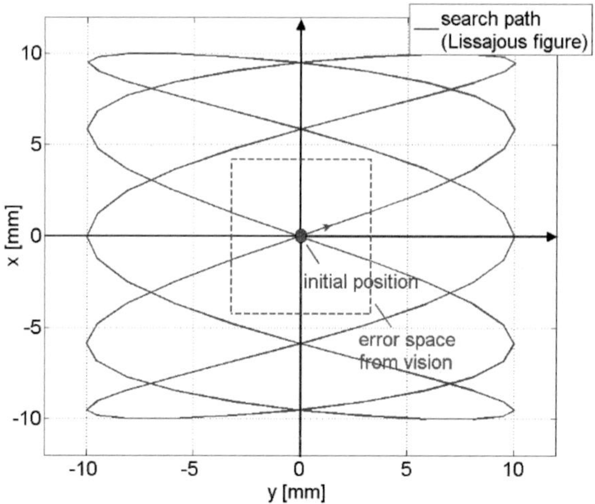

Fig. 5 Blind-Search-Path on the XY-Plane due to additional forces. The initial position is the position which comes from the socket detection by vision system. The red rectangle represents the maximum error area measured by the vision system, i.e. the socket should be inside the area

Starting from the initial pose $(x_0;\ y_0;\ z_0;\ \psi_0;\ \theta_0;\ \phi_0)$ with given parameter values, the search area covers the following region:

- $-10\,mm \leq x_0 < 10\,mm$ because of $\Delta x = 20\,mm$
- $-10\,mm \leq y_0 < 10\,mm$ because of $\Delta y = 20\,mm$
- $-9.75\,° \leq \psi_0 < 9.75\,°$ because of $\Delta\psi \approx 19.5°$

3.3 Generate Required Force to Complete the Insertion

The third step starts if the socket is found in the previous step. The criterion is the distance covered in the z-direction. If it is too small, it means that the plug remains on the surface of the socket. If the value is greater than 10 mm, we assume the socket has been found. Another indicator is the development of forces. This is explained in more detail in the next section. If the criterion is fulfilled a new impedance control will be parameterized. All DOFs stay compliant except in z-direction. The stiffness for z becomes $K_z = 4000\,N/m$. This configuration allows the generation of the required force in insertion direction and simultaneously further passive compensation of pose errors in the other DOFs. The insertion is finished when the distance *sz* has reached the depth value of the socket *sz_in*. In Fig. 6 the complete process is shown schematically.

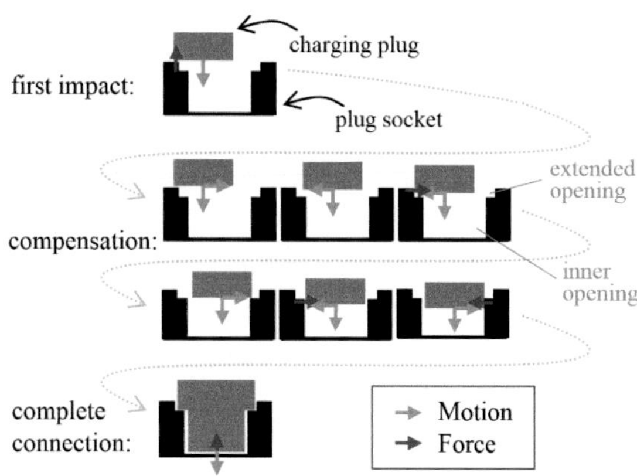

Fig. 6 Schematic Connection Process. The red arrows show the acting forces because of the blue marked motions of the robot and the plug.

The error in the z-direction is compensated during the first step of algorithm (sec. 3.1 - first impact). The rotational errors in θ and ϕ are compensated due to compliance without active motions. The search motion and the slow linear motion in z-direction will lead to collisions with the inner side of the socket. These contacts in combination with the compliant behavior of the robot effect the gradual adaption of the socket's pose and carry out the alignment with the socket pose. This phase is successful when the distance covered in z-direction *sz* is greater than the specified limit value *sz_f* (sec. 3.2 - compensation). Full connection is created in the third step.

4 Experiments and Results

A series of experiments has been carried out to evaluate the performance of the proposed algorithm. During the first test phase at laboratory set-up (Fig. 3) the positioning errors of the plug are varied up to the boundary of the search path area (Fig. 5). We consciously exceeded the expected errors from the vision system (Table 1) for testing the robustness of our approach. In a second phase, we evaluate our search algorithm together with the vision system on a real electric vehicle (Fig. 3). Additionally we tested the presented approach for another kind of plug to show the transferability.

Much more than 100 different start positions were tested in test phase one. If the initial errors are inside the actively searched area (x, y, ψ) and inside the maximum errors of θ and ϕ the insertion was always successful. The active compensation reliably corrects 8 mm in x and y respectively 8° in ψ. The passive compensation corrects up to 5° in θ and ϕ. For making it more robust it is possible to integrate an active search as it is realized for ψ. But this is not necessary because the errors will not become such high in the real application. Hence, the laboratory test phase was successful and has demonstrated robustness of the algorithm. After these tests it was evaluated together with the vision system at a real car, i.e. the initial position was determined by image processing. As it was expected, there were no failures because the positioning by the vision system is much more accurate than our test positions during laboratory evaluation. We tested it again more than 100 times and if the socket was detected by vision the following insertion was always successful. Thus, we proved the functionality of the suggested insertion algorithm. In the following we will explain the process of one trial, representative for all of them.

In the time response of the position (Fig. 7, left), the blind-search phase is visible because before the robot is moving forward in z-direction it only moves within the xy-plane. The size of the scanned area (blue curve in the xy-plane in Fig.7, left) is much smaller than it is shown in Fig. 5, because at the beginning of step 2 (compensation of the pose errors) the plug is already inside the extended opening of the socket (Fig. 6). The right plot of Fig. 7 shows a slow increase in *Fz* after three seconds, while *Fx* and *Fy* still alternate but now with higher amplitudes. The higher maximum values for *Fx* and *Fy* indicates that the search motion is even more limited. Looking for this in the plot of position (Fig. 7, left), we see the limitation really increases when the robot moves forward in z-direction. At this moment the plug enters the inner opening of the socket. The increase of *Fz* results from the slow linear motion in z-direction and the straining of the virtual spring-damper-system.

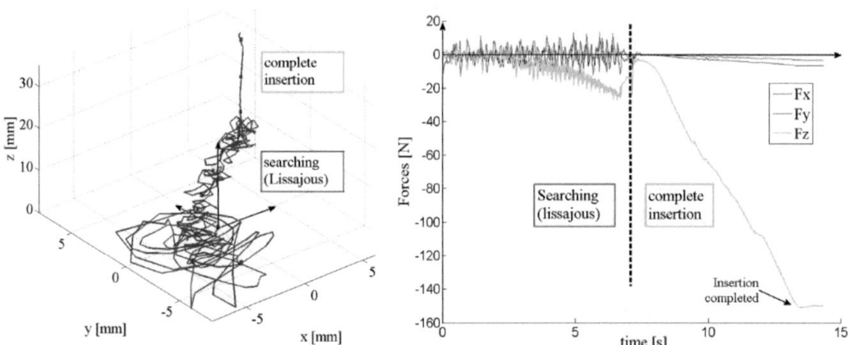

Fig. 7 Left plot: 3D position of the plug during insertion process. Right plot: Forces in translational DOFs at the plug during the insertion process

When *Fz* or the distance in z-direction reach their limits the search phase ends. In the presented test, the distance *sz* triggered the stop condition, i.e. the searching was successful. After that the final insertion phase is started. As described in previous sections the algorithm sets the robot very stiff in z-direction and thus it is able to build up the required force. This happens gradually to prevent the robot and the environment from damage. Now *Fz* becomes slowly greater while the other forces stay nearly at zero level (Fig. 7, right)). Transition from step 2 to step 3 is also visible in the position plot of Fig. 7. This is when it becomes nearly the simple straight line parallel to the z-axis. Presently, the algorithm stops in step 3 when the specified distance sz_i is reached. In the future, there will be a digital signal generated when the insertion is completed.

During the whole insertion process the joint torques of the robot are checked for excessive values. This is very important for safety and indispensable for detection of problems and prevention from damage. Additionally, in the final version of the charging station there will be a camera for controlling the global workspace. It will be able to stop the robot if any danger for human or machine is detected, when e.g. a human moves into robot's workspace.

To show the adaptability to other kinds of plugs and other application we test the algorithm for insertion of a heavy current plug (Fig. 8). As expected it works when we adjust the amplitudes of the search paths and the value of the generated force during the insertion phase. The adjustment mainly affects the frequency of the searching path around the z-axis and the Lissajous oscillation in xy-plane. It should be decreased because the notch is comparatively hard to find due to the small clearance (< 0.5mm).

Fig. 8 Socket (left) and plug (right) for a heavy current connection. It is visible that a small notch guarantees the suitable connection, i.e. the absolutely correct pose has to be found by the algorithm. There is only a very small clearance value (< 0.5 mm) between plug and socket.

5 Conclusion

In this paper we presented a method for a peg-in-hole task where the pegs are charging plugs for electric vehicles. The plug has 7 pins, which is very challenging. We used a robot with programmable compliant behavior. Based on this possibility we implemented a compliant blind-search strategy. The algorithm starts after the robot receives a coarse position of the socket from the vision system. The functionality was verified by numerous experiments. They prove, our approach is able to compensate the expected position errors between the plug and the socket and the robustness of the algorithm is much higher than actually requested by means of Table 1. Thus, the success rate for realistic initial error values was up to 100%. If the error values are different in other applications, the search area can be adjusted by changing the parameters of the additional forces.

We suggest verifying this approach in the future for other settings because it can be easily transferred to other tasks when the estimation of coarse pose is available. The developed algorithm can be easily adapted to solve similar problem which frequently occur in assembly or service robotics. One of them can be the establishing of a heavy current connection which was successfully tested during the presented research project.

References

1. Fuelmatics Systems AB, July 5, 2015. http://fuelmatics.com/
2. RotecEngineering, July 5, 2015. http://rotec-engineering.nl/en/automation/refueling-robot/
3. GINKO, June 28, 2015. https://www.tu-chemnitz.de/etit/sse/Forschung/Projekte/ginko.html
4. Whitney, D.E.: Quasi-static-assembly of compliantly supported rigid parts. In: Journal of Dynamic Systems, Measurement and Control Division, pp. 65–77 (1981)
5. Sharma, K., Shirwalkar, V., Pal, P.K.: Intelligent and environment-independent Peg-In-Hole search strategies. In: 2013 International Conference on Control, Automation, Robotics and Embedded Systems (CARE), pp. 1–6, December 16–18, 2013
6. Chhatpar, S.R., Branicky, M.S.: Localization for robotic assemblies with position uncertainty. In: Proceedings of the Intelligent Robots and Systems, (IROS 2003), vol. 3, pp. 2534–2540 (2003)
7. Newman, W.S., Zhao, Y., Pao, Y.-H.: Interpretation of force and moment signals for compliant peg-in-hole assembly. In: Proceedings of the 2001 IEEE International Conference on Robotics and Automation, ICRA, vol. 1, pp. 571–576 (2001)
8. Thomas, U., Molkenstruck, S., Iser, R., Wahl, F.M.: Multi sensor fusion in robot assembly using particle filters. In: 2007 IEEE International Conference on Robotics and Automation, pp. 3837–3843 (2007)
9. Yamamoto, Y., Hashimoto, T., Okubo, T., Itoh, T.: Task analysis of ultra-precision assembly processes for automation of human skills. In: Proceedings of the 2001 IEEE/RSJ International Conference on Intelligent Robots and Systems, vol. 4, pp. 2093–2098 (2001)
10. Savarimuthu, T.R., Liljekrans, D., Ellekilde, L.-P., Ude, A., Nemec, B., Kruger, N.: Analysis of human peg-in-hole executions in a robotic embodiment using uncertain grasps. In: 2013 9th Workshop on Robot Motion and Control (RoMoCo), pp. 233–239 (2013)
11. Goto, T., Takeyasu, K., Inoyama, T.: Control Algorithm for Precision Insert Operation Robots. Transactions on Systems, Man and Cybernetics **10**(1), 19–25 (1980)
12. Stemmer, A., Albu-Schaffer, A., Hirzinger, G.: An analytical method for the planning of robust assembly tasks of complex shaped planar parts. In: 2007 IEEE International Conference on Robotics and Automation, pp. 317–323 (2007)
13. Park, H., Bae, J.-H., Park, J.-H., Baeg, M.-H., Park, J.: Intuitive peg-in-hole assembly strategy with a compliant manipulator. In: 2013 44th International Symposium on Robotics (ISR), pp. 1–5 (2013)
14. Liu, S., Liu, C., Liu, Z., Xie, Y., Xu, J., Chen, K.: Laser tracker-based control for peg-in-hole assembly robot. In: 2014 IEEE Annual International Conference on Cyber Technology in Automation, Control, and Intelligent Systems (CYBER), pp. 569–573 (2014)

15. Song, H.-C., Kim, Y.-L. Song, J.-B.: Automated guidance of peg-in-hole assembly tasks for complex-shaped parts. In: Intelligent Robots and Systems (IROS 2014), pp. 4517–4522 (2014)
16. Albu-Schaffer, A., Ott, C., Frese, U., Hirzinger, G.: Cartesian impedance control of redundant robots: recent results with the DLR-light-weight-arms. In: Proceedings of the IEEE International Conference on Robotics and Automation, ICRA 2003, vol. 3, pp. 3704–3709 (2003)
17. Wu, Y., Tarn, T., Xi, N., Isidori, A.: On robust impact control via positive acceleration feedback for robot manipulators. In: Proceedings of the 1996 IEEE International Conference on Robotics and Automation, vol. 2, pp. 1891–1896 (1996)
18. Bdiwi, M., Winkler, A., Suchy, J., Zschocke, G.: Traded and shared vision-force robot control for improved impact control. In: 2011 8th International Multi-Conference on Systems, Signals and Devices (SSD), pp. 1–6 (2011)
19. KUKA Sunrise.Workbench1.1, Manual, KUKA Laboratories GmbH 2014, pp. 304–305 (2014)
20. Wikibooks, September 17, 2015. http://en.wikibooks.org/wiki/Trigonometry/For_Enthusiasts/Lissajous_Figures

Double A* Path Planning for Industrial Manipulators

Pedro Tavares, José Lima and Pedro Costa

Abstract The scientific and technological development, together with the world of robotics, is constantly evolving, driven by the need to find new solutions and by the ambition of human beings to develop systems with increasingly efficiency. Consequently, it is necessary to develop planning algorithms capable of effectively and safely move a robot within a given non structured scene. Moreover, despite of the several robotic solutions available, there are still challenges to standardise a development technique able to obviate some pitfalls and limitations present in the robotic world. The Robotic Operative System (ROS) arise as the obvious solution in this regard. Throughout this project it was developed and implemented a double A* path planning methodology for automatic manipulators in the industrial environment. In this paper, it will be presented an approach with enough flexibility to be potentially applicable to different handling scenarios normally found in industrial environment.

Keywords Path planning · A* · ROS · Configuration space · Kinematics · Industrial environment

1 Introduction

Robotic manipulators based solutions have attracted substantial interest from the engineering research community, as they present key characteristics for streamlining automated processes. There are numerous applications for these robots such as transportation of equipment, pick and place operations, dangerous and inaccessible tasks [8, 11]. Robots currently used in industrial environment are commonly associated

P. Tavares(✉)
FEUP - Faculty of Engineering of University of Porto, Porto, Portugal
e-mail: pedro.tavares@gmail.com

J. Lima · P. Costa
INESC TEC - INESC Technology and Science Formely INESC Porto, Porto, Portugal

L.P. Reis et al. (eds.), *Robot 2015: Second Iberian Robotics Conference*,
Advances in Intelligent Systems and Computing 418,
DOI: 10.1007/978-3-319-27149-1_10

with the optimization of manufacturing processes [3]. Moreover those same robots easily enable assembly operations or surveillance in the same non structured environments. There have also been developed robots to a wide range of other applications such as medical practice or military missions.

Currently the organizational growth is associated to the development of small and medium enterprises (SMEs) [20]. Nowadays the development of robots is based on the ability of those robots being autonomously adaptable to several applications and environments. Furthermore, there have been efforts towards the possibility of such robots becoming highly efficient operators. In that regard robots are developed in order to be able to complete high demanding task in terms of strength or precision. Moreover to keep up with the recent trend of industrial environment and SMEs, there is the need develop robots able to be profitable in small areas and with small work loads [2, 15].

The ultimate goal in the development of present and future robots is focused in the ability of the robot be fully independent. There have been some works addressing intelligent tasks planning intending to recreate the human behaviour. In that dimension is also inserted the path planning of a given task. This step is transversal to all robotic approaches. The path planning essentially is the description of the sequence of configurations of the robot at each moment in order to reach a desired destination. When considering robotic manipulator arms, that configuration is associated with the state of each joint. A correct path planning allows one to reduce the distance between points (origin and destination), minimizing the execution time, and the effort required to the robot.

Despite the utility associated with robotic solution, there is still room for improvement. Actual solutions are focused in one previously defined task and lack some flexibility to autonomously react to changes. Moreover, most of those solutions are implemented without using a modular approach, which means that a minor change in the required task may imply a completely restructuring of the source code that rules the robot.

When considering this problem there is a recent and growing solution, the Robotic Operative System (ROS). In short, ROS is a set of software libraries and tools that can be used to build robot applications. This methodology includes drivers and state-of-the-art algorithms, and has the advantage of being open source. Moreover ROS allows the decomplexation of any given problem. This way, instead of a single complex algorithm, the idea behind ROS is to develop multiple simple algorithms that put together assure a modular solution to any complex problem, while adding flexibility to the overall system.

An important aspect to consider when developing a novel robot is the selection of the methodology for handling of the objects of interest. In this paper, it will be presented a generic approach with enough flexibility to be potentially applicable to different scenarios of object handling in robotics. Moreover it will be presented a new path planning algorithm. The main aim is for this application to be applied to pick and place routines in robotics allowing present and future robots that use this approach to be completely adaptive to any given task.

The current paper is structured in six main sections. Section 2 aims to provide an overview of the state of the science in robotic systems optimization approaches, specially in path planning techniques. Section 3 (System Architecture) will describe our approach to solve typical robotics problems based on the use of industrial manipulators. Section 4 (Configuration Space and Kinematics) intends to demonstrate the importance of discretize the robot's space. Section 5 aims to describe the new double A* planner developed throughout the project. Section 6 (Experimental Validation) synthetizes the results obtained by that planner as well as a comparison with the state-of-the-art algorithm currently used. Section 7 (Discussion and Future Perspectives) reviews the contribution of this project to the scientific world and the future option of it.

2 Related Work

The robotic development is associated with the scientific and technological development. Nowadays robotic solutions based on manipulator arms in the industrial environment are frequently associated to pick and place operations. These operations require the robot to handle objects in a given environment, while safely move around that same environment. Examples of applications of these robots include the packing of medical drugs, automation systems for the handling of products in the food industry [4] and the automation of production lines in the automotive industry [10].

Pick and place operation comprehends the handling of a object of interest using a robotic manipulator. This handling vary from a simple transportation process to the assembly of parts in a production line. In order to clarify and simplify this operation, in 2000, Mattone *et. al* described the two fundamental problems associated with pick and place techniques, dividing them into two steps: (1) - sensing, detecting and classifying the objects to be sorted and (2) - gripping [12, 13].

In 2005, Gecks and Henrich developed an algorithm to perform pick and place operations with complete safety and real time behavior by using image processing of several stationary cameras [7]. Last year, Daoud *et al.* developed a method that allowed the coordination between several robots in order to maximize the production rate [5]. There have also been efforts towards the improvement of the gripping strategy. For instance, Wang et al proposed a three step strategy that allowed to align the tool responsible for the gripping operation with the object of interest ensuring a correct pick and place operation [19].

When considering a physical implementation, there is the need to understand the types of manipulators that can be used. Nowadays there are several with enough flexibility such as the Delta Robot or the DexTAR robot (see fig 1).

Furthermore, when designing a robotic system using manipulator arms, it is necessary to consider a control tier that controls the system movement and minimizes and compensate the errors throughout the operation. In 2002, Son suggest a learning algorithm that associated to a fuzzy optimal process facilitates pick and place approaches such as part insertion. Moreover Son also demonstrated that by using

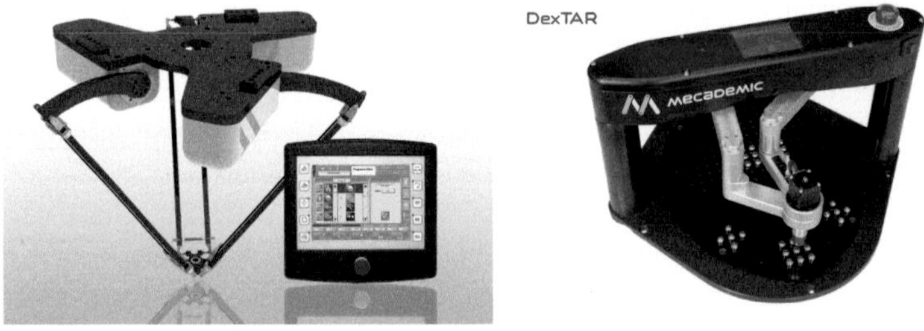

Fig. 1 At the right - The Delta robot; at the left - The DexTAR robot

a fuzzy control based on measured force and moment information it is possible to avoid jamming related to a quasi-static part insertion [17, 18].

Another important step to be considered when developing the robotic system is the motion planning. This is one of the most important areas of robotics research. In 1992, Hwang and Ahuja surveyed the work on gross-motion planning, explaining the key steps in motion planning [9]. Moreover Barraquand and Latombe, in 1991, proposed a new approach to robot path planning that consisted of building and searching a graph connecting the local minima of a potential function defined over the robot's configuration space [1]. This new approach was proposed considering robots with multiple degrees of freedom. Later Ralli and Hirzinger refined that same algorithm accelerating the system, calculating solutions with a lower estimated executing time [16]. The path planning becomes complexer when there are inserted obstacles in a given environment. Although the previous mention approaches took into consideration that case, in 2008, Yao *et al.* developed an obstacles avoidance algorithm, inserting those in the configuration space of the robot [21].

Lately, ROS has also present crucial developments in that regard, mostly with the software *MoveIt!*. This is a state of the art software for mobile manipulators that covers motion planning, manipulation, 3D perception, kinematics, control and navigation [14]. Therefore *MoveIt!* presents itself as a utility tool for all robotic system that include motion planning.

Although all strategies presented above have been implemented and validated in the industrial world they have limitations and thus consideration should be given on which approach to use for a given application.

3 System Architecture

Most robotic applications using mobile manipulators can be defined as a multi-tier problem. Typically, there are considered three main tiers (see table 1).

The objective behind this architecture is to develop a simple and universal method for all solutions that require mobile manipulators. Moreover this architecture allows

Table 1 Typical Tiers in a Mobile Manipulator based Solution

Tier	Tier Specification	Application Examples
1	Environment Recognition	Image Processing; Camera-Laser Triangulation; Sonar; Others.
2	Motion and Strategy Planning	Path Planning; Task Planning; Heuristics; Robot Movement; Joint Control.
3	Control and Actuation	*Gripping* and *Grasping*; Error Correction; Adaptive Algorithms; Others.

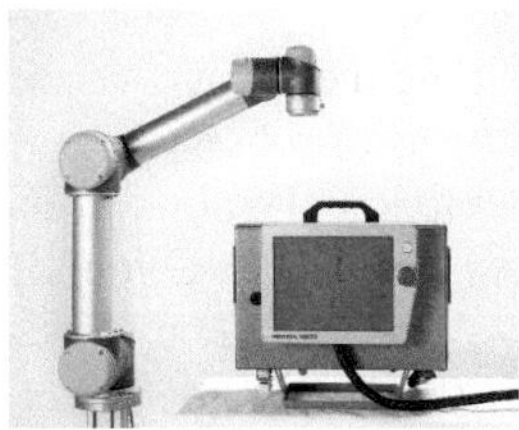

Fig. 2 The robot UR5 used in the project (at the left); Robotiq 3-finger Hand (at the middle); VERSABALL (at the right).

to consider several robots and adds enough flexibility so that those can be inserted in any given environment.

Throughout this project it was followed a three tiers approach stated in table 1 so that the overall project could autonomously adapt to any given task and space. Moreover it was selected a specific robot, the universal robot 5 (UR5) and multiples tools such as ROBOTIQ 3-finger hand, Festo Grippers or VERSABALL (see fig 2).

4 Configuration Space and Kinematics

Once detected the surrounding environment of the robot, there is need to define a configuration space for the same robot. Thus, initially, we have to consider a way to transform the 3D space into a discrete space of configurations. Another aspect to consider is the kinematics associated to the selected robot. This aspect allows one to associate the current state of the robot with a Cartesian pose.

For a robotic manipulator, a configuration space q can be defined as a set of values that define the current position of the robot. These values are typically associated to the set of joint values of the manipulator of interest. This way, when considering manipulators with N joints and therefore N degrees of freedom (such as the UR5), we easily can transform a redundant three dimensions problem to a N dimensional but well defined problem.

When considering the kinematics of a robot, we can use the DH method proposed by Denavit and Hartenberg [6]. Essentially by using this method we can define a transformation from the origin of the robot to the end-effector of the already mention robot as a product of N−1 elementary transformations between the N joints of the manipulator (see equation 1).

$$T_N^0 = T_1^0 * T_2^1 * T_3^2 * ... * T_N^{N-1} \tag{1}$$

Each elemental transformation can be defined as a translation and a rotation between each joint. Thus, we can define a homogeneous transformation matrix, based on the four DH parameters (a_i; α_i; d_i; q_i) as presented below (see equation 2).

$$\begin{bmatrix} R_y^x & \mathbf{P_y^x} \\ 0 & 1 \end{bmatrix} = \begin{bmatrix} cos(\theta_i) & -sin(\theta_i) * cos(\alpha_i) & sin(\theta_i) * cos(\alpha_i) & a_i * cos(\theta_i) \\ sin(\theta_i) & cos(\theta_i) * cos(\alpha_i) & -cos(\theta_i) * cos(\alpha_i) & a_i * sin(\theta_i) \\ 0 & sin(\alpha_i) & cos(\alpha_i) & d_i \\ 0 & 0 & 0 & 1 \end{bmatrix} \tag{2}$$

As mentioned before, the robot of choice for this project was the UR5. For this particular robot the DH parameters can be found in table 2.

Table 2 UR5 DH Parameters

Joint	a_i	α_i	d_i	θ_i
1	0	$\pi / 2$	d_1	θ_1
2	$-a_2$	0	0	θ_2
3	$-a_3$	0	0	θ_3
4	0	$\pi/2$	d_4	θ_4
5	0	$-\pi/2$	d_5	θ_5
6	0	0	d_6	θ_6

5 Double A* - Path Planning Algorithm

Path planning algorithms try to minimize the path between two points. However in order to do so there is the need to define a configuration space of all possible outputs that the robot can assume (see fig 3).

Since that the robot is controlled by joint position it is beneficial to define a configuration space as a joint space. Currently manipulator robots generally have six degrees of freedom. Thus, the discretization of the configuration space is frequently a problem due to the lack of memory to store that configuration space. Supposing that

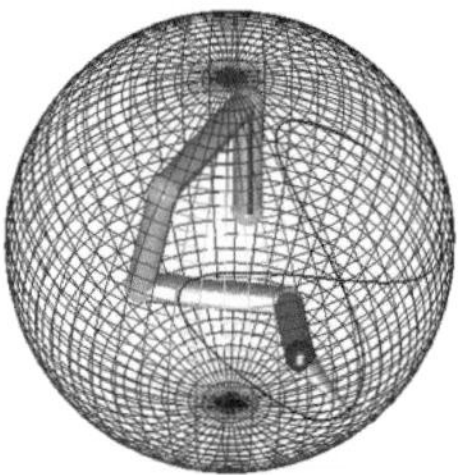

Fig. 3 Configuration Space Diagram

we want a definition of one degree and considering the interval $[-\pi,\pi]$, there would be approximately $2.2 * 10^{15}$ possible configurations. Therefore it is not viable to store all those configurations and the properties associated to each one (availability, index, the joints state and others).

When considering this memory problem, the precision of the discretization had to be reduced. As such it was implemented a double planner, with two phases (1) - approach phase; (2) - precision phase. In the first phase the configuration space was divided in fragments of 20º each. Thus, the storage space require of all configurations becomes significantly lower. The second planner allows to add precision to the system. This phase is based in the same fundamental ideas. However it is only applied when near the destination of interest. Thus, the range of configuration viable for the planning of a path are less and, based on that, the division of each configuration can be preciser. This way it is assure that the position can be rapidly achieve and the final pose is precise and coherent with the goal position.

Once defined the discretization strategy, there is the need to define a data structure to use throughout the planner algorithm. Considering that this double planner is based on the A* family algorithms, there is the necessity of creating an unique index that can identify all possible configurations. Moreover it is included a property that classifies each configuration as free space or obstacle. There are also fields that allow to store the previous configuration while performing the algorithm. This allows one to identify the correct sequence of configurations at the end of the planner algorithm. Based on those configurations is applied a heuristic rule which identifies the next probably best step.

Moreover the property that describes the space as free or occupied allows to classify each configuration regarding their availability. As such, when an object is inserted in the robot's space, it is possible to associate its position to a configuration and set that property as impossible to reach.

To save all this intermediate steps was used a heap based structure named *Open List* in which there were inserted every configuration in the neighbourhood to the current that may be beneficial for the path planning. These configurations are denominated as neighbours and they are ordered in the list in terms of relevance obtained from the application of the heuristic.

The process of finding neighbours runs as a cycle where there is identified and stored all configurations near to the current one. This cycle is stopped when found the goal configuration. At that point, there is withdrawn the intermediate configurations that allowed to reach the desired goal storing them in a new heap named *Closed List*.

Finally, when conclude this all process we just have to use the indexes stored in that *Closed List* and transform them to robot joints states so that the robot can be command to reach those states and reach the final configuration required.

6 Experimental Validation

In order to evaluate the performance of the path planner developed it was considered multiple poses around the robot's space. As such, it was defined poses within distinct quadrants. All the tests performed presented similar results to those described below. In table 3 and 4 there is stated the joint evolution of the robot while trying to reach a pre defined pose. To this particular pose ((x,y,z) = (0.100, 0.075, 0.725)) the final resulting configuration after applying inverse kinematics is (167.5 ; -53.0 ; -49.5 ; 107.8 ; -100.0 ; 140.6).

Table 3 Approximation Step: Joints Evolution - Pose 1

Iteration	θ_1	θ_2	θ_3	θ_4	θ_5	θ_6
1	0	0	-140	-80	-100	-60
2	20	-20	-120	-60	-120	-40
...	...	...	...	...	...	...
12	160	-60	-60	100	-120	140

Table 4 Precision Step: Joints Evolution - Pose 1

Iteration	θ_1	θ_2	θ_3	θ_4	θ_5	θ_6
1	161.1	-58.9	-58.9	100	-101.1	140
2	162.2	-57.8	-57.8	101.1	-101.1	140
...	...	...	...	...	...	...
9	166.7	-53.3	-50	107.8	-101.1	140

Analysing the joint evolution and the configuration goal it is possible to identify a error lower than 1º per joint resulting in a total error of 3.3º when considering all joints. Moreover, by applying the path determinated by the double A* planner the ultimate position would be (0.095, 0.073, 0.724) which represent a deviation of 0.5cm. All other tests followed similar patterns with the deviation being limited to 1cm and a total of 4º (less than 1º per joint).

Furthermore there was tested the ability to avoid obstacles. As such it was inserted an obstacle in the middle of a previously calculated trajectory requiring a new calculation to the path planner. The results are expressed in tables 5, 6, 7 and 8. The first two represent the trajectory calculated without any obstacles and the last two correspond to the case where those obstacles were inserted.

Table 5 Approximation phase: Index Evolution (Without Obstacles) - Pose 2

Iteration	Index	Configuration
1	17964258	[0 ; 0 ; -140 ; -80 ; -100 ; -60]
2	19754340	[20 ; -20 ; -120 ; -100 ; -120 ; -60]
3	19759830	[20 ; -20 ; -100 ; -120 ; -140 ; -60]
4	19765662	[20 ; -20 ; -80 ; -120 ; -140 ; -60]

Table 6 Precision Phase: Index Evolution (without obstacles) - Pose 2

Iteration	Index	Configuration
1	1988369	[21.1 ; -20 ; -62.2 ; -102.2 ; -122.2 ; -41.1]
2	3976738	[22.2 ; -18.9 ; -63.3 ; -103.3 ; -123.3 ; -42.2]
...	...	...
10	19808654	[31.1 ; -11.1 ; -66.7 ; -110 ; -124.4 ; -44.4]

Table 7 Approximation Phase: Index Evolution (with several obstacles)) - Pose 2

Iteration	Index	Configuration
1	17964582	[0 ; 0 ; -140 ; -60 ; -100 ; -60]
2	19754664	[20 ; -20 ; -120 ; -80 ; -120 ; -60]
3	17870586	[0 ; -20 ; -100 ; -100 ; -140 ; -60]
4	19765662	[20 ; -20 ; -80 ; -120 ; -140 ; -60]

Table 8 Precision Phase: Index Evolution (with several obstacles)) - Pose 2

Iteration	Index	Configuration
1	1988369	[21.1 ; -20 ; -62.2 ; -102.2 ; -122.2 ; -41.1]
2	3976738	[22.2 ; -18.9 ; -63.3 ; -103.3 ; -122.2 ; -42.2]
...	...	...
10	19808654	[31.1 ; -11.1 ; -66.7 ; -110 ; -124.4 ; -44.4]

Reviewing tables 7 and 8 it is easy to understand that the algorithm continues completely operational as it avoids from such said obstacles and adapts the remaining trajectory accordingly.

6.1 Comparison with MoveIt!

When evaluating the efficiency of a new planner is important to compare it with the state-of-the-art solutions. As such, the path obtained by the double A* was compared to the OMPL planner pre-defined by the *MoveIt!* solution for the UR5.

When comparing both algorithms it is possible to conclude that in the generality of cases the double A* planner finds the solution in less steps, determinates a shorter

Table 9 MoveIt vs Double A*: Number of iterations - Pose 1

Algorithm	Steps	Distance (m)	Execution Time (s)
MoveIt	28	1.1328	5.894
Double A*	11+9 = 20	0.8095	4.2775

Table 10 *MoveIt* vs Double A*: Number of iterations - Pose 3

Algorithm	Steps	Distance (m)	Execution Time (s)
MoveIt	—	—	—
Double A*	11+7 = 18	0.8651	4.5555

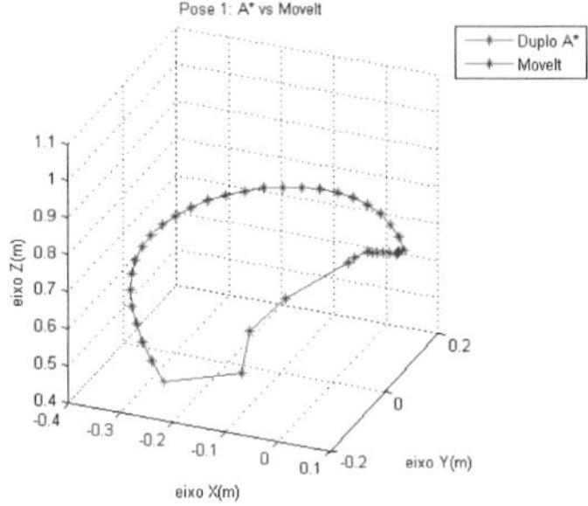

Fig. 4 MoveIt vs Double A* - Pose 1

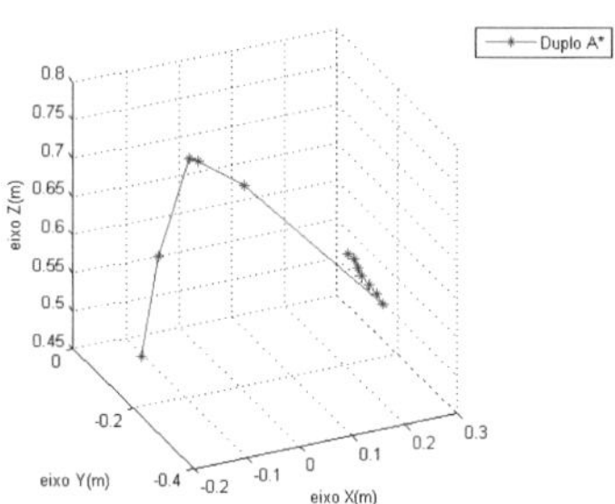

Fig. 5 MoveIt vs Double A* - Pose 3

path and the estimated executing time is also lower. Moreover the A* planner is an optimal and complete algorithm. As such there is always a solution as long as if the required pose is reachable by the robot. This fact can be highly valuable since that cases like pose 3 (fig 5) do not have a defined trajectory using *MoveIt!*, however it has solution using the double A* planner.

7 Discussion and Future Perspectives

Throughout this project, it was developed several solution that can be valuable by both scientific and technological communities.

Kinematics is a frequent problem associated with all robotic challenges that evolve robot movement and control of such said movement. During the project already presented, the robot chosen was the universal 5 (UR5). Focusing this robot and the kinematics problem associated to it, there was identified and described the interference of each joint in the Cartesian movement produced by the robot. Moreover equations were defined to understand the robot state at each moment.

Another aspect to consider when analysing robotic solution using manipulators is the path planning. The primordial objective of this project was the development of a planner able to compete with the state of the art algorithms used currently. In that regard it was developed a double A* planner.

In general, this planner produces better results than other planners present in both industry and research areas. Furthermore, the execution time and the number of steps require to perform the path produced by such planner are significantly lower. Another advantage is the division of the path in two main phases. The first one (the approach phase) allows to accelerate the process by guiding the robot to a position near the destination required in a couple of steps. Then the second phase (the precision phase) allows one to refine the path done in the previous stage, while guiding the robot to the final position with an insignificant error. This way the goal is achieve faster and in a precise fashion.

Although this algorithm shows good result, the approach phase can produce a overly direct trajectory that may require too much effort to the robot joints. This is due to the fact of the configuration space being to imprecise and that a iterative correction may provoke too sharp Cartesian variations. As such, as future work it would be advantageous to cross the solution produced by the double A* planner with a mitigation algorithm that considers the Cartesian space.

In conclusion, the work present formalizes a planner able to be inserted in most of the robotic solution that use manipulators. This planner can be valuable in several fields of robotics based on its flexibility as it produces interesting results for any required pose as long as it is in the robot range. This planner was tested also in two application cases: EuRoC and Amazon Picking Challenge, showing good results and proving to be a viable solution for future applications.

Acknowledgments The research leading to these results has received funding from the European Union Seventh Framework Programme FP7/2007-2013 under grant agreement no287787.

The full project title: SMErobotics, the European Robotics Initiative for Strengthening the Competitiveness of SMEs in Manufacturing by integrating aspects of cognitive systems.

References

1. Barraquand, J., Latombe, J.C.: Robot motion planning. a distributed representation approach. International Journal of Robotics Research **10**(6), 628–649 (1991)

2. Bischoff, R., Kurth, J., Schreiber, G., Koeppe, R., Albu-Schaeffer, A., Beyer, A., Eiberger, O., Haddadin, S., Stemmer, A., Grunwald, G., Hirzinger, G.: The KUKA-DLR lightweight robot arm - a new reference platform for robotics research and manufacturing. In: 2010 41st International Symposium on Robotics (ISR) and 2010 6th German Conference on Robotics (ROBOTIK), pp. 1–8 (2010)
3. Chen, H.P., Wang, J.J., Zhang, G., Fuhlbrigge, T., Kock, S.: High-precision assembly automation based on robot compliance. International Journal of Advanced Manufacturing Technology **45**(9–10), 999–1006 (2009)
4. Chua, P.Y., Ilschner, T., Caldwell, D.G.: Robotic manipulation of food products - a review. Industrial Robot-an International Journal **30**(4), 345–354 (2003)
5. Daoud, S., Chehade, H., Yalaoui, F., Amodeo, L.: Efficient metaheuristics for pick and place robotic systems optimization. Journal of Intelligent Manufacturing **25**(1), 27–41 (2014)
6. Foundation, O.S.R.: Why gazebo? (2014). http://gazebosim.org/
7. Gecks, T., Henrich, D., Ieee: human-robot cooperation: safe pick-and-place operations. In: 2005 IEEE International Workshop on Robot and Human Interactive Communication. Ieee, New York (2005)
8. Hvilshoj, M., Bogh, S.: "little helper" - an autonomous industrial mobile manipulator concept. International Journal of Advanced Robotic Systems **8**(2), 80–90 (2011)
9. Hwang, Y.K., Ahuja, N.: Gross motion planning - a survey. ACM Computing Surveys **24**(3) (1992)
10. Jiang, X., Koo, K.M., Kikuchi, K., Konno, A., Uchiyama, M., Ieee: robotized assembly of a wire harness in car production line. In: Ieee/Rsj 2010 International Conference on Intelligent Robots and Systems (Iros 2010), pp. 490–495 (2010)
11. Kim, J.H.: Automated medicine storage and medicine introduction/discharge management system, October 9, 2012. US Patent 8,281,553
12. Mattone, R., Campagiorni, G., Galati, F.: Sorting of items on a moving conveyor belt. part i. a technique for detecting and classifying objects. Robotics and Computer-Integrated Manufacturing **16**(2–3), 73–80 (2000)
13. Mattone, R., Divona, M., Wolf, A.: Sorting of items on a moving conveyor belt. part 2: performance evaluation and optimization of pick-and-place operations. Robotics and Computer-Integrated Manufacturing **16**(2–3), 81–90 (2000)
14. Merlet, J.P.: Direct kinematics of parallel manipulators. IEEE Transactions on Robotics and Automation **9**(6), 842–846 (1993)
15. Mikael, H., Erik, H., Mats, J.: Robotics for SMEs - investigating a mobile, flexible, and reconfigurable robot solution. In: 39th International Symposium on Robotics, ISR 2008, pp. 56–61
16. Ralli, E., Hirzinger, G.: Fast path planning for robot manipulators using numerical potential fields in the configuration space, vol. 3, pp. 1922–1929 (1994)
17. Son, C.: Intelligent robotic path finding methodologies with fuzzy/crisp entropies and learning. International Journal of Robotics and Automation **26**(3), 323–336 (2011)
18. Son, C.M.: Optimal control planning strategies with fuzzy entropy and sensor fusion for robotic part assembly tasks. International Journal of Machine Tools & Manufacture **42**(12), 1335–1344 (2002)
19. Wang, L.D., Ren, L., Mills, J.K., Cleghorn, W.L.: Automated 3-d micrograsping tasks performed by vision-based control. Ieee Transactions on Automation Science and Engineering **7**(3), 417–426 (2010)
20. Westhead, P., Wright, M., Ucbasaran, D.: The internationalization of new and small firms: A resource-based view. Journal of Business Venturing **16**(4), 333–358 (2001)
21. Yao, L., Ding, W., Chen, Y., Zhao, S.: Obstacle avoidance path planning of eggplant harvesting robot manipulator. Nongye Jixie Xuebao/Transactions of the Chinese Society of Agricultural Machinery **39**(11), 94–98 (2008)

Mobile Robot Localization Based on a Security Laser: An Industry Scene Implementation

Héber Sobreira, A. Paulo Moreira, Paulo Gomes Costa and José Lima

Abstract Usually the Industrial Automatic Guide Vehicles (AGVs) have two kind of lasers. One for navigation on the top and others for obstacle detection (security lasers). Recently, security lasers extended its output data with obstacle distance (contours) and reflectivity, that allows the development of a novel localization system based on a security laser. This paper addresses a localization system that avoids a dedicated laser scanner reducing the implementations cost and robot size. Also, performs a tracking system with precision and robustness that can operate AVGs in an industrial environment. Artificial beacons detection algorithm combined with a Kalman filter and outliers rejection method increase the robustness and precision of the developed system. A comparison between the presented approach and a commercial localization system for industry is presented. Finally, the proposed algorithms were tested in an industrial application under realistic working conditions.

Keywords AGV · Mobile robotic · Localization · Artificial beacons · Kalman filter · Outliers rejection · Security laser

1 Introduction

One of the most important requirement of an industry mobile robot is to own a robust self-localization. Basically, it can be defined as the task of estimating the robot's pose in a map of the environment. This task had captured the attention of researches, developers and technology transfers of mobile robots over the last years.

H. Sobreira(✉) · A.P. Moreira · P.G. Costa · J. Lima
NESC TEC (formerly INESC PORTO) - Robotics and Intelligent Systems, Porto, Portugal
e-mail: heber.m.sobreira@inescporto.pt, {amoreira,paco}@fe.up.pt, jllima@ipb.pt

H. Sobreira · A.P. Moreira · P.G. Costa
Faculty of Engineering of University of Porto, Rua Dr. Roberto Frias, 4200-465 Porto, Portugal

L.P. Reis et al. (eds.), *Robot 2015: Second Iberian Robotics Conference*,
Advances in Intelligent Systems and Computing 418,
DOI: 10.1007/978-3-319-27149-1_11

There are some solutions adopted that solves the localization problem but bring some disadvantages [1] [2].

The aim of this work is to develop a localization system, based on reflector detection using a security laser, with precision and robustness that can operate in an automatic guided vehicle (AGV) in an industrial environment.

The motivation for this approach is related to the fact that the security laser is a mandatory equipment in most applications using AGVs. Therefore, reusing this equipment for localization purposes avoids the use of a second navigation laser at the top of the robot, which leads to lower equipment costs. Furthermore, in some applications, it is not possible to use a navigation laser due to limitations of maximum height AGV, as can be seen in the particular case represented in figure 1.

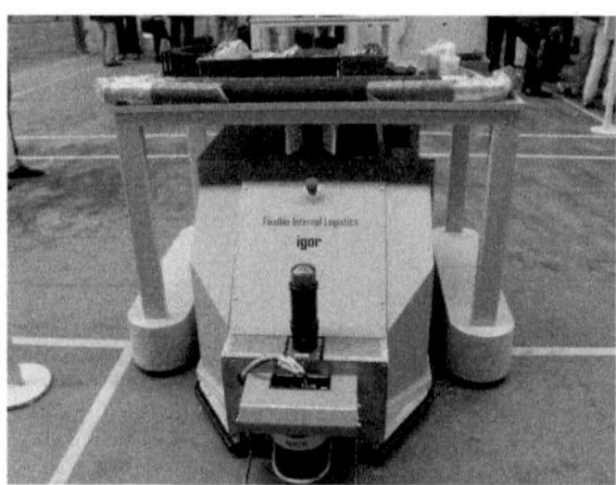

Fig. 1 Robot in the shop floor of an industrial environment

In this case, the AGV is carrying tables above it, and being a robot short enough to be able to pass beneath the table, does not allow the use of a top navigation laser. This approach intends to develop smaller AGV's instead of the AGV's that use a navigation laser (reflectors triangulation), and with more flexibility than the traditional small industrial AGVs which navigation is based on the tracking of magnetic bands.

A possible indicator of the market value for such a solution, location with reflectors and security laser, is that there are already manufacturers providing models of security lasers, with the ability to detect reflectors.

However, there are many problems associated to this type of solution. The first one is related to the angular opening of the security laser, which is smaller than the one of the navigation laser (typically a navigation laser is 360, while a safety laser is 190° or 270° depending from where it is installed on the AGV). Secondly, we also have the problem where the security laser is located. Being in a lower position, the possibility of beacons occlusion is higher as well as the possibility of false positive detection that is typically because there is a greater quantity / variety of objects at a short distance from the floor. It is clear that in this case we can not use the same algorithm used in navigation laser. These facts emphasize the need for new algorithms much more robust and efficient/reliable supporting more outliers and smaller amount of information without significantly degrading the accuracy of the location. The system is developed in ROS [3].

This paper is organized as follows: after the introduction, section 2 addresses the state of the art where related work is described. Section 3 presents the algorithm whereas section 4 addresses the practical results. Finally, section 5 rounds up with conclusions and points out some future work.

2 Related Work

Over the last two decades researches have been working in the mobile robot localization [7]. There are a huge variety of solutions based on several approaches. Unfortunately, commercial solutions do not include the localization system and just a few ones work in well-known and controlled environments. Using lasers it is possible to apply algorithms that find the match between the information from the laser and the map [4] [11]. But, the related localization systems should be based on indistinguishable beacons. With this criteria, there is the well-known Thrun et al. of book [5] as an extended Kalman filter approach.

The core of this localization algorithm is the identification of the beacons that are indistinguishable between them to find the correspondence position in the map. Several approaches exist on the literature, such as Thrun [5] that uses the maximum likelihood data estimation to compute for each iteration and for each beacon the probability density function. The desired solution is the one that maximizes that function. There are other approaches, like Ronzoni et al. [8] that reaches the global localization based on the distance of the reflectors. By this approach encoders odometry data is not used and global positioning is computed without previous information of robot localization.

3 Algorithm

Problem Definition. The problem of beacon based localization is defined as the estimation of the AGV absolute pose $X_v = [x_v y_v \theta_v]^T$ in an external referential W (figure 2) being W the world referential and R the AGV relative referential).

This estimation is performed from:

- The reflectores map $M_B = [M_{B,1} \ ... \ M_{B,numB}]$. It consists in a set of $numB$ fixed positions where reflectors are installed. So $M_{B,i} = [x_{B,i} \ y_{B,i}]$ is the position of reflector i in the external referential W. The reflectors have a cylinder form with fixed radius (B_{radius}) and are indistinguishable between them. In figure 2 is represented a reflector by the yellow circle.
- The laser measures $Z_L(k) = Z_{L,i}(k), C_{L,i}(k) : i \in [1 \ numL]$. Where $Z_{L,i}(k)$ corresponds to relative position i at instant k in polar coordinates ($r_{L,i}$ distance, $\varphi_{L,i}$ angle) of the detected obstacle. $c_{L,i}$ is a Boolean related with the reflexivity of the target (a reflector presents high reflexivity). The dashed lines in figure 2 represent

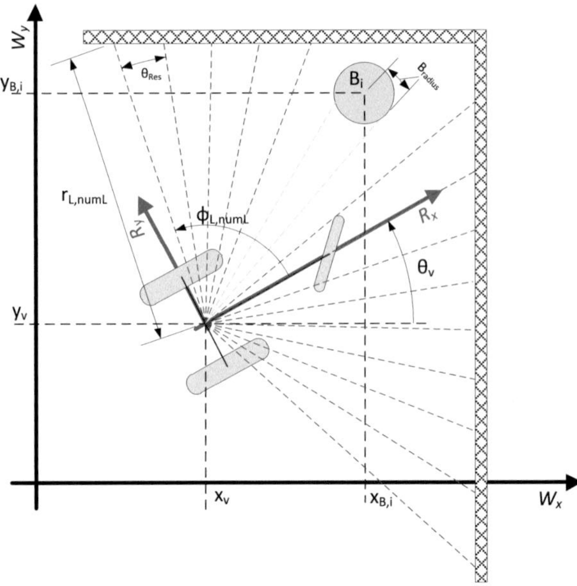

Fig. 2 AGV and world referentials with beacon map

the laser measures, the yellow ones have a true $c_{L,i}$ (high reflexivity) while the red ones have a false $c_{L,i}$ (low reflexivity).

- The odometry data $u(k)$. It is provided by the robot encoders and measure the relative displacement of AGV in R referential.

Data acquired by robot sensors are corrupted by external factors and the system is exposed to outliers like the false detections of reflectors (other materials with high reflectivity) and its occlusions. The map (M_B) acquisition is a subject to be address in other paper.

Problem Approach. The problem of localization can be expressed as a block diagram as presented in figure 3. Each block presents inputs (red), outputs (blue) and parameters (green). The gray area refers the sensor fusion algorithm known as the Extended Kalman Filter (EKF).

Next subsection presents the EKF application to perform the localization task based on reflectors and then further subsections address the reflectors detection, association and outlier filtering.

Extended Kalman Filter. The EKF is a well-known algorithm applied to sensor fusion in mobile robotics area. In that case it computes the statistical data related to the state estimation (estimated pose and its covariance) from fusion of the distance and angles measured from reflectors and odometry data. It also deals with sensors errors modulated as gaussian noise. Despite the linearization issues of EKF could

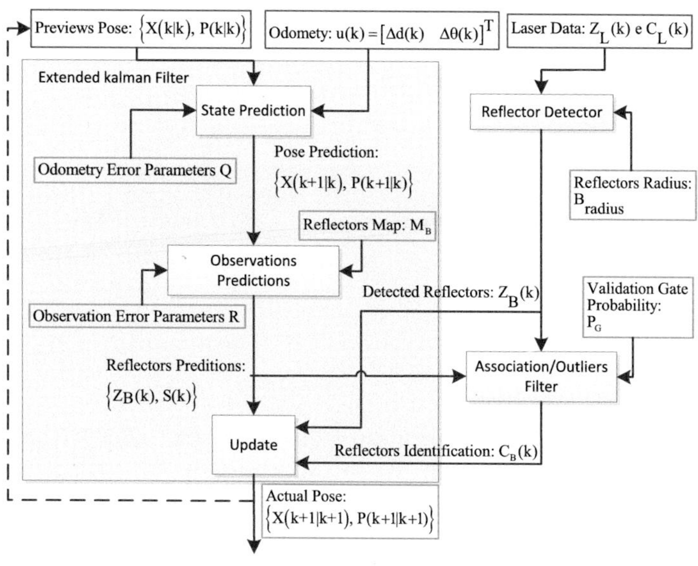

Fig. 3 System architecture

be a problem, it fits well in our problem because industrial application usually requires high precision solutions (low covariance) and the effect of linearization error is attenuated. Besides, industrial applications usually requires high frequency rate solutions where Kalman filter fits better in opposition to other approaches such as particle filters [6].

In order to apply an EKF, it is necessary to define the models: state transition $f(.)$ and observation $h(.)$.

As it can be expressed through equation 1, $f(.)$ models the evolution of the robot pose based on last state and odometry data $u(k)$. $Q(k)$ is the state noise variance and it depends on the values of $u(k)$. Odometry model was based on [10].

$$X_v(k+1) = \begin{bmatrix} x_v(k+1) \\ y_v(k+1) \\ \theta_v(k+1) \end{bmatrix} = f\left(X_v(k), u(k)\right) + N\left(0, Q(k)\right) \tag{1}$$

$h(.)$ states robot pose (X_V) with expected reflectors detections (Z_B). The observation model assume that measurements are affected by additive gaussian noise with zero mean and covariance R. In the presented work, R is assumed constant and a parameter of the system. For further details see [9].

$$\begin{bmatrix} r_{B,i} \\ \phi_{B,i} \end{bmatrix} = h\left(M_{B,i}, X_v(k)\right) + N(0, R) \tag{2}$$

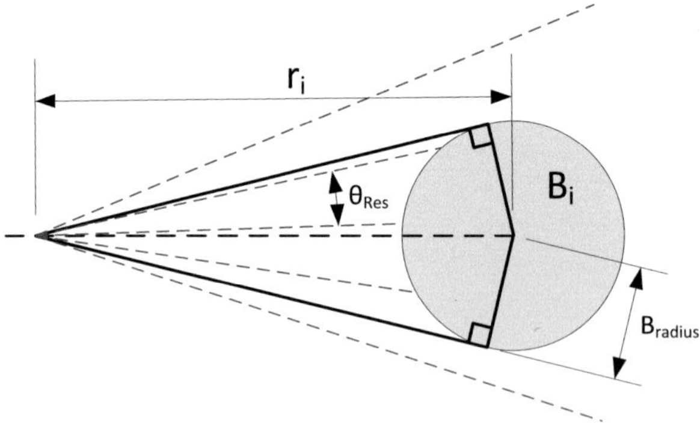

Fig. 4 Laser beams intercepting the reflector beacon.

At this point we are able to present our application of EKF filter. It is an adaptation of [9].

The used Kalman filter algorithm is described in previous work [11]. As input there is previous pose estimation and its covariance $(\widehat{X}(k|k)P(k|k))$, laser data (Z_L) and odometry data $u(k)$. As output the filter presents the pose estimation with its uncertainty characterized by the co-variance matrix $P(K+1|K+1)$. More details about algorithm can be found in [9].

Reflector Detection. The reflector detector module processes the acquired data from laser Z_L detecting reflectors and measures its positions Z_B related to the robot referential. Z_B provides a set of polar positions represented by $[r_i, \varphi_i]^T$.

The Reflector Detector performs the following steps:

* Splits Z_L in clusters based on reflexivity $C_{L,i}$. One clusters is sequence of measures with high reflexivity.

* For each cluster is calculated polar coordinates Z_B. For now it is only considered the central measure, so we also consider a constant R in the equation 2. We intend to refine this model in the future.

* It is applied a filter "detector filter" to eliminate possible outliers.

The detector filter is based on the geometric relationship of figure 4, that allows to model the number of the beams measuring object (number of elements in a cluster) according to its distance r. With this, and knowing the radius of the installed reflectors (B_{radius}) and the laser resolution (θ_{Res}) the model presented in equation 3 is defined.

$$M_{num}(r) = floor\left(\frac{2 * \arcsin\left(\frac{B_{radius}}{r}\right)}{\theta_{\text{Res}}}\right) \tag{3}$$

By this way it is possible to ignore objects with dimensions other than reflectors. The error between the number of measurements and the model allows to reject or accept each cluster.

Association/Outliers Filter. This module has two functions: identify the detected reflectors and filter outliers that the previous filter (reflector detector) could miss. This filtering takes into account the current estimation of the robot pose, $\widehat{Z}_B(.)$ and $P(.)$, rejecting measures according to the probability of their occurrence.

This algorithm is presented in algorithm 1, where Z_B are the detected reflectors and $\widehat{Z}_B$ and $S(.)$ is the predicted observations and its covariance. As result of this algorithm C_B consists in an array (with the same size of Z_B) where each element $C_{B,i}$ shows the reflectors map index of $M_{B,i}$ associated to detection $Z_{B,i}$.

Data: $Z_B(k),\ \widehat{Z}_B(k),\ S(k)$
1 **for** *all beacons* $Z_{B,i}$ **do**
2 Association:
3 $C_{B,i}(k) = \underset{j}{\operatorname{argmax}} \det\left(2\pi * S_j(k)\right) * \exp -\frac{1}{2} * \left(Z_{B,i}(k) - \widehat{Z}_{B,j}(k)\right)^T * \left[S_j(k)\right]^{-1} * \left(Z_{B,i}(k) - \widehat{Z}_{B,j}(k)\right)$
4 Association filter:
5 **if** $\left(Z_{B,i}(k) - \widehat{Z}_{j=C_{B,i}(k)}(k)\right)^T * \left(\mathrm{S}_{j=C_{B,i}(k)}(\mathrm{k})\right)^{-1} * \left(Z_{B,i}(k) - \widehat{Z}_{j=C_{B,i}(k)}(k)\right) > \chi^2_{df=2}$ **then**
6 $C_{B,i}$ = INVALID_BEACON_ID
7 **end**
8 **end**
Result: $C_B(k)$

Algorithm 1. Association Outliers Filter.

A brief description of Algorithm 1 is presented:

Line 3: $Z_{B,i}$ observation is associated with the element $M_{B,i}$, that maximizes the probability of their occurrence. For each element j of the map M_B is computed the value of probability distribution function of $Z_{B,i}$ observation occurrence (likelihood). The observation i is associated with the element j of the map that presents a higher value (maximum likelihood). For more details see [9].

Line 5: $\chi^2_{df=2}$ is a constant value function of P_G (figure 3 Validation Gate Probability). It is define, in observations space, a zone where the probability of an observation falls inside it is greater or equal to P_G. $\chi^2_{df=2}$ corresponds to the inverse of the density function of a chi-square distribution with two degrees of freedom. In this point we enhance the outliers statistical description of [9], where is suggested a filtering only based on likelihood value of the observations.

Later in figure 8 we can see the representation of the validation area in Cartesian space. The blue ellipses corresponding to the area in which, taking into account the AGVs estimated pose, the probability of being detected a reflector is P_G, which in this particular case is 95%.

As conclusion, Kalman filter to merge information of the detected reflectors and odometry data was used. The robustness of this algorithm relies on outliers filtering performed by two filters in series: detector filter and association filter.

4 Practical Results

As a way to validate our approach with the ground truth it was used another AGV which was equipped with a commercial navigation laser based on triangulation (SICK NAV350). This laser does not only provides the robot pose but also the reflectors relative position (Z_B of figure 3).

In figure 5 we marked the position of the four reflectors installed (blue) and the path (red) made by the AGV, that begins to move forward from P0 and when reaches P1 reverses the direction moving to P2, go forward to P3 and finally it reverses again to P0. Along this path it is compared the pose provided by our system with the pose provided by laser triangulation system. We point out that in this experiment we use a navigation laser and not a security laser, and thus the four reflectors are installed in the walls so they are always visible along the path for laser has Field of View (FOV) of 360°.

As expected we obtained low positioning and orientation errors as seen in the blue lines of figure 6, in which our algorithm worked with a 360° field of view. But with this we can address the response of the system with less sensory information. With this purpose, our systems field of view was reduced to 180° by software (ie measures exceeding 90 are ignored). This situation is analogous to the use of security lasers in which the field of view is typically reduced and positioned at a lower high. By imposing this restriction to our system we obtain the results shown by the red lines of figure 6. The green line concerns the visible reflectors for the reduced F.O.V.

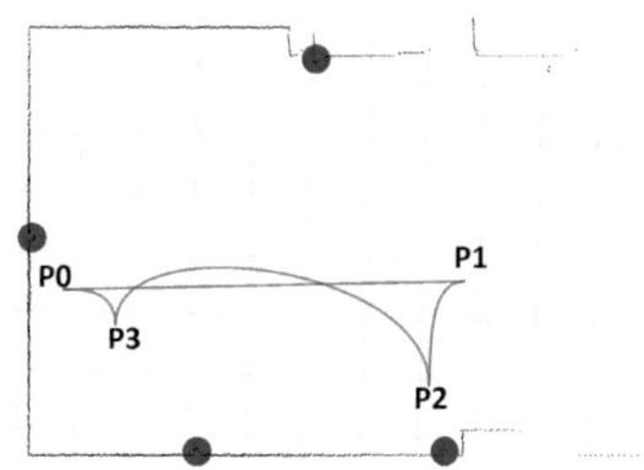

Fig. 5 Map of the AGV path testing. Grid size: 1m.

It is noted that the accuracy of our system is not significantly affected by the reduction of the F.O.V. (except for situations where less than two visible beacons are available and only odometry is used for long period of time). These results show the advantage of using Kalman filter. The system used here as ground truth is based on laser triangulation, and therefore only works if they have at least three reflectors visible simultaneously (the recommended value is five visible reflectors). On the other hand, the Kalman filter, merging angles and distances measures to reflectors with odometry, can estimate the robot pose having even less than three reflectors line of sight.

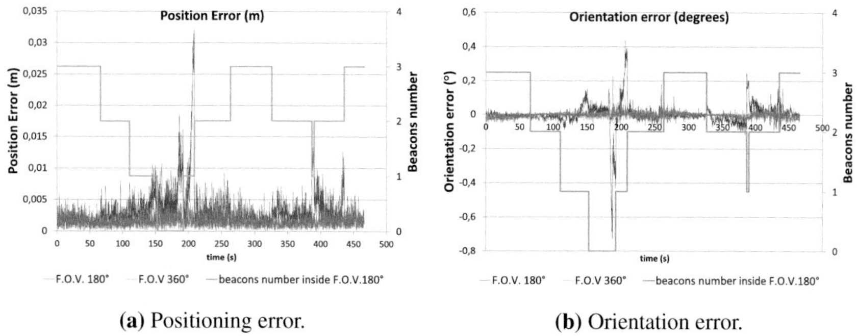

(a) Positioning error. **(b)** Orientation error.

Fig. 6 Comparison of error between the 360° FOV triangulation laser (blue line) and 180° FOV (red line) from security laser.

Robustness Analisys-Outliers Rejection. In the previous section we analyzed the system's accuracy using a navigation laser whereas this sections addresses the safety laser and the platform presented in figure 1. Here we analyze the system response when exposed to adversity in a real manufacturing environment.

In the scope of Project PRODUTECH PSI PPS3 it was done a public demonstration of the developed system which was used to validate our developed localization system in industry environment. Figure 7 (left) shows the demonstration scenario in manufacturing environment. Under this demonstration it was possible to collect data, and validate the system in the sense that the robot successfully accomplishes its mission for a period of 6 consecutive hours without failures. The AGV navigated in an area of 24m x 13m at a speed of 0.5m/s, and its mission was to carry tables between working stations, and to do so it is required an accuracy of about 1 cm in position and 3 degrees in orientation. This demonstration was especially demanding for the localization system because there were about fifty people walking around in the AGV navigation area and the floor was uneven. Therefore, at certain points of its trajectory, the laser range was reduced to 4m because the laser was pointing to the floor. All together it increases the number of reflectors occlusions and increases the amount of outliers that the system is exposed.

In order to give an idea of the system exposure to outliers in this demonstration we present the figure 7 right. Here we have marked the position of the installed reflectors through the dark blue circles, and the considered path in the generation of this image. The AGV has moved backwards from P0 to P1 and then followed straight ahead to P2. During this path the position of the reflectors detected by the safety laser is marked with yellow and red dots. The yellow dots correspond to inliers and the red dots are the outliers rejected by the detector filter and the association filter.

We also show a screenshot of the navigation system interface (figure 8) taken during path described above. In this we have represented: The reflectors map M_B (dark blue dots), the covariance of the AGV estimated position through the red ellipse (right upper corner), and the corresponding covariance observations prediction represented by blue ellipses. This area around the reflectors position corresponds to the validation area in which is based on the association filter. The observations that

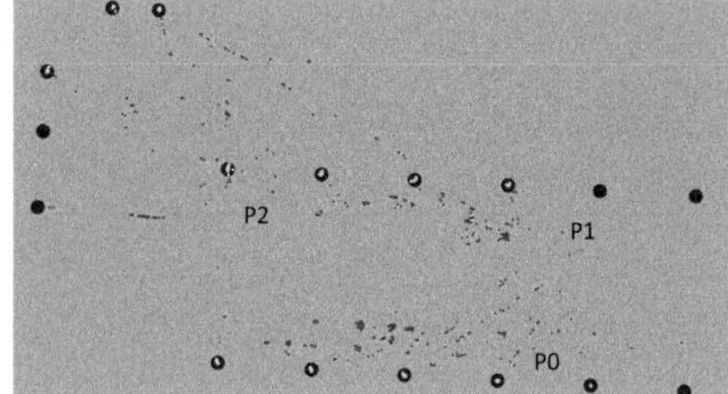

Fig. 7 Right: Industrial environment scenario where localization tests were developed. Reflector instalation in the ADIRA real industry environment (dark blue). Left: Red points are outliers, yellow points are the reflectors detection during one run. Grid size: 1m.

fall out this area means that are less likely to 5%, are marked as outliers and ignored by the system. As we see in the figure 8 the detected reflectors are represented with yellow circles, inliers, and red the measures rejected, outliers (left bottom corner). The black dots represent the measurements of the distances provided by the safety laser (Z_L).

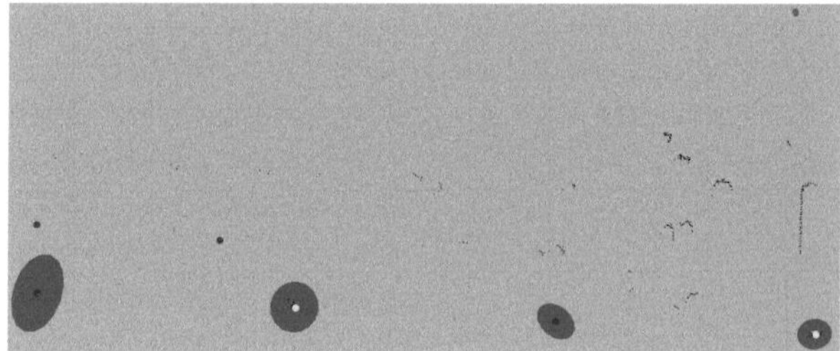

Fig. 8 Screenshot of the application during localization tests. Red points are rejected outliers, yellow points are accepted reflectors, blue ellipses show accepted area to inliers, red ellipse (upper right corner) represents the AVG covariance positioning and black points represents the remaining acquired data by security laser scanner. Grid size: 1m.

At this point it becomes evident that there are a large number of false positives in reflectors detection by the safety laser. The correct outliers filter in applications such as these take a central role. In the present approach, this filtering is performed on two points of the system, on the reflector detector (detector filter) and after the association process (association filter). In order to highlight the importance of the filtering process we select a part of the collected data and once again test the system with three different settings. Figure 9 presents the result for positioning and orientation respectively without any filter (blue line), only with detection filter (green line) and finally with both (detector and association filter) in red line. As it is easily to understand, the absence of filtering system produces noisily, mistakenly and unstable results. The detector filter improves the result of positioning but it still lacks stability and introduces some error. Finally, the use of both filters allows to perform a stable measure where all the outliers are correctly identified, without unexpected variations in position and orientation.

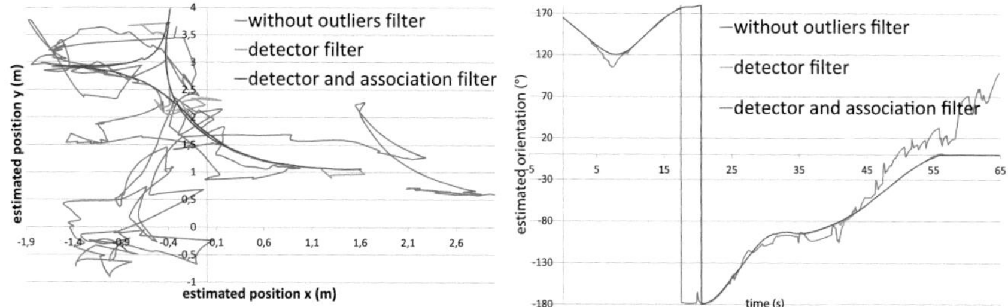

Fig. 9 Comparison of filtering rejection outliers. Right: Position, left: Orientation.

Videos containing the localization system interface during the tests above mentioned and in transporting mission can be downloaded at these links:

- Without outliers filter: http://youtu.be/wCv9qVTSICg
- Detector filter: http://youtu.be/iTCb5UR6CRE
- Detector and association filter: http://youtu.be/4_Io52ORvOE
- Laboratory: http://youtu.be/6SQ3llbTSFk
- Adira industrial environment: https://youtu.be/cyokKOBxcY0

5 Conclusion and Future Work

The developed localization system based on a security laser scanner was applied in an industrial AGV. The experiments allow to confirm that in real industrial environment there are advantages in the proposed localization algorithm over the laser positioning state of the art systems.

The Extended Kalman Filter was applied as a multi fusion sensor system in order to combine the odometry information and the result of the developed system. It allows to perform the localization task with less number of sensors than commercial solutions. As final remark the presented algorithm ensures precision (10mm or best depending on the number of reflectors) and fast computation time (5ms), without which it would not be possible to use the localization system in an industrial environment.

As a future direction, enhancing the beacon detection system (increasing the number of detected features) will allow to benefit of better robustness and precision. Moreover, implementing redundant methods will improve the good results of the presented system.

Acknowledgment This work is financed by the ERDF European Regional Development Fund through the COMPETE Programme (operational programme for competitiveness) and by National Funds through the FCT Fundao para a Cincia e a Tecnologia (Portuguese Foundation for Science and Technology) within project FCOMP-01-0124-FEDER-037281.

References

1. Wulf, O., Lecking, D., Wagner, B.: Robust self-localization in industrial environments based on 3D ceiling structures. In: Proceedings of the 2006 IEEE/RSJ International Conference on Intelligent Robots and Systems (2006)
2. Liu, J., Yin, B., Liao, X.: Robot Self-localization with Optimized Error Minimizing for Soccer Contest. Journal of Computers **6**(7) (2011)
3. Quigley, M., Gerkey, B., Conley, K., Faust, J., Foote, T., Leibs, J., Berger, E., Wheeler, R., Ng, A.Y.: ROS: an open-source robot operating system. In: Proc. Open-Source Software workshop of the International Conference on Robotics and Automation, Kobe, Japan, May 2009
4. Lauer, M., Lange, S., Riedmiller, M.: Calculating the perfect match: an efficient and accurate approach for robot self-localization. In: RoboCup Symposium, Osaka, Japan, July 13–19, 2005, pp. 142–53 (2005)
5. Thrun, S., Burgard, W., Fox, D.: Probabilistic Robotics. Massachusetts Institute of Technology (2006)
6. Grisetti, G., Stachniss, C., Burgard, W.: Improved Techniques for Grid Mapping with Rao-Blackwellized Particle Filters. IEEE Transactions on Robotics **23**(1), 34–46 (2007)
7. Borenstain, J., Everett, H.R., Feng, L., Wehe, D.: Mobile Robot Positioning and Sensors and Techniques. Journal of Robotic Systems, Special Issue on Mobile Robots **14**(4), 231–249 (1997)
8. Ronzoni, D., Olmi, R., Secchi, C., Fantuzzi, C.: AGV global localization using indistinguishable artificial landmarks. In: IEEE International Conference on Robotics and Automation, Shanghai, pp. 287–292. IEEE (2011). doi:10.1109/ICRA.2011.5979759
9. Thrun, S., Burgard, W.: Probabilistic Robotics (Intelligent Robotics and Autonomous Agents series). The MIT Press (2005)
10. Eliazar, A.I., Parr, R.: Learning probabilistic motion models for mobile robots. In: Proceedings of International Conference on Machine Learning (2004)
11. Sobreira, H., Pinto, M., Moreira, A.P., Costa, P.G., Lima, J.: Robust robot localization based on the perfect match algorithm. In: Proceedings of the 11th Portuguese Conference on Automatic Control Lecture Notes in Electrical Engineering 2014, Portugal, vol. 321, pp. 607–616 (2015)

Part III
Legged Locomotion Robots

Energy Efficient MPC for Biped Semi-passive Locomotion

C. Neves and R. Ventura

Abstract Traditional methods for robotic biped locomotion employing stiff actuation display low energy efficiency and high sensitivity to disturbances. In order to overcome these problems, a semi-passive approach based on the use of passive elements together with actuation has emerged, inspired by biological locomotion. However, the control strategy for such a compliant system must be robust and adaptable, while ensuring the success of the walking gait. In this paper, a Model Predictive Control (MPC) approach is applied to a simulated actuated Simplest Walker (SW), in order to achieve a stable gait while minimizing energy consumption. Robustness to slope change and to external disturbances are also studied.

Keywords Model predictive control · Biped locomotion · Simplest walker · Energy efficiency

1 Introduction

In the past years, robotics has evolved from mechanisms working in closed environments like factories to toys and devices available in every house, interacting with humans of all ages. The aim of anthropomorphic robotics is to create mechanisms that are able to blend in our human oriented world and handle locomotion challenges such as uneven terrain or stairs as easily as humans.

One of the main challenges of these mechanisms is the high energy consumption required to control the high number of servos typically used in robotics. Aiming at the creation of energy efficient walkers, McGeer proposed a passive dynamic walking mechanism[10, 11] that is capable of walking down a slope actuated merely by gravity. The fact that the gait that naturally rose from this setup was so human like has encouraged others to develop and study passive dynamic walkers and their natural gaits.

C. Neves(✉) · R. Ventura
Institute of Systems and Robotics, Instituto Superior Técnico, Lisbon, Portugal
e-mail: {cneves,rodrigo.ventura}@isr.tecnico.ulisboa.pt

L.P. Reis et al. (eds.), *Robot 2015: Second Iberian Robotics Conference*,
Advances in Intelligent Systems and Computing 418,
DOI: 10.1007/978-3-319-27149-1_12

However, passive mechanisms are not controlled and therefore are not able to deviate from obstacles or walk on horizontal ground. A semi-passive approach has been extensively studied [4, 15, 16, 17, 18], which incorporates typical actuators alongside compliant elements, making use of their natural dynamics. Energy Storage and Return (ESAR) elements are usually elastic mechanisms, like springs, rubber bands or compliant sheets, that are able to deform and store energy that can be returned to power the joints, drastically reducing the power consumption of the actuators. Compliant actuation also transforms the joint into a low bandwidth system, which reduces the instability and absorbs vibrations from the foot-ground impacts.

Given the natural instability of biped locomotion and in order to make the best use of these elements, the control methods applied to these mechanisms need to be robust to handle compliance and external disturbances while ensuring energy efficiency. The main challenges to be faced are (1) the mix of continuous dynamics and discrete jumps, (2) handling model errors that emerge from the passive elements and (3) handling external disturbances. Model Predictive Control techniques are a general method to control these systems, by minimizing a constrained cost function over a sliding horizon, based on the model of continuous flow and discrete jumps[3].

In the next Section, the Simplest Walker model is presented, as well as the equations of motion and transition used in the simulation. A cell-mapping method for determining a stable gait for a given slope is explained at the end of the section. In Section 3, the Model Predictive Control approach is explained and the details about the application to an actuated SW model are given. The results of this approach are shown in Section 4. Finally, some conclusions are taken from these results and future work is detailed.

2 The Simplest Walker

The Simplest Walker mechanism model used for this work is presented in [6] and several variations have been proposed and analysed in [2, 7, 8, 9]. It consists on a compass like mechanism, formed by two massless legs with length l connected at the hip by a frictionless joint, with mass M at the hip and two masses m at the feet, as shown in Fig. 1. The model is defined by the angle between the perpendicular to the slope and the stance leg θ and the angle between the stance and swing legs ϕ.

2.1 Equations of Motion

The Simplest Walker can be modelled as an hybrid system, since it exhibits both continuous and discrete dynamics. A typical step is divided into single support and double support stage. During the single support stage, the back leg swings to the front, while the center of mass of the mechanism moves forward. The double support stage starts at heel strike, when the weight shifts from a leg to another and the swing leg

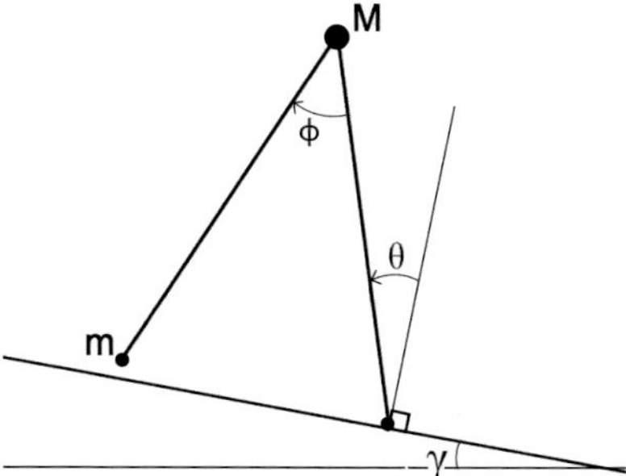

Fig. 1 Simplest Walker Model

becomes the stance leg, and stops at foot liftoff. For this model, the walking step starts with both feet on the ground and the swing phase is assumed to be slipless and the heel-strike impact is also slipless, inelastic and instantaneous. The swing foot scuffing the ground when the swing leg meets the stance leg is ignored, since it is unavoidable for a model with straight legs.

Model State. In cartesian coordinates, the simplest walker configurations is defined by the feet and hip positions, as

$$x = \begin{bmatrix} x_{stance} \\ y_{stance} \\ x_{hip} \\ y_{hip} \\ x_{swing} \\ y_{swing} \end{bmatrix} = \begin{bmatrix} 0 \\ 0 \\ -l\sin(\theta-\gamma) \\ l\cos(\theta-\gamma) \\ -l\sin(\theta-\gamma)+l\sin(\theta-\phi-\gamma) \\ l\cos(\theta-\gamma)-l\cos(\theta-\phi-\gamma) \end{bmatrix} \tag{1}$$

In order to reduce dimensionality, generalized coordinates were employed and the state of the system is given by

$$q = \begin{bmatrix} \theta \\ \dot{\theta} \\ \phi \\ \dot{\phi} \end{bmatrix} \tag{2}$$

Single Support. The single support stage, or swing phase, is characterized by continuous dynamics. The equations of motion are obtained using Lagrangian Mechanics, as it is shown in [7]. By application of the $M \gg m$ approximation and rescaling time with $\sqrt{\frac{l}{g}}$, the continuous equations of motion are given by:

$$\ddot{\theta} = \sin(\theta - \gamma) \tag{3}$$
$$\ddot{\phi} = \sin(\theta - \gamma) + \left\{\dot{\theta}^2 - \cos(\theta - \gamma)\right\} \sin(\phi) \tag{4}$$

For the application of the Model Predictive Control, the hip and stance ankle joints are actuated. The non-dimensional torque vector u is obtained by dividing the torque U by Mgl and is added to the previous equations, and we get

$$\ddot{\theta} = \sin(\theta - \gamma) + u_{(1)} \tag{5}$$
$$\ddot{\phi} = \sin(\theta - \gamma) + \left\{\dot{\theta}^2 - \cos(\theta - \gamma)\right\} \sin(\phi) + u_{(2)} \tag{6}$$

Double Support. The double support stage, for this model, is instantaneous and represents a jump in the hybrid model at heel strike. In this case, making use of the conservation of angular momentum, we obtain the following general coordinates q^+ after heel strike from q^- before heel strike:

$$q^+ = \begin{bmatrix} \theta^+ \\ \dot{\theta}^+ \\ \phi^+ \\ \dot{\phi}^+ \end{bmatrix} = \begin{bmatrix} \theta^- \\ -2\theta^- \\ cos(2\theta^-)\dot{\theta}^- \\ cos(2\theta^-)(1 - cos(2\theta^-))\dot{\theta}^- \end{bmatrix} \tag{7}$$

It is also worth noticing that the initial conditions of a new step are only dependent on θ and $\dot{\theta}$. This means that it is possible to analyse the gait merely by looking at these two variables.

2.2 Step Failure

For this work, a step is considered to fail if:

- Forward fall is detected when $\theta < -\pi/4$
- Backward Fall is detected when $\theta > \pi/4$
- Running is detected when $Q_v = -\cos(q_{(1)})(q_{(3)}^2 - \cos(q_{(1)} - \gamma)) < 0$
- Backward walking is detected when $\phi_{(k-1)} < 0$ and $\phi_{(k)} > 0$ or $\dot{\theta} > 0$
- Swing foot goes under ground while scuffing
- Swing foot touches ground behind stance foot

2.3 Gait Stability

A passive walking gait is highly unstable and only a narrow set of initial conditions will result in a successful step and only a subset of these will result in a stable gait[19]. A step can be represented by a stride function $S_{(\theta,\dot{\theta})}$, which maps the evolution of the state space from the beginning of a step at foot liftoff to the end of the step at heel strike. The analysis of the stride function would traditionally be made by analysing the trajectory of each variable of the state space. Given the cyclic behaviour of a walking gait, a Poincaré map was used [13, 14], where the evolution of the state is intersected by a 2D Poincaré section at heel strike, which allows the evaluation of the stability of the gait based on the discrete values of $[\theta, \dot{\theta}]$ at heel strike.

Like so, if the mechanism is released and the evaluation of the state space results in a sequence S_q, it is considered a stable gait if the values remain in a bound interval or unstable otherwise. If the stride function shows a fixed point $S_{(\theta^*, \dot{\theta}^*)} = [\theta^*, \dot{\theta}^*]$, the walker starts a cyclic walking gait. If this gait is stable, there will be a range of initial conditions that leads to it, and this point is called the limit cycle. There may be other $n-$periodic solutions, where the steps repeat themselves after n steps.

In order to discover the basin of attraction of a passive mechanism, the cell mapping method explained in [19] is used as follows: a cell matrix is populated with the initial conditions of a step and the cell closest to the finish conditions after a step. Afterwards, an analysis is performed to each cell to check if the step is successful and if a stable gait raises from it. If a step fails, the corresponding starting cell points to a sink cell.

The plot shown in Fig. 2 is obtained using starting conditions $\theta \in [0, 1]$ and $\dot{\theta} \in [-1, 0]$ with $\Delta\theta = \Delta\dot{\theta} = .005$, while in Fig. 3 the conditions used are $\theta \in [.1, .2]$ and $\dot{\theta} \in [-.2, -.1]$ with $\Delta\theta = \Delta\dot{\theta} = .0005$. The light grey area represents unsuccessful steps (39528 instances), the dark grey represents successful steps that result in unstable gaits (818 instances) and the black area represents stable passive gaits (55 instances).

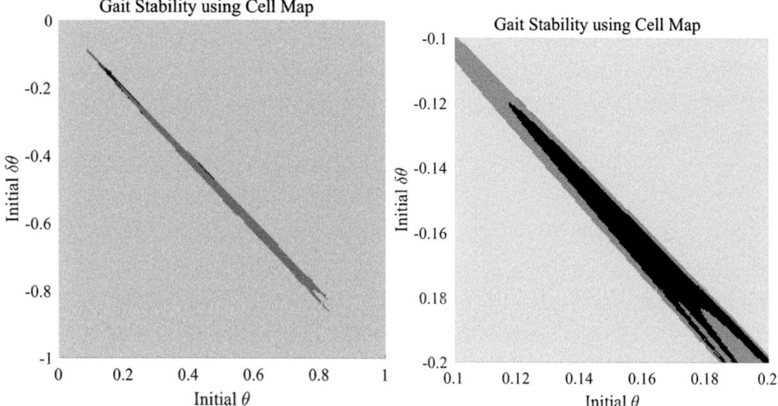

Fig. 2 Stability analysis of SW on a $\gamma = 0.004$ slope

Fig. 3 Zoom of stability analysis of SW on a $\gamma = 0.004$ slope around Limit Cycle conditions

The Limit Cycle is obtained by using a Newton-Raphson iterative method based on the linearization of the step function [19]

$$S_{(v+\Delta v)} \approx S_{(v)} + J\Delta v \tag{8}$$

where $v = [\theta, \dot{\theta}]$ and $J = \frac{\delta S}{\delta v}$. The Jacobian is obtained numerically, by simulating a step with small perturbations from the initial point. The NR algorithm is as follows:

repeat

Compute $J = \begin{bmatrix} \frac{\Delta S_1}{\Delta v_1} & \frac{\Delta S_1}{\Delta v_2} \\ \frac{\Delta S_2}{\Delta v_1} & \frac{\Delta S_2}{\Delta v_2} \end{bmatrix}$;

$\Delta v = [I - J]^{-1}(S_{(v)} - v)$;

$v = v + \Delta v$

until $|\Delta v| < \epsilon$;

The resulting conditions are a fixed point of the step function, but it's stability still must be evaluated. If the fixed point correspond to limit cycle conditions, then both eigenvalues of the Jacobian verify $|eigen(J)| < 1$. The basin of attraction for the limit cycle conditions is obtained by calculating the Jacobian as before and, if both eigenvalues verify the above condition, the basin of attraction is at least as large as the perturbation. The obtained Limit Cycle for these conditions is $[0.1534 - 0.1561]$, and is shown in both figures as a black cross. These results are consistent with previous works in the area [2, 7, 8, 9, 19].

3 Model Predictive Control

Model Predictive Control (MPC) [1, 12] is a strategy where constrained optimization of a cost function is used to generate inputs for a dynamic system. The cost function is evaluated at every time step and over a discrete sliding time horizon of N time slots.

A linear MPC has been used for biped locomotion control in [5], but for this work, two non-linear MPC based approaches were explored: a gait generator and a trajectory follower. Both were implemented using a Sequential Quadratic Programming (SQP) algorithm.

3.1 Gait Generator

The gait generator approach will run an event based MPC, where the cost function is computed over all time steps until a heelstrike is detected. The goal is to achieve initial conditions for the next step that reduce the need of actuation, which means to aim at the Limit Cycle conditions or at least to start within the Basin of Attraction. Therefore, given a slope estimation γ, the cost function will penalize deviations from the step length L_{step} and the step period T_{step} corresponding to the Limit Cycle conditions. In order to reinforce that the new start conditions should start a stable gait, the cost function will also include a term for the $|eigen(J)|$.

The optimized input vector $\bar{u}^*$, which controls the actuated stance ankle and hip joint at each timestep, is obtained from an initial input vector $\bar{u}_0 = \left[\bar{u}_{0(1)}\bar{u}_{0(2)} \dots \bar{u}_{0(m)}\right] = 0_{2m\times 1}$, where $m = ceil\left(\frac{\frac{6}{5}T_{step}}{\Delta t}\right)$, by running the SQP algorithm. The cost function terms associated with the desired step length, step period and the eigenvalues of the Jacobian are plotted in Fig. 4 and are of the form

$$J_{term} = C\|x - x_d\|^2 + \frac{1}{1 + \exp^{-2C(x - tol*x_d)}} + \frac{1}{1 + \exp^{2C(x + tol*x_d)}} \tag{9}$$

where x is the value obtained by the current simulation, x_d is the desired value and C is a constant factor. This function was designed in order to be continuous and to create a valley for acceptable values for each term. The cost function terms associated with spent energy and change of actuation are, respectively,

$$J_u = W \sum_{k=1}^{m} \|\bar{u}_{(k)}\|^2 \tag{10}$$

$$J_{\Delta u} = W \sum_{k=1}^{m} \|\Delta\bar{u}_{(k)}\|^2 \tag{11}$$

where W is a weight constant. In addition to the function cost, the optimizer must verify non-linear constraints that ensure the step is successful and that the physical limitations of the actuator are respected (maximum amplitude and maximum rate of change).

This process will output an optimized input $\bar{u}^*$ and a state trajectory for a step. The main disadvantage of this approach is the execution time of the optimization process, due to the large number of elements of the optimized vector and the highly non-linear step function.

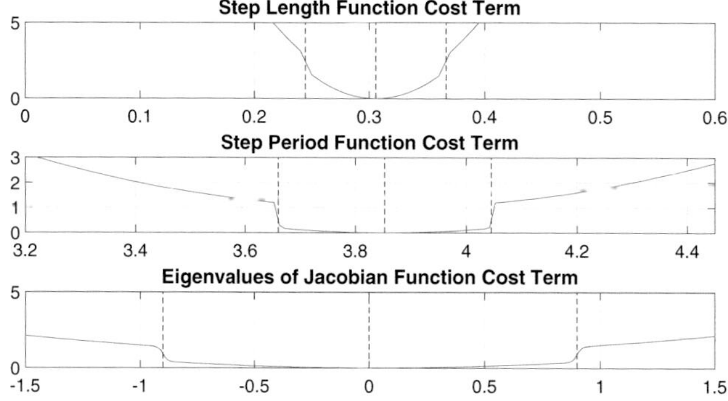

Fig. 4 Cost Function Terms for Step Lenght, Step Period and Eigenvalues of Jacobian

3.2 Trajectory Follower

In this approach, the controller will attempt to follow the given state trajectory for the next m timesteps, by optimizing the initial input vector. Let $q_P = \left[q_{P(1)} \ldots q_{P(N)}\right]$ be a state trajectory, provided either from the Gait Generator or a simulated passive step, $\bar{u}_P = \left[\bar{u}_{P(1)} \ldots \bar{u}_{P(N)}\right]$ be the initial guess vector provided to the optimizer, $q_{(k)}$ be the current state and m the sliding window horizon.

The cost function to be optimized for this method has a term to penalize deviations from the trajectory and an energy term, as in:

$$J = \sum_{i=1}^{m} w_q \|q_{(k+i)} - q_{P(k+i)}\|^2 + \sum_{i=1}^{m} w_u \|u_{(k+i)}\|^2 \tag{12}$$

where k is the current time step, w_x and w_u are weight factors. At each timestep, the input vector for the following m timesteps is optimized in order to follow the given trajectory while saving energy if possible and reacting to disturbances, and only the first element of the optimized vector is applied to the robot.

4 Simulation and Results

The MPC based controller designed for this work aims at controlling a Simplest Walker simulated mechanism on different slopes, starting with different initial conditions, until it reaches the basin of attraction, while maintaining energy consumption to a minimum and being robust to external disturbances. If the starting conditions of a step are within the basin of attraction or that given slope, the predicted trajectory will be that of a passive step and no actuation is predicted to be necessary. Otherwise, the SQP optimizer will output the best actuation and the predicted trajectory is computed.

Afterwards, the mechanism is controlled to follow the given trajectory. At each time step, the best actuation is recalculated for a sliding window horizon and the next state is estimated. If a small disturbance is detected, the optimizer will adapt the actuation outputed from the generator. If a large disturbance is detected, the gait generator is ran again.

4.1 Increased Stability

Given the unstable nature of biped walking, one of the main goals of actuation is to ensure a stable gait. For the first simulation, the mechanism is placed in a slope with $\gamma = 0.004$ and is given as goals the step length $L_{step} = 0.3057$ and step period $T_{step} = 3.8517$, starting with different initial conditions. In Fig. 5, the successful steps

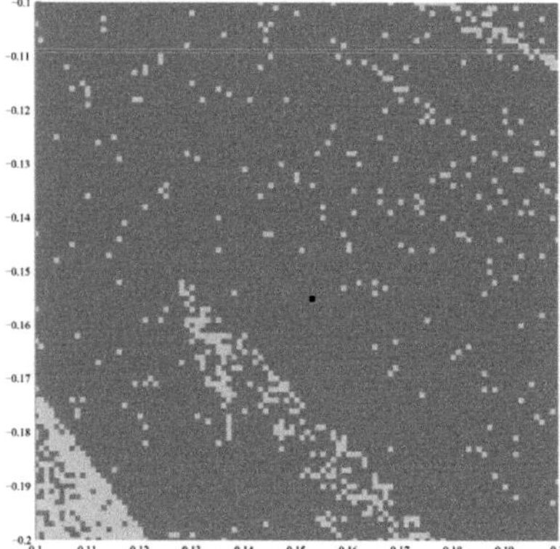

Fig. 5 Actuated Simplest Walker conditions leading to stable passive gait

are shown in grey, while the unsuccessful are shown in light grey. When compared to Fig. 3, we can see the main goal is achieved, since the amount of successful starting conditions increases.

4.2 *Robustness to External Disturbances*

In order to test the robustness to external disturbances during the trajectory following stage, a constant external disturbance was applied to both joints during the third step. This actuated response is shown in Fig. 6, as well as the step evolution, with a straight line connecting points before the disturbance and dashed line connecting spots afterwards. The 7th step is already part of the basin of attraction for that slope and it will reach a stable gait. Larger disturbances will trigger the Gait Generator to choose the best reaction.

4.3 *Execution Time*

The main downside of this method is the execution time of the gait generator, both when starting a step or a large disturbance is detected. Fig. 7 shows the step evolution and execution time of each step. The first iteration takes about 6 minutes to evaluate the instability of the starting conditions, generate a trajectory and follow it. From that point forward, the mechanism enters the basin of attraction and it will simply follow the trajectories of a passive mechanism, with iterations taking about 15 seconds, for a window horizon of $m = 10$.

When a slope change is introduced mid gait, the necessary time to project a new gait trajectory is also very lengthy. In Fig. 8, the mechanism is started in the basin

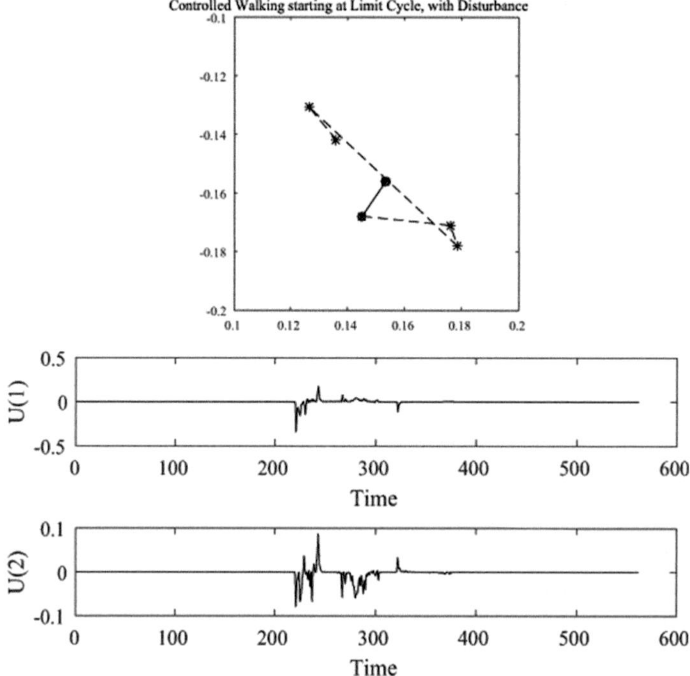

Fig. 6 Step Evolution when facing external disturbances

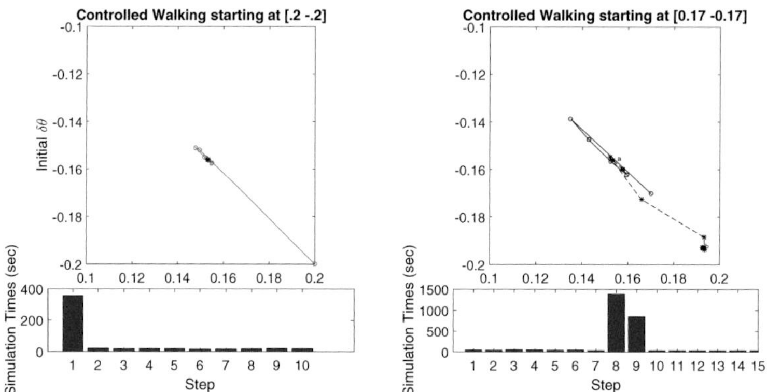

Fig. 7 Step Evolution and Execution Time starting at $[.2 - .2]$ on $\gamma = .004$ slope

Fig. 8 Step Evolution and Execution Time starting at $[.17 - .17]$ on changing slope

of attraction of a $\gamma = .004$ slope and the initial iterations take about 5 seconds to generate a trajectory and 15 seconds to follow it. At the 7th step, the slope is changed to $\gamma = .008$ and the Gait Generator takes about 21 minutes to generate and follow a new trajectory, and for the 8th step it takes about 14 minutes. After that, the mechanism enters the basin of attraction of the new slope, and all iterations return to execution times of about 25 seconds.

5 Conclusions and Future Work

In this paper, an MPC approach was used to control a semi-passive simplest walker mechanism. This method showed good results, as it is able to generate stable gaits that tend to the Limit Cycle conditions for different slopes and is robust to external disturbances and slope changes, while attempting to keep low energy consumption.

The main downside of this method is the time it takes to provide a gait trajectory and the optimized input vector, both at startup and when a strong external disturbance is detected. Other optimization methods or cost functions must be explored in order to improve efficiency and the use of approximate models must be taken in consideration.

For future work, a learning algorithm is to be implemented to provide the MPC with the most energy efficient step length and period, based on the slope and desired speed of the mechanism. In addition, a comparison study must be performed between the presented gait generator and other methods, such as CPG's, and Energy Save and Return elements should be added to the model.

Acknowledgements This work was supported by the project FCT [UID/EEA/ 5009/2013] and was partially funded with grant SFRH/BD/80044/2011, also from Fundação para a Ciência e a Tecnologia.

References

1. Bemporad, A., Morari, M.: Robust model predictive control: a survey. In: Robustness in Identification and Control, pp. 207–226. Springer (1999)
2. Bhounsule, P., et al.: Foot placement in the simplest slope walker reveals a wide range of walking solutions. IEEE Transactions on Robotics **30**(5), 1255–1260 (2014)
3. Camacho, E., Ramirez, D., Limon, D., de la Pena, D.M., Alamo, T.: Model predictive control techniques for hybrid systems. Annual Reviews in Control **34**(1), 21–31 (2010). http://www.sciencedirect.com/science/article/pii/S1367578810000040
4. Daerden, F., Lefeber, D.: Pneumatic artificial muscles: actuators for robotics and automation. European Journal of Mechanical and Environmental Engineering **47**, 10–21 (2000)
5. Diedam, H., Dimitrov, D., Wieber, P.B., Mombaur, K., Diehl, M.: Online walking gait generation with adaptive foot positioning through linear model predictive control. In: IEEE/RSJ International Conference on Intelligent Robots and Systems, IROS 2008, pp. 1121–1126. IEEE (2008)
6. Garcia, M., Chatterjee, A., Ruina, A., Coleman, M.: The simplest walking model: stability, complexity, and scaling. Journal of Biomechanical Engineering **120**(2), 281–288 (1998)
7. Goswami, A., Thuilot, B., Espiau, B.: Compass-like biped robot part i: Stability and bifurcation of passive gaits. INRIA Research Report No. 2996 (1996)
8. Goswami, A., Thuilot, B., Espiau, B.: A study of the passive gait of a compass-like biped robot: symmetry and chaos. International Journal of Robotics Research **17**(12), 1282–1301 (1998)

9. Hobbelen, D.G.: Limit cycle walking. TU Delft, Delft University of Technology (2008)
10. McGeer, T.: Passive dynamic walking. The International Journal of Robotics Research **9**(2), 62–82 (1990)
11. McGeer, T.: Passive walking with knees. In: Proceedings. 1990 IEEE International Conference on Robotics and Automation, vol. 3, pp. 1640–1645 (1990)
12. Morari, M., Lee, J.H., Garcia, C., Prett, D.: Model predictive control. Preprint (2002)
13. Morimoto, J., Nakanishi, J., Endo, G., Cheng, G., Atkeson, C.G., Zeglin, G.: Poincare-map-based reinforcement learning for biped walking. In: Proceedings of the 2005 IEEE International Conference on Robotics and Automation, ICRA 2005, pp. 2381–2386. IEEE (2005)
14. Morris, B., Grizzle, J.: A restricted poincaré map for determining exponentially stable periodic orbits in systems with impulse effects: application to bipedal robots. In: 44th IEEE Conference on Decision and Control and 2005 European Control Conference, CDC-ECC 2005, pp. 4199–4206. IEEE (2005)
15. Neves, C., Ventura, R.: Survey of semi-passive locomotion methodologies for humanoid robots. In: 15th International Conference on Climbing and Walking Robots and the Support Technologies for Mobile Mechanics, pp. 393–400 (2012)
16. Vanderborght, B., Ham, R.V., Lefeber, D., Sugar, T.G., Hollander, K.W.: Comparison of mechanical design and energy consumption of adaptable, passive-compliant actuators. International Journal of Robotics Research **28**(1), 90–103 (2009)
17. Vanderborght, B., Verrelst, B., Ham, R.V., Damme, M.V., Lefeber, D., Duran, B.M., Beyl, P.: Exploiting natural dynamics to reduce energy consumption by controlling the compliance of soft actuators. International Journal of Robotics Research **25**, 343–358 (2006)
18. Williamson, M.M.: Series elastic actuators. Tech. rep., Massachusetts Institute of Technology - Artificial Intelligence Laboratory (1995)
19. Wisse, M.: Essentials of dynamic walking; Analysis and design of two-legged robots. Ph.D. thesis, TU Delft, Delft University of Technology (2004)

Monte-Carlo Workspace Calculation of a Serial-Parallel Biped Robot

Adrián Peidró, Arturo Gil, José María Marín, Yerai Berenguer, Luis Payá and Oscar Reinoso

Abstract This paper presents the Monte-Carlo calculation of the work-space of a biped redundant robot for climbing 3D structures. The robot has a hybrid serial-parallel architecture since each leg is composed of two parallel mechanisms connected in series. First, the workspace of the parallel mechanisms is characterized. Then, a Monte-Carlo algorithm is applied to compute the reachable workspace of the biped robot solving only the forward kinematics. This algorithm is modified to compute also the constant-orientation workspace. The algorithms have been implemented in a simulator that can be used to study the variation of the workspace when the geometric parameters of the robot are modified. The simulator is useful for designing the robot, as the examples show.

Keywords Climbing robots · Hybrid robots · Monte-Carlo · Redundant robots · Workspace

1 Introduction

The workspace of a manipulator can be defined as the set of positions and orientations that can be attained by the end-effector, and it plays a crucial role when designing the robot or planning its movements. Methods for determining the workspace can be classified as analytic or numerical. Analytic methods obtain closed-form descriptions of the boundaries of the workspace, they are more efficient but limited to specific classes of manipulators [5, 16]. Numerical methods [4, 11] can be applied to wider classes of robots and are more flexible. Amongst the numerical methods, Monte-Carlo algorithms [1] are specially interesting for complex and redundant robots, such as humanoid robots [9].

A. Peidró(✉) · A. Gil · J.M. Marín · Y. Berenguer · L. Payá · O. Reinoso
Systems Engineering and Automation Department, Universidad Miguel Hernández, Elche, Spain
e-mail: {apeidro,arturo.gil,jmarin,yberenguer,lpaya,o.reinoso}@umh.es

L.P. Reis et al. (eds.), *Robot 2015: Second Iberian Robotics Conference*,
Advances in Intelligent Systems and Computing 418,
DOI: 10.1007/978-3-319-27149-1_13

In this paper, we apply a well-known Monte-Carlo algorithm [1] to compute the workspace of a robot designed to climb and explore 3D structures, with the purpose of studying how this workspace is affected by the geometric design parameters of the robot. 3D structures, such as metallic bridges or power transmission lines, require periodic maintenance and inspection tasks that are dangerous for human workers due to risks such as falling from height. To avoid these risks, many climbing robots have been developed to execute these tasks. Climbing robots can have serial [3, 8, 12, 14, 17], parallel [2] or hybrid [7, 15] architecture. Serial architectures have smaller load capacity than parallel robots, but have larger workspaces, which is useful for exploring complex 3D structures. Parallel robots have a limited workspace, but a high load-to-weight ratio that is useful for climbing robots since they must carry their own weight. Finally, hybrid robots have the advantages of both architectures, which makes them very interesting for climbing 3D structures.

The robot studied in this paper is shown in Figure 1a. The robot is biped and hybrid, since each leg is composed of two serially connected parallel mechanisms. Its legs are specially designed to facilitate the execution of the typical movements that are necessary to explore 3D structures, such as movements along beams or columns, or transitions between planes with different spatial orientation. Also, the robot is redundant, because it has 10 degrees of freedom between its feet. Due to the complexity of this robot, it is difficult to obtain an analytic description of its workspace, hence we decided to use a simple Monte-Carlo method.

This paper is organized as follows. Section 2 briefly describes the architecture of the robot and analyzes the workspace of the parallel mechanisms of the legs. In Section 3, the solution to forward kinematics is used with a Monte-Carlo method to compute the workspace of the robot. Then, Section 4 presents a simulation tool developed to study the relation between the design parameters of the robot and its workspace. Finally, the conclusions are exposed in Section 5.

2 Robot Architecture and Workspace of the Parallel Mechanisms

Figure 1a shows the studied robot. It has two legs $\{A, B\}$ connected to a hip H through revolute joints (angles θ_A and θ_B). Each leg j is composed of a core link C_j and two platforms P_{1j} and P_{2j}. Each platform is connected to C_j through a passive slider and two linear actuators in parallel, constituting the parallel mechanism of Figure 1b. Thus, each leg is composed of two parallel mechanisms of this type connected in series. The robot has 10 degrees of freedom: eight linear actuators in the legs and two revolute joints in the hip. The reference frames E_A and E_B of Figure 1a are attached to the platforms P_{1A} and P_{1B}, which are the feet. The reference frames H_A and H_B are fixed to the hip. All these reference frames have their origins in the middle plane of the legs, as shown in Figure 1a.

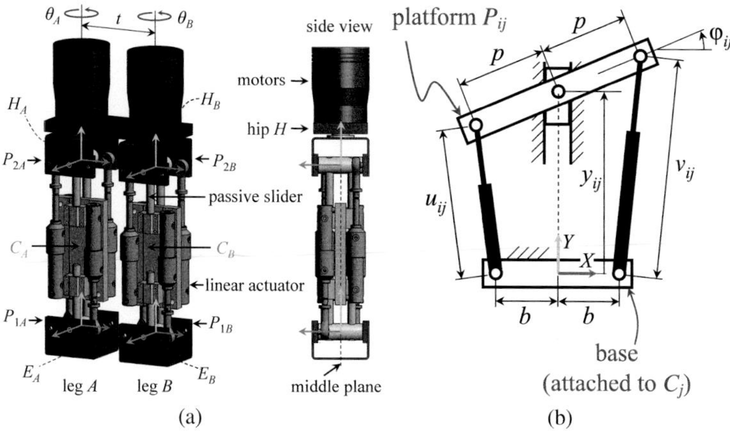

Fig. 1 (a) CAD model of the studied climbing robot. The X, Y, and Z axes of the reference frames are indicated with red, green, and blue colors, respectively. (b) 2-DOF parallel mechanism of the legs of the robot.

Next, we will analyze the workspace of the parallel mechanisms. Figure 1b shows the i-th parallel mechanism of the leg j ($i \in \{1, 2\}$, $j \in \{A, B\}$), which has a platform P_{ij} connected to the core link C_j through a passive slider and two linear actuators of lengths u_{ij} and v_{ij}. The forward kinematics of this mechanism consists in computing $\{y_{ij}, \varphi_{ij}\}$ in terms of $\{u_{ij}, v_{ij}\}$, and this problem was solved in [10]. The inverse problem can be easily solved analyzing Figure 1b:

$$u_{ij} = \sqrt{(p\cos\varphi_{ij} - b)^2 + (y_{ij} - p\sin\varphi_{ij})^2} \tag{1}$$

$$v_{ij} = \sqrt{(p\cos\varphi_{ij} - b)^2 + (y_{ij} + p\sin\varphi_{ij})^2} \tag{2}$$

In practice, the linear actuators have a minimum length $\rho_0 > 0$ and a stroke $\Delta\rho > 0$, which means that $u_{ij}, v_{ij} \in [\rho_0, \rho_0 + \Delta\rho]$. Thus, the workspace can be defined as the set of pairs (y_{ij}, φ_{ij}) for which the right-hand side of both Eqs. (1) and (2) is in $[\rho_0, \rho_0 + \Delta\rho]$. For example, Figure 2 shows the workspace for $b = p = 4$, $\rho_0 = 19.5$, and $\Delta\rho = 5$ (all in cm). This workspace is composed of four regions R_i enclosed by the curves where u_{ij} or v_{ij} equal ρ_0 or $\rho_0 + \Delta\rho$. The configuration of the mechanism is different in each region, as shown in Figure 2. Only the configurations of R_1 are valid, since the configurations of the other regions require mechanical interferences: y_{ij} cannot be negative (regions R_3 and R_4), and the linear actuators cannot interfere with the passive slider (regions R_2 and R_4). Thus, for this example the workspace is defined as follows:

$$WS = \left\{(\varphi_{ij}, y_{ij}) \in \mathbb{R}^2 : \varphi_{min} \le \varphi_{ij} \le \varphi_{max}, \underline{y_{ij}}(\varphi_{ij}) \le y_{ij} \le \overline{y_{ij}}(\varphi_{ij})\right\} \tag{3}$$

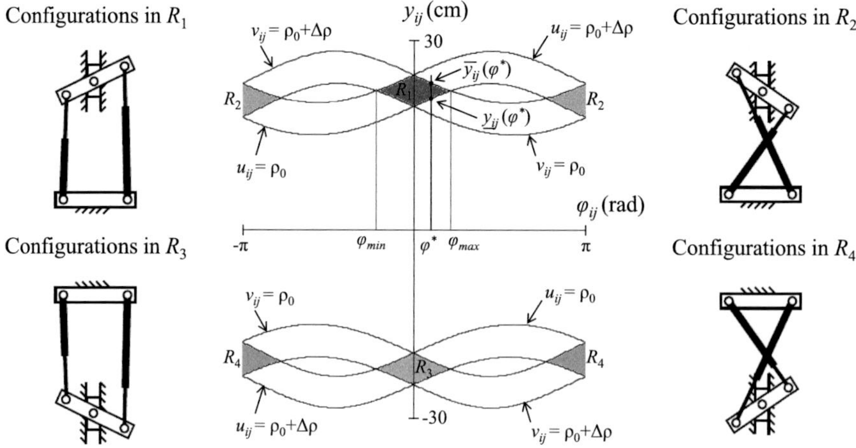

Fig. 2 Workspace of a parallel mechanism with $b = p = 4$, $\rho_0 = 19.5$, $\Delta\rho = 5$ (cm)

where $\underline{y_{ij}}(\varphi^*)$ and $\overline{y_{ij}}(\varphi^*)$ are the lower and upper bounds of the variable y_{ij} for $\varphi_{ij} = \varphi^*$, respectively (see Figure 2). In the following, we will assume that the workspace of the parallel mechanisms has the form of Eq. (3), which defines a more general set. Although this type of workspace has been derived from a particular geometry, the workspace of the parallel mechanisms will be similar to the region R_1 of Figure 2 if b, p, and $\Delta\rho$ are similar and small compared to ρ_0.

3 Monte-Carlo Workspace Calculation

In this section, we will use the equations of forward kinematics with a Monte-Carlo method to compute the workspace. The workspace considered here is the set of points that can be attained by one foot of the robot (free foot) when the other foot is fixed. Since the robot is symmetric, we can consider, without loss of generality, foot A as the fixed foot and foot B as the free one.

3.1 Forward Kinematics

Next, we will compute the position and orientation of the foot B relative to the foot A in terms of the rotations of the hip (angles θ_A and θ_B) and the rotations φ_{ij} and translations y_{ij} of the parallel mechanisms ($i \in \{1, 2\}$, $j \in \{A, B\}$). First, we will obtain the relative position and orientation between the hip and the foot of a generic leg j. According to Figure 3, the position and orientation of the hip relative to the

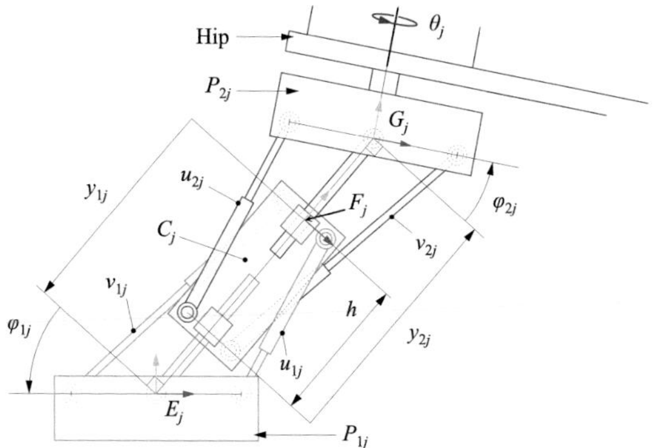

Fig. 3 Kinematics of the generic leg $j \in \{A, B\}$.

foot of the leg j can be obtained multiplying the following matrices, which relate the reference frames E_j, F_j, G_j, and H_j:

$$
{}^{E_j}\mathbf{T}_{H_j} = {}^{E_j}\mathbf{T}_{F_j}\,{}^{F_j}\mathbf{T}_{G_j}\,{}^{G_j}\mathbf{T}_{H_j} = \begin{bmatrix} c_{\varphi_{1j}} & s_{\varphi_{1j}} & 0 & y_{1j}s_{\varphi_{1j}} \\ -s_{\varphi_{1j}} & c_{\varphi_{1j}} & 0 & y_{1j}c_{\varphi_{1j}} \\ 0 & 0 & 1 & 0 \\ 0 & 0 & 0 & 1 \end{bmatrix} \times
$$

$$
\begin{bmatrix} c_{\varphi_{2j}} & -s_{\varphi_{2j}} & 0 & 0 \\ s_{\varphi_{2j}} & c_{\varphi_{2j}} & 0 & y_{2j}-h \\ 0 & 0 & 1 & 0 \\ 0 & 0 & 0 & 1 \end{bmatrix} \begin{bmatrix} c_{\theta_j} & 0 & s_{\theta_j} & 0 \\ 0 & 1 & 0 & 0 \\ -s_{\theta_j} & 0 & c_{\theta_j} & 0 \\ 0 & 0 & 0 & 1 \end{bmatrix} = \begin{bmatrix} c_{\theta_j}c_{\Phi_j} & s_{\Phi_j} & s_{\theta_j}c_{\Phi_j} & y_j s_{\varphi_{1j}} \\ -c_{\theta_j}s_{\Phi_j} & c_{\Phi_j} & -s_{\theta_j}s_{\Phi_j} & y_j c_{\varphi_{1j}} \\ -s_{\theta_j} & 0 & c_{\theta_j} & 0 \\ 0 & 0 & 0 & 1 \end{bmatrix} \tag{4}
$$

where $\Phi_j = \varphi_{1j} - \varphi_{2j}$, $y_j = y_{1j} + y_{2j} - h$, and h is the size of the core link of the leg j. The symbols s_x and c_x denote respectively $\sin(x)$ and $\cos(x)$, and θ_j is the angle that the frame G_j must be rotated about its Y axis to align that frame with the frame H_j of the hip (shown in Figure 1a). Particularizing the previous matrix for the two legs of the robot ($j = A$ and $j = B$), we obtain the position and orientation of the foot B with respect to the foot A as follows:

$$
{}^{E_A}\mathbf{T}_{E_B} = {}^{E_A}\mathbf{T}_{H_A}\,{}^{H_A}\mathbf{T}_{H_B}\,{}^{H_B}\mathbf{T}_{E_B} = {}^{E_A}\mathbf{T}_{H_A} \begin{bmatrix} \mathbf{I} & [t, 0, 0]^T \\ \mathbf{0}_{1\times 3} & 1 \end{bmatrix} \left({}^{E_B}\mathbf{T}_{H_B}\right)^{-1} \tag{5}
$$

where $\mathbf{I}$ is the 3×3 identity matrix, $\mathbf{0}_{1\times 3} = [0, 0, 0]$, and ${}^{H_A}\mathbf{T}_{H_B}$ encodes the position and orientation of the frame H_B relative to the frame H_A (see Figure 1a). Performing the products of Eq. (5), the matrix that encodes the position and orientation of the frame E_B relative to the frame E_A can be written as follows:

$$
{}^{E_A}\mathbf{T}_{E_B} = \begin{bmatrix} {}^{E_A}\mathbf{R}_{E_B} & \left[p_x\ p_y\ p_z\right]^T \\ \mathbf{0}_{1\times 3} & 1 \end{bmatrix} \tag{6}
$$

The position vector of the previous matrix has the following expression:

$$\begin{bmatrix} p_x \\ p_y \\ p_z \end{bmatrix} = y_A \begin{bmatrix} s_{\varphi_{1A}} \\ c_{\varphi_{1A}} \\ 0 \end{bmatrix} + y_B \begin{bmatrix} -c_\Theta c_{\Phi_A} s_{\varphi_{2B}} - s_{\Phi_A} c_{\varphi_{2B}} \\ c_\Theta s_{\Phi_A} s_{\varphi_{2B}} - c_{\Phi_A} c_{\varphi_{2B}} \\ s_\Theta s_{\varphi_{2B}} \end{bmatrix} + t \begin{bmatrix} c_{\theta_A} c_{\Phi_A} \\ -c_{\theta_A} s_{\Phi_A} \\ -s_{\theta_A} \end{bmatrix} \tag{7}$$

where $\Theta = \theta_A - \theta_B$. The rotation submatrix of ${}^{E_A}\mathbf{T}_{E_B}$ has the following form:

$$^{E_A}\mathbf{R}_{E_B} = \begin{bmatrix} s_{\Phi_A} s_{\Phi_B} + c_\Theta c_{\Phi_A} c_{\Phi_B} & s_{\Phi_A} c_{\Phi_B} - c_\Theta c_{\Phi_A} s_{\Phi_B} & s_\Theta c_{\Phi_A} \\ c_{\Phi_A} s_{\Phi_B} - c_\Theta s_{\Phi_A} c_{\Phi_B} & c_{\Phi_A} c_{\Phi_B} + c_\Theta s_{\Phi_A} s_{\Phi_B} & -s_\Theta s_{\Phi_A} \\ -s_\Theta c_{\Phi_B} & s_\Theta s_{\Phi_B} & c_\Theta \end{bmatrix} \tag{8}$$

Next, Eqs. (7) and (8) will be used to compute the following workspaces:

- Reachable workspace: the set of points $\mathbf{P} = [p_x, p_y, p_z]^T$ that can be reached by the free foot with at least one orientation
- Constant-orientation workspace: the set of points $\mathbf{P} = [p_x, p_y, p_z]^T$ that can be reached by the free foot with a specific orientation

3.2 *Computation of the Reachable Workspace*

Once the solution to forward kinematics of the complete biped robot is available, the reachable workspace can be easily generated using a Monte-Carlo method [1]. This approach consists in varying randomly the following variables in their ranges: $\{\varphi_{1A}, \varphi_{2A}, \varphi_{1B}, \varphi_{2B}, y_A, y_B, \theta_A, \theta_B\}$, generating a point $\mathbf{P} = [p_x, p_y, p_z]^T$ for each value of these variables. The generated points form a 3D cloud point in space that constitutes a discrete approximation of the workspace.

To apply this method, we must find the variation ranges of the previous variables, for each leg j. It will be assumed that $\theta_j \in [-\pi, \pi]$ (the legs can perform complete revolutions about the axes of the hip). The angles φ_{ij} must belong to the workspace of the parallel mechanisms of the legs. It will be assumed that such workspaces are of the type studied in Section 2, which implies that $\varphi_{min} \leq \varphi_{ij} \leq \varphi_{max}$. Finally, the valid ranges for y_j can be found as follows: given $\varphi_{ij} \in [\varphi_{min}, \varphi_{max}]$ ($i \in \{1, 2\}$), then according to Eq. (3), y_{ij} must satisfy:

$$\underline{y_{ij}}(\varphi_{ij}) \leq y_{ij} \leq \overline{y_{ij}}(\varphi_{ij}) \tag{9}$$

Since $y_j = y_{1j} + y_{2j} - h$, then the variables y_j must verify:

$$\underline{y_{1j}}(\varphi_{1j}) + \underline{y_{2j}}(\varphi_{2j}) - h \leq y_j \leq \overline{y_{1j}}(\varphi_{1j}) + \overline{y_{2j}}(\varphi_{2j}) - h \tag{10}$$

Once the variation ranges of all the variables are known, the Monte-Carlo algorithm to compute the reachable workspace can be summarized in Algorithm 1. In this algorithm, N_r is the number of sampled random points.

Algorithm 1. Monte-Carlo calculation of the reachable workspace

1: $WS = \emptyset \rightarrow$ The reachable workspace is initialized as an empty set.
2: **for** $k = 1$ to N_r **do**
3: Randomly sample θ_A and θ_B in $[-\pi, \pi]$
4: Randomly sample $\varphi_{1A}, \varphi_{2A}, \varphi_{1B}$, and φ_{2B} in $[\varphi_{min}, \varphi_{max}]$
5: Compute the lower and upper limits for y_j ($j \in \{A, B\}$):
6: $\underline{y_j} = \underline{y_{1j}}(\varphi_{1j}) + \underline{y_{2j}}(\varphi_{2j}) - h$
7: $\overline{y_j} = \overline{y_{1j}}(\varphi_{1j}) + \overline{y_{2j}}(\varphi_{2j}) - h$
8: Randomly sample y_j in $[\underline{y_j}, \overline{y_j}]$ ($j \in \{A, B\}$)
9: Compute the position $\mathbf{P} = [p_x, p_y, p_z]^T$ of the free foot using Eq. (7)
10: Add the point $\mathbf{P}$ to WS
11: **end for**

To sample the variables θ_A and θ_B in line 3 of Algorithm 1, a uniform distribution can be used. However, the variables y_j and φ_{ij} (whose limits define the limits of the workspace) should be sampled using a beta distribution with parameters $\alpha, \beta \in (0, 1)$. Using this non-uniform distribution for these variables favors the generation of random points close to the boundaries of the workspace, which results in a better definition of these boundaries [6].

3.3 Computation of the Constant-Orientation Workspace

The constant-orientation workspace is the set of points of the space that can be reached with a desired orientation, which can be specified as:

$$^{E_A}\mathbf{R}_{E_B} = \begin{bmatrix} r_{11} & r_{12} & r_{13} \\ r_{21} & r_{22} & r_{23} \\ r_{31} & r_{32} & r_{33} \end{bmatrix} \tag{11}$$

where r_{ij} are known quantities. Algorithm 1 can still be used to generate random points in the constant-orientation workspace. However, unlike in Algorithm 1, not all the angles $\{\varphi_{ij}, \theta_j\}$ can be sampled independently now: these angles must satisfy certain relations to guarantee that the generated random points have the desired orientation. Two cases are distinguished:

Case 1: $r_{33}^2 \neq 1$. Equating the element (3,3) of matrices (8) and (11) permits computing the angle Θ as follows:

$$c_\Theta = r_{33} \longrightarrow s_\Theta = \sigma\sqrt{1 - r_{33}^2} \longrightarrow \Theta = \theta_A - \theta_B = \text{atan2}(s_\Theta, c_\Theta) \tag{12}$$

where $\sigma \in \{-1, 1\}$. Once s_Θ is known, Equating the elements (1,3), (2,3), (3,1) and (3,2) of Eqs. (8) and (11) allows for the calculation of Φ_A and Φ_B:

$$c_{\Phi_A} = r_{13}/s_\Theta, \quad s_{\Phi_A} = -r_{23}/s_\Theta \longrightarrow \Phi_A = \varphi_{1A} - \varphi_{2A} = \text{atan2}(s_{\Phi_A}, c_{\Phi_A}) \quad (13)$$

$$c_{\Phi_B} = -r_{31}/s_\Theta, \quad s_{\Phi_B} = r_{32}/s_\Theta \longrightarrow \Phi_B = \varphi_{1B} - \varphi_{2B} = \text{atan2}(s_{\Phi_B}, c_{\Phi_B}) \quad (14)$$

Note that Eqs. (12), (13), and (14) fix the differences $\theta_A - \theta_B$, $\varphi_{1A} - \varphi_{2A}$, and $\varphi_{1B} - \varphi_{2B}$, respectively. Thus, we cannot give random values to the six angles $\{\varphi_{1A}, \varphi_{2A}, \varphi_{1B}, \varphi_{2B}, \theta_A, \theta_B\}$ simultaneously. Instead, we can give values to the angles $\{\theta_B, \varphi_{2A}, \varphi_{2B}\}$ and compute the other three angles using the previous equations to guarantee that the generated points have the desired orientation. Note that, after calculating $\{\varphi_{1A}, \varphi_{1B}\}$, these angles may not be in $[\varphi_{min}, \varphi_{max}]$, in which case the point must be discarded since it does not satisfy the joint limits.

Case 2: $r_{33}^2 = 1$. In this case, Θ can be calculated from Eq. (12), but Φ_A and Φ_B cannot be computed from Eqs. (13) and (14) since $s_\Theta = 0$. To compute these angles, we substitute $c_\Theta = r_{33}$ into the elements (1,2) and (2,2) of Eq. (8) and equate these elements to r_{12} and r_{22}:

$$\begin{bmatrix} r_{12} \\ r_{22} \end{bmatrix} = \begin{bmatrix} s_{\Phi_A} c_{\Phi_B} - r_{33} c_{\Phi_A} s_{\Phi_B} \\ c_{\Phi_A} c_{\Phi_B} + r_{33} s_{\Phi_A} s_{\Phi_B} \end{bmatrix} = \begin{bmatrix} \sin(\Phi_A - r_{33}\Phi_B) \\ \cos(\Phi_A - r_{33}\Phi_B) \end{bmatrix} \quad (15)$$

where the last equality is true because $r_{33} = 1$ or $r_{33} = -1$. In this case, Algorithm 1 can also be used with the following modification: $\{\varphi_{1B}, \varphi_{2B}, \varphi_{2A}\}$ are randomly sampled, whereas φ_{1A} is computed as $\varphi_{1A} = \varphi_{2A} + r_{33}\Phi_B + \text{atan2}(r_{12}, r_{22})$, discarding the point if $\varphi_{1A} \notin [\varphi_{min}, \varphi_{max}]$.

Finally, the previous methods compute discrete approximations of the solid workspace. However, for practical purposes (e.g., for visualization) it is sufficient to know the surfaces that delimit these solids. The boundaries of the computed workspaces can be extracted using the algorithm described in [1], which defines a 3D grid composed of N_g boxes in each dimension. The boxes that contain workspace points are marked with "1", whereas the remaining boxes are marked with "0". Then, the workspace boundary is composed of the boxes that are marked with "1" and have at least one neighboring box marked with "0".

4 Simulation Tool and Examples

This section presents a simulation tool developed in Java to study the workspace of this robot. The tool can be downloaded from http://arvc.umh.es/parola/climber.html and may require the latest version of Java.

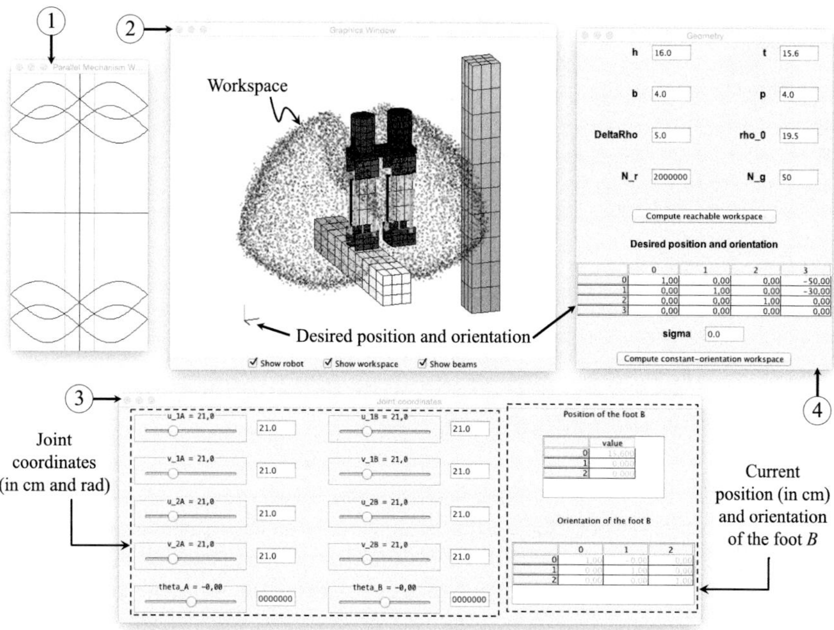

Fig. 4 Simulator developed to study the workspace of the biped robot.

The tool, which has four windows, is shown in Figure 4. Window 1 shows the workspace of the parallel mechanisms, as defined in Section 2, so that the user can see if this workspace has the form of Eq. (3). Window 2 shows the biped robot in a virtual environment composed of two beams, along with the workspace calculated using the previous Monte-Carlo method. In the simulator, the foot A is fixed to the horizontal beam, and the foot B is free. Window 3 has some sliders and numeric fields to modify the value of the joint coordinates. When a joint coordinate is modified, the forward kinematics is solved and the posture of the robot in Window 2 is modified accordingly. Window 3 also shows the current position and orientation of the foot B (vector $[p_x, p_y, p_z]^T$ and matrix ${}^{E_A}\mathbf{R}_{E_B}$). Finally, Window 4 can be used to design the robot: in it, the six design parameters $\{h, b, t, p, \Delta\rho, \rho_0\}$ (in cm) can be modified to study how the shape and size of the workspace varies with these parameters.

Next, we will analyze some examples that show how this tool can be used to design the robot. In the following examples, the shown workspaces have been obtained using $N_r = 2 \cdot 10^6$ random points and $N_g - 50$.

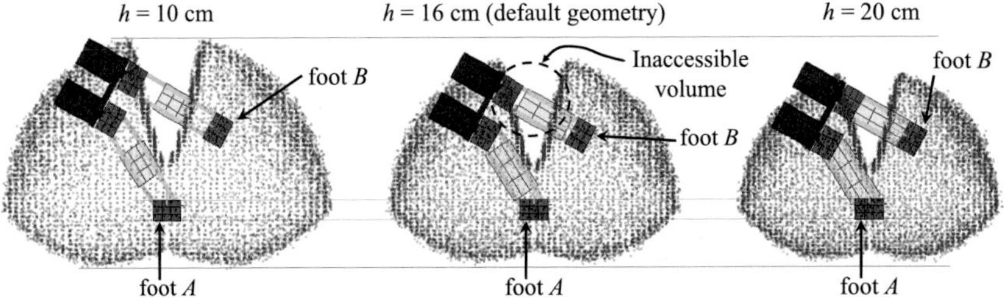

Fig. 5 Variation of the reachable workspace when h is modified.

4.1 Example 1: Sensitivity of the Reachable Workspace with Respect to the Design Parameters

In this example, we will study the changes in the shape and size of the reachable workspace when perturbing the design parameters from their default values: $\rho_0 = 19.5$, $\Delta\rho = 5$, $b = p = 4$, $t = 15.6$, $h = 16$ (all in cm). The reachable workspace for this geometry is shown in Figure 5 (center), which shows that the points above the fixed foot A cannot be reached by the foot B. Next, we will vary the design parameters (one at a time, keeping the rest at their default values) to obtain a larger workspace in which the region above the foot A is accessible.

Figure 5 shows that increasing h reduces the size of the reachable workspace, leaving its shape practically unaffected. If the parameters t and ρ_0 are respectively varied in the intervals [10, 20] cm and [15, 25] cm in the simulator, it can be checked that the size of the workspace increases with these parameters, but its shape hardly varies with them. Also, it can be checked that varying b in (0, 10] cm hardly affects the shape or size of the reachable workspace. Thus, varying these four parameters generates workspaces where the points above the foot A are still inaccessible. However, varying the parameter p modifies noticeably the shape of the workspace, as shown in

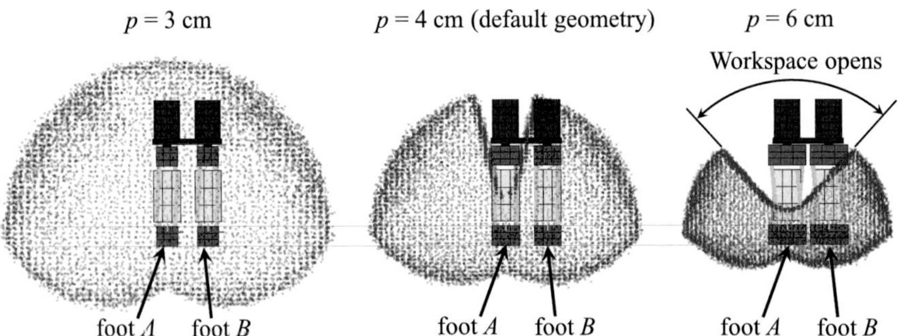

Fig. 6 Variation of the reachable workspace when p is modified.

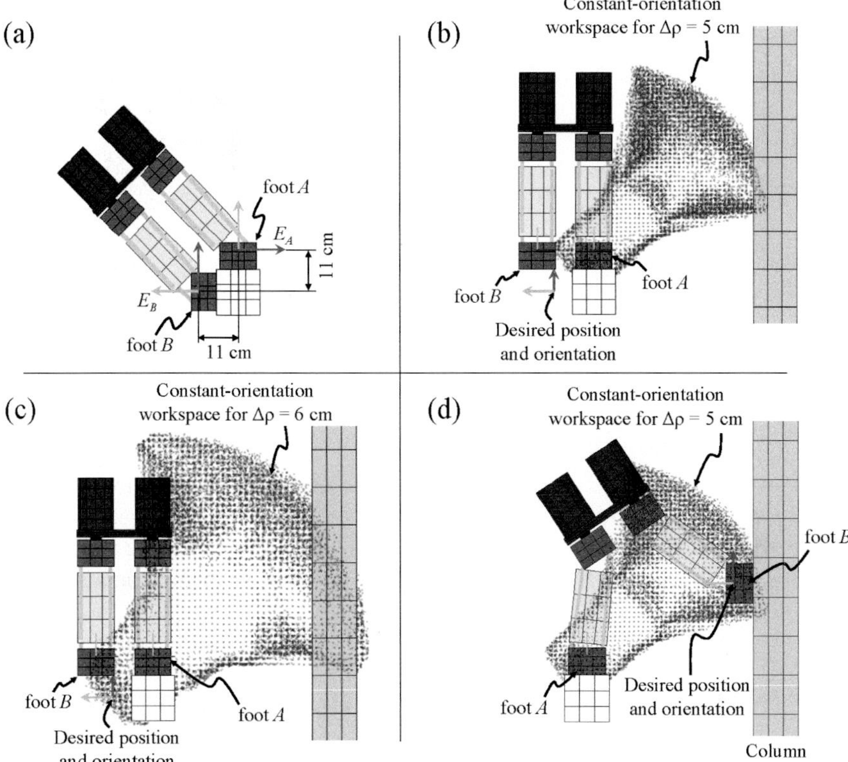

Fig. 7 (a) Desired position and orientation to change between different faces of a beam. For $\Delta\rho = 5$ cm (b), the constant-orientation workspace for the desired orientation does not contain the desired point, but it contains the point for $\Delta\rho = 6$ cm (c). (d) Performing transitions between different beams using the default geometry.

Figure 6. This figure shows that the reachable workspace opens as p increases. Thus, it is convenient to reduce p as shown in Figure 6 in order to eliminate the inaccessible region above the foot A. It can be checked that varying $\Delta\rho$ in [3, 6] cm produces a similar effect in the opposite direction: the workspace opens as $\Delta\rho$ decreases.

4.2 *Example 2: Transition Between Different Faces of a Beam*

In this example, we assume that all the design parameters are fixed at their default values except $\Delta\rho$, whose value must be chosen so as to permit the robot to perform a transition between different faces of a beam, as shown in Figure 7a. According to this figure, the desired position and orientation for the foot B relative to the fixed foot A are given by the following matrices:

$$^{E_A}\mathbf{R}_{E_B} = \begin{bmatrix} 0 & -1 & 0 \\ 1 & 0 & 0 \\ 0 & 0 & 1 \end{bmatrix}, \quad \begin{bmatrix} p_x \\ p_y \\ p_z \end{bmatrix} = \begin{bmatrix} -11 \\ -11 \\ 0 \end{bmatrix} \text{cm} \tag{16}$$

Introducing these matrices in the simulator, and using $\Delta\rho = 5$ cm (default stroke) yields the constant-orientation workspace of Figure 7b. Note that, for the default stroke, the desired point cannot be attained with the desired orientation because it lies outside the computed constant-orientation workspace. However, if the workspace is recalculated for $\Delta\rho = 6$ cm, we obtain the workspace of Figure 7c, which contains the desired point. Thus, choosing a linear actuator with a stroke of 6 cm would permit the robot to change between different faces of the beam in this example.

Note that the orientation defined in Eq. (16) is also necessary to attach the foot B to the column, as indicated in Figure 7d. As shown in this figure, the constant-orientation workspace for $\Delta\rho = 5$ cm contains points that are near the surface of the column. Thus, it is possible to attach the foot B to the column using the default geometry.

5 Conclusions and Future Work

This paper has presented a Monte-Carlo workspace analysis of a serial-parallel climbing robot. First, the workspace of the parallel mechanisms has been calculated to use it in the calculation of the workspace of the complete robot. Then, the solution to forward kinematics has been used together with a Monte-Carlo method to compute the reachable and constant-orientation workspaces of the robot. These calculations have been implemented in a simulator, which has been used to manually and visually study the sensitivity of the shape and size of the workspace with respect to the design parameters. In the future, an algorithm to automatically perform the sensitivity analysis and optimization of the design of the robot will be devised, similar to [13]. Also, we will compute other types of workspace (such as the orientation workspace) and study the singularities.

Acknowledgements This work has been supported by the Spanish Ministry of Education, Culture and Sport through an FPU grant (Ref: FPU13/00413), by the Spanish Ministry of Economy and Competitiveness through Project DPI2013-41557-P, and by the Generalitat Valenciana through project AICO/2015/021.

References

1. Alciatore, D., Ng, C.: Determining manipulator workspace boundaries using the Monte Carlo method and least squares segmentation. ASME Robotics: Kinematics, Dynamics and Controls **72**, 141–146 (1994)
2. Aracil, R., Saltaren, R.J., Reinoso, O.: A climbing parallel robot: a robot to climb along tubular and metallic structures. IEEE Robotics & Automation Magazine **13**(1), 16–22 (2006)
3. Balaguer, C., Giménez, A., Pastor, J.M., Padrón, V.M., Abderrahim, M.: A climbing autonomous robot for inspection applications in 3D complex environments. Robotica **18**(3), 287–297 (2000)
4. Bohigas, O., Manubens, M., Ros, L.: A complete method for workspace boundary determination on general structure manipulators. IEEE Transactions on Robotics **28**(5), 993–1006 (2012)
5. Bonev, I.A., Gosselin, C.M.: Analytical determination of the workspace of symmetrical spherical parallel mechanisms. IEEE Transactions on Robotics **22**(5), 1011–1017 (2006)
6. Cao, Y., Lu, K., Li, X., Zang, Y.: Accurate numerical methods for computing 2D and 3D robot workspace. Int. J. of Advanced Robotic Systems **8**(6), 1–13 (2011)
7. Figliolini, G., Rea, P., Conte, M.: Mechanical design of a novel biped climbing and walking robot. In: Parenti-Castelli, V., Schiehlen, W. (eds.) ROMANSY 18 Robot Design, Dynamics and Control, pp. 199–206. Springer, Vienna (2010)
8. Guan, Y., Jiang, L., Zhu, H., Zhou, X., Cai, C., Wu, W., Li, Z., Zhang, H., Zhang, X.: Climbot: a modular bio-inspired biped climbing robot. In: Proceedings of the 2011 IEEE/RSJ Int. Conf. on Intelligent Robots and Systems, pp. 1473–1478, September 2011
9. Guan, Y., Kazuhito, Y., Zhang, X.: Numerical methods for reachable space generation of humanoid robots. Int. J. of Robotics Research **27**(8), 935–950 (2008)
10. Kong, X., Gosselin, C.M.: Generation and Forward Displacement Analysis of R$\underline{P}$R-PR-R$\underline{P}$R Analytic Planar Parallel Manipulators. ASME J. Mech. Design **124**(2), 294–300 (2002)
11. Merlet, J.P.: Determination of 6D workspaces of Gough-type parallel manipulator and comparison between different geometries. Int. J. of Robotics Research **18**(9), 902–916 (1999)
12. Shvalb, N., Moshe, B.B., Medina, O.: A real-time motion planning algorithm for a hyper-redundant set of mechanisms. Robotica **31**(8), 1327–1335 (2013)
13. Silva, V.G., Tavakoli, M., Marques, L.: Optimization of a Three Degrees of Freedom DELTA Manipulator for Well-Conditioned Workspace with a Floating Point Genetic Algorithm. Int. J. Natural Computing Research **4**(4), 1–14 (2014)
14. Tavakoli, M., Marques, L., De Almeida, A.T.: 3DCLIMBER: Climbing and manipulation over 3D structures. Mechatronics **21**(1), 48–62 (2011)
15. Tavakoli, M., Zakerzadeh, M.R., Vossoughi, G.R., Bagheri, S.: A hybrid pole climbing and manipulating robot with minimum DOFs for construction and service applications. Industrial Robot: An International Journal **32**(2), 171–178 (2005)
16. Wang, Q., Wang, D., Tan, M.: Characterization of the analytical boundary of the workspace for 3–6 SPS parallel manipulator. In: Proceedings of the 2001 IEEE Int. Conf. on Robotics and Automation, pp. 3755–3759 (2001)
17. Yoon, Y., Rus, D.: Shady3D: a robot that climbs 3D trusses. In: Proceedings of the 2007 IEEE Int. Conf. on Robotics and Automation, pp. 4071–4076 (2007)

A Control Driven Model for Human Locomotion

Diana Guimarães and Fernando Lobo Pereira

Abstract This article concerns the modeling of human locomotion with a view to the design of advanced control systems that are capable of supporting natural mobility, and, thus, promoting inclusivity and quality of life.

The complexity of the model (i.a. degrees of freedom and motion planes taken into consideration) was carefully chosen to include the relevant features of the motion dynamics while remaining as simple as possible. The outcome is a model composed by three components (stance leg, swing leg and trunk) that are articulated to achieve balanced motion patterns in both transitory and periodic contexts. Each leg has 3 links connected by pitch joints and the trunk has a single link.

Significant attention was dedicated to the generation of natural (human-like) motion references, in order to achieve a safe and anthropomorphically correct motion that respects the human joints' constraints and can be adjusted to the multiple daily-life situations.

Keywords Assistive robotics · Human locomotion modeling · Hybrid systems · Adaptative control systems

1 Introduction

In daily life, a person has to use his/hers motor skills to overcome multiple situations and obstacles. The locomotion system will act differently in each one of these situations, adjusting, for example, walking speed, step length or torso position. People with locomotion disabilities are sometimes not able to do the required adjustments on their own with very significant negative consequences in their quality of life and their inclusion in society.

D. Guimarães(✉) · F. Lobo Pereira
Faculdade de Engenharia da Universidade do Porto, Porto, Portugal
e-mail: dlguim@fe.up.pt

L.P. Reis et al. (eds.), *Robot 2015: Second Iberian Robotics Conference*,
Advances in Intelligent Systems and Computing 418,
DOI: 10.1007/978-3-319-27149-1_14

This article is part of a investigation that will contribute towards a solid foundation for the future design of a human locomotion system. The aim is to investigate and develop models of human biped locomotion which are suited to the design of control systems capable of recognizing the present environmental conditions and adapting itself to ensure stability.

First of all, the human biped motion is investigated having in mind:

- the specification of requirements for natural motion,
- the fulfillment of such requirements with a simple mechanical structure controlled by an appropriate control architecture.

This work differs from the current state-of-the-art in the fact that:

- it preserves the proportions and main segments of the human body and motion features, although studying it only in the sagittal plane (where more significant changes occur); and
- the modeling and control strategies were based not only in the state-of-the-art for robots but also in a detailed analysis of human motion and the most common needs of people with balance difficulties.

The presented model is a good trade-off between simplicity (computationally important due to the real-time requirements) and the complexity needed to represent properly such complex system as it is the human locomotion system and its dynamic behavior. The outcome of the current work provides the base of the control architecture, which has been conceptualized to be able to provide assistance and ensure the user's safety, thus contributing to the development of new medical devices that help promote an active lifestyle.

This article is organized as follows: sections 1 to 4 describe motion modeling (kinematics, dynamics and impact mechanics). Section 5 presents the consistency verification mechanism developed to ensure that the sequence of inputs over time leads to the expected human-like pattern generation. Finally, some simulation results of the implementations are presented and conclusions are taken.

2 State-of-the-Art

Given the critical dependence of control systems on locomotion models, this section concerns not only the key issues arising in modeling but it also addresses control systems and tools for locomotion.

Humanoid robots have multiple purposes, such as entertaining or helping the elderly in daily tasks, but, above all, they allow better understanding of the influence in locomotion of the variation of some parameter – foot shape, weight distribution and posture.

To reach the performance humanoid robots have today the inertial forces resulting from body acceleration and matching external disturbances need to be taken into consideration and controlled, which led the to the use of ZMP (Zero Moment

Point) instead of COG (center of gravity) as stability criterion. This dynamic walking idea can be combined with models going from the single inverted pendulum up to many links, establishing a trade-off between complexity- real-time performance and stability- speed of motion.

In the most complex models, different simplifications were done, such as not considering the feet[25][23][7] (thus working with a 5 links model), considering a constant height for the hip [25], analyzing only the sagittal plan[25][23][7] or "imposing holonomic constraints on the robot's configuration parameterized by a monotonically increasing function of the robot's state" in order to reduce the stability analysis problem "from a 5-dimensional to a scalar Poincaré return map", [23].

According to [23], a pre-computed ZMP trajectory tracking using PID and a computed torque or sliding mode was compared by Tzafestas, Park and Kim combined computed torque with gravity compensation, while Fujimoto combined it with foot force control, and Mitobe et al. tried using computed torque to regulate swing leg and COM' position.

A controller based on a 3D gait prediction is suggested in [3]. The authors of this paper combined the "state-of-the-art walking controllers from the robotics field and state-of-the-art musculoskeletal models in the biomechanics field" to predict the consequences of lower limb surgeries.

In [6], the trunk position in the walking direction is used to help maintain stability (if COG is too ahead of the support foot, the trunk will lean backward), and the influence of the arms movement is analyzed in [8].

Unlike [25] and [23], there are authors that consider feet as the main concern in robot motion. The authors of [16] propose the implementation of force sensors in the sole of the feet to calculate the real ZMP position and compare it with a database of gaits with different parameters that will be adjusted online by a neuro-Fuzzy controller. The paper [21] tests a foot positioning compensator (FPC) "to adaptively modify the robot's foot positioning". In the same year, [2] tried to make their robot walk more like a human by adding a toe joint and focusing in its behavior during the support and swing phase and adding the possibility of stretching the support leg's knee.

By abandoning the walk on flat floor problem, an interesting comparative study between analytical method (DH based), Neural Networks and Fuzzy Logic performance in finding optimal gait for ditch crossing by a 7 links Humanoid was conducted [22].

Based on the network of online controllers, the same authors present a solution for stairs in [13].

Curiously, [1] refrained from the use of techniques such as computed torque or adaptative control, because from their point of view, they were effective for the trajectory tracking but complex, time-consuming and required a precise and accurate model. Therefore, they chose an iterative learning control method.

Regarding the control architectures, it is clear that the most used architecture has three main blocks: an (offline) trajectory generator, a (network of) controller(s) and a stabilizer. The function of the stabilizer is to react in real-time to external disturbances, giving robustness to the system.

Some works use approaches or address the locomotion topic with a perspective that worth being mentioned, such as: the usage of PD controller with gravity compensation in [5]; the Neural Network used in [22]; in paper [1], locomotion is seen as a coordinated control effort and a prosthetic leg is designed based on an hierarchical (master-slave) architecture.

More recently, [11] has chosen to, based on the inverted pendulum model, abdicate of the motion periodicity to work with unit step. On a different approach, [20] optimized the hip height movement to be able to walk faster.

All the analyzed works make an interesting approach to the biped robots theme. Naturally, each author makes the simplifications that allow a lower workload without damaging their goals.

3 The Proposed Approach

The goal of the developed work is to contribute to the design of a device that can provide a more inclusive mobility service to people with a locomotion disability and at the same time bring the medical assistance and R&D community closer.

Therefore, the modeling and control strategies were based, not only, in the state-of-the-art for robots, but also in a detailed analysis of human motion and the most common needs of people with balance difficulties. Also the control architecture was conceptualized to be able to provide assistance and ensure the user's safety.

The human locomotion was perceived as a hybrid system and the impact mechanics was taken into consideration in the model. The existence of feet (and all the other human leg segments) and a correct but versatile human walking pattern are the key characteristics in the developed model.

The generation of natural motion references is also key, since it is responsible for ensuring the consistency between the reference signal in each local controller and simultaneously adapt them to robustly preserve the motion characteristics despite external perturbations or modeling inadequacies.

4 Mechanical Modeling of Human Locomotion

The central problem in modeling human locomotion is to find a simple, but still complete and realistic model, because overly strong assumptions can rule out important dynamic issues, but doing none leads to a very heavy model, which is bad for real time processing. In the developed work, the strategy chosen was to start with a very simple model and adding complexity along the way, Hence, in section 4.1 a one degree of freedom (DOF) model is presented. At first the model is linearized and later the constraints are removed and the model's properties and behavior are analyzed in the non-linear domain. Then, a more complex model is built to get the model closer to the configuration of the human body. Such model has 6 DOF and is, therefore, still far from the detailed configuration of the human locomotion system but presents a good trade-off between functionality and complexity, allowing the analysis of the locomotion's main features without being excessively heavy computationally. This multi-link model is presented in sections 4.2 and 4.3 since they exhibit two different layers: the kinematics and the Dynamics, respectively.

4.1 Single Inverted Pendulum

The most basic model of locomotion is the planar single inverted pendulum approach and has been used by several authors.

Such model, illustrated below, allows the representation of two fundamental human activities: standing still and walking. To study the walking motion in further detail, a hybrid automaton ([9]) was adopted.

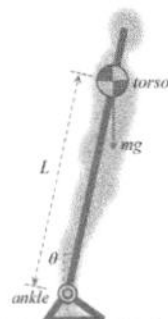

Fig. 1 Single inverted Pendulum approach

The system respects Newton's Second Law, yielding the general equations below

$$ma = \sum \overrightarrow{F} \tag{1}$$

$$I\ddot{\theta} = \sum \overrightarrow{T} \tag{2}$$

Regarding rotation, equation (3) can be written

$$I\ddot{\theta} = mgLsin(\theta) + u \tag{3}$$

where u, I and m are the ankle torque, inertia and mass of the body, respectively. L and θ are as depicted in Figure 1.

- **Standing Still**

To remain still, the right side of conditions (1) and (2) have to go simultaneously to zero.

On level ground, the reference value for θ is zero degrees. Any deviation is expected to be corrected quickly, so the approximation $sin\theta = \theta$ is valid and equation

(3) can be linearized around its (unstable) equilibrium point and written in the state-space form:

$$\dot{X} = \begin{bmatrix} 0 & 1 \\ \frac{g}{L} & 0 \end{bmatrix} X + \begin{bmatrix} 0 \\ \frac{1}{mL^2} \end{bmatrix} u \tag{4}$$

with

$$X = \begin{bmatrix} \theta \\ \dot{\theta} \end{bmatrix} \tag{5}$$

- **Walking**

The linearization performed above is still valid if the steps are very small (implying θ is not much larger than 5 degrees).

Analyzing walking motion based on the single pendulum model without the small-steps constraint enters the nonlinear domain. Equation (3) can be rewritten in the generic form of a second order system.

The physical intuition provided by the inverted pendulum reveals that it has an unstable equilibrium point in the interval of $[-\pi/2, \pi/2]$ and a stable equilibrium point in the interval of $[\pi/2, 3 * \pi/2]$ (remember that the upright position was considered to be in $\theta = 0$). A way to confirm this is through a phase portrait of the system (see [9] for detais).

4.2 *Multi-link Model: Kinematics*

Constructing a more complex model enables the study of other features. By observing human locomotion, it is possible to conclude that the same joint motion does not always have the same effect on body behavior. It is important that the model captures these details. So, we will start by having legs with 3 links with one pitch degree of freedom (DOF) and one link trunk also with a pitch DOF.

The DH method ([10]) is a systematic way to define the motion of a system composed by rigid bodies connected by single or multiple degrees of freedom joints each. It associates a local frame to each single degree of freedom joint. Following specific rules, the relationship between adjacent local reference frames is described by only four parameters (a, α, d and θ), which enable the construction of the homogeneous matrix H, which is the product of rotations and translations.

Thus, any transformation R between two frames is found multiplying all the ${}^{i-1}H_i$ matrices separating their local frames,

$${}^{0}R_i = {}^{0}H_1 \, {}^{1}H_2 ... {}^{i-2}H_{i-1} \, {}^{i-1}H_i = {}^{0}R_{i-1} \, {}^{i-1}H_i \tag{6}$$

with H denoting the transformation between two consecutive frames.

This transformations allows us to describe the coordinates of the origin of a frame in a second Cartesian System.

Table 1 DH parameters for the support leg

DH parameters	Frame				
	1	2	3	4	5
a_i	$x_0 + L_{foot}$	$-L_{foot}$	$-L_{heel}$	$-L_{shin}$	$-L_{thigh}$
α_i	- 90	0	0	0	0
d_i	0	0	0	0	0
θ_i	0	θ_{toe-st}	90	$\theta_{ankle-st}$	$\theta_{knee-st}$

Table 1 contains the resulting DH parameters for the support leg.

The variable x_0 stands for the distance in x from the global frame to the heel of the support leg, which is initially zero and increases at each step executed. The letters "st" in the angles represent "stance" (synonymous for support) and helps to distinguish from the angles of the other leg.

- **Forward Kinematics**

The forward kinematics is obtained by expressing behaviors (motion) of joint space into the Cartesian space.

By using equations (6) and table 1, it is possible to express, for example, the knee position in the global frame

$$x_{knee} = x_0 + L_{foot} - L_{foot}\, cos(\theta_{toe}) + L_{shin}\, sin(\theta_{toe} + \theta_{ankle}) + L_{heel}\, sin(\theta_{toe})$$
$$z_{knee} = L_{shin}\, cos(\theta_{toe} + \theta_{ankle}) + L_{heel}\, cos(\theta_{toe}) + L_{foot}\, sin(\theta_{toe})$$

4.3 *Multi-link Model: Dynamics*

The equations associate the configurations in space over time with the forces or torques needed to originate them. The Euler- Lagrange equation is used, because this approach is beneficial in systems with higher dimension, that are submitted to holonomic constraints and where a good perception of the body behavior as a whole is needed.

Under the assumption that each link behaves as a rigid body, we determine the Lagrangian and the Euler-Lagrange equation ((7)).

$$u_i = \frac{\mathrm{d}}{\mathrm{d}t}\left(\frac{\partial \mathcal{L}}{\partial \dot{q}_i}\right) - \frac{\partial \mathcal{L}}{\partial q_i} \tag{7}$$

$q_i \in R^n$ represents the generalized coordinates and u_i is the generalized torque/force performing work on q_i, $(i = 1, ..., N)$. Note that u_i are non-conservative and N is the number of links of the robot.

The motion equations can be written in the generic vector format

$$u = B(q)\,\ddot{q} \;+\; C(q,\dot{q}) \;+\; G(q) \tag{8}$$

- **Ground Impact Forces**

In the heel-strike stage, a significant impact between the swing leg and the ground occurs.

These contact forces bring a new term into expression (8), which are now written as

$$u + J^T\,F_c = B(q)\,\ddot{q} \;+\; C(q,\dot{q}) \;+\; G(q) \tag{9}$$

In a real situation, the impact is instantaneous (when compared with the gait cycle duration) and the foot does not rebound. Due to these constraints, velocity has a discontinuity (although there is no discontinuity in configuration). Mathematically,

$$B(\dot{q}^{\,+} - \dot{q}^{\,-}) = J^T \lim_{\Delta t \to 0} \int_{\Delta t} F_c \Leftrightarrow B(\dot{q}^{\,+} - \dot{q}^{\,-}) = J^T\,\lambda \tag{10}$$

where λ and J denote the external impulses at collision points and the Jacobian.

Furthermore, the velocity of the foot along the y axis will also be always positive, once we consider that macroscopically the ground has no elasticity.

Thus, combining it with (10) we obtain

$$B\,\dot{q}^{\,+} - J^T\lambda = B\,\dot{q}^{\,-} J^T\,\dot{q}^{\,+} = 0 \tag{11}$$

Solving the upper equation in system (11) depends on whether J is full rank or not.

In the first situation, we can multiply both sides of the equation by $J\,B^{-1}$ and since J and B are full-ranked, $J\,B^{-1}\,J^T$ is invertible. Note that the rate of displacement previous to impact is related with the joint velocity by the Jacobian, this is, $\Delta \dot{pos}^{\,-} = (J\,\dot{q}^{\,-})$, resulting

$$\begin{aligned}
\lambda &= (J\,B^{-1}\,J^T)^{-1}\,(-\,J\,\dot{q}^{\,-})\\
\dot{q}^{+} &= \dot{q}^{\,-} \;+\; B^{-1}\,J^T\,(J\,B^{-1}\,J^T)^{-1}\,(-\,J\,\dot{q}^{\,-})
\end{aligned} \tag{12}$$

allowing us to know the magnitude of the impact force and the velocity afterwards.

When J is not full rank, a different procedure is needed. See [9] for details.

5 Generation of Natural Motion References

In this section, is described the generation and consistency verification of trajectories that "feed" the developed model. It is one of the main features of this concept in relation to the state-of-the-art and plays a very important role in the overall system. The generation of reference signals is a key issue in the design of a control system because such signals are what determine if model described (in spite of its simplicity relatively to the human locomotion system) will display behaviors close to the expected human behaviors.

The idea, central to the chosen approach, is to define appropriate references that are capable to generate an anthropomorphically correct walk, that will be adapted to each one of the locomotion circumstances person's height or walking speed but always without exceeding each joint's limits and, at the same time, is robust to the multiple perturbations that naturally arise in any more or less structured environment.

The first possibility taken into consideration was using experimental data collected from humans, as the one published in [19]. However, the presence of significant standard deviation and the absence of its variation explicitly with the person's height, age or gender for each joint does not enable a human-like pattern when introduced in the model.

This problem was solved by resorting to the analysis of human motion in order to extract qualitative features that are always present in the human locomotion gait cycle independently of the subject's body characteristics or speed of motion. Then, having these features and using the human body constraints, the motion references can be adjusted to the user's biometric information and the desired walking style. The remaining data is automatically adjusted and the waveforms can be artificially generated.

To realize this complex task, this module will, together with others in the architecture top layer, check the motion specification feasibility before generating a new reference signal.

6 Results

In this section are presented the results together with a brief description.

All the simulations presented were implemented using Matlab.

The ability to generate motion references that are perceived as natural is, as detailed earlier, extremely important in the overall system, since they are the controllers' reference signal. Thus, the goal here is to guarantee the satisfaction of the constraints (if not all, at least the strongest ones) and the generation of a human-like motion pattern. The only input required is the person's/robot's height. The length of each segment is then automatically obtained. Nevertheless, the length of each segment can be manually introduced, in case a simulation with unusual body proportions is intended.

Figure 2 show the region of admissibility (support leg) for two individuals of different height: individual 1 is 1,60m height and individual 2 is 1,90m height.

The back parallelepiped represents the usual range of each human joint when walking.

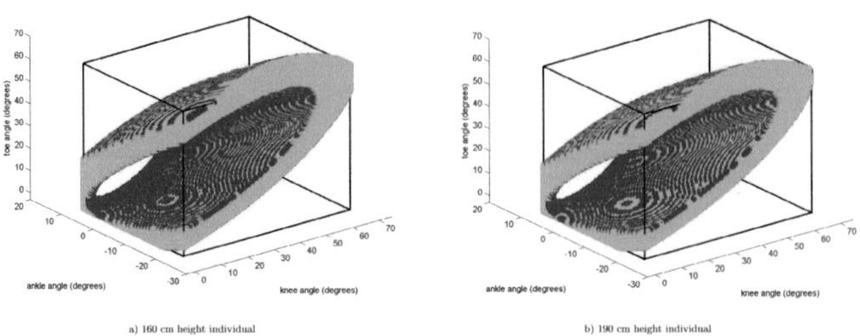

Fig. 2 Feasibility Verification results - region of admissibility for the support leg in individuals with different heights (left- 160cm; right 190cm)

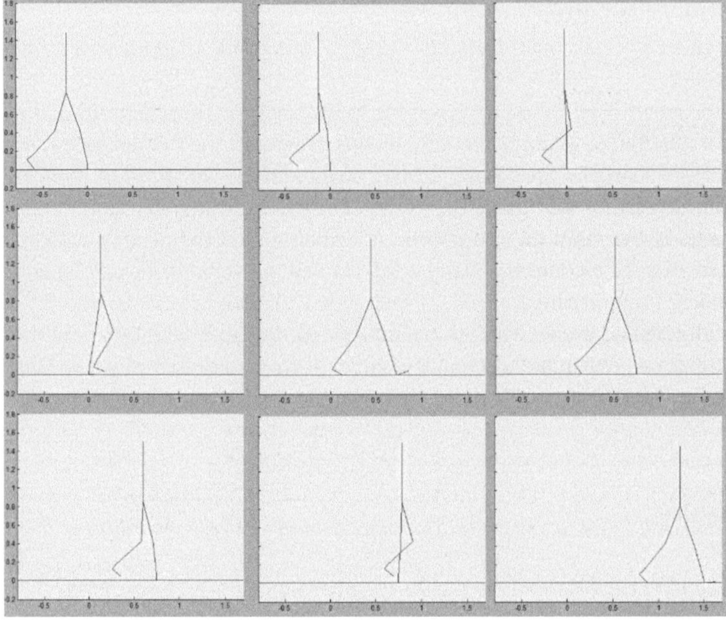

Fig. 3 Simulation results of the implemented model and designed trajectories - entire body in 2D

Specialized literature describes the hip (and all other joints) displacement over the gait cycle as waves with a characteristic shape, amplitude range and instants for the occurrence of minimum and maximum. Then, from this huge number of combinations, just some produce a hip displacement over time that resembles the human motion pattern. The region of admissibility is colored in green with the minimum colored in red and the maximum colored in blue.

Note that stepping out of the region of admissibility, would not only compromise the anthropometry of the walking pattern but also have impact on the region of admissibility of the swing leg.

A similar process was followed for the swing leg. This time, the constraint being on the foot (heel and toe) height rather then the hip's.

At this point, given the obtained region of admissibility and the information taken from the literature (trend analysis and a rough idea of the value of each angle in some meaningful moments of the gait cycle), the trajectory of each joint can be generated.

The resulting references proved their ability to generate a correct motion when introduced into the mathematical model, which was the ultimate goal (in figure 3 can be seen the displacement (in meters) of a standard height individual). Additionally, it is able to walk at different speeds, for a configurable distance always respecting the periodicity of the motion and, at the same time, without of the constraints imposed by other authors.

Furthermore, the accomplishment of this goal allows the research to progress with a solid base for the development and tests of the control system.

7 Conclusion

This article describes the investigation and development of human locomotion model that preserves the main particularities of the human body and is the base for the implementation of a control architecture.

This work brought a huge variety of challenges, such as obtaining reference signals that could generate a anthropomorphic walking pattern, which turned out to be an obstacle that made us rethink the system's architecture, so that we the system could ensure safety, versatility in multiple situations.

Also, comparatively to some works reviewed in the state-of-the-art, it presents a new focus and does not rely on constraints that damage the motion pattern characteristic of humans. The obtained model has also some degree of configurability (walking speed/ walking distance are input parameters, body proportions can be changed if intended,...).

The constraints imposed by the range of motion of the human joints and the periodicity of the motion are always respected. Moreover, the accomplished model is a solid base for future developments, either in terms of modeling or in terms of the development of the control system, in the sense that the model complexity is the outcome of a requirements analysis and a well thought overall system, in which the model is just a piece. On the other hand, the intention to model dynamics and impacts is the bring the model closer to reality, so that it doesn't need to be redone later.

More that the simulation results themselves, the impact of this work lies on the vision it's sharing: when developing robotic systems that aim to promote inclusivity, modeling aspects as the degrees of freedom/ segments considered, dynamics and impacts modeling, respect for human constraints motion periodicity and real-time requirements, or the generation of anthropomorphic motion references and its relation the human balance metrics, all of them gain tremendous relevance.

Given the complexity of the overall system, there is still a long way to go. The next steps would still be in simulation context furthering the modeling of dynamics, impacts and environment; ensure the robustness of the controllers to various situations and total system integration. Upon conclusion of these stages a hardware implementation has to be planned.

References

1. Wang, F., et al.: A coordinated control strategy for stable walking of biped robot with heterogeneous legs. Industrial Robot: An International Journal **36**(5), 503–512 (2009)
2. Miura, K., et al.: Human-like walking with toe supporting for humanoids. In: 2011 IEEE/RSJ International Conference on Intelligent Robots and Systems, pp. 4428–4435, September 2011
3. Fluit, R., et al.: A simple controller for the prediction of three-dimensional gait. Journal of Biomechanics **45**, 2610–2617 (2012)
4. Azevedo, C., the BIP team: Control architecture and algorithms of the anthropomorphic biped robot bip2000 (2000)
5. Hernández-Santos, C., Rodriguez-Leal, E., Soto, R., Gordillo, J.: Kinematics and dynamics of a new 16 dof humanoid biped robot with active toe joint. International Journal of Advanced Robotic Systems **9**, August 2012
6. Douat, L.R.: Estabilização do caminhar de um robot bipede de 5 elos com compensação do movimento dorsal. Tech. rep., Universidade Federal de Santa Catarina, June 2008
7. Zonfrilli, F., Oriolo, G., Nardi, D.: A biped locomotion strategy for the quadruped robot sony ers-210, September 2001
8. Gonçalves, A.Q.: Locomoçã bípede. Tech. rep., Universidade do Minho, December 2011
9. Guimaraes, D.L.: Biped locomotion systems analysis, modeling and control. Dissertation Work (2013)
10. Hartenberg, R.S., Denavit, J.: Kinematic synthesis of linkages. McGraw-Hill (1965)
11. Heo, J.W., Oh, J.H.: Biped walking pattern generation using an analytic method for a unit step with a stationary time interval between steps. IEEE Transactions on Industrial Electronics **62**(2), 1091–1100 (2015)
12. Kim, J.-Y., Park, I.W., Oh, J.H.: Walking control algorithm of biped humanoid robot on uneven and inclined floor
13. Kim, J.-Y., Park, I.W., Oh, J.H.: Realization of dynamic stair climbing for biped humanoid robot using force/torque sensors. J. Intell. Robot. Syst. **56**, 389–423 (2009)
14. Suwanratchatamanee, K., Matsumoto, M., Hashimoto, S.: Walking on the slopes with tactile sensing system for humanoid robot. In: International Conference on Control, Automation and Systems 2010, Korea, pp. 350–355, October 2010
15. Li, Y.: An optimal control model for human postural regulation. Tech. rep., University of Maryland (2010)
16. Manuel Crisóstomo, J.P.F., Coimbra, A.P.: Controlo de um robô bípede com base em sensores de força nos pés. In: Asociación Española para el Desarrollo de la Ingeniería Eléctrica (eds.) 9th Spanish Portuguese Congress on Electrical Engineering, Marbella, Spain, June 30–July 2, 2005. ISBN 84-609-5231-2
17. Spong, M.W., Hutchinson, S., Vidyasagar, M.: Robot Modeling and Control, 1st edn. John Wiley and Sons, Inc. (2006)
18. Monteiro, R.: Desenvolvimento de um controlador dinâmico para rôbos humanoides nao. Tech. rep., Faculdade de Engenharia da Universidade do Porto, Julho 2012
19. Levangie, P., Norkin, C.: Joint Structure and function - A Comprehensive analysis, 4th edn. MacLennan and Petty (2005)
20. Shafii, N., Lau, N., Reis, L.: Learning a fast walk based on zmp control and hip height movement. In: 2014 IEEE International Conference on Autonomous Robot Systems and Competitions (ICARSC), pp. 181–186, May 2014

21. Tao, X., Qijun, C.: A simple rebalance strategy for omnidirectional humanoids walking by learning foot positioning. In: 2011 8th Asian Control Conference, Taiwan, pp. 1340–1345, May 2011
22. Vundavilli, P.R., Pratihar, D.K.: Dynamically balanced optimal gaits of a ditch-crossing biped robot. Robotics and Autonomous Systems, 349–361 (2010)
23. Westervelt, E.R.: Toward a coherent framework for the control of planar biped locomotion. Tech. rep., University of Michigan (2003)
24. Li, Y., Levine, W.S., Loeb, G.E.: A two-joint human posture control model with realistic neural delays. IEEE Transactions on Neural Systems and Rehabilitation Engineering **20**(5), 738–748 (2012)
25. Zonfrilli, F.: Theoretical and experimental issues in biped walking control based on passive dynamics. Tech. rep., Universita degli Studi di Roma "La Sapienza", December 2004

Biped Walking Learning from Imitation Using Dynamic Movement Primitives

José Rosado, Filipe Silva and Vítor Santos

Abstract Exploring the full potential of humanoid robots requires their ability to learn, generalize and reproduce complex tasks that will be faced in dynamic environments. In recent years, significant attention has been devoted to recovering kinematic information from the human motion using a motion capture system. This paper demonstrates the use of a VICON system to capture human locomotion that is used to train a set of Dynamic Movement Primitives. These DMP can then be used to directly control a humanoid robot on the task space. The main objectives of this paper are: (1) to study the main characteristics of human natural locomotion and human "robot-like" locomotion; (2) to use the captured motion to train a DMP; (3) to use the DMP to directly control a humanoid robot in task space. Numerical simulations performed on V-REP demonstrate the effectiveness of the proposed solution.

Keywords Biped locomotion · Motion capture · Movement primitives · Nonlinear oscillators · Inter-limb coordination · Learning by demonstration

---

J. Rosado(✉)
Department of Computer Science and Systems Engineering,
Coimbra Institute of Engineering, IPC, Coimbra, Portugal
e-mail: jfr@isec.pt

F. Silva
Institute of Electronics Engineering and Telematics of Aveiro, Department of Electronics, Telecommunications and Informatics, University of Aveiro, Aveiro, Portugal
e-mail: fmsilva@ua.pt

V. Santos
Institute of Electronics Engineering and Telematics of Aveiro,
Department of Mechanical Engineering, University of Aveiro, Aveiro, Portugal
e-mail: vitor@ua.pt

L.P. Reis et al. (eds.), *Robot 2015: Second Iberian Robotics Conference*,
Advances in Intelligent Systems and Computing 418,
DOI: 10.1007/978-3-319-27149-1_15

1 Introduction

Programming robots to perform complex tasks and extend its repertoire can be extremely tedious and time consuming. Learning from demonstration is a promising methodology that offers a more intuitive approach to teach a robot how to generate its own motor skills [1-2]. To this end, the robot should be able to estimate human poses when performing a desired task, as well as to translate the skeleton data into appropriate motor commands. In the last years, a large body of work has studied the use of computer vision-based human motion capture for extracting poses as input to robotic applications [3-7].

In this work we present a study of human gait and human like-robot gait captured using a marker system. By human like-robot gait we intend that the movement performed by the demonstrator should be done as similar as the movement performed by current biped robots is done: slightly bent legs and foot always parallel to the ground. It is our intention to try to use this kind of movement to modulate a set of Dynamic Movement Primitives and try to adapt them, so they can be used to control a full-humanoid robot model in the V-REP simulation software [8]. The remainder of the paper is organized as follows: Section 2 reviews the mathematical formulation of rhythmic movement primitives and their modulation. Section 3 does a brief description of the capture system and an analysis on the captured movement is done. Section 4 talks about the adaption of the captured movement to the biped robot and shows an example of the result obtained.

2 Dynamic Movement Primitives

Robot learning from demonstration is a powerful approach promoting movements that look natural and predictable [9-12]. However, there are several challenges when transferring skills from humans to robots since the simple reproduction of the task by the robot may be of limited interest. A common challenge is to identify policy representations well suited for robot learning. Nonlinear dynamical and nonparametric regression techniques are promising candidates to achieve continuous learning and on-line adaptation. This section reviews the mathematical formulation of dynamic movement primitives applied in periodic tasks and some of their useful properties.

2.1 Mathematical Formulation

Dynamical system motor primitives have become a robust policy representation, for both discrete and periodic movements, that facilitates the process of learning and improving the desired behavior [13]. The basic idea behind DMP is to use an analytically well-understood dynamical system with convenient stability properties and modulate it with nonlinear terms such that it achieves a desired point or limit cycle attractor. The approach was originally proposed by Ijspeert *et al.* [14] and, since then, other mathematical variants have been proposed [15,16].

In the case of rhythmic movements, the dynamical system can be defined in the form of a linear second order differential equation that defines the convergence to the goal g (baseline, offset or center of oscillation) with an added nonlinear forcing term f that defines the actual shape of the encoded trajectory. This model can be written in first-order notation as follows:

$$\begin{aligned} \tau\dot{z} &= \alpha_z[\beta_z(g-y)-z]+f \\ \tau\dot{y} &= z \end{aligned} \tag{1}$$

Here, τ is a time constant and the parameters $\alpha_z, \beta_z > 0$ are selected and kept fixed, such as the system converge to the oscillations given by f around the goal g in a critically damped manner, y is the position and z is the speed. The forcing function f (nonlinear term) can be defined as a normalized combination of fixed basis functions:

$$\begin{aligned} f(\phi) &= \frac{\sum_{i=1}^{N}\psi_i\omega_i}{\sum_{i=1}^{N}\psi_i} r \\ \psi_i(\phi) &= \exp(-h_i(\cos(\phi-c_i)-1) \end{aligned} \tag{2}$$

where ω_i are adjustable weights, r characterizes the amplitude of the oscillator, ψ_i are von Mises basis functions, N is the number of periodic kernel functions, $h_i > 0$ are the widths of the kernels and c_i equally spaced values from 0 to 2π in N steps (N, h_i and c_i are chosen a priori and kept fixed). The phase variable ϕ bypasses explicit dependency on time by introducing periodicity in a rhythmic canonical system. This is a simple dynamical system that, in our case, is defined by a phase oscillator:

$$\tau\dot{\phi} = \Omega \tag{3}$$

where Ω is the frequency of the canonical system. In short, there are two main components in this approach: the transformation system that provides the shape of the trajectory patterns and the canonical system that provides the synchronized timing.

In order to encode a desired demonstration trajectory y_{demo} as a DMP, the weight vector has to be learned with, for example, statistical learning techniques such as locally weighted regression (LWR). Instead of using a single global model to fit all of the training data, local models attempt to fit the data only in a region around the location of the query using a distance weighted regression. The motivation for exploring locally weighted techniques came from their suitability for online robot learning, because of their fast incremental learning.

The input to the learning algorithm is the demonstration trajectory defined by the triplets of position, velocity and acceleration: $y_{demo}(t)$, $\dot{y}_{demo}(t)$ and $\ddot{y}_{demo}(t)$. Re-arranging equation (1) and equating y to y_{demo}, z to $\tau\dot{y}_{demo}$ and $\dot{z}$ to $\tau\ddot{y}_{demo}$, we obtain:

$$f_{target} = \tau^2 \ddot{y}_{demo} - \alpha_z\left[\beta_z\left(g - y_{demo}\right) - \tau\dot{y}_{demo}\right] = \frac{\sum_{i=1}^{N} \psi_i \omega_i}{\sum_{i=1}^{N} \psi_i} r \tag{4}$$

LWR finds, for each kernel function ψ_i, the weight vector ω_i that minimizes the following quadratic error criterion:

$$J_i = \sum_{k=1}^{M} \psi_i^k \left[f_{target}^k - \omega_i r\right]^2 \tag{5}$$

where k represents an index associated to the discrete time steps. This is a weighted linear regression problem that can be solved using a batch or an incremental regression. The incremental solution can be derived using recursive least squares with a forgetting factor λ [24]. Given the target data f_{target}, the update of the weights ω_i is performed for each time-step k as follows:

$$\begin{aligned} w_i^{k+1} &= w_i^k + \psi_i P_i^{k+1} r e^k \\ P_i^{k+1} &= \frac{1}{\lambda}\left(P_i^k - \frac{{P_i^k}^2 r^2}{\frac{\lambda}{\psi_i} + P_i^k r^2} \right) \\ e^k &= f_{target}^k - w_i^k r \end{aligned} \tag{6}$$

On the one hand, when the forgetting factor λ is one, batch and incremental learning regressions provide identical weights ω_i. On the other hand, when the forgetting factor is less than one, the differences may appear since the incremental regression tends to forget older data and give more weight to recent ones.

2.2 Extension to Multiple Degrees-of-Freedom

It is worth noting that a DMP contains one independent dynamical system per dimension of the space in which it is learned. The extension of the previous concepts to multiple degrees-of-freedom (DOF) is commonly performed by sharing one canonical system among all DOFs, while maintaining a set of transformation systems and forcing terms for each DOF. Therefore, the canonical system provides the temporal coupling among DOFs, the transformation system achieves the desired attractor dynamics for each individual DOF and the respective forcing terms modulate the shape of the produced trajectories.

Typically, the encoded variables represent different quantities such as joint angles or the three-dimensional Cartesian coordinates of the robot's end-effector. However, the adaptation of learned motion primitives to new situations becomes difficult when the demonstrated trajectories are available in the joint space. For low-DOF robots, for which inverse kinematics can be easily solved, a solution is to learn DMPs in task space [17] where DMP parameters directly relates to task variables. Other authors [18] have suggested the use of dimensionality reduction to infer new spaces, from joint-space trajectory demonstrations, in which the DMP parameters can be related to task variables.

3 Demonstration Signal Capture

3.1 VICON System Description

The VICON system used on our captures uses a total of nine infrared cameras with Full HD resolution up to 2000fps and two video cameras. Combined with this system, two force platforms capable of measuring the forces and moments on the 3 axis directions are available. Non invasive EMG sensors on a total of thirty are also available. The system works by using a total of thirty-six reflective markers disposed as depicted in Fig. 1.

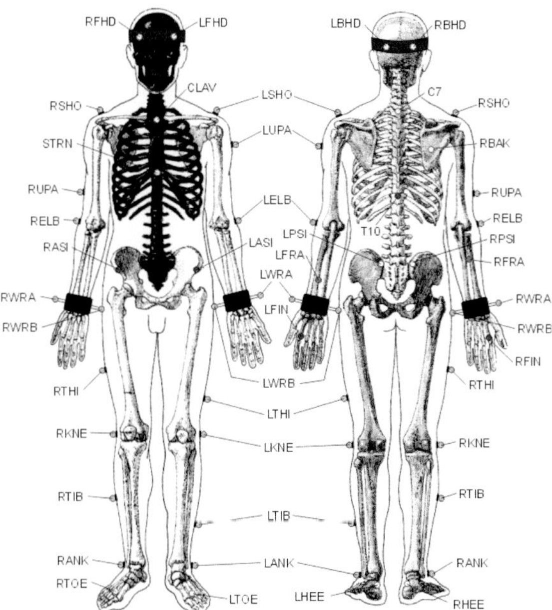

Fig. 1 Placement of the reflective markers on the VICON system.

By combining the information captured by the infrared cameras for each marker together with the provided software, the Cartesian position of each marker at each frame can be computed and retrieved having as base a referential with the origin on the center of the room.

Fig. 2 Example of "robot-like" gaiting.

As said before, it was asked to the subject to perform a natural human walking and a robot-like gait and Fig. 2 illustrates this concept.

3.2 VICON Data Analysis

Across the experience, it was requested that the subject would walk always on the direction of the y axis of the VICON referential. This requirement allows extracting directly some of the walking parameters (Step length, walking period, Foot clearance, Hip height) from the VICON data, without the need of extra computations.

The first value we observe is the Step length in boths walking modes. Since the movement was performed on y axis direction, the Step Length (Sl) can be directly obtained from y data for any of the markers placed on the foot (Fig. 3). A change in these values means the foot is moving and of course, no variation means that the foot is stopped. Analysis of these values gives an average step length of 118.7 cm for human natural walking and 50.3 cm for robot-like walking, which is not surprising, since the bent knees no robot-like gait do not allow for the leg to extend so much, as it happens on human natural walking movement.

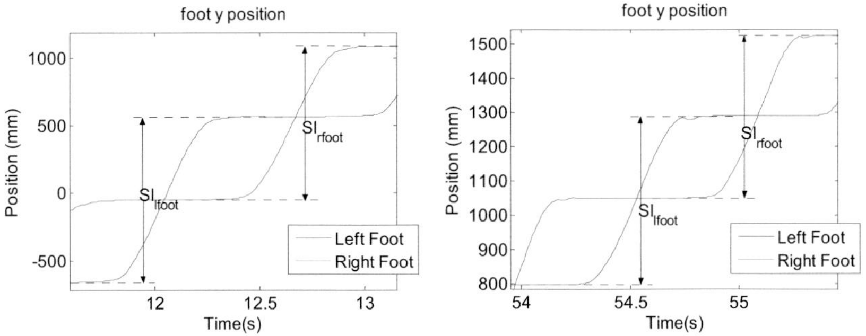

Fig. 3 Left and Right foot displacement on the y axis for human walking (left) and robot-like walking (right). Values taken from LTOE and RTOE markers (timings are different, because the two type of movements was performed in succession).

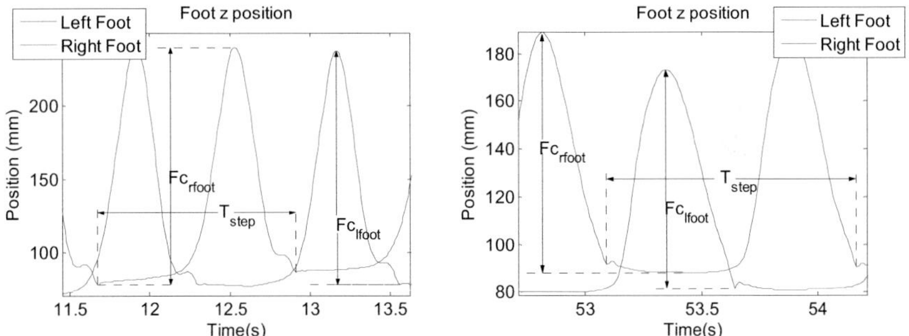

Fig. 4 Left and Right foot vertical position for human (left) and robot-like (right) walking modes. Values taken from LANK and RANK markers.

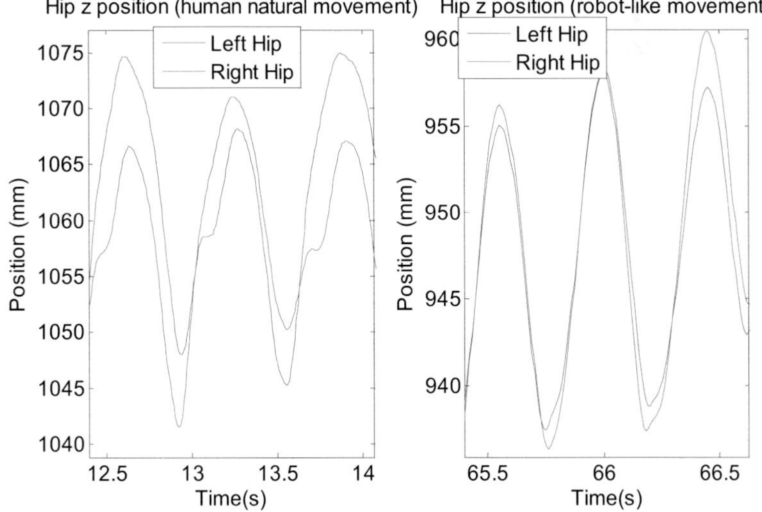

Fig. 5 Vertical position of the left and right hips for both walking modes.

Since the *z* axis represents the vertical axis on our referential, values like Foot clearance (Fc), walking period and Hip height can be obtained by looking at any of the foot markers coordinates on this axis. These trajectories are represented in Fig. 4 and the average Foot clearance is 21.1 cm for human walking and 13.3 cm for robot-like movement. Notice that there's a slight difference between the left and right foot tops (and bottoms), but this is due to the marker placement not be exactly the same). As for the walking period (also marked in Fig. 4), this varies between 1.188 s and 1.003s for each walk method respectively. As for Hip height (Fig. 5), we have the value taken from the average of the four markers (LASI, RASI, LPSI, and RPSI), it has an sinusoidal aspect, with an average of 105.9 cm and a small value of 95.7cm for robot-like walking mode, as expected.

In legged locomotion, each leg goes to one of two phases: swing or stance phase, but in human walking, the stance phase is composed of several stages:

- Heel Strike: it's the time the heel touches the ground, after the swing phase;
- Early Flat Foot: after the heel touches the ground, the toes start descending and all the foot is in touch with the ground. This happens at this stage.
- Heel Rise: before the swing phase, the heel starts rising. This is also known as the Late Flat Foot phase.
- Toe Off: This marks the end of stance phase and the beginning of swing phase. It's the moment the toes leave the ground.

All these stages can be seen by looking at the z trajectory of the toe and the heel markers (Fig. 6). Analysis of these graphics gives a 60% of stance phase, which agrees with the value referred in literature [19]. In a perfect robot-like walking, none of these stages should be present.

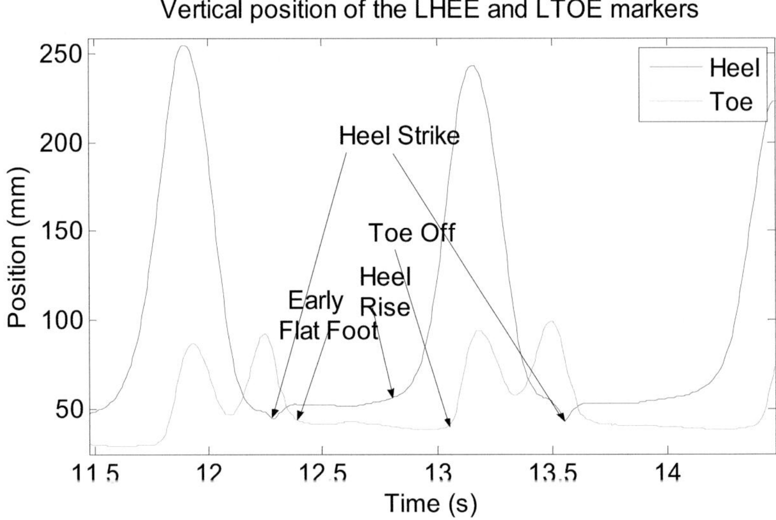

Fig. 6 Stages of the human gait

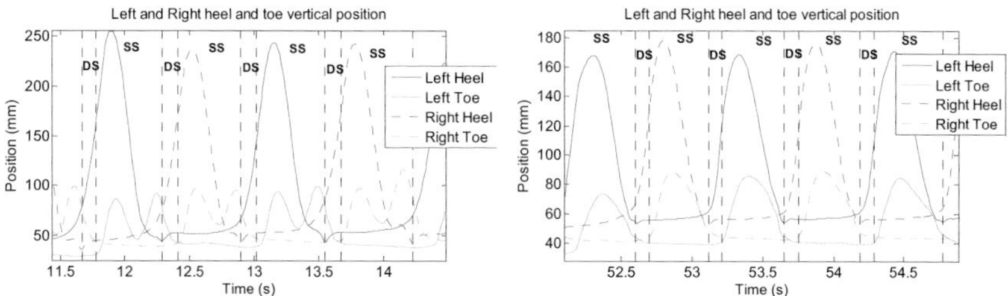

Fig. 7 Single and Double Support for the human natural gait (left) and robot-like gait (right).

In any biped locomotion there's always a Double Support phase (when both feet are in the ground) and a Single Support (when only one foot is on the ground). Having in mind the mentioned phases above and with both feet heel and toe marker, we can identify these Single Support (SS) and Double Support (DS) stages (Fig. 7). Notice that a perfect robot-like gait would not have the any of the stages mentioned above for human locomotion, but the stages Heel Rise and Toe Off are still present in our experiments.

4 V-REP Simulation

In order to extract a valid signal that can be used to train a DMP and control our robot, we face essentially two problems: (1) what values use, Cartesian space or Join Space? (2) Scale problem: the human used to capture data has not the same dimensions as the robot in simulation. In our previous works we have successfully used the Cartesian space by assuming a referential for each leg on the correspondent hip. We have also seen that the Cartesian space simplifies tasks like increasing Foot clearance or Step length, since these values relate directly to this space (we have already seen that when examining the gait patterns on previous sections). Choosing this option also easily solves the second problem, since it's just a question of direct scale (relate the step length to those possible by the robot and the hip distance to the ground to the one present on the robot). In Fig. 8 we have the Cartesian space coordinates for the left foot tip obtained, after some processing (scaling and filtering to remove some noise). By using the DMP properties, the right foot tip is obtained by simply using a oscillator for that foot that is out of phase 180° in respect to the other foot.

These signals are then converted, through an inverse kinematics algorithm, to the desired joint trajectories used as reference input to a low-level feedback controller. Since they relate directly to the foot trajectory, changing these signals will affect parameters of the locomotion like the Hip Height, the Step Length, the Foot Clearance or even the forward velocity. Fig. 9 shows a sequence of images of the robot successful walking using the signal of Fig. 8.

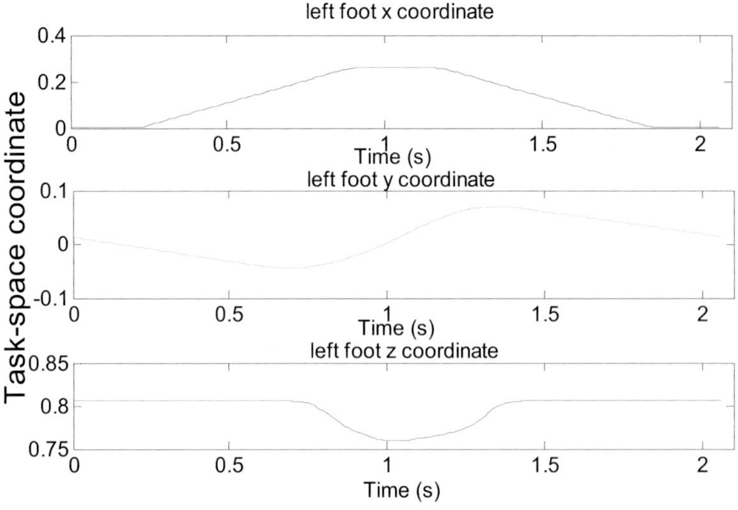

Fig. 8 Reference Cartesian left foot obtained from the VICON data

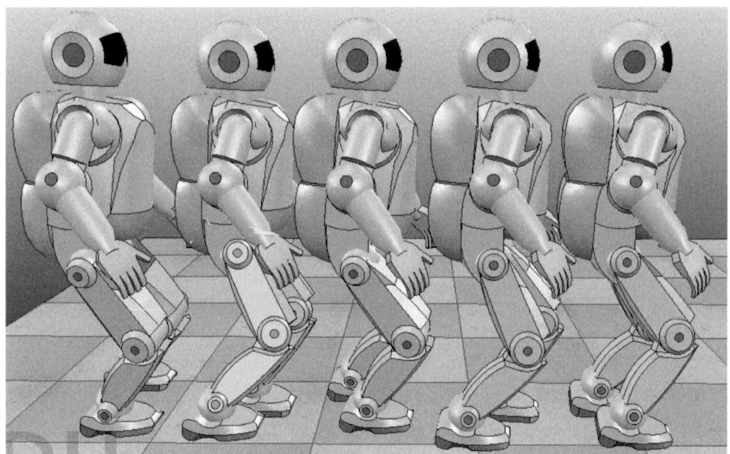

Fig. 9 Sequence of images of the robot walking using the DMP trained with the VICON data

5 Conclusions

We have performed a study of the human locomotion and robot-like locomotion based on motion data captured using a VICON platform and the results are in line with what was expected. Next we created a reference signal to train a DMP that can be used to control a biped robot. Numerical simulations shown that the methodology works and it can be used to successfully control the robot. Ongoing work involves the analysis of other situations, like climbing stairs, slopes and overcome obstacles, in order to extract information that can be used by the robot when faced with the same situations. Analysis of the EMG data will also be performed, in order to study the energy differences when performing a human natural gait vs. a robot-like gait.

References

1. Billard, A., Callinon, S., Dillmann, R., Schaal, S.: Robot programming by demonstration. In: Siciliano, B., Khatib, O. (eds.) Handbook of Robotics. Springer, New York (2008)
2. Argall, B.D., Chernova, S., Veloso, M., Browning, B.: A Survey of Robot Learning from Demonstration. Robotics and Autonomous Systems **57**(5), 469–483 (2009)
3. Dasgupta, A., Nakamura, Y.: Making feasible walking motion of humanoid robots from human motion capture data. In: IEEE International Conference on Robotics and Automation, pp. 1044–1049 (1999)
4. Elgammal, A., Lee, C.-S.: Tracking People on a Torus. IEEE Transactions on Pattern Analysis and Machine Intelligence **31**(3), 520–538 (2009)
5. Inamura, T., Toshima, I., Tanie, H., Nakamura, Y.: Embodied Symbol Emergence Based on Mimesis Theory. International Journal of Robotics Research **23**(4–5), 363–377 (2004)
6. Kulic, D., Takano, J.W., Nakamura, Y.: I"ncremental Learning, Clustering and Hierarchy Formation of Whole Body Motion Patterns using Adaptive Hidden Markov Chains. International Journal of Robotics Research **27**(7), 761–784 (2008)
7. Shon, A.P., Grochow, K., Hertzmann, A., Rao, R.P.: Learning Shared Latent Structure for Image Synthesis and Robotic Imitation. In: Weiss, Y., Schlkopf, B., Platt, J.C. (eds.) Advances in Neural Information Processing Systems. MIT Press, Cambridge (2005)
8. Rohmer, E., Singh, S., Freese, M.: V-REP: a versatile and scalable robot simulation framework. In: IEEE/RSJ International Conference on Intelligent Robots and Systems, pp. 1321–1326 (2013)
9. Argall, B., Chernova, S., Veloso, M., Browning, B.: A survey of robot learning from demonstration. Robotics and Autonomous Systems **57**(5), 469–483 (2009)
10. Billard, A., Callinon, S., Dillmann, R., Schaal, S.: Robot programming by demonstration. In: Siciliano, B., Khatib, O. (eds.) Handbook of robotics. Springer, New York (2008)
11. Schaal, S., Ijspeert, A., Billard, A.: Computational approaches to motor learning by imitation. Philosophical Transaction of the Royal Society of London: Series B, Biological Sciences **358**, 537–547 (2003)
12. Breazeal, C., Scassellati, B.: Robots that imitate humans. Trends in Cognitive Science **6**(11), 481–487 (2002)
13. Kober, J., Peters, J.,: Policy search for motor primitives in robotics. Machine Learning (2010)
14. Ijspeert, A., Nakanishi, J., Schaal, S.: Movement imitation with nonlinear dynamical systems in humanoid robots. In: Proceedings of the 2002 IEEE International Conference on Robotics and Automation, pp. 1398–1403 (2002)

15. Gams, A., Ijspeert, A.J., Schaal, S., Lenarcic, J.: On-line learning and modulation of periodic movements with nonlinear dynamical systems. Autonomous Robots **27**(1), 3–23 (2009)
16. Ijspeert, A., Nakanishi, J., Hoffmann, H., Pastor, P., Schaal, S.: Dynamical movement primitives: learning attractor models for motor behaviors. Neural Computation **25**, 328–373 (2013)
17. Pastor, P., Hoffmann, H., Asfour, T., Schaal, S.: Learning and generalization of motor skills by learning from demonstration. In: Proceedings of the IEEE International Conference on Robotics and Automation (2009)
18. Bitzer, S., Havoutis, I., Vijayakumar, S.: Synthesising Novel Movements through Space Modulation of Scalable Control Policies. From Animals to Animats. Springer (2008)
19. Umberger, B.R.: Stance and Swing Phase Costs in Human Walking. Journal of the Royal Society Interface **7**(50), 1329–1340 (2010)

Reconfiguration of a Climbing Robot in an All-Terrain Hexapod Robot

Lisbeth Mena, Héctor Montes, Roemi Fernández, Javier Sarria and Manuel Armada

Abstract This work presents the reconfiguration from a previous climbing robot to an all-terrain robot for applications in outdoor environments. The original robot is a six-legged climbing robot for high payloads. This robot has used special electromagnetic feet in order to support itself on vertical ferromagnetic walls to carry out specific tasks. The reconfigured all-terrain hexapod robot will be able to perform different applications on the ground, for example, as inspection platform for humanitarian demining tasks. In this case, the reconfigured hexapod robot will load a scanning manipulator arm with a specific metal detector as end-effector. With the implementation of the scanning manipulator on the hexapod robot, several tasks about search and localisation of antipersonnel mines would be carried out. The robot legs have a SCARA configuration, which allows low energy consumption when the robot performs trajectories on a quasi-flat terrain.

Keywords Hexapod robot · Walking and climbing robot · SCARA configuration · Antipersonnel landmines · Control architecture

1 Introduction

Locomotion of legged robots has been of great interest by many research groups, for several decades, with the objective to reproduce movements carried out by animals when they move. A historical perspective of walking robot technology

L. Mena · H. Montes · R. Fernández · J. Sarria · M. Armada
Centro de Automática Y Robótica CSIC-UPM, Ctra. Campo Real Km 0.220,
Arganda Del Rey, 28020 Madrid, Spain

H. Montes(✉)
Facultad de Ingeniería Eléctrica, Universidad Tecnológica de Panamá,
Via UTP, Panama City, Panama
e-mail: hector.montes@csic.es

L.P. Reis et al. (eds.), *Robot 2015: Second Iberian Robotics Conference*,
Advances in Intelligent Systems and Computing 418,
DOI: 10.1007/978-3-319-27149-1_16

can be found in the literature, for example in [1-3], among other. In these references, the evolution of the legged robots, besides other concepts, are described. On the other hand, the development of new technologies has allowed the design and construction of an innovative generation of legged robots (climbing and walking) in the last decade, e.g., Tri-Athlete Rover developed by Jet Propulsion Lab [4], the RiSE, the BigDog, the LittleDog, the LS3 and the Cheetah [5-9] developed by Boston Dynamics in cooperation with other institutions, and partially funded by DARPA.

In just over two decades the Centre for Automation of Robotics – CAR (CSIC-UPM) has designed and developed some kinds of legged robots (climbing and walking robots) for several applications. Examples of some of them are TRACMINER, RIMHO, RIMHO2, ROWER, REST, REST-II, SILO4, SILO2, SILO1, SILO6, ROBOCLIMBER [2, 10-19], which have been used as robot prototypes (with one, two, four and six legs) from educational purposes until carrying out consolidation works of rocky mountainsides.

Firstly, in this paper, it is presented a brief description of the previous climbing robot, whose name is REST (from the acronym in Spanish: *Robot Escalador de Soldadura a Tope*) [13, 20]. This brief description consists of the explanation of the main parts of the robot, focusing in the subsystems that have been reconfigured.

Secondly, the modifications carried out in some parts/subsystems of the climbing robot to convert it in a walking robot will be presented. Fundamentally, the modifications were carried out in the feet, in the electronics, in the power supply, the rear control panel, the wireless communication, and, evidently, in the design of the control algorithms.

The general description of the all-terrain hexapod robot will be presented, taking into account the leg configuration, the subsystems that form the hexapod robot, the working area of the leg, and its control architecture.

Finally, in this article some experimental results are shown when the hexapod robot executes a gait to carry out a task established. In this case, currently, the hexapod walking robot performs gaits for humanitarian demining tasks. However, this walking machine could be prepared to complete other kind of works in indoor and outdoor environments, according its size and versatility.

2 Brief Description of the Previous Climbing Robot

REST climbing hexapod robot was designed with the purpose to move high payloads on vertical ferromagnetic surface. The main areas of applications of this robot comprise inspection, maintenance, and intervention tasks in industrial environment. Specifically, the REST could perform autonomous welding tasks on steel sheets used in hulls of civil ships [13, 20].

SCARA configuration is used in the legs of the REST, constructed with a hard structure of aluminum 7075, in order to avoid deformations when the robot climbs on a vertical surface. Each leg has three degree of freedom (DOF), two rotational, and one prismatic. In the last DOF (prismatic), it is installed the electromagnetic foot.

The electromagnetics feet can be compared with grasping's devices due to the feet must be "gripped" to the ferromagnetic surface, to move on the vertical surface. The electromagnetic foot consists of an array of eight special electromagnets (permanent magnets + electromagnets), which are attached to the wall by default (without electric energy), and unattached when the electric current is applied. This guarantees the robot does not fall down when the electric power supply is cut, unexpectedly.

Fig. 1(a) shows the REST climbing hexapod robot on a ferromagnetic surface with the power cord attached at the front of its body, and a steel rope connected between the robot body and a fixed point in the upper part of the wall, for security purpose. For more details of this robot please refer to [13, 20].

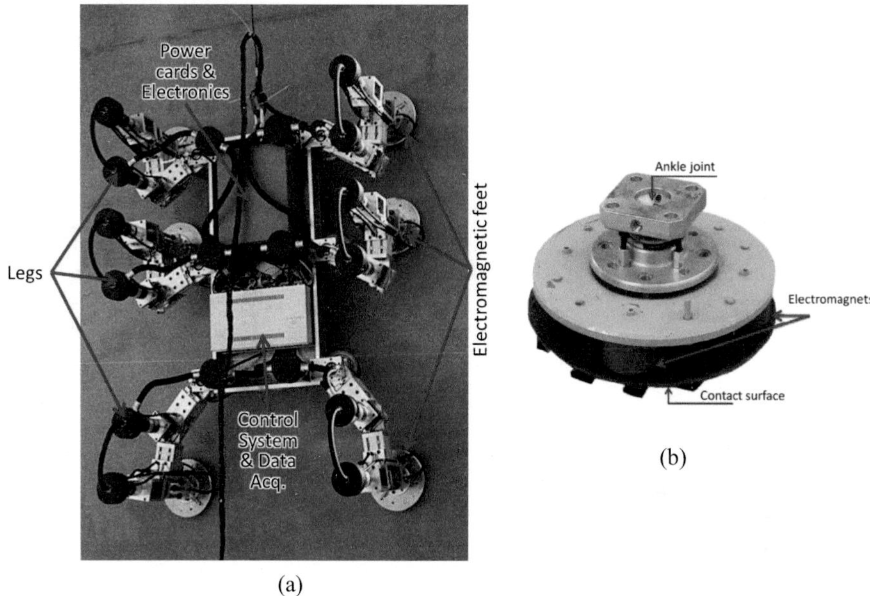

Fig. 1 (a) REST climbing robot on a ferromagnetic surface; (b) Electromagnetic foot of the REST.

3 Reconfiguration of the Climbing Robot in a Walking Robot

It is evident that one of the first modifications realised in the climbing robot to convert it to a walking robot must be the new design of the feet. However, some details must be considered in the realization of this design, how much load must support each foot depending the terrain where the robot moves, must be considered rotational passive joint to avoid some inaccuracy in the positioning of the robot legs, is it necessary a position sensor in order to know when the foot contacts with the terrain, and force sensors, etc.

In addition, to the modifications on the robot feet, other changes or inclusion of new subsystems were carried out. Some modifications or other add-on have been realized/installed in the robot, for example, in the electronics, in the power supply, in the rear control panel on the robot, the wireless communications, and in the control algorithms.

The new design of each foot consists of the some parts, one of them is the internal axis of the foot (of steel plated with nickel) ending in a nylon cone. This foot supports loads up to 2000 kg, which it is important for the new reconfigured robot. The internal axis of the foot must move inside the other parts that form the foot, in order to activate the inductive proximity sensor installed in the up part of the foot. This indicates that the foot has made contact with the terrain. With the use of a spring the axis of the foot returns to the original position, in this case the proximity sensor send to the control system an off signal. Fig. 2 depicts the schematic diagram and one picture of the new foot installed on the walking robot.

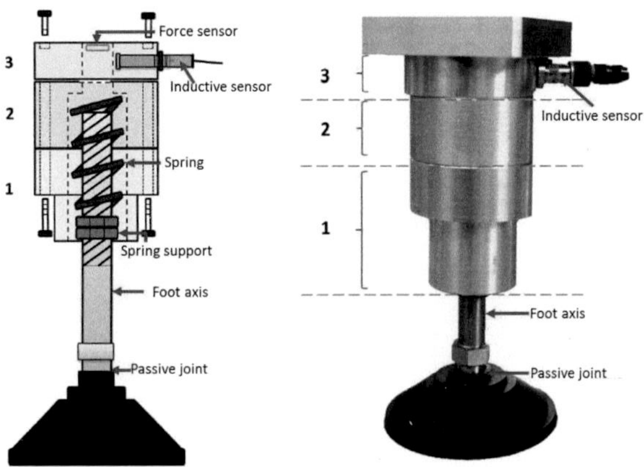

Fig. 2 Schematic diagram (to the left) and a picture (to the right) of the new foot designed for the all-terrain walking robot.

The previous robot had a tethered cable, which supplied of electric energy to the robot, and Ethernet connection between on-board computer and an external computer. This tethered cable was eliminated in order to provide autonomy to the new reconfigured robot. The power supply was replaced by two batteries of 12 VDC connected in series and two DC-DC converters with the power and characteristics required for the control stage and the power stage of the system. On the other hand, the communication between the on-board PC and the remote PC has been realized by mean of a wireless connection. Two types of standard wireless connections have been installed in the robot. One of them is a wireless connection using the local area network (LAN) of any institution where the robot is working. The other wireless connection is a point-to-point connection between the on-board computer and a remote portable computer. In Fig. 3 the two wireless connections implemented in the reconfigured robot are shown.

The first option is very useful in laboratory and in indoor environments (see Fig. 3(a)), since it is easy connection. However, in other environment where LAN connections are impossible that exist (e.g., at some external environment as land infested by antipersonnel landmines), then this option is not useful. On the other hand, the second option of wireless communications will facilitate the connection between the robot and remote PCs in outdoor environments, where not exists a local network provider. In this case a point-to-point network is configured by directly connecting the access point to one or more remote computers, assigning them fixed IPs addresses (see Fig. 3(b)). Off-the-shelf devices are used in both standard wireless communications, which will depend on the bandwidth required for the tasks that will be carried out in different applications.

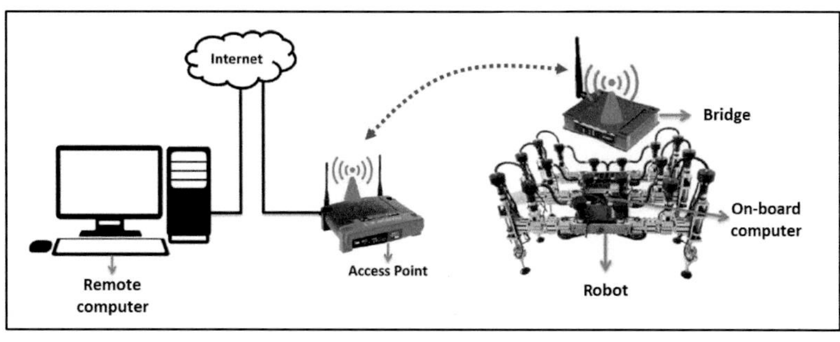

(a)

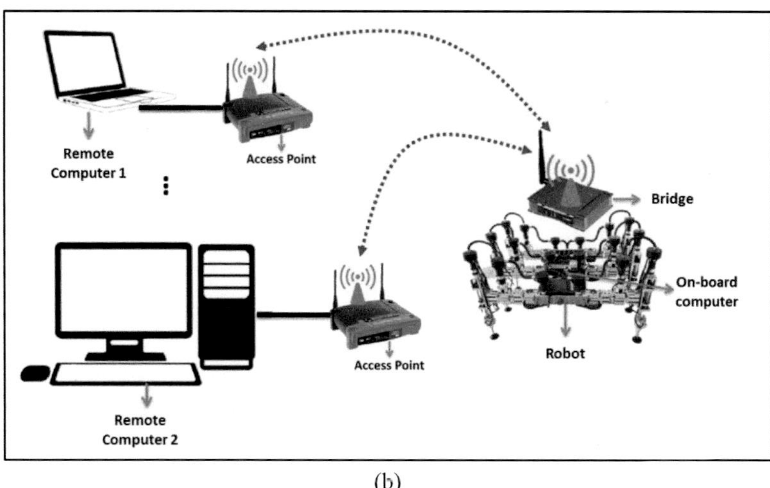

(b)

Fig. 3 Wireless communication between the robot and the remote computer. (a) Communication through the LAN; (b) Communication through point-to-point connection.

4 Description of the All-Terrain Hexapod Robot

The all-terrain hexapod robot has six legs in SCARA configuration of 3 DOF each one. The first two joints, the shoulder and the elbow, are rotational and the third joint, where the foot is implemented, is prismatic. Each leg has five inductive proximity sensors in order to initialize each leg and subsequently to verify the position of the leg during the locomotion process, besides to detect the contact between the feet with the soil. Fig. 4 shows the configuration of one of the legs.

The four external legs installed on the robot have similar working areas among them, but are different that the working area of the legs assembled in the middle of the robot. All working areas of the legs are restricted by the body robot and the collocation of the respective adjacent legs. Fig. 5 shows the working area of the external legs of the robot. In the figure it is possible to see the three joints (q_1, q_2 and q_3) and the top view (left) and frontal view (right) of the working area of the leg.

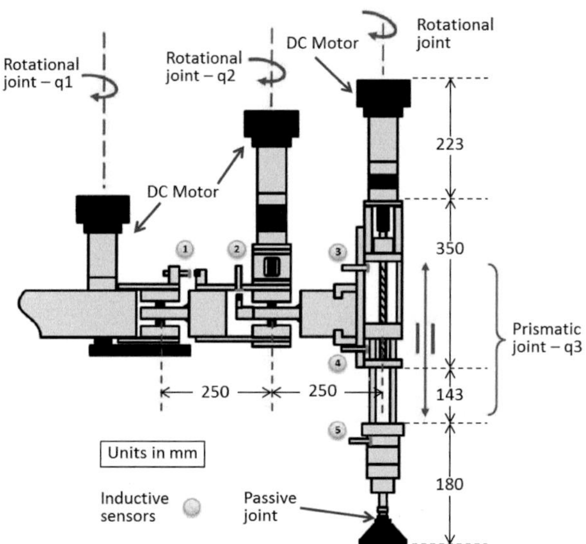

Fig. 4 Configuration of the leg.

The system architecture of this all-terrain robot consist of an on-board computer, control cards, data acquisition boards, power cards, signals conditioner cards, positioning sensors, DC motors, Wi-Fi communication system, DGPS, batteries, and other devices and accessories. This system architecture provides a reliable starting point for developing several control strategies in order to carry out several tasks in outdoor environment, e.g. in humanitarian demining tasks [21-22]. Fig. 6 shows the main subsystems implemented in the hexapod robot.

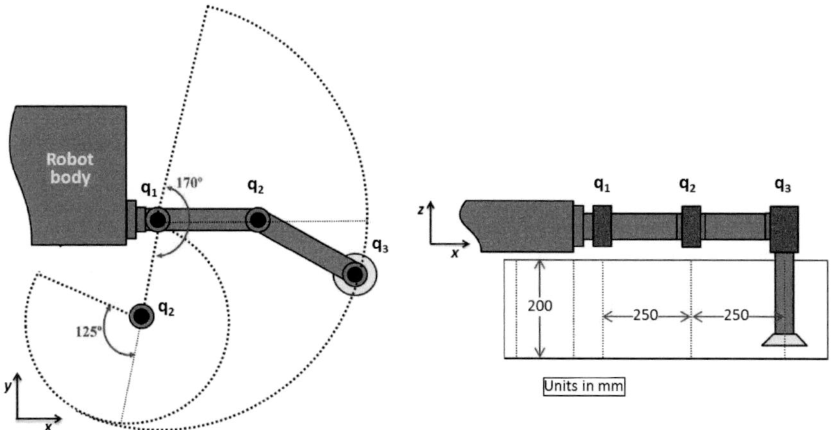

Fig. 5 Working area of one leg of the robot.

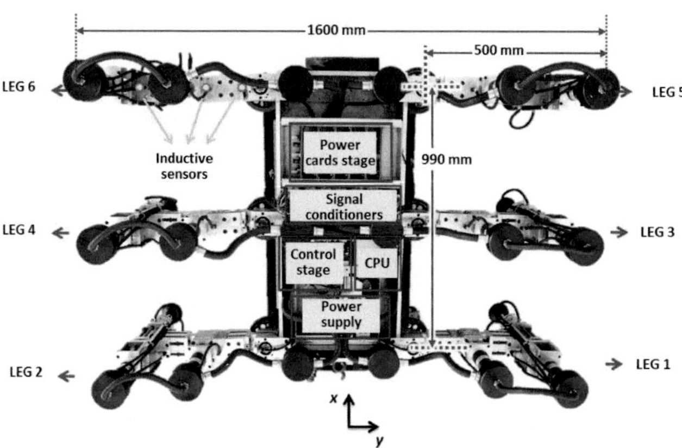

Fig. 6 Top view of the hexapod robot with its main subsystems.

This walking robot is able to load different tools on-board in order to carry out several tasks in outdoor environment. Up to 300 kg of mass can to support this robot, for this reason it could perform several kind of works. One example is carry out stable locomotion in order that a scanning manipulator arm can perform suitable movements of its end effector, where is installed the metal detector head, in humanitarian demining applications [21-22]. Fig. 7 shows the general control architecture of the hexapod robot.

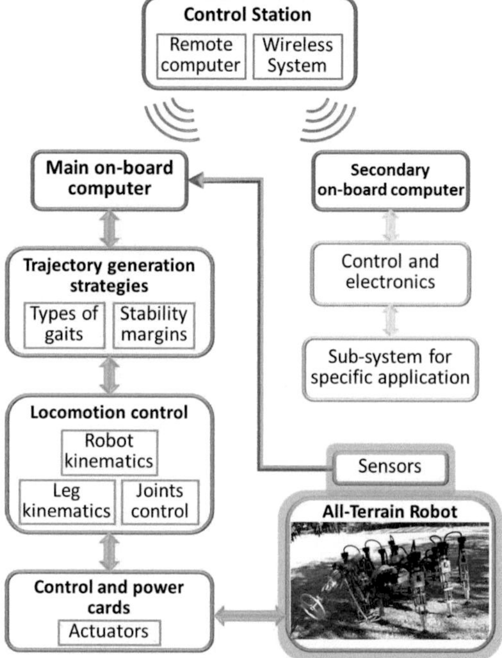

Fig. 7 General control architecture of the hexapod robot.

The general control architecture shown in Fig. 7 presents the main parts implemented in the all-terrain hexapod robot, and other stages that may have the tools installed over this robot in order to perform specific tasks. The communication between the base station computer and the on-board computer is by means of Wi-Fi. High level instructions are only sent by the remote computer, besides through this computer the teleoperation tasks could be carried out. The locomotion control of the robot with different gaits is performed with the control of on-board computer. On the other hand, a secondary on-board computer can be installed on the hexapod robot, if an external tool is installed and if required by this external system. Other control and electronics card must be installed in order to drive the external tool. One example of an external tool would be a scanning manipulator arm to detect and localize antipersonnel mines [21-22].

5 Experimental Results

Several experimental tests in indoor and outdoor environment with different locomotion patterns have been carried out with the reconfigured all-terrain hexapod robot. This robot has demonstrated high stability during the execution of several trajectories. It has been able to move extra loads installed at the body front during some locomotion tests, and the additional power consumption has been insignificant with respect to the execution of gait patterns without extra payload.

Consequently, it will be feasible to load on-board different tools and other additional sensors, without to contribute to significant energy consumption.

Fig. 8 shows some experimental results during the execution of a discontinuous gait using the alternating tripod mode. In this case only the Cartesian positions of the leg 1 of the robot are shown. Only two steps are presented in Fig. 8 in order to see, in best way, the details of the experimental results.

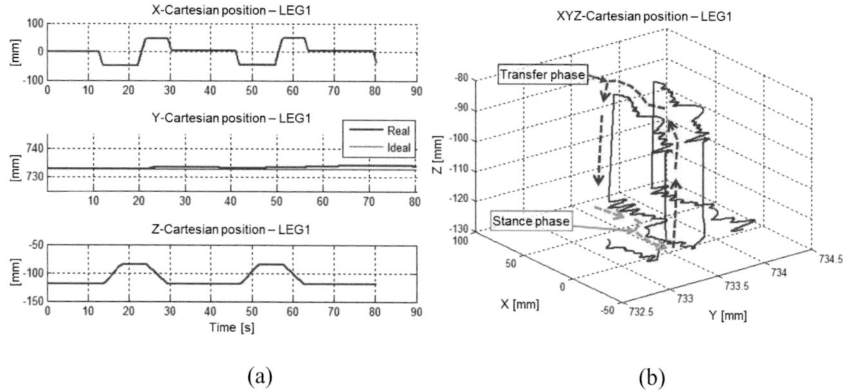

Fig. 8 Some experimental results during performing of an alternating tripod discontinuous gait. (a) Cartesian position of Leg 1; (b) Phases of transfer and stance of leg 1.

The Y-axis drift of the leg 1 position is appreciated (see Fig. 8(a)). However, this drift is about 1 mm in two steps, therefore, it can be corrected easily. This minor deviation is due to some backlashes there are in the leg joints. In Fig. 8(b) the transfer and stance phase of the leg 1 during two steps are shown. The vibrations observed in Y-axis, besides, are related with the intrinsic forces in the robot structure when the machine moves-forward its body.

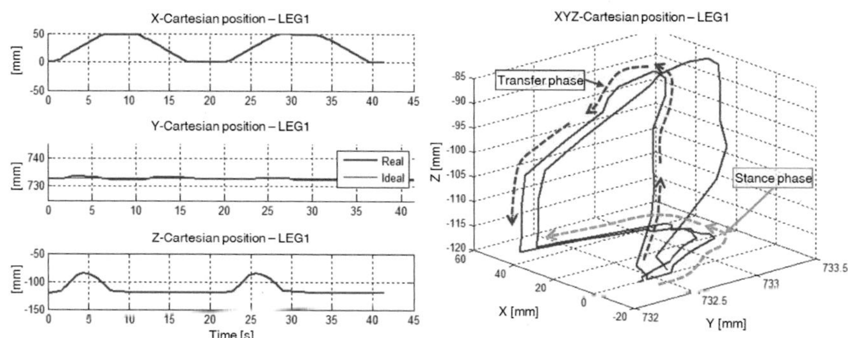

Fig. 9 Some experimental results during performing of an alternating tripod continuous gait. (a) Cartesian position of Leg 1; (b) Phases of transfer and stance of leg 1.

Fig. 9 shows some experimental results during the execution of a continuous gait using the alternating tripod mode. As in the previous figure only the positions of the leg 1 are shown. The other leg positions that form the same tripod are similar. Evidently, each tripod has a phase of 180º respect to each other. In this mode the transfer and stance phase are carried out suitable way considering the experimental results presented in Fig. 8. The backlashes problems diminish when the hexapod robot executes a continuous gait. This is because during continuous forward-motion of the robot the backlashes are compensated.

Fig. 10 shows a photographic sequence of a specific trajectory followed by the all-terrain hexapod robot in outdoor environment. Several steps were carried out by the robot, which loads a scanning manipulator arm installed in front of the robot. During this test the manipulator was not used. However, the hexapod robot and the scanning manipulator are used to carry out detection and localization tasks of antipersonnel landmines in order to humanitarian demining applications.

Fig. 10 Photographic sequence (left to right and up to down) of a gait pattern of the all-terrain hexapod robot.

6 Conclusions

This work presented some details of the reconfiguration of a climbing robot named REST in an all-terrain hexapod robot. This walking robot must be able to carry out different kind of works in external environment. However, this machine could perform several tasks in indoor environments, too. One example is to perform specific trajectories with a scanning manipulator loaded on its structure in order to perform humanitarian demining tasks.

Since this all-terrain hexapod robot can sustain high payloads (up to 300 kg), it could carry out several kinds of tasks, with a good stability, bearing different tools/sensors installed on it. These characteristics will be very important to carry out future projects for different types of applications. Therefore, the reconfiguration made to the original climbing robot to an all-terrain walking robot has been successful. Future works with this hexapod robot will show important results related with the energy saving, while the robot performs several gaits with different loads.

Acknowledgments The authors acknowledge funding of this work under TIRAMISU Project - Grant Agreement Nº 284747 of the 7FP, and partial funding from the RoboCity2030-III-CM project (Robótica aplicada a la mejora de la calidad de vida de los ciudadanos. Fase III; S2013/MIT-2748), funded by Programas de Actividades I+D en la Comunidad de Madrid and cofunded by Structural Funds of the EU. Dr. Héctor Montes also acknowledges support from Universidad Tecnológica de Panamá.

References

1. Bekey, G.A.: Autonomous Robots: From Biological Inspiration to Implementation and Control. MIT Press, USA (2005)
2. Gonzalez de Santos, P., Garcia, E., Estremera, J.: Quadrupedal locomotion: an introduction to the control of four-legged robots. Springer, Germany (2006)
3. Tenreiro, J.A., Silva, M.: An overview of legged robots. In: Proc. International Symposium on Mathematical Methods in Engineering, Ankara, Turkey (2006)
4. Heverly, M., Matthews, J., Frost, M., Quin, C.: Development of the Tri-ATHLETE lunar vehicle prototype. In: Proceedings of the 40th Aerospace Mechanisms Symposium, NASA, May 12, 2010
5. Saunders, A., Goldman, D.I., Full, R.J., Buehler, M.: The RiSE Climbing Robot: Body and Leg Design. Unmanned Systems Technology VIII, Proc. of SPIE **6230**(17), 1–13 (2006)
6. Raibert, M., Blankespoor, K., Nelson, G., Playter, R.: The BigDog team: BigDog, the rough-terrain quaduped robot. In: Proc. The IFAC Workshop on Navigation, Guidance and Control of Underwater Vehicles, IFAC NGCUV 2008, Killaloe, Ireland, April 8–10, 2008
7. Boston Dynamics: LittleDog Robot 1.0 User Guide. Cambridge, MA, p. 64 (2006)
8. Mandelbaum, R.: Legged Squad Support System, Industry Day. DARPA (2008)
9. Massachusetts Institute of Technology: Run, cheetah, run: New algorithm enables cheetah robot to run and jump, untethered, across grass. ScienceDaily, September 15, 2014. www.sciencedaily.com/releases/2014/09/140915114524.htm
10. Gonzalez de Santos, P., Armada, M., García, E., Akinfiev, T., No, J., Prieto, M., Nabulsi, S., Cobano, J.A., Ponticelli, R., Sarria, J., Reviejo, J., Salinas, C. y Ramos, A.: Desarrollo de robots caminantes y escaladores en el IAI-CSIC, Departamento de Control Automático Instituto de Automática Industrial - CSIC, Congreso Español de Informatica – CEDI (2007)
11. Gonzalez de Santos, P., and Jimenez, M.A.: Generation of Discontinuous Gaits for Quadruped Walking Machines. Journal of Robotic Systems **12**(9), 599–611 (1995)
12. Gonzalez de Santos, P., Armada, M.A., Jimenez, M.A.: Ship building with ROWER. IEEE Robotics and Automation Magazine **7**(4), 35–43 (2000)

13. Grieco, J.C., Prieto, M., Armada, M., Gonzalez de Santos, P.: A six-legged climbing robot for high payloads. In: Proc. 1988 IEEE International Conf. on Control Applications, Trieste, Italy, pp. 446–450 (1998)
14. de Santos, Gonzalez: P., Garcia, E., Estremera, J. and Armada, M.A.: DYLEMA: Using walking robots for landmine detection and location. The International Journal of System Science **36**(9), 545–558 (2005)
15. Montes, H.: Análisis, diseño y evaluación de estrategias de control de fuerza en robots caminantes. Ph.D. Thesis, Universidad Complutense de Madrid (2005)
16. Akinfiev, T., Armada, M., Montes, H.: Vertical movement of resonance hopping robot with electric drive and simple control system. In: Proc. of IEEE, 11th Mediterranean Conference on Control and Automation MED 2003, Rhodes, Greece (2003)
17. Armada, M., Caballero, R., Akinfiev, T., Montes, H., Manzano, C., Pedraza, L., Ros, S., González de Santos, P.: Design of SILO2 humanoid robot. In: Proc. of The Third IARP International Workshop on Humanoid and Human Friendly Robotics, Tsukuba, Ibaraki, Japan, pp. 37–42 (2002)
18. Nabulsi, S., Montes, H., Armada, M.: ROBOCLIMBER: control system architecture. In: Proc. Int. Conf. of Climbing and Walking Robots, CLAWAR 2004, Madrid, Spain, pp. 943–952 (2004)
19. Nabulsi, S., Sarria, J., Montes, H., Armada, M.: High Resolution Indirect Feet-Ground Interactions Measurement for Hydraulic Legged Robots. IEEE Transactions on Instrumentation and Measurement **58**(10), 3396–3404 (2009)
20. Grieco, J.C.: Robots Escaladores. Condiciones de Diseño, estabilidad y Estrategias de Control. Ph.D. Thesis, Universidad de Valladolid (1997)
21. Montes, H., Mena, L., Fernández, R., Sarria, J., González de Santos, P., Armada, M.: Hexapod robot for humanitarian demining. In: Proc. RISE 2015, 8th IARP Workshop on Robotics for Risky Environments [CD], January 28–29, 2015, Naval Academy, Lisbon, Portugal (2015)
22. Montes, H., Mena, L., Fernández, R., Sarria, J., Armada, M.: Inspection platform for applications in humanitarian demining. In: 18th International Conference on Climbing and Walking Robots and the Support Technologies for Mobile Machines. Assistive Robotics, September 6–9, 2015, HangZhou, China, pp. 446–453 (2015)

Review of Control Strategies for Lower Limb Prostheses

César Ferreira, Luis Paulo Reis and Cristina P. Santos

Abstract Each year thousands of people lose their lower limbs, mainly due to three causes: wars, accidents and vascular diseases. The development of prostheses is crucial to improve the quality of millions of people's lives by restoring their mobility. Lower limb prostheses can be divided into three major groups: passive, semi-active or variable damping and powered or intelligent. This contribution provides a literature review of the principal control strategies used in lower limb prostheses, i.e., the controllers used in energetically powered transfemoral and transtibial prostheses. We present a comparison of the presented literature review and the future trends of this important field. It is concluded that the use of bio-inspired concepts and continuous control combined with the other control approaches can be crucial in the improvement of prosthesis controllers, enhancing the quality of amputee's lives.

Keywords Prosthesis · Impedance · Central Pattern Generator (CPG) · Electromyography (EMG) · Finite state machine

1 Introduction

The possibility of moving from one place to another using our locomotive system is synonymous of freedom to humans [6]. However, each year thousands of people

C. Ferreira(✉) · C.P. Santos
Department of Industrial Electronics, University of Minho, Azurém Campus, Guimarães, Portugal
e-mail: cesarferreira1990@gmail.com, cristina@dei.uminho.pt

L.P. Reis
Department of Information Systems, University of Minho, Azurém Campus, Guimarães, Portugal
e-mail: lpreis@dsi.uminho.pt

C. Ferreira · L.P. Reis · C.P. Santos
ALGORITMI Research Center, Guimarães, Portugal

L.P. Reis
LIACC - Artificial Intelligence and Computer Science Laboratory, Porto, Portugal

L.P. Reis et al. (eds.), *Robot 2015: Second Iberian Robotics Conference*,
Advances in Intelligent Systems and Computing 418,
DOI: 10.1007/978-3-319-27149-1_17

lose that freedom, mainly due to three causes: wars, accidents and vascular diseases (especially diabetes) [6, 11]. Currently, there are more than 30 million amputees in the world [2] and the development of prostheses is crucial to improve the quality of millions of people's lives by restoring their mobility.

During the last decades, the prostheses have evolved significantly. The researchers are creating prostheses that function as extensions of the human body (structurally, neurologically and dynamically) [19] and not lifeless mechanisms. Some important technological advances are responsible for the prostheses improvement. We highlight the creation of energy-storing ankle-foot complexes, the integration of microprocessors in the prostheses control, the development and use of more efficient actuator technologies (like series elastic actuators) and the improvement of control frameworks that exploit principles of bio-inspired movements [19, 25]. However, current prostheses still present some shortcomings mainly related with the control strategies and, therefore, the development of lower limb prostheses capable of restoring the quality of life of many amputees is still a problem to be solved.

The main goals of lower limb prostheses are to provide a comfortable walking and to emulate the natural movements of human joints [7, 5]. Mimicking the functions of human limbs, several tasks of daily living can be restored, such as walking, climbing stairs, walking up and down a slope or simply balancing their legs.

Lower limb prostheses can be divided into three major groups: passive, semi-active or variable damping and powered or intelligent [27, 11].

Passive prostheses are mechanically designed to store and release energy throughout each stance period of the step cycle, but they are not capable of providing more power than the one that was stored [16]. Consequently, these prostheses are incapable of restoring some activities of daily living, like walking up stairs or slopes, jumping or running, since they require positive power at the knee and ankle joints [11, 9]. Furthermore, lower limb amputees using passive devices present asymmetric gait patterns, they expend 60% more energy than healthy people and their walking speed is slower than normal [9, 3].

Semi-active or variable damping prostheses have a power source but only to modulate damping levels. These devices have advantages over passive prostheses, since they improve the knee stability and adaptation to different walking speeds [27]; however, they cannot produce positive net power and, therefore, these prostheses cannot replicate the activities of daily living.

The human ankle generates positive net power and performs a peak power over the stance period, mainly at moderate to fast walking speeds. None of the above types of prostheses are capable of mimicking the human knee and/or ankle behavior. Active or intelligent devices can produce the necessary positive net work and consequently restore locomotive functions, including running, jumping and walking up and down stairs [26]. These kind of prostheses are considered intelligent because they are coupled to the human's gait through an electronics-based system capable of recognizing the user's intent and then performing the desired limb movement [38].

The current active prostheses controllers reveal some problems [18, 29] and, for that reason, they should be improved to apply robotic technology in powered prosthetic knees and ankles with greater preponderance. This contribution provides a literature review of the main control strategies used in lower limb prostheses, i.e., the controllers used in energetically powered transfemoral and transtibial prostheses.

The paper is organized as follows. In section 2 we describe the most commonly control strategies applied in powered lower limb prosthetic devices. In section 3 we present a brief comparison of the presented literature review and the future trends of this field. Finally, in section 4, the conclusions are presented.

2 Control Strategies

The control approaches applied in the majority of powered active prosthesis can be divided into four categories: echo control [15], finite state impedance control (FSI) [27, 36], electromyography (EMG) based control [41] and Central Pattern Generator (CPG) based control [37]. At the end of the section we also present some works that do not fit in the four categories aforementioned.

Hereinafter, the more important works related with these control strategies are presented.

2.1 *Echo Control*

The initial developments of active lower limb prostheses occurred in the 70's decade [15]. The authors presented an echo control applied in an active transfemoral prosthesis. This control approach coordinates the prosthesis' joint position based on the movement of the sound leg, i.e, the recorded trajectory of the healthy limb is played back on the prosthesis. This control strategy presents a relevant problem since the control is not executed on real-time, i.e, requires a half-step delay [41].

2.2 *Finite State Impedance Control*

The FSI-based control has been the commonly used strategy in the design of intelligent prosthesis control [26]. In this kind of controllers, the step cycle is divided into different states (states of the finite state machine) and, in each gait state, a different joint impedance is modeled using virtual spring-damper systems.

Henceforth, it is presented the literature review of the more important works made in the past years that aimed to create FSI controllers capable of controlling lower limb transfemoral/transtibial prostheses.

Sup et al. [35] introduced a transfemoral prosthesis with pneumatically powered knee and ankle joints. The control strategy used by the authors was the finite state impedance control. Most of the prior works control the prosthetic devices generating the desired knee and/or ankle joint position trajectory instead of using impedance control. Position-based control brings two relevant problems when it comes to the control of lower limb prostheses: the computed position trajectories are based on measurements of the healthy limb trajectory, such as echo control presented in subsection 2.1 and the position tracking demands high output impedance, forcing the amputee to react to the device instead of interacting with it. The implemented impedance based control generates different joint torques (knee and ankle) to be applied in each finite state of the gait. The impedance model of each joint was calculated using a virtual nonlinear spring and damper. In 2007, this prosthesis enhanced the functionality of transfemoral prostheses when compared to existing passive devices.

Two years following, Lambrecht and Kazerooni [24] presented an innovative semi-active knee prosthesis. This prosthesis combines the safety of a passive damped hydraulic device and the better mobility of an active transfemoral prosthesis, reducing hip hike and torque and occasional stair climbing capability. The control system was based on a finite state machine that applies the necessary level of active power or passive impedance, dependent on the gait's phase of the device.

Sup et al. [36] presented a new prosthesis. Compared to their previous work [35], the authors developed an electrically powered active knee and ankle prosthesis. The control framework is divided into three levels: high, middle and low. The high level controller recognizes the user's intent and transmits this information to the middle level controller. This action is responsible for generating the joints' torque using a finite state machine that defines what should be the necessary impedance in each phase of the step cycle. Finally, the low level controllers are force controllers that ensure that the commanded torques (knee and ankle) are respected. This work showed that this prosthesis was capable of enabling amputees to have similar biomechanics to those of healthy humans and to produce human-scale power. However, the prosthesis would have to be tested in multiple amputee subjects.

Martinez-Villalpando and Herr [27] introduced a powered transfemoral prosthesis with two series-elastic actuators (SEA) in an agonist-antagonist disposition. A finite state controller was implemented and, like the previous works, the controller defined the torques to be applied in the knee joint. The hardware configuration of this prosthesis allowed to obtain a more efficient prosthesis in terms of electrical energy cost (directly compared with [36]) without losing the capability of mimicking the knee biomechanics of humans. This prosthesis was tested for level-ground walking. As future work, the authors proposed to include a musculoskeletal reflex controller to improve the system's robustness to irregular terrains.

Eilenberg et al. [9] presented an adaptive muscle-reflex controller used to provide torque commands to a transtibial prosthesis. The device is actuated through a series-elastic actuator, also used in the work of Martinez-Villalpando and Herr [27]. The control architecture is composed by a finite state machine, a dorsiflexor model, a plantar flexor model and the low level torque control. The finite state machine determines when the transition from one state to another occurs, and then the dorsiflexor and plantar flexor models define the correct torque to be applied in the ankle joint, depending on the phase in which the leg entered (stance or swing). The dorsiflexor and plantar flexor models are two virtual actuators. The former represents biological dorsiflexor muscles and is implemented as a rotary spring-damper, while the plantar flexor model is implemented with a Hill type muscle. The low level torque control ensures that the command torques generated by the dorsiflexor and plantar flexor models are applied in the ankle-foot prosthesis. Results showed that this prosthesis is capable of mimicking the human ankle's behavior in level-ground walking. Furthermore, the neuromuscular controller was able to intrinsically adapt to ground slope.

Recently, Lawson et al. [25] presented the second version of the prosthesis presented in their previous work [36], commonly known as Vanderbilt leg. As in the previous work [36], the controller was based on finite state machines that generate the necessary knee and ankle joints to be applied in the device. In this contribution, the authors presented two different state machines. One state machine for the stair ascent controller and another for the stair descent controller. The controllers were tested by a 23 years old amputee subject. Results showed that during stair ascent the powered prosthesis provided knee and ankle joint kinematics more similar to healthy joint kinematics than passive devices. In the stair descent experiments, only the ankle kinematics of the powered prosthesis are more similar to healthy joint kinematics than passive prostheses.

Liu et al. [26] presented a more robust finite state impedance control for transfemoral prostheses. The authors used Dempster-Shafer theory (DST) to improve the design of the transition rules. The intrinsic controller of the prosthesis is a finite state impedance controller (similar to Sup et al. work [36]). As stated before, in this kind of controllers, a finite state machine is necessary to distinguish the different gait phases to apply the correct torque command. In this contribution, the authors showed that using the DST theory can improve some aspects in the controller: (a) the transition rules ensured reliable transition across subjects; (b) the system presented reduced control errors and (c) initial tuning process simplified and shortened training time. These advantages make the prosthesis more practical for the daily living.

2.3 EMG-Based Control

In surface EMG technique (non invasive technique) electrodes are placed on the skin of the residual limb to record the electrical activity produced by muscles.

Hereinafter, it is presented the literature review of the most important works made in the past years that aimed to develop EMG-based controllers capable of controlling powered lower limb prostheses.

Huang et al. [21] presented a classifier of user's locomotion modes using surface electromyography (EMG) signals and a pattern recognition (PR) system. The system was tested with EMG data collected from two transfemoral amputees and eight healthy subjects. The classifier was capable of recognizing seven task modes: level ground walking, stepping over obstacles, climbing stairs, descending stairs, contralateral turning, ipsilateral turning and standing still.

Wu et al. [41] presented a proportional EMG-based controller for an active transfemoral prosthesis, enabling the user to control the prosthesis with his or her muscle signals. The surface EMG signals are used to calculate the control inputs for the implemented active-reactive controller. The authors presented a virtual agonist-antagonist structure combined with the linear muscle model to calculate the torque to be applied in the knee joint. Thereby, the muscle activation measured by the surface EMG electrodes are directly used (proportional control) in the generation of the necessary torque to be applied in the prosthesis.

Recently, Wang et al. [40] also presented a proportional EMG-based controller for an active ankle foot prosthesis, also used in other work [3]. The implemented controller can be divided into three modules: finite state machine, plantar flexor module and EMG module. The finite state machine is used to always have the prosthesis synchronized with the gait cycle. The powered plantar flexor module is responsible for generating the adequate torque within each phase of the gait cycle. Finally, the EMG controller is used to modulate the gain parameter of the torque generator (plantar flexor module). Like in the previous work [41], the EMG signal also influences directly the joint torque generation. This kind of controller allows the user to control directly the applied torque during the walking stance phase.

2.4 CPG-Based Control

The control strategies aforementioned reveal problems [29], mainly due to the application of control by stages, unlike continuous biological control [10, 37].

Central Pattern Generators (CPGs) are biological neural networks, regulated by simple sensory signals, which are being modeled and extensively studied [37, 22]. They can produce coordinated rhythmic patterns, such as walking, running, swimming and flying [29, 37]. CPGs' studies have suggested novel robot control techniques and their application to different robotic platforms [28], exoskeletons [33, 32] and prostheses [37].

CPG-based control strategies seem to constitute a new and full of potential research avenue for lower limb prosthesis control, mainly due to their properties [37]: intrinsic rhythmic behavior, limit cycle stability, smooth online trajectory modulation by parameters change, low computational cost, easy feedback integration, robustness, provide for coupling/synchronization and entrainment phenomena when coupled to mechanical systems. Considering all these factors, several works explored the use of CPGs applied to energetically powered transfemoral prostheses.

Ryu et al. [34] presented a CPG-based controller for a lower limb prosthesis implemented in a simulated humanoid robot (HOAP-3). In the biped robot, the healthy limb (right leg) was controlled by planned motion trajectories and the prosthesis or robotic limb (left leg) was controlled by the CPG controller. The CPG was implemented by a frequency-adaptive Matsuoka neural oscillator and had two functions: to generate the trajectory of the joints and to coordinate both limbs. The robotic limb or prosthesis was equipped with a spring damper component between the hip and ankle joints to provide compliant behavior. Both controllers were synchronized by the touch signal and joint angles of the healthy limb.

Guo et al. [17] presented a controller that combined CPGs and EMG. The prosthesis controller was divided into two levels and was designed to control both knee and ankle joints. The high level control was responsible for recognizing the human's intent using surface EMG signals through the Support Vector Machine (SVM) method. The low level controller (MCU controller) was built to drive the actuator using CPGs. The CPG network was implemented through non-linear Rayleigh oscillators, one in each joint.

Torrealba et al. [38] presented a biomechatronic knee prosthesis. The authors developed two control strategies: a bio-inspired controller and an adaptive proportional controller. The former was constituted by three modules: (a) Accelerometry-based Events Detection Algorithm (AEDA); (b) Amplitude Controlled Phase Oscillators (ACPOs); (c) Knee Angle Generator (KAG). The AEDA algorithm detects seven events along the gait cycle, based on accelerometers data. The CPG is constituted by one ACPO and a KAG to control the prosthesis knee joint. The function of the ACPO is to track the gait cycle phase of the prosthesis and the KAG is to generate the knee angle reference to be followed by the knee joint. The adaptive proportional controller is active only during extension and consists of a control action proportional to the measured knee angle. The prosthesis was tested in able bodies and amputee subjects. The authors concluded that for these particular prosthesis, the adaptive proportional controller fits better since the device does not have the optimal architecture to deal with the CPG-based control. The authors claimed that the CPG controller seemed to be a very promising approach to control active prostheses.

In another relevant work, Geng et al. [12] presented an active transfemoral prosthesis with a four-bar linkage architecture. The controller was designed to control the knee joint based on biological concepts of CPGs. The CPG was modeled using a nonlinear oscillator. The Hopf oscillator learned pre-defined knee trajectory and replicated it on the prosthesis.

Duvinage et al. [7] introduced a biologically inspired approach divided into three modules: a Brain-Computer Interface (BCI) system, a gait model and an orthosis. The presented orthosis was designed for people suffering from foot drop problems. The BCI system was developed to detect the user's intent and to define some high level commands for the gait model. Later, it was developed using biological concepts of CPGs. The authors developed a PCPG (Programmable Central Pattern Generator) capable of generating periodic patterns after a learning step. Specifically, the PCPG only controlled the swing phase of the gait cycle and, therefore, the orthosis tracked the position pattern similar to a healthy foot generated by the PCPG. The stance phase of the gait cycle was driven in passive mode, i.e., the foot experienced a free motion around an equilibrium point.

Torrealba et al. [37] claimed that we are facing the beginning of a new era of prostheses, the cybernetic prostheses. The authors stated that applying finite state control (non-continuous approach) is wasting part the capability of controllable actuators to modulate the prosthesis response along the gait. Thereby, the authors proposed a continuous control approach based on biological concepts of Central Pattern Generators applied on a prosthetic knee. The CPG was capable of providing a continuous knee angle reference for the control loop. The authors improved their previous work [38] by adding a second ACPO in the adaptive CPG control. This second ACPO allowed to have a smooth and continuous control response during the walking gait, consequently accomplishing the main goal of this work - to have a continuous control of a prosthetic knee. The authors stated that this controller was the first step in the development of biological controllers that can mimic human locomotion functions.

2.5 Other Control Strategies

In this literature review we present other control strategies that cannot be included in the main control strategies presented above.

Au et al. [3] presented a powered transtibial prosthesis named MIT Powered Ankle-Foot Prosthesis. This prosthesis had a SEA actuator and presented the capability of improving the amputee metabolic economy (from 7% to 20%) when compared to a passive-elastic prosthesis. A finite-sate controller was designed to implement two different control strategies, depending on the prosthesis step phase: stance or swing. If the prosthesis was in stance phase, the controller generates torque commands, but if the prosthesis was in swing phase, the controller generates position commands.

Holgate et al. [20] presented a new control algorithm based on phase plane invariants. The authors defended the same idea of Torrealba et al. [37], since they claimed that the prosthesis should be controlled by a continuous control, preventing the device controller from being misguided in some situations. Thereby, the researchers proposed a tibia-based controller that had a mathematical relationship between the tibia and ankle angles. In this controller, the authors used the tibia angular position to obtain the desired ankle position that would be applied in a prosthesis. The controller was tested in an active robotic prosthesis named Sparky [4]. This control algorithm has the advantage of being continuous and never getting stuck on state of operation and it cannot be misguided by an erratic transition.

Parsan and Tosunoglu [30] introduced a control algorithm to drive the ankle foot of the prosthesis. In this controller, the authors used the knee angle to calculate the desired ankle joint to be applied in the transtibial prosthesis, i.e., the functions governing the ankle angle are directly dependent on the knee angle. The researchers constructed a Matlab Simulink model to test the controller. The authors concluded that this control algorithm had potential to be applied in active prostheses, but still needed to be improved.

Recently, Gregg and Sensinger [14] presented a novel control strategy for powered transtibial prostheses. The author used a planar 6-link biped model to test the developed controller. The researchers defined the Center of Pressure (CoP) as the point on the foot sole where the resultant ground reaction force is transmitted. The CoP is independent over walking speeds, heel heights and body weights and therefore, it presents an invariant circular rocker shape. Thereby, the authors used this invariant shape to control the knee and ankle angles of the prosthesis, i.e., the knee and ankle trajectories are calculated as a function of the CoP. This control strategy showed an important feature: tuning few parameters (when compared to the works of Sup et al. [36]; Eilenberg et al. [9] and Holgate et al. [20]). This feature can possibly enhance the clinical viability of powered devices.

3 Comparative Analysis and Future Trends in Lower Limb Prostheses Control

All the control strategies presented above have advantages and disadvantages. They have potential to increase the prosthesis performance, especially when compared with passive prosthesis.

The echo control approach is the least used controller in active prosthesis because of the intrinsic problem related with the necessity to require a half step delay [41], that makes impossible real time control.

Generally, the finite state impedance control is the most used approach because it's close to the type of actuation of humans, i.e., the joints are actuated in form of impedance [36, 9, 27]. However, the use of finite state machines is a bottleneck of this kind of controllers, since the system is not using all the capability of current controllable actuators [37]. Other problem related with this controllers is the necessity of tuning a lot of parameters for each subject [1], which makes the prostheses less viable.

The EMG-based control allows the incorporation of input from the user's nervous system in the controller, enabling a feedforward intentional control of the prosthesis. Consequently, a more natural and physiological control can be designed [2]. This control approach also has some disadvantages, like motion artifacts, fatigue, misalignment of the electrode [39].

The designing of continuous control applying biological concepts of CPGs [37] can lead to cybernetic prostheses. The continuous control is more similar to human control. However, in this continuous control, the prosthesis was position controlled, which departs from the actuation type of humans.

A set of recent ideas may bring new insights into the development of new and more effective control strategies for the locomotion problem. First, there is a growing consensus that both intrinsic and sensory feedback signals play a crucial role [23, 8, 13]. Second, multiple inputs to spinal CPG accommodate its locomotor output to different mechanical and task demands. In particular, separate speed signals for each limb compose the velocity control signal that drives the CPG for asymmetric gait control. Third, there is evidence of independence of neural control of intra- versus interlimb parameters during walking [31]. Finally, the CPG could model low dimensional representations (motor primitives), thus reducing the control problem to the modulation of their amplitude, duration and timing.

4 Conclusion

This literature review has established the state of the art of control strategies applied in powered lower limb prostheses. The control approaches vary depending on the available prosthetic device, on the structure of the implemented control strategy and on the instrumentation of the device necessary to detect the gait events or the user's intent.

Actually, the main bottleneck of lower limb prostheses is the controller, since the mechanical systems have been very refined. It is not yet defined what is the best type of control that should be used to replicate more closely the biomechanics of human lower limbs. In fact, prostheses need to be improved to enable people to perform all the activities of daily living, restoring their mobility and dignity. It is expected that the powered devices will be able to have such a performance, but, at the same time, they will reduce the metabolic energy expended during walking, they will restore symmetric gaits and they will decrease problems such as osteoarthritis in the joints of the intact limb.

The development of cybernetic prostheses seems to be a way forward to obtain prostheses capable of mimicking the human locomotor system. However, the use of bio-inspired concepts and continuous control combined with the other control approaches can be crucial in the development of prosthesis controllers. The combination of the advantages of several control strategies will enable the improvement of the prosthesis performance, enhancing the quality of millions of people's lives.

Acknowledgments This work has been supported by FCT - *Fundação para a Ciência e Tecnologia* in the scope of the project: PEst-UID/CEC/00319/2013 and the Portuguese Science Foundation (grant SFRH/BD/102659/2014).

References

1. Aghasadeghi, N., Zhao, H., Hargrove, L.J., Ames, A.D., Perreault, E.J., Bretl, T.: Learning impedance controller parameters for lower-limb prostheses. In: 2013 IEEE/RSJ international conference on Intelligent robots and systems (IROS), pp. 4268–4274. IEEE (2013)
2. Alcaide-Aguirre, R.E., Morgenroth, D.C., Ferris, D.P.: Motor control and learning with lower-limb myoelectric control in amputees. J. Rehabil. Res. Dev. **50**(5), 687–698 (2013)
3. Au, S.K., Herr, H., Weber, J., Martinez-Villalpando, E.C.: Powered ankle-foot prosthesis for the improvement of amputee ambulation. In: 29th Annual International Conference of the IEEE Engineering in Medicine and Biology Society, EMBS 2007, pp. 3020–3026. IEEE (2007)
4. Bellman, R.D., Holgate, M., Sugar, T.G., et al.: Sparky 3: design of an active robotic ankle prosthesis with two actuated degrees of freedom using regenerative kinetics. In: 2nd IEEE RAS & EMBS International Conference on Biomedical Robotics and Biomechatronics, BioRob 2008, pp. 511–516. IEEE (2008)
5. Borjian, R., Lim, J., Khamesee, M.B., Melek, W.: The design of an intelligent mechanical active prosthetic knee. In: 34th Annual Conference of IEEE Industrial Electronics, IECON 2008, pp. 3016–3021. IEEE (2008)
6. Chen, M.Y., Lin, Y., Xiong, H., Torrealba, R.R., Fernández-López, G., Grieco, J.C.: Towards the development of knee prostheses: review of current researches. Kybernetes **37**(9/10), 1561–1576 (2008)

7. Duvinage, M., Castermans, T., Hoellinger, T., Reumaux, J.: Human walk modeled by PCPG to control a lower limb neuroprosthesis by high-level commands. In: Proceedings of the 2nd International Multi-Conference on Complexity, Informatics and Cybernetics (IMCIC2011). Citeseer (2011)
8. Dzeladini, F., Van Den Kieboom, J., Ijspeert, A.: The contribution of a central pattern generator in a reflex-based neuromuscular model. Frontiers in Human Neuroscience **8** (2014)
9. Eilenberg, M.F., Geyer, H., Herr, H.: Control of a powered ankle-foot prosthesis based on a neuromuscular model. IEEE Transactions on Neural Systems and Rehabilitation Engineering **18**(2), 164–173 (2010)
10. El-Sayed, A.M., Hamzaid, N.A., Abu Osman, N.A.: Technology efficacy in active prosthetic knees for transfemoral amputees: A quantitative evaluation. The Scientific World Journal 2014 (2014)
11. Geng, Y., Xu, X., Chen, L., Yang, P.: Design and analysis of active transfemoral prosthesis. In: Proceedings of the IECON, pp. 1495–1499 (2010)
12. Geng, Y., Yang, P., Xu, X., Chen, L.: Design and simulation of active transfemoral prosthesis. In: 2012 24th Chinese Control and Decision Conference (CCDC), pp. 3724–3728. IEEE (2012)
13. Gossard, J.P., Dubuc, R., Kolta, A.: A hierarchical perspective on rhythm generation for locomotor control. Breathe, Walk and Chew; The Neural Challenge: Part II, p. 151 (2011)
14. Gregg, R.D., Sensinger, J.W.: Biomimetic virtual constraint control of a transfemoral powered prosthetic leg. In: American Control Conference (ACC 2013), pp. 5702–5708. IEEE (2013)
15. Grimes, D., Flowers, W., Donath, M.: Feasibility of an active control scheme for above knee prostheses. Journal of Biomechanical Engineering **99**(4), 215–221 (1977)
16. Grosu, S., Cherelle, P., Verheul, C., Vanderborght, B., Lefeber, D.: Case study on human walking during wearing a powered prosthetic device: Effectiveness of the system human-robot. Advances in Mechanical Engineering **6**, 365265 (2014)
17. Guo, X., Chen, L., Zhang, Y., Yang, P., Zhang, L.: A study on control mechanism of above knee robotic prosthesis based on cpg model. In: 2010 IEEE International Conference on Robotics and Biomimetics (ROBIO), pp. 283–287. IEEE (2010)
18. Hargrove, L.J., Simon, A.M., Young, A.J., Lipschutz, R.D., Finucane, S.B., Smith, D.G., Kuiken, T.A.: Robotic leg control with emg decoding in an amputee with nerve transfers. New England Journal of Medicine **369**(13), 1237–1242 (2013)
19. Herr, H.M., Kornbluh, R.D.: New horizons for orthotic and prosthetic technology: artificial muscle for ambulation. In: Smart structures and materials. pp. 1–9. International Society for Optics and Photonics (2004)
20. Holgate, M., Sugar, T.G., Böhler, A.W., et al.: A novel control algorithm for wearable robotics using phase plane invariants. In: IEEE International Conference on Robotics and Automation, ICRA 2009, pp. 3845–3850. IEEE (2009)
21. Huang, H., Kuiken, T., Lipschutz, R.D., et al.: A strategy for identifying locomotion modes using surface electromyography. IEEE Transactions on Biomedical Engineering **56**(1), 65–73 (2009)
22. Ijspeert, A.J.: Central pattern generators for locomotion control in animals and robots: a review. Neural Networks **21**(4), 642–653 (2008)
23. Kuo, A.D.: The relative roles of feedforward and feedback in the control of rhythmic movements. Motor Control Champaign **6**(2), 129–145 (2002)
24. Lambrecht, B.G., Kazerooni, H.: Design of a semi-active knee prosthesis. In: IEEE International Conference on Robotics and Automation, ICRA 2009, pp. 639–645. IEEE (2009)
25. Lawson, B.E., Varol, H.A., Huff, A., Erdemir, E., Goldfarb, M.: Control of stair ascent and descent with a powered transfemoral prosthesis. IEEE Transactions on Neural Systems and Rehabilitation Engineering **21**(3), 466–473 (2013)
26. Liu, M., Zhang, F., Datseris, P., Huang, H.H.: Improving finite state impedance control of active-transfemoral prosthesis using dempster-shafer based state transition rules. Journal of Intelligent & Robotic Systems **76**(3–4), 461–474 (2014)
27. Martinez-Villalpando, E.C., Herr, H.: Agonist-antagonist active knee prosthesis: A preliminary study in level-ground walking. J. Rehabil. Res. Dev. **46**(3), 361–374 (2009)

28. Matos, V., Santos, C.P.: Towards goal-directed biped locomotion: Combining cpgs and motion primitives. Robotics and Autonomous Systems **62**(12), 1669–1690 (2014)
29. Nandi, G.C., Ijspeert, A., Nandi, A.: Biologically inspired CPG based above knee active prosthesis. In: IEEE/RSJ International Conference on Intelligent Robots and Systems, IROS 2008, pp. 2368–2373. IEEE (2008)
30. Parsan, A., Tosunoglu, S.: A novel control algorithm for ankle-foot prosthesis. In: Florida conference on recent advances in robotics, Boca Raton (2012)
31. Reisman, D.S., Wityk, R., Silver, K., Bastian, A.J.: Split-belt treadmill adaptation transfers to overground walking in persons poststroke. Neurorehabilitation and neural repair **23**(7), 735–744 (2009)
32. Ronsse, R., Vitiello, N., Lenzi, T., van den Kieboom, J., Carrozza, M.C., Ijspeert, A.J.: Human-robot synchrony: flexible assistance using adaptive oscillators. IEEE Transactions on Biomedical Engineering **58**(4), 1001–1012 (2011)
33. Ronsse, R., Vitiello, N., Lenzi, T., Van Den Kieboom, J., Carrozza, M.C., Ijspeert, A.J.: Adaptive oscillators with human-in-the-loop: Proof of concept for assistance and rehabilitation. In: 2010 3rd IEEE RAS and EMBS International Conference on Biomedical Robotics and Biomechatronics (BioRob), pp. 668–674. IEEE (2010)
34. Ryu, J.K., Chong, N.Y., You, B.J., Christensen, H.: Adaptive cpg based coordinated control of healthy and robotic lower limb movements. In: The 18th IEEE International Symposium on Robot and Human Interactive Communication, RO-MAN 2009, pp. 122–127. IEEE (2009)
35. Sup, F., Bohara, A., Goldfarb, M.: Design and control of a powered knee and ankle prosthesis. In: 2007 IEEE International Conference on Robotics and Automation, pp. 4134–4139. IEEE (2007)
36. Sup, F., Varol, H.A., Mitchell, J., Withrow, T.J., Goldfarb, M.: Preliminary evaluations of a self-contained anthropomorphic transfemoral prosthesis. IEEE/ASME Transactions on Mechatronics **14**(6), 667–676 (2009)
37. Torrealba, R.R., Cappelletto, J., Fermín, L., Fernández-López, G., Grieco, J.C.: Cybernetic knee prosthesis: application of an adaptive central pattern generator. Kybernetes **41**(1/2), 192–205 (2012)
38. Torrealba, R.R., Pérez-D'Arpino, C., Cappelletto, J., Fermín-Leon, L., Fernández-López, G., Grieco, J.C.: Through the development of a biomechatronic knee prosthesis for transfemoral amputees: mechanical design and manufacture, human gait characterization, intelligent control strategies and tests. In: 2010 IEEE International Conference on Robotics and Automation (ICRA), pp. 2934–2939. IEEE (2010)
39. Tucker, M.R., Olivier, J., Pagel, A., Bleuler, H., Bouri, M., Lambercy, O., del R. Millán, J., Riener, R., Vallery, H., Gassert, R.: Control strategies for active lower extremity prosthetics and orthotics: a review. Journal of Neuroengineering and Rehabilitation **12**(1), 1 (2015)
40. Wang, J., Kannape, O., Herr, H.M., et al.: Proportional emg control of ankle plantar flexion in a powered transtibial prosthesis. In: 2013 IEEE International Conference on Rehabilitation Robotics (ICORR), pp. 1–5. IEEE (2013)
41. Wu, S.K., Waycaster, G., Shen, X.: Electromyography-based control of active above-knee prostheses. Control Engineering Practice **19**(8), 875–882 (2011)

Part IV
Machine Learning in Robotics

Visual Inspection of Vessels by Means of a Micro-Aerial Vehicle: An Artificial Neural Network Approach for Corrosion Detection

Alberto Ortiz, Francisco Bonnin-Pascual, Emilio Garcia-Fidalgo and Joan P. Company

Abstract Periodic visual inspection of the different surfaces of a vessel hull is typically performed by trained surveyors at great cost, both in time and in economical terms. Assisting them during the inspection process by means of mechanisms capable of automatic or semi-automatic defect detection would certainly decrease the inspection cost. This paper describes a defect detection approach comprising: (1) a Micro-Aerial Vehicle (MAV) which is used to collect images from the surfaces under inspection, particularly focusing on remote areas where the surveyor has no visual access; and (2) a coating breakdown/corrosion detector based on a 3-layer feed-forward artificial neural network. The success of the classification process depends not only on the defect detector but also on a number of assistance functions that are provided by the control architecture of the aerial platform, whose aim is to improve picture quality. Both aspects are described along the different sections of the paper, as well as the classification performance attained.

Keywords Corrosion detection · MAV · Artificial neural network

1 Introduction

The movement of goods by ships is today one of the most time and cost effective methods of transportation. The safety of these vessels is overseen by Classification Societies, who are continually seeking to improve standards and reduce the risk of

A. Ortiz(✉) · F. Bonnin-Pascual · E. Garcia-Fidalgo · J.P. Company
Department of Mathematics and Computer Science, University of Balearic Islands, Palma, Spain
e-mail: {alberto.ortiz,xisco.bonnin,emilio.garcia,joanpep.company}@uib.es

A. Ortiz—This work is partially supported by the European Social Fund through grant FPI11-43123621R (Conselleria d'Educacio, Cultura i Universitats, Govern de les Illes Balears) and by the EU FP7 project INCASS (GA 605200). This publication reflects only the authors' views and the European Union is not liable for any use that may be made of the information contained therein. The authors also thank Fabricio Ardizon for his involvement in project INCASS.

L.P. Reis et al. (eds.), *Robot 2015: Second Iberian Robotics Conference*,
Advances in Intelligent Systems and Computing 418,
DOI: 10.1007/978-3-319-27149-1_18

maritime accidents. Structural failures are a major cause of such accidents, which can be usually prevented through timely maintenance. As such, vessels undergo annual inspections, with intensive Special and Docking Surveys every five years, what ensures that the hull structure and related piping are all in satisfactory condition and are fit for the intended use over the next five years.

An important part of vessels maintenance has to do with the visual inspection of the hull, where the surveyor is expected to be within arm's reach to the structure under inspection. As it is well known, the hull can be affected by different kinds of defective situations typical of steel surfaces, such as coating breakdown, corrosion and, ultimately, cracks. These defects are indicators of the state of the metallic surface and, as such, an early detection prevents the structure from buckling and / or fracturing.

To carry out this task, the vessel has to be emptied and situated in a dockyard where scaffolding and / or cherry-pickers must be used to allow the human inspectors to reach the area under inspection. For some vessels (e.g. Ultra Large Crude Carriers, ULCC), this process can mean the visual assessment of more than 600,000 m^2 of steel. Besides, the surveys are on many occasions performed in hazardous environments for which the operational conditions turn out to be sometimes extreme for human operation. Moreover, total expenses involved by the infrastructure needed for close-up inspection of the hull can reach up to $1M once you factor in the vessel's preparation, use of yard's facilities, cleaning, ventilation, and provision of access arrangements. In addition, vessel owners experience significant lost opportunity costs while the ship is inoperable. It is therefore clear that any level of automation of the inspection process is to lead to a reduction of the inspection time, a reduction of the financial costs involved and/or an increase in the safety of the operation.

One of the main goals of the already concluded EU FP7 project MINOAS was to develop a fleet of robotic platforms with different locomotion capabilities with the aim of teleporting the human surveyor to the different vessel structures to be inspected. Given the enormity of these structures and the requirement for vertical motion as part of the inspection process, a multirotor platform, due to their small size, agility and fast deployment time, was selected as one of the members of the robot fleet, and later adapted to the inspection application [11]. In accordance to some constructive advice from end-users at the end of project MINOAS, a re-design of this platform has been undertaken within the EU FP7 follow-up project INCASS, which will be described in this paper.

As an additional contribution in this line, this paper presents a novel solution for detecting coating breakdown / corrosion in images as a support for surveyors during vessel inspection. The solution here described adopts an approach based on a 3-layer feed-forward neural network which detects pixels suspected to correspond to defective areas. To improve the success rate, the aforementioned aerial platform has been fitted with a number of autonomous functionalities for enhanced image capture. This is achieved by means of extensive use of behaviour-based high-level control.

The paper is organized as follows: Section 2 reviews previous work on the subject, Section 3 describes the aerial platform that is used for image collection, Section 4 describes the corrosion detection algorithm, Section 5 reports on the result of a number of experiments using vessel surface images, and, finally, Section 6 concludes the paper.

2 Previous Work

Referring to automated visual defect detection, the scientific literature contains an important number of proposals. Among other possibilities, these can be roughly classified in two categories, depending on whether they look for defects specific of particular objects or surfaces —e.g. LCD displays [7], printed circuit boards [18], copper strips [28], ceramic tiles [6], etc.— or, to the contrary, they aim at detecting general and unspecific defects –e.g. [1, 4, 8, 15, 19].

Within the first category, one can find a large collection of contributions for automatic vision-based crack detection, mainly for concrete surfaces —e.g. [12, 21, 26]. However, regarding corrosion, apart from some contributions of part of the authors of this paper [3, 5], to the best of our knowledge, the number of works which can be found is rather reduced [16, 17, 23, 25, 27]. First of all, [16] makes use of color wavelet-based texture analysis algorithms for detecting corrosion, while [17] utilizes the watershed transform applied over the gradient of gray-level images, [23] uses wavelets for characterizing and detect corrosion texture in airplanes, [25] adopts an approach based on the fractal properties of corroded surfaces and [27] also focuses on corrosion texture using the standard deviation and the entropy as the discriminating features. In our previous works, we have used texture and colour descriptors through a cascade of weak classifiers [3] and Law's energy filters and Adaboost [5], among others.

3 The Aerial Platform

3.1 General Overview

In line with the robotic platform developed for the MINOAS project, the new micro-aerial vehicle is based on a multi-rotor configuration. The control software has been configured to be hosted by any of the research platforms developed by Ascending Technologies (the quadcopters Hummingbird and Pelican, and the hexacopter Firefly), although it could be adapted to other systems. The AscTec vehicles are equipped with an inertial measuring unit (IMU) that comprises a 3-axis gyroscope, a 3-axis accelerometer and a 3-axis magnetometer. Attitude stabilization control loops linked to the onboard IMU and thrust control run over the main ARM7 microcontroller as part of the platform firmware. The manufacturer leaves almost free an additional secondary ARM7 microcontroller which can execute onboard higher-level control loops.

Fig. 1 A Hummingbird platform featuring the visual inspection sensor suite: (green) front-looking and bottom-looking optical flow sensors, (red) ultrasound sensor, (orange) height sensor, (yellow) embedded PC, (blue) camera.

Figure 1 shows a Hummingbird platform configured for the visual inspection application, whose sensor suite comprises:

- Two optical flow sensors, one looking to the ground and the other pointing forward, to estimate the vehicle speed with regard to, respectively, the ground and the front wall (i.e. the surface to be inspected). To this end, we make use of the PX4Flow sensor developed within the PX4 Autopilot project [14]. This sensor comprises a CMOS high-sensitivity imaging sensor, an ARM Cortex M4 microcontroller to compute the optical flow at 250 Hz, a 3-axis gyroscope for angular rate compensation, and a MaxBotix ultrasonic (US) sensor HRLV-EZ4, with 1 mm resolution, used to scale the optical flow to metric velocity values. The distance measured by the US sensor is an additional output supplied by this device.
- Range sensors pointing in different directions for obstacle detection and collision prevention. Currently, we use two Maxbotix HRLV-EZ4 sensors oriented to the left and to the right. This kind of sensor provides information at 10 Hz and its detection range is up to 5 m.
- An additional range sensor which provides height estimation when flying above the 5 m covered by the PX4Flow oriented downwards. To this end, we use the Teraranger One [22], which consists in an infrared time-of-flight measurement sensor that provides an extended range of 14 m.
- A set of cameras that collect the requested images from the vessel structures under inspection. The specific configuration depends on the inspection to perform and the payload capacity of the platform. For instance, the Hummingbird shown in Fig. 1 features a minimalistic configuration comprising a single forward-looking uEye UI-1221LE camera.

The vehicle carries an additional processing board which avoids sending sensor data to a base station, but process them onboard and, thus, prevent communications latency inside critical control loops. The configuration shown in Fig. 1 comprises a Commell LP-172 Pico-ITX board featuring an Intel Atom 1.86 GHz processor and 4 GB RAM.

3.2 Control Software

The aerial platform implements a control architecture that follows the *supervised autonomy* (SA) paradigm [9]. This is a human-robot framework where the robot implements a number of autonomous functions, including self-preservation and other safety-related issues, which make simpler the intended operations for the user, so that he/she, which is allowed to be within the general platform control loop, can focus in accomplishing the task at hand. Within this framework, the communication between the robot and the user is performed via qualitative instructions and explanations: the user prescribes high-level instructions to the platform while this provides instructive feedback. In our case, we use a joystick to introduce the qualitative commands and a *graphical user interface* (GUI) to receive the robot feedback. Joystick commands and the GUI are handled at a *base station* (BS) linked with the MAV via a WiFi connection.

In more detail, the control software has been organized around a layered structure distributed among the available computational resources. On the one hand, as said above, the *low-level control* layer implementing attitude stabilization and direct motor control executes over the main microcontroller as the platform firmware provided by the manufacturer [13]. On the other hand, *mid-level control*, running over the secondary microcontroller, comprises height and velocity controllers which map input speed commands into roll, pitch, yaw and thrust orders. Lastly, the *high-level control* layer, which executes over the embedded PC, implements a reactive control strategy coded as a series of ROS nodes running over Linux Ubuntu, which combine the user desired speed command with the available sensor data —x, y, z velocities, height z and distances to the closest obstacles—, to obtain a final and safe speed set-point that is sent to the speed controllers.

Speed commands are generated through a set of robot behaviors organized in a hybrid competitive-cooperative framework [2]. That is to say, on the one hand, higher priority behaviors can overwrite the output of lower priority behaviors by means of a suppression mechanism taken from the *subsumption* architectural model. On the other hand, the cooperation between behaviors with the same priority level is performed through a *motor schema*, where all the involved behaviors supply each a motion vector and the final output is their weighted summation. An additional flow control mechanism selects, according to a specific input, between the output provided by two or more behaviours.

Figure 2 details the behavior-based architecture, grouping the different behaviors depending on its purpose. A total of four general categories have been identified for the particular case of visual inspection: (a) *behaviors to accomplish the user intention*, which propagate the user desired speed command, attenuating it towards zero in the presence of close obstacles, or keeps hovering until the WiFi link is restored after an interruption; (b) *behaviors to ensure the platform safety within the environment*, which prevent the robot from colliding or getting off the safe area of operation, i.e. flying too high or too far from the reference surface that is involved in optical flow measurements; (c) *behaviors to increase the autonomy level*, which

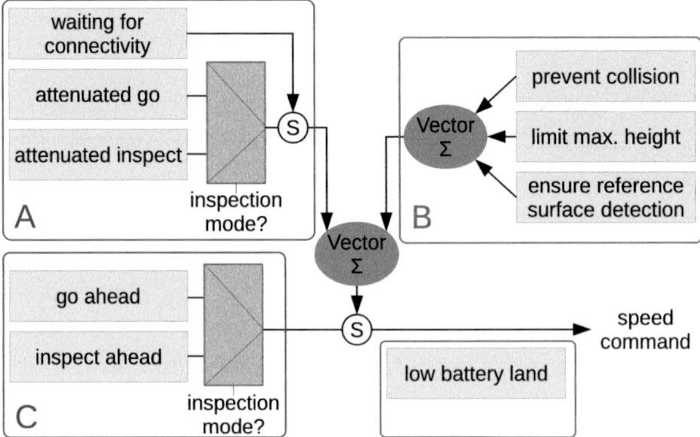

Fig. 2 MAV behaviors: A – behaviors to accomplish the user intention, B – behaviors to ensure the platform safety within the environment, C – behaviors to increase the autonomy level, and D – behaviors to check flight viability.

provide higher levels of autonomy to both simplify the vehicle operation and to introduce further assistance during inspections; and (d) *behaviors to check flight viability*, which checks whether the flight can start or progress at a certain moment in time. Some of the behaviors in groups (a) and (c) can operate in the so-called *inspection mode*. While in this mode, the vehicle moves at a constant and reduced speed (if it is not hovering) and user commands for longitudinal displacements or turning around the vertical axis are ignored. In this way, during an inspection, the platform keeps at a constant distance and orientation with regard to the front wall, for improved image capture.

4 Artificial Neural Network for Corrosion Detection

This section describes a coating breakdown/corrosion (CBC) detector based on a *multi-layer perceptron* configured as a *feed-forward neural network* (FFNN), which discriminates between the CBC and the NC (non-corrosion) classes.

4.1 Background

An *artificial neural network* (ANN) is a computational paradigm that consists of a number of units (neurons) which are connected by weighted links (see Fig. 3). This kind of computational structure learns from experience (rather than being explicitly programmed) and is inspired from the structure of biological neural networks and their way of encoding and solving problems. An FFNN is a class of ANN which organizes neurons in several layers, namely one input layer, one or more hidden layers, and one output layer, in such a way that connections exist from one layer

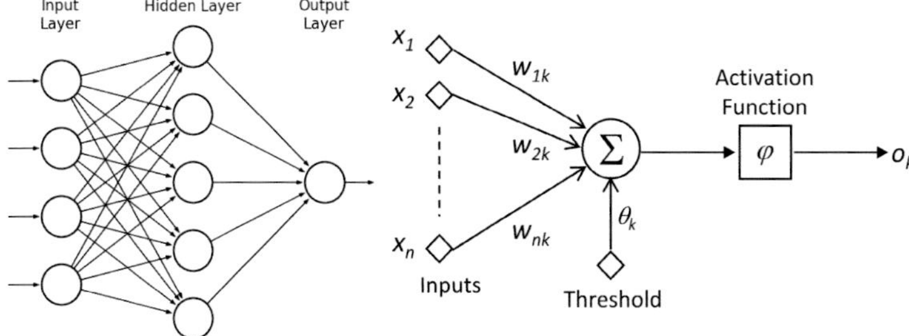

Fig. 3 (left) Topology of a simple FFNN comprising one single hidden layer. (right) Structure of an artificial neuron.

to the next, never backwards [24], i.e. recurrent connections between neurons are not allowed. Arbitrary input vectors propagate forward through the network, finally causing an activation vector in the output layer. The entire network function, which maps input vectors onto output vectors, is determined by the connection weights of the net w_{ij}.

Every neuron k in the network is a simple processing unit that computes its activation output o_k with respect to its incoming excitation $x = \{x_i \mid i = 1, \ldots, n\}$, in accordance to $o_k = \varphi\left(\sum_{i=1}^{n} w_{ik} x_i + \theta_k\right)$, where φ is the so-called activation function, which, among others, can take the form of e.g. the hyperbolic tangent $\varphi(z) = 2/(1 + e^{-az}) - 1$. Training consists in tuning weights w_{ik} by optimizing the summed square error function $E = 0.5 \sum_{p=1}^{N} \sum_{j=1}^{r} (o_j^p - t_j^p)^2$, where N is the number of training input patterns, r is the number of neurons at the output layer and (o_j^p, t_j^p) are the current and expected outputs of the j-th output neuron for the p-th training pattern. Taking as a basis the *back-propagation algorithm*, a number of alternative training approaches have been proposed through the years, such as the *delta-bar-delta rule*, *QuickpPop*, *Rprop*, etc [10].

4.2 Network Configuration

In order to determine an optimal setup for the classifier, we have considered a number of plausible combinations and performed tests accordingly in order to adopt the best configuration. First of all, we have defined both color and texture descriptors to characterize the neighbourhood of each pixel. Next, we have also considered different structures for the NN varying the number of hidden neurons. In detail:

– As for color, we have checked the *hue* and *saturation* channels of the HSV color space, as well as the r and g channels of normalized RGB. Furthermore, two kinds of descriptors have been considered for both cases: (1) the average value of each channel within a neighbourhood, and (2) stacked 8-bin histograms for downsampled intensity values for each channel in the same neighbourhood. In the former

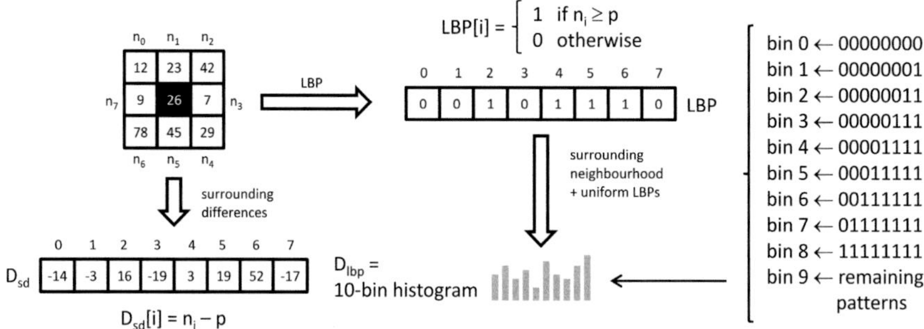

Fig. 4 Illustration of the *surrounding differences* (D_{sd}) and *uniform LBP* (D_{lbp}) texture descriptors (in the latter case, to achieve rotational invariance, every bin accounts not only for the indicated pattern but also for the corresponding rotations).

case, the descriptor comprises two components, while the latter configuration results in a total of 8 + 8 = 16 components.

- Regarding texture, center-surround differences have been considered in the form of: (1) signed *surrounding differences* (SD) between a central pixel and its neighbourhood, and (2) 10-bin histograms of *uniform local binary patterns* (LBP) [20]. See Fig. 4 for an illustration of both descriptors for a distance of one pixel between the neighbourhood $\{n_i\}$ and the central pixel p. During testing, the 2-pixel distance case has also been taken into account.
- Finally, the number of hidden neurons have been varied from 0.5 to 2.5 times the number of components of the input pattern (in 0.1 increments).

All in all, a total of 2 (texture descriptors) $\times$ 2 (1-pixel & 2-pixel distances) $\times$ 2 (color descriptors) $\times$ 2 (average/histogram) $\times$ 21 (hidden neuron quantities) $=$ 336 combinations have been considered. Since every configuration was trained with 5 different initializations (to avoid dependence in this regard), the total number of tests finally performed was 1680. All layers make use of the hyperbolic tangent as activation function.

5 Experimental Results

Figure 5 represents, in TPR-FPR space, the performance of the full set of 336 configurations, where TPR is the *true positive rate* [TPR = TP/(TP+FN)] and FPR is the *false positive rate* [FPR = FP/(TN+FP)], where a *positive* represents membership to the CBC class. In the TPR-FPR space, the perfect classifier lies at the (0,1) point.

Among all classifiers, those whose performance lie closer to the (0,1) point are clearly preferrable to those ones that are farther. Table 1 shows, in this regard, the average and standard deviations of those distances for every combination of descriptors, measured over a total of $21 \times 5 = 105$ trainings, varying the number of hidden

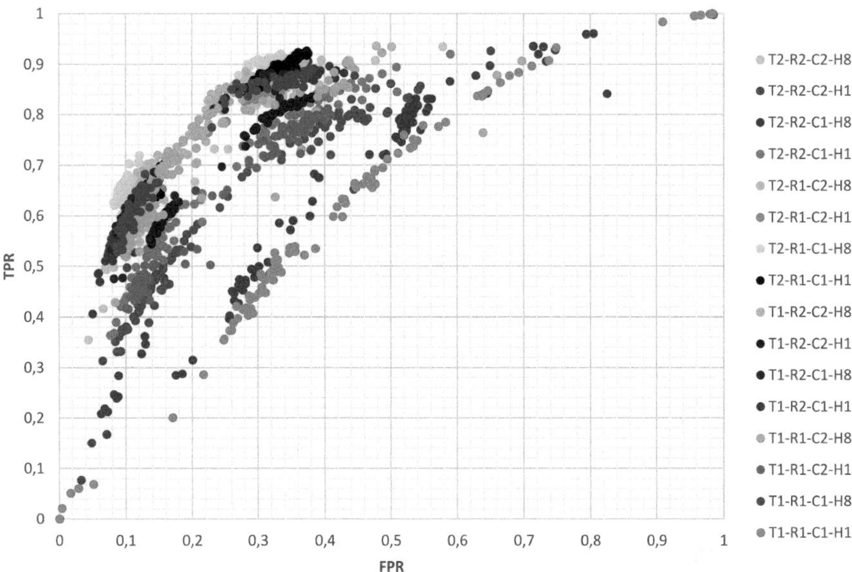

Fig. 5 FPR *versus* TPR for all combinations of descriptors. [T1 = SD, T2 = LBP, Ri = i-pixel distance, C1 = (h,s) color channels, C2 = (r,g) color channels, H1 = average, H8 = 8-bin histogram.]

Table 1 Average and standard deviation of distances to point (0,1).

Combination	C1 / H1	C1 / H8	C2 / H1	C2 / H8
T1 / R1	0.73 (0.16)	0.40 (0.09)	0.47 (0.06)	0.39 (0.07)
T1 / R2	0.64 (0.12)	0.36 (0.05)	0.41 (0.03)	0.35 (0.06)
T2 / R1	0.35 (0.02)	**0.33 (0.02)**	0.38 (0.04)	0.34 (0.02)
T2 / R2	0.47 (0.07)	0.43 (0.07)	0.51 (0.07)	0.42 (0.08)

Table 2 Performance details for the best combination of descriptors.

T2/R1/C1/H8		truth			
		CBC	NC	CBC	NC
classification output	CBC	656,680	3,661,374	644,853	1,769,771
	NC	91,092	9,841,962	110,165	11,726,319
success rate		0.74		0.87	
TPR		0.88		0.85	
FPR		0.27		0.13	
distance to (0,1)		0.30		0.20	

neurons as specified above. As can be observed, the best combination results to be LBP histograms at 1-pixel distance for texture and 8-bin histograms of (h,s) channels for color (case T2/R1/C1/H8), which exhibits the shortest average distance to (0,1) as well as the smallest standard deviation, what indicates little variation in global

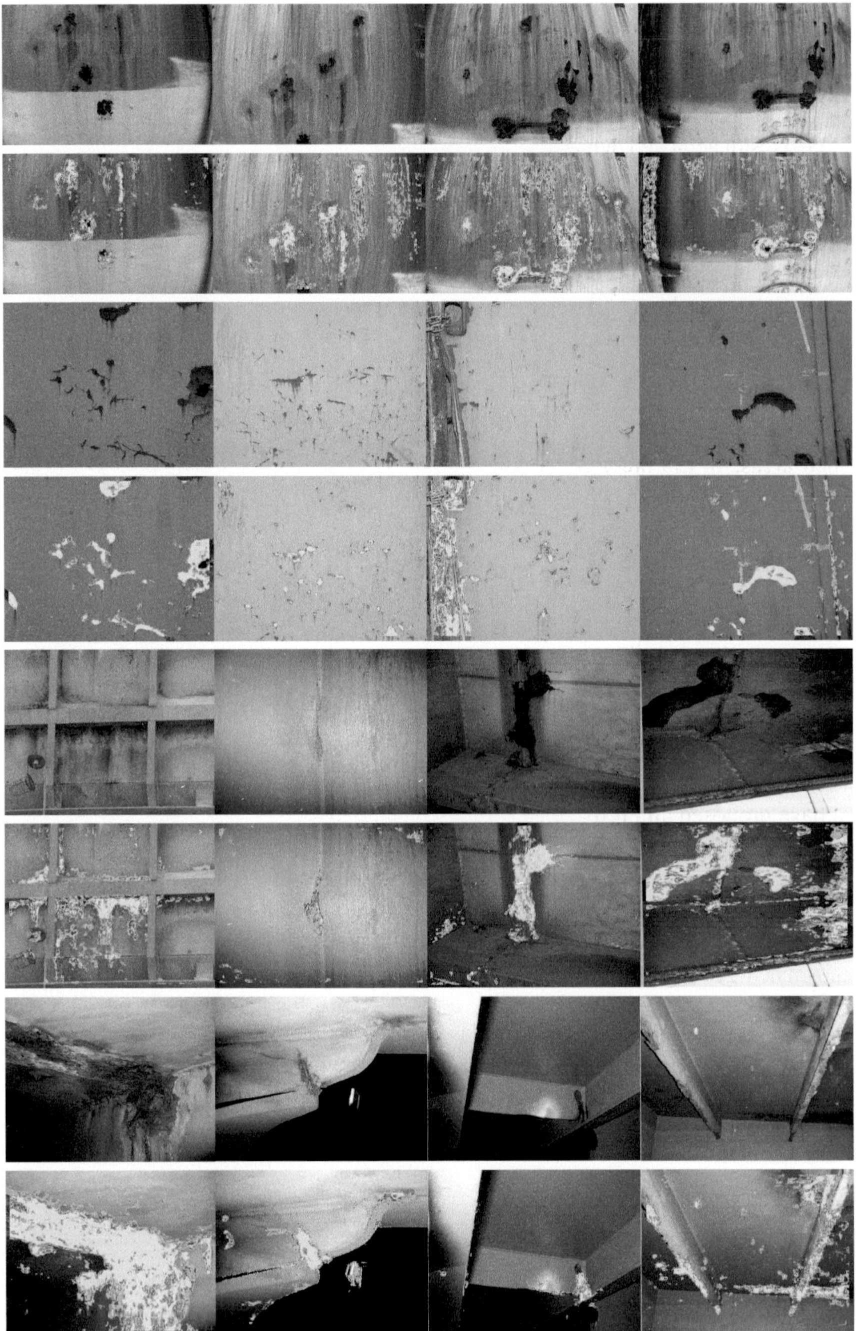

Fig. 6 Classification results for images containing corroded areas.

performance among the different trainings and amounts of hidden neurons (from $0.5 \times (10 + 8 + 8) = 13$ to $2.5 \times 26 = 65$ neurons). During this selection, all neighbourhoods were set to 15×15 pixels for all descriptors.

Table 2(left) provides the details for the previous combination of descriptors and the best result achieved, which corresponds to the case of 37 hidden neurons, while the results contained in Table 2(right) are for the same configuration but after tuning the descriptors with additional training rounds. Improved performance was attained after combining LBP and SD descriptors, and after varying the neighbourhood sizes. The final configuration has resulted to be: 10-bin histograms of uniform LBPs at 1-pixel distance measured over 17×17-pixel windows, 8 surrounding differences at 1-pixel distance, 8-bin histograms of hue values and 8-bin histograms of saturation values, both measured over 11×11-pixel values, and a 3-layer FFNN, with 10+8+8+8=34-component input vectors and 37 neurons in the hidden layer.

To finish, Figure 6 shows classification results for some images containing corroded areas. In the pictures, yellow and red pixels are deemed to be quite likely affected by corrosion, while for green pixels the likelihood is much lower.

6 Conclusions and Future Work

A Micro-Aerial Vehicle to be used for vessel visual inspection has been presented, together with a corrosion detection approach based on an artificial neural network. The MAV control approach is based on the SA paradigm, and hence the user is introduced in the platform control loop. For the specific problem of visual inspection, we have proposed and fully described a behaviour-based control architecture which include among its functionalities the enhancement of image capture. Regarding the corrosion detection approach, the classifier building process has been described and successful detection results have been reported. Next steps for improving corrosion detection performance include enhancing the navigation capabilities of the MAV (and hence get higher quality images), and reduce false positives by means of a general defects pre-detection stage, so that the ANN corrosion detector is mostly fed with defective area images.

References

1. Amano, T.: Correlation based image defect detection. In: Proc. ICPR, pp. 163–166 (2006)
2. Arkin, R.C.: Behavior-based Robotics. MIT press (1998)
3. Bonnin-Pascual, F., Ortiz, A.: Combination of weak classifiers for metallic corrosion detection and guided crack location. In: IEEE Intl. Conf. Emerging Tech. & Factory Automat. (2010)
4. Bonnin-pascual, F., Ortiz, A.: A probabilistic approach for defect detection based on saliency mechanisms. In: IEEE Intl. Conf. Emerging Tech. & Factory Automat. (2014)
5. Bonnin-Pascual, F., Ortiz, A.: Corrosion Detection for Automated Visual Inspection. In: Aliofkhazraei, D.M. (ed.) Developments in Corrosion Protection, pp. 619–632. InTech (2014)

6. Boukouvalas, C., Kittler, J., Marik, R., Mirmehdi, M., Petrou, M.: Ceramic tile inspection for colour and structural defects. Technical Report CS-EXT-1995-052 (1995), also: Proc. AMPT, pp. 390–399 (1995)
7. Chang, C.-L., Chang, H.-H., Hsu, C.P.: An intelligent defect inspection technique for color filter. In: Pro. IEEE Intl. Conf. Mechatronics, pp. 933–936 (2005)
8. Castilho, H., Pinto, J., Limas, A.: An automated defect detection based on optimized thresholding. In: Proc. Intl. Conf. Image Analysis & Recognition, pp. 790–801 (2006)
9. Cheng, G., Zelinsky, A.: Supervised Autonomy: A Framework for Human-Robot Systems Development. Auton. Robot. **10**, 251–266 (2001)
10. Duda, R., Hart, P., Stork, D.: Pattern Classification. Wiley (2001)
11. Eich, M., Bonnin-Pascual, F., Garcia-Fidalgo, E., Ortiz, A., Bruzzone, G., Koveos, Y., Kirchner, F.: A Robot Application to Marine Vessel Inspection. J. Field Robot. **31**(2), 319–341 (2014)
12. Fujita, Y., Mitani, Y., Hamamoto, Y.: A method for crack detection on a concrete structure. In: Proc. ICPR, pp. III: 901–904 (2006)
13. Gurdan, D., Stumpf, J., Achtelik, M., Doth, K.M., Hirzinger, G., Rus, D.: Energy-efficient Autonomous Four-rotor Flying Robot Controlled at 1 kHz. In: Proc. IEEE ICRA, pp. 361–366 (2007)
14. Honegger, D., Meier, L., Tanskanen, P., Pollefeys, M.: An open source and open hardware embedded metric optical flow CMOS camera for indoor and outdoor applications. In: Proc. IEEE ICRA, pp. 1736–1741 (2013)
15. Hongbin, J., Murphey, Y., Jinajun, S., Tzyy-Shuh, C.: An intelligent real-time vision system for surface defect detection. In: Proc. ICPR, pp. III: 239–242 (2004)
16. Jahanshahi, M., Masri, S.: Effect of color space, color channels, and sub-image block size on the performance of wavelet-based texture analysis algorithms: An application to corrosion detection on steel structures. In: ASCE Intl. Workshop on Computing in Civil Eng. (2013)
17. Ji, G., Zhu, Y., Zhang, Y.: The corroded defect rating system of coating material based on computer vision. In: Trans. Edutainment VIII, LNCS, vol. 7220, pp. 210–220. Springer (2012)
18. Jiang, B.C., Wang, C.C., Chen, P.L.: Logistic regression tree applied to classify PCB golden finger defects. Intl. Jour. Advanced Manufacturing Technology **24**(7–8), 496–502 (2004)
19. Kumar, A., Shen, H.: Texture inspection for defects using neural networks and support vector machines. In: Proc. IEEE ICIP. vol. III, pp.353–356 (2002)
20. Ojala, T., Pietikäinen, M., Harwood, D.: A comparative study of texture measures with classification based on featured distributions. Pattern Recognit. **29**(1), 51–59 (1996)
21. Oullette, R., Browne, M., Hirasawa, K.: Genetic algorithm optimization of a convolutional neural network for autonomous crack detection. In: Proc. IEEE CEC, pp. I: 516–521 (2004)
22. Ruffo, M., Castro, M.D., Molinari, L., Losito, R., Masi, A., Kovermann, J., Rodrigues, L.: New Infrared Time-of-flight Measurement Sensor for Robotic Platforms. In: IMEKO TC4 Int. Symposium and Int. Workshop on ADC Modelling and Testing, pp. 13–18 (2014)
23. Siegel, M., Gunatilake, P., Podnar, G.: Robotic assistants for aircraft inspectors. IEEE Instrumentation & Measurement Magazine **1**(1), 16–30 (1998)
24. Theodoridis, S., Koutroumbas, K.: Pattern Recognition. Academic Press (2006)
25. Xu, S., Weng, Y.: A new approach to estimate fractal dimensions of corrosion images. Pattern Recogn. Lett. **27**(16), 1942–1947 (2006)
26. Yamaguchi, T., Hashimoto, S.: Fast Crack Detection Method for Large-size Concrete Surface Images Using Percolation-based Image Processing. Mach. Vision Appl. **21**(5), 797–809 (2010)
27. Zaidan, B.B., Zaidan, A.A., Alanazi, H.O., Alnaqeib, R.: Towards corrosion detection system. Intl. Jour. Computer Science Issues **7**(1), 33–35 (2010)
28. Zhang, X., Liang, R., Ding, Y., Chen, J., Duan, D., Zong, G.: The system of copper strips surface defects inspection based on intelligent fusion. In: Proc. IEEE Intl. Conf. Automation and Logistics, pp. 476–480 (2008)

Analyzing the Relevance of Features for a Social Navigation Task

Rafael Ramon-Vigo, Noe Perez-Higueras, Fernando Caballero and Luis Merino

Abstract Robot navigation in human environments is an active research area that poses serious challenges in both robot perception and actuation. Among them, social navigation and human-awareness have gained lot of attention in the last years due to its important role in human safety and robot acceptance. Several approaches have been proposed; learning by demonstrations stands as one of the most used approaches for estimating the insights of human social interactions. However, typically the features used to model the person-robot interaction are assumed to be given. It is very usual to consider general features like robot velocity, acceleration or distance to the persons, but there are not studies on the criteria used for such features selection.

In this paper, we employ a supervised learning approach to analyze the most important features that might take part into the human-robot interaction during a robot social navigation task. To this end, different subsets of features are employed with an AdaBoost classifier and its classification accuracy is compared with that of humans in a social navigation experimental setup. The analysis shows how it is very important not only to consider the robot-person relative poses and velocities, but also to recognize the particular social situation.

Keywords Human-robot interaction · Supervised learning · Social robot

1 Introduction

Telepresence systems allow a human controller (the visitor) to interact remotely with people. Called by some "Skype on a stick", in such systems the visitor pilots a

---

R. Ramon-Vigo(✉) · N. Perez-Higueras · L. Merino
Pablo de Olavide University, Seville, Spain
e-mail: rramvig@upo.es

F. Caballero
University of Seville, Seville, Spain

L.P. Reis et al. (eds.), *Robot 2015: Second Iberian Robotics Conference*,
Advances in Intelligent Systems and Computing 418,
DOI: 10.1007/978-3-319-27149-1_19

Fig. 1 The picture depicts a telepresence robot with partial autonomy in terms of navigation and body pose control, considering social feedback, in elderly centres. A typical situation is presented here.

remotely located robot that results in a more physically presence than with standard teleconferencing. One of potential problems of telepresence systems is the cognitive overload that arises by having to take low (navigation commands) and high level decisions (interaction) at the same time. This may lead to mistakes at low level and to give less attention to the high level tasks [13]. To allow the visitor focusing in the interaction with other people, we aim to enhance the autonomy of the telepresence robot to perform low-level decisions for the controller regarding navigation and body pose in social settings (see Figure 1).

Actually, partial autonomy in terms of navigation is a feature requested by telepresence users [4]. Enhancing the autonomy of the telepresence robot in terms of navigation involves not only ensuring a safe and efficient navigation but also social interaction and social awareness when performing the robot tasks. For instance, approaching a person should be performed in a socially appropriate manner. In addition, when accompanying a person, some social rules must be maintained.

To this end, novel approaches are based on learning socially acceptable behaviors from real data collected under various social situations, avoiding manual explicit formulation of the behaviors. This is particularly interesting in the setup of telepresence robots, as there is a controller from which we can obtain information. In the last years, several contributions have been presented in this direction: supervised learning is used in [12] to learn appropriate human motion prediction models that take into account human-robot interaction when navigating in crowded scenarios. Unsupervised learning is used by Luber et al., [9] to determine socially-normative motion prototypes, which are then employed to infer social costs when planning paths. In [5], a model based on social forces is employed. The parameters for the social forces are learnt from feedback provided by users.

An additional approach is learning from demonstrations [2]: an expert indicates the robot how it should navigate among humans. One way to implement it is through Inverse Reinforcement Learning (IRL) [1], in which a reward (or cost) function is recovered from the expert behavior, and then used to obtain a corresponding robot policy. In [7], a path planner based on inverse reinforcement learning is presented. As the planner is learned from exemplary trajectories involving interaction, it is also aware of typical social behaviors. Inverse reinforcement learning for social navigation is also considered in [10]. However, while in [7] the costs are used to path plans, in [10] the authors employ these techniques to learn local execution policies, thus providing direct control of the robot. This can be combined with other planning techniques at higher levels, while alleviating the complexity associated to learning.

Most of those works assume that the learned cost function depends on a set of predefined and hand-coded features of the state, like person distance and others. This paper presents a procedure to discriminate between features in order to choose those ones that could better describe the task of navigating among other people. Closest to this work, in [14] the authors present a software framework to select the features in the design choice in IRL by means of investigating the effect of selecting several feature sets in the evaluation of two different IRL approaches. However, while they compare different IRL methods for different set of features, in this work we leverage feedback provided by the telepresence users to employ a supervised learning method to determine the importance of the features.

The paper presents an analysis of the feature importance, as well as the data used for learning. A dataset of the robot navigating through other persons in different social configurations is employed here to learn how to classify different state/action pairs as socially normative behavior or not, using AdaBoost. This approach has been used by [11] to determine the features to take into account when the robot selects and follows a human leader to take advantage of their motion. Here we analyze a different task, and a different set of features. A similar approach has been used by [3] in order to determine the most important features on which a person 2D range-based classifier could rely on.

The structure of the paper is as follows: next section describes the experimental set up followed to retrieve both the human demonstrations of the robot navigating and the feedback signal from an user observeing that. The features considered are explained in Section 3 and the results of the evaluations are showed at Section 4. Finally, the conclusions and future developments are detailed in Section 5.

2 Experimental Setup

As a robot social behavior is very difficult to describe mathematically, we aim to learn adequate behaviors in social situations by observing real demonstrations of the task to be accomplished. In particular, in this paper we analyze a social navigation task consisting on approaching a person, called interaction target, with a telepresence robot. Thus, we perform a set of social navigation episodes in which we can create and control specific social situations for navigation. This data will be used to associate the robot and its environment context, such as the position of people in the room, to various types of direct and indirect feedback.

The experiments carried out involve the use of two rooms (see Figure 2): the interaction room and the visitor room. The interaction room is where all the pre-defined social interactions between the robot and the persons present at the scene

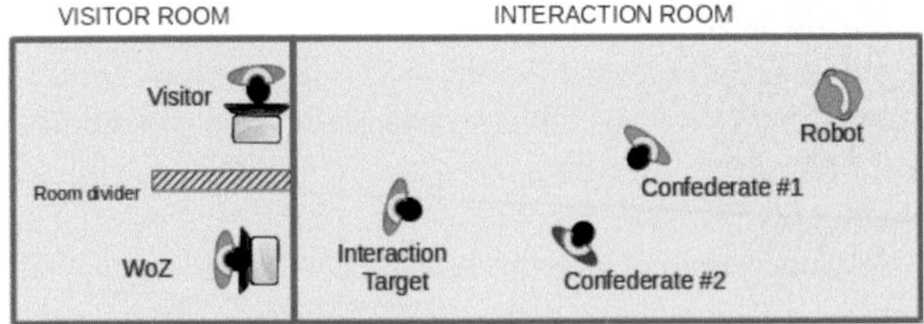

Fig. 2 Overview of the visitor and interaction rooms.

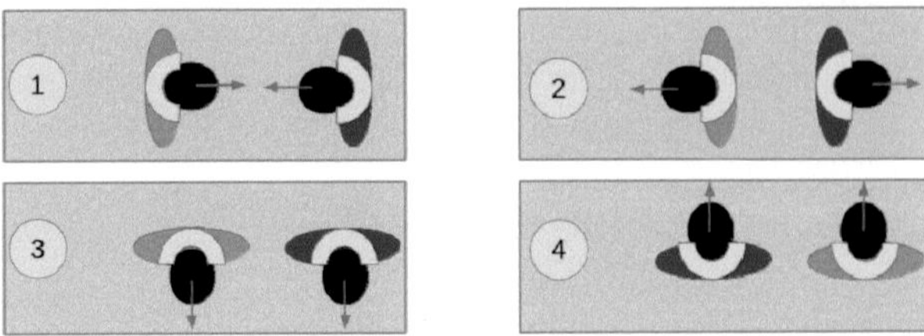

Fig. 3 Different social scenarios regarding to the pose adopted by the confederates.

take place. Due to control and repeatability issues, we propose to use confederates[1] that can conduct the social situation on an established plan. Among the confederates, we have an interaction target, which is the subject that the visitor intends to interact with using the telepresence robot. Finally, we also have the telepresence robot itself.

In the visitor room we have a subject (the visitor) that is observing, through the telepresence robot, what is going on in the interaction room. The visitor was instructed to provide an instantaneous feedback signal based on the behavior played by the robot in terms of what could be a normative socially behavior or not. This feedback signal will be detailed in the next sections.

Also present in the visitor room is a Wizard of Oz (WoZ) [8]. The WoZ is responsible of the behavior and low level control of the robot, but the visitor is misinformed that the robot is autonomous. In fact, the WoZ is physically separated from the visitor by a room divider and all the time conducts the experiment, not only driving the robot, but also informing the visitors about the current attempt of the robot and carrying the execution timing.

2.1 Experiments

The task of the robot was to reach the interaction target while dealing with the social scenarios (depicted in Figure 3) performed by the other two confederates, and then return to its starting position. Although the scenarios performed include

[1] A confederate in this context is any person who takes part in the experiment but is not a subject. Even if the subjects themselves are aware of this fact.

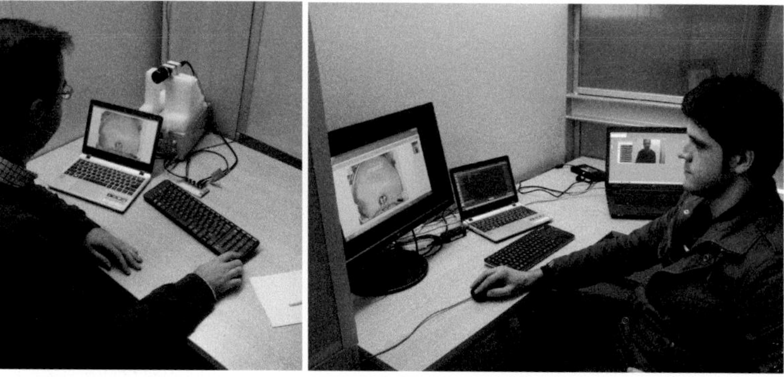

Fig. 4 Left: the visitor is depicted. He observes the scene through the telepresence robot interface in a laptop; he has a keyboard to enter his feedback by pressing a key. Right: the WoZ performing the navigating teleoperating the robot through the telepresence interface. There was no visual contact between them.

static and dynamic configurations, this work deals only with the static ones as a preliminar study. In the static scenarios, the social obstacles and the interaction target remain standing in the scene at the same place. The robot has to navigate towards the interaction target (and then moving back), performing a trajectory that avoids in some way the social obstacles. As mentioned before, the different configurations that the confederates could adopt are depicted in Figure 3.

In the adjacent room, the WoZ was controlling the robot while the visitor was observing and evaluating its actions, thinking that it was autonomous. Figure 4 describes the realization of the visitor room.

2.2 Data Gathering

The effective area of the interaction room is $6x4$ meters, and it is covered by a motion capture system, in particular the OptiTrack[2] system. This tool allowed us to collect detailed information about the positions and orientations of all the elements present in the interaction room, i.e. the robot, the confederates and the interaction target. Sensing the robot environment during the experiments is necessary in order to derive the states and features that can be later used for learning.

Another important source of data is the feedback signal that the visitor provides. During the experiments, the visitor gives direct feedback to the robot's instantaneous behavior using timestamped button presses (as showed in Figure 4 (left)) whenever he/her feels that the robot is behaving wrongly. The visitor was instructed to press a button when he observed such an action and keep pressing it until he thought that

[2] https://www.naturalpoint.com/OptiTrack/

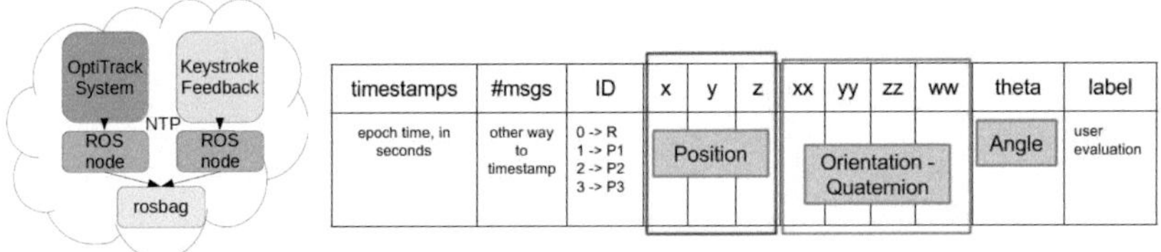

Fig. 5 Left: Integrating the state capture data and labeling information into rosbag log utility. The NTP ensures the time synchronization. Right: Formatting of the extracted data for each trajectory into the txt files from bags. For each ID (robot, R; interaction target, P1; confederates, P2 and P3) the position and orientation are stored.

the robot had returned to normal behavior. During the trajectories, the WoZ would deliberately execute socially unacceptable actions sometimes.

Before the integration and collection of the data from the motion capture system Optitrack and the labeling information provided by the visitor subject through the keystroke system (see Figure 5 (left)), a data post-processing step was performed by applying a smoothing interpolation. These allowed us to deal with small drop outs and tracking errors while performing the experiments.

Based on this data, several features can be derived and employed to represent states used in the learning process (see Section 3). For all the features, additional Gaussian filters were used in order to suppress any residual noise. This was required because we are also interested in some features that are obtained by differentiating previous ones.

It is important to notice that the observed feedback signal in the experiments is very sparse. In addition we have very limited knowledge about the meaning of the feedback and its duration. Thus, the dataset was pre-processed in order to increase the duration of the feedback signal, making it less sparse while keeping its interpretation meaningful. This is done by extending the feedback duration forward and backwards. This filter is reasonable under the assumption that the evaluator had some delay time in their reaction (backwards justification) and that the robot actions are smooth, i.e., the robot does not escape from a situation instantaneously (forward justification). The chosen extended duration was a second, centered at the exact time in which the visitor stamped his label. This value was settled empirically after analyzing the total time duration of each single trajectory and the usual length of 'train of keystrokes' observed when a person evaluate a bad behavior.

Despite the feedback signal extension performed, the number of bad examples and good examples are clearly unbalanced on behalf of good ones. However, the Adaboost algorithm implemented [3] deals with such kind of unbiassed datasets.

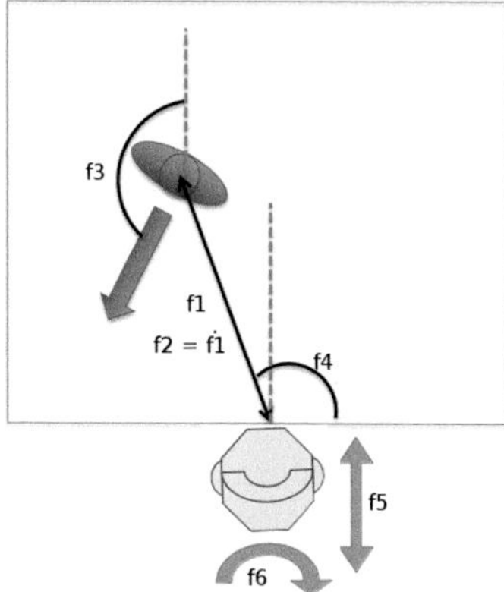

Fig. 6 Features considered in this work. All of them are average values (moving average with a 1-second window) and with respect to the closest person.

3 Features Considered

The objective is finally to transfer the good behaviors of the WoZ into the robot navigation stack, so that the robot is able to execute the task by itself in a socially adequate manner. The first important issue is to determine which information about the state is relevant for such task.

We will use the feedback given by the visitor. The idea is to use a supervised machine learning approach to reproduce the labels given by the visitor using as inputs different sets of features from the state. In general, features do not contribute equally to predict the target response; in many situations the majority of the features are in fact irrelevant.

There are some machine learning resources, like individual decision trees, that intrinsically perform feature selection by selecting appropriate split points. This information can be used to measure the importance of each feature; the basic idea is: the more often a feature is used in the split points the more important that feature is. In this work, we use AdaBoost [6], whose core principle is to fit a sequence of weak learners (i.e., models that are only slightly better than random guessing, such as small decision trees) on repeatedly modified versions of the data. The input to the algorithm is a set of labeled training data (e_n, l_n) , $n = 1, ..., N$, where each e_n is an example and $l_n \in \{+1, -1\}$ indicates whether e_n is positive or negative respectively.

It is interesting to choose a weak learner that is fast to be trained, as usually a large number of them is required. Examples of weak learners are decision trees, multi-layer perceptron and radial basis function. In this work, decision stumps are used, which is a one-level decision tree, making predictions based on a single threshold over a single feature [15].

From the features point of view, one requirement was to keep the model as simple as possible: the level of complexity must be such that the task description is not trivial, which would make learning redundant, or too complex, which would make learning impossible. So thus, the features considered were (see Figure 6):

- f_1: Average Distance To Closest Person
- f_2: Average Relative Velocity To Closest Person
- f_3: Average Relative Orientation To Closest Person
- f_4: Average Angle To Closest Person
- f_5: Average robot's linear velocity
- f_6: Average robot's angular velocity

All the features are computed from the recorded data using a moving average with a 1-second window. Although these features are quite basic, they are sufficient for the initial static scenario described in 2.1. All of the features used here are also employed in the literature, as in [14] and [7], but in this case we do not use a discrete approximation. It should be pointed out that, from the point of view of the approach used in this work, it is plausible to add whatever feature we can measure while the robot is performing the navigation task.

4 Results

A total of 4 sessions of the task described above were carried out. Each session involved 12 different executions, by using different trajectories, locations of the interaction target and configuration of the confederates. That results in 48 trajectories. For each session, a unique visitor evaluated the robot behavior into bad examples (by keystroke) or good examples (by default, no keystroke needed).

Each experiment was tagged with information related with the type of trajectory, target layout, etc. This classification allows us to analyze the feature importance over different set of trajectories, since we can take into account the whole set of trajectories or some subset of them according to the three last tags indicated before. We can also choose between different social navigation tasks, being more specific with respect to what kind of navigation task we attempt to, i.e. navigating towards a target while two persons are interacting (confederates' pose 1 of Fig. 3) or not.

The tags allow distinguishing between approaching the interaction target and getting away from him, called *go* and *return* trajectories respectively. This division is made because the type and nature of the interaction might be different when the robot approaches a person than when the robot leaves him away.

In addition to this, different subsets of features are considered at the learning process of the AdaBoost from those suggested at section 3. Thus, 6 different evaluations regarding to the set of features employed for the AdaBoost learning process will be considered:

- $\mathcal{F}_1 = (f_1, f_2, f_3, f_4)$
- $\mathcal{F}_2 = (f_1, f_2, f_3)$
- $\mathcal{F}_3 = (f_1, f_2, f_4)$

- $\mathcal{F}_{1*} = (f_1, f_2, f_3, f_4, f_5, f_6)$
- $\mathcal{F}_{2*} = (f_1, f_2, f_3, f_5, f_6)$
- $\mathcal{F}_{3*} = (f_1, f_2, f_4, f_5, f_6)$

The evaluations $\mathcal{F}\_*$ consider the last two features described at Section 3, i.e. f_5 and f_6. Both of them could be described as action-based features, because they enclose the robot actions. Thus, the idea is that these features could enclose robot behaviors in terms of bad (or good) maneuvers related to natural or smooth movements that may not disturb the people surrounding the robot but that they can be perceived as strange or unusual. On the other hand, sets $\mathcal{F}_2$ and $\mathcal{F}_3$ (and their $*$ extensions) are intended to study the effects of suppressing some features in the classification task. We focused in f_3 and f_4 in order to determine which of them would be more descriptive for the navigation tasks.

Next paragraphs present the evaluation of the different features. In the following, only the experiments with configuration 1 for the confederates (see Fig. 3) were considered. The procedure followed consists on cross validation: 4 random sets of training samples have been selected with other 4 sets of testing samples. The ratio of samples was 80% for each set of training and 20% for each set of testing. In the following, the details of the specific trajectories and features employed will be provided.

4.1 Evaluation 1

Table 1 takes into account the whole set of trajectories gathered at the experiments. This case represents the most global description of the task, since no distinction was done between the type of the trajectory, targets' layout and confederates' poses. We intended to classify all the social configurations with a single classifier. We do that with and without considering the action-based features.

Table 1 All trajectories and configurations are considered from the executed examples. Evaluating the overall classification with respect all f_1, f_2, f_3, f_4 features, with and without action-based features.

	All trajectories; All poses			
	$\mathcal{F}_1$		$\mathcal{F}_{1*}$	
	Detected Label		Detected Label	
True Label	No-Social	Social	No-Social	Social
No-Social	46%	**81%**	59%	**85%**
Social	**54%**	19%	**41%**	15%

The results shown in Table 1 indicate that it is not possible to correctly classify all the trajectories. Next, we considered ways to alleviate the complexity of the model being learned, and proposed some easier models by reducing the variance across the examples. During the experiments, the WoZ always introduced the visitor the individual trajectories, differentiating between *go* and *return* trajectories (explained above). Thus, the next natural step was making this distinction. Table 2 evaluates the construction of two different classifiers for both *go* and *return* types of trajectories, taking into account the sets of features $\mathcal{F}_1$ and $\mathcal{F}_{1*}$.

Table 2 *go* and *return* trajectories with confederates' pose 1: Evaluating the effects of adding action-based features.

	Type: *go*				Type: *return*			
	Set $\mathcal{F}_1$		Set $\mathcal{F}_{1*}$		Set $\mathcal{F}_1$		Set $\mathcal{F}_{1*}$	
	Detected Label		Detected Label		Detected Label		Detected Label	
True Label	No-Social	Social	No-Social	Social	No-Social	Social	No-Social	Social
No-Social	**93%**	10%	**93%**	3%	**67%**	13%	**94%**	7%
Social	7%	**90%**	7%	**97%**	33%	**87%**	6%	**93%**

It can be seen that the introduction of the action-based features improves the classification task in all the cases (see Tables 1 and 2). Other important conclusion is the fact that training a single classifier with all the examples performs poorly with respect to training different classifiers for some specific configurations of the task. Performance differences observed between types *go* and *return* may not be as obvious at a first glance. This could be produced by the lack of some unknown features, like context information, or by a significant difference between both trajectories due to the proximity of the goal to a target in the case of the *go* task.

4.2 Evaluation 2

This evaluation focuses on the analysis of the impact of some features. The main objective is to study how the introduction (or removal) of a specific feature into the classification could affect the performance obtained in $\mathcal{F}_1$ and $\mathcal{F}_{1*}$. Tables 3 and 4 presents the confusion matrix of sets $\{\mathcal{F}_2, \mathcal{F}_3\}$ and $\{\mathcal{F}_{2*}, \mathcal{F}_{3*}\}$, respectively.

Table 3 *go* and *return* trajectories with confederates' pose 1: Evaluating the effects of extracting a feature. No action-based features considered.

	Type: *go*				Type: *return*			
	Set $\mathcal{F}_2$		Set $\mathcal{F}_3$		Set $\mathcal{F}_2$		Set $\mathcal{F}_3$	
	Detected Label		Detected Label		Detected Label		Detected Label	
True Label	No-Social	Social	No-Social	Social	No-Social	Social	No-Social	Social
No-Social	**84%**	6%	**76%**	21%	**90%**	38%	**67%**	20%
Social	16%	**94%**	24%	**79%**	10%	**62%**	33%	**80%**

Table 4 *go* and *return* trajectories with confederates' pose 1: Evaluating the effects of extracting a feature. Action-based features considered here.

	Type: *go*				Type: *return*			
	Set $\mathcal{F}_{2*}$		Set $\mathcal{F}_{3*}$		Set $\mathcal{F}_{2*}$		Set $\mathcal{F}_{3*}$	
	Detected Label		Detected Label		Detected Label		Detected Label	
True Label	No-Social	Social	No-Social	Social	No-Social	Social	No-Social	Social
No-Social	**99%**	3%	**93%**	11%	**88%**	12%	**87%**	10%
Social	1%	**97%**	7%	**89%**	12%	**88%**	13%	**90%**

If we compare the Tables 1, 2, 3 and 4, it can be observed that the best performance is obtained by using the set $\mathcal{F}_{2*}$, which performs only slightly better than using the set $\mathcal{F}_{1*}$. Comparing the Tables 3 and 4, we also conclude that the feature f_3 (average relative orientation to closest person) is more relevant than the feature f_4 (average angle to closest person) for the description of the task in both situations (*go* and *return* trajectories).

5 Conclusions and Future Work

This paper presented an analysis of the features used to model the interaction between persons and robots when the latter are performing a particular social navigation task. An experimental setup have been conceived and implemented in order to gather enough data for the validation of the technique. The experiment also included feedback from the user so that supervised learning approaches can be applied to learn such features.

Two main conclusions arise from the data evaluation: First, the type of trajectory followed by the robot plays an important role in the learning phase. A classifier trained with *go* and *return* trajectories performs worst than two different classifiers, one per type of trajectory. We guess this is produced by the different perspectives when the robot needs to approach a goal close to a person (*go*) with respect to a less constrained goal in terms of gaze and body pose expectations (*return*), which basically are considered different navigation tasks. Furthermore, the social situation of other persons in the environment (represented in the experiments by the confederates) is also important, even in such a simple task. Finally, the introduction of features related with the robot action clearly improve the classification. The evaluations showed that considering the robot actions (in terms of velocities and accelerations) help to identify if the robot is behaving normative or not

Future work considers extending this study to more complex escenarios, including dynamic confederates and targets. In addition, the set of possible features will be increased, considering also some high level information as groups of persons or social situation.

Acknowledgements This work is partially funded by the European Commission, FP7, under grant agreement no. 611153 (TERESA).

References

1. Abbeel, P., Ng, A.Y.: Apprenticeship learning via inverse reinforcement learning. In: Proceedings of the twenty-first international conference on Machine learning, ICML 2004, p. 1. ACM, New York (2004). http://doi.acm.org/10.1145/1015330.1015430
2. Argali, B., Chernova, S., Veloso, M., Browning, B.: A survey of robot learning from demonstrations. Robotics and Autonomous Systems **57**, 469–483 (2009)
3. Arras, K.O., Mozos, O.M., Burgard, W.: Using boosted features for the detection of people in 2d range data. In: Proc. International Conference on Robotics and Automation, ICRA (2008)
4. Desai, M., Tsui, K., Yanco, H., Uhlik, C.: Essential features of telepresence robots. In: 2011 IEEE Conference on Technologies for Practical Robot Applications (TePRA), pp. 15–20, April 2011
5. Ferrer, G., Garrell, A., Sanfeliu, A.: Robot companion: a social-force based approach with human awareness-navigation in crowded environments. In: 2013 IEEE/RSJ International Conference on Intelligent Robots and Systems (IROS), pp. 1688–1694, Nov 2013
6. Freund, Y., Schapire, R.E.: A decision-theoretic generalization of on-line learning and an application to boosting. Journal of Computer and System Sciences **55**(1), 119–139 (1997). http://www.sciencedirect.com/science/article/pii/S002200009791504X
7. Henry, P., Vollmer, C., Ferris, B., Fox, D.: Learning to navigate through crowded environments. In: ICRA 2010, pp. 981–986 (2010)
8. Knox, W., Spaulding, S., Breazeal, C.: Learning social interaction from the wizard: a proposal. In: Workshops at the Twenty-Eighth AAAI Conference on Artificial Intelligence (2014)
9. Luber, M., Spinello, L., Silva, J., Arras, K.: Socially-aware robot navigation: a learning approach. In: IROS, pp. 797–803. IEEE (2012)
10. Perez-Higueras, N., Ramon-Vigo, R., Caballero, F., Merino, L.: Robot local navigation with learned social cost functions. In: 2014 11th International Conference on Informatics in Control, Automation and Robotics (ICINCO), vol. 02, pp. 618–625, Sept 2014
11. Stein, P., Spalanzani, A., Santos, V., Laugier, C.: On leader following and classification. In: 2014 IEEE/RSJ International Conference on Intelligent Robots and Systems (IROS 2014), pp. 3135–3140. IEEE (2014)
12. Trautman, P., Krause, A.: Unfreezing the robot: navigation in dense, interacting crowds. In: IROS, pp. 797–803. IEEE (2010)
13. Tsui, K., Desai, M., Yanco, H., Uhlik, C.: Exploring use cases for telepresence robots. In: 2011 6th ACM/IEEE International Conference on Human-Robot Interaction (HRI), pp. 11–18, March 2011
14. Vasquez, D., Okal, B., Arras, K.O.: Inverse reinforcement learning algorithms and features for robot navigation in crowds: an experimental comparison. In: IEEE/RSJ Int. Conf. on Intelligent Robots and Systems (IROS 2014). Chicago (2014)
15. Viola, P., Jones, M.: Robust real-time face detection. International Journal of Computer Vision **57**(2), 137–154 (2004). http://dx.doi.org/10.1023/B:VISI.0000013087.49260.fb

Decision-Theoretic Planning with Person Trajectory Prediction for Social Navigation

Ignacio Pérez-Hurtado, Jesús Capitán, Fernando Caballero and Luis Merino

Abstract Robots navigating in a social way should reason about people intentions when acting. For instance, in applications like robot guidance or meeting with a person, the robot has to consider the goals of the people. Intentions are inherently non-observable, and thus we propose Partially Observable Markov Decision Processes (POMDPs) as a decision-making tool for these applications. One of the issues with POMDPs is that the prediction models are usually handcrafted. In this paper, we use machine learning techniques to build prediction models from observations. A novel technique is employed to discover points of interest (goals) in the environment, and a variant of Growing Hidden Markov Models (GHMMs) is used to learn the transition probabilities of the POMDP. The approach is applied to an autonomous telepresence robot.

Keywords Markov decision processes · Social robot navigation · GHMM

1 Introduction

Social robots are becoming a strong trend in the last years. It is clear that future robotic applications will require robots to coexist with human beings. In scenarios populated with people, robots should behave in a *social* manner [2]. This includes not only considering humans in a different way as other "obstacles", but also reasoning about people intentions and reacting to them. For instance, in applications like robot guidance or meeting with a person, robots have to consider people's goals and commitment in order to actuate in advance [6]. Moreover, robots may want to avoid certain places when humans intend to go not to disturb them [14].

I. Pérez-Hurtado(✉) · L. Merino
Pablo de Olavide University Seville, Seville, Spain
e-mail: iperde@upo.es

J. Capitán · F. Caballero
University of Seville, Seville, Spain

L.P. Reis et al. (eds.), *Robot 2015: Second Iberian Robotics Conference*,
Advances in Intelligent Systems and Computing 418,
DOI: 10.1007/978-3-319-27149-1_20

In this paper, we consider the application of telepresence robots in scenarios like meeting a person at a particular place. The objective is to increment the social intelligence and autonomy of the telepresence robot, so that the robot can execute the low-level navigation tasks and the user can concentrate in the interaction with his/her peers. This is relevant, as it has been observed that one of the problems of telepresence systems is the cognitive overload that arises by having to take low-level (navigation commands) and high-level decisions (interaction) at the same time. This may lead to mistakes at low level and to give less attention to the high-level tasks [15].

These scenarios are uncertain by nature: robot actions are not always deterministic; the environment may be dynamic; and sensors are noisy. Furthermore, the main source of uncertainty comes from people intentions, which are non-observable and should be modeled probabilistically. For this reason, Partially Observable Markov Decision Processes (POMDPs) are proposed in this paper for decision making in these setups. POMDPs provide a sound mathematical framework for decision-theoretic problems in uncertain domains, and have already been used for social applications [2, 3, 6]. Even though they have traditionally faced scalability issues and can be computationally costly, recent advances in online [12, 13] and offline [8] solvers are making POMDPs increasingly practical for robot planning in large domains.

POMDPs use prediction models to infer the state. For instance, people motion models are required in most social tasks. However, in most works, basic or hand-crafted models for people movement are used [2, 6]. However, motion patterns and people intentions are affected by points of interest in the environment, and follow repetitive patterns: people move between doors and corridors following common trajectories; places of interest are common goals or affect people behavior, attracting them (e.g., vending machines) or repelling them (e.g., grass lawns); etc. Therefore, machine learning techniques are considered in the literature to estimate such *interesting* points and learn models for people intentions [3, 17].

In this paper, we contribute by using a POMDP model for a social task of a robot meeting with a person, where human motion and intentions are modeled automatically by an extension of Growing Hidden Markov Models (GHMMs) [16]. This GHMM model is first learnt from data observed by the robot and then integrated within the POMDP in order to predict the person intentions or goals. Later, an offline solver is used to compute an approximate optimal policy for the robot. During the execution phase, as the robot interact with the person, a probability distribution over the person position and intention is estimated using the learnt GHMM. That *belief* is also used to feed the POMDP policy that selects best actions for the robot in order to meet the person or avoid him/her depending on his/her goal.

We show results to prove the feasibility of the method in an indoor scenario where a telepresence robot [11] has to navigate autonomously around meeting people at certain points of interest (e.g., coffee machine) and not bothering them at others (e.g., toilets).

The remainder of the paper is as follows: Section 2 defines the problem as a social task for robot navigation; Section 3 describes our approach and models for decision making; Section 4 provides experimental results; and Section 5 discusses conclusions and future work.

Fig. 1 The telepresence robot is located in a meeting area, and the task consists of approaching people that go to typical interaction points (like coffee machines). These points of interest are learnt previously from data.

2 Problem Definition

In this work, we focus on a social task where a telepresence robot needs to interact with people in the environment. The main objective of a telepresence robot is to act as an *avatar* for a remote user, carrying a video-conference system on board and allowing that remote user to *sense* and *interact* with the environment from the distance.

As commented, the objective is that the robot carries out the low-level navigation tasks, allowing the user to concentrate on the interaction through the robot. In particular, the task we consider here allows the user to connect to the robot, which operates autonomously in a certain area (for instance, a meeting room or coffee area), waiting for people to appear (see Fig. 1). Then, the robot automatically should go and catch the person at their destination so that the remote user can establish a conversation. For that, the robot must reason about the possible intentions of the person and distinguish between two types of destinations: adequate and inadequate spots for having a conversation. For example, there are some places where people go to interact with others, like a coffee machine or a rest area. However, the robot should not disturb people when they intend to go to the toilet or exit the area.

The robot can detect and track people nearby, but its only available information for the task is a map of the scenario. The intention of each person entering in the operational area is not observable, so the robot needs to plan where to go taking into account uncertainties in people's positions and intentions. Moreover, we want the robot to discover automatically which are the locations of the *hot* spots of the scenario, where people intend to go, either to interact with others or to leave the scenario.

We propose a POMDP to model and solve this social task, since it allows the robot to deal with the uncertainties associated with its sensors, actions and the people around in a compact manner. POMDPs are also adequate for different multi-objective problems if their reward and cost functions are designed properly. Furthermore, we aim to learn the spatial structure of human trajectories in a specific environment by building an Instantaneous Topological Map (ITM) that can be viewed as a dynamic occupancy grid map. A Hidden Markov Model (HMM) is then built over this ITM and used as the transition model for the former POMDP. The details are described in the next section.

3 A POMDP for Social Navigation

The problem in Section 2 can be modeled as a POMDP where the robot maintains a belief over a person detected and its intentions. Here, we assume that the robot can only reason about a person at once, so when there are multiple people around, it only focuses on a particular one. We use a GHMM to model person motion patterns and discover automatically the points of interest in the scenario by observing people around.

3.1 POMDP Preliminaries

Formally, a discrete POMDP is defined by the tuple $\langle S, A, Z, T, O, R, h, \gamma \rangle$ [5].

- The *state space* is the finite set of possible states $s \in S$, for instance robot and people poses.
- The *action space* is defined as the finite set of possible actions that the robot can take, $a \in A$.
- The *observation space* consists of the finite set of possible observations $z \in Z$ from the onboard sensors.
- After performing an action a, the state transition is modeled by the conditional probability function $T(s', a, s) = p(s'|a, s)$, which indicates the probability of reaching state s' if action a is performed at state s.
- The observations are modeled by the conditional probability function $O(z, a, s') = p(z|a, s')$, which gives the probability of getting observation z given that the state is s' and action a is performed.
- The reward obtained for performing action a at state s is $R(s, a)$.

The state is not fully observable; at every time instant the agent has only access to observations z which give incomplete information about the state. Thus, a belief function b is maintained by using the Bayes rule. If action a is applied at belief b and observation z is obtained, a new belief b' is given by:

$$b'(s') = \eta O(z, a, s') \sum_{s \in S} T(s', a, s) b(s), \qquad (1)$$

where the normalization constant:

Fig. 2 The space is discretized by learning a topological map from people tracks obtained with the robot sensors (in blue). Some of the nodes are identified during the learning phase as goals (in red).

$$\eta = p(z|b, a) = \sum_{s' \in S} O(z, a, s') \sum_{s \in S} T(s', a, s) b(s) \tag{2}$$

gives the probability of obtaining a certain observation z after executing action a for a belief b.

The objective of a POMDP is to find a policy that maps beliefs into actions in the form $\pi(b) \rightarrow a$, so that the *value* is maximized. This value function represents the expected total reward earned by following π during h time steps starting at the current belief b: $V^{\pi}(b) = E\left[\sum_{t=0}^{h} \gamma^t r(b_t, \pi(b_t)) | b_0 = b\right]$, where $r(b_t, \pi(b_t)) = \sum_{s \in S} R(s, \pi(b_t)) b_t(s)$. Rewards are weighted by a discount factor $\gamma \in [0, 1)$ to ensure that the sum is finite when $h \rightarrow \infty$. Therefore, the optimal policy π^* is the one that maximizes that value function: $\pi^*(b) = \arg\max_{\pi} V^{\pi}(b)$.

3.2 States

The state s of our POMDP consists of three factors: the robot position, the person position and the person goal. As we employ a discrete POMDP, the scenario is divided into non-overlapping regions in order to discretize the robot and person positions (see Fig. 2). Each region has a centroid and all regions are combined into a topological map that is discovered automatically, as it will be described in the next section. Also, there is a finite set of goals (each goal corresponds to a region) where the person can go, which are discovered automatically too, as explained later.

We assume that the localization system of the robot is good enough to be able to determine its region with high certainty. Therefore, the robot position is assumed observable, being just necessary to keep a belief over the person position and intention. Note that the most relevant uncertainty of the problem comes from the person intentions which are non-observable by nature.

3.3 State Transitions: A Growing Hidden Markov Model

As described in Section 3.1, the POMDP planner needs a probabilistic transition function of the state $T(s', a, s)$, which models the dynamics of people locations and motion intentions. Instead of handcrafting this transition model (a Markov model), we have developed an extension of GHMMs [16] to learn this transition function from data.

In a GHMM, there is a discrete representation of the space, which is divided into regions. Transitions are only allowed between neighboring regions. The learning process consists of estimating the best space discretization, and identifying neighboring regions and transition probabilities from observed data. Thus, first a topological map is built with the ITM algorithm [4]; then, an HMM is built from the ITM, and its transition (and prior) probabilities are trained with the incremental Baum-Welch technique [7]. Once the GHMM has been trained, it is integrated with the POMDP before starting the task execution.

Learning Phase. The ITM algorithm structures the space in a graph whose nodes represent the centroid of Voronoi regions where people have been observed and edges represent connections between adjacent/neighboring regions (in the mathematical sense). In order to enforce an average geometrical distance between nodes, a threshold τ to insert new nodes is defined. Each node maintains and updates a Gaussian distribution in relation with the observations (2D positions of people) within its corresponding region. We propose a variant of the original ITM algorithm where a bivariate Gaussian distribution $\mathcal{N}(\mu_n, \Sigma_n)$ is updated for each region n after each new observation, where μ_n and Σ_n are respectively the mean and the covariance matrix of all the observations (x, y) related to n. Thus, instead of using a fixed covariance matrix for all the nodes, each node stores and updates a specific covariance matrix, so the observation model can adapt to the characteristics of the different parts of the scenario. The centroids are also updated in a different manner as in the original ITM algorithm, and each centroid is computed as the mean of its associated observations.

People goals are automatically discovered by applying hypothesis testing (t-test), there are two types of goals: entry/exit points where people appear or disappear and standing points where people stop longer than usual in the scene. The algorithm is adaptive, i.e, nodes, edges and goals are created, erased and updated dynamically as more people are observed. A $p - value$ threshold is defined in order to accept or refuse new goals.

Finally, a HMM is used to model all the transition state probabilities, being the state of a person its node of the graph and its goal or intention. In particular, prior and transition probabilities are computed by applying the Baum-Welch algorithm, including people positions and velocities as observations. A sampling ratio parameter T_s is used to sample the observed trajectories at a constant rate and feed the Baum-Welch algorithm (hence, each state transition corresponds with a time T_s). In the GHMM framework, the HMM can be trained several times during the learning phase. For this paper, the HMM has been generated and trained once after the creation of the topological map and goal discovery, since we are using an offline approach. For more details about the learning phase, please refer to [9].

Belief Estimation. Once the transition probabilities of the GHMM have been learnt (for the person position and goal), the belief over the person position and intention can be updated each T_s with Equation 1. Note that the robot actions do not affect people positions nor intentions in our model. Moreover, the robot positions are considered observable and its transition probabilities for each action are hand-coded.

3.4 *Observations, Actions and Rewards*

The robot has sensors onboard to measure its own pose and estimate the person position. At each moment, it can either determine the region where the person is or not detect anything. Sensors are noisy, and the probability of non-detecting the person (false negative) is p_f. Moreover, if the person is in a certain region, it could be detected in the adjacent regions.

This probability of erroneous detection depends on the distance between the centroids of the actual region and the observed region. The probability of detecting a person for each region is modeled by a Gaussian centered in the centroid of the region. Those Gaussian distributions vary for each region and are learnt together with the GHMM [9].

In addition, the robot can take movement actions at each iteration of the planner. In particular, the robot can decide either to stay where it is or to move to an adjacent region. Those transitions are not modeled as deterministic and there is certain probability that the robot may end up in an erroneous region. Moreover, there is a cost associated with moving to an adjacent region, whereas there is no cost associated with staying in the same region.

The objective of the robot is to come across the person in order to have a conversation, therefore the reward function is designed with this purpose. For that, two different types of goals are considered: adequate or inadequate. *Adequate* goals are those where the robot can go and have an interaction with the person. *Inadequate* goals are those where the robot should not bother the person and go to its *home* position, defined beforehand. Thus:

- If the person intends to go to an adequate goal and the robot is there, it gets a positive reward R_{pos}.
- If the person intends to go to an inadequate goal and the robot is at *home* position, it also gets a positive reward R_{pos}.
- If the person intends to go to an adequate goal and the robot is not there when the person arrives, it gets a negative penalty R_{neg}.
- If the person intends to go to an inadequate goal and the robot is there, it gets a negative penalty R_{neg}.

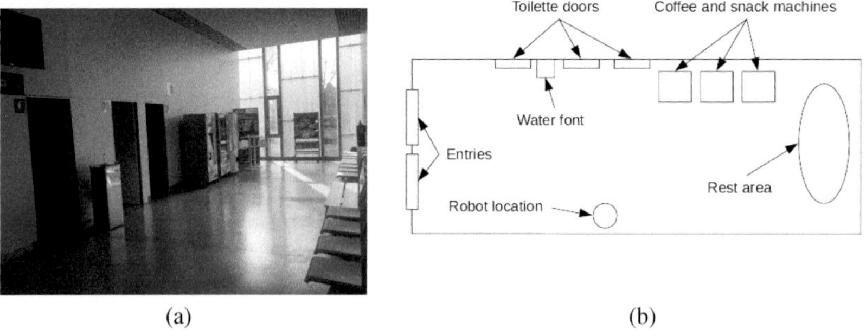

(a) (b)

Fig. 3 (a) Experimental area at university. (b) Schematic view with the main points of interest and the *home* position for the robot.

The *home* position is a region more or less centered in the scenario where the robot can wait for people to arrive. The reward function tries to encourage the robot to catch the people who go to adequate goals and to arrive there before them. It also forces the robot to go back to *home* if the person intends to go to an inadequate goal. The fact that the person comes across the robot in an inadequate place is penalized because it may be considered disturbing.

4 Experiments

In this section, we present some experimental results to show the feasibility of our approach. We implemented our decision-making algorithm in a real telepresence robot.

4.1 Experimental Setup

The scenario used for our social task is one of the rest areas in Pablo de Olavide University (Fig. 3), which is a space of 4.30 × 11.80 meters with a single entry/exit point at one side and several points of interest: a spot with a coffee and a snack machine, a door to the toilets, a water font and a rest area with magazines.

We implemented our methods for people motion model and decision making in C++ under the Robot Operating System (ROS) framework. We used for the experiments the TERESA robot [11], which is equipped with two laser-scanners (front and back) and a video-conference system. The robot had a map of the scenario and was able to localize itself and navigate between waypoints thanks to the ROS navigation stack (`amcl` and `move_base` packages).

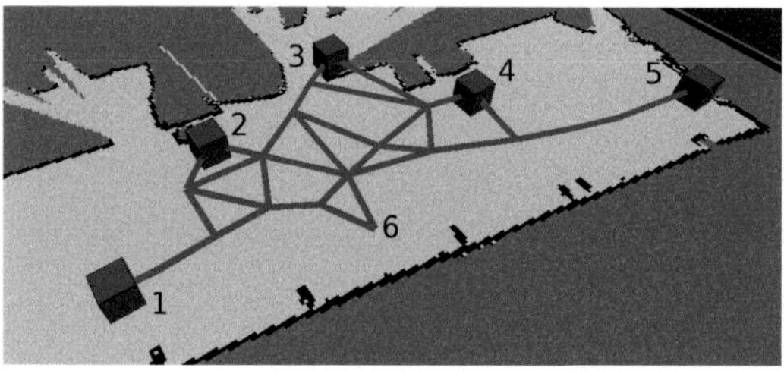

Fig. 4 Topological map (blue) and discovered goals (red) for the scenario. Node 6 of the topological map is used as *home* position for the robot.

First, we placed the robot in the scenario without moving from the *home* position, just observing trajectories of people passing by. With that information, we ran the algorithm described in Section 3.3 to learn a topological map of the scenario with the centroids of the regions, the possible goals for people and a GHMM with the transition probabilities.

The robot used the two laser-scanners for person detection and tracking, applying the algorithm in [1] and a Kalman Filter for temporal tracking and velocity estimation. More than 200 people trajectories were recorded in a dataset and used to train the models[1], generating a topological map with 21 nodes, 32 edges and 5 discovered goals (see Fig. 4). The algorithm was able to discover as goals all the points of interest in the scene: (1) entry/exit door; (2) water font, (3) toilets, (4) coffee/snack machines, (5) rest area. The corresponding GHMM was trained by sampling the people trajectories at $T_s = 1$ Hz, resulting in 105 states and 1,705 transition probabilities.

Once the GHMM was learnt, we implemented our POMDP model[2] to obtain a policy for the robot. In this case, we used an offline POMDP solver, Symbolic Perseus [10]. During the experiments, we ran two different modules: a module using the GHMM to estimate the belief of the person position and intention, and a module to determine the best action for the robot at each time given the current belief. The estimator module is executed at 1 Hz whereas the decision-maker at 0.33 Hz. Moreover, the decision-maker commands the robot to stay at the same region or to go to adjacent ones, which means sending to the `move_base` navigator the corresponding waypoint (centroid of the destination region).

4.2 Results

In order to evaluate the behavior of the robot with the computed policy, we ran different trials where people were appearing at the scenario and going to different places. In general, we observed a common behavior: the robot waits before moving

[1] The ITM algorithm was executed with $\tau = 1$ meter for node insertion and $p - value = 10^{-4}$ for hypothesis testing.

[2] The parameters were set as $p_f = 0.1$, $R_{pos} = 10$ and $R_{neg} = -10$.

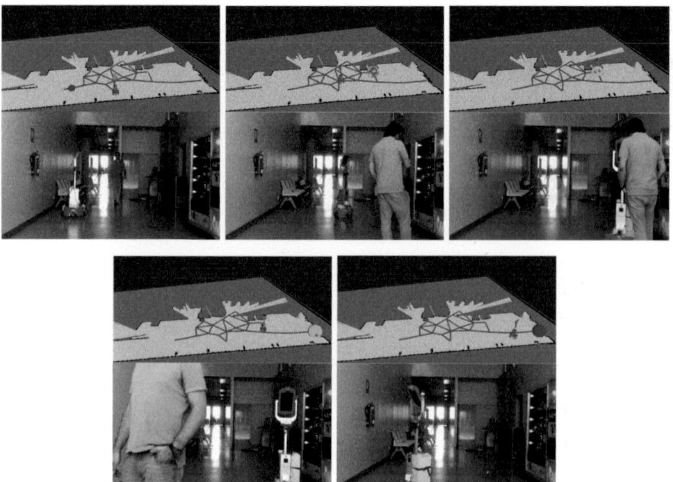

Fig. 5 Several snapshots with the person going to the coffee machine and later to the rest area.

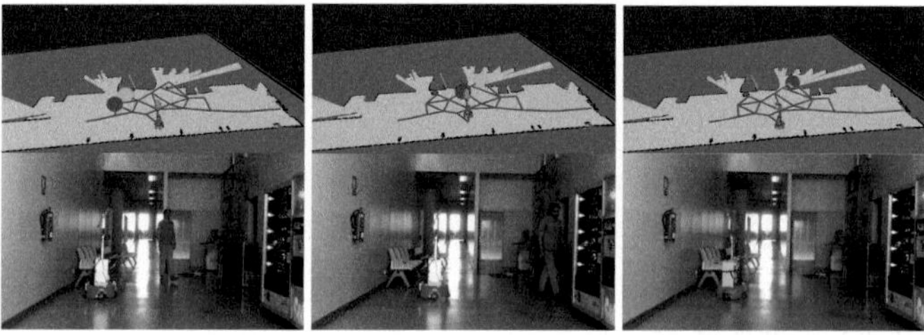

Fig. 6 Several snapshots with the person going to the toilet.

when it is not very sure about the person intention (high uncertainty in the goal belief); however, when the certainty on the person intention increases, the robot goes and perform its task, i.e., meeting the person if he/she goes to an adequate destination or coming back to the *home* position if he/she goes to an inadequate destination.

In Figures 5, 6 and 7, three sequences of snapshots of an experiment are depicted. Due to space limitations, only a few examples of the robot behavior are shown. Their corresponding representations with the RVIZ visualization tool are also shown, where the belief over the person positions is depicted with red spheres at the nodes of the topological map (the bigger the sphere, the higher the probability), and the belief over the intentions with green spheres at the goal positions.

In particular, it can be seen in Figure 5 how the robot waits until it knows with certainty where the person goes, and then it goes to the coffee machine to meet him. Later, it also follows him to the rest area. In Figure 6, the robot beliefs at first that the person goes to the water font and starts moving there, but then it changes its belief to the toilet and decides not to go back to the *home* position. Finally, in Figure 7, the

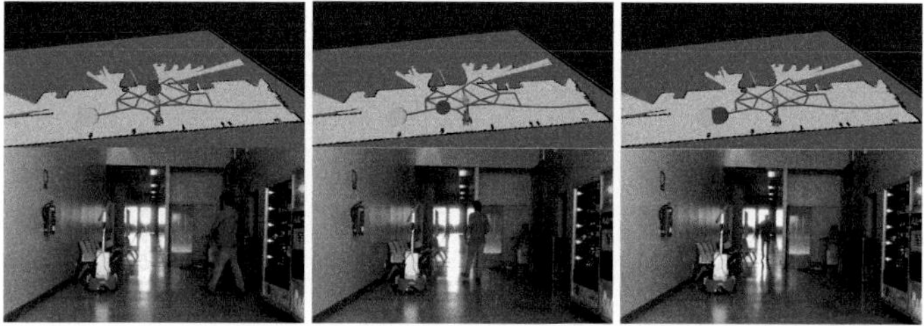

Fig. 7 Several snapshots with the person going out of the toilet and leaving the room. Real images and RVIZ representations are combined together.

person gets out of the toilet and its intention is clear from the beginning, so the robot does not move and let the person exit the room.

5 Conclusions

This paper has presented a decision-theoretic approach for social navigation under uncertainties. The social task (where a robot needs to interact with people) and its associated uncertainties are modeled using a POMDP, which allows the robot to reason in a principled manner about the possibilities in the future, and about the uncertain (non-observable) intentions of people. Moreover, a prediction model for people motion is learnt from observed data and used to train the POMDP policy. In particular, an extension of the GHMM framework is proposed, which allows the robot to discover automatically points of interest in the scenario and the transition probabilities for people movements.

The paper applies the concepts to a social task with a telepresence robot, and it presents results obtained with a real system. These results show how the robot is able to accomplish its task in a satisfactory way even with noisy sensors and reasoning on future people intentions. We believe that our approach should behave in a more robust manner than other greedy policies where the planner does not consider uncertainties in the future. If the robot only takes into account the current belief on the person intentions, it will be more prone to erroneous decisions based on incorrect beliefs.

As future work, we plan to perform a benchmarking comparison of our policy against other simpler (greedy) policies using systematic simulations. Also, we would like to explore the use of online approaches, where the POMDP prediction model and policy are learnt during the execution of the task.

Acknowledgements This work is partially funded by the EC-FP7 under grant agreement no. 611153 (TERESA) and by the Junta de Andalucia through the project PAIS-MultiRobot (TIC-7390). I. Pérez-Hurtado is also supported by the Postdoctoral Junior Grant 2013 co-funded by the Spanish Ministry of Economy and Competitiveness and the Pablo de Olavide University.

References

1. Arras, K.O., Mozos, O.M., Burgard, W.: Using boosted features for the detection of people in 2D range data. In: Proceedings of the IEEE International Conference on Robotics and Automation (ICRA), pp. 3402–3407. Roma, April 2007
2. Bai, H., Cai, S., Ye, N., Hsu, D., Lee, W.S.: Intention-aware online POMDP planning for autonomous driving in a crowd. In: 2015 IEEE International Conference on Robotics and Automation (ICRA), pp. 454–460, May 2015
3. Foka, A., Trahanias, P.: Probabilistic autonomous robot navigation in dynamic environments with human motion prediction. International Journal of Social Robotics **2**(1), 79–94 (2010). http://dx.doi.org/10.1007/s12369-009-0037-z
4. Jocksuch, J., Ritter, H.: An instantaneous topological map for correlated stimuli. In: Proceedings of the International Joint Conference on Neural Networks, vol. 1, pp. 529–534. Washington (1999)
5. Kaelbling, L.P., Littman, M.L., Cassandra, A.R.: Planning and acting in partially observable stochastic domains. Artificial Intelligence **101**, 99–134 (1998)
6. Merino, L., Ballesteros, J., Pérez-Higueras, N., Ramón-Vigo, R., Pérez-Lara, J., Caballero, F.: Robust person guidance by using online POMDPs. In: Armada, M.A., Sanfeliu, A., Ferre, M. (eds.) ROBOT 2013: First Iberian Robotics Con-ference, Advances in Intelligent Systems and Computing, vol. 253, pp. 289–303. Springer International Publishing (2014)
7. Neal, R., Hinton, G.: A view of the EM algorithm that justifies incremental, sparse, and other variants. In: Jordan, M. (ed.) Learning in Graphical Models (1998)
8. Ong, S.C., Png, S.W., Hsu, D., Lee, W.S.: POMDPs for robotic tasks with mixed observability. In: Proceedings of the Robotics: Science and Systems Conference. Seattle (2009)
9. Perez-Hurtado, I., Capitan, J., Caballero, F., Merino, L.: An extension of GHMMs for environments with occlusions and automatic goal discovery for person trajectory prediction. In: European Conference on Mobile Robots. p. To appear. Lincoln (2015)
10. Poupart, P.: Exploiting Structure to Efficiently Solve Large Scale Partially Observable Markov Decision Processes. Ph.D. thesis, University of Toronto (2005)
11. Shiarlis, K., Messias, J., van Someren, M., Whiteson, S., Kim, J., Vroon, J., Englebienne, G., Truong, K., Evers, V., Perez-Higueras, N., Perez-Hurtado, I., Ramon-Vigo, R., Caballero, F., Merino, L., Shen, J., Petridis, S., Pantic, M., Hedman, L., Scherlund, M., Koster, R., Michel, H.: Teresa: a socially intelligent semi-autonomous telepresence system. In: Workshop on Machine Learning for Social Robotics at ICRA-2015. Seattle (2015)
12. Silver, D., Veness, J.: Monte-Carlo planning in large POMDPs. In: Advances in Neural Information Processing Systems (2010)
13. Somani, A., Ye, N., Hsu, D., Lee, W.S.: DESPOT: Online POMDP planning with regularization. In: Burges, C., Bottou, L., Welling, M., Ghahramani, Z., Weinberger, K. (eds.) Advances in Neural Information Processing Systems, pp. 1772–1780. Curran Associates, Inc. (2013)
14. Tipaldi, G.D., Arras, K.O.: Planning problems for social robots. In: Proc. International Conference on Automated Planning and Scheduling (ICAPS 2011). Freiburgu (2011)
15. Tsui, K., Desai, M., Yanco, H., Uhlik, C.: Exploring use cases for telepresence robots. In: 2011 6th ACM/IEEE International Conference on Human-Robot Interaction (HRI), pp. 11–18, March 2011
16. Vasquez, D., Fraichard, T., Laugier, C.: Growing Hidden Markov Models: An Incremental Tool for Learning and Predicting Human and Vehicle Motion. The International Journal of Robotics Research **28**(11–12), 1486–1506 (2009)
17. Xie, D., Todorovic, S., Zhu, S.: Inferring "dark matter" and "dark energy" from videos. In: IEEE International Conference on Computer Vision (ICCV), pp. 2224–2231 (2013)

Influence of Positive Instances on Multiple Instance Support Vector Machines

Nuno Barroso Monteiro, João Pedro Barreto and José Gaspar

Abstract This work studies the influence of the percentage of positive instances on positive bags on the performance of multiple instance learning algorithms using support vector machines. There are several studies that compare the performance of different types of multiple instance learning algorithms in different datasets and the performance of these algorithms with the supervised learning counterparts. Nonetheless, none of them study the influence of having a low or high percentage of positive instances on the data that the classifiers are using to learn. Therefore, we have created a new image dataset with different percentages of positive instances from a dataset for pedestrian detection. Experimental results of the performance of mi-SVM and MI-SVM algorithms on an image annotation task are presented. The results show that higher percentages of positive instances increase the overall accuracy of classifiers based on the maximum bag margin formulation.

Keywords Multiple instance learning · mi-SVM · MI-SVM

1 Introduction

In supervised learning, the classifier is provided with a training set that consists of instances and the corresponding labels. The training set is then used to obtain a classifier that can predict the labels for novel instances [4]. Nonetheless, the correspondence requirement between instances and labels is difficult or even prohibitive

N.B. Monteiro(✉) · J. Gaspar
Institute for Systems and Robotics (ISR/IST), LARSyS, University of Lisbon, Lisbon, Portugal
e-mail: {nmonteiro,jag}@isr.ist.utl.pt

N.B. Monteiro · J.P. Barreto
Institute for Systems and Robotics, University of Coimbra, Coimbra, Portugal
e-mail: jpbar@isr.uc.pt

L.P. Reis et al. (eds.), *Robot 2015: Second Iberian Robotics Conference*,
Advances in Intelligent Systems and Computing 418,
DOI: 10.1007/978-3-319-27149-1_21

for some applications like object detection. Annotating whole images is easier and faster than annotating and identifying relevant image regions.

Multiple instance learning appeared as a more flexible paradigm assuming that there is some ambiguity in how the labels are assigned. Namely, in multiple instance learning, the instances are grouped into bags and the labels are assigned to the bags instead of being assigned to each of the instances. The labels are then learned using a multiple instance learning assumption like the weighted collective assumption [10] or the standard multiple instance learning assumption. The standard multiple instance learning assumption is the one that is considered in this work, and states, for a binary classification problem: a bag is positive if at least one of the instances in that bag is positive, a bag is negative if all the instances in that bag are negative. Therefore, the true input labels are not known during training, i.e., the true input labels are latent variables. Note also that a positive bag can contain negative instances.

In order to understand better the multiple instance learning problem, let us consider an example adapted from [7]. Consider that we have a classroom and we know some professors that have access to that classroom, and others who do not have access. Each professor has a key chain with a few keys that can be differentiated, for example, by color. The goal is to predict if a given key or a given key chain allows to open the door of the classroom. Using the multiple instance learning framework, the bags will correspond to the key chains that are labeled as positive or negative according to the access that the professor has to the classroom. The instances are the keys contained in the key chains. Using the assumption, we know that keys on key chains from professors that do not have access to the classroom do not open the door. Thus, to solve this problem we have to find the key that is common in all positive key chains. If we consider the color as a feature, the keys colors that appear in negative key chains could be ruled out. Hopefully, there is one key color that remains and this will correspond to the key that opens the door to the classroom. If the classifier can correctly identify this key, it can predict if a key or key chain is able to give access to the classroom. From this simple example, we can see that multiple instance learning algorithms can have several formulations because they can aim at designing classifiers for better discriminating bags or instances.

The performance of multiple instance learning algorithms have been compared between themselves and with the supervised learning counterparts in several application domains. But none of them focused on what is the influence of a different percentage of positive instances on the classifiers. Therefore, this work intends to study this influence on the performance of the classifiers. We focus on mi-SVM and MI-SVM which are support vector machines adapted to the multiple instance learning framework.

The paper has two major contributions: a new dataset that allows to evaluate the influence of different percentages of positive instances, and experimental evidence that increasing this percentage has a positive impact in the performance of MI-SVM.

In terms of structure, we will first present a brief review of the state of the art on multiple instance learning algorithms in Section 2. A more detailed formalism of multiple instance learning frameworks with focus on algorithms using support vector machines is provided in Section 3. The methodology used to evaluate the performance of the classifiers and the dataset used is described in Section 4. The results and major conclusions are presented in Section 5 and Section 6, respectively.

Notation: The notation followed throughout this work is the following: non-italic letters correspond to functions, italic letters correspond to scalars, lower case bold letters correspond to vectors (for example, to represent instances), and upper case bold letters correspond to matrices (for example, to represent bags).

2 Related Work

Multiple instance learning algorithms appeared to overcome some limitations of supervised learning regarding the training sets for some applications like object detection or drug activity prediction. The limitation is associated with the difficulty of providing an accurate and correctly labeled training set at the instance level.

The term multiple instance learning appeared with Dietterich *et al.* [7] in the context of drug activity prediction. Dietterich *et al.* [7] developed the Axis-Parallel Rectangle algorithm for predicting some property of the molecule based on the molecule's shape statistics. Since than, several algorithms have been proposed for solving the multiple instance learning problem. Maron *et al.* [13] proposed the Diverse Density (DD) algorithm to learn Gaussian concepts for representing the positive regions. This algorithm was then extended by Zhang *et al.* [17] to use the expectation-maximization (EM) has the optimization technique.

Other authors have proposed algorithms for multiple instance learning by adapting the support vector machine framework. The support vector machine framework aims in finding an hyperplane that is capable of separating the training data with the maximum margin possible. Andrews *et al.* [2] proposed two algorithms, mi-SVM and MI-SVM, that differ in the margin definition. The first considers the margin between the positive and negative instances while the second considers the margin between the bags. In the MI-SVM, the bags are not represented by multiple instances but rather by the most positive instance for the positive bags and by the least negative instance for the negative bags. Bunescu *et al.* [5] extended the MI-SVM algorithm by adding constraints to ensure that at least one of the instances in a positive bag is positive. This algorithm is shown to work well for sparse positive bags.

The previous approaches modify the objective function for adapting the support vector machines to the multiple instance learning framework. There are other approaches that modify the kernels used. Gartner *et al.* [11] proposed two kernels: statistic kernel and the normalized set kernel. In the statistic kernel, the bag is transformed into a feature vector by selecting the minimum and maximum values for each feature from all instances in the bag. In the normalized set kernel, the bag is represented as the sum of all instances that belong to the bag normalized by the 1 or 2-norm.

These algorithms have been applied in different application domains like classification of molecules [7], content based image retrieval [2], text classification [2], among others. These algorithms have different formalisms and a more detailed review can be found in the following articles [1, 3, 10].

Since there are a large number of applications and a high number of algorithms for multiple instance learning, several studies have concentrated in finding the best multiple instance classifier. Ray and Craven [16] concluded that there is no optimal multiple instance learning algorithm since their performance depend on the data. Furthermore, the multiple instance learning algorithms have been compared to the correspondent supervised learning algorithms using different datasets [5, 7, 16]. These studies concluded that the multiple instance learning algorithms are consistently superior, although this observation also depends on the data. Nonetheless, none of these studies have considered the rate of positive instances in their data while evaluating the performance of the multiple instance learning algorithms. The rate of positive instances on positive bags is directly related with the standard multiple instance learning assumption. Therefore, in this work we want to evaluate if the rate of positive instances on positive bags influence the performance of multiple instance learning algorithms. This will allow to determine if this should be considered as a variable when analyzing the performance of these algorithms.

3 Multiple Instance Learning

In this section we will present the formalism associated with the multiple instance learning problem. Remember from Section 1 that in supervised learning, the training set consists of example pairs: an input and a corresponding label. In multiple instance learning, the inputs are grouped into bags and the labels are assigned to the bags of inputs. Thus, we do not know which of the inputs or pair of inputs is responsible for the label. In this sense, the multiple instance learning framework makes weaker assumptions about the labeling information.

In order to formulate the multiple instance problem, let us consider a training data Φ with N pairs of examples:

$$\Phi = \{(\mathbf{X}_1, y_1), (\mathbf{X}_2, y_2), \ldots, (\mathbf{X}_N, y_N)\} \tag{1}$$

where $\mathbf{X}_i$ is the bag of the i-th example, and y_i is the corresponding bag label of the i-th example. A bag $\mathbf{X}_i$ is composed of M_i instances:

$$\mathbf{X}_i = \left\{\mathbf{x}_{i1}, \mathbf{x}_{i2}, \ldots, \mathbf{x}_{iM_i}\right\} \tag{2}$$

where $\mathbf{x}_{ij} \in \chi_j$ is the j-th instance of the bag $\mathbf{X}_i$ and $\chi_j = \mathbb{R}^{D_j}$ is the D_j-dimensional Euclidean space (dimension of the j-th instance). For simplifying the notation, assume from now on that all bags in the N pairs of examples have the same number of instances $M_i = M$ and all instances belong to the same D-dimensional Euclidean space $\chi_j = \chi = \mathbb{R}^D$.

A bag label $y_i \in \Upsilon$ is the result of the labels given to each of the instances $y_{ij} \in \upsilon$ that compose the bag $\mathbf{X}_i$. For simplifying this exposure, consider a binary classification problem where $\Upsilon = \{-1, 1\}$ and $\upsilon = \{-1, 1\}$. Using the standard multiple instance learning assumption for binary classification, mentioned in Section 1, the bag label y_i corresponds to:

$$y_i = \max_j y_{ij} = \begin{cases} 1 & \exists_j : y_{ij} = 1 \\ -1 & \forall j : y_{ij} = -1 \end{cases} \tag{3}$$

The goal of the multiple instance learning algorithm is to train an instance classifier $\mathrm{f}(\mathbf{x}) : \chi \rightarrow \upsilon$ or a bag classifier $\mathrm{F}(\mathbf{X}) : \chi^M \rightarrow \Upsilon$. From (3), a bag classifier can be obtained from a correct instance classifier by $\mathrm{F}(\mathbf{X}_i) = \mathrm{sign}\left(\max_j \mathrm{f}\left(\mathbf{x}_{ij}\right)\right)$. Hence, most of the multiple instance learning algorithms aim to learn instance classifiers instead of bag classifiers.

The focus of this work is to evaluate the performance of support vector machines in the multiple instance learning framework in an image classification task, therefore we will now present the formalism of the two classifiers used: Maximum Instance Margin (mi-SVM) and Maximum Bag Margin (MI-SVM). But first, let us introduce the support vector machine framework for supervised learning.

3.1 Support Vector Machines

Consider a training set Ω with K examples:

$$\Omega = \{(\mathbf{x}_{i1}, y_{i1}), (\mathbf{x}_{i2}, y_{i2}), \ldots, (\mathbf{x}_{iK}, y_{iK})\} \tag{4}$$

in a binary classification problem $y_{ij} = \{-1, 1\}$. The objective of a support vector machine framework is to find the hyperplane that separates the examples with the biggest margin possible. The margin is defined as the smallest distance between the hyperplane and a positive and a negative example.

The support vector machine soft-margin formulation is:

$$\begin{aligned} &\min_{\mathbf{w}, w_0, \xi_i} \frac{1}{2} \|\mathbf{w}\|^2 + C \sum_{j=1}^{K} \xi_{ij} \\ &s.t. \quad \xi_{ij} \geq 0 \\ &\qquad y_{ij}\left(\mathbf{w} \cdot \mathbf{x}_{ij} + w_0\right) \geq 1 - \xi_{ij} \end{aligned} \tag{5}$$

where $\mathbf{w}$ is the hyperplane normal, w_0 is the hyperplane offset, and the ξ_{ij} are the slack variables that allow to apply the support vector machine framework to data that is not separable. This means that we are allowed some mislabeled examples. This leads to a quadratic programming problem that is convex and easily solvable. The examples that are nearest to the hyperplane are called the support vectors.

In this work we do not intend to detail the support vector machine formalism. Therefore, for a detailed introduction to support vector machines the reader should refer to [12, 14].

3.2 Maximum Instance Margin: mi-SVM

The maximum instance margin formulation of the support vector machines aims to recover the instance labels of the positive bags.

In support vector machines for supervised learning, the labels y_{ij} of each instance $\mathbf{x}_{ij}$ in a training set are known. In multiple instance learning this is not the case, only the labels y_i of a bag $\mathbf{X}_i$ are known. Considering the standard multiple instance learning assumption (3), we can see that the labels for each instance of a negative bag are also known and therefore the margin could be defined as in a regular support vector machine. However, the labels for each instance of a positive bag are unknown and therefore computing the margin is more complicated. Andrews *et al.* [2] treated the instance labels y_{ij} as unknown integer variables:

$$
\begin{aligned}
\min_{y_{ij}} \ & \min_{\mathbf{w}, w_0, \xi_i} \ \tfrac{1}{2} \|\mathbf{w}\|^2 + C \sum_{i,j} \xi_{ij} \\
s.t. \ & \xi_{ij} \geq 0 \\
& y_{ij} \left(\mathbf{w} \cdot \mathbf{x}_{ij} + w_0 \right) \geq 1 - \xi_{ij} \\
& y_{ij} = -1, \ \forall i : y_i = -1 \\
& \sum_j \tfrac{y_{ij}+1}{2} \geq 1, \ \forall i : y_i = 1
\end{aligned}
\tag{6}
$$

From (6) we can see that in mi-SVM multiple negative or positive instances in positive bags can be support vectors. The problem leads to a mixed integer program that is hard to solve. Nonetheless, the integer variables, the hidden labels y_{ij}, reduce the problem to a quadratic programming problem. Andrews *et al.* [2] proposed an heuristic that consists of two steps: first train a SVM classifier considering a given value for the instance labels, then use the new classifier to update the instance labels y_{ij} of positive bags. This process is computed until no changes occur in the instance labels.

3.3 Maximum Bag Margin: MI-SVM

The maximum bag margin formulation of the support vector machines aims to recover the key positive instance for every positive bag.

In this formulation, the margin of a bag corresponds to the maximum distance between the hyperplane and all of the instances that belong to a bag:

$$y_i = \max_j \left(\mathbf{w} \cdot \mathbf{x}_{ij} + w_0 \right) \tag{7}$$

From (7) we can see that the margin of a positive bag is determined by the most positive instance while the margin of a negative bag is determined by the least negative instance. The bag margin formulation is:

$$\begin{aligned} &\min_{\mathbf{w}, w_0, \xi_i} \frac{1}{2} \|\mathbf{w}\|^2 + C \sum_{i,j} \xi_{ij} \\ &s.t. \;\; \xi_{ij} \geq 0 \\ &\quad\;\; y_i \max_j \left(\mathbf{w} \cdot \mathbf{x}_{ij} + w_0 \right) \geq 1 - \xi_{ij} \end{aligned} \tag{8}$$

This optimization is not convex, therefore Andrews *et al.* [2] introduced an extra variable $s(j)$ for each bag. This variable denotes the instance that is selected as the witness instance of a positive bag. The formulation (8) is now given by:

$$\begin{aligned} &\min_{s(j)} \min_{\mathbf{w}, w_0, \xi_i} \frac{1}{2} \|\mathbf{w}\|^2 + C \sum_{i,j} \xi_{ij} \\ &s.t. \;\; \xi_{ij} \geq 0 \\ &\quad\;\; \mathbf{w} \cdot \mathbf{x}_{ij} + w_0 \leq -1 + \xi_{ij}, \forall i : y_i = -1 \\ &\quad\;\; \mathbf{w} \cdot \mathbf{x}_{is(j)} + w_0 \geq 1 - \xi_{ij}, \forall i : y_i = 1 \end{aligned} \tag{9}$$

In this formulation, each positive bag is represented by only one positive instance. All the negative instances in the positive bags are disregarded. Like the formulation of mi-SVM, this corresponds to a mixed integer program. Nonetheless, the integer variables, the variables $s(j)$, reduce the problem to a quadratic programming problem. Andrews *et al.* [2] also proposed an heuristic that consists in two steps: first train a classifier like a regular supervised learning considering a given witness instance, then using the new classifier select new witness instances for the positive bags. The optimization process ends when the selected witness stops to change.

4 Methodology

The objective of this work is to compare the performance of multiple instance support vector machines' learning algorithms with the percentage of positive instances on positive bags. This performance is evaluated in an image annotation task using a dataset for pedestrian detection. Image annotation task consists on identifying if an image has a person and if a given region of that image contains a person or a part of a person.

Normally, datasets in multiple instance learning do not report the percentage of positive instances included on the positive bags and there is no study evaluating the influence of positive instances on this type of classifiers. Therefore, a new multiple instance learning dataset has been created based on a real dataset for pedestrian detection, the HDA Person Dataset [9, 15].

Fig. 1 Examples of positive and negative instances obtained during the transformation of the HDA Person Dataset. A and C: Negative instances (cyan) drawn from negative bags. B and D: Negative (cyan) and positive (red) instances drawn from positive bags.

The HDA dataset is a high resolution image sequence dataset for research on high definition surveillance, pedestrian detection and re-identification. The dataset comprises information from 18 cameras with different resolutions. A total of 13 image sequences are labeled. The labeled data includes 64.028 annotations from 85 persons in a total of 75.207 frames. The annotations include information about the person bounding box position and unique identification, occlusion and type of detection (person or crowd). Additionally, the annotations have information about the camera and frame number.

In order to transform this dataset, we adopted a similar strategy to Andrews *et al.* [2]. This takes in consideration the standard multiple instance learning assumption for binary classification: an image is positive if at least there is one person or a part of a person on the image, and an image is negative if there is no person or part of a person on the image. Therefore, the positive images are randomly drawn from the annotated frames of the cameras while the negative images are sampled from the non-annotated frames. Notice that frames annotated with occlusion are not considered, in order to not provide erroneous features to the classifier. Remember that the bounding box for an occluded person is drawn by estimating the whole body extent in the HDA Person Dataset.

In a positive image (Figure 1.B and Figure 1.D), the positive image regions are obtained by sampling randomly the bounding box area while the negative image regions are drawn from the remaining area of the annotated frames. In a negative image (Figure 1.A and Figure 1.C), the negative image regions are randomly sampled from the entire area of the non-annotated frame. In either of the cases, the maximum overlapping area between image region of the same image is 40%, and the window considered for each image region is 60 x 60. Furthermore, in order to

Table 1 Number of instances for each training and testing set obtained from the HDA Person Dataset.

Set	Positive Instance Rate	Positive Bags		Negative Bags	Total Bags	
		Total Instances	Positive Instances	Negative Instances	Total Instances	Positive Instances
Testing	0.469	409	192	500	909	192
Training-10	0.100	750	75	750	1500	75
Training-30	0.300	750	225	750	1500	225
Training-50	0.500	750	375	750	1500	375
Training-70	0.699	747	522	750	1497	522
Training-90	0.899	739	664	750	1489	664

comply with a given percentage of positive image regions on positive images, only annotated frames that satisfied this percentage have been considered for each of the training datasets obtained. Like the MIL datasets mentioned previously, the number of instances per bag is also variable. The features are extracted from the image regions using the integral channel features defined by Dollar *et al.* [8] and that have been used for pedestrian detection. These features include color (LUV color space), gradient magnitude, and gradient histograms. Remember from Section 3 that each instance belongs to a D-dimensional Euclidean space, a feature corresponds to each of the D components of that Euclidean space. The number of features per image region is 600.

The training and testing sets have been obtained using the approach defined above. A total of 5 training sets were created, each with a different percentage of positive instances on positive bags: 0.1, 0.3, 0.5, 0.7, and 0.9. In order to have an unbiased comparison, the testing set consists of sets with different percentage of positive instances: 0.1, 0.2, 0.3, 0.4, 0.5, 0.6, 0.7, 0.8, and 0.9. Each of these sets are composed of approximately the same number of bags. Notice also that the images for the training and testing sets were obtained from different cameras. The training set was obtained from cameras 53, 56, 57 and 58, while the testing set was obtained from cameras 50 and 59. The images used correspond to cameras of equal resolution 1280 x 800.

In conclusion, the HDA Person dataset was transformed into a new dataset with training sets that consist of 150 bags (75 positive and 75 negative bags), and into a testing set of 100 bags (50 positive and 50 negative bags). The summarized statistics of each dataset obtained can be found in Table 1.

Similarly to the previous datasets, this dataset is used for training and testing the mi-SVM and MI-SVM classifiers. The parameters for each of the kernels (linear, polynomial and RBF [6]) were analyzed and optimized using 3-fold cross-validation using a training set of 150 bags (75 positive and 75 negative bags) obtained similarly to the testing set described above.

5 Results

The classifier and kernels accuracies depend on the data and on the application domain [16]. Therefore, we started by evaluating the performance of each of the classifiers and kernels on a specific training set (Training-CL) before analyzing the influence of the positive instances on positive bags. This training set corresponds to a mix of bags with different percentages of positive instances in order to get training data that is not biased towards one of the training sets with a specific percentage of positive instances. This analysis will allow us to select the best classifiers and kernels to the problem of pedestrian detection. The results obtained are presented in Table 2.

From Table 2, we can conclude that the mi-SVM algorithms are consistently better than the MI-SVM counterparts. This means that the features present at the instance level are sufficient to learn a classifier that can discriminate between the positive and negative bags. Furthermore, we can conclude that the mi-SVM classifier with RBF kernel has the highest bag and instance classification accuracies for the adapted HDA Person dataset. Notice also that the instance accuracies for mi-SVM with a linear kernel are similar to the ones obtained using a RBF kernel. This analysis allows us to determine that the classifiers for this data should use RBF kernels. Therefore, the remaining analysis were made using the mi-SVM classifier with RBF kernel and the MI-SVM classifier with RBF kernel.

As mentioned in Section 4, the performance of the mi-SVM and MI-SVM classifiers is evaluated for different percentages of positive instances on positive bags. The classification accuracies obtained are reported in Table 3. From Table 3, the increasing percentage of positive instances has higher influence on the maximum bag margin algorithms than on the maximum instance margin algorithms. In the mi-SVM, the instance classification accuracies are very similar for the different percentages of positive instances (maximum difference between the minimum and maximum classification accuracies is 1.8%), and the bag classification accuracy reaches a maximum when the training set has 50% of positive instances and then starts to decrease. In the MI-SVM, the bag and the instance classification accuracies increase significantly with the increasing percentage of positive instances.

The results in Table 3 give information about the overall accuracy of the classifier. In Figure 2, we present the performance of the classifier on positive and negative instances. From Figure 2, we can reinforce that the impact of an increasing percentage of positive instances is higher on MI-SVM classifier. Namely, when the rate of positive instances increases the classification accuracy of positive bags and instances increases. Nonetheless, this increase is followed by a decrease in the classification accuracy of negative bags and instances. This decrease occurs at a significantly less extent on instances than on bags. Regarding the mi-SVM classifiers, the influence of positive instances is more noticeable at the bag level. The change in the classification accuracy is driven by the change on the classification accuracy of negative bags (maximum difference between the minimum and maximum classification accuracies is 11%). The results show that the classifier is biased towards negative bags when there is a low and high percentage of positive instances. At the instance level, the accuracy does not have significant changes with the increase of positive instances.

In conclusion, for maximum bag margin algorithms, the increase of positive instances allow to obtain classifiers with higher accuracy (Figure 3). Nonetheless, at the bag level, the overall accuracy of the classifier does not benefit when the percentage of positive instances is too high (90%). These findings can be explained

Table 2 Bag and instance classification accuracies on an adapted HDA Person dataset. The higher classification accuracies are presented in bold.

Training Set	Bag Classification Accuracy						Instance Classification Accuracy					
	mi-SVM			MI-SVM			mi-SVM			MI-SVM		
	Linear	Polynomial	RBF	Linear	Polynomial	RBF	Linear	Polynomial	RBF	Linear	Polynomial	RBF
Training-CL	0.680	0.510	**0.770**	0.600	0.500	0.710	0.878	0.677	**0.879**	0.804	0.789	0.816

Table 3 Bag and instance classification accuracies for the mi-SVM and MI-SVM with RBF kernel obtained from training in datasets with different rates of positive instances.

Training Sets	mi-SVM.RBF		MI-SVM.RBF	
	Bag Accuracy	Instance Accuracy	Bag Accuracy	Instance Accuracy
Training-10	0.730	0.882	0.500	0.789
Training-30	0.760	**0.889**	0.500	0.789
Training-50	**0.790**	0.871	0.580	0.795
Training-70	0.780	0.882	**0.760**	0.843
Training-90	0.750	0.879	0.750	**0.864**

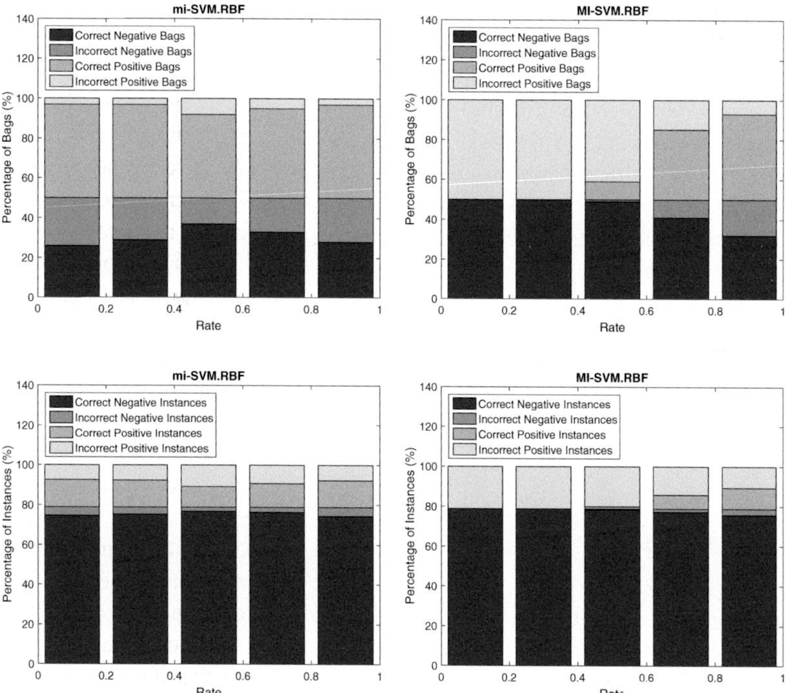

Fig. 2 Percentages of correctly and incorrectly classified bags (top) and instances (bottom) with increasing percentage of positive instances on positive bags for mi-SVM (left) and MI-SVM (right) classifiers.

by the fact that these multiple instance algorithms use the most positive instance of each positive bag and the least negative of each negative bag to find the hyperplane that best separate the two types of instances. Thus, the higher percentage of positive instances allows the classifier to span more hypothesis for finding the instance that is most representative of the positive class. Nonetheless, after finding the witness instance of the positive class, adding more instances would not help increase the accuracy of the classifier. These results suggest that is better to use images whose random sampling has higher likelihood of providing positive instances.

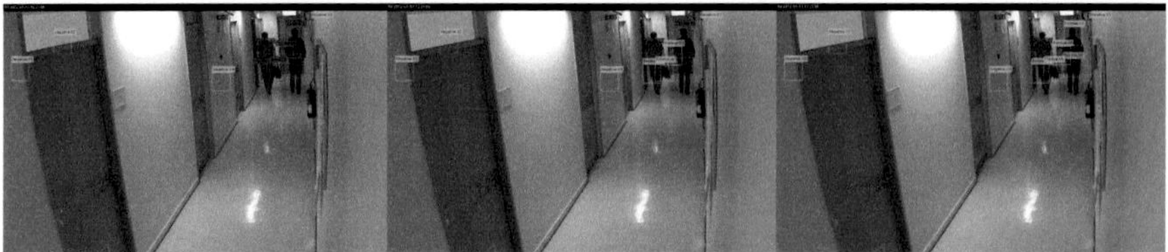

Fig. 3 Example of an instance classification using the MI-SVM classifier for increasing rate of positive instances on positive bags: 0.5, 0.7, and 0.9. The image regions in green correspond to instances correctly labeled while the image regions in red correspond to mislabeled instances.

On the other hand, for maximum instance margin algorithms, the increase of positive instances do not exhibit any influence on the classifier accuracy at the instance level. This may be due to the fact that these algorithms use multiple positive or negative instances to determine the hyperplane. Although the percentage of positive instances is increasing, the negative instances are present in a high percentage in all training sets. Therefore, these classifiers are highly capable of determining the negative instances from the remaining instances. This could justify why the bag and instance accuracy of these classifiers is high throughout the several rate of positive instances. Nonetheless, at the bag level, the classifier shows signals of overfitting for high and low rate of positive instances.

6 Conclusions

In this work, the influence of positive instances on positive bags is analyzed on support vector machines adapted for the multiple instance learning framework: mi-SVM and MI-SVM. The algorithms were evaluated using a new dataset obtained from an existing pedestrian detection dataset.

The results show that the increasing percentage of positive instances have a positive impact in the overall accuracy of the MI-SVM classifier as a result of the maximum bag margin formulation. For the mi-SVM classifiers, no relevant changes occur at instance level. At bag level, the classifier is biased towards negative bags at low and high rates of positive instances affecting the overall performance of the classifier. Therefore, the rate of positive instances should be considered while evaluating the performance of mi-SVM and MI-SVM classifiers.

As future work, we would like to extend this analysis to more multiple instance learning algorithms and consider algorithms that are based on other multiple instance learning assumptions.

Acknowledgements This work has been partially supported by the Portuguese Foundation for Science and Technology (FCT) project [UID / EEA / 50009 / 2013], by the CMU-Portugal Project AHA [CMUP-ERI / HCI / 0046 / 2013], and by the EU Project POETICON++ EU-FP7-ICT-288382. Nuno Barroso Monteiro is funded by FCT PhD grant PD/BD/105778/2014.

References

1. Amores, J.: Multiple instance classification: Review, taxonomy and comparative study. Artificial Intelligence **201**, 81–105 (2013)
2. Andrews, S., Tsochantaridis, I., Hofmann, T.: Support vector machines for multiple-instance learning. In: Advances in Neural Information Processing Systems, pp. 561–568 (2002)
3. Babenko, B.: Multiple instance learning: algorithms and applications. View Article PubMed/NCBI Google Scholar (2008)
4. Bishop, C.M.: Pattern recognition and machine learning. Springer (2006)
5. Bunescu, R.C., Mooney, R.J.: Multiple instance learning for sparse positive bags. In: Proceedings of the 24th International Conference on Machine Learning, pp. 105–112. ACM (2007)
6. Burges, C.J.: A tutorial on support vector machines for pattern recognition. Data Mining and Knowledge Discovery **2**(2), 121–167 (1998)
7. Dietterich, T.G., Lathrop, R.H., Lozano-Pérez, T.: Solving the multiple instance problem with axis-parallel rectangles. Artificial Intelligence **89**(1), 31–71 (1997)
8. Dollár, P., Tu, Z., Perona, P., Belongie, S.: Integral channel features. In: BMVC, vol. 2, p. 5 (2009)
9. Figueira, D., Taiana, M., Nambiar, A., Nascimento, J., Bernardino, A.: The hda+ data set for research on fully automated re-identification systems. In: 2014 Workshops Computer Vision-ECCV, pp. 241–255. Springer (2014)
10. Foulds, J., Frank, E.: A review of multi-instance learning assumptions. The Knowledge Engineering Review **25**(01), 1–25 (2010)
11. Gärtner, T., Flach, P.A., Kowalczyk, A., Smola, A.J.: Multi-instance kernels. In: ICML, vol. 2, pp. 179–186 (2002)
12. Hastie, T., Tibshirani, R., Friedman, J.: The elements of statistical learnin (2009)
13. Maron, O., Lozano-Pérez, T.: A framework for multiple-instance learning. Advances in Neural Information Processing Systems, 570–576 (1998)
14. Murphy, K.P.: Machine learning: a probabilistic perspective. MIT press (2012)
15. Nambiar, A., Taiana, M., Figueira, D., Nascimento, J.C., Bernardino, A.: A multi-camera video dataset for research on high-definition surveillance. International Journal of Machine Intelligence and Sensory Signal Processing **1**(3), 267–286 (2014)
16. Ray, S., Craven, M.: Supervised versus multiple instance learning: an empirical comparison. In: Proceedings of the 22nd International Conference on Machine Learning, pp. 697–704. ACM (2005)
17. Zhang, Q., Goldman, S.A.: EM-DD: an improved multiple-instance learning technique. In: Advances in Neural Information Processing Systems, pp. 1073–1080 (2001)

A Data Mining Approach to Predict Falls in Humanoid Robot Locomotion

João André, Brígida Mónica Faria, Cristina Santos and Luís Paulo Reis

Abstract The inclusion of perceptual information in the operation of a dynamic robot (interacting with its environment) can provide valuable insight about its environment and increase robustness of its behaviour. In this regard, the concept of Associative Skill Memories (ASMs) has provided a great contributions regarding an effective and practical use of sensor data, under a simple and intuitive framework [2, 13]. Inspired by [2], this paper presents a data mining solution to the fall prediction problem in humanoid biped robotic locomotion. Sensor data from a large number of simulations was recorded and four data mining algorithms were applied with the aim of creating a classifier that properly identifies failure conditions. Using Support Vector Machines, on top of sensor data from a large number of simulation trials, it was possible to build an accurate and reliable *offline* fall predictor, achieving accuracy, sensitivity and specificity values up to 95.6%, 96.3% and 94.5%, respectively.

J. André · C. Santos
DEI/EEUM - Dep. de Electrónica Industrial, Esc. de Engenharia da Universidade do Minho, Guimarães, Portugal
e-mail: {joaocandre,cristina}@dei.uminho.pt

B.M. Faria(✉) · L.P. Reis
ESTSP/IPP – Esc. Sup. de Tecnologia da Saúde do Porto, Instituto Politécnico do Porto, Porto, Portugal
e-mail: btf@estsp.ipp.pt, lpreis@dsi.uminho.pt

L.P. Reis
DSI/EEUM - Dep. de Sistemas de Informação, Esc. de Engenharia da Universidade do Minho, Guimarães, Portugal

J. André · C. Santos · L.P. Reis
Centro ALGORITMI, Guimarães, Portugal

B.M. Faria
LIACC – Laboratório de Inteligência Artificial e Ciência de Computadores, Porto, Portugal

B.M. Faria
INESC TEC – INESC Tecnologia e Ciência, Porto, Portugal

L.P. Reis et al. (eds.), *Robot 2015: Second Iberian Robotics Conference*,
Advances in Intelligent Systems and Computing 418,
DOI: 10.1007/978-3-319-27149-1_22

1 Introduction

Learning in a robotic environment can bring about many advantages and performance improvement, the same way it is a necessary skill in human behaviour [12], but nonetheless it comprises a demanding task for every robot. With humans, learning is an on-going process, and even after grasping the basic motions necessary for a specific movement, there is continuous improvement on that skill, either by trial-and-error or inferences from sensor data and past experiences. Humans perform this type of learning almost unconsciously, predicting their actions' impact on their surroundings and improving them in order to achieve better results. At this point, however, robots do not use sensor information not nearly as automatically, leading to time-consuming parameterization for reasonable performance [11].

What makes robotic control such a challenging task is the sheer amount of variables involved in the interaction between a robot and its surroundings. The amount of information to be processed at each fraction of a second is unmeasurable and it as such it is not realistically possible to account in advance for all potential disruptive interferences in the way of movement execution. There is a need for alternative ways to increase robustness in robotic motions, either at hardware level (passive adaptation with compliant joints [13]) or at control level (extensive calibration, exploration though vision systems [13]). Popular solutions are often based on a feedback loop that corrects movement plans as needed in response to perceptible changes in the environment [13].

An attractive feature in robotic systems is the ability to manipulate sensor information to improve its walking performance. Ideally, a robot would be able to identify potential failure/uncharacteristic/unfamiliar conditions and learn from past experiences only based on the perceptual information its hardware provides. This can prompt the robot to take counter-measures in time to ensure task completion, reacting to sudden fluctuations in the environment that otherwise would have a disruptive effect in movement execution. Several failure detection systems include particle [14] filters and neural-network based approaches [5, 6], however these are highly focused in industrial and wheeled robots, and up to the author's knowledge such a framework in humanoid/legged robots is lacking.

Associative Skill Memories (ASMs) [13] have paved the way towards a more elegant and practical way to solve this problem. They entail a simple and intuitive way to look at perceptual information and its relation with how individual skills or basic motions are performed. The basic principle of the *skill memories* [13] is not in the sensor data itself but on the *association* (hence the term *Associative*) of that same data with the specific skill or motion it represents - based on the premise that *stereotypical movements* tend to leave similar *sensor footprints* with each execution, even if the environment is dynamically changing.

A framework was proposed in [2] to properly construct an ASM, as well as compress and adapt the learning process to periodic movements such as biped locomotion, the main focus of this work. A failure predictor was then implemented through simple statistical processes on top of the compressed ASM. However, in this work we opted for a different approach, with the goal of improving the prediction procedure. From the ASM premise that stereotypical skills leave unique sensor footprints, instead of processing the raw sensor data into a phase-indexed ASM, we executed several Data Mining algorithms on the raw sensor data, creating a classifier aimed at *offline* detection. While without several advantages of a phase-indexed memory, this analysis might help identify specific issues regarding sensor selection, and achieve better results. This way, four data mining algorithms were run in similar conditions and with the same input data: Support Vector Machines, Nearest Neighbour, Naive Bayes, and Artificial Neural Networks.

The robotic motion of interest in this work is in biped locomotion which is complex, periodic, and heavily influenced by the characteristics of the surrounding environment (type of terrain, actuator noise, etc) [16]. We make use of an already established Central Pattern Generator (CPG) architecture that generates effective and stable biped locomotion for the Robotis DARwIn-OP robot [16]. Using the simulation software *Webots*, an extensive number of simulation trials - simple walking motions of preset duration - was conducted and their sensor data was stored, along with a cost metric previously used in other works[2, 3]. With this cost metric and the definition of a cost threshold, simulation trials were labeled as either *successful* or *unsuccessful*, a boolean value added to the mining dataset.

This procedure assumes there is variability among the simulation trials - white noise was injected directly in the joint CPG dynamics. The magnitude of this noise was 1% in half of the simulation trials and 0.1% in the other half. This ensures there is a significant number of successful and unsuccessful trials (excessive perturbations leads most of the time to a robot fall). However, it should be noted that these values were determined empirically with objective of achieving a similar number of successful and unsuccessful trials - in real world conditions, it is expected that noise levels are even greater.

There should also be made the distinction between what is considered a simple *failure* and robot fall. The CPG controlled running in the DARWiN-OP humanoid robot leads the robot in a forward trajectory, and as such any deviation from the supposed path caused by external perturbations is taken as *undesirable* and considered a *failure*. In fact, as will be shown in the following sections, the cost metric helps not only identify the trials here the robot fell, but also quantifies performance in regards to the walking distance and direction. Additionally, there is also a cost penalty in the case of robot fall.

The paper's structure is as follows: Section 2 offers a brief description of the locomotion system adopted in this work by the humanoid robot, Section 3 details the procedure through which the simulation trials were conducted, followed by Section 4 that presents the main results and their discussion.

2 Humanoid Biped Walking

The simulation trials performed in this work were identical to the ones in previous works[2]. A robust and versatile dynamic controller based on *Central Pattern Generators (CPGs)* and improved with a sensor feedback loop [2, 9] generates the walking motions for the 6 joints in each leg. Variability is achieved by injecting white noise on top of these joint dynamics. As a detailed description of this controller can be found on[1–3, 8] and it falls out of scope of this work, only a brief overview of the locomotion system will be presented.

2.1 Central Pattern Generators

Central Pattern Generators (CPGs) are neural circuits in vertebrates' spinal cord, generating rhythmic activation patterns for several rhythmic motor actions, such as locomotion. It has been show that a similar approach can be effective when designing bioinspired robotic controllers. Walking motor patterns in such controllers are organized as coupled unit-burst elements, consisting of a motion pattern generator driven by a rhythmic generator [8, 16] - a phase oscillator controls the pace in the generation of rhythmic motions in a given leg i, setting the current position in the gait cycle:

$$\dot{\phi}_i = \omega + k \sin(\phi_i - \phi_o + \pi) \tag{1}$$

where the phase $\phi_i \in [-\pi, \pi]$ (rad.s^{-1}) increases monotonically and linearly with rate ω, and k is the coupling strength between right and left leg oscillators, which defines the anti-phase relationship between contra-lateral motion generators on each leg - keeping both legs coordinated during locomotion. This type of phase coupling has evident advantages, such as the straightforward maintenance of phase relationships and entrainment, used to achieve interlimb coordination among the unit generators.

Motion generators create rhythmic trajectories through the sum of motion primitives, encoded as a set of non-linear dynamical equations with well-defined attractor dynamics and easily modulated in amplitude, frequency, and offset. Joint position $z_{i,j}(t)$ is generated according to the current phase ϕ_i:

$$\dot{z}_{i,j} = \sum_{n=1}^{N} f_j^n(z_{i,j}, \phi_i) - \alpha(z_{i,j} - O_{i,j}) \tag{2}$$

where i, j specifies the jth joint of the ith leg. $O_{i,j}$ is the rhythmic motion offset and α the relaxation parameter for this offset. Each function $f_j^n(z_{i,j}, \phi_i)$ specifies a single motion primitive, and the final joint trajectories result from the sum of all N motion primitives actuating in each joint j, which for the DARwIn-OP are: *hip roll*

Table 1 CPG Parameters for straight walking.

Amplitude	DARwIn	Offset	DARwIn
$A_{\text{balancing}}$	14	O_{hYaw}	0
$A_{\text{flex,hip}}$	15	O_{hRoll}	1
$A_{\text{flex,knee}}$	30	O_{hPitch}	-25
A_{yield}	5	O_{knee}	40
A_{rotation}	5	O_{aRoll}	-1
A_{compass}	11	O_{aPitch}	22
$\omega(\text{rad.s}^{-1})$	9.81	k	7

(hRoll), *hip yaw* (hYaw), *hip pitch* (hPitch), *knee* (knee), *ankle roll* (aRoll) and *ankle pitch* (aPitch), as can be seen in Figure 1.

In order to maintain the feet parallel to the ground at all times, motions assigned to the ankle are symmetric to those performed by the hip and the knee joints. A general schematic of the CPG system is shown in Figure 1, and a summary of the contribution of each motion to the joint trajectories $\dot{z}_{i,j}$ is presented in expressions (3)-(8):

$$\dot{z}_{i,\text{hRoll}} = f_{\text{hRoll}}^{\text{balancing}} \tag{3}$$

$$\dot{z}_{i,\text{hYaw}} = f_{\text{hYaw}}^{\text{rotation}} \tag{4}$$

$$\dot{z}_{i,\text{hPitch}} = f_{\text{hPitch}}^{\text{flex}} + f_{\text{hPitch}}^{\text{compass}} \tag{5}$$

$$\dot{z}_{i,\text{kPitch}} = f_{\text{kPitch}}^{\text{flex}} + f_{\text{kPitch}}^{\text{yield}} \tag{6}$$

$$\dot{z}_{i,\text{aPitch}} = f_{\text{aPitch}}^{\text{flex}} + f_{\text{aPitch}}^{\text{yield}} + f_{i,\text{aPitch}}^{\text{compass}} \tag{7}$$

$$\dot{z}_{i,\text{aRoll}} = f_{\text{aRoll}}^{\text{balancing}} \tag{8}$$

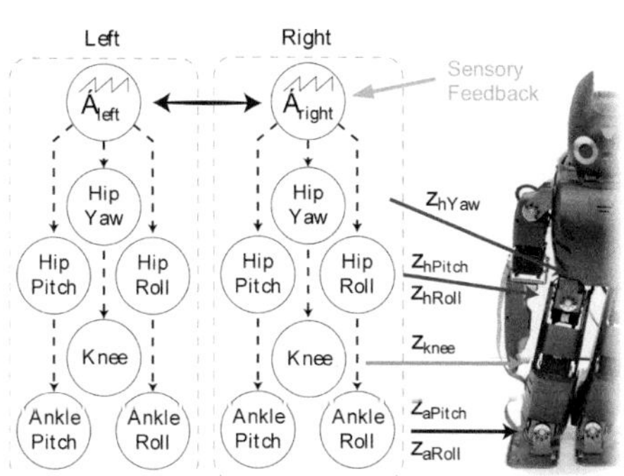

Fig. 1 General Schematic of CPG locomotion.

Biped walking typically exhibits a duty factor of $\beta = 0.6$, achieving two types of support during a step cycle: single and double support. During the first the robot's body is propelled forward the body forward using one leg, while the other swings frontwards until the foot is completely grounded. Double support implies the phase where both legs support the robot simultaneously, in the transition between alternate leg swings. A detailed description as well as an extensive analysis of this CPG solution is provided in [8].

2.2 Perceptual Feedback

Since we consider only natural that locomotion should be adapted considering physical interactions of the robot with its surroundings and body changes, it becomes of utmost importance, when designing controllers for legged locomotion [4, 8], to be able to include reactive behaviors, achieved e.g. with differential equations in movement generation. Phase regulation mechanisms (such as in [4, 8]) have shown reasonable performance, however they do not deal with quantitative information, that describes the extent to which the body and environment interactions are *felt*. Herein, we build upon the idea described in [9] and instruct our biped humanoid robot to adapt its gait proportional to the foot contact with the ground.

We implemented a simple sensory-driven feedback loop from sensor information recorded by the foot force sensors, to adjust oscillator phase in a quantitative way (Figure 1), as follows:

$$\dot{\phi}_i = \omega + k \sin(\phi_i - \phi_o + \pi) + \sigma \text{GRF}_i \sin(\phi_i). \tag{9}$$

This expression adds a third element to (1),responsible for sensory feedback and its role in the computation of $\dot{\phi}_i$. σ is the feedback strength, and GRF_i (the *Ground Reaction Force (GRF)* on the ith leg) is the sum of the values measured by the four force sensors on each foot.

This feedback mechanism is based on the premise that leg oscillators should 1) slow down proportionally to GRF during *stance* phase and 2) speed up similarly during the *swing* phase. Higher GRF values during stance imply that weight is primarily distributed on a single leg, thus the opposite leg is assumed to still be in swing. When considering stance-swing transition, in order to maintain the stability of the biped robot, the standing leg is *forced* to slow down by decreasing $\dot{\phi}_i$ and waiting for the opposite leg to touch the ground (initiate stance) before proceeding to swing. This delay is proportional to the GRF_i value measured by the foot force sensors - greater weight on a single leg slows down the walking cycle proportionally on the same leg until that weight is relieved when the opposing leg enters stance phase. Similarly, during swing-stance transition, phase variation is accelerated if GRF_i has a non-zero value, implying that the foot touched the ground ahead of time (the leg should still be in swing), and $\dot{\phi}_i$ is increased. There, the feedback term on (9) has a negative value during stance-swing transition, and inversely during swing-stance transition. Increasing or decreasing $\dot{\phi}_i$ has a similar effect on the opposite leg oscillator due to the coupling term in (9). It is important to note that the value of σ was hand tuned (by trial-and-error) to this particular application ($\sigma = 0.005$).

When compared to the feedback system described in [9], in addition to the obvious distinction in robot morphology (quadruped vs biped), the previous work discarded intralimb coordination due to single-joint limbs, while in this work the DARwIn-OP robot has 6 DoF per leg, addressed in the oscillators and the CPG system. Additionally, closed loop CPG dynamics use coupled oscillators to achieve synchronization between both legs (no type of interlimb coordination system was implemented in [9]). Despite these relevant differences between the quadruped robot used in [9] and the DARwIn-OP upon which our work was focused, the basic motivation and methodology still applies.

3 Experiments and Simulation

Each trial consists of 10 CPG periods of stable walking (after an initial balancing and stepping phase, roughly 25 seconds). Data from a set of 32 preselected sensors is recorded at a rate identical to the simulation integration step, $t = 0.008\ s$. This leads to approximately 1600 data values for each sensor along the 10 movement periods. Having data mining in mind, this amount of information should be reduced in order to achieve a tractable dataset. Therefore these values are filtered by a mean filter of variable length/window $w = 10, 20, 30, 40, 50$ leading to 5 datasets with variable size.

3.1 Simulation Cost

In order to proper evaluate performance of the robot in each trial, a cost C_t is computed for each simulation trial, allowing quantification of the walking performance according to the movement objective - linear forward displacement, and actively providing a reliable performance measure, illustrating how close is the final behavior in a particular trial to the movement objective. Therefore am instantaneous cost function was established that provides the cost C_n of trial t at the nth instant of the simulation, based on the frontal distance traveled by the DARwIn-OP in that instant - favoring frontal displacement but penalizing lateral movement:

$$C_n(t) = -\frac{|\Delta z_n|}{D_n}\frac{|\Delta z_n|}{\|d_n\|}\left(z_n - z_{n-1}\right), \tag{10}$$

where $\Delta z_n = z_n - z_0$, $\Delta x_n = x_n - x_0$ and $\|d_n\| = \sqrt{\Delta z_n^2 + \Delta x_n^2}$ are the frontal, lateral and total displacement respectively (with z_0 and x_0 as the initial frontal and lateral positions). The total distance traveled is defined as

$D_n = \sum_{k=0}^{n} \sqrt{(z_{k+1} - z_k)^2 + (x_{k+1} - x_k)^2}$. In order the compute the total cost $C(t)$ of a trial t, instantaneous costs $C_n(t)$ are summed over all N simulation instants:

$$C(t) = \sum_{n=1}^{N} C_n(t) + f, \quad f = 10 \vee f = 0. \tag{11}$$

Lower costs imply better walking performance, and as such have a greater weight on the final ASM values. Expression (10) is enough to identify falls since after falling there is no more forward displacement and thus final cost is lower than when fall does not occur. Additionally, f acts as a *fall penalty* when the robot falls, taking the value of 10 if the DARwIn-OP falls at any point during the trial (boolean classification based on the height of the robot's center of mass), and 0 otherwise (empirical, hand-tuned value). This increases the difference between success and failure among the trials.

In addition to this, a cost threshold value C_s was introduced because, as previously explained, the occurrence of fall is not enough to define a *successful* or an *unsuccessful* trial. The noise in the joint dynamics can cause enough perturbation to the walking patterns that the robot deviates from its intended path and moves in undesirable directions, while maintaining its posture, or lead to extremely unstable but upright locomotion. Since in both cases this penalty is not applied and the sensor informations holds no interest to the problem, C_s was established with the value of -0.35 (found empirically), in order to separate successful experiences from *bad* trials. In practical terms, a trial is only considered successful with a minimal frontal displacement without falling.

3.2 Sensor Selection and Dataset Construction

A proper selection of the simulation sensor data to store can have real impact on the outcome of the learning. We are looking for sensors with unique, unambiguous characteristics about the movement, without the risk of being associated to more than a single robotic skill - illustrating the motion in its most *canonical* form. Additionally, periodic movements should be represented by periodic sensory data, and as such absolute or non-periodic components should be removed from the signals, either with a high-pass filter or by simply avoiding absolute sensory information. Among these sensor values, CPG oscillators are included to supply the phase evolution during trial execution.

We have presented several signal processing steps in order to optimize data specificity, minimize redundancy and bulk size, and adjust for periodic motions (such as locomotion), in other works[2]. However, in this paper we opted to use the raw sensor data from the 10 locomotion periods. This allows the data mining algorithms to use all the data at their disposal, and ultimately tells us what sensor data can be considered irrelevant (no contribution to failure detection - no significant variation

Table 2 Sensor data selected to be part of the ASM.

Description	# sensors
Gyroscope readings	3
Center of Mass (CoM) position	3
Accelerometer readings	4
Left Foot Touch Sensors	4
Right Foot Touch Sensors	4
$\Theta_{\text{left},j}$	6
$\Theta_{\text{right},j}$	6
$\phi_{\text{left,right}}$	2
Total	32

between successful and unsuccessful trials) right from the start, which can lead to future reduction of the dataset and amount of data. The sensors selected in Table 2 (a total of 32) were deemed most relevant within the context of biped locomotion.

However, the number of data values for the whole 10 locomotion periods can prove to be an obstacle. At a 8 ms integration step, each sensor has around 1600 values for a trial. Considering 32 sensors, each trial would contribute with 51200 values. Since we need a large number of trials for the results to be reliable, and around 1000 trials were conducted, the dataset would have 51 million values, which is computably intractable. For lack of time and computation resources, the raw data was filtered in order to build a smaller dataset.

Therefore these values are filtered by a mean filter of variable length/window $w = 10, 20, 30, 40, 50$ leading to 5 datasets with variable size. The minimum window size leads to the largest dataset, which comprises a matrix of about 500 by 512 values. The dataset is constructed by placing filtered sensor data on the same matrix line, with each line representing a trial. Different values of joint noise were adopted - 1% for half of the 1000 samples and 0.1% for the other half. Experiments were organized with these different datasets comparing results and trying to understand which algorithms produce the best model. The cost variable was also added in order to discover if there was any real gain or improvement in the results with quantitative performance information.

3.3 Data Mining

A comparison analysis was performed using several classification algorithms [10], such as:

- **Support Vector Machines (SVM)** - is a technique based on statistical learning theory which works very well with high-dimensional data and avoids the curse of dimensionality problem. The objective is to find the optimal separating hyperplane between two classes by maximizing the margin between the classes' closest points.

- **Nearest Neighbour (NN)** - represents each example as a point in a d-dimensional space, where d is the number of attributes. Given a test example the proximity to the other data points in the training set is computed, using a measure of similarity or dissimilarity, such as the Euclidian distance or its generalization, the Minkowski distance metric, the Jaccard Coefficient or Cosine Similarity.
- **Naive Bayes (NB)** - is a Bayesian classifier which presents the probability of an object and which applies the Bayes Theorem to produce the classification. The Naïve Bayes classifier assumes that the presence of a particular feature of a class is unrelated to the presence of any other feature. An advantage is that it requires a small amount of training data to estimate the parameters necessary for classification.
- **Artificial Neural Networks (ANN)** - is a mathematical/computational model that attempts to simulate the structure of biological neural systems. The ANN consists of an interconnected group of artificial neurons and processes information using a connectionist approach. In most cases an ANN is an adaptive system that changes its structure based on external or internal information that flows through the network during the learning phase. The interconnections are used to send signals from one neuron to the other. The concept of weights between nodes is also present since it is used to establish the importance from one connection to the other. The network may contain several intermediary layers between its input and outputs layers. The intermediary layers are called hidden layers and the nodes embedded in these layers are called hidden nodes. In a feed-forward neural network the nodes in one layer are connected only to the nodes in the next layer. The Perceptron is the simplest model since it does not use any hidden layers.

The experiments with the classifiers were performed under the assumption of the *k-fold* method. K-fold cross validation is used to determine how accurately a learning algorithm will be able to predict data that it was not trained with. When using the k-fold method, the dataset is initially randomly partitioned into k groups [7, 10]. The learning algorithm is then trained k times, using all of the training set data points except those in the kth group. K-fold cross validation is extremely useful, if the correct value of k is chosen, because provides the best estimate cross validation error. Using $k = 10$ was assumed to be a good rule of approximation although the true best value differs for each algorithm and each dataset. Performance of the models was also achieved using the accuracy, sensitivity and specificity values and all the algorithms were tested using *Rapidminer Studio 6.4* software [15]. Accuracy is the proportion of true results (both true positives and true negatives) among the total number of cases.

The SVM classifier used a Radial Basis Function (RBF) kernel with $\epsilon = 0.001$, the type of ANN used was a 3 rounds perceptron, with a learning rate of 0.05, and the NN algorithm used had a k value of 3 and was based on euclidean distance measurement.

4 Results and Discussion

The overall results about the experiments with the algorithms described in Section 3, using different temporal window and with or without the variable cost are presented in Table 3.

Table 3 Accuracy results (%) achieved by the classifiers with several window sizes and with/without cost metric. K-fold of 1000 trials separating in test and train subsets.

	10		20		30		40		50	
	without	with	without	with	without	with	without	with	without	with
LIBSVM	93.8	94.2	95.6	96.5	95.1	97.7	95.0	98.0	94.3	98.8
NN	91.6	92.9	91.3	94.6	93.0	96.3	92.3	96.7	92.4	97.6
NB	93.2	98.2	93.0	98.3	93.1	98.4	93.0	98.3	93.6	98.6
ANN	91.1	91.7	92.2	93.6	91.6	92.4	91.5	91.6	91.6	94.9

Without using simulation cost as part of the dataset, best performance was consistently achieved by Support Vector Machines (SVMs). With the cost variable as part of the dataset, however, Naive Bayes reached the highest accuracy values. There is no significant improvement with the size of the mean filter window, with only slightly higher values for Naive Bayes with the cost metric. This is as expected because the greater the filter window, more information is lost in the process since there is greater downsampling.

On the other hand, the presence of the cost variable provides significantly better results, achieving accuracy values up to 98.6%. This was also expected as the cost function allows to differentiate among the *good* trials, and extract higher contributions from the best trials.

Even so, it is noteworthy that all values are well above the 90% mark, which proves that the approach taken, as well as the data selection, has real value in such applications.

Since fall prediction was performed *offline* with the bulk of the raw data as an input to the classifier, it is not yet possible to quantify the detection advance - how early does the classifier detects failure conditions before a robot fall or failure. Moreover, an improved sensor selection could also have been performed with the results achieved - which sensors offer a greater contribution to the final results. Both these steps will be taken into account in future work.

5 Conclusion and Future Work

Overall, very good results were obtained regarding the classification of new trials, with all values after data mining correctly predicting whether or not a certain sensor footprint is associated with potential failure conditions. Although the work was developed in its entirety with *offline* methods, its value should still hold when performing the experiments in real-time (using current data as the input). In this case it would be interesting to quantify the advance that failure prediction occurs - which has already been done in similar works [2], albeit with different classification methods.

There are however limitations to this approach. Particularly regarding the size of data to be used as input to the classification system. In real-time, walking is a process executed continuously or during extensive periods. As such, the amount of sensor information can quickly become intractable. Such is the case of periodic movements - several alternative methods to reduce bulk data size were proposed in [2].

Another valuable work line is to introduce the notion of ASM in the process. Instead of using the raw data, the distance to the ASM value [2] could be used as the input to the classifiers. This has the advantage of supplying *a priori* information to the process and might yield satisfactory results. There could also be gains in including the moment of failure/fall instead of a boolean success/failure value - quantitative information that might give a greater contribution to trials where failure occurred in later stages of the simulation. Additionally, it would be interesting to perform similar procedures with other types of robotic skills, and to ideally save each memory data and/or classifier parameters as a specific memory of each skill. This will eventually lead to a library o skills that the robot can access at any time, and helps it identify current context and conditions in order to properly respond and adapt to its environment.

Acknowledgments This work has been supported by FCT - Fundação para a Ciñcia e Tecnologia in the scope of projects PEst-UID/CEC/00319/2013 and PEst-UID/CEC/00027/2013, and the Portuguese Science Foundation (grants SFRH/BD/107891/2015 and SFRH/BPD/102411/2014)

References

1. André, J., Santos, C., Costa, L.: Path integral learning of multidimensional movement trajectories. In: Proceedings of the International Conference of Numerical Analysis and Applied Mathematics 2013 (ICNAAM 2013), Rhodes, Greece (2013)
2. André, J., Santos, C., Costa, L.: Skill memory in biped locomotion. Journal of Intelligent & Robotic Systems, 1–19 (2015). doi:10.1007/s10846-015-0197-z, http://dx.doi.org/10.1007/s10846-015-0197-z
3. André, J., Teixeira, C., Santos, C., Costa, L.: Adapting biped locomotion to sloped environments. Journal of Intelligent & Robotic Systems, 1–16 (2015). doi:10.1007/s10846-015-0196-0, http://dx.doi.org/10.1007/s10846-015-0196-0
4. Aoi, S., Ogihara, N., Funato, T., Sugimoto, Y., Tsuchiya, K.: Evaluating functional roles of phase resetting in generation of adaptive human bipedal walking with a physiologically-based model of the spinal pattern generator. Biological Cybernetics **102**(5), 373–387 (2010)
5. Christensen, A.L.: Fault Detection in Autonomous Robots. Phd, Université Libre de Bruxelles (2008)
6. Dev Anand, M., Selvaraj, T., Kumanan, S.: Fault detection and fault tolerance methods for industrial robot manipulators based on hybrid intelligent approach. Advances in Production Engineering and Management **7**(4), 225–236 (2012)

7. Holte, R.C.: Very simple classification rules perform well on most commonly used datasets. Machine Learning **11**(1), 63–90 (1993). doi:10.1023/A:1022631118932
8. Matos, V., Santos, C.P.: Central pattern generators with phase regulation for the control of humanoid locomotion. In: IEEE-RAS International Conference on Humanoid Robots, Osaka, Japan (2012)
9. Owaki, D., Morikawa, L., Ishiguro, A.: Listen to body's message: quadruped robot that fully exploits physical interaction between legs. In: 2012 IEEE/RSJ International Conference on Intelligent Robots and Systems, pp. 1950–1955. IEEE, Vilamoura (2012)
10. Tan, P.N., Steinbach, M., Kumar, V.: Introduction to data mining. In: Pearson Education, Inc. (2006)
11. Pastor, P., Kalakrishnan, M., Chitta, S., Theodorou, E., Schaal, S.: Skill learning and task outcome prediction for manipulation. In: 2011 IEEE International Conference on Robotics and Automation, pp. 3828–3834 (2011)
12. Pastor, P., Kalakrishnan, M., Meier, F., Stulp, F., Buchli, J., Theodorou, E., Schaal, S.: From dynamic movement primitives to associative skill memories. Robotics and Autonomous Systems **61**(4), 351–361 (2013)
13. Pastor, P., Kalakrishnan, M., Righetti, L., Schaal, S.: Towards associative skill memories. In: 12th IEEE-RAS International Conference on Humanoid Robotics (Humanoids), Osaka, Japan, pp. 309–315 (2012)
14. Plagemann, C., Stachniss, C., Burgard, W.: Efficient failure detection for mobile robots using mixed-abstraction particle filters. In: European Robotics Symposium 2006, pp. 93–107 (2006)
15. Rapidminer: Predictive analytics reimagined. http://support.robotis.com/en/product/darwin-op.htm (accessed July 2015)
16. Santos, C.P., Matos, V.: CPG modulation for navigation and omnidirectional quadruped locomotion. Robotics and Autonomous Systems **60**(6), 912–927 (2012)

Part V
Rehabilitation and Assistive Robotics

User Intention Driven Adaptive Gait Assistance Using a Wearable Exoskeleton

Vijaykumar Rajasekaran, Joan Aranda and Alicia Casals

Abstract A user intention based rehabilitation strategy for a lower-limb wearable robot is proposed and evaluated. The control strategy, which involves monitoring the human-orthosis interaction torques, determines the gait initiation instant and modifies orthosis operation for gait assistance, when needed. Orthosis operation is classified as assistive or resistive in function of its evolution with respect to a normal gait pattern. The control algorithm relies on the adaptation of the joints' stiffness in function of their interaction torques and their deviation from the desired trajectories. An average of recorded gaits obtained from healthy subjects is used as reference input. The objective of this work is to develop a control strategy that can trigger the gait initiation from the user's intention and maintain the dynamic stability, using an efficient real-time stiffness adaptation for multiple joints, simultaneously maintaining their synchronization. The algorithm has been tested with five healthy subjects showing its efficient behavior in initiating the gait and maintaining the equilibrium while walking in presence of external forces. The work is performed as a preliminary study to assist patients suffering from incomplete Spinal cord injury and Stroke.

Keywords Adaptive control · Exoskeleton · Gait initiation · Gait assistance · Wearable robot

1 Introduction

Human centered rehabilitation is essential for ensuring the user involvement in a therapy. Several human centered strategies, such as patient cooperative and support motor function assessment, oriented to the development of robot

V. Rajasekaran(✉) · J. Aranda · A. Casals
Universitat Politécnica de Catalunya, Barcelona-Tech, Barcelona, Spain
e-mail: rajasekaran.vijaykumar@gmail.com, {joan.aranda,alicia.casals}@upc.edu

J. Aranda · A. Casals
Institute for Bioengineering of Catalonia, Barcelona, Spain

L.P. Reis et al. (eds.), *Robot 2015: Second Iberian Robotics Conference*,
Advances in Intelligent Systems and Computing 418,
DOI: 10.1007/978-3-319-27149-1_23

behaviors have been widely studied [1]. These strategies support the assist-as-needed concept by determining the level of robotic assistance provided to the user. Referring to gait, assistance must be dynamically adapted to the patient's needs and thus, it is necessary to develop a personalized assistance in function of the user intentions and movements, as well as the therapy, which involves the knowledge of assistance to be perceived. Other factors have a direct influence on the quality of assistance, such as the availability of mechanical support, control strategies, combination of assistive devices etc. One of the widely used approaches to detect and evaluate the need of assistance is by evaluating the position errors. However, the use of a predefined trajectory pattern, without other inputs, imposes a complete assistance which might induce slacking and harm the patient. Thus, it is necessary to measure the human-orthosis interaction torques, to evaluate the user performance and status, to design a hybrid combination of force-position control. Assistance in robotic rehabilitation can be achieved using an effective control strategy [2] such as impedance or adaptive control, which act based on the subject's performance. Such control strategies operate under the principle of assistance-as-needed, in which assistive forces increase as the participant deviates from the desired trajectory. This deviation can also be used as input to generate a trigger to initiate the movement or the assistance in accordance to the user's performance.

The detection of the best instant for gait initiation and termination has been studied by many researchers in order to design a volitional control based robotic rehabilitation [3, 4]. One of the widely used approaches for monitoring the human intention relies on the use of brain machine interfaces such as in XoR [5]. These systems are efficient in monitoring the user intention, because a real displacement of the joint position is not needed to initiate the gait. Instead, gait initiation in MINDWALKER is based on the displacement of the CoM (center of mass) which is calculated heuristically [6]. HAL (Hybrid Assistive limb) adapts to the user movements by sensing the muscle signals to measure the muscle forces and to support voluntary motion of the patient [7, 8], as well as to determine the joint stiffness to be applied.

Gait assistance using wearable robots is challenging in terms of determining the suitable assistance for dynamic stability, considering the ground reaction forces. Several wearable robots have succeeded in providing dynamic stability, such as BLEEX [9], XPED2 [10], Ekso (earlier eLegs) [11], Rex (Rex Bionics) and Re-Walk [12], and have proven to be efficient in providing assistance on a passive range of motion and using complex systems.

The objective of this work is to develop an assistive control strategy for a wearable robot to perform user-dependent gait initiation and assistance. The gait initiation approach ensures the user involvement in the therapy and their motivation in performing the task. The control strategy implies the implementation of an impedance based approach for gait assistance, without using neither treadmill nor body weight support. The absence of weight compensation carries with it the challenge of maintaining the equilibrium in presence of ground reaction forces. The goal is to develop an efficient control model for a low-cost wearable robot and to validate the assistive behavior of the robot for patients with neurological disorders. A hybrid position and interaction torques based control strategy is presented to continuously adapt the user movements to the desired gait pattern in real time. This real time adaptation also ensures synchronization among the joint trajectories to maintain dynamic stability.

2 Human Centered Gait

The human-orthosis interaction torques are essential in defining the dynamic analysis of a human centered control strategy. Hence, the mathematical model for the dynamic analysis of an exoskeleton can be represented as

$$M_{ort}(q)\ddot{q} + C_{ort}(q,\dot{q}) + G_{ort}(q) = \tau_a + \tau_{pat} + \tau_d \tag{1}$$

where $q, \dot{q}, \ddot{q}$ are the vectors of joint positions, velocities and accelerations. $M_{ort}(q)$ is the inertia matrix, $C_{ort}(q,\dot{q})$ is the centrifugal and Coriolis vector and $G_{ort}(q)$ represents the gravitational torques. τ_a and τ_{pat} are the orthosis and patient torques respectively and τ_d corresponds to the external disturbances acting on the subject. These actuator and patient torques are influenced by the human-orthosis interaction, while the external disturbances can be due to any assistive or external sources which can affect the dynamic stability of the robot. In the present work, τ_{pat} is used for gait initiation and to determine the level of assistance to be exerted by the orthosis.

2.1 *Gait Initiation*

The gait initiation can be defined as the time *t* in which the user intends to perform a movement and can be determined from the human-orthosis interaction torques. This type of user dependent initiation is efficient in influencing or motivating the user to provide an input movement. The user motivation is necessary to improve the therapy and avoid slacking. Since human-orthosis interaction torques based gait initiation can be a drawback in differentiating between a tremor and the user intention, the joint position and torques must be considered too to trigger the therapeutic procedure. Additionally, this strategy permits the user to initiate the therapy with the leg he or she feels more comfortable, fig. 1.

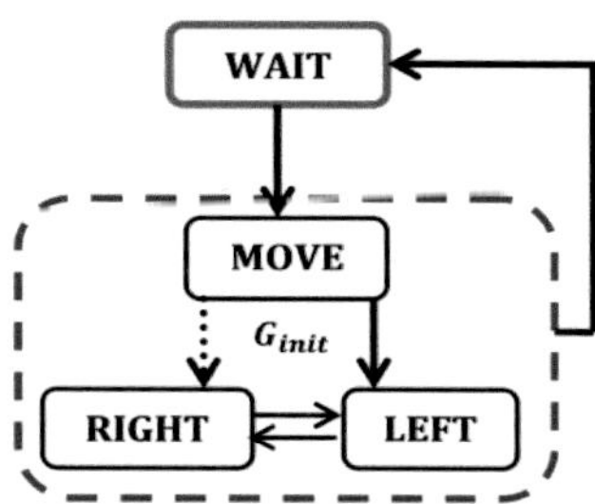

Fig. 1 State transition of the gait initiation algorithm

The gait initiation strategy considers the joint positions and the human-orthosis interaction torques of both legs. This can be represented as,

$$G_{init} = Leg_x \,, x \in [R, L] \tag{2}$$

where, G_{init} is the gait initiation trigger obtained from the user for Leg_x. θ_R and θ_L are the joint angles of right and left leg respectively; τ_R and τ_L corresponds to the measured interaction torques in right and left leg respectively.

Algorithm 1: Gait initiation

Case $\theta_R(t) \neq \theta_R(t-1), \tau_R(t) \neq \tau_R(t-1)$ && $\theta_L(t) = \theta_L(t-1)$

$x = R$

Else $x = L$

2.2 Gait Assistance

In the present work, for each joint, the actuator works in collaboration with the patient. The actuator torque can be modified by varying the joint stiffness parameter, which invariably modifies the corresponding joint trajectory and force compensation. This stiffness variation alters the actuator torque which determines the degree of control transferred from the orthosis to the human or vice versa. Such an impedance control scheme has been widely used for its compliant behavior, which results in an adaptive walking pattern and a more natural interaction between patient and orthosis. Thus the impedance control can be determined by equation (3)

$$F = Ma + Cv + K.\left(\theta_{ref} - \theta_{act}\right) \tag{3}$$

where, θ_{ref} and θ_{act} are the reference and actual joint positions respectively, K is the stiffness parameter of the joint and F represents the applied force to the joint. M represents the mass, C is the damping constant and *a* and *v* represent the acceleration and velocity of the robot. Here, the input sample rate to the system is maintained constant and the damping coefficient is kept small, therefore the velocity of the orthosis is not modified by the user and thus, it does not induce any significant effect on the applied force. Hence the force equation, influenced by the position error, is modified as:

$$F = K(\theta_{ref} - \theta_{act}) \tag{4}$$

The value K can be determined dynamically based on the performance of the user and the level of assistance to be exerted by the orthosis.

$$K_{t+1} = K_t \pm \Delta K \tag{5}$$

$$\Delta K = \left|\left(\frac{\theta_{ref} - \theta_{act}}{s * \tau_{pat}}\right)\right| \tag{6}$$

where, *s* is a confidence factor which is used to determine the stiffness value at time *t+1*. The confidence factor is a variable which shall be defined by the therapist in function of the capabilities of the patient. This confidence factor can be varied according to the progress of the user, in order to modify the time instant at which assistance is to be initiated. A low confidence factor means that the level of assistance should be high and a higher confidence factor indicates that the subject is capable of walking without or with little assistance. A variation of the K value results in a change in the force acting at the joint level what is perceived as assistance or resistance by the patient.

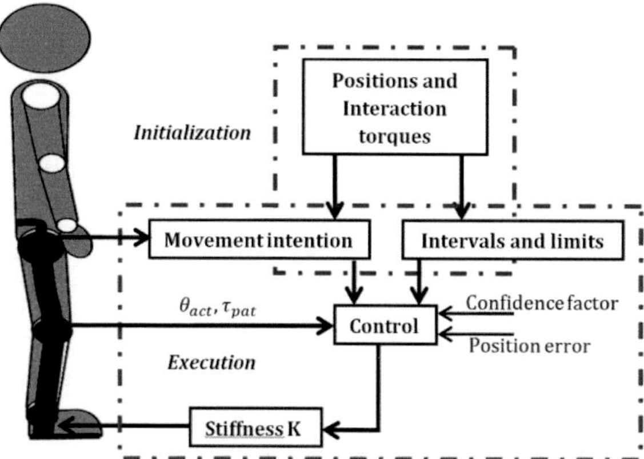

Fig. 2 Schematic *representation of the user-intention based adaptive control strategy*

The stiffness variation module, as shown in fig. 2, is responsible for incrementing, decrementing or maintaining the stiffness parameter of each joint. Within a given range of trajectory errors, stiffness is computed from position errors, but when the defined error thresholds are surpassed, stiffness should be modified according to the measured interaction torques. This condition takes place in general with the presence of external perturbations. The following are the parameters involved in defining this stiffness variation:

θ_e- *Position error (deg)*
$\theta_e Th_{up}$ – *Upper threshold of position error (deg)*
$\theta_e Th_{lo}$ – *Lower threshold of position error (deg)*
τ_{pat} - *Human-orthosis interaction torques (Nm)*
$\tau_{pat} Th_{lo}$- *Lower threshold of interaction torques (Nm)*
$\tau_{pat} Th_{up}$- *Upper threshold of interaction torques (Nm)*
ΔK - *Stiffness variation (N/m)*

Algorithm 2 : Gait assistance

Case $\theta_e > \theta_e Th_{up}$

$$K = K + \Delta K$$

Case $\theta_e < \theta_e Th_{lo}$ && $\tau_{pat} Th_{lo} < \tau_{pat} < \tau_{pat} Th_{up}$

$$K = K - \Delta K$$

Else
Maintain K

An average walking pattern generated from a set of recorded walking patterns of healthy individuals is used as reference input. Based on this pattern and the patient contribution, the stiffness K establishes the operating mode at each joint, assistive or resistive. In a first set of trials, for each subject the maximum interaction torques are obtained by applying a low stiffness value at each joint. From these maxima, we extract the upper and lower thresholds of interaction torques to dynamically define the operation mode. These thresholds are obtained by multiplying these maxima interaction torques by the confidence factor (s).

3 Experimental Procedure

The proposed intention driven adaptive strategy is based on the position error and interaction torques. The strategy needs an initial study about the user adaptation to determine the gait initiation and assistance scenario. Hence, the experimentation consists of two phases: initialization and execution. The initialization phase involves monitoring the interaction torques and joint positions with no-assistance provided by the orthosis in order to be able to define s. This initialization phase is used to parameterize the user intentions and adaptations to the movement. In the execution phase, the changes in movement and interaction torques are used as a trigger for gait initiation. The interaction torques, limited by the confidence factor, are used to determine the time instants of stiffness variation. Both these phases are performed and evaluated using a lower-limb exoskeleton.

3.1 Exoskeleton

H1 is a 6 DoF (degree of freedom) wearable lower limb orthosis with an anthropomorphic configuration to assist individuals with incomplete Spinal cord injury (SCI) or Stroke. The exoskeleton, shown in fig. 3, has been built within the framework of the Hyper* project. H1 has three joints for each leg: hip, knee and ankle, each joint is powered by a DC motor coupled with a harmonic drive gear. The exoskeleton is equipped with potentiometers and strain gauges to measure the joint angles and human-orthosis interaction torques on the links respectively. A detailed description about the exoskeleton structure and communication parameters is detailed in [13]. The variable stiffness control ensures a safe therapeutic experience [14, 15]. The exoskeleton permits a stiffness value within the range of 1-100 N/m. A low stiffness value (<10N/m) will not cause any significant effect on the user's behavior. Similarly, a high stiffness value (> 80N/m) will provide a completely assisted movement, with few or no input from the user. The initial stiffness value must be defined taking into account the user's capability and the degree of assistance to be applied by the orthosis.

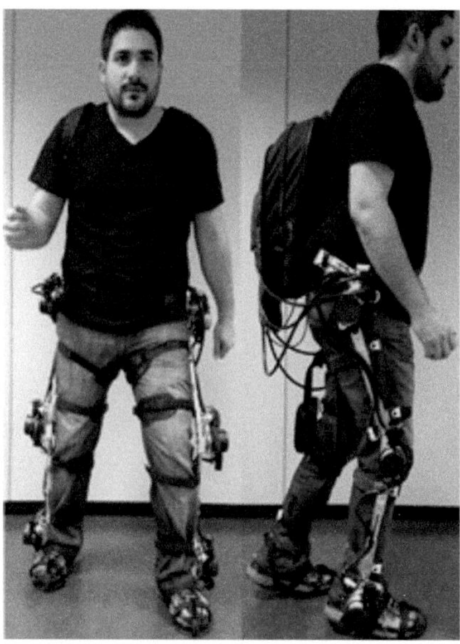

Fig. 3 Participant wearing the HYPER exoskeleton H1

3.2 Experimentation

Initially the walking pattern is generated from tests on subjects applying a low stiffness value (20N/m). This phase is necessary to obtain the pattern of interaction torques and to allow the user to adapt to the orthosis. This initialization also determines the minimum interaction torque observed in the users along with the deviation in position. The evolution of these interaction torques are used to determine the initiation of stiffness variation, determining the adaptive behavior to be exerted by modifying stiffness. This variable stiffness results in either an assistive or resistive behavior. The confidence factor is used to determine the initiating time of the gradual actuation of the joints stiffness functions, thusachieving a smooth performance without affecting the joints trajectories. The gradual increase or decrease of the stiffness value smoothens the interaction torques.

The setup includes a recorded gait pattern obtained from healthy users and optimized after some repetitions of gait cycles. The values of stiffness and confidence factor are defined based on the subject's health condition. Since the strategy is tested with healthy individuals, the initial stiffness and confidence factor are assumed to be 50 N/m and 0.9 respectively. High interaction torques are found in healthy subjects, so a higher confidence factor is needed to define their thresholds.

The evaluation of the control strategy is performed by following a protocol and considering an intermediate pause between consecutive trials to allow identifying the user intention after recovering from fatigue. An interval of 10-20 seconds is introduced at the beginning of each trial, for gait initiation algorithm, with an auditory cue to notify the subject to initiate the movement. Similarly for evaluating the gait assistance strategy pause time of 1-2 minutes is considered at the end of every 10 minutes walking test. Thus the study involves two types of walking experiments, gait initiation for 3 minutes and gait assistance for 10 minutes. The gait assistance experiment is performed for 30 minutes, i.e. 3 sets of 10 minutes walking test.

4 Results and Discussion

The proposed adaptive control strategy has been tested and evaluated with five healthy individuals of the age group 37±9, weight 80±8kg, and height 1.75±0.05m. The results section has been divided into two parts in order to explain the gait initiation and the gait assistance scenarios.

4.1 Gait Initiation

In a gait cycle, the knee joint plays a key role for both the initialization of the movement and the swing state. Hence, the gait initiation strategy is evaluated by monitoring the deviation in the knee joint movement with respect to the expected pattern, along with the interaction torques. The flexion and extension movement of the knee joint is monitored to differentiate between the user's intention and tremor movement. As shown in fig. 4, the right leg of the user showed gait movement initiation in most of the trials. This can be seen from the shift in the interaction torques of the right and left leg, both in hip and knee joint. The initiation of the gait is characterized by the flexion movement of the knee joint. For instance, at time 97 seconds the right knee joint shows a little displacement in the movement which initiates the gait cycle. The hip joint trajectory appears after a few seconds, immediately followed by the transition to the left leg.

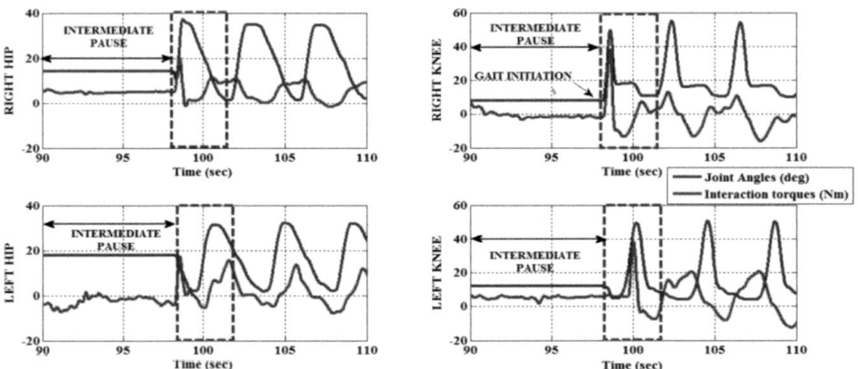

Fig. 4 Gait initiation sequence of a healthy user, blue curve signifies the joint trajectories and red curve indicates the interaction torques

4.2 Gait Assistance

The efficiency of the adaptive gait assistance provided by the control model is evaluated in comparison with a reference gait pattern. Fig. 5 shows the reference gait patterns and the resulting mean gait cycles of the five healthy subjects. The subjects performed a free normal walking movement, with low stiffness. In this case, the deviation from the desired trajectory was found to be high. After a series of trials (10) this error decreased gradually due to the effect of the adaptive stiffness acting on the joints. The stiffness variation helped to maintain this error within a specified range and following a similar pattern of incrementing and decrementing K at every joint, thus resulting in an assistive or resistive behavior.

The results of one of the subjects are used to show the response of the control strategy. The gait performance of a healthy user, as shown in fig. 6, demonstrates the influence of the stiffness variation proposed in this work. The initial walking with low stiffness value is presented as the 'no-assistance' mode. In comparison with the reference pattern, the no assisted walking is found to produce a maximum deviation. After the application of a variable stiffness, the user is able to walk within a predefined error limits. The stiffness variation also converges with respect to the movement at the end of 10 trials. At the end of 20 trials, the user is following a movement which is quite similar to the reference pattern.

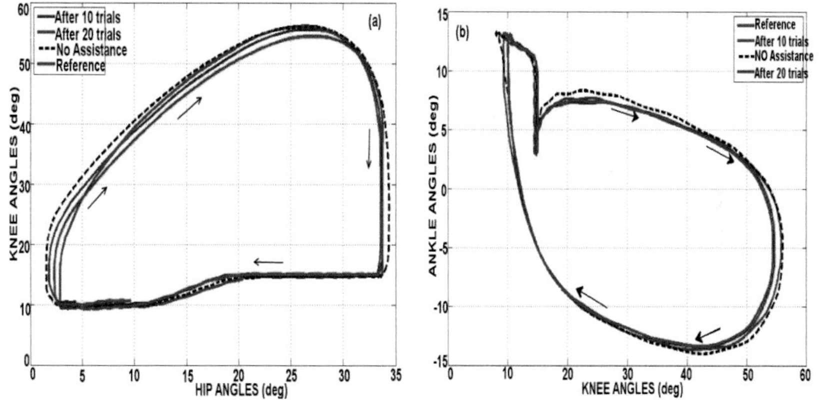

Fig. 5 Changes in gait pattern of a healthy subject due to effect of adaptive stiffness: (a) Hip (deg) - Knee (deg), (b) Knee (deg) - Ankle (deg)

The error was found to be within the defined limits for all the users. The hip joint showed a little variation and more adaptable behavior in terms of stiffness changes in real time. Since the exoskeleton is a planar robot, the lateral hip movement cannot be monitored. However, this orthosis limitation does not affect the proposed control strategy. A significant variation of the stiffness is found in both ankle and knee joints. The hip joint stiffness varied in a short range which is evident from the interaction torques in fig. 7. This can be due to the lateral movement of the user's hip joint which compensates the joint trajectory.

The interaction torques of the ankle joint is in the limits of 12 Nm to -3 Nm, as shown in fig 7, and with the application of the confidence factors the threshold is limited to 10 Nm to -1 Nm. This threshold limit is used to initiate the stiffness increment when the position error threshold is reached. Similarly, in the knee joint, the interaction torques are within the limits of 14 Nm to -14 Nm and after the application of the confidence factor the thresholds are 12 Nm and -12 Nm. The interaction torques are bounded within the limits even in the presence of maximum stiffness.

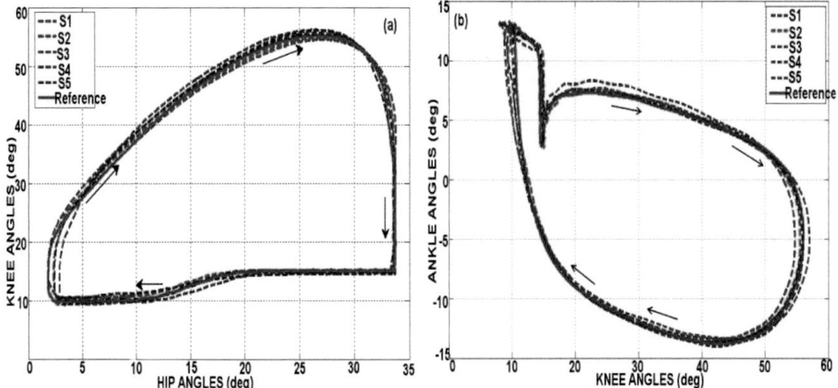

Fig. 6 Reference gait pattern and the resulting mean gait pattern of each subject: (a) Hip (deg) – Knee (deg), (b) Knee (deg) - Ankle (deg)

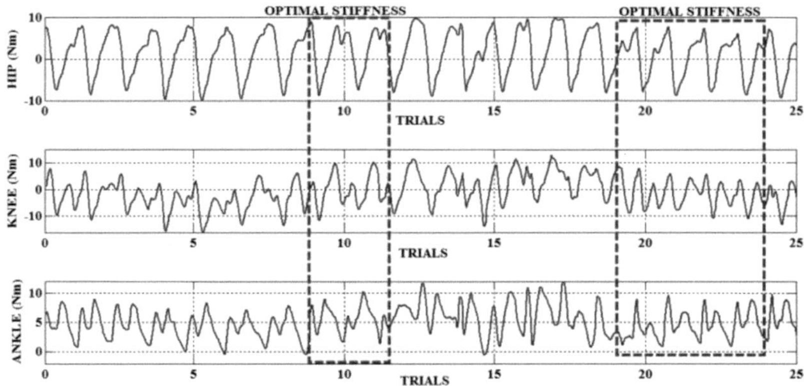

Fig. 7 Interaction torques of each joint showing the change in behavior while stiffness converges (highlighted region) to an optimal value

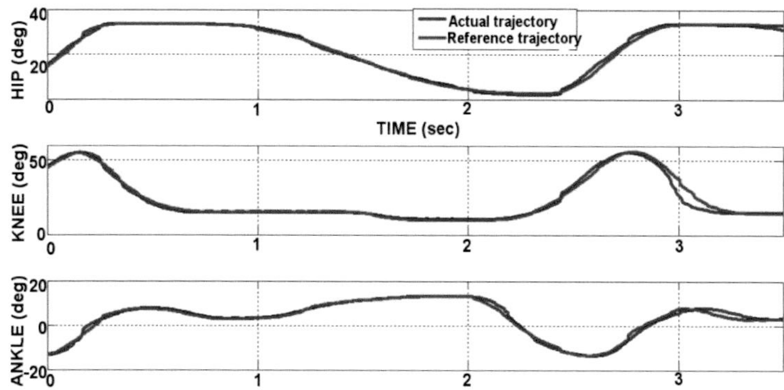

Fig. 8 Trajectory deviation of a healthy subject between trials 10 and 11

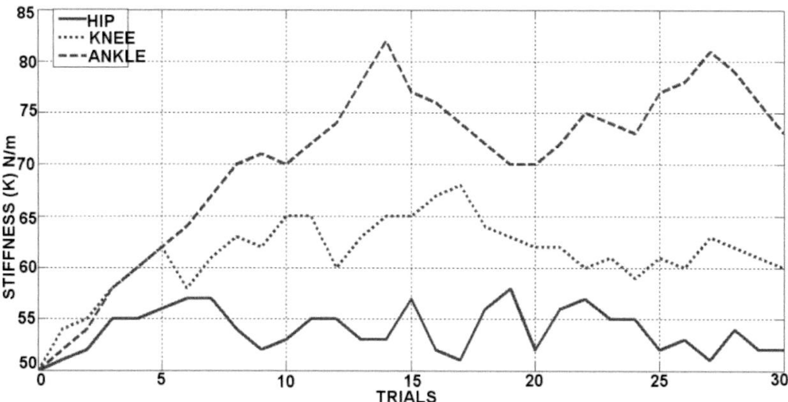

Fig. 9 Stiffness variation of a healthy subject

Since the flexion and extension movement of the knee joint is essential in walking for maintaining the transition between gait phases, the stiffness variation converges, after a few gait trials, as shown in fig. 8. The trajectory deviation is found to be within a small range, but with a delay, in the knee joint, as shown in fig. 9. In case of the ankle joint the stiffness behavior shows a different pattern due to the compensation of ground reaction forces. *In the ankle joint, the deviation from the reference position is found to be higher, which explains the pattern of stiffness variation.*

The confidence factor is used to act on the joint gradually, by varying its stiffness. The consequence is the relax intervals that appear as negative slope (decreasing stiffness), which results in achieving a smooth behavior of the system. Lower confidence factors will result in few and shorter steps of stiffness variation, so the increment will be faster. On the contrary, a higher confidence factor will limit the increase of stiffness. The gradual increase in the stiffness value is due to the permanent difference in position error. The error in position of the joint in combination with the change in interaction torques results in a high stiffness value.

5 Conclusion

A user intention based adaptive walking strategy has been evaluated in function of the position error and human-orthosis interaction torques, thus ensuring an effective and safe therapy. The gait initiation is determined at the beginning of each gait cycle from the input orders received from the user. The stiffness value of each joint adapts dynamically to the user needs and keeps the position error bounded within the specified limits in real time. The wearable robot was tested with no body weight compensation, which demonstrates the reliability of the control strategy in terms of ensuring dynamic stability in presence of ground reaction forces. The results of the proposed control method were evaluated by comparing the resulting trajectory with a predefined gait trajectory. The experimental results showed that the user's gait intention was observed effectively with no delay and followed by the leg movement. The results demonstrated that the evolution of the stiffness value does not follow a similar pattern for all the joints. The stiffness value converges to be within a given range after a series of trials. The stiffness variation was in coordination with the flexion and extension movements. This demonstrates the efficiency of the proposed method for a real time process involving multiple joints.

This work has been the basis for a further study combining this gait initiation strategy with a brain machine interface that considers neurological signals and for the evaluation of the performance of the adaptive control in the assistance of incomplete SCI and Stroke patients, including the presence of muscle stimulation (FES), which, from the control perspective, acts as external disturbances.

Acknowledgment This research is supported by project HYPER (Hybrid Neuroprosthetic and Neurorobotic devices for functional compensation and rehabilitation of motor disorders), grant CSD2009-00067 CONSOLIDER INGENIO 2010 from MINECO (Spanish Ministry for Science and Education).

References

1. Yan, T., Cempini, M., Oddo, C.M., Vitiello, N.: Review of assistive strategies in powered lower-limb orthoses and exoskeletons. Robotics and Autonomous Systems **64**, 120–136 (2015)
2. Marchal-Crespo, L., Reinkensmeyer, D.: Review of control strategies for robotic movement training after neurologic injury. Journal of NeuroEngineering and Rehabilitation **6**(20) (2009)
3. Novak, D., et al.: Automated detection of gait initiation and termination using wearable sensors. Medical Engineering and Physics **35**, 1713–1720 (2013)
4. Hasani, W., Mohammed, S., Rifaï, H., Amirat, Y.: Powered orthosis for lower limb movements assistance and rehabilitation. Control Engineering Practice **26**, 245–253 (2014)

5. Noda, T., et al.: Brain-controlled exoskeleton robot for BMI rehabilitation. In: 12th IEEE RAS International Conference on Humanoid Robots, Osaka, Japan, pp. 21–27 (2012)
6. Wang, S., et al.: Design and Control of the MINDWALKER exoskeleton. IEEE Transactions on Neural Systems and Rehabilitation Engineering **23**(2), 277–286 (2015)
7. Kawamoto, H., et al.: Voluntary motion support control of robot suit HAL triggered by bioelectrical signal for hemiplegia. In: 32nd International Conference of the IEEE EMBS, Buenos, Aires, pp. 462–466 (2010)
8. Aach, M., et al.: Voluntary driven exoskeleton as a new tool for rehabilitation in chronic spinal cord injury: a pilot study. The Spine Journal **14**, 2847–2853 (2014)
9. Kazerooni, H., Racine, J.L., Huang, L., Steger, R.: On the control of the Berkeley lower extremity exoskeleton (BLEEX). In: IEEE International Conference on Robotics and Automation, ICRA, pp. 4353–4360 (2005)
10. van Dijk, W., van der Kooij, H.: XPED2: A passive exoskeleton with artificial tendons. IEEE Robotics and Automation Magazine, 56–61, December 2014
11. Strausser, K.A., Kazerooni, H.: The development and testing of a human machine interface for a mobile medical exoskeleton. In: IEEE/RSJ International Conference on Intelligent Robots and Systems, San Francisco, CA, USA, pp. 4911–4916 (2011)
12. Talaty, M., Esquenazi, A., Briceño, J.E.: Differentiating ability in users of the rewalk powered exoskeleton. In: IEEE International Conference on Rehabilitation Robotics, Seattle, USA, pp. 1–5 (2013)
13. Bortole, M., Del-Ama, A.J., Rocon, E., Moreno, J.C., Brunetti, F., Pons, J.L.: A robotic exoskeleton for overground gait rehabilitation. In: IEEE International Conference on Robotics and Automation, Karlsruhe, Germany, pp. 3356–3361 (2013)
14. Rajasekaran, V., Aranda, J., Casals, A.: Recovering planned trajectories in robotic rehabilitation therapies under the effect of disturbances. International Journal of System Dynamics Applications **3**(2), 34–49 (2014)
15. Rajasekaran, V., Aranda, J., Casals, A., Pons, J.L.: An Adaptive control strategy for postural stability using a wearable exoskeleton. Robotics and Autonomous Systems **73**, 16–23 (2015)

Control of the E2REBOT Platform for Upper Limb Rehabilitation in Patients with Neuromotor Impairment

Juan-Carlos Fraile, Javier Pérez-Turiel, Pablo Viñas, Rubén Alonso, Alejandro Cuadrado, Laureano Ayuso, Francisco García-Bravo, Felix Nieto, Laurentiu Mihai and Manuel Franco-Martin

Abstract In this paper, the most significant aspects of the new robotic platform E2REBOT, for active assistance in rehabilitation work of the upper limbs for people with neuromotor impairment, are presented. Special emphasis is made on the characteristics of their control architecture, designed based on a three level model, one of which implements a haptic impedance controller, developed according to the "assist as needed" paradigm, looking to dynamically adjust the level of assistance to the current situation of the patient, in order to improve the results of the therapy. The two modes of therapy that supports the platform are described, highlighting the behavior of the control system in each case and

J.-C. Fraile · J. Pérez-Turiel(✉)
ITAP – Instituto de Tecnologías Avanzadas de la Producción, Univ. de Valladolid, Valladolid, Spain
e-mail: {jcfraile,turiel}@eii.uva.es

P. Viñas · R. Alonso · A. Cuadrado
Centro Tecnológico Fundación CARTIF, Valladolid, Spain
e-mail: {pabvin,rubalo,alecua}@cartif.es

L. Ayuso · F. García-Bravo
Aplifisa S.L., Salamanca, Spain
e-mail: {lauri,fgarciabravo}@aplifisa.es

F. Nieto · L. Mihai
IDECAL, Barcelona, Spain
e-mail: {felnie,laumih}@idecal.es

M. Franco-Martin
Instituto Ibérico de Investigación en Psicociencias, Zamora, Spain
e-mail: mfm@intras.es

L.P. Reis et al. (eds.), *Robot 2015: Second Iberian Robotics Conference*,
Advances in Intelligent Systems and Computing 418,
DOI: 10.1007/978-3-319-27149-1_24

describing the criteria used to adapt the behavior of the robot. Finally, we describe the ability of the system for the automatic recording of kinematic and dynamic parameters during the execution of therapies, and the availability of a management environment for exploiting these data, as a tool for supporting the rehabilitation tasks.

Keywords Robotics · Rehabilitation · Neuromotor impairment · "assist as needed" · Haptic controller · Impedance control

1 Introduction

Acquired brain injury (ABI) is the result of a sudden injury in brain structures. It is a major cause of disability that impairs the quality of life of people who have suffered, since it involves both physical and cognitive difficulties, and even emotional and relational disorders. An acquired brain injury is defined as damage to the brain, which occurs after birth and is not related to a congenital or a degenerative disease. These impairments may be temporary or permanent and cause partial or functional disability or psychosocial maladjustment. The most common causes of ABI are head trauma, stroke, brain tumors, cerebral anoxia and brain infections. Acquired brain injury is more common after the age of 55, and their risk increases proportionally with age. The World Health Organization estimates that nearly half of the world population could suffer ABI damage.

Recent advances have resulted in promising new strategies for the rehabilitation of stroke patients, including drug therapies, early and very early mobilization, virtual reality technologies in combination with conventional rehabilitation, and robotics. The use of robotic assistive devices for rehabilitation therapy has proven to be an excellent tool that helps in managing physical therapy for patients who suffered an acquired brain injury. Several studies have shown that rehabilitation therapies with assistive robots have positive effects on the improvement of patients with ABI, since with these therapies there is some evidence of improved neuronal plasticity [1]. There are many examples of robotic devices for rehabilitation [2] [3]. A review of robotic systems for rehabilitation of the upper limb in patients with ABI from a historical perspective, and with a focus on the most innovative methods to encourage patient involvement in therapy can be seen in [4].

Clinical evidence regarding the relative effectiveness of different types of robotic therapy controllers is limited. There are several published studies [5, 6, 7] in order to assess the effectiveness of the use of assistive robots for patients with brain damage. These studies indicate that patients undergoing therapy with robots do not show significant improvements in the movement of the upper limbs for most activities of daily living (ADLs), but they do show a significant improvement in their motor function. In a more recent review [8] it shows that patients receiving robot-assisted therapy after stroke are more likely to improve their generic activities of daily living.

E2REBOT is a low cost robotic platform for active assistance in upper limbs rehabilitation tasks, aimed at people with disabilities due to an acquired brain injury. This article describes this platform indicating the design requirements in section 2, and with special emphasis on the characteristics of their control architecture in section 3. This architecture includes a haptic impedance controller developed according to the "assist as needed" paradigm that tries to adapt on-line the robot level of assistance to the patient's condition during the course of therapy. This is reflected in the two modes of therapy that supports the platform, as described in section 4. Finally, in section 5, the capacity of the system for automatic registration of kinematic and dynamic parameters during the execution of therapies is shown, together with the availability of a management environment for exploiting these data.

2 Design Requirements of E2REBOT

To ensure that E2REBOT could be used for rehabilitation of the upper limbs, the following requirements were defined for its development [9]:

- The robot should allow the patient to perform flexion-extension tasks of the elbow and abduction-adduction of the glenohumeral joint.
- The robot must be intrinsically safe, with transparent dynamics (backdrivability), this implies a simple mechanical design, looking to maintain the robot dynamics at the lowest possible levels.
- The workspace of the robot must allow for the completion of the natural movements of the human arm. The range of motion of this in a planar surface is approximately 800 mm × 500 mm. This allows the patient to perform full shoulder abduction and a complete elbow extension, provided that during flexion-adduction the medial plane is not exceeded.
- The robot may incorporate a torque/force sensor, integral with the gripper, with the aim of measuring the dynamic parameters of interaction between the user and the robot.
- The software of the platform must provide adequate multimedia interfaces for representing virtual environments and analyzing data collected during therapy sessions.
- The system must include a database to store information about patients, therapies, and kinematic (position, velocity) and dynamic (forces) parameters for further analysis.

3 E2REBOT Features

E2REBOT is an "end-effector" robot designed to assist in the rehabilitation of the upper limbs in patients with neuromotor impairment. Figure 1 shows a E2REBOT prototype during the trials period. It has a two-Cartesian axes (XY) mechanical structure, and is equipped with a force/torque JR3 sensor mounted on the terminal element, where the patient grabs the robot during the rehabilitation therapy. This sensor measures the reaction force between the robot and the patient. Useful dimensions of the work surface where the patient performs the rehabilitation tasks are 975 x 600 mm.

Since the main purpose of this design is to ensure intrinsic safety by the reversibility of the actuator-transmissions assembly, two high performance Maxon DC brushless motors were selected to move the end effector in the X and Y directions. In order to obtain a good weight/ torque relation a transmission through gears was chosen (4.3: 1). The consequences of using low rate transmissions are very beneficial as it allows to increase the output torque and, at the same time, preventing no mechanical nonlinearities, as the resisting torque due to the permanent magnet of the brushless motors and discontinuities of surface friction.

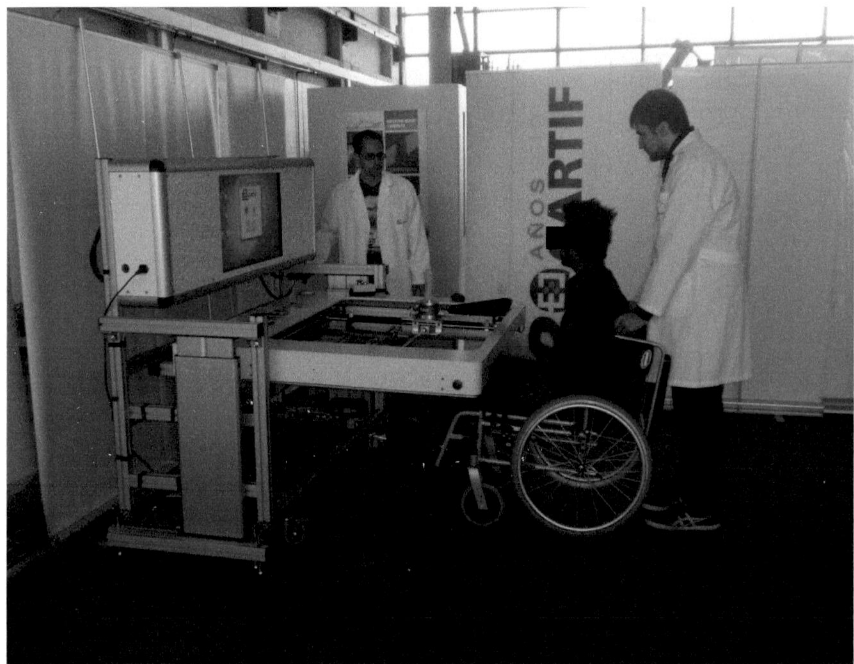

Fig. 1 Built prototype of the E2REBOT platform

The architecture developed for the supervision and control of E2REBOT, and the corresponding software implementation, are modular and are organized at three levels: the upper runs on a Personal Computer, the intermediate is implemented in a CompactRIO (National Instruments) controller, and the lower two Motor Control Units (MCUs) are used to control the robot axes.

E2REBOT implements safety measures through redundant hardware and software. There are four limiters which cut the motor current in the vicinity of the mechanical limits in each lane. This contributes not only to security but avoids degradation of hardware under potential impacts. The software security layer includes a speed/force observer that provides a control signal that buffers the robot motion to avoid damage to the patient when it detects a dangerous situation (speed or force above a preset maximum value).

Further details on the design, features, control and operation of this robotic platform can be found in references [9] and [10].

3.1 Kinematics

E2REBOT kinematics is based on two prismatic joints (Figure 2). The first joint moves the X axis (0 mm ÷ 975 mm) and the second one displaces the Y axis (0 mm ÷ 600 mm).

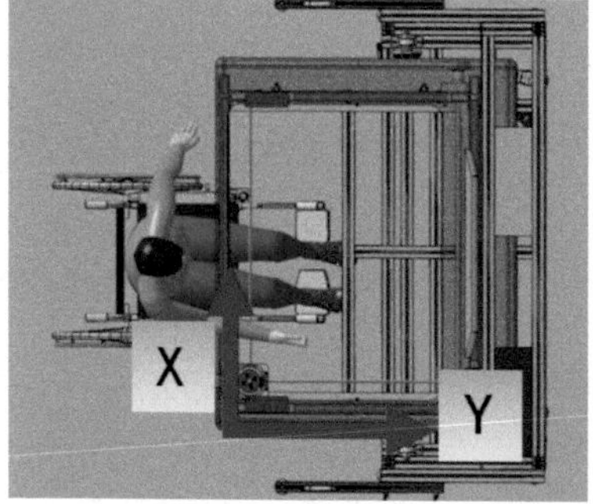

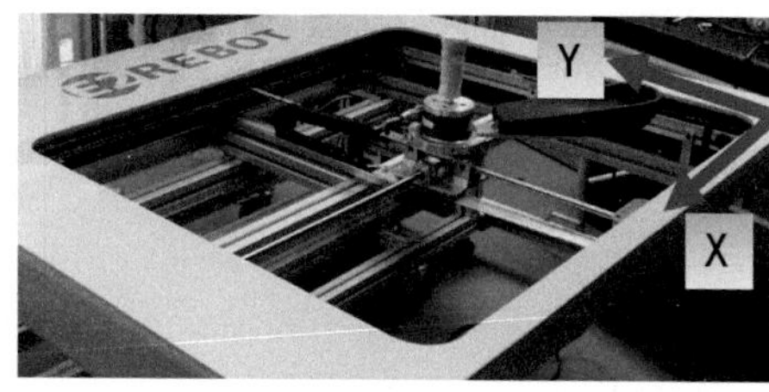

Fig. 2 Linear axes and workspace of E2REBOT

These two joints are connected to the gripper, which has an associated force/torque JR3 sensor. This kinematic arrangement ensures that the handling element performs only linear motions, and has a fixed orientation in space during the movement. This kinematic structure was selected because it offers the patient a working area suitable to the mobility needs of the human arm, and allows therapies of elbow flexion-extension and abduction-adduction of the glenohumeral joint.

3.2 Control Architecture

Level 1 - higher level: this level software was developed by using the C # language, and runs on a Personal Computer. It includes the following software modules:

- Virtual rehabilitation therapies that the patient must perform. In order to implement the visual environment associated with the therapies, Microsoft's Visual Studio 2010 XNA framework has been used.
- Man-machine interface that allows to select among predefined therapies and modify its settings.
- Recording of parameters related to the therapy session: date and time values, position and velocity of the end effector, force exerted by the patient on the robot, and assistive force that brings the robot to the patient. This data is stored in the PC every 30 ms, and will be used for off-line analysis, so that the therapist can analyze the patient's progress over time.
- On-line computation of the "degree of support", K value, that the robot provides the patient during the course of therapy. Its value depends on the difference between the actual and desired positions of the end-effector. This information is sent to the intermediate level by TCP sockets.

Level 2 - Intermediate: This level is implemented in the FPGA (field-programmable gate array) of the CompactRIO controller. At this level the software was developed by using LabVIEW. This level is responsible for:

- This level must run the haptic impedance controller and determine the value of the desired force, associated with the degree of assistance provided by the robot to the patient.
- Recording and processing of data from the JR3 torque-force sensor.
- It must perform the necessary calculations in order to obtain the torque set points for the axes motors. This set point is an analog signal in the 0 - 5V DC range, which is sent to the lower level.

Level 3 - lower level: This level is implemented in two ESCON 50/5 Maxon Motor Control Units (MCUs) controlling the two axes of E2REBOT. The software implements a PID control loop in order to control the motors torque.

The E2REBOT control scheme is shown in the block diagram of Figure 3. An impedance based haptic controller with force feedback has been developed, allowing the non-linear dynamics of the robot (mechanical high coefficients of friction, viscosity and Coulomb phenomena appearing in the dynamics of the robot X,Y axis) to be "transparent". In addition, it has been necessary to apply a "feedforward" compensation because the hardware where the control is implemented has a low bandwidth, and we could not apply an excessive increase in the gain of the KD error (measured as the difference between the desired and the real forces).

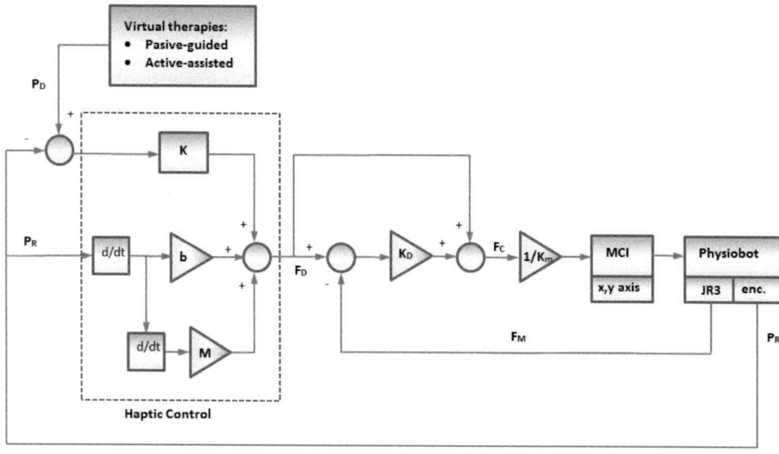

Fig. 3 Block diagram of the control implemented in E2REBOT

Our impedance model is a mass-spring-damper system with a mass M, K is a spring, and B is a damping impedance. The impedance controller computes the desired force (FD) in accordance with the desired position (PD) of the grip, obtained from the data of the virtual therapies implemented in the computer (high level control). This desired force is compared with the force (FM) measured by the JR3 sensor (force exerted by the patient during the therapeutic exercises). The difference FD - FM, is multiplied by the error gain KD, and joins the feedforward loop to generate the force Fc. Using the motor constant Km, the force Fc is transformed into the intensity of current Ic, which is applied to each MCU to control the position and velocity of the robot axes.

Giving appropriate values to the parameters of virtual impedance model, characterized by K, B and M, it is possible to generate anisotropic force fields ("force tunnels") that make easier to execute certain movements, and help patients to move in the vicinity of the planned trajectory.

4 Neuro-Motor Rehabilitation Therapies

The rehabilitation therapies implemented in the E2REBOT platform have been developed based on the "assist as needed" methodology, through a haptic impedance controller, which calculates "on-line" the magnitude of the assistance that the robot gives the patient. The "assist as needed" approach tries to increase patient involvement, providing assistance only when necessary, depending on the degree of interaction between the patient and the system. We have developed two ways of neuro-motor therapy: passive-guided mode and active-assistive mode. Both modes include a graphical environment in which different games, designed to be used as therapeutic tools, can be executed, since patients interact with the game by using the robot as a joystick.

4.1 Passive-Guided Mode

The passive-guided mode is aimed at implementing a conventional rehabilitation therapy in which the patient is guided by the therapist to complete a given task. In a similar way, the robot guides the patient by creating a field of isotropic radial force acting in the same way in all directions, preventing the patient to move the system too slow or too fast. It is important to provide physical training, not only to teach the patient to reach a goal in his/her workspace, but to try to control the way that he/her achieves the goal. This requires a deeper cognitive effort, which can promote the process of neuroplasticity.

In order to implement this passive-guided mode the therapist moves the end effector of the robot to follow the path he/she want the patient subsequently to execute. This path is stored in the CompactRIO controller. Then the therapist selects a level of assistance that he/she wants the robot "helps" the patient, and the patient performs, with the robot, the previously stored trajectory. The level of assistance selected by the therapist allows the haptic control "generate" a viscoelastic force tunnel along each point of the recorded path, which keeps the patient in a "very close" path to the desired trajectory.

Three levels: easy, medium and hard, related to the level of assistance that the robot gives the patient, have been implemented in the therapies based on the passive-guided mode. For each of these three levels the impedance controller parameters (K, B, and M) have predefined constant values, which remain fixed throughout the execution of the rehabilitation task. In Table I the values of these parameters for each level are given.

Table 1 Impedance controller settings for all three levels of assistance in the passive-guided mode

level	K (N/m)	B (N/m/s)	M (Kg)
easy	110	18	1
medium	80	18	1
hard	60	18	1

The higher the value of K, the robot gives more support to the patient during the execution of therapy, and therefore the less the error resulting between the actual measured position and the desired position. The values shown in the table were obtained empirically by tests performed by therapists working with E2REBOT. We consider that the value of the force exerted by the subjects during therapy will be between 0 and 50 N. From this we found that the parameters in Table 1 provide a "nice" behavior and keep the system stable, without the appearance of resonance phenomena or noise sensitivity reactions.

The interface developed for the patient to perform this type of therapy is shown in Figure 4 (left screenshot). The path to be followed by the patient is shown on the screen by moving a "white hand". The "blue hand" represents the point in the path where the set patient-robot is moving. If the patient cannot "follow" the path, the robot controller, depending on the level of assistance scheduled, "guides" the patient "up to" the desired path through the forces virtual tunnel generated by the controller.

4.2 Active-Assisted Mode

This mode offers a kind of task-oriented therapy, complemented by an "assist as needed" algorithm to modulate the amount of assistance (Assistance Force) provided by the robot to the patient, based on his/her performance. Active-assisted mode uses the robot to help patients in the execution of movements. This technique is routinely applied by physiotherapists in the rehabilitation clinic practice.

The active-assisted mode regulates "on-line", by applying the "assist as needed" paradigm, the amount of help (assisting force) that the robot gives the patient at every moment during execution of the path, detecting the force exerted by the patient to move the robot. A software module on the computer calculates on-line, every 300 ms., the value of controller "K" parameter (its value determines the level of assistance provided). This value is sent to the haptic impedance controller in the CompactRIO (low level control), which generates the desired set point force to assist the patient in moving the robot.

In this mode of operation the patients can move freely a hand icon on the screen. The aim is to reach a small virtual mouse that runs away from them. The mouse behavior, such as speed or visual range, is highly customizable so the therapist can adjust the degree of difficulty of his capture. Three types of behavior (easy, medium and hard) have been implemented, depending on the speed of movement and difficulty of mouse capture.

Fig. 4 Virtual interfaces developed to execute passive-guided tasks (left) and active-assisted tasks (right)

To implement this active-assisted therapy a "cat and mouse" type game has been developed on the computer (see Figure 4, right screenshot). Each time the mouse is reached, the game score is increased in a given amount, and the mouse runs away from his position across the virtual field of repulsive forces, centered in the patient-controlled hand. This is the way pseudorandom paths are generated.

5 Registration and Management of Therapeutic Parameters

The evaluation of the effectiveness of the rehabilitation therapy after an ABI is a need recognized and addressed in different guidelines and clinical recommendations [11]. However, there is no consensus on the Outcome Measures to be used to evaluate specific aspects of the therapy, which makes it difficult in many cases a direct comparison of the results between different studies. Although there are studies that have pointed out that the measure of kinematic parameters in patients with ABI are adequate to evaluate their functional performance [12], it has not been devoted sufficient attention to studying the wide variety of kinematic parameters used in studies of assisted rehabilitation with robots and in particular, the validity of these parameters to reflect significantly the changes that try to induce in the patients.

Therefore, in E2REBOT we have developed an information management system, to analyze the bio-kinematics and bio-dynamics of the movements performed during therapy. We believe that it is possible to determine the state of a patient (diagnostic) and its evolution in a simple and economical way, from the point of view of the time clinicians have to invest, since these analysis can be performed without performing specific test to measure the execution of therapies, because the registration is performed simultaneously with the execution of the task by the patient.

The variables recorded by the E2REBOT platform when a patient performs a therapy, are, among other parameters, the target position point in the desired trajectory and the current position of the robot-patient set, target and speeds for the patient-robot set and the forces exerted by the patient on the robot (F*pacient*) and the assistive one provided by the robot to the patient (F*asistiv*).

In Figure 5 and in red color, the evolution of the desired trajectory that has been recorded by the physiotherapist in the passive-guided mode is shown. In the left graph the set of paths performed by a healthy person is shown in blue. With no impairment most of the paths reach almost the same points. However, in the chart at right, the set of trajectories performed by a patient with motor disabilities it is shown in blue. The great dispersion of the paths reached by the patient during therapy is observed, and can also be seen that the patient is not able to follow the path recorded by the physical therapist in the farthest zone (y-axis values between 300 ÷ 400 mm) . These data These data provide us with information that facilitates the diagnosis and severity of the possible paresis of the patient, which could be an instrument, not only for rehabilitation but it opens a field of work for the evaluation of motor paresis of the upper limb, and even to be able to establish degrees of severity depending on the paths followed. Obviously this proposal would be subject to validation of the platform as an evaluation tool.

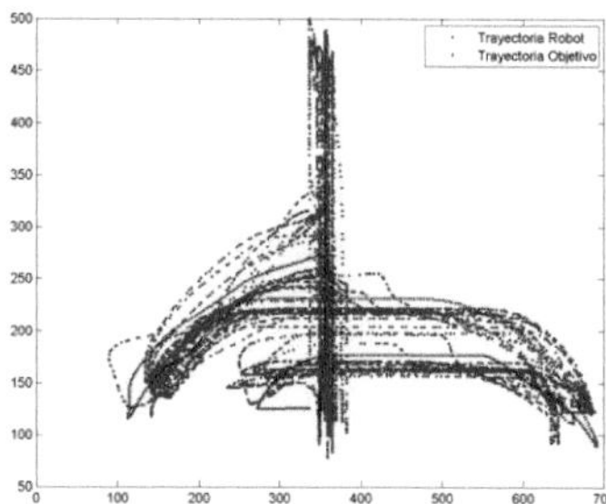

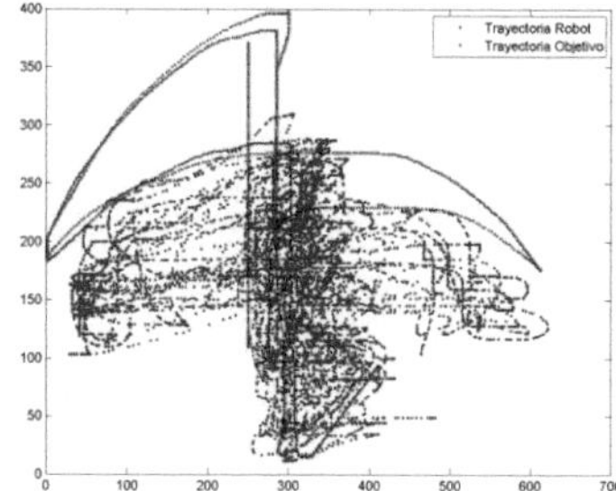

Fig. 5 Representation [mm] of the data recorded about the desired path (red) and the set of trajectories performed by the patient (blue) during therapy in "passive-guided" mode. The data in the graph on the left correspond to a healthy person. On the right, the trajectories of a patient with motor disabilities are shown.

6 Conclusions

In this work, the development of E2REBOT, a new prototype of intelligent robotic device for neuro-rehabilitation of the upper limb in patients with acquired brain damage, is shown. This robotic platform allows the application of therapies that are adapted dynamically on-line to the needs of each patient ("assist as needed" paradigm).

The haptic control strategy implemented in E2REBOT, by means of an impedance controller with force feedback, allows the patient robot interaction while performing rehabilitation therapies. In addition, a "feedforward" compensation has been necessary because the hardware where the control has been implemented has a low bandwidth, and we could not apply an excessive increase in the gain of the Kd error (measured as the difference between the desired and actual forces). This control strategy creates a "force field" that holds the patient's hand in a "tunnel", whose form is determined by the values of the parameters characterizing the K, B and M tensors of the haptic controller.

Including a management information system in the robotic platform allows storing and analyzing a large number of bio-kinematic and bio-dynamic variables while performing therapies. We believe that this analysis provides relevant information that will enable clinicians to make decisions to improve the efficiency and effectiveness of the functional rehabilitation of patients. In addition, the possibility of using the system as a tool for evaluating motor impairments of the upper limb and establish differentiated degrees of gravity is proposed, although it will be a future project whose results must be validated.

Acknowledgements This work has been partially funded by the Spanish CDTI, by reference project IDI-20130740, and by the Spanish Ministry of Economy and Competitiveness, through DPI2013-47196-C3-3-R project.

References

1. Warraich, Z., Kleim, J.A.: Neural plasticity: the biological substrate for neurorehabilitation. PM&R **2**(12), S208–S219 (2010)
2. Scott, S.H., Dukelow, S.P.: Potential of robots as next-generation technology for clinical assessment of neurological disorders and upper-limb therapy. J. Rehabil. Res. Dev. **48**(4), 335–353 (2011)
3. Bartenbach, V., Sander, C., Pöschl, M., Wilging, K., Nelius, T., Doll, F., Burger, W., Stockinger, C., Focke, A., Stein, T.: The BioMotionBot: A robotic device for applications in human motor learning and rehabilitation. Journal of Neuroscience Methods **213**(2), 282–297 (2013)
4. Blank, A.A., French, J.A., Pehlivan, A.U., O'Malley, M.K.: Current trends in robot-assisted upper-limb stroke rehabilitation: promoting patient engagement in therapy. Current Physical Medicine and Rehabilitation Reports **2**(3), 184–195 (2014)
5. Hesse, S., Heß, A., Werner, C., Kabbert, N., Buschfort, R.: Effect on arm function and cost of robot-assisted group therapy in subacute patients with stroke and a moderately to severely affected arm: a randomized controlled trial. Clinical Rehabilitation **28**(7), 637–647 (2014)
6. Hogan, N., Krebs, H.I.: Interactive robots for neurorehabilitation. Restorative Neurology and Neuroscience **22**, 349–358 (2004)
7. Liao, W.W., Wu, C.Y., Hsieh, Y.W., Lin, K.C., Chang, W.Y.: Effects of robot-assisted upper limb rehabilitation on daily function and real-world arm activity in patients with chronic stroke: a randomized controlled trial. Clinical Rehabilitation **26**(2), 111–120 (2012)
8. Timmermans, A.A., Lemmens, R.J., Monfrance, M., Geers, R.P., Bakx, W., Smeets, R.J., Seelen, H.A.: Effects of task-oriented robot training on arm function, activity, and quality of life in chronic stroke patients: a randomized controlled trial. J. Neuroeng. Rehabil. **11**(45), 0003–11 (2014)
9. Guerrero, C.R., Marinero, J.C.F., Turiel, J.P., Muñoz, V.: Using "human state aware" robots to enhance physical human–robot interaction in a cooperative scenario. Computer Methods and Programs in Biomedicine **112**(2), 250–259 (2013)
10. Fraile Marinero, J.C., Pérez Turiel, J.; Rodríguez Guerrero, C., Oliva, P.: Evolución de la plataforma robotizada de neuro-rehabilitación physiobot. In: VII Congreso Iberoamericano de Tecnologías de Apoyo a la Discapacidad (Iberdiscap 2013), Santo Domingo (República Dominicana), pp. 287–292 (2013)
11. Mehrholz, J., Haedrich, A., Platz, T., Kugler, J., Pohl, M.: Electromechanical and robot-assisted arm training for improving generic activities of daily living, arm function, and arm muscle strength after stroke. The Cochrane Library (2012)
12. Sivan, M., O'Connor, R.J., Makower, S., Levesley, M., Bhakta, B.: Systematic review of outcome measures used in the evaluation of robot-assisted upper limb exercise in stroke. Journal of Rehabilitation Medicine **43**(3), 181–189 (2011)

Design and Development of a Pneumatic Robot for Neurorehabilitation Therapies

Jorge A. Díez, Francisco J. Badesa, Luis D. Lledó, José M. Sabater, Nicolás García-Aracil, Isabel Beltrán and Ángela Bernabeu

Abstract This paper presents a new robotic system for upper limb rehabilitation. It is designed to assist the upper limb in therapies for both sitting and supine position, helping patients to carry out the required movements when they could not perform them. In the first part of the paper, the mechanical design and the development of the first prototype is exposed in detail. In the second part, new control strategy that modify the behavior of the rehabilitation robot according to different potential and force fields has been presented. Then, some experimental results of the performance of the implemented control with healthy subjects are reported.

Keywords Rehabilitation robotics · Robot design · Mechanical design · Control

1 Introduction

Along the next four decades it is expected that the population older than 60 years will increase in about a 50%, from 274 million in 2011 to 418 million foreseen in 2050 [1]. This fact implies the rising of age related diseases cases, specially cerebrovascular accidents or strokes, which may suppose the first cause of disability in many countries [2].

In order to recover their independence during daily living activities of people that suffered this kind of accident, mobility recovery rehabilitation therapies play an essential role. Studies in neuroplasticity conclude that active performance of motions in a coherent environment may help to restore motor skill of the affected limb [3].

J.A. Díez(✉) · F.J. Badesa · L.D. Lledó · J.M. Sabater · N. García-Aracil ·
I. Beltrán · Á. Bernabeu
Biomedical Neuroengineering Research Group, Miguel Hernandez University, Av. Universidad s/n, 03202 Elche, Spain
e-mail: {jdiez,fbadesa,llledo,j.sabater,nicolas.garcia}@umh.es, ibeltran@clinicabenidorm.com
http://nbio.umh.es

L.P. Reis et al. (eds.), *Robot 2015: Second Iberian Robotics Conference*,
Advances in Intelligent Systems and Computing 418,
DOI: 10.1007/978-3-319-27149-1_25

1.1 State of the Art

The use of robotic devices, as a possible rehabilitation strategy to achieve motor recovery, can be justified because of its potential impact on better therapeutic treatment and motor learning [4]. In the last years, several research groups have developed different robotic devices for upper-limb robot-aided neurorehabilitation. Most of these devices are actuated by electric drives [5–7], while there are only a few examples of rehabilitation robots based on pneumatic actuation systems. Some examples of pneumatic robots are iPam robot [8], which is actuated by pneumatic cylinders, and Rupert exoskeleton [9], which is actuated by artificial pneumatic muscles.

According to some recent reviews on the state of art in upper-limb rehabilitation robotics [10–12], there are no evidences of any pneumatic device for early delivering of rehabilitation therapy in supine position, being NEREBOT [13] the only electrical driven device designed to deliver this kind of therapies.

1.2 Goals

As an alternative to the current trend on this field, an arm rehabilitation robot with three active degrees of freedom (DOF), actuated by pneumatic technology, is being developed. It is designed to assist the upper limb rehabilitation therapies for both siting and supine position, helping patients to carry out the required movements when they could not perform them. In addition, the device is able to be moved freely by the patient as well as it can even oppose to the subject's movement in later rehabilitation stages.

In this paper, the development of a first functional prototype will be exposed, including its mechanical design and the statement of a new control strategy for this kind of robot. Some tests with healthy subjects are presented in order to validate the robot's behavior, so that the design can be checked out and refined before clinical tests with real patients.

2 Mechanical Design

The device which is being designed must have 6 DOFs in all: 3 of them will be active so the robot can position the patient's wrist inside its workspace, and the 3 remaining will be passive to allow the upper limb to orientate freely.

Since working conditions among different degrees of freedom may be very uneven, the mechanical design will endeavor to uncouple them. This way the different mechanical subsystems can be modeled and studied separately to the remaining ones in addition to simplify control task. As a result, the device has been divided into three main independent subsystems, which will be described separately:

- Planar DOF: Two first active degrees of freedom which control wrist position in planes parallel to the ground. These degrees of freedom must overcome resistance of patient's arm.
- Vertical DOF: Third active degree of freedom which control wrist height respect to the ground. This single degree of freedom works in different conditions than the previous ones, since it must balance arm's weight aside from the resistance that the patient can exert.
- End-Effector: This subsystem comprises three passive rotational degrees of freedom that will uncouple the orientation of the arm to the position of the wrist.

These three subsystems are assembled together on a wheeled frame that allows to transport easily the whole devices. This frame contains a lifting column in order to be able to adjust robot height so it can be used with any kind of wheelchair or stretcher. The prototype that has been built is shown in Fig. 1.

It is important to emphasize that one of the main features to be taken into account in the design of rehabilitation robotic devices is safety. For this reason, pneumatic actuators have been selected to drive the robot. They are considered safer than electric or hydraulic ones for robots interacting with patients, as the compressibility of the air makes them more backdrivable, specially rotary drives. Nevertheless, pneumatic

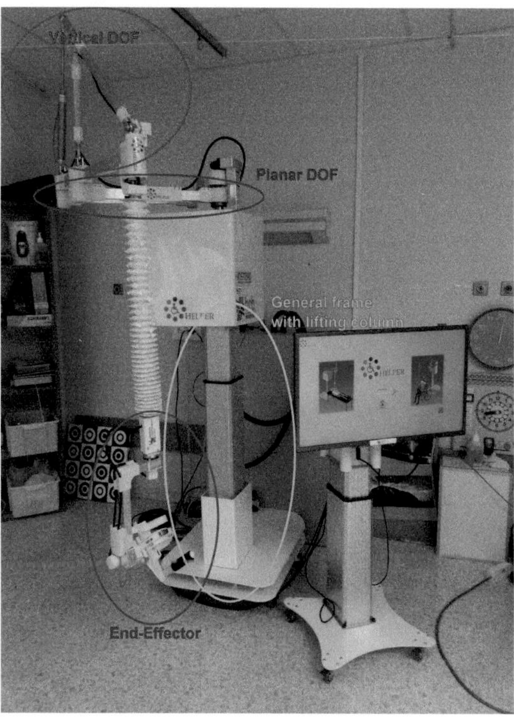

Fig. 1 First prototype of the Pneumatic Rehabilitation Robot.

actuators present some limitations which set out some implementation difficulties, due to their low response speed and small range of movements. This choice will condition greatly the kind of mechanical solutions implemented.

2.1 Planar Degrees of Freedom

For the first two degrees of freedom, an articulated parallelogram based on a previous work [14] has been developed. This previous work, PUPArm robot, is a rehabilitation robot design to deliver rehabilitation therapies in siting position, performing assisted movements of the patient's arm in a horizontal plane. As the main objective is different for the presented robotic system, some analysis have been re-computed:

- Kinematic design. Since this new robotic device will be able to administrate rehabilitation therapies for both siting and supine position, the kinematic analysis has been computed again in order to be able to achieve the user's arm workspace for these two positions.
- Dynamic design. Once the length of the links are changed, the torques exerted by the actuators must be higher in order to apply the same forces as the PUPArm robot in the human-robot interaction. In this regard, the immediately higher pneumatic swivel module has been selected, which can exert double the torque than the previous one.
- Structural design. As the structure of the planar degrees of freedom is going to be the frame of the vertical one, a structural analysis has been done. For this purpose, using results of Section 2.2 and maximum shear-stress theory, the structure has been designed to have a failure safety factor of 10.

2.2 Vertical Degree of Freedom

This degree of freedom presents some additional problems to the planar ones since the gravity acts against the robot when lifting patient's arm; thereby, the mechanism must balance out patient's weight in addition to the muscular tone.

Furthermore, rehabilitation exercises require a wide amplitude of movement, so a trade-off must be reached between maximum force and range of movement, as the mechanical advantage is inversely proportional to the amplification of the displacement.

In order to achieve a satisfactory solution, a collection of requirements has been clearly stated:

- Mechanism must perform pure one-dimensional straight trajectories.
- Vertical displacement of the end-effector must be about 55 cm or greater, so that the patient's arm can be fully extended, which defines the minimum workspace needed.

- Net static lifting force must be over 50 N as a reference value.
- This mechanism must be backdrivable enough, so the patient can perform free movements of the end-effector with minimum resistance.
- It must be as lightweight as possible in order not to increase excessively robot inertia, making easy the control tasks.
- Finally, the designed device must be completely out of the space occupied by patient and therapist avoiding to interfere or hit them.

Among the wide variety of linkages and mechanisms that might perform the required motion, a five-bar linkage actuated by a pneumatic cylinder (Fig.2 left) has been chosen since it has parameters enough to achieve a good compromise solution between range of motion (Δa_b) and maximum force over the end-effector. These parameters are five link dimensions (l_1,l_2,l_3,d_1,d_2) and two cylinder extreme positions (a_{c0}, a_{c1}). The device will be set up on the long link of the articulated parallelogram, that will act as a frame, as shown in Fig.1.

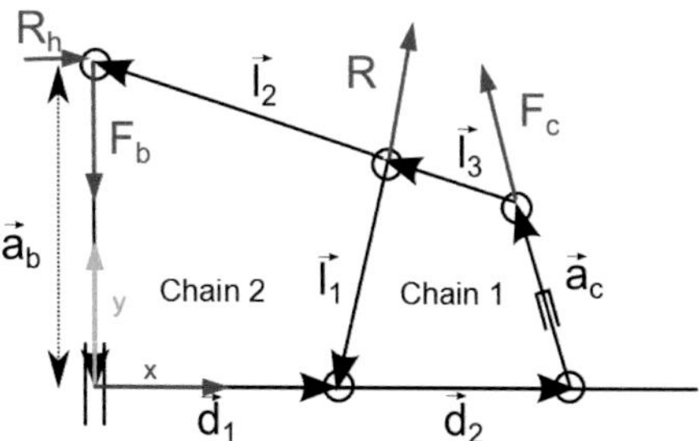

Fig. 2 Vectorial representation of vertical degree of freedom, where $\mathbf{a}_c$ represents the pneumatic cylinder, and $\mathbf{a}_b$ correspond to the vertical displacement.

The kinematics of this mechanism have been modeled by using the equations of the closed kinematic chains (1 to 9) shown in Fig.2.

$$a_{cx} + l_{3x} + l_{1x} + d_2 = 0 \quad (1)$$

$$a_{cy} + l_{3y} + l_{1y} = 0 \quad (2)$$

$$-l_{1x} + l_{2x} + d_1 = 0 \quad (3)$$

$$-l_{1y} + l_{2y} + a_b = 0 \quad (4)$$

$$l_1 = \sqrt{l_{1x}^2 + l_{1y}^2} \quad (5)$$

$$l_2 = \sqrt{l_{2x}^2 + l_{2y}^2} \quad (6)$$

$$l_3 = \sqrt{l_{3x}^2 + l_{3y}^2} \quad (7)$$

$$a_c = \sqrt{a_{cx}^2 + a_{cy}^2} \quad (8)$$

$$l_{2y}/l_{2x} = l_{3y}/l_{3x}. \quad (9)$$

Moreover a static forces model (Fig.2 right) has been obtained using static equilibrium equations (10 to 16) on the link composed by $\mathbf{l}_2$ and $\mathbf{l}_3$.

$$F_{cx} + R_x + R_h = 0 \tag{10}$$

$$F_{cy} + R_y - F_b = 0 \tag{11}$$

$$-F_{cy} * l_{3x} + F_{cx} * l_{3y} - R_h * l_{2y} - F_b * l_{2x} = 0 \tag{12}$$

$$F_{cx} = F_c * \frac{a_{cx}}{a_c} \tag{13}$$

$$F_{cy} = F_c * \frac{a_{cy}}{a_c} \tag{14}$$

$$R_x = R * \frac{-l_{1x}}{l_1} \tag{15}$$

$$R_y = R * \frac{-l_{1y}}{l_1}. \tag{16}$$

Where $\mathbf{R}_h$ is the reaction force of the linear guide, $\mathbf{F}_b$ is the load on the end effector, $\mathbf{R}$ is the reaction force of the bar $\mathbf{l}_1$ and $\mathbf{F}_c$ the force exerted by the cylinder $\mathbf{a}_c$.

Design parameters are subjected to several external restrictions imposed by requirements, working environment and mechanical issues such as:

- Commercial pneumatic cylinder size. Concretely, a 20 cm stroke cylinder is used, with initial length of 36.5 cm. Which define the values of a_{c0} and a_{c1}.
- Maximum length of horizontal projection of the linkage equal to 80 cm, which is the span of the frame.
- Maximum height lower than 70 cm so it does not collide with the ceiling.
- Stroke of the end-effector about 55 cm.
- Frame joints must be as close to the planar DOF joints as possible. Therefore, forces are not applied directly to frame link, but in the nodes.

With the purpose of obtaining a satisfactory linkage sizing, a combination of graphical and analytic methods has been used. First of all, a pre-sizing is achieved by representing graphically a mechanism that fulfill the external restrictions in both extreme position and adjusting visually the link dimensions so a reasonable linkage is obtained. Then, the mechanism is simulated using the mathematical models stated before, checking link trajectories and forces over it. If the result is not the desired one, link dimensions are modified. This process is repeated until a satisfying solution is reached.

One of the key parameters is the relation between the forces exerted on the end-effector F_b and the required force in the cylinder F_c. This way, the factor $\frac{F_c}{F_b}$ shows the force needed to exert with the actuator to lift 1 Newton of weight applied on the end-effector; this value varies along the height so it is imperative to obtain a curve (Fig.3) that allow the control system to lift patient's arm in any position. A polynomial will be fit to that curve in order to obtain a handy expression to compute the required force in each position.

The most satisfactory link sizes correspond to: $l_1 = 43cm$, $l_2 = 60.5cm$, $l_3 = 16.35cm$, $d_1 = 55cm$, $d_2 = 23.5cm$. It presents the force curve and polynomial fitting shown in Fig.3, which means that with a cylinder that can exert a maximum force of 550 N it might lift a total weight of about 117 N. Note that the mechanical advantage of the linkage lever is on the patient, oppositely to the usual trend that gives advantage to the actuator. This effect is desired to make the system more backdrivable since the user will be always in the most favorable position to overcome the actuator, but this one will have still force enough to help the patient when needed.

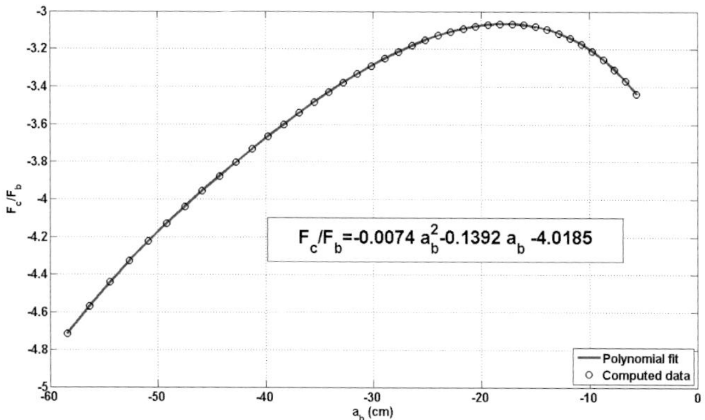

Fig. 3 Static forces on the linkage.

2.3 *End-Effector*

The end-effector is a device that consists on three rotary joints, which allows the arm to orient freely to the wrist position so that the patient can be in the most comfortable posture. The axes of these joints are perpendicular between each other and converge in a single point that will be considered the center of the wrist. These DOFs are sensed in order to know which is the orientation so a virtual reality system can perform an accurate representation of the patient's arm.

As shown in Fig.4 the x-axis is placed along the direction of the vertical DOF and it will be responsible of decoupling the orientation of the planar DOF to the patient's arm. The y-axis is located perpendicular to the former one and contains the wrist center, this will allow to rotate the forearm in a plane perpendicular to the ground. Finally the z-axis is placed along the pronation-supination axis of the arm, enabling this movement. The designed device and a prototype are shown in Fig.4

X and Y axes are sensed by using precision potentiometers attached to their respective shafts, but measuring the rotation of the Z-axis requires a more complex system. Since this DOF is made of several pieces that can be detached while setting it up on the arm, it presents some discontinuities in its mechanical design that preclude the use of potentiometers or other similar devices. As a solution, an incremental encoder has been built with optical sensors and 3D-printed parts (Black parts in Fig.4). This device has a resolution of 2 degrees, which is enough to get a smooth representation of the arm in a virtual environment.

3 Control

A recent review of control strategies for robotic movement training [15], shows that the most commonly used control strategy is impedance control, and there are only a few studies that use methods based on potential fields, some for lower-limb rehabilitation [16], [17], and only a few for upper-limb rehabilitation [18].

For the presented rehabilitation robotic system, a new approach for the control of the robot based on artificial potential fields has been developed. This control strategy tries to modify the behavior of the rehabilitation robot according to different potential and force fields.

The main objective of the developed control is to modify the robot behavior in accordance with a force field defined along its workspace, trying to mimic the corrective actions done by the therapists in the rehabilitation therapy session. In order to define these suitable force field responses, different parameters, such as the precision (p) to reach the target, the static friction (F_f) and the maximum force exerted by the robot (F_{max}) have been taken into account.

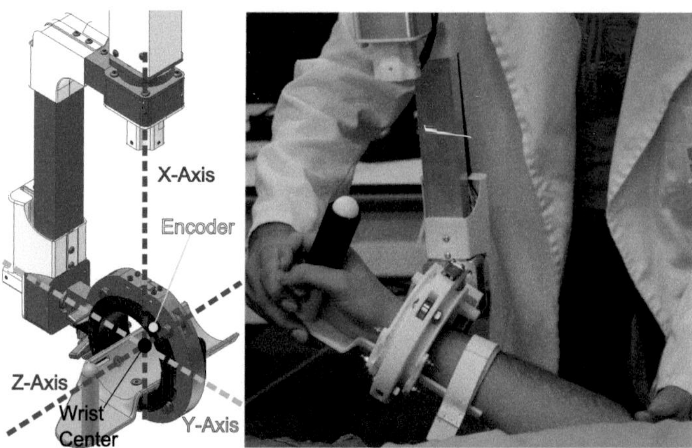

Fig. 4 End-effector CAD design (left) and end-effector prototype with an attached subject (right).

3.1 Implementation

As pointed out above, the active degrees of freedom of this robotic system are divided, and decoupled, in horizontal plane movements and vertical axis movements. In this respect, the implemented controller has been divided in two separates control loops.

Horizontal Plane Movements Control. For this first control loop, a force field with a Gaussian profile has been defined. With this field shape, we have a different force response which changes slowly near the target. In (17) a general definition of a Gaussian is shown, while equations (18) and (19) are the rectification of the Gaussian to have an adequate force field for our requirements: zero force in the target position and maximum force far from the target. In these expressions, a is the maximum amplitude of the Gaussian, b is the position of the center of the peak and c is the parameter that define the width of the Gaussian (is the x value where the inflection point occurs).

$$f(x) = a\exp\left(\frac{-(x-b)^2}{2\cdot c^2}\right) \tag{17}$$

$$\|\mathbf{F_g}\| = a - a\exp\left(\frac{-(x-b)^2}{2\cdot c^2}\right) \tag{18}$$

$$\mathbf{F_g} = -sign(x)\left(a - a\exp\left(\frac{-(x-b)^2}{2\cdot c^2}\right)\right)\cdot\hat{\mathbf{x}} \tag{19}$$

In order to adjust the force field response to the restrictions of precision, static friction and maximum force, parameters a, b and c have been selected as follows:

$$\begin{aligned} a &= F_{max} \\ b &= 0 \\ c &= \sqrt{\frac{-p^2}{2\cdot\left(1-\left(\frac{F_f}{F_{max}}\right)\right)}} \end{aligned} \tag{20}$$

Vertical Axis Movements Control. For this control loop, and taking into account the effects of the gravity, a quadratic potential field, defined by (21), has been selected due to its shape is more restrictive than the Gaussian one.

The force field derived from this potential field increases linearly with the distance from the target position, as is shown in (22). The parameter ζ exposed in these equations determine the width of the quadratic potential shape.

$$U_q = \frac{1}{2}\zeta\,(x-x_0)^2 \tag{21}$$

Table 1 Path following RMSE (cm)

	X-axis	Y-axis	Z-axis	Path
User 1	2.38	1.70	2.49	3.84
User 2	1.64	1.57	1.91	2.97
User 3	2.01	1.57	3.27	4.14
User 4	1.45	1.76	2.55	3.43
User 5	1.80	1.29	2.89	3.64
User 6	1.67	1.28	3.16	3.80
User 7	2.41	2.10	4.27	5.37
User 8	1.67	1.33	2.47	3.26
User 9	1.96	1.60	2.04	3.26
Mean	1.89	1.58	2.78	3.75
STD	0.32	0.25	0.68	0.67

$$\mathbf{F_q} = -\nabla U_q = -\zeta\ (x - x_0) \cdot \hat{\mathbf{x}} \tag{22}$$

The parameter ζ has been defined in order to obtain a resultant force of value F_f in positions at distance p from the target (23). Also, taken into account the maximum force exerted by the robot (F_{max}), a threshold has been achieved by combining both quadratic and conic potentials, as is shown in (24).

$$\zeta = F_f/p \tag{23}$$

$$U = \begin{cases} \frac{1}{2}\,\zeta\ (x - x_0)^2 & |x - x_0| < x^* \\ x^*\ \zeta\ (x - x_0) - \frac{1}{2}\,\zeta\ (x^*)^2 & |x - x_0| \geq x^* \end{cases} \tag{24}$$

where $x^* = F_{max}/\zeta$ is the position where the force due to (24) is equal to F_{max}.

3.2 Experimental Results

In order to show the performance and accuracy of the developed control system, an experimental trial has been carried out. In this experiment, nine students and staff members of the Biomedical Neuroengineering Group of Miguel Hernandez University participated in the experiment. All were healthy, with no major cognitive or physical deficits.

In this experiment, a therapist performs a guided movement of the user's arm, which is recorded by the robotic system. Then, the robot has to reply twice this movement in order to guide the user's arm along the desire path.

Table 1 shows the root mean square error (RMSE) along the trajectory as well as the RMSE for each axis of the 3D space. Mean and standard deviation of these results can be seen graphically in Fig. 5.

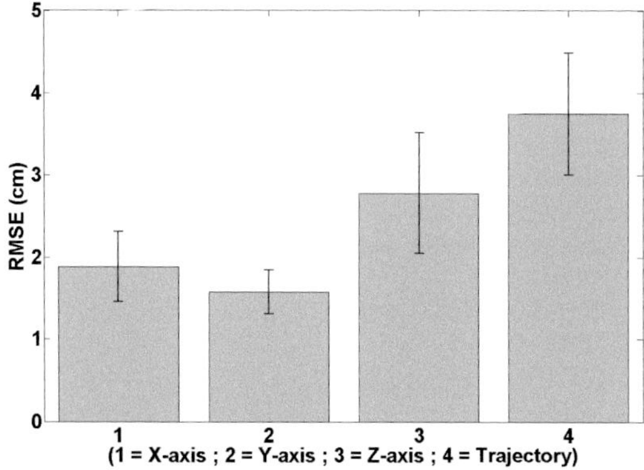

Fig. 5 Path following RMSE.

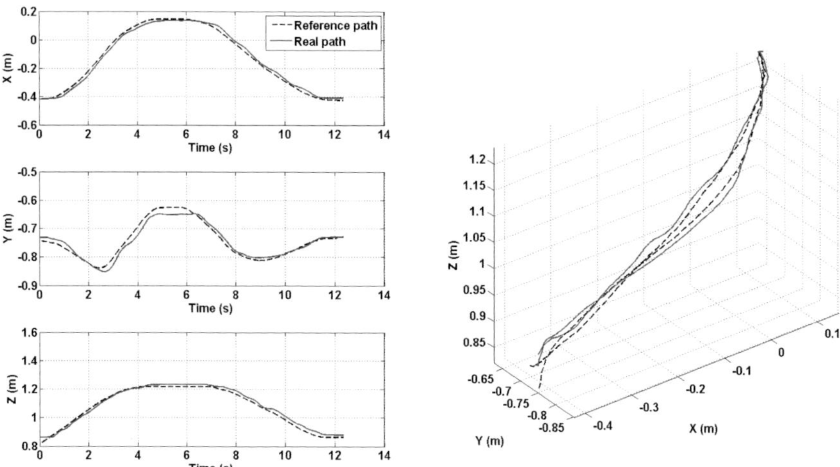

Fig. 6 Path following.

4 Conclusion

Despite of the strict and conflicting requirements imposed on the design, the chosen alternatives have proven to be congruent and work together correctly. This results in a robust system that can resist and exert the required forces and torques during the rehabilitation therapy.

Finally, results of the experimental trial shows that the controller based on potential fields achieve a stable and safety behavior of the robot with an an acceptable accuracy. In Fig. 6 an example of a trajectory perform by one user is shown. These results give rise to start a wide clinical trial with post-stroke patients in order to compare the benefits of the presented rehabilitation robotic system with traditional therapies and other alternative technologies.

References

1. U. Nations: World population prospects. The 2010 revision (2011)
2. Prange, G., Jannink, M., Groothuis-Oudshoorn, C., Hermens, H., Ijzerman, M.: Systematic review of the effect of robotaided therapy on recovery of the hemiparetic arm after stroke. J. Rehab. Res. Develop. **43**(2), 171–184 (2006)
3. Schaechter, J.D.: Motor rehabilitation and brain plasticity after hemiparetic stroke. Prog. Neurobiol. **73**, 61–72 (2004)
4. Krebs, H.I., Hogan, N., Aisen, M.L., Volpe, B.T.: Robot-aided neurorehabilitation. IEEE Trans. Rehabil. Eng. **6**(1), 75–87 (1998)
5. Lum, P.S., Burgar, C.G., Van der Loos, M., Shor, P.C., Majmundar, M., Yap, R.: The MIME robotic system for upper-limb neuro-rehabilitation: results from a clinical trial in subacute stroke. In: 9th International Conference on Rehabilitation Robotics ICORR 2005, pp. 511–514 (2005)
6. Nef, T., Guidali, M., Klamroth, V., Riener, R.: ARMin-Exoskeleton for Stroke Rehabilitation. IFMBE Proceedings **25**(9), 127–130 (2009)
7. Loureiro, R., Amirabdollahian, F., Topping, M., Driessen, B., Harwin, W.: Upper Limb Robot Mediated Stroke Therapy:GENTLE/s Approach. Auton. Robots **15**(1), 35–51 (2003)
8. Jackson, A., Holt, R., Culmer, R., Makower, S., Levesley, M., Richardson, R., Cozens, J., Williams, M., Bhakta, B.: Dual robot system for upper limb rehabilitation after stroke: the design process. Journal of Mechanical Engineering Science- Proceedings of the Institution of Mechanical Engineers, Part C **221**, 845–857 (2007)
9. Balasubramanian, S., Ruihua, W., Perez, M., Shepard, B., Koeneman, E., Koeneman, J., Jiping, H.: Rupert: an exoskeleton robot for assisting rehabilitation of arm functions. In: Virtual Rehabilitation, pp. 163–167 (2008)
10. Masiero, S., Armani, M., Rosati, G.: Upper-limb robot-assisted therapy in rehabilitation of acute stroke patients: focused review and results of new randomized controlled trial. J. Rehabil. Res. Dev. **48**(4), 355–366 (2011)
11. Morales, R., Badesa, F.J., Garca-Aracil, N., Sabater, J.M., Prez-Vidal, C.: Pneumatic robotic systems for upper limb rehabilitation. Medical & Biological Engineering & Computing **49**(10), 1145–1156 (2011). ISSN 0140-0118
12. Nordin, N., Xie, S.Q., Wnsche, B.: Assessment of movement quality in robot- assisted upper limb rehabilitation after stroke: a review. J. Neuroeng. Rehabil. **12**(11), 137 (2014). doi:10.1186/1743-0003-11-137
13. Masiero, S., Celia, A., Rosati, G., Armani, M.: Robotic-assisted rehabilitation of the upper limb after acute stroke. Arch. Phys. Med. Rehabil. **88**, 142–149 (2007). doi:10.1016/j.apmr.2006.10.032. [PMID: 17270510]
14. Badesa, F.J., Llinares, A., Morales, R., Garcia-Aracil, N., Sabater, J.M., Perez-Vidal, C.: Pneumatic planar rehabilitation robot for post-stroke patients. Journal of Biomedical Engineering: Applications, Basis and Communications **26**(02), 1450025 (2014)
15. Marchal-Crespo, L., Reinkensmeyer, D.J.: Review of control strategies for robotic movement training after neurologic injury. Journal of NeuroEngineering and Rehabilitation **6**, 20 (2009). doi:10.1186/1743-0003-6-20
16. Banala, S.K., Agrawal, S.K., Scholz, J.O.: Active leg exoskeleton (alex) for gait rehabilitation of motor-impaired patients. In: IEEE 10th International Conference on Rehabilitation Robotics (ICORR), pp. 401–407 (2007)
17. Banala, S.K., Agrawal, S.K., Kim, S.H., Scholz, J.O.: Novel gait adaptation and neuromotor training results using an active leg exoskeleton. IEEE/ASME Transactions on Mechatronics **99**, 1–10 (2010)
18. Mihelj, M., Nef, T., Riener, R.: A novel paradigm for patient-cooperative control of upper-limb rehabilitation robots. Adv. Robotics **21**, 843–867 (2007)

An Active Knee Orthosis for the Physical Therapy of Neurological Disorders

Elena Garcia, Daniel Sanz-Merodio, Manuel Cestari, Manuel Perez and Juan Sancho

Abstract This paper presents the design of a new robotic orthotic solution aimed at improving the rehabilitation of a number of neurological disorders (Multiple Sclerosis, Post-Polio and Stroke). These neurological disorders are the most expensive for the European Health Systems, and the personalization of the therapy will contribute to a 47% cost reduction. Most orthotic devices have been evaluated as an aid to in-hospital training and rehabilitation in patients with motor disorders of various origins. The advancement of technology opens the possibility of new active orthoses able to improve function in the usual environment of the patient, providing added benefits to state-of-the-art devices in life quality. The active knee orthosis aims to serve as a basis to justify the prescription and adaptation of robotic orthoses in patients with impaired gait resulting from neurological processes.

Keywords Rehabilitation robots · Exoskeletons · Knee orthoses · Gait rehabilitation · Neuro-rehabilitation

1 Introduction

Multiple Sclerosis (MS) is the most common neurological disease among young adults, it is usually diagnosed between 18 and 35 years. MS is a chronic, degenerative central nervous system disease. The symptoms of MS are very diverse and vary from

E. Garcia(✉) · M. Perez · J. Sancho
Centre for Automation and Robotics, CSIC-UPM, 28500 Arganda del Rey, Madrid, Spain
e-mail: elena.garcia@csic.es

D. Sanz-Merodio
Marsi Bionics SL, 28500 Arganda del Rey, Madrid, Spain

M. Cestari
Department of Electrical and Computer Engineering, University of Houston, Houston, TX 77004-4005, USA

L.P. Reis et al. (eds.), *Robot 2015: Second Iberian Robotics Conference*,
Advances in Intelligent Systems and Computing 418,
DOI: 10.1007/978-3-319-27149-1_26

person to person depending on the areas affected: fatigue, visual disturbances, balance problems, walking difficulty and coordination, etc. It is not possible to predict the disease course. It is more common among women and young adults. The disease leads to consequences that limit quality of life, so it has a great social impact. It is the second leading cause of disability in young people, behind traffic accidents [1].

Stroke is the leading cause of permanent disability in Europe and USA [2]. Neurological impairment after stroke frequently leads to hemiparesis or partial paralysis of one side of the body that affects the patient's ability to perform activities of daily living (ADL) such as walking and eating. One-third of surviving patients from stroke do not regain independent walking ability and those ambulatory, walk in a typical asymmetric manner. Neurological impairment after stroke can lead to reduced or no muscle activity around the ankle and knee causing the inability to lift the foot (drop foot).

Poliomyelitis (or simply polio) is an infectious disease transmitted by a virus (poliovirus). After World War II, the first polio vaccine was able to control the disease. Polio mainly affects children under three years. About 95% of cases are completely asymptomatic, but 0.5-1% are paralytic. The key of disability in patients with chronic sequelae of poliomyelitis is the emerging muscular weakness resulting from the combination of the initial effects in acute attack of polio and subsequent neuronal damage. This muscle deterioration often affects the ability for ambulation of patients with sequelae of poliomyelitis (PPS) and they need technical aids and orthotics for ADL.

Rehabilitation addresses these neurological diseases from a multidisciplinary approach. The goal of rehabilitation therapy is to achieve the highest level of physical, mental, emotional and social independence. If the rehabilitation treatment begins in the early stages of the disease, it can improve the general condition of the patient, quality of life, and prevent complications. The goal of rehabilitation exercises is to perform specific movements that provoke motor plasticity to the patient and therefore improve motor recovery and minimize functional deficits. Movement rehabilitation is limb dependent, thus the affected limb has to be exercised [3].

Traditional rehabilitation therapies are very labor intensive especially for gait rehabilitation, often requiring more than three therapists together to assist manually the legs and torso of the patient to perform training. This fact imposes an enormous economic burden to any country's health care system thus limiting its clinical acceptance. All these factors stimulate innovation in the domain of rehabilitation in such a way it becomes more affordable and available for more patients and for a longer period of time. The variability of symptoms, unpredictability of the illness course, and degenerative character of these three neurological disorders described above make it difficult to find a solution that can assist in the physical therapy by providing walking assistance. The requirement is for a general-purpose robotic orthosis, reconfigurable based on the illness progression, and provided with advanced perception-action capabilities that self-adapt to the disease variable symptoms. This is the main objective of our approach in the development of a robotic orthosis that provides gait assistance and rehabilitation to this group of patients. This paper presents the MB-ActiveKnee, its design and preliminary functinality tests.

2 The Enabling Technology

2.1 *Current State of the Technology*

Although research on joint-specific orthoses is in progress, at present two types of braces are widely used in the clinical domain for gait rehabilitation of neurological disorders (see Fig.1):

- AFO (Ankle Foot Orthosis), which controls the foot and ankle, and indirectly the knee.
- KAFO (Knee Ankle Foot Orthosis) for the control of foot, knee and ankle.

The control of the knee joint is critical in the rehabilitation of neurological diseases. Conventional KAFOs use some type of knee joint close (sealing ring, Swiss close), which may or may not lock the knee in extension while walking, and unlock for sitting. The locking of the knee during walking is indicated for: Instability of the knee in the sagittal plane, quadriceps weakness, weakness of hip extensors, inability to balance the trunk over the extremities.

These conventional passive orthoses, although less complex and cheaper, cannot supply energy to the affected limbs, hence are limited compared to active devices. Proprioception and skills are required to maintain a stable walk, controlling the body position with the center of mass always ahead of the knee.

To approach the active orthosis field, robotics is an emerging field which is expected to grow as a solution to automate rehabilitation and training. Robotic orthoses can (i) replace the physical training effort of a therapist, allowing more intensive repetitive motions and delivering therapy at a reasonable cost, (ii) assess quantitatively the level of motor recovery by measuring force and movement patterns, (iii) potentially assist in daily life activities providing power to walk up slopes or stairs. Many robotic systems have been developed to enforce or restore ankle and knee motion specifically. These systems can be grouped into stationary or active orthoses.

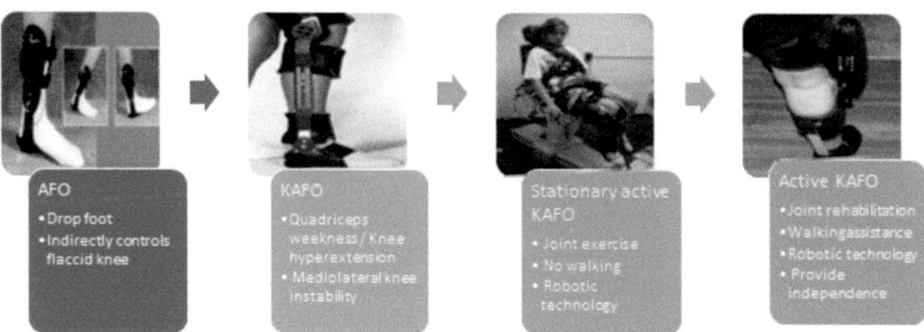

Fig. 1 State of the art orthoses

Stationary System: Stationary systems are those robotic mechanism designed to exercise the human ankle and knee motions without walking. The patient is positioned always in the same place, and only the target limb is exercised. Devices like the Rutgers Ankle[4], the High Performance Ankle Rehabilitation Robot[5] by the IIT, based on a Stewart Platform. A more recent system, the Active Knee Rehabilitation Orthotic Devices (AKROD)[6], provides variable damping at the knee joint, controlled in ways that can facilitate motor recovery in post stroke and other neurological disease patients.

Active Orthoses: On the contrary to stationary systems, active foot orthoses are actuated exoskeletons that the user wears while walking overground or in a treadmill. They are intended to control position and motion of the ankle, compensate for weakness, or correct deformities. They are an evolution of traditional passive lower limb orthoses, with additional capabilities to promote appropriate gait dynamics for rehabilitation. They have the potential of providing tools to facilitate functional recovery, reducing cost of treatment and providing patients with adequate level of independence.

- Active Foot Orthoses: Currently, the only commercialized system for rehabilitation of the ankle joint is the Anklebot (Interactive Motion Technologies, Inc.), developed at the Massachusetts Institute of Technology (MIT) to rehabilitate the ankle after stroke[7]. It allows normal range of motion in all 3 DOF of the foot relative to the shank while walking overground or on a treadmill. Pilot controlled trials with such device showed a general improvement in the walking distance covered and time. The MIT also developed an Active Ankle-Foot Orthosis (AAFO) where the impedance of the orthotic joint is modulated throughout the walking cycle to treat drop-foot gait[8].
- Active Knee Orthoses are mostly dissipative by combining an electro-rheological fluid-based variable damper with a modified commercial knee brace[6]. This device is intended to provide resistive torques to the user for rehabilitation purposes.
- Active Knee-Ankle-Foot-Orthosis (AKAFO): Commercial devices like C-Brace by Ottobock are a combination of AFO and KO, having a dissipative variable damper at the knee. Quasi-passive devices are being tested in clinical setting for stroke, MS, PPS, and other neurological affections[9]. Active KAFOs having a powered knee joint are mainly at a research stage powered by artificial pneumatic muscles [10] or by electrical motors [11].

For many patients, a programmable actuated orthosis could guide and facilitate the recovery of a more efficient and clinically desirable gait pattern via retraining sessions. Current clinical practice is generally restricted to brief periods of less than 1 hour of gait training provided a few times per week. In between these sessions, patients continue to walk using their typical gait pattern, and likely reinforce compensatory gait patterns. A wearable training orthosis could be used by patients to guide them through a targeted gait pattern while undertaking daily activities. This strategy of reinforced therapy in a real-world environment has the potential to provide more effective gait retraining, improving one's ability to ambulate.

Finally, clinical studies conducted still show little evidence for a superior effectiveness of the robotic therapy, although a clear benefit is shown in reduced therapist effort, time, and costs [12]. It has been shown that robotic rehabilitation can be as effective as manually assisted training for recovery of locomotor capacity, but a higher benefit should be desirable to spread its use in clinics worldwide.

This paper presents the design and preliminary tests of an orthotic device configured as an active KAFO(see Fig. 2). Joint actuation is based on the ARES technology [13] developed at the Centre for Automation and Robotics and currently being commercialized by Marsi Bionics. The ARES technology provides power and controllable compliance to each joint.

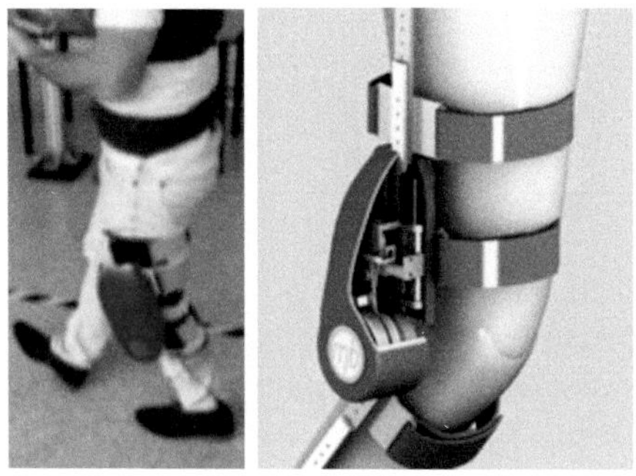

Fig. 2 Active knee orthosis using ARES technology

3 Biomechanics Analysis of the Human Knee Motion

Let us distinguish six phases of the knee motion along the gait cycle for regular walking on level ground. Figure 3 shows the knee angle vs. torque and power consumed at the knee during regular walking on level ground, and distinguishes these six phases:

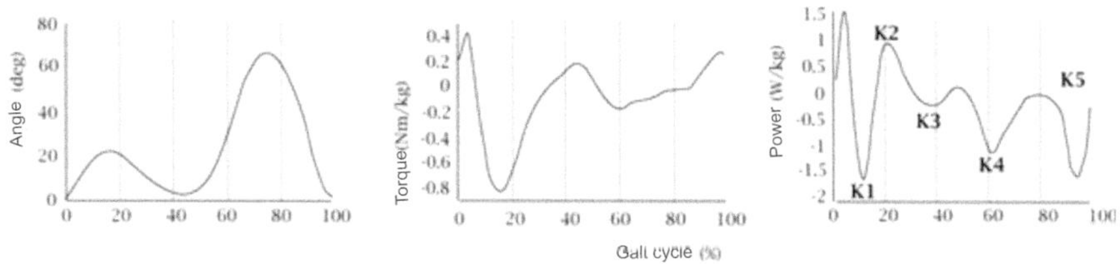

Fig. 3 Joint angle trajectory, torque and power consumption at the knee when walking on level ground

Phase 1 Load response at stance (0% to 15% of the gait cycle): Immediately after heel strike the supporting leg slightly flexes. This slight bending allows accommodating the body weight on the supporting foot. Drawing the knee angle versus torque (see Figure 4) it can be observed that in this phase the knee acts as a spring which elastic constant is the slope of the curve mentioned.

Phase 2 Knee extension at stance (15% to 42% of the gait cycle): After a maximum knee flexion of near 20 degrees at stance, the knee is then extended, reaching the maximum extension at 42% of the walking cycle, although the knee is not completely stretched (not reaching 0 degrees).

Phase 3 Fast knee bending (42% to 62% of the gait cycle): At toe off, between the end of leg stance and the beginning of leg swing, a fast knee flexion occurs preparing for leg swing. Here again the leg acts as a spring but featuring a lower stiffness than in the load response phase.

Phase 4 Knee flexion at swing (from 62% to 73% of the gait cycle): While the hip transfers the foot forwards, the knee remains flexed (near 60 degrees) to avoid tripping over the floor. At this phase the knee does negative work against gravity by braking to avoid the fall of the calf.

Phase 5 Knee extension at swing (from 73% to 95% of the walking cycle): The knee extends to prepare for support again.

Phase 6 Knee flexion for heel strike (95% to 100% of the walking cycle): Just before leg stance begins again, the knee flexes minimally to minimize the foot speed relative to the ground, this substantially reduces the impact with the ground, and thus improves stability and energy consumption is reduced.

Based on these biomechanics characteristics of the human knee, the MB-ActiveKnee, an active knee brace has been designed. Particularly relevant, the human knee load response properties will be mimicked by a spring with the adequate stiffness, while an actuator provides the necessary energy demanded in those knee phases of high consumption.

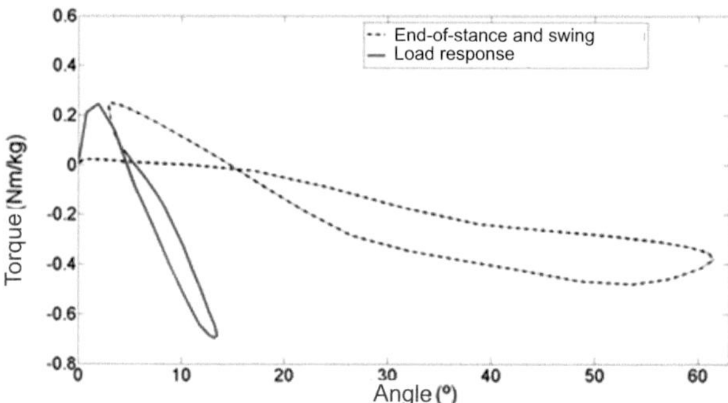

Fig. 4 Torque versus knee angle. An almost linear behavior is observed in the load response phase

4 Design of the MB-Active Knee

The MB-ActiveKnee is a joint work between the Centre for Automation and Robotics and Marsi Bionics. It has been conceived to provide active knee control in the leg stance providing the necessary rigidity, and active power during swing. Therefore this active brace compensates for the lack of knee mobility or quadriceps weakness, present in MS, Stroke and PPS.

Figure 5 presents the CAD design of the MB-ActiveKnee. It is an AKAFO (Active Knee Ankle Foot Orthosis), where the ankle joint responds passively. The brace extends down to the foot to provide better grip and stability on the subject. The shank link is adjustable in extension from 20 cm to 70 cm.

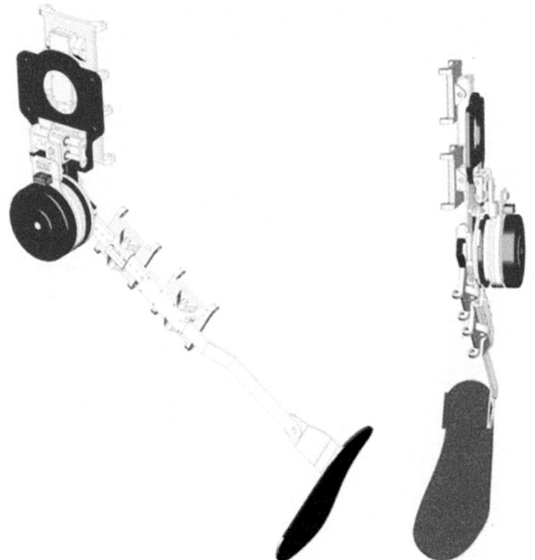

Fig. 5 CAD detail design of the Active Knee

The knee actuator is based on the ARES [14] patented technology. It can be conceptually divided into two parts, a rigid actuation set and a transmission based on elastic elements. A gear motor combination is used to provides speed and torque ratings above 2.5 rad/s and 30 Nm respectively, sufficient for the requirements of this application.

As shown in Figure 6, the passive elastic elements in series to the transmission allows for a safe user-robot interaction, intrinsically absorbing undesired perturbations. Furthermore, this design allows also to absorb jerky movements usual in some patients. A position sensor measures the spring deflection, thus providing a measure of the torque being applied by the following relationship:

$$\tau = \frac{2\Delta x K_{eq} L_{arm}}{\cos(\theta)} \tag{1}$$

Where Δx is the measured deflection of the elastic elements, K_{eq} is the equivalent elastic constant of the springs and L_{arm} is the radial distance to the joint axis.

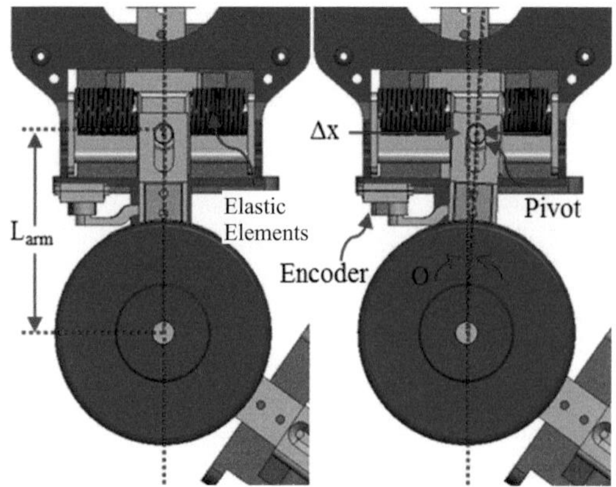

Fig. 6 ARES spring deflection at torque transmission

5 Hardware

Power to weight and power to volume ratios have been optimized during the selection of hardware components. Lithium-ion batteries provide more than three hours of continuous operation.The orthosis includes a magnetic position sensor for measuring the joint angle. As mentioned, the joint torque is computed from the measured spring deflection, by which good parallel force-position control can be performed.

Foot plantar pressure sensors are used for real-time detection of the gait phase. An insole with two pressure sensors, one at the heel and one in the toe is integrated in the orthosis. Figure 7 shows how the sensors are very effective in differentiating the stance and swing phases. It can be observed how the beginning of the stance is detected by the heel sensor (in red line) while the end of the stance is detected by the toe sensor (in green line).

6 Control Approach

The control approach has been designed to exploit the advantages of the mechanical design to mimic the biomechanics of the human knee. The innovative mechanical design and underlying technology confer great versatility to the MB-ActiveKnee brace, facilitating adaptation to the symptomatology of the patient. To offer such versatility the following control schemes have been implemented:

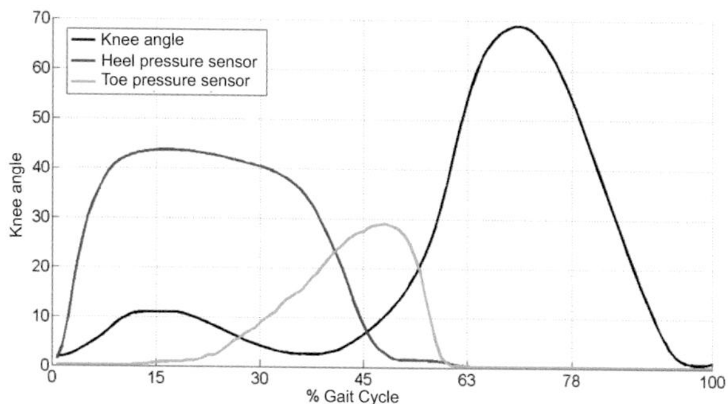

Fig. 7 Relationship between knee angle and plantar pressure

Zero force control: Based on a purely force control scheme, the knee brace "gets out of the away" when the user initiates the motion. This allows the user and the therapist to manipulate the brace effortlessly or to programm rehabilitation exercises through demonstration of the therapeutical movements. The brace is made transparent to the user, who does not perceive resistance to his/her own motion.

Position control: The knee is able to repeat a number of joint trajectory patterns that the user or the therapist programmed.

The combination of these control schemes yields the following operation modes:

Gait training by learn and repeat: The patient having the brace adequately adjusted, the therapist performs a series of movements which the MB-ActiveKnee learns and then repeats cyclically.

Gait assistance: MB-ActiveKnee assist in gait performance to patients having weak quadriceps. MB-ActiveKnee detects gait phase and provides the additional power required to complete the phase.

As discussed in Section 3, where the biomechanics behavior of the knee is studied, in the load response phase the human knee behaves like a spring. The elastic elements of the MB-ActiveKnee knee show a elastic constant similar to that obtained in the human knee load response phase (see Figure 4). During operation the MB-ActiveKnee reaches a 40% reduction of energy consumption and provides a more natural movement of the knee[14] by taking advantage of this intrinsic impedance during the load response, while power is only supplied during swing to flex and extend the knee.

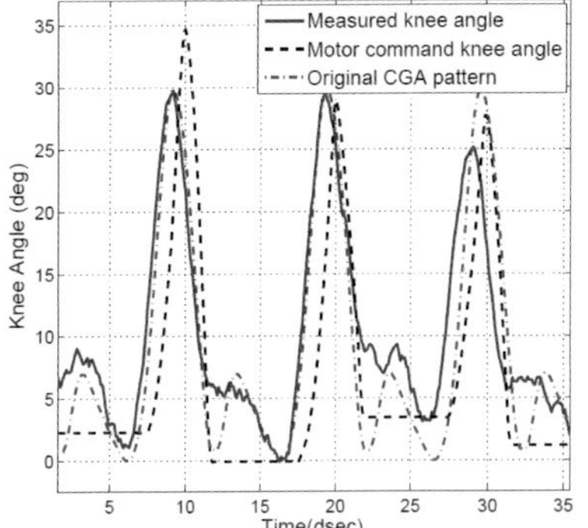

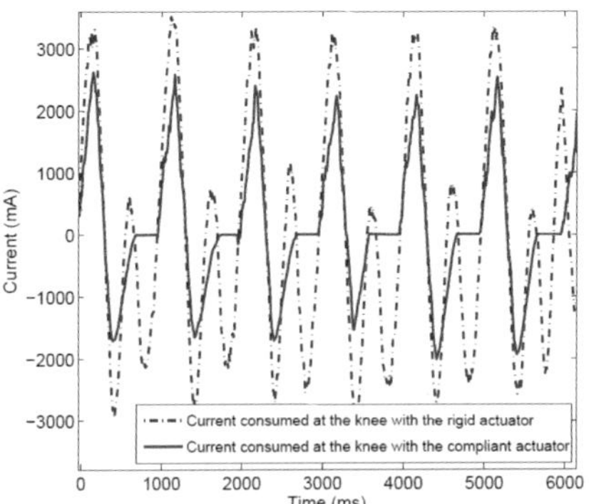

Fig. 8 Knee joint trajectory along three gait cycles using ARES elastic actuation vs. CGA pattern. Notice that the power during stance is provided by the elastic elements while the actuator is not commanded

7 Results

Figure 8 shows in black line the motor trajectory, in blue line the resulting knee joint trajectory provided by the MB-ActiveKnee brace, while it is compared to the Clinical Gait Analysis (CGA) pattern for the human knee. During the load response the motor is not commanded, not providing power to the brace, which is simply powered by the elastic elements, able to provide sufficient adaptation to body weight during stance. Therefore a reduction in energy consumption is resulting in longer battery life.

Functional tests were performed in healthy subjects. Gait speed and maximum knee flexion/extension angles are configurable for each patient via a user interface. During the tests the MB-ActiveKnee adapts naturally to the user detecting gait phases and providing the required power. A video of this performance is shown in https://www.youtube.com/watch?v=oDd7dI3lZc0.

8 Conclusions and Future Work

This paper describes the design and functionality of the MB-ActiveKnee brace developed by Marsi Bionics SL (www.marsibionics.com) in collaboration with the Center for Automation and Robotics (CSIC-UPM). It is an AKAFO for use in the therapy of multiple sclerosis, stroke, post-polio syndrome or spinal cord injury.

MB-ActiveKnee results in a light device (1 Kg), allowing portability with a range of more than three hours in continuous operation. It incorporates the ARES technology, a geared motor assembly which allows to provide the necessary power in swing phases but also provides energy-efficient weight bearing during stance based on the inherent elastic behaviour of the ARES joint. As a result the user interaction is safe throughout time as these springs absorb shocks.

The functionality of this AKAFO brace has been tested in a healthy user with significant results. Work is in progress to test the device on a number of voluntary patients suffering from quadriceps weakness, in different types of ground profiles, uphill, downhill and flat, also up and down stairs. The clinical evaluation will provide understanding of the potential, limitations and challenges of the robotic technology applied to the orthotics field.

Acknowledgements This work has been partially funded by the Spanish Ministry for Economy and Competitiveness through grant DPI2013-40504-R and by EU FP7 Echord++ Exp 401 EXOTrainer.

References

1. European Commission Health and Consumer Protection Directorate: Some elements on the situation of multiple sclerosis in the european union (2013)
2. Adams, R.J., Lloyd-Jones, D., Brown, T.M., et al.: Heart disease and stroke statistics-2010 update: a report from the american heart association. Circulation **121**(7), e46–e215 (2010)
3. Schmidt, H., Werner, C., Bernhardt, R., Hesse, S., , Kruger, J.: Gait rehabilitationmachines based on programmable footplates. Journal of NeuroEngineering and Rehabilitation **4**(2) (2007)
4. Girone, M., Burdea, G., Bouzit, M., Popescu, V., Deutsch, J.E.: Stewart platform-based system for ankle telerehabilitation. Autonomous Robots **10**(2), 203–212 (2001)
5. Saglia, J.A., Tsagarakis, N.G., Dai, J.S., Caldwell, D.G.: A high-performance redundantly actuated parallel mechanism for ankle rehabilitation. International Journal of Robotics Research **28**(9), 1216–1227 (2009)
6. Nikitczuk, J., Weinberg, B., Canavan, P.K., Mavroidis, C.: Active knee rehabilitation orthotic device with variable damping characteristics implemented via an electrorheological fluid. IEEE/ASME Transactions on Mechatronics **15**(6), 952–960 (2010)
7. Roy, A., Krebs, H.I., Patterson, S.L., et al.: Measurement of human ankle stiffness using the anklebot. In: IEEE International Conference on Rehabilitation Robotics, Noordwijk, The Netherlands, pp. 356–363 (2007)
8. Blaya, J.A., Herr, H.: Adaptive control of a variable-impedance ankle-foot orthosis to assist drop-foot gait. IEEE Transactions on Neural Systems and Rehabilitation Engineering **12**(1), 24–31 (2004)
9. Shamaei, K., Napolitano, P., Dollar, A.: Design and functional evaluation of a quasi-passive compliant stance control knee-ankle-foot orthosis. IEEE Transactions on Neural Systems and Rehabilitation Engineering **22**(2), 258–268 (2014)
10. Sawicki, G.S., Ferris, D.P.: A pneumatically powered knee-ankle-foot orthosis (KAFO) with myoelectric activation and inhibition. Journal of Neuroengineering and Rehabilitation **6**(1), 23–39 (2009)
11. Arazpour, M., Chitsazan, A., Bani, M., Rouhi, G., Ghomshe, F., Hutchins, S.: The effect of a knee ankle foot orthosis incorporating an active knee mechanism on gait of a person with poliomyelitis. Prosthet. Orthot. Int. **37**(5), 411–414 (2013)
12. Diaz, I., Gil, J.J., Sanchez, E.: Lower-limb robotic rehabilitation: Literature review and challenges. Journal of Robotics **2011**, 11 (2013)
13. Cestari, M., Sanz-Merodio, D., Garcia, E.: Articulation with controllable stiffness and force-measuring device.Patent ES P201330882, WO 2014/198979 A1 (2013)
14. Cestari, M., Sanz-Merodio, D., Arevalo, J., Garcia, E.: An adjustable compliant joint for lower-limb exoskeletons. IEEE/ASME Transactions on Mechatronics **20**(2), 889–898 (2015)

Part VI
Robotic Applications in Art and Architecture

LSA Portraiture Robot

Bruno Rodrigues, Eduardo Cruz, André Dias and Manuel F. Silva

Abstract This paper describes the development of an application that allows an ABB robot arm to automatically perform the portrait of people. The Portraiture Robot performs the picture of a human face on paper. The developed system consists of 4 steps: (i) image acquisition through a webcam, (ii) image processing to retrieve the contours and features of the person's face, (iii) vectorization of the coordinates in the image plane, and (iv) conversion of the coordinates to the RAPID programming language. To get only the person's face, is performed a background subtraction and to obtain only the necessary information from the image are used filtering techniques to remove the features and contours of the person's face. To convert these points into x, y coordinates, the contours are vectorised and sent to a file, saved according to a defined protocol, and allowing to create a program for the robot. The developed application allows processing of all blocks listed above in real-time and in a robust manner, having the ability to adapt to any environment and allowing continued use. The work was validated through the participation in the 2014 Portuguese Robotics Open, and in an ISEP exhibition that occurred in Maia, always with good results.

Keywords Computer vision · Industrial robot manipulator · Drawing · OpenCV · ROS · QtCreator

1 Introduction

The field of artificial vision and its integration with robotics has gained increasing prominence. In view of the development of skills in this area, as well as in the area

B. Rodrigues · E. Cruz · A. Dias · M.F. Silva(✉)
Department of Electrical Engineering, LSA - Autonomous Systems Laboratory,
ISEP-IPP - School of Engineering of the Polytechnic of Porto, Porto, Portugal
e-mail: {1110403,1110200,apd,mss}@isep.ipp.pt
http://lsa.isep.ipp.pt/

B. Rodrigues · E. Cruz · A. Dias · M.F. Silva
INESC-TEC - INESC Science and Technology, Rua Dr. António Bernardino de Almeida,
Porto, Portugal

L.P. Reis et al. (eds.), *Robot 2015: Second Iberian Robotics Conference*,
Advances in Intelligent Systems and Computing 418,
DOI: 10.1007/978-3-319-27149-1_27

of programming of industrial robots, it was decided to develop a robot that can draw pictures of humans, and serve as a demonstration platform for study visits to ISEP's control laboratory and exhibitions.

The main objective of this project is to develop a system that makes a robot doing a portrait of a human face on paper. The system developed is separated into two parts: the first related to the field of artificial vision and responsible for collecting information of the human face, and the second related to robotics and in charge of representing the information obtained previously. This system should be as functional and versatile as possible. The process that takes place from collecting the information of the human face until the completion of the picture should be as automated as possible, within a few security parameters.

The objectives of this project included the development of (i) an open-source application for the image processing and its drawing using an ABB industrial robot arm, (ii) a system for the detection of contours and features of a human face adopting image processing techniques, (iii) a graphical user interface (GUI) allowing to control the entire image acquisition and processing process, (iv) the optimization of the robot drawing time / quality of the drawing relation, (v) a system robust enough in order to adapt to different environments, and (vi) perform the system validation in distinct environments.

Bearing these ideas in mind, the sequel of the paper is organized as follows. Section two presents a brief state of the art and section three introduces the high level system architecture. Next, sections four and five are devoted to describing the computer vision and the robotics applications developed, respectively. Section six presents several tests performed to validate the system, and the results obtained. Finally, section seven summarizes the main conclusions of the work and states some ideas for future improvements.

2 Related Work

A few systems with somehow similar objectives have already been presented before and others are described in the literature.

The "Robotlab Autoportrait" project uses a KUKA robot arm with a tool that allows it to use a pen and draw people's faces [9]. The arm is equipped with a camera that takes a picture, after which the robot starts drawing the face. In the end the robot shows the drawing to the audience. The level of automation is an interesting aspect because the robot is able to put the arm in a position to take the photo and, in the end, is able to pick up the screen, show it to people and then clean it. The picture is taken in a controlled environment where the focus of light is facing the person, leading to a good image quality.

The work "Human portrait generation system for robot arm drawing" was developed for a humanoid robot [5]. After being captured the image of the person, is performed the face detection using a learning algorithm for faces, being used the skin color to define the region of interest (RoI). For defining the representative lines of the

RoI is used the Sobel technique and an algorithm is used to remove unwanted noise from these lines. To optimize the drawing process, the image undergoes a process of thinning of the face lines, as well as a process of removing unnecessary points. After this, all coordinates of key facial lines are recorded and sent to the robot arm. One point to emphasize of this algorithm is that the image has a good process of the face identification. However, since there is the need for a learning process by the algorithm (which is computationally heavy), its use is limited for different features without prior knowledge.

The Aikon project resulted in several works related to robotics and artificial vision, the most known being "Paul and Pete The Sketching Robots" [13]. The initial version of this project had only a robotic arm (Paul) to which was later added another arm (Pete). In this project there is a a webcam (installed on a servomotor which allows its movement during the drawing process) that takes a photo, being created a drawing and generated the necessary commands to the robots perform the picture. The end result is a picture done on paper, on which are present the shadows of the person's face [8]. This project is developed in the Robotic Operation System (ROS) platform and uses the Open Computer Vision (OpenCV) library. The robot draws the protruding lines that are captured by a Gabor filter. After this, an algorithm intended to represent shadows is used, to give a more realistic appearance to the drawing. The technique in question has too much unnecessary noise, eventually covering the face features. Another disadvantage is that, not doing any pretreatment of the image so that it only sketches the face of the person, is dependent on having a white panel as background, eventually stop being robust to adapt to any environment.

The "Portrait Bot" is a project developed at the University of Freiburg, Germany, which uses a PR2 robot, along with the ROS platform and the OpenCV library [11]. The robot is controlled via a GUI that runs on a remote computer. After the image acquisition, a learning algorithm is used to detect people's faces, creating a mask. This way, just the person's face is obtained, removing any background objects that could also be drawn. The detection of the contours and features is performed through the Canny technique. To improve the picture, and optimize the design process, two algorithms are used: one that removes lines too short and another that allows linking the lines that are close. In the result of this project it is noted that there is a good detail of a person's face, but there is also a lot of information lost what ends up losing definition. The process for getting only a person's face uses face learning, therefore not being robust for anyone.

In one of the latest projects on portraitists robots, Google developed an application that captures a user's face image through the computer's camera. This is converted into a drawing which is then used to create the robot commands. This information is sent to a museum where a robot (the sketchbot robot used was developed by Google) draws the portraits in a litter box [4].

Recently Markiewicz developed a work entitled "Robotic Portrait Drawer" [6]. This system uses a Webcam to take a picture and an ABB IRB 140 robot to draw it, being divided into four phases: (i) the acquisition of the camera image, (ii) the image vectorization, (iii) the transfer process data to the robot controller and (iv) the design of the vectorized image using the robotic arm. The main problems with this work were the absence of autonomous working of the system and its lack of robustness to different environments.

3 System Architecture

As depicted in Figure 1, the general architecture of the system is divided into two parts: the artificial vision part and the robot one. The first part is responsible for the image acquisition, its treatment and vectorization. Since a robot program consists of a set of points/targets in space (set of coordinates) that define the paths that robots should travel, the points resulting from the image vectorization are saved in a file. The second part is tasked with the conversion of the relevant image points to a program in the robot programming language (RAPID). After this, the program is sent to the robot (via File Transfer Protocol (FTP)), which is constantly checking the communication line waiting for the file to be downloaded.

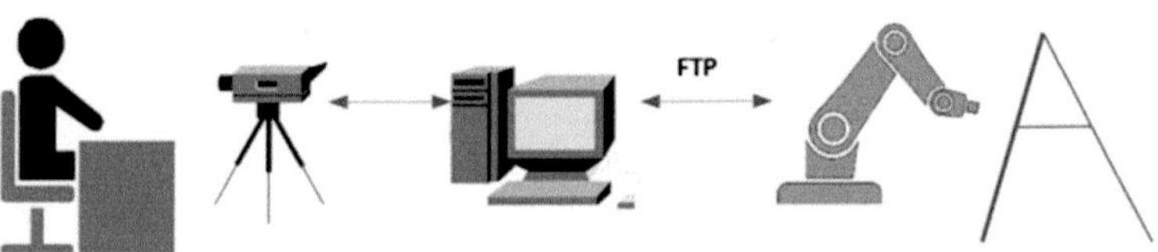

Fig. 1 System main architecture.

To acquire a person's image, it is needed a camera with an USB connection to the computer. The camera used in this project is a NGS Xpress Cam-300, with a resolution of 5 megapixels. A computer is necessary to take a picture with the webcam, and also to process the image and send the resulting file with the robot program to the robot, via FTP, so that it is able to draw the portrait. In order for these two parts to communicate with each other, and make the whole system work, the file shared between them follows a defined protocol.

4 Generation of the Face Contours

Based on this architecture, for the generation of the face contours were followed the next three steps:

– image acquisition and processing: at this phase there are requirements on the identification and extraction of features in individuals presenting different faces,

with varying brightness issues, and the need to perform a background removal, to ensure that only useful information is extracted from the image;
- evaluation of state of art image filtering techniques;
- conversion of the points obtained in the image plane for a RAPID language data structure, based on the need to resize the image to different sizes of paper sheets, as well as the translation to the robot zero (reference of coordinates system).

The software was developed in the ROS [7] platform (version Fuerte, installed on a computer with the Ubuntu 12.04 operating system) together with the OpenCV library, thereby achieving one of this project objectives (the use of open-source software). ROS allows a connection between the interface and the webcam (Figure 2) and, together with the OpenCV, performs the image processing, in order to extract the features in the face and then the coordinates in the image plane.

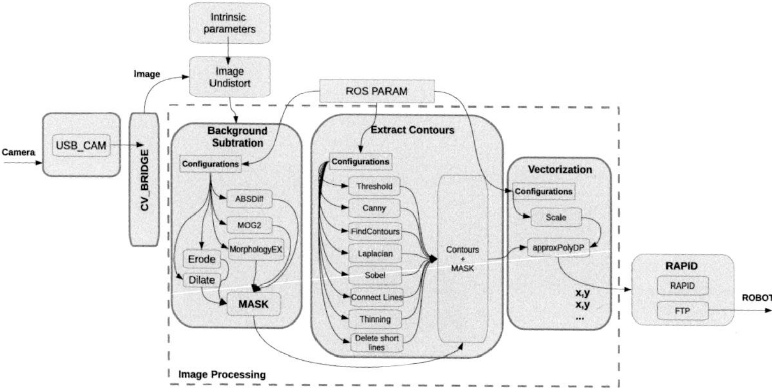

Fig. 2 ROS architecture.

To be able to receive the image from the webcam, an usb_cam module is used. This module publishes the images in a topic, in order to make them usable, but the OpenCV doesn't recognize an image from this topic. As such, is used the CVBridge, which connects the topic and the library [10].

A problem inherent in most cameras is that they capture images that exhibit a distortion created by the lens. This phenomenon can be modelled as radial or tangential, and is possible to compute its coefficients using calibration algorithms. In this work, to remove the camera distortion and have a clean image were used the intrinsic parameters obtained to rectify it. These parameters are always constant in the camera, unless there is a change of its focus [2].

To process the image is created a module. The first step of this module is to remove the background, through a background subtraction operation – this way the robot doesn't draw anything except the face. Its possible to use two algorithms for this purpose: Absdiff [3] and MOG2 [12]. Both create a mask based on the difference between the image with the face and the image with the background. With the aim of improving this mask, a morphology algorithm (morphologyEX), an erode and a dilate operation are applied to the image. The next step is to detect the minimum area of the mask, and cut it out of the image. The mask is then applied after the image has been treated, to avoid that the applied algorithms for contour detection find the contour of the masks edge.

The contours detection techniques identify sudden changes in brightness levels and colors, based on the use of filters. To detect the contours, a group of algorithms are applied such as Threshold, Canny, Findcontours, Laplacian and Sobel. Its also possible to manipulate the brightness and the contrast as to accentuate the contours. To optimize and improve the process of drawing, the module connects near lines, eliminates small lines and straightens them.

Before the vectorization of the contours, an image resize is done, so that it may be possible to draw the image in a sheet of paper. A translation is applied to the image to match that reference with the robot workobject reference. The points are saved in a file, and a point with the coordinates $(-1; -1)$ is created in the end of every line.

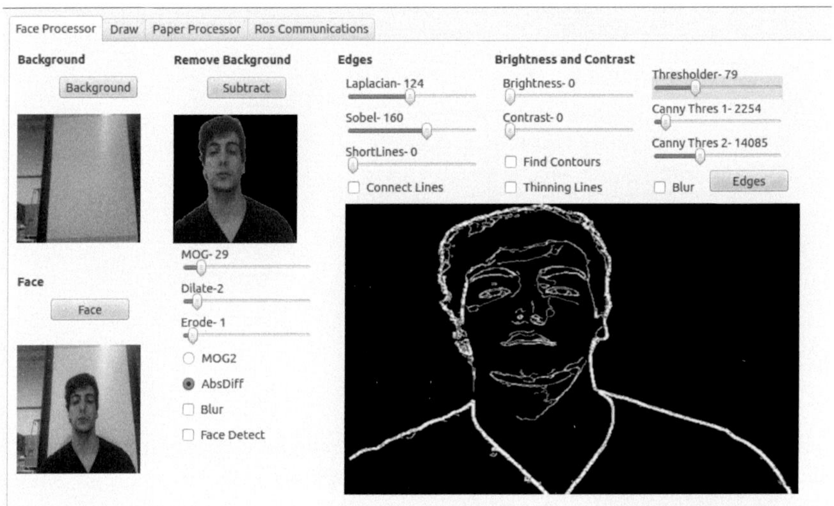

Fig. 3 Graphical user interface.

All these procedures can be done in an interface (developed in the QT programming language) to facilitate the allocation of the various parameters which exist in the different referenced algorithms and also visualize the resulting aspect of the image (Figure 3). The interface is individualized from the image processing module and publishes the values of the parameters in the ROSPARAM, as can be seen in Figure 4. The module responsible for the processing will subscribe the values of the parameters and use them in the filters.

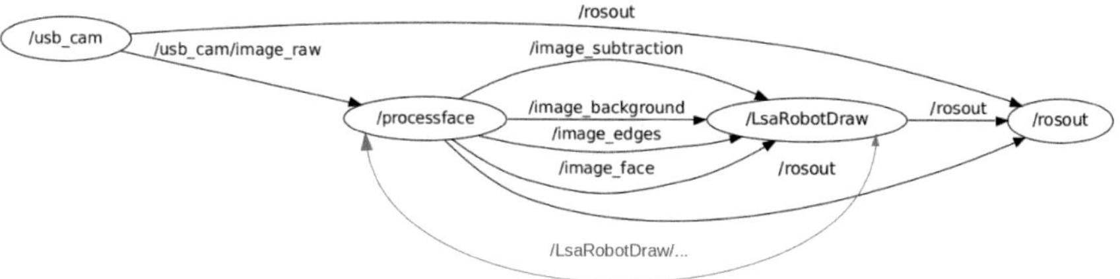

Fig. 4 Comunication middleware architecture.

5 Robotic Drawing

After producing the file with the relevant image points, a C++ program was created to read those points (coordinates, as depicted in Figure 5) and "translate" the code into the ABB robot RAPID programming language [1]. The RAPID program that was created makes the robot draw the portrait. The reading of the points and the conversion to RAPID code is done in three main steps:

- read the file with the coordinates;
- save the coordinates in dynamic memory;
- generate the code with the information needed.

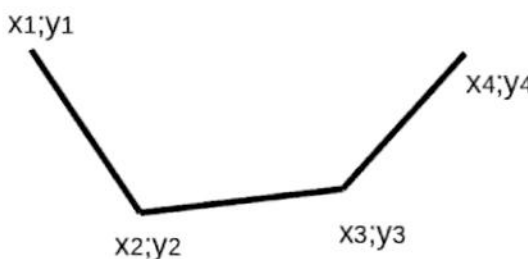

Fig. 5 Decomposition of lines and curves into points with given coordinates.

The reading of the points is done in the same order the points are written in the file by the artificial vision part. The points defined in the file are part of the drawing and will become the targets to the robot movements. When the lines ends a target is created with the last coordinates but with a higher position – this way the pen stops writing. The approach to the next line is done creating a target with the same coordinates but with a higher position – this way the pen approaches the paper vertically. Figure 6(a) presents a flowchart with the steps, executed by the code.

After the execution of this code, a file describing all the movements that the robot needs to perform to execute the drawing is created. The file describes the movements required to the robot's pencil execute a trajectory equal to the sketch of the face, obtained by the vision part. The file mentions two elements that must be predefined in the robot – the tool and the work space. The tool used in the project was a mechanism that makes it possible for the robot to use a pen or a marker. The work space adopted

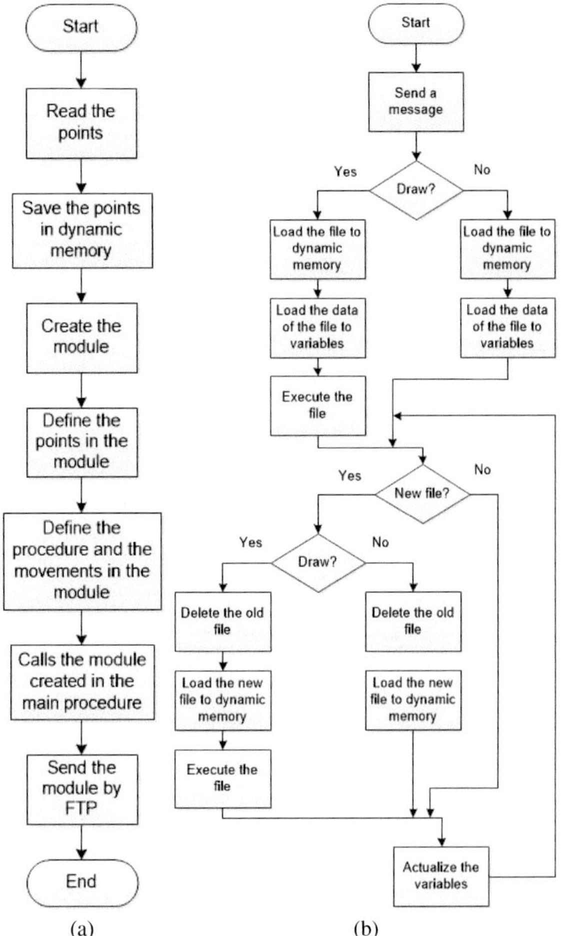

Fig. 6 C++ program (a), and robot program (b) flowcharts.

was a support for the sheet of paper that fixes it while the portrait is being executed. To send the file to the robot an Ethernet connection was created between the robot and the computer, through the interface. The file is sent to a specific localization in the robots memory (defined by the programmer). In the robot, a program is running waiting for the file in the location defined earlier.

The flowchart depicted in Figure 6(b) shows how this scheme works. It is possible to notice that the verification is done by the record of the date when the file was placed in the robot memory. If the date of the file is more recent than the date of the previous file, a message is sent to the robot controller asking if the user wants to draw or not: if the user chooses "yes" the robot loads the file and starts executing it; if the user presses "no" the robot waits for a new file. In both cases the date of the file is saved.

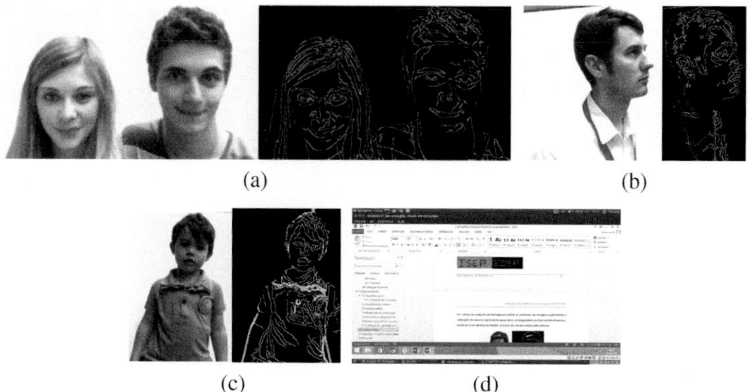

Fig. 7 Resulting image with two people standing together (a), a person standing in profile (b), a full body person (c), and a hand written message (d).

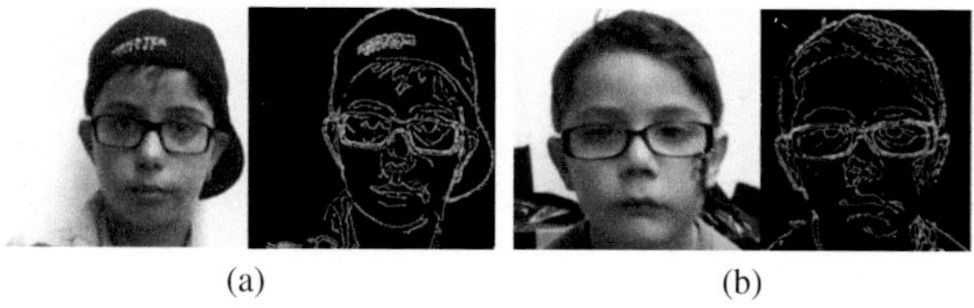

Fig. 8 Resulting image of a person using a cap (a), and using eyeglasses (b).

When the robot starts to execute the file, it consequently executes all movements defined. Simultaneously, the portrait begins to be done with the work space defined and the tool prepared.

6 Tests Performed and Results Achieved

With the work completed, this system was on display at the 2014 Portuguese Robotics Open (FNR'2014), that took place in Espinho, and at an ISEP dissemination action in Maia, where it was possible to carry out a set of tests and get the results of this project.

6.1 Tests and Results on the Image Generation

Having the ability to detect the foreground, and not using any learning algorithm for facial recognition, this system allows drawing more than just a face. It permits to draw

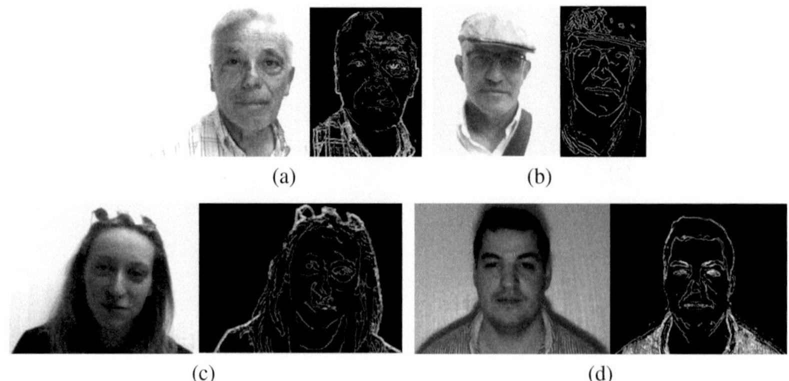

Fig. 9 Resulting image of a person with baldness (a), with a beret on his head (b), with sunglasses on the head (c), and with checkerboard pattern on his shirt (d).

people together (Figure 7(a)), people in profile (Figure 7(b)), full body (Figure 7(c)) and even anything standing in front of the camera (Figure 7(d)).

By using a set of techniques to withdraw the outlines of images and allowing the calibration of the various required parameters, it is possible to get a good picture of the person with a good detail of the features, including the caps (Figure 8(a)) and eyeglasses (Figure 8(b)) used by them.

However, one of the major problems of artificial vision is the instability against lighting variations. It was verified at FNR'2014, which took place in a sports hall containing numerous sources of light, that shadows were easily created on the faces of several people, mainly on people with whiter skin or who have baldness (Figure 9(a)), which causes the light to be reflected and negatively contributing to the image detail. The same happens when people use accessories that reflect light, such as berets (Figure 9(b)) and sunglasses (Figure 9(c)) on their heads. Another point to mention is the fact that the drawing of people who use clothes with a chessboard pattern is not affected, as shown in Figure 9(d), but the drawing process becomes quite time consuming.

6.2 *Tests and Results on the Robot Drawing*

After developing the work were performed several tests, to determine which of the options led to a picture with better quality. In the various options, the component that changes is the tool used. In the case of the IRB120 there were two cases: in the first was used a marker with a spring and in the second a marker without a spring. In the case of the IRB140 was used just a common pen.

In the tests performed with the marker loaded with a spring it was found that the spring exerted excessive force on the paper sheet. This made the portrait traces got too

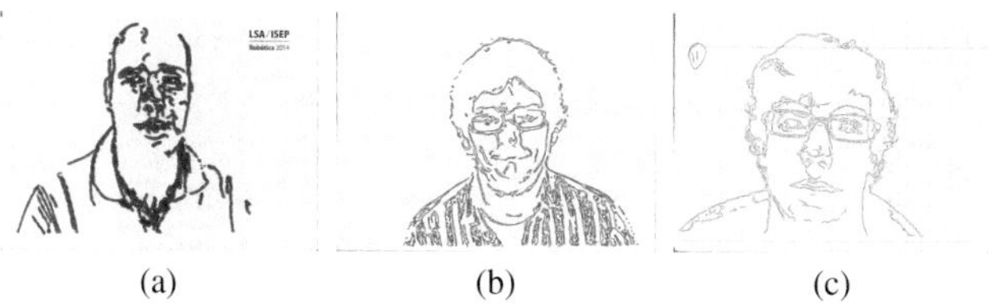

(a) (b) (c)

Fig. 10 Picture of a portrait done with the marker loaded with a spring in the IRB120 robot (a), of a portrait done with the marker without a spring in the IRB120 robot (b), and of a portrait done with a common pen in the IRB120 robot (c).

thick, removing some details from the pictures (Figure 10(a)). In the tests performed with the marker without spring it was found that only the force made by the marker mass was enough to make a thin line. Pictures of this test show a higher quality when compared with portraits performed in the previous test since, as the trace is thin, are shown more details of the human face (Figure 10(b)). Finally, in the tests conducted with the pen it was found that the trace is thinner than the trace obtained in the tests carried out with the marker. In these tests was obtained a higher quality in pictures due to the trace. In the pictures it was possible to obtain almost all the same details as in the image obtained by the artificial vision system (Figure 10(c)).

It can be concluded that the appropriate tool is the pen. However, due to a lack of an appropriate adaptor, this tool is only used in the ABB IRB 140 robot.

7 Conclusions and Future Developments

This paper presented the development of an open-source solution that allows performing the image processing of a person's face to be portrayed by an ABB robot arm, being all the stated objectives achieved. Accordingly, were applied together a set of image processing techniques to obtain a robust system for detection of features and contours of human faces. In order to allow a better parameter adjustment associated with each technique, a GUI was developed in QT, able to communicate with the ROS framework. To optimize the drawing process, an algorithm to join the near lines and remove outliers was implemented. One of the problems encountered relates to the connection process with the robot arm. In this sense, an FTP connection was incorporated, that allows sending the sequence of points in the RAPID language, and initialize the drawing process.

This system is very versatile and easy to use, and can work on multiple workspaces and be easily used by a user with general knowledge of the system. The process that occurs from the collection of the image until the completion of the picture was automated, taking into account the safety standards required for the intended purpose. As a result of this project, was developed a robust solution that can serve as a demonstration system in outreach events of ISEP. The system was on display at the FNR'2014 and also at an ISEP dissemination action in Maia, with evidence of a huge interest in people.

Concerning ideas for future improvements, among them are the detection of shadows in order to allow more realistic portraits, the detection and drawing of different line widths and the use of colors in the drawings. It is also being considered to use charcoal or graphite pencils on the robot to do the portraits. These types of pencils would make the portrait look more artistic and more appealing to the public. Finally, it is also being considered to change the order in which lines and dots are drawn so that these are designed in an order according to its length: first the line of greater length, then the line with the second largest length and so forth. This would make the public first visualize the main features of the face and only latter the details.

Acknowledgments This work is financed by the ERDF-European Regional Development Fund through the COMPETE Programme (operational programme for competitiveness) and by National Funds through the FCT-Fundação para a Ciência e a Tecnologia (Portuguese Foundation for Science and Technology) within project "FCOMP-01-0124-FEDER-037281". The authors acknowledge the support of ISEP, in order to be able to show this robot at the FNR'2014, and from the Mechanical Department of ISEP, who borrowed an ABB IRB 120 robot to be present at the several demonstrations performed.

References

1. ABB: Introduction to rapid programming - operating manual. http://developercenter.robotstudio.com/BlobProxy/manuals/IntroductionRAPIDProgOpManual/Custom/IntroRAPIDProgOpManual.html (acessed June 5, 2015)
2. Budhiraja, R.: Camera calibration tool. http://rahulbudhiraja.com/project/camera-calibration-tool/ (acessed May 30, 2015)
3. Code, O.: Simple webcam intruder alarm part 1. http://opencv-code.com/tutorials/simple-webcam-intruder-alarm-part-1/ (acessed May 30, 2015)
4. the Interface, T.: Google chrome web labs sketchbots: bits to atoms, zen-style. http://through-the-interface.typepad.com/through_the_interface/2013/01/google-chrome-web-labs-sketchbots-bits-to-atoms-zen-style.html (acessed May 30, 2015)
5. Lin, C.Y., Chuang, L.W., Mac, T.T.: Human portrait generation system for robot arm drawing. In: IEEE/ASME International Conference on Advanced Intelligent Mechatronics, AIM 2009, pp. 1757–1762, July 2009
6. Markiewicz, L.M.: Robotic Portrait Drawer. Msc. thesis, ISEP-IPP
7. Martinez, A., Fernández, E.: Learning ROS for Robotics Programming. Packt Publishing (2013). http://www.packtpub.com/learning-ros-for-robotics-programming/book
8. Project, T.C.: Meet paul and pete, the sketching robots. http://thecreatorsproject.vice.com/blog/meet-paul-and-pete-the-sketching-robots-2 (acessed June 10, 2015)
9. robotlab: Portraitzeichnungen aus der hand eines roboters. http://www.robotlab.de/index.htm (acessed September 20, 2015)
10. ROS. Org: cv_bridge. http://wiki.ros.org/cv_bridge?distro=fuerte (acessed June 10, 2015)
11. ROS.Org: portrait_bot. http://wiki.ros.org/portrait_bot (acessed June 10, 2015)
12. Study, M.C.V.: (opencv study) background subtractor MOG, MOG2, GMG example source code (backgroundsubtractorMOG, backgroundsubtractorMOG2, background-subtractorGMG). http://feelmare.blogspot.pt/2014/04/opencv-study-background-subtractor-mog.html (acessed May 30, 2015)
13. Tresset, P., Leymarie, F.F.: Portrait drawing by paul the robot. Computers & Graphics **37**(5), 348–363 (2013). http://www.sciencedirect.com/science/article/pii/S0097849313000149

Human Interaction-Oriented Robotic Form Generation

Reimagining Architectural Robotics Through the Lens of Human Experience

Andrew Wit, Daniel Eisinger and Steven Putt

Abstract Within the discipline of architecture, the exploration and integration of robotics has recently become an area of rapid development and investment. But with the current majority of architectural robotics research focused primarily around the realms of digital fabrication and biologic form/material optimization, there are few examples of direct translation from human generated data to form and processes, particularly as it pertains to the human experience of, and the interaction with architectural artifacts. Through a series of three case studies each building upon the previous, this paper investigates how the interconnection of secondary, smaller data harvesting/translating robotic systems in collaboration with larger industrial systems can be integrated within the conceptual design workflow to allow for the creation of unique/interactive tools for the materialization of human interaction through design, robotic control, and fabrication.

Keywords Robotic fabrication · Big data · Computation · Robotic manipulation

1 Introduction

Within the discipline of architecture, the exploration and integration of robotics has recently become an area of rapid investment and development [3], [5], [7].

A. Wit(✉)
Tyler School of Art, Temple University, Philadelphia, USA
e-mail: andrew.wit@temple.edu

D. Eisinger · S. Putt
College of Architecture and Planning, Ball State University, Muncie, USA
e-mail: {dmeisinger,stputt}@bsu.edu

L.P. Reis et al. (eds.), *Robot 2015: Second Iberian Robotics Conference*,
Advances in Intelligent Systems and Computing 418,
DOI: 10.1007/978-3-319-27149-1_28

This exploration in robotics is often limited by the narrow framework of exploration focused around the areas of digital and robotic fabrication. Although robotic fabrication allows for the creation of unique and innovative solutions for more complex construction problems, the appearance of the final resulting artifacts can lack a human touch and can often be easily traced back to a specific tool methodology, material or structural exploration.

Rather than initiating the design process with a desired end resultant (artifact) or set of known robotic processes in mind, this paper asks, "Can conceptual form, tools and processes be generated through the harvesting, translation and utilization of unmediated, human interaction-based data sources?" Through a series of case studies, this paper examines:

1. Human-robot interactive data generation and collection.
2. One-to-one, real-time data translation into 2D artifacts.
3. The conversion of collected data into 3D spatial potentials.
4. The harnessing and translating of collected data in the creation of unique, human interaction-derived industrial robot movements and end-effectors that create a large-scale wound-composite interior installation to be displayed at Ball State University.

Through these case studies, this paper will examine the potential of utilizing interactive robotic data collection and fabrication systems as a design tool that could be utilized to create novel, experience driven design solutions for the initiation of the design process.

2 Data

With the volume of data available at our fingertips growing on a daily basis, understanding innovative and meaningful methods for its utilization within architecture has become an important yet difficult undertaking. Data can enable architects and researchers to comprehend and predict local and global trends based on continuously updated information, but deciphering and filtering which data is relevant and meaningful can become a complex problem. In addition, data selection does not necessarily lend itself immediately to a physical manifestation. Rather, data may suggest the appropriateness of a certain type of solution for a given region, which can then be implemented through other means.

As opposed to investigating the use of such a process to solve specific global problems, this paper explores novel processes for the generation and collection of localized data through human-robot interaction and the direct translation of that data into the production of 2D and 3D tangible artifacts. In order to minimize the number of initial variables, case studies utilized local data generated by uninformed human participants interfacing a robotic drawing machine, rather than harnessing "Big Data".

2.1 Data Types & Collection Methodologies

With the vast amounts of existing data available and the increasing ease of collecting new data digitally, it can be difficult to choose a specific aspect to inform the design process. Since a direct translation to physical artifacts was a desired outcome for the case studies featured in this project, it was important that the chosen data type could be easily translated into a format understood by various types of robotic systems that function in two or three dimensions. The exact data type and mapping (translation) process employed in each case is discussed with the respective case below.

3 Case Studies

Throughout the course of this research, three distinct case studies were developed as a means of testing the of feasibility of translating global data into a design language. Employing local data arising from simple human-robot interaction, each test built upon the previous and developed the concept of translation as a design tool. Each successive test added complexity and scale resulting in the translation of collected data into:

1. 2D physical artifacts in the form of robotic drawings.
2. Industrial-scale robotic movements with the potential to define visual spatial constructs.
3. 3D physical artifacts in the form of a large-scale, composite-based, industrial robot-fabricated interior installation.

Following, this paper discusses the processes and artifacts created throughout each of these case studies.

4 The Lean Mean Data Harvesting Drawing Machine

The drawing machine was the starting point of this research and became the platform upon which subsequent case studies were developed. The initial focus of development was on the creation of a framework for the direct collection and translation of data arising from an unmediated human-robot interface. While human interaction data could be collected in any number of ways, a simple option was to collect data with the ability to be scaled and mapped onto a spatial coordinate system. Since color data is often represented with three values (e.g. red, green, blue), it lends itself to XYZ coordinate mapping.

A method was developed to collect color data and map it to a Cartesian coordinate system based on an interchangeable translation algorithm. Using webcams and video processing tools including Firefly (for Grasshopper, a visual programming environment for Rhino) and OpenCV (a standard computer vision library), an average color value was calculated for a region of interest at the center of each analyzed frame. To keep the amount of data collected manageable, only one or two frames were processed per second. Subsequently either Grasshopper or Python was used to perform the mapping for each project.

The Lean Mean Data Harvesting Drawing Machine was created to implement this methodology and consisted of a three-part system:

1. A webcam that collected color data from the surrounding environment and from human-robot interaction.
2. A single-board computer (Raspberry Pi) that translated the collected color data into 2D coordinates for the integrated drawing machine that also stored the color data for use in subsequent tests.
3. A small-scale LEGO and custom component based robot that created 2D artifacts while attracting participants to interact with the machine.

Placed in unassuming public spaces, no explanation of the machine or its translation algorithms was given. Individuals approached and interacted of their own accord while the machine recorded the experiences through drawn lines in 2D space and as recorded data points within an external hard drive.

4.1 Motion Translation Algorithm

During initial data collection sessions the translation algorithm employed involved the following steps:

1. RGB color values were converted to HSB (hue, saturation, brightness) values.
2. The hue spectrum was mapped to a circle such that hue values translated to degrees of rotation, or the orientation of a vector.
3. Saturation and brightness values were summed and mapped to distances from the center of the drawing surface to the edges of the drawing area, providing a magnitude for the vector.
4. The vector, plotted from the center of the drawing surface, pointed to a specific XY coordinate for each color.

4.2 The Drawing Machine

The drawing machine was designed as a simplistic robotic tool that would allow for intuitive human-robot interaction (Figure 1). Constructed of laser cut acrylic, foam core and Plexiglas sheets, the design was concerned with simplicity over precision, providing the opportunity for uncontrolled variables to appear. LEGO Mindstorms robotics components were employed to generate and control movement. LEGO NXT motors were mounted on the upper corners of the acrylic frame and LEGO wheel hubs were used as spools. The monofilament wound around the spools was connected to a drawing "puck", which was designed to slide across the drawing surface with minimum contact aside from the pen attached through its center.

To convert the XY coordinates provided by the Raspberry Pi to pen movement, a LEGO NXT microprocessor running the LeJOS NXJ open source operating system and a Java-based control program performed another translation, calculating (based on the geometry of the drawing surface) the degrees of rotation that each motor (and associated spool) needed to rotate to result in the correct amount of unspooled filament. The calculations were simple, avoiding the calculus that might be employed in a more precise system.

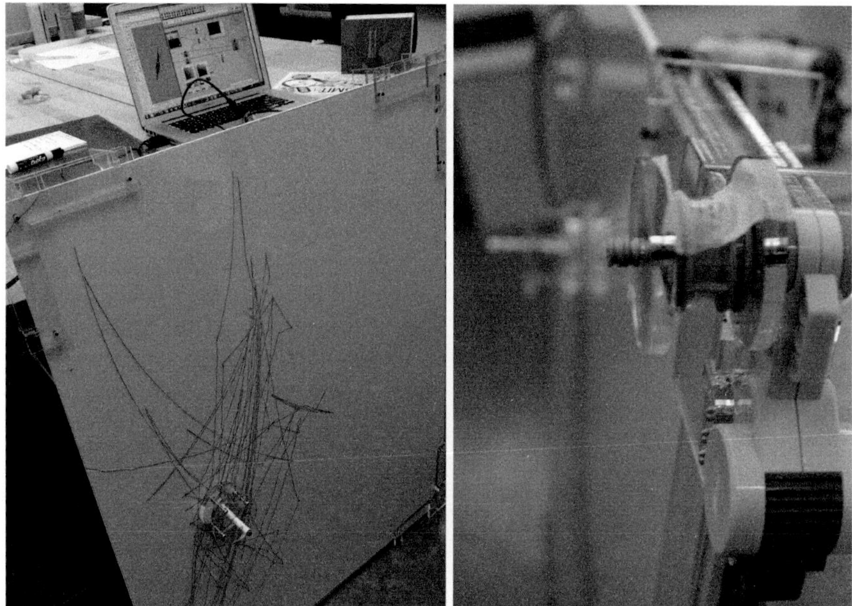

Fig. 1 The drawing machine & Detail of motor and Spool

4.3 User Interaction

Because the drawing machine was constantly receiving and processing input (whether or not any participants were present) and because the webcam picked up subtle variations in light from frame to frame as lighting conditions changed, the machine was almost constantly in motion. Even though most of its movements were small when no one was interacting with the machine, it generated enough noise and movement to catch the attention of many passersby. The machine's response to the presence of people was often dramatic due to the significant changes in color that the camera observed during such periods of interaction. This created additional interest, and at times a large number of people gathered around the machine, some gesticulating wildly to see what kinds of movement they could stimulate. The absence of explanation combined with a clear connection between interaction and movement stimulated curiosity and created a memorable experience for the individuals interacting with the machine.

4.4 Visual Output

The drawings produced by the drawing machine reflect qualities of the environment in which it was placed and of the user interaction that transpired. The example below (Figure 2)was the first drawing generated over an extended period of time, from late afternoon until after dusk. The changing quality of light in the space was tracked by the machine and is evident in the drawing—the location that the pen idled during periods without interaction (producing the densest markings) started out towards the left and migrated to the right over the course of the time the machine was active. The dramatic swoops above and below the center are the direct result of occasions where human interaction was taking place.

Upon completion of a series of tests in several locations, it was deemed that the initial system created for the translation of the data created by human/robot interaction directly into a physical artifact was successful and should be developed further into a 3D system.

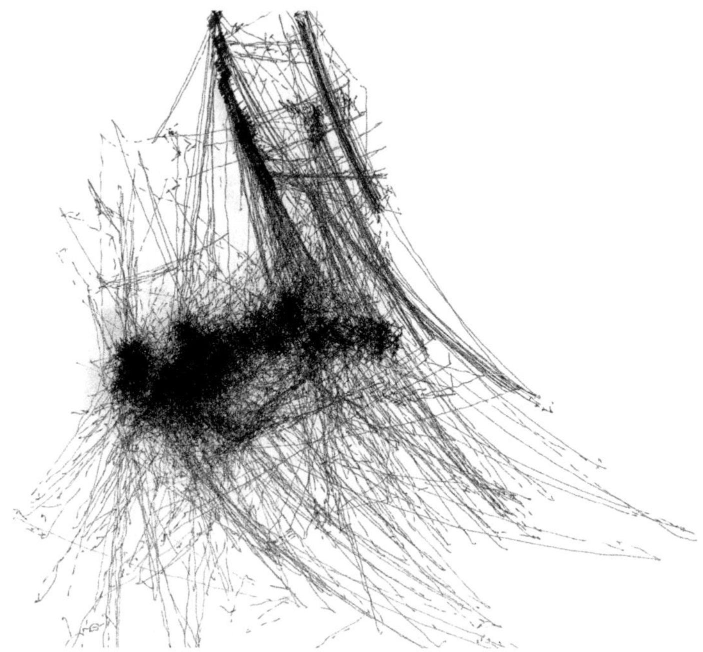

Fig. 2 Output from drawing machine depicting human interaction

5 Connect-The-Dots: Large-Scale Robotic Spatial Generation

Building on the successful completion of the first case study, a series of 3D tests were then initiated. Rather than initially focusing on the creation of physical artifacts as in the first case study, Connect-the-Dots focused on the linkages and processes necessary to directly convert the previously collected human interaction data into controlled, industrial robot motion. In addition to the data collection network established for the first case study, Connect-the-Dots consisted of a four-part system:

1. A Rhino 3D and Grasshopper-based translation system that mapped the data onto a 3D coordinate space.
2. Grasshopper plug-ins Robots.IO and KUKA|prc simulated robot motion and generated control code.
3. A single-board computer (Raspberry Pi) monitored KUKA control program indicators at run time and notified the end-effector of any status changes.
4. A custom, Arduino-based robot end-effector altered the state and color of an LED based on received notifications synced with the robot's movements.

The visualization of the previously collected human interaction data was then translated into 3D motions and visualized through long-exposure photography.

5.1 Translation and Robot Simulation

The algorithm employed to translate color data to 3D coordinates was incredibly simple: red, green and blue (RGB) values were mapped directly to X, Y and Z coordinates and scaled according to the working envelope of the university's KUKA industrial robot arm. The result was a series of 3D points, which were then visualized by connecting the dots in sequence with an interpolated curve.

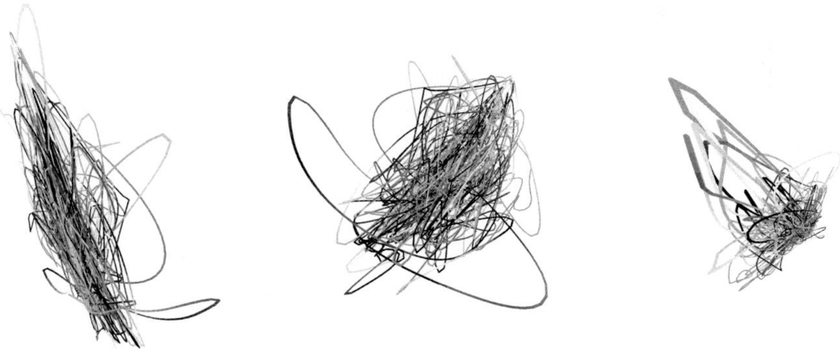

Fig. 3 Simulations of 3Dspatial onstructs created through human data inputs

The Robots.IO Grasshopper plug-in provided the means to simulate and test the viability of the robot motion before any physical test. This involved modeling the robot's working environment, specifying the position of the camera that would be capturing the output, designating the data-generated points as waypoints for the robot and its end-effector, using vectors from the waypoints to the planned camera location to dictate the orientation of the end-effector at each point, and parametrically determining the location of the points in space (as a group) to result in a tool path without singularities or unreachable waypoints.

The simulation process made it clear that new or modified tools are necessary when it comes to the generation of industrial robot motion instructions in the context of a data driven design process, particularly where high precision and repeatability are not required.

5.2 End-Effector and Robotic Output

To render the output of data-generated robotic motion, a simple solution was developed which allowed for the generation of spatial potential through light. A secondary system consisting of a single-board computer (Raspberry Pi) listening to the KUKA controller outputs and an Arduino-based end-effector ran simultaneously during the robot's movement. The Raspberry Pi listened to output signals from the generated robot control code and relayed them via XBee radio to the end-effector that controlled the state (on/off) and color of an LED light mounted to the end of the robot. Through the use of long exposure photography, these controlled lighting motions became traces in 3D space and suggested a potential spatial reality.

Fig. 4 Long exposure spatial output test

The Connect-the-Dots case study allowed for the project's first successful tests in human interaction data-generated, large-scale robotic movements that generated a physical spatial construct (Figure 4). By creating variations in the translation algorithm, any number of variations in form, size and color could now be generated based on minimal amounts of input data and could be used to suggest spaces virtual and/or potential. In addition, like the drawing machine this approach constructs feedback loop. While latency is a factor in its current manifestation (due to the manual steps required between interaction and rendering), the technology exists to explore real-time generation of spatial potentials in response to human interaction [9].

6 Module Maker: Composite Winding End-Effector and Aggregate Module Generation

Whereas the previously discussed data collection and translation systems have demonstrated the potential for the translation of human interaction data sources into both 2D drawn artifacts as well as visually recorded robotic movements, the Module Maker was created to show that these processes could be utilized for the creation orfaddition to the design of large-scale artifacts or spaces, realized through robotic fabrication. A commissioned installation served as the vehicle for developing these processes for 3D artifacts.

As with the first test, this phase of research set up a framework for the production of human interaction data-determined artifacts within certain constraints. These included:

1. The installation would be created using a series of globally unique aggregating modules.
2. Donated 1/8" pre-impregnated carbon fiber tow would be the production material.
3. The module size would be limited by the size of an available on-site oven to be used for the curing process.
4. The end-effector material would be required to have a high melting point and/or combustion temperature.

These constraints led to a unique design solution and opportunities for further research.

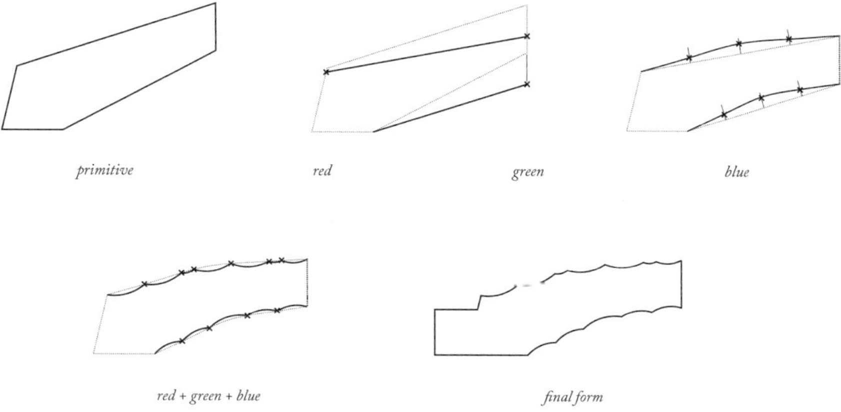

Fig. 5 Translation of color data into end-effector shape

6.1 Translation and End-Effector Generation

This project once again drew on the color data collected using the drawing machine. The translation system was developed using Grasshopper and again made use of RGB values, but instead of mapping to XYZ coordinates, these values were tied to specific parametric elements controlling the angles and inflections of the end-effector edges as illustrated in Figure 5 and superimposed in Figure 6.

Typically significant effort is required in the design and fabrication of unique end-effectors. As the timeline was short for this phase of the project and over one hundred unique modules were necessary, it was imperative to make the fabrication phase as quick and seamless as possible. For this reason and due to the time required for the composite to cure in an oven, cardboard was chosen for the end-effectors. Cardboard could be laser cut quickly using the generated end-effector shape paths, could withstand the 260° F curing temperature, and could be easily removed and recycled once curing was complete. In addition, to minimize the time required in swapping out end-effectors, a laser cut acrylic attachment plate was developed that simultaneously received three separate units and secured them with simple locking mechanisms.

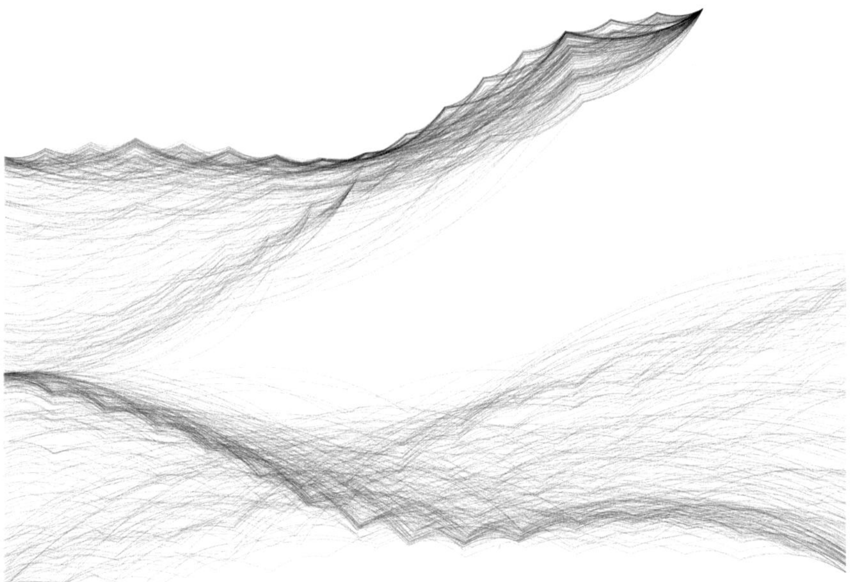

Fig. 6 Human experience-generated end-effector variations

The generated cardboard end-effectors required some assembly once laser cut. Using a numbering and labeling system implemented in Grasshopper and etched into the cardboard during laser cutting, two end-effector shapes were paired for each end-effector and connected through slots in the cardboard with laser cut bridging pieces. These could easily be collapsed after curing to allow the larger side shapes to be extracted with minimal effort. The pre-impregnated carbon fiber tow being used did not closely adhere to the cardboard during curing and as a result the final modules required little to no cleanup.

6.2 Module Creation Through Robotic Winding

Once a few of the end-effectors were cut and assembled the actual winding of the carbon fiber tow around the end-effectors could proceed while more end-effectors were being cut. The winding process incorporated the college's KUKA KR60-3 industrial robot arm with the attached custom end-effectors as well as a stationary tube through which the tow was fed. Due to time constraints and in order to provide a consistent aesthetic, the winding pattern for this exercise was identical for each end-effector, leaving the variation in the end-effectors to allow for the expression of the data. This variation led to varied points of contact between the tow and cardboard while the density of the winding and the degree of intersection was generally consistent.

The methods described for this case study and the previous ones could be applied to other aspects of this design and fabrication solution. It would be possible, for example, to use the data to control parameters related to the winding pattern (and associated robotic movements) for each end-effector, which would reduce the opportunity for error during winding and allow structural concerns to be addressed precisely [8].

Unlike the previous studies, this particular methodology does not lend itself to a real-time feedback loop; however, the result of this process does create the opportunity for a different sort of experience. Like the drawings produced by the drawing machine, there is a mystery related to the form of the modules. They are moments frozen in time and space expressing latent qualities of the initiating interactions. The installation constructed from these modules leverages the latency in this process to both preserve and create human experience.

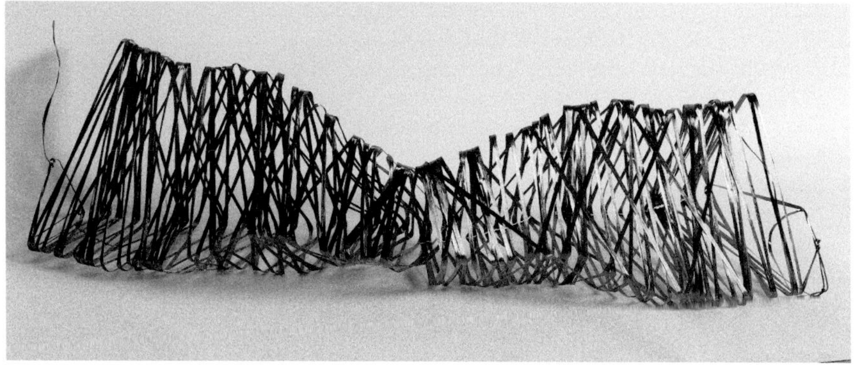

Fig. 7 Two data-driven, wound-composite modules

7 Conclusion

Robotics, including the forms employed in this paper's case studies and beyond, have not only a growing place in architecture, but also in the human experience as technologies develop and become more accessible. Architecture has the opportunity to leverage these new technologies to connect to its users in ways that highlight and value the human experience in the built environment. Although this research is still in its early phases, the case studies in this paper demonstrate the potential of using the data captured by human/robot interaction as a means to create form or as a means to add a level of uniqueness to existing designs. These novel methodologies for translating human interaction data directly into physical artifacts through robotics suggest a means for architects to rethink the initiation and process of design.

The potentials for design found within human experience and interaction are worth further exploration. Continuing this strand of research, forthcoming case studies will examine other forms of human interaction data as well as employing large-scale robotics in real-time interaction.

Acknowledgements We would like to thank the following for their support throughout this research's process:

Ball State University department of architecture for their continued grant funding. Robots.IO for the extended trial usage of their robot simulation and code generation software. The CAP Fab Lab for their use of space and continued support. And finally, TCR Composites for their kind donation of pre-impregnated composite tow.

References

1. Braumann, J., Brell-Cokcan, S.: Real-time robot simulation and control for architectural design. In: Proceedings of eCAADe Conference, Prague, Czech Republic, pp. 469–476 (2012)
2. Brown, B., Bughin, J., Byers, A.H., Chui, M., Dobbs, R., Manyika, J., Roxburgh, C.: Big Data: The Next Frontier for Innovation. Competition and Productivity. McKinsey & Company, N.Y. (2011)
3. Daas, M.: Towards a taxonomy of architectural robotics. In: Proceedings from the SIGRADI Conference, Montevideo, Uruguay, pp. 623–626 (2014)
4. Fox, M., Kemp, M.: Interactive Architecture. Princeton Architectural Press, N.Y. (2009)
5. Gramazio, F., Kohler, M.: Made by Robots: Challenging Architecture at the Large Scale. Wiley, London (2014)
6. Gross, J., Green, K.E.: Architectural Robots Inevitably. Interactions, vol. XIX, N.Y. (2012)
7. Knippers, J., La Mangna, R. Waimer, F.: Integrative numerical techniques for fibre reinforced polymers – forming process and analysis of differentiated anisotropy. In: Proceedings from the IASS Symposium, Wroclaw, Poland (2013)
8. Partridge, K.: Robotics. The Reference Shelf, vol. 81, N.Y. (2010)
9. Sanfilippo, F., Hatledal, L.I., Zhang, H., Fago, M., Pettersen, K.Y.: JOpenShowVar: an open-source cross-platform communication interface to kuka robots. In: Proceedings of the IEEE conference on Information and Automation (ICIA). Hailar, China (2013)

Robot-Aided Interactive Design for Wind Tunnel Experiments

Maider Llaguno Munitxa

Abstract The objective of this study is to investigate the effect of architectural geometry and materiality on airflow around buildings. For this purpose it is relevant to look for interactive design and analysis platforms that enable the analysis of architectural form and material variations while promoting the participation of designers in the analysis process. Today wind tunnel experiments are mostly deployed for design post-rationalization purposes, complicating the interaction between designers and the experimental environment, and constraining the number of design tests to be performed. The following research proposes to collapse the modeling and sensing processes within the wind tunnel with the aid of a robotic arm, to enable a real time design feedback informed by airflow analysis. Building geometry and surface studies have been conducted aided by robotic modeling and sensing, in a low speed and turbulence open circuit wind tunnel for a single building array and street canyon configuration. The recorded velocity profile variations reveal that mean flow statistics are sensitive to the texture variations.

Keywords Air quality · Architecture · Interactive design · Robot · Sculpting · Sensing · Wind tunnel

1 Introduction

Airflow plays an important role in urban comfort and pollutant dispersion. Regular arrays of cubic buildings have been widely investigated to understand the effect of the morphological parameters of the plan area densities λ_T and frontal area densities λ_F in the street canyon airflow. These parameters have been used to identify different flow regimes [1, 2] and although to a lesser extent, the building length to

M.L. Munitxa(✉)
ETH Zurich, Institute of Technology in Architecture, Chair of Structural Design,
HIL E 43.3, Wolfgang-Pauli-Str. 15, 8093 Zurich Hoenggerberg, Switzerland
e-mail: llaguno@arch.ethz.ch

L.P. Reis et al. (eds.), *Robot 2015: Second Iberian Robotics Conference*,
Advances in Intelligent Systems and Computing 418,
DOI: 10.1007/978-3-319-27149-1_29

height ratio influence has also been researched as well as the influence of the angle of attack of the wind flow to the main street axis. Modeling of street canyon pollution was the major topic of the European Research network TRAPOS [3].Within this network, street-canyon flow and dispersion characteristics for idealized street canyons were investigated with numerical models and atmospheric boundary layer wind tunnel.

Most of the numerical and wind tunnel studies developed to model street canyon flow and pollutant dispersion, have been devoted for the exploration of idealized building configurations, i.e. buildings of rectangular shapes and flat roofs. However prior literature has demonstrated that the influence of roof geometries on street airflow patterns and pollutant dispersion is not negligible [4, 5, 6, 7, 8, 9, 10, 11]. These findings show that airflow and turbulence characteristics inside the urban canopy are highly variable and highly dependent on architectural features and building arrangements. Therefore to seek for a more detailed understanding of the effect of architecture in the airflow and turbulence characteristics, and to develop more accurate building geometry urban canopy parameterizations, further building geometry and material studies are to be developed. For this purpose a collaboration between the design and scientific community would be most productive.

To date, the physics and computational expertise required to develop fluid dynamic simulations and wind tunnel experiments have constrained the study of the phenomena of airflow to the field of civil and aeronautical engineering remaining dissociated from the discipline of architecture and as a result from the processes of design and planning. Furthermore, the economical and time constraints are also to be taken into account. This is true specially when dealing with design problems at the building and urban scale, where at the initial stages of the design development sophisticated Computational Fluid Dynamic (CFD) simulation platforms or Wind Tunnel (WT) tests are often unable to follow the rapidity of the design changes and decision making. This is the case especially for WT experiments, which are usually deployed as part of design post rationalization processes lacking to inform formal and material decisions.

The ambition of this research is twofold. On the one hand, it seeks to explore the effect of small-scale architectural features such as building surface details and texture variations on local airflow. On the other, it aims to propose a design/analysis platform to enable a higher engagement of the design community on the study of the relationship between formal and material considerations and airflow around buildings.

Usually CFD simulation platforms and WT experiments follow a linear i) model ii) measure iii) analyze cycle. This workflow complicates the interaction between the designer and the experiment constraining the number of design tests to be performed and in most of the cases limiting the studies to the end of the design process. The following paper describes an alternative method. With the aid of a robotic arm, the modelling and sensing phases are collapsed to enable a real time design feedback informed by airflow analysis. This ambition is explored through WT experiments aided by robotic modeling and sensing in a low speed and turbulence open circuit WT (Fig. 1).

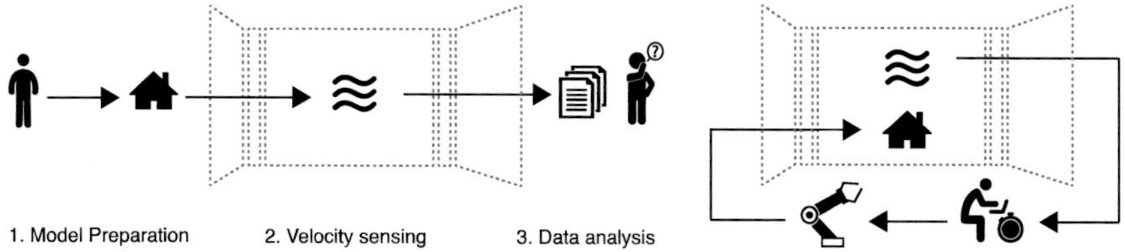

Fig. 1 WT work-flow diagram a) Existing work-flow b) Proposed work-flow.

The utilization of specialized robots for dynamic model positioning in WT environments have been proved effective to enable complex model movements and adjustments in their interaction with the approaching flow [12][13]. Industrial robots have also been used to automate probe positioning [14]. However a design-oriented approach where the robot contributes not only for the sensor positioning automation process, but also for the modelling of the building geometries is still to be explored.

With this ambition, the experiment proposes the utilization of an industrial 4-axis SCARA robot that operates in two modes: i) modeling mode and ii) sensing mode. While in modeling mode, through subtractive prototyping, the model is adjusted informed by the designers input who visualizes airflow data (mean velocity $\langle u \rangle$) collected by three static sensors located in the area under study. Once a desired geometry and airflow performance is achieved, the robot setup is changed to the sensing mode and a sensor is coupled to the robot arm to map the immediate environment of the physical model at a higher spatial resolution. This setup aims at an interactive data acquisition and data processing method for the designer to perform WT analysis within an intuitive design platform. Prior literature on design oriented WT experiments is presented in the next section. In section 3 the experimental setup is described and in section 4 a case study is presented. A summary and conclusions are included in section 5.

2 Design and Airflow Analysis

While concerns on air pollution and urban health have been one of the main focus of meteorologists and environmental scientists since the beginning of the 19th century [15, 16], the initial WT explorations on the influence of form and materiality on airflow, were initiated at the turn of the 19th century within the research on aerodynamics.

Gustave Eiffel's experimental work was pioneer to converge the research on architecture, airflow and human flight. Eiffel's experiments developed at the WT at Champ-de-Mars in 1909 and Auteuil in 1914 [17] aimed to understand the differences geometrical variations imposed on the recorded pressurized streamlines on solid objects. Eiffel not only developed novel technologies of data gathering and WT design, but through a design oriented work-flow, he engaged on systemic studies of airplane and building geometries. At a time when the sanitary conditions in Paris were critical, his work became influential for contemporary French architects and hygienists concerned with the problems of air quality and urban health.

Another interesting reference where WT experiments and design explorations converged, came from the hand of the Hungarian architects Viktor and Aladar Olgyay. The Olgyay brothers focused on the study of passive architectural notions to modulate interior and exterior microclimates. In their time in Princeton in the 1950s, the Olgyays devised the Thermoheliodon [18][19], a didactic machine that was able to model solar incidence, wind, humidity and soil conditions, while enabling iterative investigations on building principles and structures.

The experimental work developed by Gustave Eiffel and the Olgyay brothers, was successful in converging design ambitions and airflow and microclimate analysis. In both cases, their direct engagement with the analytic process enabled a design/analysis work-flow that attended to design considerations as well as to scientific enquiry. The following experimental setup aims to embrace this philosophy for the exploration of architectural geometry and surface definition in relation to airflow around buildings.

3 Experimental Setup

The experimental setup is composed by a i) low speed and turbulence open circuit WT, ii) a 4-axis SCARA robot, iii) building models, iv) a measuring system, v) a graphic user interface, vi) digitalization process for final data collection.

3.1 Wind Tunnel

The experiments have been conducted in a low speed and turbulence open circuit WT also described as the "Eiffel type" WT [20]. The flow passes through a conditioning area with a settling chamber constructed by i) a mesh to reduce the incoming turbulence intensity, ii) a flow straightening honeycomb built out with 10mm diameter and 150mm deep plastic pipes, iii) a screen, and a square 4:1 contraction area. The test section plane of the tunnel is 0.45 m (height) x 0.45 m (width) built out of plexiglass surfaces to allow full visual accessibility to the workable area. The test section walls are operable for full accessibility for the installation of the physical models. The top of the test section provides an accessible slot adapted to the movement range of the 4-axis SCARA robot, and at the bottom, 3 access points have been located in the rotting plate, for accessibility of 3 Pitot-static probes. The diffuser area is followed by a three-phase and 2.5 kilowatt motor which drives the axial flow fan with 7 blades with a blade angle of 33°, a diameter of 630 mm and with variable pitch-blades enabling a velocity range of 0.5 to 15 m s^{-1} (Fig. 2).

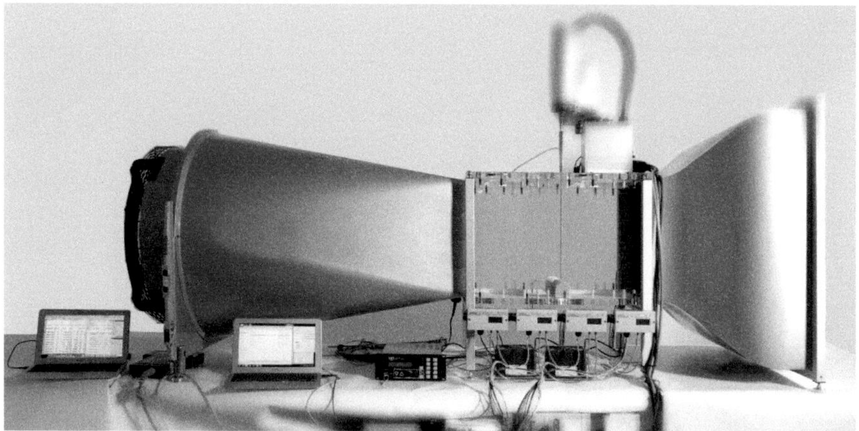

Fig. 2 WT setup. Diffuser, test section, contraction zone.

3.2 Robot Setup

A 4-axis SCARA robot with four degrees of freedom, a reach of 250 mm and a stroke (Z-Axis) of 150 mm has been mounted over the WT test section. The speed of the robot has been programmed with 1-100 %, 1% (0.04 m s^{-1}) increments. The robot performs two tasks: i) modeling building geometries and surface textures ii) positioning the sensing probe for larger spatial resolution mapping once the modeling period has been concluded.

For the modeling phase, to minimize flow obstruction while providing enough rigidity to reduce vibrations, a thin tungsten carbide bar has been clamped to the robot spindle, and spherical milling tools of ø 3 mm & 5 mm have been attached to its end. While the milling process is ongoing, the designer visualizes the mean velocity $\langle u \rangle$ collected by three Pitot-static tubes located in the area under study.

After the modeling period is concluded, the milling setup is replaced by the sensor. While in sensing mode, the data acquisition system and sensor positioning is performed with the aid of the 4-axis SCARA robot. While sensing probes are usually either stationary or manually positioned, this limits the amount of sample points that can be taken. This is one of the main reasons why methods such as the Laser-Doppler Anemometry (LDA) and Particle Imaging Velocimetry (PIV) which provide an overall image of the field flow of the WT, are increasingly more popular. However the economic and technical costs are also to be taken into account. The use of a robot for dynamic sensing, enables the data gathering at a higher spatial resolution (Fig. 3).

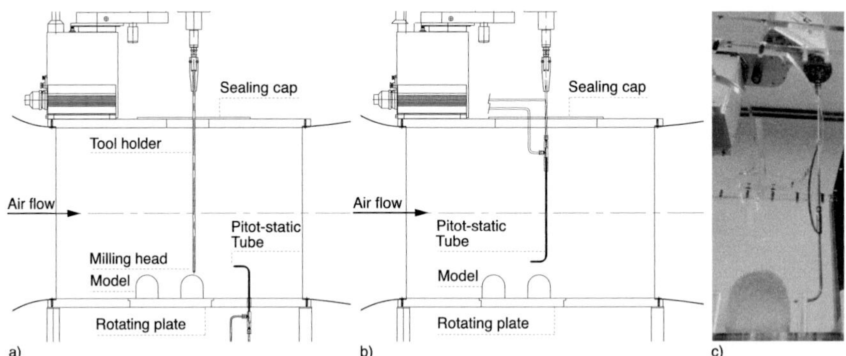

Fig. 3 Test section view. a) Milling process with Pitot-static tubes fixed to the bottom plate. b) & c) Pitot-static tubes attached to the robot arm for a high spatial resolution mapping of the area under study.

The robot movement range provides access to the full width of the test section. In order to seal the top of the test section access slot while the robot is in motion, a sliding enclosing cap is attached to the moving robot arm.

3.3 Building Models

The models have been built with pressed Silica sand to facilitate the milling process. Coarse sand, with grain sizes between 1 mm and 2 mm have been used. In order to guarantee an adequate stiffness, the sand has been mixed with resins and then pressed. The combination of large grain sizes and soft adhesion resins, enable an easier milling process minimizing the vibration problems caused by milling inside the test section. The residual sand particles are deposited in the bottom of the test section. Tests with and without the sand deposition have been performed and the $\langle u \rangle$ measurements remained consistent, therefore no further corrections were deemed necessary.

3.4 Measuring Technique

While the robot setup is arranged in the modeling mode, three Pitot-static probes are fixed to the three accessible slots in the bottom of the test section to record the mean velocity $\langle u \rangle$. A fourth Pitot-static probe is located in the free stream for velocity calibration purposes.

The Pitot-static sensors are connected to a differential capacitance manometer with a signal conditioner to collect the pressure differential ΔP. The manometer receives ± 13 V DC and delivers a pressure signal of 0-10 V DC which is directly proportional to the pressure, that is to say, the signal conditioning electronics will provide an output varying from 0 to 10 volts with a linear relationship between output voltage and pressure: 10 V DC represents 133.32 Pa. The resolution being $1\text{x}10^{-6}$ and an accuracy of 0.12% of reading ±zero/span coefficient.

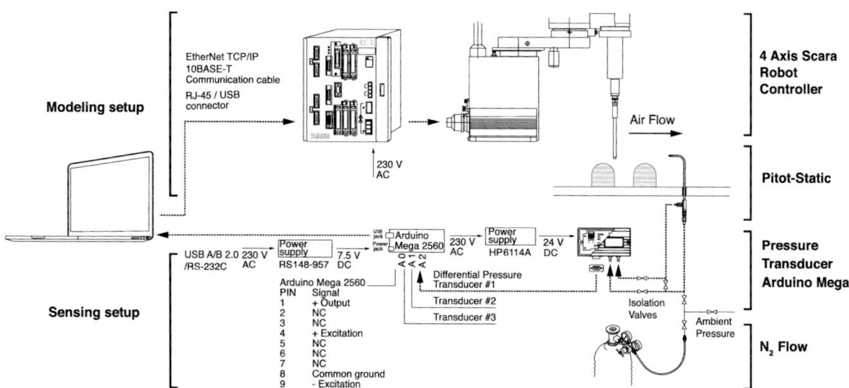

Fig. 4 Setup diagram. Milling and sensing technical setup during test studies.

An Arduino Mega 2560 microcontroller board based on 8-bit Atmel µP manages the analogue signal received from a voltage divider that is connected to the signal conditioner of the capacitance manometer. The voltage divider converts the signal from 0-10 V to 0-5 V to adapt the analogue output to the requirements of the Arduino Mega 2560. The microcontroller on the board is programmed using the Arduino programming language (based on Wiring) and the Arduino development environment (based on Processing). The Arduino board communicates through the serial line with the Matlab application in the computer. (Fig. 4).

Through Matlab code, the spike detection and removal method is performed for signal filtering following the methodology as described by Højstrup (1993) [21]; Vickers and Mahrt (1997) [22]. The method computes the mean and standard deviation for a series of moving windows of length L1. When a point in the window exceeds 3.5 times the standard deviation from the window mean, is considered a spike. The point is replaced using linear interpolation between data points. When four or more consecutive points are detected, they are not considered spikes and are not replaced.

The Pitot-static tubes provide a velocity capturing range from 0.5 m s^{-1} to 15 m s^{-1}. The sampling frequency has been set to 1 kHz or 200 Hz per channel (4x). One minute sampling time is performed after every milling sequence to guarantee that the flow is fully developed.

To avoid the sand particles from entering the Pitot-static tubes, a steady back flow of nitrogen N_2 is ejected from the Pitot-static tubes to protect the measuring circuit while the milling operation is being performed.

3.5 User Interface

Matlab scripts for statistical analysis have been written to analyze the time series of the captured data. The Matlab graphical interface is used to provide visual feedback on the statistics and enable an interactive feedback from the user end.

The Matlab plots collect the evolution of the time series in real time, providing both a qualitative and quantitative visualization of the results to facilitate the user with the necessary data to inform the next operation.

Custom milling scripts have been prepared for the user to test different texture combinations. The user can define the variables as well as the desired application area (Fig. 5).

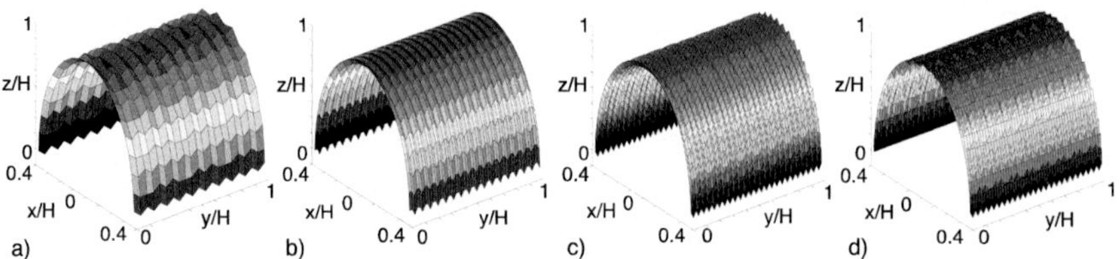

Fig. 5 Examples of milling sample algorithms a) Coarse-differential striation. b) Coarse-alternated striation. c) Fine-stepped-spiked striation. d) Fine-2D-spiked striation.

3.6 Digitalization

The robot milling trajectories are used to reproduce the final models digitally. These models are visualized together with the recorded airflow data. When a higher resolution representation of the physical model is desired to capture its material qualities, a 3D scanning of the model is performed.

4 Case Study

4.1 General Setup and Design Objective

The following case study aimed to develop an exploration of the influence of building surface textures and ornamental details on the neighboring airflow dynamics. The effect of building surface detailing on airflow separation in the interface of the Urban Canopy Layer (UCL) and the free atmosphere has been studied looking for the understanding of the sensitivity of texture variations on flow separation and reattachment, considerations that may potentially affect street ventilation. The angle of attack has been constrained to 90º, representing canyons that are perpendicular to the mean wind.

Studies on i) a single building array and ii) an urban street canyon configuration are presented. The size of the initial building models is 55 mm (width) x 450 mm (length) x 55 mm (height). The scale of the models has been constrained within these limits to minimize the blockage effect. The model proportions have been kept consistent with that of a homologous building geometry to ensure geometric similarity. The Bubble project database [23] has been used as reference for the geometrical definition. The length scale and reference velocity in the field have been considered L_{ref} = 15m and U_{ref} = 2 m s^{-1} respectively with a Re_{RH}= 2.46 10^6

while the values in the experimental setup are L_m = 0.055 m, U_m = 3.5 m s^{-1} and 7.5 m s^{-1} with Re_{ML} = 1.58 10^4 and 3.39 10^4 respectively.

4.2 Work-Flow

As described in section 3, initially the robot has been setup in modeling mode to develop the geometrical definition of the physical model. The geometries of the studied physical models are i) the flat roof geometry ii) the barrel vault geometry, both comprised within the described bounding box. Sample texture ornamentation algorithms have been developed through Matlab code and combined and applied in the desired areas of the physical models. Through combination and definition of the sample code variables, different pattern scales, depths and application areas have been defined.

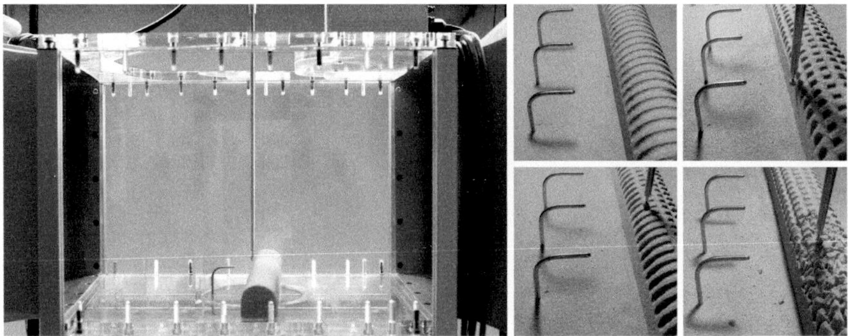

Fig. 6 Milling process. a) WT milling setup b) Instances of 3 milling sequences. Radial striation, longitudinal striation and random striation.

The geometrical sequences followed in one of the performed iterations are displayed in (Fig. 6, 7). Once the desired formal and surface definition is obtained, the milling tool attached to the robot arm is replaced by the sensor holder to perform a high spatial resolution mean velocity mapping. Finally, the final models are digitally represented together with the mean air velocity visualizations.

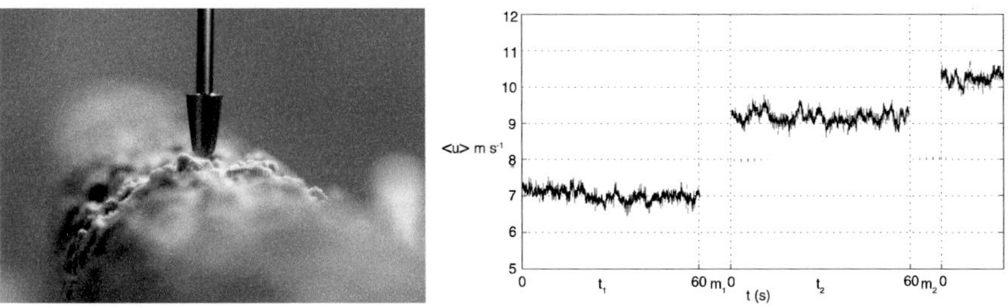

Fig. 7 Recording time series. a) Milling phase instant. b) $\langle u \rangle$ plot of sequences of time-series t_1 ,t_2 and milling phases m_1 ,m_2

4.3 Results

The results plotted in (Fig. 8) show comparisons of geometries obtained through a sequence of modeling operations and subsequent sensing phases. Once the milling processes have been concluded, the vertical rakes at 0.5W downstream have been recorded with the aid of the robot in sensing mode. (Fig. 8.a) displays the vertical profile inside a canyon and (Fig. 8.b) displays the results for a building configuration with a single building array. These results reveal that mean flow statistics are sensitive to the texture variations and therefore affect the flow attachment and separation features over building arrays. Judging from the variability of the $\langle u \rangle$ profiles, the effect of geometrical and texture variations on airflow are not negligible.

According to the performed studies, from the barrel vault iterations, the flow separation appears to be most sensitive to the roof corner radius. The configuration that shows a variable corner radius with a decreasing slope towards the leeward face has resulted in longest flow attachment. Longitudinal striation also appears to contribute for a delay of the flow attachment.

From the performed rectangular initial geometries, the ones subject to diagonally oriented striation patterns appear to contribute most to the delay of the flow separation.

The plots obtained from the single array building configuration option and the street canyon configuration display consistent tendencies.

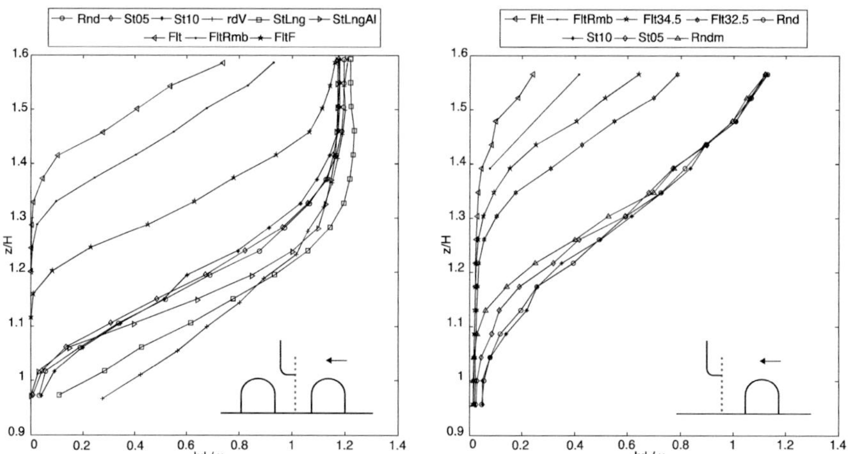

Fig. 8 Envelope texture variation $\langle u \rangle$ profiles captured with robot with sensing probe. a) Street canyon b) Single building array. Rnd, round roof. St05, r = 5 mm striation. St10, r = 10 mm striation. rdV, variable radius. StLng, longitudinal striation. StLngAl, in all envelope. Flt, flat roof. FltRmb, flat roof with romboidal striation. FltF, flat roof faceted. Flt34.5, flat roof with rounded corners r = 34,5 mm. Flt32.5, r = 32.5 mm. Rndm, random striation.

Fig. 9 Texture studies on vaulted building geometries.

5 Conclusions

A design-analysis work-flow has been proposed to develop interactive design explorations within a WT. The proposed work-flow, with the aid of a 4-axis SCARA robot, facilitates a designer friendly design/analysis environment by collapsing the modeling and sensing processes. While the proposed workflow enables a progressive building geometry definition informed by airflow studies, it requires a trial and error process guided by the designer. Performance oriented design often seeks for the most optimum result based on a I/O logic. Instead, the proposed work-flow facilitates the generation of various possible outcomes to be judged and adjusted by the designer. Reminiscent of Karl Culmann argument that favored descriptive and graphical geometrical representations as opposed to calculus [24] given their visual and designer-oriented approach, as well as the experimental work developed by Eiffel and the Olgyay brothers, this philosophy promotes a didactic approach through an iterative process for design and analysis fine-tuning.

The exploration developed in the case study presented in this paper has compared the effect of building envelope ornamentation on the flow separation over a single building array and a street canyon configuration. On flat roof and vaulted roof initial geometries, texture studies have been developed and mean velocity $\langle u \rangle$ measurements have been conducted to understand the sensitivity of neighboring airflow dynamics to the geometrical and surface variations. The results display that flow separation is sensitive to the surface condition, affecting the mean velocity $\langle u \rangle$ wind profile over the building roofs and therefore potentially influencing the ventilation potential at the street level.

Similar studies for the flat roof and barrel vault geometry as well as for the single building array and street canyon configuration have been performed through

CFD Large Eddy Simulations (LES) aiming to understand the influence of surface textures on the mean and turbulence statistics. Given the limit in extent of the present paper, this studies will be subject to other publication. Ventilation potential and Air Exchange Rate (ACH) calculations have also been performed for similar street canyon configurations which are part of another publication which is currently under development.

For the future work, apart from the mean velocity measurements, the recording of the turbulence statistics would be of interest. An understanding of the turbulence variability in relation to the texture variations would be useful to understand the variations in the ventilation potential amongst the studied geometries and textures. Data collections for $\langle u \rangle$ and u' but also the vertical component $\langle w \rangle$ and w' would be necessary for these computations.

Acknowledgements This paper would have not be possible without the support of the chair in Structural Design at the ETH in Zurich, the Princeton University Environmental Fluid Mechanics group, and the Basque Government- Dpt. of Education and Research Fund.

References

1. Oke, T.R.: The urban energy balance. Progress in Physical Geography **12**(4), 471–508 (1988)
2. Stewart, I.D., Oke, T.R.: Local climate zones for urban temperature studies. Bulletin of the American Meterological Society **93**(12), 1879–1900 (2012)
3. Berkowicz, R.: The European Research Network Trapos - Results and Achievements. Report, Loutraki, Greece (2001)
4. Rafailidis, S., Schatzmann, M.: Concentration measurements with different roof patterns in street canyons with aspect ratios B/H=1/2 and B/H1. Report, Metereology Institute, University of Hamburg (1995)
5. Rafailidis, S.: Influence of building areal density and roof shape on the wind characteristics above a town. Boundary-Layer Meteorology **85**(2), 255–271 (1997)
6. Kastner-Klein, P., Plate, E.J.: Wind-tunnel study of concentration fields in street canyons. Atmospheric Environment **33**, 3973–3979 (1999)
7. Kastner-Klein, P., Berkowicz, R., Britter, R.: The influence of street architecture on flow and dispersion in the street canyons. Meteorol. Atmos. Phys. **87**, 121–131 (2004)
8. Huang, Y., Hu, X., Zeng, N.: Impact of wedge-shaped roofs on airflow and pollutant dispersion inside urban street canyons. Building and Environment **44**(12), 2335–2347 (2009)
9. Theodoridis, G., Moussiopoulos, N.: Influence of building density and roof shape on the wind dispersion characteristics in an urban area: A numerical study. Environmental Monitoring and Assessment **65**(1), 451–458 (2000)
10. Xie, X., Huang, Z., Wang, J., Xie, Z.: The impact of solar radiation and street layout on pollutant dispersion in street canyon. Building and Environment **40**(2), 201–212 (2005)
11. Yassin, M.F.: Impact of height and shape of building roof on air quality in urban street canyons. Atmospheric Environment **45**(29), 5220–5229 (2011)

12. Bruckmann, T., Hiller, M., Schramm, D.: An active suspension system for simulation of ship maneuvers in wind tunnels. Mechanisms and Machine Science **5**, 537–544 (2013)
13. Bayati, I., Belloti, M., Ferrari, D., Fossati, F., Gilberti, H.: Design of a 6-DoF robotic platform for wind tunnel tests of floating wind turbines. Energy Procedia **53**, 313–323 (2014)
14. Kuka Industrial Robots: Robot guides probe in wind tunnel. http://www.kuka-robotics.com/usa/en/solutions/solutions_search/L_R198_Robot_guides_probe_in_wind_tunnel.htm (accessed January 15, 2014)
15. Thorsheim, P.: Inventing Pollution: Coal, Smoke, and Culture in Britain since 1800. Ohio University Press, Ohio (2006)
16. Howard, L.: The Climate of London. Joseph Rickerby Printer, London, 2nd edn., available via IAUC (1833). http://www.urban-climate.org/documents/LukeHoward_Climate-of-London-V1.pdf (accessed July 15, 2015)
17. Eiffel, G.: Nouvelles recherches sur la résistance de l'air et l'aviation, faites au Laboratoire d'Auteuil. Librarie Aeronautique, E. Chiron, Paris (1919)
18. Olgyay, V., Sorenson, A.: Thermoheliodon. Laboratory machine for testing thermal behavior of buildings through model structures. Report, Princeton University, Princeton (1956)
19. Barber, D.A.:Climate and Region, The Post-War American Architecture of Victor and Aladar Olgyay. A Journal of American Architecture and Urbanism, 68–75 (2014)
20. NASA.: Open Return Wind tunnel. https://www.grc.nasa.gov/www/K-12/airplane/tunoret.html (accessed September 15, 2015)
21. Højstrup, J.: A statistical data screening procedure. Meas. Sci. Technol. **4**, 153–157 (1993)
22. Vickers, D., Mahrt, L.: Quality control and flux sampling problems for tower and aircraft data. Journal of Atmospheric and Oceanic Technology **14**, 512–526 (1997)
23. Rotach, M.W., Vogt, R., Bernhofer, C., Batchvarova, E., Christen, A., Clappier, A., Voogt, J.A.: BUBBLE – an Urban Boundary Layer Meteorology Project. Theoretical and Applied Climatology **81**(3), 231–261 (2005)
24. Maurer, B., Ramm E.: Draw the language of the engineer, in karl culmann und die graphische statik, vol. 205. Ernst and Sohn, Berlin (1998)

Part VII
Simulation and Competitions in Robotics

A Coordinated Team of Agents to Solve Mazes

David Simões, Rui Brás, Nuno Lau and Artur Pereira

Abstract Mazes have been famously chosen as a great challenge for robots, either real or virtual, to solve, where agents have to explore the maze and fulfil goals. Mazes can be explored with greater speed by using a group of agents, as opposed to a single-agent system. There is, however, a greater degree of complexity in the implementation of a distributed team of agents that can coordinate to complete their tasks faster and more efficiently.

This paper explores the CiberMouse competition problem, where a team of virtual agents need to complete tasks within an unknown maze, with as much efficiency as possible. Their solution has shown great results in the challenge and has won the CiberMouse 2015 competition. The team can solve many complex mazes, in a smart and mostly collision-free manner. Our agents struggle with very tight paths, but compensate by having flexible high-level behaviours which allow them an efficient maze exploration.

Keywords Mobile intelligent robotics · Ciber mouse · Multi agent system · Maze problem

1 Introduction

CiberMouse is a robotics competition in a simulation environment running on a computer network. It consists in a team of five homogeneous agents that have at their disposal several sensors, up to four of which can be requested to the simulator each cycle, as well as a communication channel between them. The system simulates real-world conditions and introduces delays in messages and noise in sensor readings and in action commands. It also simulates range in sensors, speed and acceleration

D. Simões · R. Brás · N. Lau · A. Pereira(✉)
University of Aveiro, Aveiro, Portugal
e-mail: {david.simoes,nuno.carrulo,nunolau,artur}@ua.pt

L.P. Reis et al. (eds.), *Robot 2015: Second Iberian Robotics Conference*,
Advances in Intelligent Systems and Computing 418,
DOI: 10.1007/978-3-319-27149-1_30

physics. In the beginning, the maze layout is unknown. The first objective is to locate and reach a specific zone of the map, the *Beacon Area*, and gather the whole team in that area. The second and final objective is to head back to the starting area, the *Home Area*, and also gather the whole team in that area.

We propose a system where each agent explores a section of the maze, maps the maze as it explores, and tries to find the *Beacon Area*. After doing so, if its peers are not aware of the beacon's location, the agent searches for them until they all gather there. After completing the first objective, they head to the *Home Area*, where their second and final goal is completed.

The remainder of the paper is structured as follows. Section 2 presents related work. Section 3 describes the overall design ideas of our solution. Section 4 describes the architecture of our solution, namely the implemented behaviours and their workflow, while Section 5 describes the reading process of the sensors and the techniques employed to mitigate their noise. Section 6 describes the techniques used to define the path of the agents, the construction and map sharing process and the beacon triangulation methodology. Section 7 describes the communications protocol and Section 8 describes the exploration algorithm of the team. Finally, Section 9 shows performance results of our proposal and Section 10 draws conclusions and presents the future work.

2 Related Work

Maze solving robotic systems, as previously stated, are popular. There is a lot of research done in the area and lots of proposals submitted by the community [1, 2, 3, 4, 5, 6, 7, 8, 9, 10, 11, 12]. Most proposals can be classified in the categories of: mapping and obstacle detection, communication, path-finding algorithms, behaviour approach and exploration algorithms. We will now address these categories and list the most relevant options.

Most proposals use a map to store information about the maze. The map is, in most cases, a grid of same-sized cells. Notable exceptions are Brain [4] and Certo et al.'s proposal [13], which use a quadtree [14] approach. To map obstacles and clear areas, labels are used in each cell or node. Most proposals used *free* and *occupied* labels, as well as *visited* and *unexplored* labels. The most popular approach to setting the labels is a probabilistic map, where a cell has a given probability of being in a state, and after the probability reaches a defined threshold, that cell is considered to be in that state. Some solutions, like ITraders [5] or Speedy Gonzalez [6], however, prefer to use a last-read value approach, which may lead to inaccuracy problems. YetAnotherMouse [7] and Nai [9] use a weight-based system to set the probability values.

For communication contents, messages with the target positions or with map information are the most common, followed by agent positions and agent intentions.

While some proposals do not feature any path-finding algorithms, most solutions feature either the Dijkstra algorithm [15] or A* [16]. A notable exception is t-bots [3] and t-botsNG [12], which use a Tangent Bug [17] strategy. Proposals without path-finding algorithms simply move to the intended direction while trying to avoid and circumvent obstacles.

Most proposals feature high-level behaviours, that are based on the current situation of the agent, and dictate the robot's actions through low-level behaviours. However, approaches like ITraders or Speedy Gonzalez use a semi-reactive architecture, with only low-level behaviours. These tend to have problems in complex mazes. A notable exception is EvoRobert [18], which uses a machine learning system based in evolutionary algorithms to solve reinforcement learning problems.

The exploration algorithm is the most varied category, even though almost all proposals focus on spreading the agents through the maze. Brain uses a direction-based strategy, where each agent explores the zone it is facing, and solutions like ITraders and Eureka [8] assign the zones based on the agents positions. SWARM [2] and t-bots [3] divide the entire maze in a small set of large areas, and each area is assigned to one of the agents. Certo et al.'s proposal uses a corner exploration method, where each agent is in charge of exploring a corner of the maze, and the corners vary periodically to avoid stalling. CRACK [19] is a notable exception, where the team does not spread out, they move as a single unit and act like a single robot whose sensor range is longer.

There are also features in solutions that do not fall under any of the above mentioned categories. Some solutions, like Eureka and t-bots NG, have a behaviour to find agents that haven't found a target yet. Other solutions, like Jitters [1] and Brain, use noise-reduction techniques, like Kalman filters [20] for sensor readings and PID controllers [21] for wheel outputs, to increase the sensor accuracy and the movement performance. FAUbot, on the other hand, uses Gaussian distributions to reduce the sensor noise.

3 Proposed Solution

To fulfil CiberMouse's goals, the coordination of the five agents to explore exclusive zones of the map will lead to a more efficient exploration and, therefore, to a faster location of the *Beacon Area*. In the beginning of the simulation, agents will explore different parts of the map. When one of them finds the *Beacon Area*, it will alert the others, so that they all can gather there and complete the first goal. After that, the starting agent informs the remaining agents of the location of the *Home Area* and they all head there, completing their second and final goal.

At the start of the simulation, an exploration algorithm decides exclusive zones of the maze for each agent to explore. Communication and coordination protocols were defined to coordinate the robots during the simulation and, as agents explore the map, they register the sections they are exploring and broadcast that information to the remainder of their team. This way, all agents within communication range have a more complete vision of the maze. In order to fit all five robots inside any of the areas, a proper accommodation behaviour was also defined. Otherwise, in most cases, robots will be stuck outside the area, without room to actually enter it, due to the area being quite small.

4 Architecture of the Solution

Each robot follows the same behavioural pattern, where it performs the following actions: requesting sensors, reading sensors, deciding on the high-level behaviour, deciding on the low-level behaviour and, finally, communicating. Figure 1 shows an interaction diagram of our components. The agent starts by requesting the sensors it needs to read in the next cycle to the simulator (1). It then reads the current cycle's sensors (2) and maps out the maze accordingly (3). After that, the agent chooses a high-level behaviour (4) and uses a path-finding algorithm to get the best path to its destination (5), which was chosen by the high-level behaviour. The path-finding algorithm, logically, uses the map to calculate its results (6). After knowing where to head, the agent engages the low-level behaviours (7) to provide the correct input to the wheels (8) by reading the sensors (9). Finally, the agent decides which protocol to use and communicates information to its peers (10), possibly broadcasting its explored map (11).

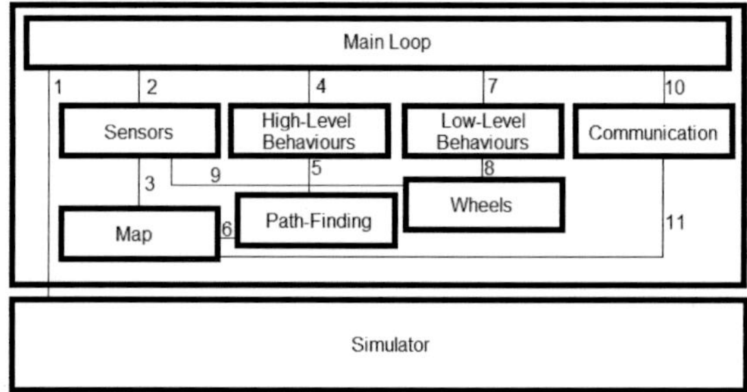

Fig. 1 Our agent's architecture

4.1 High Level Behaviours

There are four predefined high level behaviours: *Exploring*, *Chasing Beacon*, *Finding Friends* and *Going Home*.

The agent's initial behaviour is *Exploring*, where it chooses an unexplored point of the map (through the use of the algorithm described in Section 8) and tries to reach that point. Along the way, the agent maps the maze and, if it senses the beacon, marks its estimated position in order to follow it. At that moment, it changes its behaviour to *Chasing Beacon*.

The *Chasing Beacon* behaviour is further divided in three phases: (1) the agent is chasing the real beacon or a beacon estimation; (2) the agent is exploring the boundaries of the *Beacon Area*, to have a more accurate estimation of its centre; (3) the agent is waiting for its team to arrive, while moving in a circular motion in order to make room for the remaining agents. When a robot reaches the *Beacon Area*, it verifies if all robots are in contact range and, in case one or more are not, it activates the *Finding Friends* behaviour.

In the *Finding Friends* behaviour, the robot coordinates with any agents within range and they search for the robots that still do not know the location of the beacon. If possible, each agent searches for a different missing robot. The robots try to look for different robots, searching for them in the direction that they were exploring in the initial phase. In case they find missing robots along the way (by coming within communication range of each other), they inform them about the location of the beacon and return to the *Chasing Beacon* behaviour. This also occurs when they come close to the zone of the map that they decided to explore, even if they haven't found their peer, to avoid stalling.

When all the robots reached and entered the beacon area, the simulator notifies them and the *Going Home* behaviour is activated by all members. In case the path-finding algorithm is not able to find the correct path to reach that point (it may happen due to noise or erroneous readings), the robots retrace their steps since they started the maze. They do not retrace their whole path; their goal is the point closest to the *Home Area* and they take shortcuts when there are intersections in the path they are retracing. This behaviour, like the *Chasing Beacon* behaviour, is divided in phases and includes an accommodation phase. When all the robots are gathered at the *Home Area*, they end in unison.

4.2 Low-Level Behaviours

The low level behaviours are meant to allow the robot to navigate through the maze. They are semi-reactive and are based on the subsumption architecture [22]. This architecture divides an agent's behaviour into sub-behaviours, organised in a hierarchy of layers. Each layer implements a particular action, and higher levels override lower levels. In our case, the behaviours are chosen according to the readings from the obstacle sensors and to the target direction to be followed. The target direction is chosen according to high-level behaviours and to the path-finding algorithm, and that is the main interaction between high and low-level behaviours.

There are four low-level behaviours. *Follow the Wall* is the highest priority behaviour and allows an agent to "hug" a wall and follow it closely. *Follow the Beacon* is the second highest-priority behaviour and allows a robot that does not detect nearby obstacles to head in the beacon's direction, turning while moving. *Dodge Obstacle* is the second least-priority behaviour and triggers when close obstacles are detected. The robot rotates around itself to avoid them. Finally, *Wander* is the least-priority behaviour, activated by default when no other behaviour fits. The agent simply moves forward.

5 Sensors

Each virtual robot has a set of sensors it can use to gather information about external entities. The GPS provides absolute coordinates, with a random offset for each simulation. The compass provides the rotation of the agent, in relation to the virtual north of the maze. The bumper indicates whether the robot has hit something. The beacon sensor indicates the degrees of rotation, in relation to the robot's direction, of the centre of the beacon area. The obstacle sensors indicate the distance between them and the closest obstacle found in their angle of vision, and their accuracy decreases with distance. The base unit for distance measuring is the agent's diameter. All the sensors, excluding the bumper, have an additive and Gaussian noise with zero mean. The GPS has a 0.1 resolution, a 0.5 standard deviation and no delay, while the beacon sensor has a 0.1 resolution and standard deviation and a 4-turn delay. The compass sensor has a resolution of 2°, a 1° standard deviation and no delay, while the beacon sensor has a 1° resolution, a 2° standard deviation and a 4-turn delay.

The robot may only request up to four sensors per cycle. We have considered the six available sensors by the following order of priority: the *Front Sensor* and *Compass* have maximum priority and, thus, are requested in every cycle. These are considered the highest priority sensors because the robot's direction is essential to draw the map and the front sensor to detect the presence of obstacles. The *Left* and *Right Sensors* have medium priority and are requested three out of four times. Although they are also important, the impact of loosing one out of four cycles is not that meaningful and thus allows obtaining the least-priority sensors with sufficient frequency. Finally, the *Ground* and *Beacon Sensors* have minimum priority and are requested once every four cycles. They are considered the least important sensors because they do not influence the mapping process.

As described above, the readings from the sensors have noise. In order to reduce that noise, buffers are used in all the obstacle sensors, allowing the calculation of the mean of the readings, significantly reducing their noise. The mean of three readings is calculated, being enough to reduce the noise without noticeably delaying the response time of the robot. The buffers also serve the purpose of correcting delay in sensors. Because the compass has a four cycle delay, the buffers allows the storage of past values in order to obtain the sensor values corresponding with the current compass value.

The GPS and compass also have an associated noise. Because of this, a Kalman filter [23] was implemented for each one. The use of this filter is suitable because the noise from the sensors of the agents fits with Kalman assumption, that the process is linear and the noise Gaussian. The Kalman filter allows a great noise reduction and, thus, a high precision in the calculation of the location of each agent.

6 Mapping

To perform the *internal mapping* of the maze, a grid strategy was adopted, where the maze is converted into fixed-length squared cells. We chose the grid resolution as three cells being the equivalent to a robot diameter. This is the lowest resolution that allows mapping the smallest obstacles, this way allowing for a efficient path-finding algorithm and for a correct mapping of the maze. Because the GPS has a random offset for each simulation and the robot can start at any point in the maze, a direct mapping of the maze to the grid is not possible. Instead, we use private offsets for each agent that set their initial position in the maze as the centre of their grid. By having the grid being twice the size of the maximum dimensions of any given maze, we can "fit" any maze inside the grid, regardless of the GPS offsets and of the starting position of the agent.

Each cell in the map might indicate three values: *Unknown*, an unexplored zone of the map; *Clear* - a zone without obstacle; and *Blocked* - zone with obstacle. In order to reach these values, we used a weight-based system based in YAM's algorithm [7]. Cells detected and closer to the robot have a big weight, which decreases as they become distant. Each cell starts with value zero and when the robot detects that a cell is free or not, it adds or subtracts its weight. The values of the cells are limited and a cell is considered *Clear* or *Blocked* when its weight reaches a threshold. The threshold is a compromise between reaction time and certainty of the status of a cell. Higher thresholds take longer to recognise a cell but have a greater degree of certainty.

As for *external mapping*, the robot may map the maze without resorting solely to its sensors, but also using information provided by the other robots. The other robots send a type of message, *Map Update*, containing the information about the cells they have explored themselves. This information is registered in the robot's map and used in the path finding algorithms.

Although the type of cells received from the other robots are the three normal types, the robot stores the cells with different values in order to distinguish which cells it explored and which cells other robots explored, while ignoring received *Unknown* cells. It also uses its peers' positions to map out the maze: it keeps them as obstacles for a few cycles or until it receives an update on that particular peer's position. It will then mark those cells as *Clear*.

6.1 Path Finding

To reach points in the map defined by the high-level behaviours, the robot uses a path-finding algorithm, the A* [16], to find the best path to its destination. The A* is a graph traversal algorithm and it is an extension of Dijkstra's algorithm. The chosen heuristic was the euclidean distance between two cells. In order to keep the agent far from obstacles in the defined path, the algorithm checks the neighbouring cells of the path to make sure they are clear. The algorithm will always try to find a clear wide path, but will resort to tighter paths if a wide one cannot be found.

In order to reduce the computational time of A*, our algorithm does not compute the whole map. The explored limits of the map are known and, therefore, we just calculate the best path within these limits. To find possible paths through unexplored regions of the map, a margin is also added to these limits. To reach the *Beacon Area* and the *Home Area*, the robot is not searching for a single point, but an area. In those cases, we use a slightly modified version of A*, whose stopping criterion is not reaching the target cell, but reaching a given distance of the target cell. If, due to misreadings, cells in the agent's current position or in the target area are marked as occupied (which would make them impossible to reach), the algorithm randomly searches for free cells within a given radius. When found, it uses those as its starting or end points and behaves as usual. As the agent moves, the readings should stabilise and provide accurate measures for the algorithm to use.

After finding the path to the target, the algorithm will return the direction that the robot should follow, as shown in Figure 2. Low-level behaviours use this direction as their target, and act accordingly. In every cycle, the path, in blue, is recalculated and the direction may be different. The direction returned is the direction to the cell five units away from the robot, in purple, (or the last cell, in pink, if the path is shorter than five units). Choosing a cell a few units away guarantees a steady direction value is returned, even with GPS or compass noise. However, as shown in Figure 2, this cell might not represent the direction the robot should follow, due to obstacle corners. To avoid these situations, a check for obstacles between the cell and the robot is done and, if any are found, the direction of a closer cell is returned.

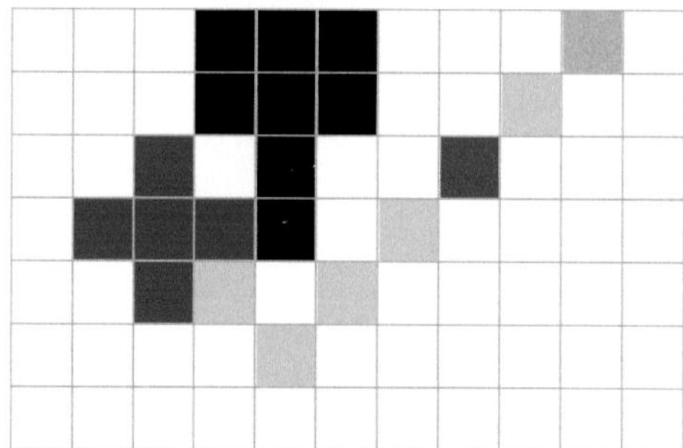

Fig. 2 Example of the corner problem of a path-finding algorithm

6.2 *Beacon Detection*

The beacon detection is based in a triangulation method, based in the direction of the robot and the beacon sensor, over time. Our algorithm traces a line every time the beacon is sighted. As new lines are added, their intersections are registered as *Beacon Points*, whose average creates a beacon position estimation. When a robot receives a *Beacon Found* message, it marks a *Beacon Point* on that location and heads towards it. When it is within the *Beacon Area*, its current position is also defined as a *Beacon Point*.

7 Communication Protocol

The communication protocol is composed by six types of messages, each message being identified by a tag in its first byte: *Hello*, *Map Update*, *Beacon Found*, *Order*, *Home Info* and *Ready to Quit*. The purpose of each message is described below.

Hello - Each agent, when it is initialised, sends a sequence of *Hello* messages where it includes its current position and the known positions of its teammates. Agents reply with *Hello* messages, forwarding information and allowing each agent to know about the location of its teammates, even if they are not within its communication radius. Our exploration protocol defines an agent to be the coordinator. While regular agents reply with *Hello* messages, if the coordinator receives the *Hello* message, it replies with an *Order* message.

Order - Its objective is to inform all the agents about the zone they have to explore, as well as the zones that their teammates will explore. This is useful for exploring faster the whole map and also to provide a "starting point" when it is necessary to search for a "lost" robot. After the reception of an *Order* message, each agent will verify if the information provided in that message is newer than the information it has. If it is (or if the agent had not received any *Order* message yet), the message is decoded and broadcast. Otherwise, it is ignored. Agents broadcasting this message allow any team members outside the coordinator's range to update their internal information.

Map Update - Its goal is to sequentially dispatch of the map and its correct incorporation into the remaining agents' local map. Values are converted from cell coordinates to GPS measures, which are common to all agents, and sent. Agents receiving the messages perform the reverse conversion, using their own private offsets, and store the information. This type of message is also sent by default; in other words, unless higher priority messages arise, *Map Update* messages are sent. The message also contains the GPS position of the agent that sent it. This way, agents maintain, at any time, the current position of peers within range. This is useful to detect a possible "false" obstacle, which is in fact another agent that is currently navigating in that zone.

Beacon Found - It is composed of a GPS position and of a list of agents. The GPS position is a random point in the *Beacon Area* or the average of the points belonging to it (depending on the high-level behaviour of the robot sending the message). The list of agents represents which agents are in range of the robot sending the message. This way, agents can infer which agents have received this message and, logically, the agents who know where the beacon is. The remaining agents will be searched for, with the purpose of informing them of the location of the beacon, according to the *Finding Friends* behaviour. This way, the team is able to gather at this area and thus conclude the first objective.

Home Info - This message is quite simple and similar to the *Beacon Found* message. Essentially, it is a GPS position of the *Home Area*, which is the starting coordinates of the coordinator. This type of message has the objective of informing all the other agents of the location of the *Home Area*. This message is only sent once, by the coordinator, when all the agents are within the *Beacon Area* and within the coordinator's range. Although the *Hello* messages carry this information, this message is necessary if the coordinator was not in range of all agents when the maze started.

Ready To Quit - This message is an empty message whose sole purpose is informing the other agents that the last objective was achieved and that the agent is waiting for the other agents to finish. If each agent finished as soon as the objective was concluded, then others would stop receiving information about its position and they would treat it as a regular obstacle, instead of registering a robot that can move and re-position itself. By using this message, agents can keep coordinating and moving until all are inside the *Home Area*, at which point they all terminate.

8 Exploration Algorithm

In order to synchronise the team of agents and coordinate each agent to search in different zones of the map, our protocol requires an agent to act as a coordinator. We chose as coordinator the agent initialised in position '1'. Initially, all agents have an array containing the team's default exploring directions. This way, isolated agents have a predefined direction to explore, known to the remainder of the team.

The coordinator broadcasts *Order* messages to update this array any time it receives *Hello* messages with updated team positions. The array is updated based on the location of the agents and the coordinator assigns the top-most left agent to the top-left corner, top-most right agent to the top-right corner, and so on. Agents will eventually reach the corner of the map, but they will not know that the maze ends there. Assuming that the corner is just an obstacle, they will try to circumvent it, and will end up exploring the entire maze. The team stops exploring when agents find the *Beacon Area* and warn the remaining members.

9 Results

Before using noise-reduction techniques, our sensor readings were quite inaccurate, with a deviation of, on average, 0.5 units in the GPS, in each axis, and 3° in the compass. However, those values oscillate a lot and may be even more inaccurate. Using Kalman filters, the variation was reduced to, on average, 0.2 units in the GPS and 2° in the compass, and the oscillation is almost nonexistent. The obstacle sensors also showed a strong deviation, increasing with the distance at which an obstacle is from the agent. With our mean filters, the deviation decreased by approximately 70%. By reducing the noise sensors, our robot's maze mapping and movement became much more reliable and, in most situations, accurate.

In order to analyse the effectiveness of our solution, the team was tested in a variety of mazes. A video with some of the test mazes is available online[1]. Through

[1] https://www.youtube.com/watch?v=f2GOFHAaApY

the analysis of our team's behaviour, we can conclude that the team can handle complex mazes and is able to explore them with great speed. The robots explore the map efficiently and also detect and find "missing" members quickly. Overall, the team was able to achieve both goals in most mazes, within a reasonable amount of time. We cannot compare our proposal with others in controlled environments, due to lack of access to other solution's binaries. However, on the CiberMouse 2015 competition, we could compare the performance of our solution with others, and our proposal showed great results, winning first place.

However, when the path-finding algorithm cannot find a path to the target (due to mapping errors and tight paths), the robot ceases to have decision-making power in the selection of the correct path and is limited to acting reactively. This may lead to looping problems and is noticeable when the robots are trying to accommodate in the *Home* and *Beacon* areas, or with tight paths in the maze. Mapping errors are mostly due to our assumption that the obstacle sensors are directional. However, they are wide-range sensors, which means that they detect obstacles up to 30° for each side of the direction they are facing. This leads to mapping inaccuracy and to the issues described above.

10 Conclusion

This paper presents the strategy and implementation of a distributed and coordinated multi-agent maze-solving team, specially adapted for the CiberMouse competition. The system makes use of a wide set of techniques that enabled noise reduction, precise mapping, fine-grained path finding and team synchronisation on the exploration of the map and the fulfilling of the competition's goals. Each technique has a specific objective and, together, they create an intelligent and synchronised team of agents that can solve a wide variety of unknown mazes. The source code is available online[2].

Our proposal can still be improved, both in the sensor problem described in the previous section, in the communication protocol, to remove the need for a coordinator, and in the team size flexibility. It has been designed with a team of five elements in mind, but supporting larger or smaller teams is possible by changing the communication protocol, where certain messages may be too big with larger teams and have to be split in two. In the end, however, our proposal has presented satisfactory results and is also a general model that can be used for most maze solving multi-agent systems with a flexible set of intelligent behaviours.

References

1. Cunha, J., Oliveira, L., Ribeiro, L., Sequeira, R.: Jitters at ciberrescue@RTSS 2009. In: RTSS - IEEE Real-Time Systems Symposium 2009, Washington DC, pp. 13–16, December 01–04, 2009

[2] https://github.com/bluemoon93/CiberMouse/tree/master/Robotics3/Cibertools

2. Sartori, J., Arefin, A.: Swarm: software automata for robotic motion. In: RTSS - IEEE Real-Time Systems Symposium 2008, Barcelona Spain, pp. 25–28 (2008)
3. Facchinetti, T.: t-bots: a coordinated team of mobile units for searching and occupying a target area at unknown location. In: RTSS - IEEE Real-Time Systems Symposium 2008, Barcelona Spain, pp. 29–32 (2008)
4. Rei, L.: Brain: an autonomous agent for cooperative navigation in a simulated environment. In: RTSS - IEEE Real-Time Systems Symposium 2008, Barcelona Spain, pp. 9–10 (2008)
5. Song, W., Zhu, H., Kulkarni, C., Kulkarni, N., Koritala, N., Belure, S., Cheng, A.M.K.: itraders: warriors of space exploration. In: RTSS - IEEE Real-Time Systems Symposium 2008, Barcelona Spain, pp. 11–12 (2008)
6. Monteiro, A., Aguiar, F., Carvalho, S.: Speedy gonzalez: a path planning and plan execution agent. In: RTSS - IEEE Real-Time Systems Symposium 2008, Barcelona Spain, pp. 21–24 (2008)
7. Ribeiro, P.: YAM (Yet Another Mouse) - Um Robot Virtual com Planeamento de Caminho a Longo Prazo. Revista do DETUA **3**(7), September 2002
8. Thekkilakattil, A., Pillai, A.S., Saravanan, V., Aysan, H.: Eureka: a team of autonomous mobile agents competing in cyberrescue 2009. In: RTSS - IEEE Real-Time Systems Symposium 2009, Washington DC, December 01–04, 2009
9. Pinto, J.: The autonomous agent Nai in RTSS 2007. In: RTSS - IEEE Real-Time Systems Symposium 2009, Tucson, Arizona, December 04–06, 2007
10. Danner, D., Kaufhold, C., Kranz, P., Muller, R., Pfaller, S., Rieß, C., Angelopoulou, E.: Faubot: purposeful navigation of a robot in a simulated environment. In: RTSS - IEEE Real-Time Systems Symposium 2009, Tucson, Arizona, December 04–06, 2007
11. Certo, J., Oliveira, J., Reis, L.: Intelligent robotic mapping and exploration with converging target localization. In: International Workshop on Intelligent Robotics (IRobot), Portuguese Conference on Artificial Intelligence (EPIA), Aveiro, Portugal (2009)
12. Vedova, M.L.D.: Introduction to t-botsng: a team of virtual robots at cyberrescue@RTSS 2009. In: RTSS - IEEE Real-Time Systems Symposium 2009, Washington DC, December 01–04, 2009
13. Certo, J., Oliveira, J., Reis, L.P.: Intelligent robotic mapping and exploration with converging target localization. In: International Workshop on Intelligent Robotics (IRobot), Portuguese Conference on Artificial Intelligence (EPIA), Aveiro, Portugal (2009). http://epia2009.web.ua.pt/onlineEdition/231.pdf
14. Samet, H.: The quadtree and related hierarchical data structures. Computing Surveys (CSUR) **16**(2), 187–260 (1984)
15. Dijkstra, E.W.: A note on two problems in connexion with graphs. Numerische Mathematik **1**(1), 269–271 (1959)
16. Hart, P., Nilsson, N., Raphael, B.: A formal basis for the heuristic determination of minimum cost paths. IEEE Transactions on Systems Science and Cybernetics **4**(2), 100–107 (1968)
17. Choset, H.M.: Principles of robot motion: theory, algorithms, and implementation. MIT press (2005). ch. Bug Algorithms, pp. 25–30
18. Alanjawi, A., Liberato, F.: Evorobert system description. In: RTSS - IEEE Real-Time Systems Symposium 2009, Tucson, Arizona, December 04–06, 2007
19. Santos, F., Corrente, G., Li, H.: Crack - ciber robots with advanced coordination kontrol. In: RTSS - IEEE Real-Time Systems Symposium 2008, Barcelona Spain, pp. 33–34 (2008)
20. Kalman, R.E.: A new approach to linear filtering and prediction problems. Journal of Fluids Engineering **82**(1), 35–45 (1960)
21. Rivera, D.E., Morari, M., Skogestad, S.: Internal model control: PID controller design. Industrial & Engineering Chemistry Process Design and Development **25**(1), 252–265 (1986)
22. Brooks, R., et al.: A robust layered control system for a mobile robot. IEEE Journal of Robotics and Automation **2**(1), 14–23 (1986)
23. Thrun, S., Burgard, W., Fox, D.: Probabilistic robotics. MIT press (2005)

Part VIII
Social Robotics: Intelligent and Adaptable AAL Systems

RFID-Based People Detection for Human-Robot Interaction

Duarte Lopes Gameiro and João Silva Sequeira

Abstract This paper discusses the use of off-the-shelf Radio Frequency Identification (RFID) detection as complementary technology to the localization of people (detection and localization relative to the robot) in social robotics scenarios. A novel model for the detection of passive RFID tags is proposed, involving the estimation in real time of a measure of the probability of the tag being detected. The method estimates the location of the tag relative to the reader with an accuracy suitable for a wide range of human-robot interactions and social robotics applications.

Keywords RFID · Social robotics · Human-Robot Interaction

1 Introduction

Social robotics growing importance requires high performance perception functionalities, namely in scenarios where human-robot interaction (HRI) is relevant. For instance, having robots behaving according to the (socially) correct proxemics requires them to perceive positioning information of the surroundings. Such information is very important for the interaction between humans and robots to be effective. This is also relevant for the robots to be accepted in human social scenarios.

Recognizing the localization of a person with respect to the robot (or vice-versa) is thus of uttermost importance if the robot is to behave according to the social norms enforced at the environment. For instance, if a robot is to greet a person that it recognizes, its movements must be adjusted in order to approach the person from a socially correct direction, e.g., from the front, using the quadrant that is most free, and carefully managing the velocity of approach.

D.L. Gameiro(✉) · J.S. Sequeira
Instituto Superior Técnico, University of Lisbon, Lisbon, Portugal
e-mail: {duarteflgameiro,joao.silva.sequeira}@tecnico.ulisboa.pt

L.P. Reis et al. (eds.), *Robot 2015: Second Iberian Robotics Conference*,
Advances in Intelligent Systems and Computing 418,
DOI: 10.1007/978-3-319-27149-1_31

This paper discusses a novel RFID based localization strategy in a mobile robotics context, where people are carrying RFID passive tags and a robot carries a reader and associated processing onboard. The application context of the proposed paradigm is that of social robotics, where the required accuracy is bounded by the size of the areas considered for proxemics purposes (see for instance [10] for a definition of such areas).

Often, the use of networked systems allow multiple sensors to be distributed throughout the environment, each extracting the adequate perception features, e.g., cameras for people detecting and tracking. State-of-the-art tracking systems such as the Vicon are accurate and fast enough to track multiple targets at 400 Hz (see www.vicon.com/products/software/tracker). This however requires a carefully controlled environment and often social environment cannot comply with such constraints, i.e., having numerous cameras may be considered too invasive and may require extensive adaptations.

Low cost vision tracking systems use low cost cameras and standard network infrastructures. State-of-the-art tracking algorithms for such systems have dynamics too slow to be compatible with natural human-robot interaction and hence tend to work best as a complement to other systems.

In this paper a people localization RFID based system is discussed. The goal is develop a system capable of providing information that can be fused with other systems, such as that from a network of cameras, and still, in case that is not possible (e.g., due to sync problems and/or occlusions) it must provide useful information for HRI purposes.

Section 2 reviews the issue of localization using RFID technology. Section 3 presents a model for the RFID reader used in the experiments in Section 5. Section 4 describes the algorithm used to estimate the location of people carrying a (passive) tag relative to the robot with the reader onboard. Final remarks are presented in section 6.

2 RFID Based Tag Localization

There is an extensive work on RFID based localization systems, including active, passive and semipassive tags and single and multiple readers. Reviews of key techniques are presented in [1], [8]. RFID detection can be used solo or as a complementary system to other localization techniques such as those based in Inertial Measurement Units (IMU) (see for instance [7] on the use of tag detection to resynchronize IMU based information).

In indoor scenarios it is common to use grids of passive tags. One or more moving readers can then be easily detected if the positions of the passive tags are known a priori. Arrays of readers can be used to provide information on the direction of the tag, [2].

Fingerprinting, i.e., recognition of tag detection patterns, has been reported to yield errors in the order of tens of centimeters for a tag density of 3.8 tag/m^2, [6].

Localization often assumes that the tag detection areas are of circular shape (see for instance [4], [9]) and tags are distributed according to regular patterns (see [5]). Simulation results are claimed to yield a localization error as low as 3 cm for a mobile robot moving up to 2 m/s and a high enough tag density.

Radio Signal Strength (RSS) has been used in low density tag distribution scenarios to yield localization errors around 1 m, [1]. Variations of the base technique include changing the power level of the reader(s) and knowing the sensitivity of specific tags to such variations, [3]. If the tags are active then by placing the tags in a carefully selected distribution, accounting for RF interferences, it is possible to use the RSS indicator generated by each tag to select the most probable regions, [2].

Scanning delays and the tag density are often referred as the most important factors generating errors, [5]. By carefully controlling the reading strategy, the effect of the scanning delay can be minimized. Multipath propagation and interference have also been referred as key factors to induce disturbances, [1].

3 RFID Reader Model

The RFID reader used in this work is a commercial SYNCO, model SR-RU-1861S, operating in the UHF 902 $\sim$ 928 MHz band. The antenna is described as omnidirectional by the manufacturer with an effective range of about 3 m, using a 8 dBi antenna (manufacturer's data). This means that, ideally, a tag placed anywhere inside the half-sphere of radius 3 m, covered by the antenna, would be detected with probability 1. The reader uses a serial RS232 connection to a laptop running software implementing the reading strategy and a simple Application Interface (API), everything implemented as a ROS[1] node. The reader only supplies the identification of the tags detected. No RSS is available.

Basic detection experiments readily show that often in indoors environments the propagation conditions are such that the half-sphere detection volume can change widely. This can happen for a variety of reasons, namely multipath, absorption, reflection, and diffraction, [8], and often cannot be controlled. For example, empirical observation has shown that in typical indoor corridor the detection volume can have the section of the corridor and be about 6–8 meters long, for the reader considered in this work. In a sense, given a generic environment it is very difficult to predict the shape of the detection volume. Often, the spherical (or half-spherical) volume assumption will not be realistic.

Additional testing also determined that a tag can be in a close neighborhood of the reader without being detected, this happening especially if there is no relative motion between the tag and the reader.

The aforementioned factors suggest that the reader can be modeled using a probabilistic-like technique by which to each point in a neighborhood of the reader is assigned a probability of a tag being detected in case it is located at that point.

[1] Robot Operating System

This probability is the quotient between the number of times a tag was detected over the total number of times the tag was placed in that position.

Figure 1 shows the model for a first model of the reader, obtained in an easy environment, i.e., wide indoors open area. The reader and tag can be seen, duly aligned. Measurements were taken along three circumferences of radius 1, 2, and 3 meters, with nine measurements each at regular angular intervals. Alternatively, non-regular spacing for the measurement points can be used, though as can be seen in the interpolated surface in Fig. 1(c) this may not have a significant impact on the results. The probabilities were estimated from 100 measurements. The model shows a central lobe, slightly deviated to the left, corresponding to the high detection probabilities.

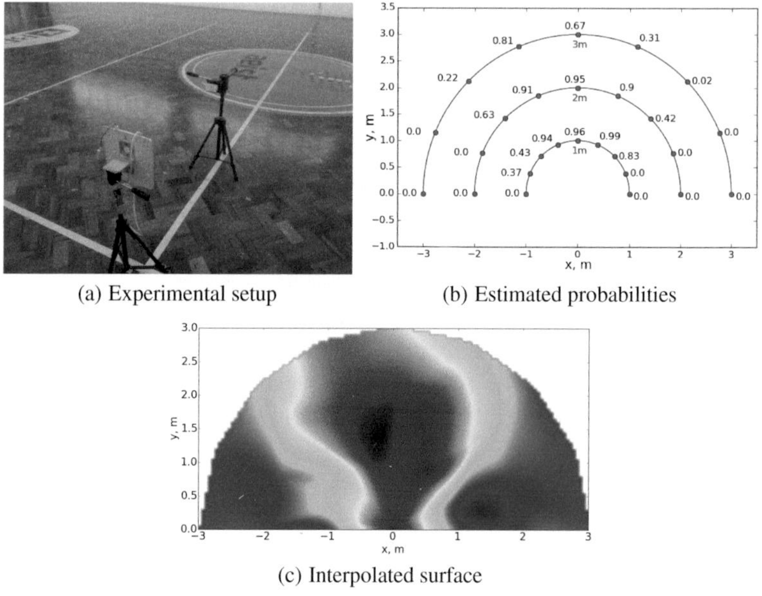

(a) Experimental setup

(b) Estimated probabilities

(c) Interpolated surface

Fig. 1 Basic probabilities model

In general, placing the RFID reader inside the robot will reduce its capabilities due to shielding and/or interference. Therefore, it can be expected that the model in Fig. 1 changes substantially when the environment around the reader changes. Figure 2 shows the RFID antenna placement in the MOnarCH robot. The antenna is located in the head of the robot and space constraints impose that it is placed facing upwards.

Figure 3 illustrates the interpolated surface, corresponding to a series of estimated probabilities at specific points in the neighborhood of the reader, placed inside the robot. The tags were kept always at the same height relative to the reader and hence the interpolated surface is two-dimensional.

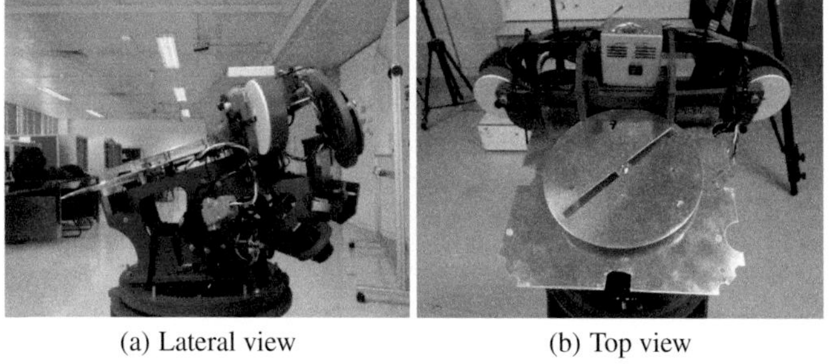

(a) Lateral view (b) Top view

Fig. 2 RFID antenna placement onboard the MOnarCH robot

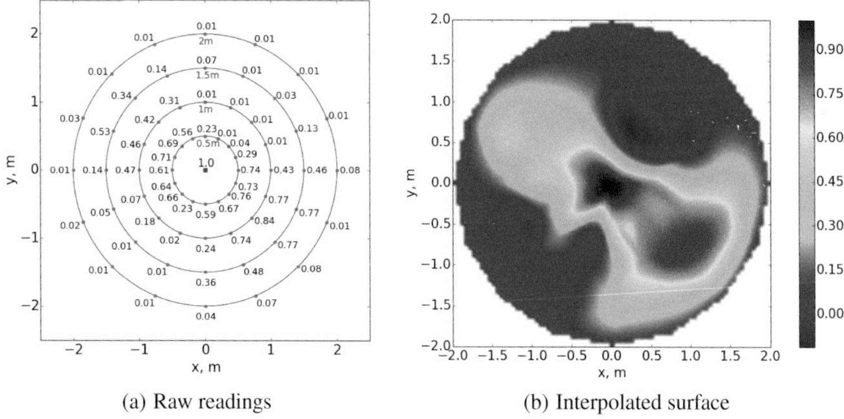

(a) Raw readings (b) Interpolated surface

Fig. 3 Probabilities model onboard the MOnarCH robot

In a more general setting a tag can change height in time, e.g., depending on the height of the person carrying it. This means that the detection conditions may vary and hence also the interpolated surface. The distribution of points obtained from the raw readings can be used to interpolate a cubic surface[2]. This surface is in fact the probabilities model of the reader.

4 Relative Location Estimation

The interpolated surface found in the previous section represents a model that embeds both reader and environment characteristics.

As the robot moves, detection probabilities are being updated. If a tag is suddenly detected then the robot may stop to better estimate the corresponding probability.

[2] Python function scipy.interpolate.griddata was used to compute the cubic surface.

In general, this will have a reduced impact on the effectiveness of the interaction with the people.

Once a probability is estimated, the interpolated model for the reader can be used to obtain a region of the most probable locations of the tag. Figure 4 illustrates a realistic situation. The person standing up is wearing a RFID tag on his shirt whereas the white robot shown carries the RFID reader inside its outer shell. The plot shows the region corresponding to the estimated probability within a $\epsilon = 0.1$ margin.

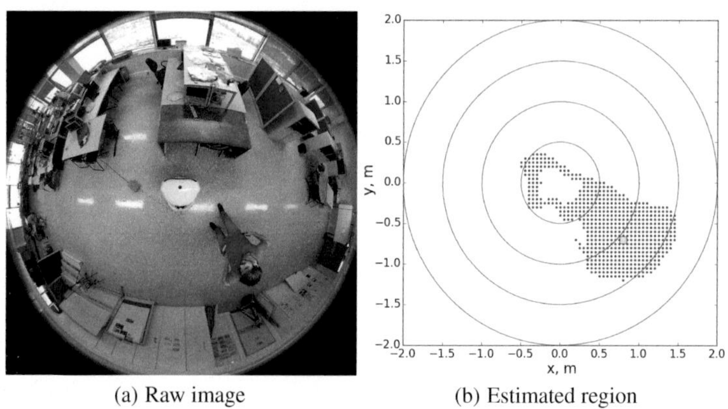

(a) Raw image (b) Estimated region

Fig. 4 Estimating the position of the RFID tag

Overall, the probability of a tag being detected is simply estimated by the counting **Algorithm 1**.

Algorithm 1. Tag detection probability estimation

Require: $total\_reads$
  $tag\_count \leftarrow 0$;
  if tag detected **then**
    for $i = 1$ **to** $total\_reads$ **do**
      re-read the tag;
      if tag is detected **then**
        $tag\_count \leftarrow tag\_count + 1$;
      end if
      wait for Δt s;
    end for
    $prob = tag\_count / total\_reads$
    publish $prob$
    return probability updated
  else
    return probability not updated
  end if

The waiting time between readings, Δt, is chosen empirically as $\Delta t = 0.05\,\text{s}$. It corresponds to the time required for the reader to return meaningful data. The detection time is then $N\Delta t$, where N is the total number of reads. By default, N is set to 100, yielding $N\Delta t = 5\,\text{s}$, which is acceptable for a number of HRI applications.

Once a probability estimate is available, **Algorithm 2** is run to select an estimate for the position of the tag, relative to the position of the robot. Figure 5 shows the result.

Algorithm 2. Tag position estimation

Require: Reader Model, ϵ

Compute the area, according to the model, with $prob \in [prob - \epsilon, prob + \epsilon]$

Compute the medial skeleton of that area

Compute the point in the skeleton with the biggest distance to the boundary of the area (this point is the position estimate)

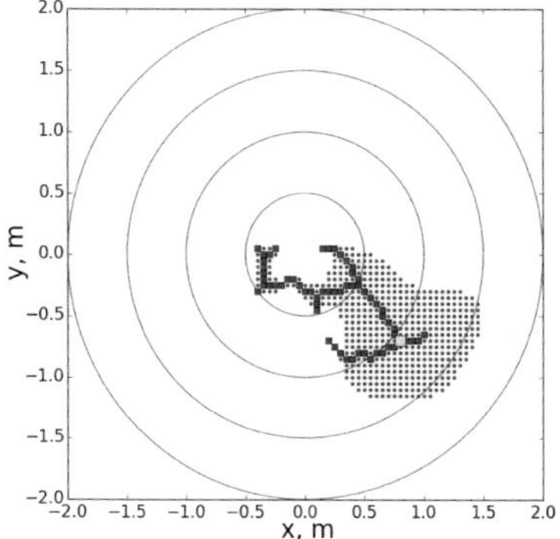

Fig. 5 Example of medial skeleton and position estimate

Algorithm 2 can return more than one point and it may even return one or more line segments (i.e. segments of the skeleton for which all points lie at the same distance of the border of the estimate region). The strategy chosen to disambiguate the output is then to mask the region around the reader (see section 5 ahead). Whenever there is not enough a priori information that can be used to select a mask then an averaging of the sets of points is returned.

5 Experiments

This section presents preliminary results obtained with the MOnarCH robot (www.monarch-fp7.eu). Figure 6 shows two views of the robot, designed for social interaction with inpatient children at a hospital. The right hand image shows the internals of the robot. The RFID reader is located in the head (see also Fig. 2).

(a) At the hospital (b) Internal structure

Fig. 6 The social robot MOnarCH

The experiments are grouped in two classes, differing on the amount of time used for detection. In the first group $N = 100$, resulting, as aforementioned, in a detection time of 5 s. The second group $N = 30$, resulting in a detection time of 1.5 s. The objective is to assess the influence of the time taken for measurement of the detection probability on the probability estimate.

To increase the flexibility of the detection one additional parameter is introduced, namely, prior information on the location of the person. This parameter masks the area around the reader such that only part of it is considered for detection. Acceptable values are (i) no information, meaning that no mask is applied and the full area around the reader is used to search for a solution, (ii) person in front of the robot, meaning that only an area of angular with 180° centered with the longitudinal axis of the reader is used, and (iii) person behind the robot, meaning that a 180° area behind the reader is used.

For the purpose of analysis, the measurements are grouped as follows. The objective of this grouping is to provide a confidence measure that a tag is detected at some angular direction that is meaningful from the perspective of HRI. In fact, for most applications involving HRI, a high accuracy in the tag position estimate it is not required. Therefore, an angular interval around the real direction of the tag is defined and a counting of the number of detections lying inside that interval provides a rough estimate of the probability that a tag approaching some direction is correct. Two widths for this interval were considered, namely 90° and 180°.

The experiments evolved as follows ($\epsilon = 0.1$ was used in all experiments). The robot (and reader) is placed at the center of a circle with 3 m radius. A person walks from outside the circle towards the robot, keeping the same angle relative to the longitudinal axis of the reader. **Algorithms 1 and 2** compute the tag angle estimate 10 times for each angle. A total of 8 angles (or 5 in the cases where a mask is applied) is used for the person approaching the robot.

The experiments were performed with the person carrying a tag at approximately the height of the robot.

Figures 7 to 12 represent the probability of the computed angle being inside the angular width region of the real angle (corresponding to the person angular position). Each experiment is shown for the two angular width values considered.

In the first experiment, Fig. 7, the probabilities that the detection is correct when the person approaches the robot by the front are clearly much higher than approaching from behind. The higher values for the 180° threshold are natural as the region considered for the counting is higher and can catch a larger number of detections.

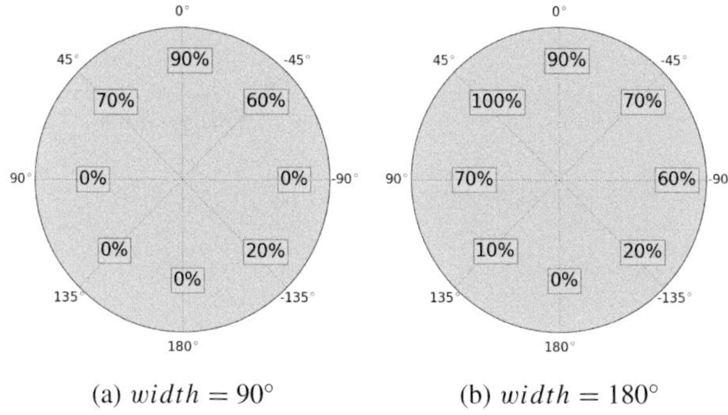

(a) $width = 90°$ (b) $width = 180°$

Fig. 7 Results for 5 s detection time and no prior information

The same experiment for detection time 1.5 s yields similar results. Though the regions in the back get lower detections than those in the front, the regions of higher detection have similar locations. The conclusion is naturally that detection confidence gets larger as the detection time increases (the results support the 5 s maximum value a priori assumed).

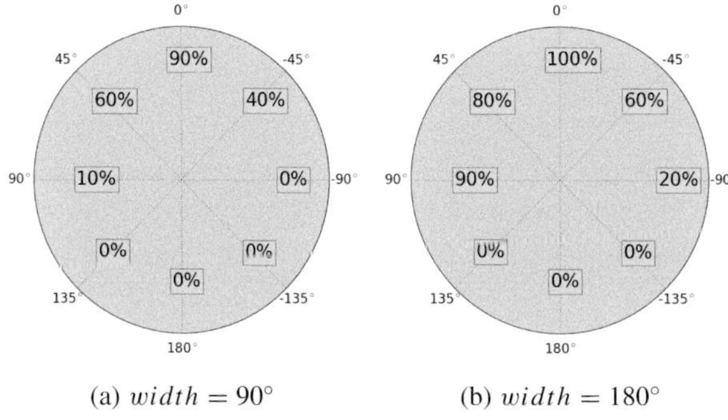

(a) $width = 90°$ (b) $width = 180°$

Fig. 8 Results for 1.5 s detection time and no prior information

Figures 9 and 10 show the second experiment masking the back part of the robot, i.e., detection is considered to be always in the front of the robot.

The results follow in the line of those of the previous experiment, namely detection confidence increases with the angular threshold and detection time.

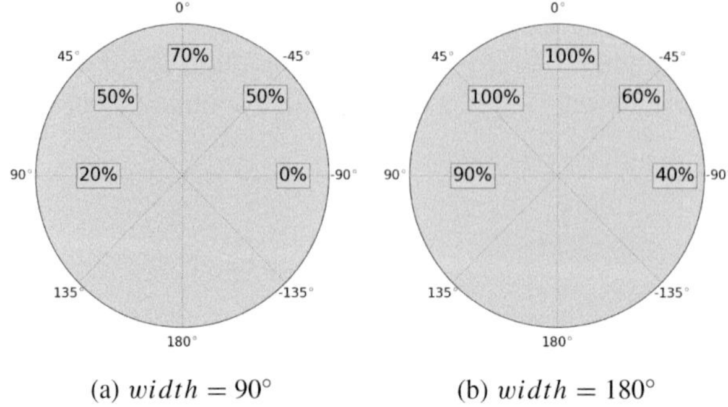

(a) $width = 90°$ (b) $width = 180°$

Fig. 9 Results for a 5 s detection time and person in front of the robot

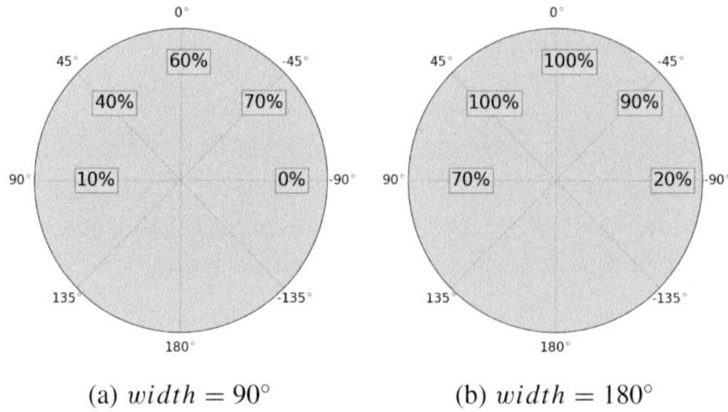

(a) $width = 90°$ (b) $width = 180°$

Fig. 10 Results for 1.5 s detection time and person in front of the robot

In the last experiment, Figs. 11 and 12, the detection mask is on the frontal area of the robot. The experiment shows much higher probabilities, suitable for HRI applications. In addition, results are fully consistent with the conclusions drawn from the two previous cases.

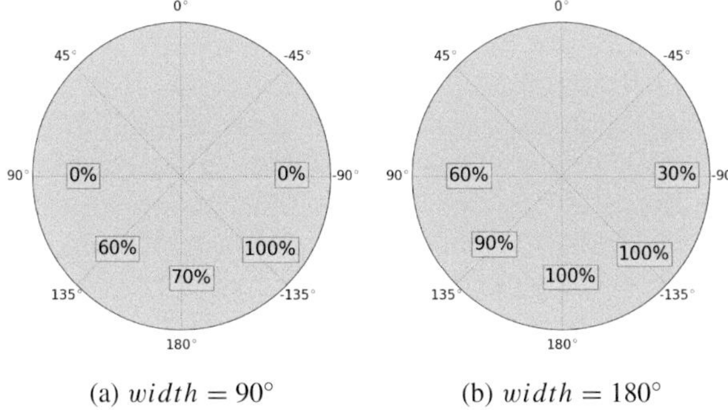

(a) $width = 90°$ (b) $width = 180°$

Fig. 11 Results for 5 s detection time and person behind the robot

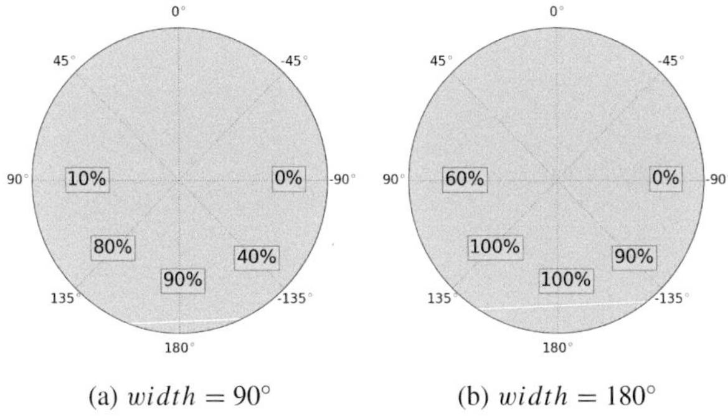

(a) $width = 90°$ (b) $width = 180°$

Fig. 12 Results for 1.5 s detection time and person behind the robot

6 Conclusions

The technique present was shown to have a dynamics compatible with a usage in proxemics related HRI, as the detection accuracy is less a limiting factor than, for instance, in estimation of the direction of movement of a person relative to a robot. Even though more accurate techniques do exist from an absolute perspective, e.g., vision, it is often the case that there are situations when they are unable to operate, e.g., in case of occlusions. Therefore, we envisage the use of the technique presented either isolated or as complement to other techniques.

Moreover, in the case of unstructured environments, where it may be difficult to install more accurate sensing, e.g., static cameras and the respective network infrastructure, the approach presented presents a clear advantage. The main limitation of this approach seems to be related to environments with difficult radio frequency propagation conditions.

Experiments show that the angle estimate is better for the first group, with 5 s detection time, but both groups have good results, namely if the output is to be used for HRI purposes. In terms of the prior information of the person position, experiments demonstrate that in the first case, i.e., the full area around the robot is used, the model only performs acceptably if the person is in front of the robot. However, if the robot is only interested in detecting people from behind, the quality of the results improves significantly. Higher probabilities appear when the angular width of the detection region is 180°, as expected. In sum, experiments yield a performance that is acceptable for most HRI applications, namely speech interaction where there is no need to know accurately the orientation of the person relative to the robot.

Future work includes (i) the tuning of parameters in the probability estimation algorithm, namely the time between readings and assessing the system in a wide variety of environments and with the people wearing the tags at a range of different heights (children are usually smaller than adults), and (ii) extensive tests in the real hospital environment. It is worth to remark that the placement of the tags does not constrain the application of this technique. In fact it is enough to run multiple instance of the detection model, tuned for different and use a voting strategy in case multiple detections arise.

Acknowledgements This work was supported by projects FP7-ICT-2011-9-601033-MOnarCH and FCT [UID/EEA/50009/2013].

References

1. Ni, L.M., Zhang, D., Souryal, M.R.: RFID-based localization and tracking technologies. Wireless Communications, IEEE **18**(2), 45–51 (2011)
2. Zhao, Y., Liu, Y., Ni, L.M.: VIRE: active RFID-based localization using virtual reference elimination. In: International Conference on Parallel Processing, ICPP 2007. IEEE (2007)
3. Chawla, K., Robins, G., Zhang, L.: Efficient RFID-based mobile object localization. In: 2010 IEEE 6th International Conference on Wireless and Mobile Computing, Networking and Communications (WiMob). IEEE (2010)
4. Munishwar, V.P., et al.: On the accuracy of RFID-based localization in a mobile wireless network testbed. In: IEEE International Conference on Pervasive Computing and Communications, PerCom 2009. IEEE (2009)
5. Nejad, K.K., Jiang, X., Kameyama, M.: RFID-based localization with Non-Blocking tag scanning. Ad Hoc Networks **11**(8), 2264–2272 (2013)
6. Jingwangsa, T., Soonjun, S., Cherntanomwong, P.: Comparison between innovative approaches of RFID based localization using fingerprinting techniques for outdoor and indoor environments. In: 2010 The 12th International Conference on Advanced Communication Technology (ICACT), vol. 2. IEEE (2010)
7. House, S., et al.: Indoor localization using pedestrian dead reckoning updated with RFID-based fiducials. In: 2011 Annual International Conference of the IEEE Engineering in Medicine and Biology Society, EMBC. IEEE (2011)
8. Bouet, M., Dos Santos, A.L.: RFID tags: positioning principles and localization techniques. In: Wireless Days, WD 2008, 1st IFIP. IEEE (2008)
9. Han, S., Lim, H., Lee, J.: An efficient localization scheme for a differential-driving mobile robot based on RFID system. IEEE Transactions on Industrial Electronics **54**(6), 3362–3369 (2007)
10. Hall, E.T.: The Hidden Dimension. Doubleday & Co. (1966)

Gaze Tracing in a Bounded Log-Spherical Space for Artificial Attention Systems

Beatriz Oliveira, Pablo Lanillos and João Filipe Ferreira

Abstract Human gaze is one of the most important cue for social robotics due to its embedded intention information. Discovering the location or the object that an interlocutor is staring at, gives the machine some insight to perform the correct attentional behaviour. This work presents a fast voxel traversal algorithm for estimating the potential locations that a human is gazing. Given a 3D occupancy map in log-spherical coordinates and the gaze vector, we evaluate the regions that are relevant for attention by computing the set of intersected voxels between an arbitrary gaze ray in the 3D space and a log-spherical bounded section defined by $\rho \in (\rho_{min}, \rho_{max}), \theta \in (\theta_{min}, \theta_{max}), \phi \in (\phi_{min}, \phi_{max})$. The first intersected voxel is computed in closed form and the rest are obtained by binary search guaranteeing no repetitions in the intersected set. The proposed method is motivated and validated within a human-robot interaction application: gaze tracing for artificial attention systems.

Keywords Human-Robot Interaction (HRI) · Artificial attention · Gaze tracing · Log-spherical · Voxel traversal algorithm

1 Introduction

With the forthcoming social or assisted living robotic platforms, interlocutor gaze has become a valuable cue to interpret intentionality [3, 11]. The next generation of robotic platforms should be able to interpret and generate social signals [17] by means of non-verbal communication. This emotional, body and attentional language

B. Oliveira · P. Lanillos · J.F. Ferreira(✉)
AP4ISR Team, Institute of Systems and Robotics (ISR), University of Coimbra,
Pinhal de Marrocos, Polo II, 3030-290 Coimbra, Portugal
e-mail: {beatriz,planillos,jfilipe}@isr.uc.pt
http://ap.isr.uc.pt, http://mrl.isr.uc.pt/projects/casir

L.P. Reis et al. (eds.), *Robot 2015: Second Iberian Robotics Conference*,
Advances in Intelligent Systems and Computing 418,
DOI: 10.1007/978-3-319-27149-1_32

Fig. 1 Representational stage in joint attention. The artificial system should be able to build an internal representation of the environment and infer the object that the interlocutor is gazing during gaze following.

will help roboticists to develop low cost robots with high user acceptance. In this sense, the direction of other's attention is crucial for mastering social interaction [16]. In fact, in the case of joint attention [3, 7], where two individuals perform a triadic relationship between them and an object, gaze direction provides useful information to know if the human is looking at the robot as well as about the direction of the shared object [2]. This location is potentially important for the correct social interaction and can also play a significant role in the task being effectuated.

Figure 1 describes a generic and simple example of social interplay within human-robot interaction (HRI) in assisted living applications. An old lady is inside the social space and engages the robot meaning that he wants to initiate interaction. Afterwards she switches the focus of attention towards somewhere outside the field of view (e.g., a computer). In order to infer intention, and therefore construct the correct attentional behaviour, we need to discover the location or the object that the interlocutor is gazing because it could be important for the interaction being performed. In this case she could want to call by internet to her daughter.

In the study presented in [7], where an artificial attention system is tested without integrating the gaze interpretation ability, it is clear that the robot cannot infer the complete non-verbal message intended by the interlocutor. In that experiment, the interlocutor shows an object to the robot by deictic fixation (e.g. gazing or grabbing the object) and then the machine should search another object with similar characteristics. They show that just using the preatentive scene segmentation the robot is not able to understand the user needs, as the robot sometimes does not figure out which object is being showed. We hypothesise that enabling the machine with a full attentional system[1] will improve reciprocity, expectation fit and interaction [6].

[1] The majority of the works in attention lacks from the integration of all needed functionalities [3]. They usually use just visual sources, there is no top-down modulation nor emotional context and social signals are not taken into account.

This requires the implementation of the representational skill, which toddlers achieve in early stages of development [12], and involves discovering the object that the interlocutor is referring to. One of the possible approaches discussed in the literature is to provide an internal representation of the environment that works as a short-term memory and stores important information for attention [4, 8], and then use the gaze cue to modulate the potential attended objects.

In this paper we present a discrete log-spherical gaze tracing algorithm to be integrated in a full-fledged attention system [7] that establish a correspondence between a set of discretised cells with potential objects being fixated by the observer. We integrate the egocentric representation of the environment, Bayesian Volumetric Map (BVM) [4], with the gaze cue, improving the capacity of the machine to infer important regions even if they are outside the field of view and providing the robot with the representational stage, thus laying the foundation for non-verbal social HRI. Although the algorithm is focused on attention in HRI, it can be used as a general voxel traversal algorithm in bounded log-spherical space representations.

The paper is organised as follows: section 1.1 and 1.2 detail the motivation of this work and the current state-of-the art methods for gaze tracing; section 2 describes the proposed solution, formalizes it mathematically and presents the tracing algorithm; section 3 shows gaze tracing working, the computational analysis and the final results when integrating the BVM with the traversal algorithm in a real HRI scenario; finally, section 4 summarises the work.

1.1 Egocentric Representation and Gaze Tracing in Artificial Attention

Regarding to attention, the information provided by the different sensors must be subjected to a spatial correspondence [3]. For instance, a sound source should be related with its potential origin location in order to make possible the correspondent attention action [8]. In fact, in overt attention, where the scene is partially observed and changes depending on the actions (e.g., head movements), if we do not enable the machine to have an internal representation of the environment with temporal registration, its actions will become reactive for each location and angles setup. Furthermore, when placing the machines in a social or human interaction context, joint attention, a primal non-verbal communication process driven by attention, should be fulfilled [3, 6]. This mechanism, where an object or location of interest is shared just by engaging and deictic cues such as gaze, seems to be the backdrop for many social cognitive skills in humans [2]. For enabling this behaviour in machines, they must have their own spatial representation of the environment to correlate the gaze cues with the potential shared objects or locations. Therefore, any perception (e.g., visual, auditory, etc.), should be related and integrated into a single egocentric reference. Recent research works have introduced cluster [5] and spherical [4, 13] representations of egocentric space to deal with these issues. In this work we will use the 3D log-spherical representation proposed in [4] to codify the perceived environment. The method is also valid when the saliency of the scene is modelled [8].

The gaze cue provides interesting information about intention and it is crucial for social development [3, 16]. In humans' social interaction, when an interlocutor wants to share an object, he switches the focus of attention from the other's face towards the object. The other interlocutor starts a phase called gaze following [2, 12] to search the object or location that it is being shared. Therefore, first we need to estimate the interlocutor gaze direction and afterwards perform the gaze following until the object is discovered. Techniques to estimate and track the gaze are out of the scope of this paper. Nevertheless, an overview of recent approaches to infer gaze are described in [9]. For gaze following, apart from the naive and deterministic solution of redirecting the robot head according to the gaze vector, some developmental learning approaches have been researched [10].

1.2 Ray Tracing

Ray tracing in 3D has become very popular in games and simulators (e.g., PowerVR Ray or NVIDIA Iray) for light rendering and in medicine for image reconstruction [14, 15]. There are three key aspects for a good ray tracing algorithm: the computational speed, the generality to multiple inputs and the implementation simplicity. On one hand, we can find in the literature two optimal algorithms in the Cartesian space for different purposes. AmanatidesWoo's algorithm [1] is a general purpose method that computes the voxels that an arbitrary ray traverse in a bounded region (box). This algorithm computes the initial intersection point to speed up the computation. Siddon's Algorithm [14] uses ray tracing to optimize the reconstruction of medical images. Finally, Thibaudeau et al. [15] developed an algorithm that traces a ray using spherical or cylindrical coordinates. It analyses each dimension of the 3D space separately (radial, and azimuthal and elevation angles) and stores the entry and exit point of each voxel. Then the intersection points of all dimensions are sorted and the repetitions are eliminated.

2 Gaze Tracing in Log-Spherical Coordinates

We generalize and abstract the problem as depicted in Figure 2. The interlocutor and the robot are represented by the green ball and the orange ball respectively. The gaze direction is described by the dashed black line and the log-spherical region is defined by the radius ρ and two angles (θ, ϕ). The gaze ray cuts through the bounded region starting in the intersection (green cross) and traverses a set of log-spherical voxels (represented in as red shapes in the figure) until it reaches the exit point. Note that further voxels from the robot are bigger because of the logarithm influence. Besides, ρ_{min} defines the egocentric gap [4]. Whilst this is the general case, the subject will

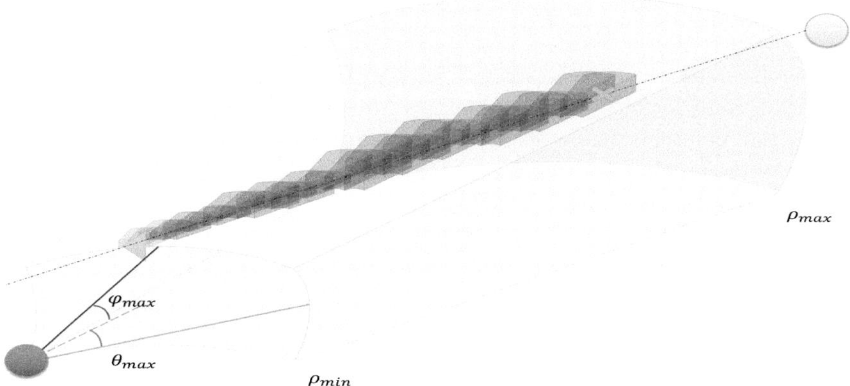

Fig. 2 Gaze tracing in a bounded log-spherical space partition. The region is compound of discrete voxels constrained by the minimum and maximum values of ρ, θ and ϕ.

usually be inside the spherical bounded region and therefore, the ray starting point will correspond to the gaze initial location[2].

2.1 Mathematical Formulation

We place the robot at the (0, 0, 0) in the Cartesian coordinate system and define the perceptive space as a bounded spherical region by its radius ρ and azimuth θ and elevation ϕ angle ranges.

$$\rho_{min} \leq \rho \leq \rho_{max} \tag{1a}$$

$$\theta_{min} \leq \theta \leq \theta_{max} \tag{1b}$$

$$\phi_{min} \leq \phi \leq \phi_{max} \tag{1c}$$

The perceptive space of the robot is modelled by means of the BVM, an egocentric and metric inference grid that encodes the spatial occupation of the environment and its dynamics [4] as well as saliency [8]. The BVM space is defined in the log-spherical coordinate systems as follows,

$$\mathcal{Y} \equiv]log_b \rho_{min}; log_b \rho_{max}] \times]\theta_{min}; \theta_{max}] \times]\phi_{min}; \phi_{max}] \tag{2}$$

where the logarithmic base is,

[2] The gaze initial location is commonly placed at the sellion: the point of the deepest depression of the nasal bones at the top of the nose.

$$b = a^{log_a(\rho_{max} - \rho min)/N}, \forall a \in \mathbb{R} \tag{3}$$

and N is the number of partitions in ρ dimension. The BVM is finally discretised into cells by defining the increment in the angles $\Delta\theta$ and $\Delta\phi$. Therefore, cell is indexed as C_{ijk} and represents the occupancy probability in the interval $[b^i + \rho_{min}, b^{i+1} + \rho_{min}] \times [\theta, \theta + \Delta\theta] \times [\phi, \phi + \Delta\phi]$. Note that i represents $\lceil log_b(\rho - \rho_{min}) \rceil$ (see Appendix). Furthermore, for a detailed explanation of the BVM see [4].

The gaze is represented as a ray expressed by the origin $p_0 = (x, y, z) \in \mathbb{R}^3$ and the director vector $\mathbf{v}_0$. Thus, given the parameter $t \in \mathbb{R}_0^+$ gaze is described by its Cartesian parametric equations,

$$y = p_0 + t\mathbf{v}_0 \tag{4}$$

The proposed method computes the collection of unique voxels traversed $\mathcal{C} = \{c_1, \ldots, c_n\}$ by the gaze ray inside the bounded region. With abuse of notation, the collection expressed in BVM indexes, as a matrix, is $\mathcal{C} = \{c_{i_1 j_1 k_1}, \ldots, c_{i_n j_n k_n}\}$.

2.2 Voxel Traversal Algorithm

Given an arbitrary gaze ray defined by Eq. (4) and the bounded region described by Eq. (1), we propose Algorithm 1 to compute the set of traversed voxels $\mathcal{C}$. The algorithm guaranties that $\mathcal{C}$ contains no repetitions and all voxels are inside the spherical region. First the logarithmic base is obtained using Eq. (3) and the initial intersection point between the ray and the region is computed as explained in section 2.3. Afterwards, we compute the next ray point using the default step. If the next ray point does not match the next voxel (i.e., the difference of indexes is not equal to 1) we apply binary search by reducing the step parameter dt to the half or incrementing by the half depending if we have fallen sort or long. In case of overshooting, when incrementing dt, in the next step we will arrive to the same cell, thus the middle point of the segment is computed. Finally, the new voxel found is added to the collection. The algorithm returns the collection $\mathcal{C}$ when the next point computed is out of bounds.

The complexity of the algorithm is in worst case $O(n \log m)$ due to the inner loop that uses binary search. In practice the number of iterations m that binary search is processed for each outer iteration is quite smaller than n and depends on the $defaultT$ parameter used. This means that the method theoretically is faster than any optimal sorting algorithm, but again, in practice this approach depends on the function $Cartesian2DiscreteLogSpherical$ that transforms the Cartesian point into the discrete log-spherical coordinates (i.e., BVM matrix indexes). The algorithm is generalizable as it works for any kind of discrete representation just by substituting the conversion function.

Algorithm 1.. Log-spherical Gaze Tracing Algorithm

Require: $p_0, \mathbf{v}_0$ ▷ Ray origin and director vector
Require: $\rho_{min}, \rho_{max}, \theta_{min}, \theta_{max}, \phi_{min}, \phi_{max}$ ▷ Region bounds
Require: $N, defaultT$ ▷ Number of partitions of radius and initial increment

$\mathcal{C} = \emptyset,\ dt = defaultT$
$b = e^{\frac{\ln(\rho_{max}-\rho_{min})}{N}}$ ▷ Eq. (3)
$t_0 = initialT(p_0, v_0)$ ▷ See section 2.3
while inside **do** ▷ Eq. (1)
  $\mathbf{p}_{xyz} = p_0 + t_0\mathbf{v}_0$
  $\mathbf{p}_{ijk} = Cartesian2DiscreteLogSpherical(\mathbf{p}_{xyz}, b)$ ▷ See Appendix
  while $\neg found$ **do**
    $t_1 = t_0 + dt$
    $\mathbf{p}'_{xyz} = p_0 + t_1\mathbf{v}_0$
    $\mathbf{p}'_{ijk} = Cartesian2DiscreteLogSpherical(\mathbf{p}'_{xyz}, b)$
    $\Delta = \left|\mathbf{p}_{ijk} - \mathbf{p}'_{ijk}\right|$
    if *block* **then** ▷ overshoot or undershoot according to previous state
      $dt = 0.5(t_{prev} + t_1) - t_0$ ▷ the middle point is selected
    else if $\Delta i > 1 \vee \Delta j > 1 \vee \Delta k > 1$ **then**
      $dt = 0.5dt$
    else if $\Delta i = 0 \wedge \Delta j = 0 \wedge \Delta k = 0$ **then**
      $dt = 1.5dt$
    else
      $found = true,\ t_0 = t_1,\ dt = defaultT$
      $\mathcal{C} = \mathcal{C} \cup C_{ijk}$ ▷ Add new voxel to the set
    end if
  end while
end while
return $\mathcal{C}$

2.3 *Computing the Initial Intersection Point*

We can rewrite the spherical region defined by inequalities (1a),(1b),(1c) into Cartesian coordinates by converting the spherical parameters (ρ, θ, ϕ) into the Cartesian system following the equations defined in (8) (see appendix). The new inequalities become:

$$\rho_{min} \leq \sqrt{x^2 + y^2 + z^2} \leq \rho_{max} \tag{5}$$

$$\theta_{min} \leq \arctan\left(\frac{x}{z}\right) \leq \theta_{max} \tag{6}$$

$$\phi_{min} \leq \arctan\left(\frac{y}{\sqrt{x^2 + y^2}}\right) \leq \phi_{max} \tag{7}$$

Then we substitute x, y, and z with Eq. (4), and solve for t as it is detailed in the appendix. The minimum t value that satisfies the equations defines unequivocally the initial intersection point t_0.

3 Results

3.1 Gaze Tracing Algorithm Example

By means of a prepared example we show the algorithm functioning. The robot is stationary at (0, 0, 0) and the gaze is defined by its initial point $p_0 = (0, 20, 20)$ and its direction $\mathbf{v}_0 = (-0.3, -0.01, 0)$. The bounded spherical region is constrained to: $\rho \in (10, 80)$, $\theta \in (-180, 180)$ and $\phi \in (-90, 90)$. Figure 3(a) shows in Cartesian coordinates the abstracted scene where the subject is gazing down and left. Figure 3(b) shows the same scene in spherical coordinates (i.e., ϕ and ρ plane). The dashed lines describe the separation between cells in the ρ axis defined by $b^i + \rho_{min}$ in the logarithmic space. We can see that the distance between the boundaries increase

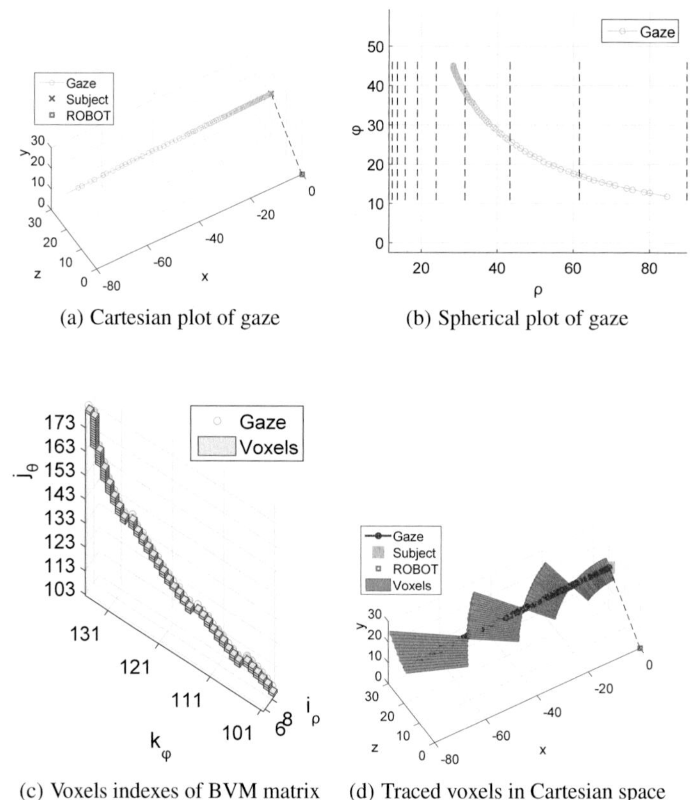

(a) Cartesian plot of gaze (b) Spherical plot of gaze

(c) Voxels indexes of BVM matrix (d) Traced voxels in Cartesian space

Fig. 3 Gaze tracing example. (a) Cartesian plot with a subject looking left and down; (b) Gaze ray converted to spherical; (c) Output of the algorithm: set of traversed voxels and their indexes in the BVM matrix; (d) Computed voxels transformed into Cartesian space overlay figure (a). Note that we represent Cartesian units in decimeters.

exponentially making closer regions to be more fine grained. The output of the algorithm, the set of traversed voxels, is shown in Fig. 3(c), where the indexes in all three dimensions are plotted as boxes. In this presented sub region of the BVM matrix (i.e., discretised log-spherical structure) we can see that as the ray goes further from the robot θ angle is increased resulting in a j decrement. Note that the angles values are positive as they are the indexes of the matrix. Finally Fig. 3(d) shows the voxels defined by the indexes converted into the Cartesian coordinate system. The egocentric nature of the log-spherical representation make the red trapezoids that represent the voxels to face the robot location.

3.2 Computational Results

In order to analyse the computational complexity of the proposed algorithm we compare it with two other state-of-the-art voxel traversal algorithms: AmanatidesWoo that works in the Cartesian space [1] and Thibaudeau that computes the intersections in the spherical space [15]. The comparison is performed by modifying the number of voxels that must be traversed. All implementations have been developed in Matlab and have been tested on an Intel Core i7-4700MQ CPU computer with 8Gb of RAM. To reduce the noise due to tasks latency we have run the algorithms 200 times for each number of voxels value.

Figure 4 shows the time comparison in *ms* of the three algorithms. We can see that AmanatidesWoo wins but it is constrained to Cartesian configurations that can exploit the mathematical properties of 3D rectangles. Although our algorithm behaves similar to Thibaudeau in terms of time computation works for log-spherical

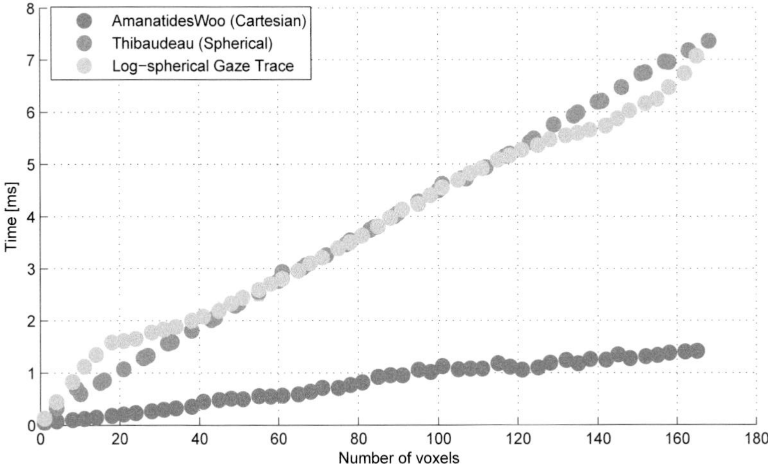

Fig. 4 Computational time comparison depending on the number of voxels traversed.

spaces, it is simpler and more generalizable. In contrast, our algorithm requires experimental tuning of the $defaulT$ parameter to minimize the computation time. For the statistical analysis we have set it to 1.5531 metres.

3.3 Integrating the Gaze Tracing with the 3D Occupancy Representation

We combine the gaze tracing proposed algorithm with the log-spherical occupancy grid (BVM) to achieve our final goal that is to discern the potential objects that the interlocutor is staring at. The BVM stores as a short-term memory the probability of occupied regions. Thus, the gaze traversed voxels that intersects the occupied ones are the locations of interest. We define the experimental set-up, as depicted in

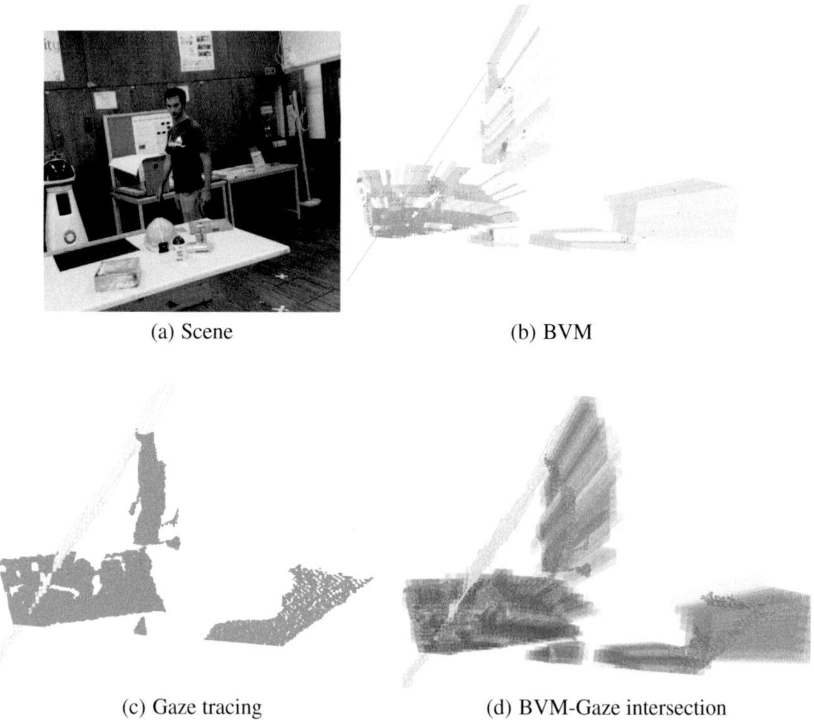

(a) Scene (b) BVM

(c) Gaze tracing (d) BVM-Gaze intersection

Fig. 5 Gaze trace combined with the BVM. (a) Robot camera; (b) Occupancy log-spherical grid (red trapezoids) with the overlaid gaze ray (yellow line); (c) voxels traversed by the gaze in the log-spherical space (yellow trapezoids); (d) Intersected voxels between the gaze ray and the occupancy grid (green trapezoids). Note that the red intensity determines the probability of a voxel being occupied and that the point cloud is overlaid in blue in (b), (c) and (d).

Fig. 5(a), with a subject that enters in the field of view of the robot and looks to an object placed on the table. The wanted voxels must satisfy that are traversed by the gaze ray and are occupied. We define an occupied voxel when its probability is higher than 0.6. The gaze tracing algorithm output is shown in Fig. 5(c) and the final pursued intersected locations are plotted as green trapezoids in Fig. 5(d). We can see that the voxels are correctly found on the table.

4 Conclusion

The penetration of social robots in assisted living applications is slowly occurring mostly due to the incapability of deploying machines with enough coherent behaviour according to human expectations. The presented results have shown how easy the proposed approach is integrated into an artificial attention system in order to identify the potential objects of interest. Thus, this proposal, as it allows to combine a social signal (gaze) and the short term-memory (egocentric representation of the environment and some stored important objects), will enable machines to understand other's attention based on his deictic gaze fixations. Furthermore, it will help the robot to recognise where the interlocutor is attending, improve subject's gaze following and enhance reciprocity and non-verbal social interplay.

We have presented a fast voxel traversal algorithm for gaze tracing in discretised log-spherical bounded spaces that works in real time. Its worst case complexity is $O(n \log m)$ and m depends on a parameter that can be experimentally optimized. The comparison of other state-of-the-art algorithms shows that the proposed approach, which is simple to implement, behaves as fast as them but allowing generalization to any bounded discretised space configuration.

When representing the subject's gaze with just one ray the uncertainty and error associated with the gaze detection measurement is not taken into account. For that purpose, we will extend the algorithm to compute a cone of rays, whose parameters will model the gaze uncertainty. Moreover, we will look for optimized implementations of the algorithm by including some of the nice properties of [15]. The extended implementation of the algorithm will be embedded in a full artificial attentional mechanism to evaluate if artificial systems can be improved when enabling the attention skill [6].

5 Appendix

Conversion from Cartesian to Spherical Coordinates.

$$\rho = \sqrt{x^2 + y^2 + z^2} \qquad \phi = \arctan\left(\frac{y}{\sqrt{x^2 + y^2}}\right) \qquad \theta = \arctan\left(\frac{x}{z}\right) \tag{8}$$

Conversion from Cartesian to Discretised Log-spherical. Applying Eq. (8), and the following equations, the discretised log-spherical indexes are found.

$$i = ceil(\log_b(\rho - \rho_{min})) \qquad j = floor(\phi) - \phi_{min} + 1 \qquad k = floor(\theta) - \phi_{min} + 1$$

Computing the Initial Intersection Point. Substituting the ray equation (4) into the inequalities presented in (5), (6) and (7) and solving for the parameter t we obtain the first intersection point of the gaze ray with the bounded spherical region.

$$t_\rho = -\frac{(x_0 v_x + y_0 v_y + z_0 v_z)}{v_x^2 v_y^2 v_z^2} \mp \mp \sqrt{\left(\frac{(x_0 v_x + y_0 v_y + z_0 v_z)}{v_x^2 v_y^2 v_z^2}\right)^2 - \frac{x_0{}^2 + y_0{}^2 + z_0{}^2 - \rho_{min}{}^2}{v_x^2 v_y^2 v_z^2}} \tag{9a}$$

$$t_\phi = -\frac{(x_0 v_x + z_0 v_z - y_0 v_y \tan(\rho)^2)}{v_x^2 + v_z^2 - v_y^2 \tan(\rho)^2} \pm \pm \sqrt{\frac{0.5(x_0 v_x + z_0 \times v_z - y_0 v_y \tan(\rho)^2)^2}{v_x^2 + v_z^2 - v_y^2 \tan(\rho)^2} - \frac{x_0{}^2 + z_0{}^2 - y_0{}^2 \tan(\rho)^2}{v_x^2 + v_z^2 - v_y^2 \tan(\rho)^2}} \tag{9b}$$

$$t_\theta = \frac{x_0 \tan(\theta) - z_0}{v_z - v_x \tan(\theta)} \tag{9c}$$

Each resultant t_ρ, t_ϕ, t_θ will be the minimum value calculated using each bound. The final t is computed by overlapping the intervals and extracting the closest one to the gaze ray origin that satisfy the conditions.

References

1. Amanatides, J., Woo, A., et al.: A fast voxel traversal algorithm for ray tracing. Eurographics **87**, 3–10 (1987)
2. Brooks, R., Meltzoff, A.N.: The development of gaze following and its relation to language. Developmental Science **8**(6), 535–543 (2005)
3. Ferreira, J.F., Dias, J.: Attentional Mechanisms for Socially Interactive Robots - A Survey. IEEE Transactions on Autonomous Mental Development **6**(2), 110–125 (2014)
4. Ferreira, J.F., Lobo, J., Bessire, P., Castelo-Branco, M., Dias, J.: A Bayesian Framework for Active Artificial Perception. IEEE Transactions on Cybernetics (Systems Man and Cybernetics, part B) **43**(2), 699–711 (2013)
5. Kuhn, B., Schauerte, B., Kroschel, K., Stiefelhagen, R.: Multimodal saliency-based attention: a lazy robot's approach. In: 2012 IEEE/RSJ International Conference on Intelligent Robots and Systems (IROS), pp. 807–814. IEEE (2012)
6. Lanillos, P., Ferreira, J.F., Dias, J.: Evaluating the influence of automatic attentional mechanisms in human-robot interaction. In: 9th ACM/IEEE International Conference on Workshop: A Bridge Between Robotics and Neuroscience Workshop in Human-Robot Interaction, Bielefeld, Germany, pp. 1–2, March 2014
7. Lanillos, P., Ferreira, J.F., Dias, J.: Designing an artificial attention system for social robots. In: 2015 IEEE/RSJ International Conference on Intelligent Robots and Systems (IROS). IEEE (2015) (to appear)

8. Lanillos, P., Ferreira, J.F., Dias, J.: Multisensory 3D saliency for artificial attention systems. In: 3rd Workshop on Recognition and Action for Scene Understanding (REACTS), 16th International Conference of Computer Analysis of Images and Patterns (CAIP), pp. 1–6 (2015)
9. Murphy-Chutorian, E., Trivedi, M.M.: Head pose estimation in computer vision: A survey. IEEE Transactions on Pattern Analysis and Machine Intelligence **31**(4), 607–626 (2009)
10. Nagai, Y.: Joint attention development in infant-like robot based on head movement imitation. In: Proc. Third Int. Symposium on Imitation in Animals and Artifacts (AISB 2005), pp. 87–96. Citeseer (2005)
11. Santos, L., Christophorou, C., Christodoulou, E., Samaras, G., Dias, J.: Development strategy of an architecture for e-health personalised service robots. IADIS International Journal on Computer Science and Information Systems **9**, 1–18 (2014)
12. Scassellati, B.: Theory of mind for a humanoid robot. Autonomous Robots **12**(1), 13–24 (2002)
13. Schillaci, G., Bodiroža, S., Hafner, V.V.: Evaluating the effect of saliency detection and attention manipulation in human-robot interaction. International Journal of Social Robotics **5**(1), 139–152 (2013)
14. Siddon, R.L.: Fast calculation of the exact radiological path for a three-dimensional CT array. Medical Physics **12**(2), 252–255 (1985)
15. Thibaudeau, C., Leroux, J.D., Fontaine, R., Lecomte, R.: Fully 3D iterative CT reconstruction using polar coordinates. Medical Physics **40**(11), 111904 (2013)
16. Vernon, D., von Hofsten, C., Fadiga, L.: A roadmap for cognitive development in humanoid robots, vol. 11. Springer (2010)
17. Vinciarelli, A., Pantic, M., Heylen, D., Pelachaud, C., Poggi, I., D'Errico, F., Schroeder, M.: Bridging the gap between social animal and unsocial machine: A survey of social signal processing. IEEE Transactions on Affective Computing **3**(1), 69–87 (2012)

Part IX
Surgical Robotics

Design of a Realistic Robotic Head Based on Action Coding System

Samuel Marcos, Roberto Pinillos, Jaime Gómez García-Bermejo and Eduardo Zalama

Abstract In this paper, the development of a robotic head able to move and show different emotions is addressed. The movement and emotion generation system has been designed following the human facial musculature. Starting from the Facial Action Coding System (FACS), we have built a 26 actions units model that is able to produce the most relevant movements and emotions of a real human head. The whole work has been carried out in two steps. In the first step, a mechanical skeleton has been designed and built, in which the different actuators have been inserted. In the second step, a two-layered silicon skin has been manufactured, on which the different actuators have been inserted following the real muscle-insertions, for performing the different movements and gestures. The developed head has been integrated in a high level behavioural architecture, and pilot experiments with 10 users regarding emotion recognition and mimicking have been carried out.

Keywords FACS · Mechatronic head · Emotional expressions · Pilot study

1 Introduction

The development of robots with social skills is one of the challenges to which more efforts has been devoted in the research centers, during the last years. These robots should be characterized by the ability of transmitting visual information through gestures, in order to achieve a natural human-to-robot communication similar to the human-to-human one.

S. Marcos(✉) · R. Pinillos
Fundación Cartif., Parque Tecnológico de Boecillo Parcela 205, 47011 Boecillo, Valladolid, Spain
e-mail: {sammar,robpin}@cartif.es

J.G. García-Bermejo · E. Zalama
Instituto de las Tecnologías de la Producción, University of Valladolid,
Paseo del Cauce 59, 47011 Valladolid, Spain

L.P. Reis et al. (eds.), *Robot 2015: Second Iberian Robotics Conference*,
Advances in Intelligent Systems and Computing 418,
DOI: 10.1007/978-3-319-27149-1_33

However, researchers have not reached a general consensus about what a robot's appearance should be. There is a large number of particular complications associated to the design of more realistic social robots [2], and although a human appearance may improve the interaction so to overcome the limitations imposed by the Uncanny Valley Theory [3], there are always inconveniences associated with the behavioral expectations that an anthropomorphic robot creates in the users [4].

Regarding the degree of realism expected for a humanoid robot, in [6] the need to use more sophisticated ways to express emotions in virtual environments was denoted. Also, the results presented in [7] show that little expressive agents had more difficulties to attain an effective interaction. On the other hand, in [8] simple drawings of faces were able to create emotional feelings on the user. The face is a highly expressive element that humans tend to interpret in face to face communication, so highly realistic artificial faces may confuse the user if they are not designed properly [7]. According to [9], there is evidence that the use of exaggerated cartoonish faces can express emotions more accurately than realistic ones.

Since the emerge of humanoid robots, the use of unrealistic or cartoonish-like humanoid robots has been justified in terms of avoiding well-known "uncanny valley". However, different studies show that user preferences in terms of appearance and behavior are not universal and may be evolving as the technology spreads. These preferences may differ depending on cultural and psychological aspects [10] [11] [12], as well as on age or gender (men vs. women, elderly vs. young people).

The first developments of expressive humanoid robots have been designed using a cartoonish simple aproach [16]. One of the first developments is the Kismet robot [17]. This robot is equipped whith an unrealistic face with large expressive eyes, eyebrows and mouth. Another example is the Robot WE-4R Waseda [18] which can output rich emotional expressions and behavior by using a simplistic face, waist, arms and neck. However, although these robots exhibit large expressiveness, they lack of the ability to express detailed emotional cues. It is necessary to incorporate movements that mimic human gestures in greater detail, as is evident in the development of robots with humanoid appearance, like the android HRP-4C [19] presented by the National Institute of Advanced Industrial Science and Technology (AIST) of Japan, or the DER2 and DER3 projects constructed by Osaka University.

There exist other approaches that try to emulate the human face structure at a greater detail. For instance, in the Geminoid robot [13] and [20] they use pneumatic actuators to display facial expressions. In [21] an interesting emotional architecture is presented, and in [22] the muscular structure of the human face is imitated with great detail. However, the obtained results either do not have a natural appearance or are very hard to implement.

The proposed approach described in this paper also looks for obtaining a realistic mechatronic face. We follow the biomechanic principles of the human face based on microexpressions. Our goal is to obtain visually realistic results, so the anatomical features of a real face are taken into account in the design of the robotic head. Our approach is similar to Hanson's work [14],[15] as the generation realistic emotional expressions is based on the human muscular model, but while Hanson's model uses 32 servo motors coupled to a mask to map the facial muscles, we have developed a

Table 1 Relevant muscles of the human face.

area	Most significant muscles
Forehead and eyebrows	Occipitalis and frontalis
Eyes and eyelids	Elevator palpebrae superioris, orbicularis oris, corrugator supercilii
Lips and mouth	Quadratus labii inferioris, Quadratus labii superioris, triangularis, zygomaticus major and minor, buccinator, orbicularis oris, caninus
Nose	Pyramidalis transverse nostril dilatator, elevator muscles, transverse
Ears	Superior and posterior auricularis

simplified model using 22 servomotors that can be engaged and disengaged easily from the facial mask, thus allowing to use different masks to the same cranial structure. Our approach also eases the integration of the robotic head with a high level control architecture developed in our lab.

2 The Human Facial Musculature

The human skull includes 14 bones to give support to the face. The facial musculature, in turn, is related directly to the different expressions the face can show. The main face muscles involved into facial gestures, along with the face areas they belong to, are shown in table 1. The synchronized movement of these muscles allows the different gestures and emotions to be performed.

2.1 FACS - Facial Action Code System

The relationship between a given muscle contraction and the corresponding shape of the face surface can be described upon the muscular anatomy. However, from the viewpoint of social interaction, the existing relationship among the facial expressions, what those expressions mean and what others can interpret from them should be established. To this end, social psychology researchers have addressed the development of standards for coding and parameterizing the facial movements objectively. This way, emotions and facial expressions can be linked upon specific patterns of facial movements.

Some recognised standars are: EMFACS [23], *Monadic Phases* [24], *the Maximally Discriminative Facial Movement Coding System* [25], and the *Facial Action Coding System* [26].

Table 2 Common AUs involved in the universal expressions.

Universal Expression	AUs
Joy	6+12
Sadness	1+4+15
Surprise	1+2+5+26
Fear	1+2+4+5+20+26
Anger	4+5+7+23
Disgust	9+15+16

The (*Facial Action Coding System*, FACS) [26] is one of the most widespread standard used for measuring and describing the face behavior.

The FACS is structured into *Action Units*, AU. An AU is a contraction or relaxation of a given set of muscles that results into a change in the face appearance. The FACS is not based on the activity of isolated muscles because certain muscles may actuate more than once during the production of a given gesture. Moreover, the face appearance changes are weakly dependent on the movement of an isolated muscle. There are more than 70 AUs involved in the movement of a human head and face. 46 AUs are linked specifically to the contraction of facial muscles, and 30 of them are linked to the contraction of specific face muscles: 12 for the upper face and 18 for the lower face.

According to Ekman, there are 6 universal expresions: happiness, disgust, sadness, anger, fear and surprise. Each expression is defined (although not completely, as we will discuss later) by a given number of AUs. The universal expresions and the most common AUs associated are listed in table 2.

3 Mechanical Design

3.1 Full Model

Figure 1 shows side and front views of the full model. A silicon skin is then added to this model, to complete the realistic look of the robotic head. For a complete description of the mechanical structure of the head please see [28].

3.2 Face Mask

First, a plaster model of the desired head has been created. As our robot is intended to interact with aged people, a mature aged face has been selected. In addition, the wrinkles ease performing the different gestures. Then, a negative mold has been generated that will be used for producing the outer surface of the mask. A mold corresponding to the inner surface of the mask has been also generated. This mold provides de required thickness to the different areas of the mask according to the

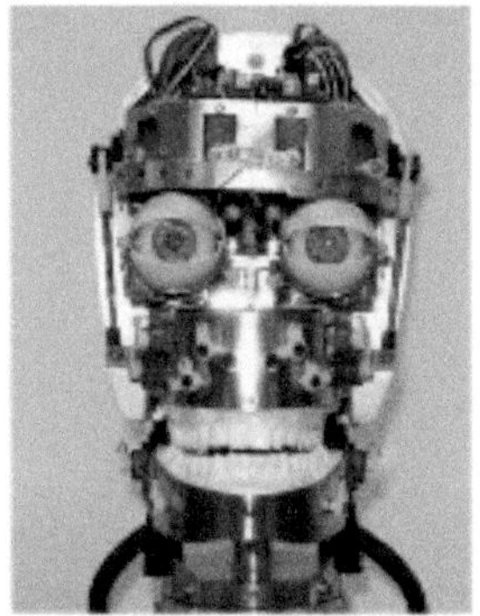
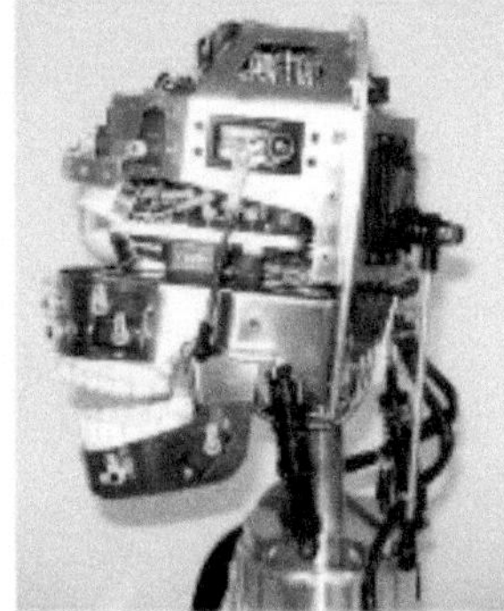

Fig. 1 Side and front view of the head.

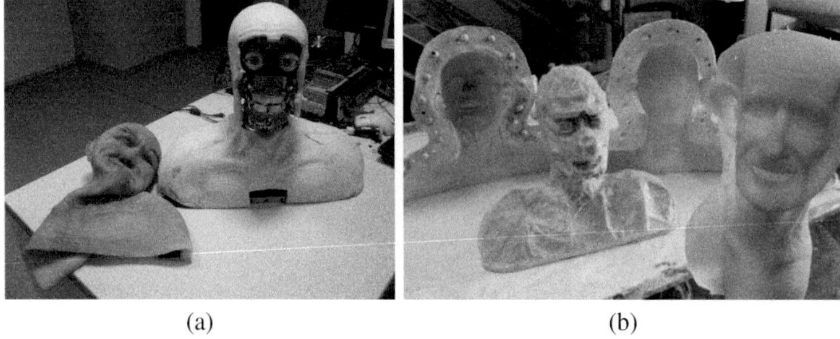

(a) (b)

Fig. 2 (a). Head with mask underlayer and silicon mask. (b) Molds employed for the mask construction.

envisaged behavior of the skin during the servomotor actions. Both outer and inner molds can be seen in figure 2.

Figure 3 shows the final look of the head, with the skin on, after the painting the outer surface and including additional elements such as eyebrows, eyelashes and facial hair. An intermediate foam layer has been added between the mechanical structure and the mask in order to fill the structure gaps and facilitate the skin movements (see figure 2).

4 Emotion Generation

The prominent biologist Charles Darwin was the first to suggest that the expressions corresponding to the main primary emotions can be easily perceived and categorized into different gesture patterns performed by the human face [1]. Following this work, Ekman [26] postulated that there are six universal expressions corresponding to

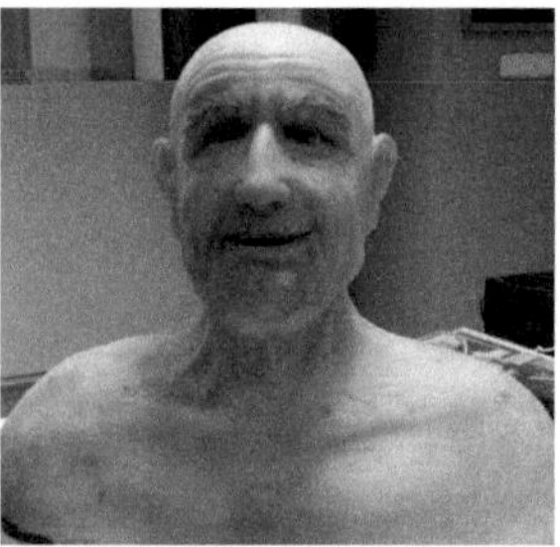

Fig. 3 Final look of the head with the silicon mask.

the six primary emotions expressions: happiness, disgust, sadness, anger, fear and surprise. In the present work we have modeled the six emotional expressions as a linear combination of the involved AUs. Each AU j is given a $weight$ value that defines its level of activation in the current gesture, according to the desired intensity of this gesture. Moreover, a gain term K_i is added which defines the intensity of the overall gesture.

$$ComplexExpression_i = K_i \sum(weight_j * AU_j) \tag{1}$$

Figure 4 shows the basic emotional expressions performed by the robot. The generation of the happiness expression involves the coordinated activation of AUs UA6+UA 12+UA25+UA26 that correspond to lifting the cheeks, pulling up the mouth corners, separating the lips and lowering the jaw. The final result can be seen in figure 5.a.

For instance, the sadness expression involves the activation of AU1+AU17 +AU54+AU55+AU63 that correspond to raising the inner part of the eyebrows, raising the chin and tilting forward the head and turning it left and moving the eyes up. (Figure 5.b).

The angriness expression shown involves the activation of AU4+AU7+AU10 +AU16+AU55+AU63 that correspond to lowering the eyebrows, narrowing the eyelid aperture, raising the upper lip, depressing the lower lip, tilting forward the head and slightly moving the eyes up. (Figure 5.c).

5 Integration in a Multimodal Architecture

Currently there is a growing interest in modular architectures and component frameworks that encourage the creation of easily modifiable and scalable open systems. This kind of initiative is of particular interest in the field of mechatronics and service robotics, where systems are specifically designed for high level interactions and sought to be easily modified and improved to meet new requirements. The main trend these days in terms of development of robotic systems is the design of architectures

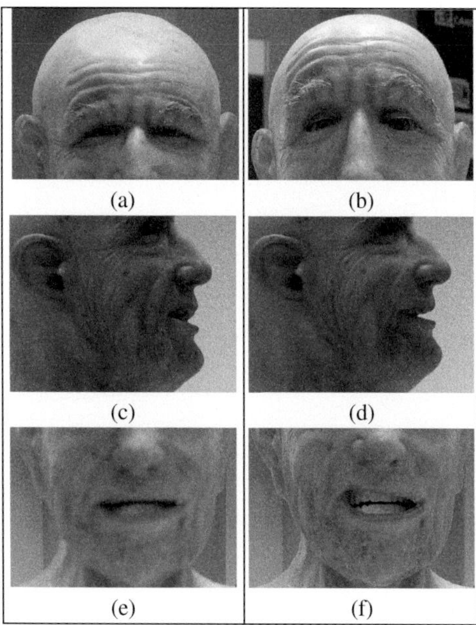

Fig. 4 Example of AUs
(a) Neutral (b) AU4 (c) Neutral (d) AU12 (e) Neutral (f) AU5

based on components or frameworks. For this purpose rules for combining components and the interaction patterns between them are defined in order to achieve a flexible system. This flexibility is understood mainly in the following aspects:

- Easy replacement of components with different functionality.

- Reuse of different modules in other projects where the functionality of those particular modules is required.

The potential of the open modular architectures is huge, and leads to reconfigurable, portable and extensible systems. This philosophy has been adopted to develop the control software of different social robots in our laboratory. Based in the ROS framework, it consists of a set of layers with different interchangeable modules (see Figure 6). The interaction between layers and modules is well defined, so a new module can easily be integrated. It has to be noted that this architecture has not only been used in robots, but also in other agents such as animated faces [5]. Thanks to the AU based emotional generation approach, both animated and robotic faces can be controlled using the same upper level components.

Making use of this architecture, the mechatroinc head and its low level control drivers have been integrated within the down layer, taking advantage of the upper layer services such as voice recognition, expression recognition or behavioural planner.

As can be seen in the video linked to this article (see Section 7 below), as an example the robotic head has been easily connected to our previously developed facial expression recognition system. This system si able to perform both: facial

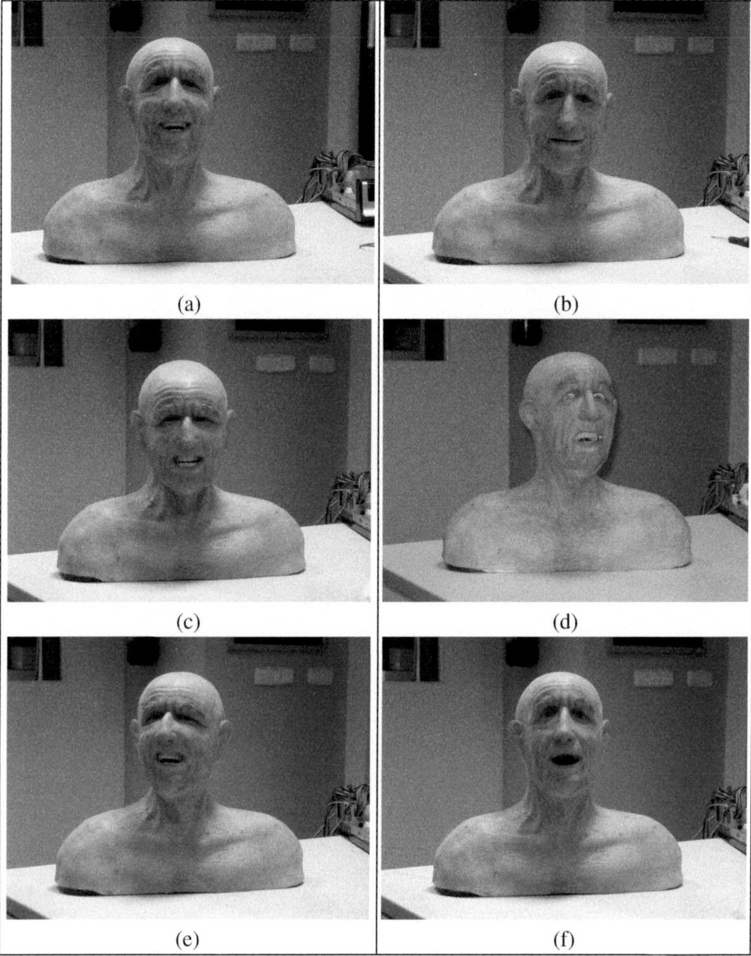

Fig. 5 Basic Emotions
(a) Joy (b) Sadness (c) Anger (d) Fear (e) Disgust (f) Surprise

expression recognition and facial emotion recognition. Expression recognition refers to the detection and analysis of facial movements, and changes in facial features from visual information. On the other hand, emotional expression recognition also associates these facial displacements to their emotional significance, given that there will be certain movements associated with more than one emotion. The proposed method [5] is divided into two different stages: the first one focuses on the detection and recognition of small groups of facial muscle actions using a combination of Gabor filters and Active Shape Models. Based on the output of this first stage, the second determines the associated emotional expression using an habituation based network plus a competitive based network. Thanks to the implemented control architecture, it

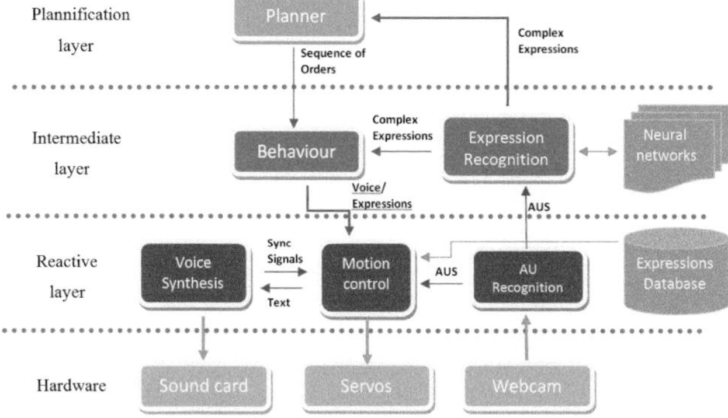

Fig. 6 Modular multimodal architecture

can be observed in the video how the mechatronic head is able to mimic the emotional expressions of the user using a regular webcam.

6 Pilot Experiments in Emotion Recognition

The mechatronic head expression generation capabilities were tested in a pilot study with 10 participants which ages ranged between 28 and 40 years old. In order to see the influence of the intensity parameter k_i (see ecuation 1) in the generation of recognisable expressions, users were asked to recognize the six universal expressions under two conditions: half- intensity ($k_i = 0.5$), and full-intensity ($k_i = 1$). Expressions were randomly generated for each condition and users were asked to anotate which expressions they were able to recognize, as well as the order the mechatronic head had displayed them. The subjects' answers in labeling an expression were scored as correct or wrong and results are shown in figure 7.

From the results it can be observed that the intensity has a great impact in the recognition of the majority of the expressions, specially in the anger expression. Apart from the joy expression, a greater intensity results in a higher recognition rate, although in expressions such as sadness or disgust the difference is not highly significant. It has to be noted that figure 7 just shows the correct answer ratio and not those answers in which users doubted between two expressions. This happened between the anger and disgust expressions and between the fear and disgust expressions. Users found it difficult to distinguish between them, and labeled the shown expressions as both anger and disgust or fear and surprise. It can be seen that this confusion is even more accentuated in the case of the half-intensity expressions. It has to be noted, however, that the small number of participants make some of the obtained results far from being significant. For instance, recognition rates of 3 out

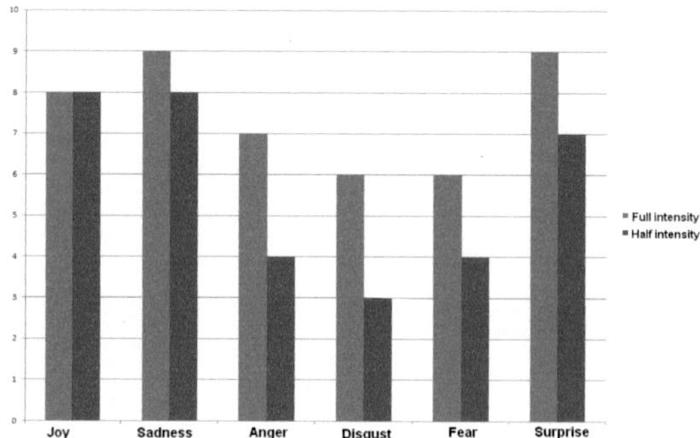

Fig. 7 Recognition rate out of ten for the six universal expressions

of 10 are close to randomness meaning that users were no able to recognize the expression. In order to that obtained results are in fact due to a lack of expressivity of our robotic head, a bigger experiment with more samples should be performed.

7 Video

A video of the head construction, and the head performing different gestures and mimicking can be seen at:

http://www.eii.uva.es/%7Eeduzal/videos/HeadMask.mp4

8 Conclusions

In this paper a methodology for developing a robotic head with large realism and expressiveness has been presented. Many researchers have suggest that the development of simplified facial animation models, even further away from human appearance in order to avoid creating false expectations on the interaction ability of the robots, should be preferred. Moreover, this would allow avoiding the adverse implications of the Mori's uncanny valley theory. However, the development of realistic robots that create the illusion of life is a challenge that contributes to raising the emotional communication ability of robots over that of simplified models.

We have addressed the development of a design methodology of realistic robotic heads, that is based on the neuromuscular human system and the FACS. On the one hand, we have undertaken a deep study of the different AUs and an analysis on which AUs are the most relevant for generating movement and emotions. This research has resulted into a refined model with 30 AUs, which can be implemented physically. On the other hand, we have designed an arrangement of servomotors that allow the facial muscles to be emulated. Also, a silicon skin has been designed and manufactured, on which the different actuators has been inserted in a muscle-insertion way, after a detailed analysis of the human anatomy. Also, the presence of wrinkles in this skin provides increased expressiveness.

An outstanding characteristic of the proposed design is that the skull, with the corresponding actuators and the associate electronics, is a separate element. In particular, the silicon skin can be easily removed away from the mechanical structure, which simplifies the assembly/ disassembly operations and thus, the robot tuning and subsequent maintenance. Even, silicon masks other than that designed in the present work could be mounted onto the same mechanical structure.

The current research work opens up some important horizons. One future research line will consist in evaluating more extensively how accurately (and pleasantly) the robot emotions are perceived by humans. Also, as there is evidence that men and women display differences in both cognitive and affective functions [29], our future experimentation should take into account those individual differences. Recent studies have examined the processing of emotions in males and females, Another interesting next-future line is to evaluate the communication skills of the robot with respect to previous researches that use virtual avatars [30] or non-realistic mechatronic heads [31].

References

1. Darwin, C.: The Expression of Emotions in Man and Animals. (872). John Murray, reprinted by University of Chicago Press (1965)
2. Duffy, B.R.: Anthropomorphism and the social robot. Robotics and Autonomous Systems **42**, 177–190 (2003)
3. Mori, M.: The uncanny valley. Energy **7**(4), 33–35 (1970)
4. Dautenhahn, K., Werry, I.: Towards interactive robots in autism therapy: background, motivation and challenges. Pragmatics and Cognition **12**(1), 1–35 (2004)
5. Marcos, S., Gmez, J., Zalama, E., Lpez, J.: Dynamic Facial Emotion Recognition Oriented to HCI Applications. Interacting with Computers **27**(2), 99–119 (2015)
6. Saugis, G., Chaillou, C., Degrande, S., Viaud, M.L., Dumas, C: A 3-D interface for cooperative work. In: Proceedings of the Conference on Collaborative Virtual Environments, pp. 17–19 (1998)
7. Paiva, A., Machado, I.: The child behind the character. IEEE Journal on Systems Man and Cybernetics, Part A **31**(5), 361–368 (2001)
8. Nass, C., Reeves, B.: The Media Equation, How People Treat Computers, Television and New Media Like Real People and Places. Cambridge University Press (1996)
9. Bartneck, C.: An Embodied Emotional Character for the Ambient Intelligent Home. Ph.D. thesis, Technische Universiteit Eindhoven (1982)
10. Koda, T., Ishida, T.: Cross-cultural study of avatar expression interpretations. In: International Symposium on Applications on Internet, pp. 130–136 (2006)
11. Deng, Z., Hiscock, M., Yun, C.: Can local avatars satisfy a global audience? A case study of high-fidelity 3D facial avatar animation in subject identification and emotion perception by us and international groups. Computers in Ententairment **7**(2), 1–26 (2009)
12. Kanda, T., Ishiguro, H., Hagita, N., Bartneck, C: Is the uncanny valley an uncanny cliff? In: Proceedings of the 16th IEEE International Symposium on Robot and Human Interactive Communication, RO-MAN 2007, pp. 368–373 (2007)

13. Nishio, S., Hiroshi Ishiguro, I., Hagita, N.: Humanoid Robots, New Developments, pp. 343–352. I-Tech Education and Publishing (2007)
14. Mazzei, D., Lazzeri, N., Hanson, D., De Rossi, D.: Hefes: an hybrid engine for facial expressions synthesis to control human-like androids and avatars. In: 4th IEEE RAS and EMBS International Conference on Biomedical Robotics and Biomechatronics (BioRob), pp. 195–200 (2012)
15. Hanson, D.F.: U.S. Patent No. 8,594,839. Washington, DC: U.S. Patent and Trademark Office (2013)
16. Beira, R., Lopes, M., Praga, M., Santos-Victor, J., Bernardino, A., Metta, G., Becchi, F., Saltaren, R.: Design of the robot-cub (iCub) head. IEEE International Conference on Robotics and Automation **51**, 94–100 (2006)
17. Breazeal, C.: Sociable Machines: Expressive Social Exchange Between Humans and Robots. Ph.D. thesis, MIT (2000)
18. Itoh, K., Miwa, H., Nukariya, Y., Zecca, M., Takanobu, H., Roccella, S., Carrozza, M. C., Dario, P., Takanishi, A. : Mechanisms and functions for a humanoid robot to express human-like emotions. In: Proceedings of the 2006 IEEE International Conference on Robotics and Automation (2006)
19. Kaneiro, F., et. al.: Cybernetic human HRP-4C: A humanoid robot with human-like proportions. Springer Tracts in Advanced Robotics, vol. 70, pp. 301–314 (2011)
20. Kobayashi, H., Hashimoto, T., Hiramatsu, S.: Development of face robot for emotional communicaion between human and robot. In: Proceedings of IEEE International Conference on Mechatronics and Automation, vol. 54, pp. 25–30 (2006)
21. Berns, K., Hirth, J., Schmitz, N.: Emotional architecture for the humanoid robot head roman. In: Proceedings of IEEE International Conference on Mechatronics and Automation, vol. 53, pp. 2150–2155 (2007)
22. Sadoyama, T., Sugahara, T., Hashimot, M., Yokogawa, C.: Development of a face robot imitating human muscle structures. Journal of Robotics and Mechatronics **19**, 324–330 (2007)
23. Friesen, W.V., Ekman, P.: Facial Action Coding System. Consulting Psychologists Press, Palo Alto (1978)
24. Tronick, E., Als, H., Brazelton, T.B.: Monadic phases: A structural descriptive analysis of infant-mother face to face interaction. Merrill-Palmer Quarterly of Behavior and Development **26**(1), 3–24 (1980)
25. Izard, C.: Innate and universal facial expressions: Evidence from developmental and cross-cultural research. American Psychological Association **115**, 288–299 (1994)
26. Ekman, P., Friesen, W.V.: Facial action coding system: A technique for the measurement of facial movement. Consulting Psychologists Press (1978)
27. Marcos, S., Gómez-García-Bermejo, J., Zalama, E.: A realistic, virtual head for human-computer interaction. Interacting with Computers **22**, 176–192 (2010)
28. Loza, D., Marcos, S., Gómez-García-Bermejo, J., Zalama, E.: Application of the FACS in the Design and Construction of a Mechatronic Head with Realistic Appearance. Journal of Physical Agents **7**(1), 30–37 (2013)
29. Montagne, B., Kessels, R.P.C., Frigerio, E., de Haan, E.H.F., Perrett, D.I.: Sex differences in the perception of affective facial expressions: Do men really lack emotional sensitivity? Cognitive Processing **6**(2), 136–141 (2015)
30. Pierce, B., Kuratate, T., Vogl, C., Cheng, G.: Mask-Bot 2i: An active customisable Robotic Head with Interchangeable Face. In: 12th IEEE-RAS International Conference on Humanoid Robots, pp. 520–525 (2012)
31. Cid, F., Moreno, J., Bustos, P., Nez, P.: Muecas: A Multi-Sensor Robotic Head for Affective Human Robot Interaction and Imitation. Sensors **14**(5), 7711–7737 (2014)

A Comparison of Robot Interaction with Tactile Gaming Console Stimulation in Clinical Applications

Jainendra Shukla, Julián Cristiano, Laia Anguera, Jaume Vergés-Llahí and Domènec Puig

Abstract Technological advancements in recent years have encouraged lots of research focus on robot interaction among individuals with intellectual disability, especially among kids with Autism Spectrum Disorders (ASD). However, promising advancements shown by these investigations, about use of interactive robots for rehabilitation of such individuals can be questioned on various aspects, e.g. is effectiveness of interaction therapy because of the robot itself or due to the sensory stimulations? Only few studies have shown any significant comparison in remedial therapy using interactive robots with non-robotic visual stimulations. In proposed research, authors have tried to explore this idea by comparing response of robotic interactions with stimulations caused by a tactile gaming console, among individuals with profound and multiple learning disability (PMLD). The results show that robot interactions are more effective but stimulations caused by tactile gaming consoles can significantly serve as complementary tool for therapeutic benefit of patients.

Keywords Robot interaction · Profound and multiple disability · PMLD · Tactile gaming console · Non-robotic stimulation · ARMONI

1 Introduction

Development of intellectual and social functioning skills of individuals with profound & multiple learning disability (PMLD) is severely slow which restrains their communication ability. They can express themselves only with the help of a

J. Shukla(✉) · J. Cristiano · J. Vergés-Llahí · D. Puig
Intelligent Robotics and Computer Vision Group, Universitat Rovira i Virgili, Tarragona, Spain
e-mail: jainendra.shukla@estudiants.urv.cat

J. Shukla · L. Anguera · J. Vergés-Llahí
Instituto de Robótica para la Dependencia, Sitges, Spain
e-mail: jshukla@institutorobotica.org

L.P. Reis et al. (eds.), *Robot 2015: Second Iberian Robotics Conference*,
Advances in Intelligent Systems and Computing 418,
DOI: 10.1007/978-3-319-27149-1_34

familiar person. Moreover, often associated medical conditions in form of neurological problems, and physical or sensory impairments limits their optimum potential and they require continuous customized assistance [1].

Around one-fourth of the world population is being affected by mental or behavioral disorders at certain stages of their life [2]. More than 780 million children, under the age of 15, are affected worldwide by certain type of intellectual disabilities [3]. It is also worth noticing that clinical experts observed more than one type of disability among 20% of all the intellectually disabled patients worldwide [2]. It is estimated that global burdens caused by these disabilities will increase to 15%, by 2020 [2]. Treatment of intellectual disabilities demands a huge physical, economical and emotional support by worldwide community.

Although much of the earlier research has been centered in patients with autism spectrum disorders (ASD) (e.g., [4, 5, 6, 7, 8]), new learning scenarios allowing the interaction between patients with PMLD and non robotics platforms (e.g., [9, 10, 11, 12, 13]) or robotics platforms (e.g., [14, 15]) are being studied. In [10], interactive tactile games were developed in IPad to promote independent living for intellectually disabled people. A system to help children with Intellectual Disability to understand healthy eating habits with an IPad based game is proposed in [11]. In the case of the robotic platforms, a humanoid robot is commonly used to directly interact with the patient in a different scenario. In [14], specific learning objectives were identified and a NAO humanoid robot was programmed to help teachers to achieve these objectives among students with PMLD. The aim of these new learning scenarios is to foster personalized therapies involving the use of new technologies taking advantage of the positive impact of these in the patient's response.

Previous researches in patients with ASD suggest that non-robotic stimulations should be used and compared with robot interactions in identical therapeutic programs, to analyze the effect of humanoid appearance and motion of the robot along with its audio and visual stimulation abilities [4]. In our previous research, we presented a case study of robot interaction among individuals with PMLD in different possible categories of clinical applications [16]. Results from this study suggested that robotic interactions can help to induce target behavior, to teach and to encourage individuals with PMLD. Above findings have inspired the proposed research to make an ideal comparison of robot interactions with non-robotic stimulations caused by ARMONI. ARMONI is a computer based tactile gaming console, developed to rehabilitate different cognitive functions among individuals with intellectual disabilities [12]. Figure 1 shows the ARMONI workstation. ARMONI activities have been found strongly associated with different aspects of cognition among individuals with intellectual disabilities.

Specifically, robot interactions in different categories of clinical applications have been compared in same program with stimulations caused by tactile gaming console, ARMONI. This will enable better-decision making about use of interactive robots in therapeutic benefits. Rest of the paper is organized as follows. Section 2 discusses the method for this research, explaining details of experiments and participants. Section 3 presents and discusses the findings of this work. Finally, section 4 concludes the paper by summarizing this work and highlighting the future directions.

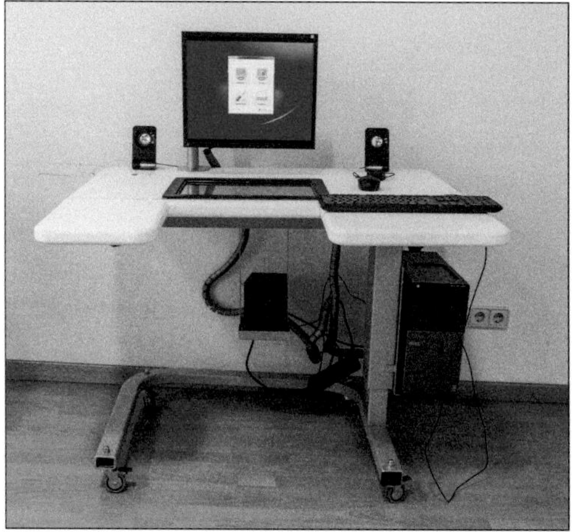

Fig. 1 ARMONI Workstation

2 Method

2.1 Approach

The case study was performed over a period of four months at a trial room in Ave Maria Foundation [1]. Ave Maria Foundation is the residential and clinical facility of the participants hence it provides a familiar environment to all the participants. All trials have been performed satisfying the medical ethics requirements.

In earlier work, authors performed a case study of robot interactions in different possible categories of robot applications among individuals with PMLD [16]. In a similar manner, four unique activities were identified from available set of activities in ARMONI, representing similar categories as used with robot interaction study. The activities were chosen based upon following criteria:

1. Similarity with the activities used in earlier robot interaction study.
2. Relevant representation of the category by selected activity.
3. Simplicity for participants at the execution level.
4. Availability in ARMONI.

These activities are listed in table 1. Details of each activity are described as below:

1. Dance Video: This activity was aimed to observe the response of the participants towards a dance video of the robot. In this activity, a video of NAO performing a dance composition while singing a song was showed to the participants. The dance routine was exactly similar to as used in previous robot interaction study.

[1] Ave Maria Fundació, http://www.avemariafundacio.org/inici.html

Table 1 Activities

Activity used in ARMONI	Activity used in Robot Interaction Study[16]
Dance Video	Dance choreography
Identify the Sounds	Touch my head
Tale Account	Learn the senses
Coordination	Guess emotions

2. Identify the Sounds: The aim of this activity was to induce a target behavior in participants. In this activity, participants were presented with a human figure on the screen and name of the body parts are played one after another. Participants are supposed to touch the respective body parts of the figure on tactile screen.
3. Tale Account: This activity is aimed to provide feedback and encouragement to the participant to achieve a certain target behavior. ARMONI prompts the participants to touch an answer image corresponding to a specific query related to a festival which they are aware with. If the participant is not able to answer within a certain time period, they are encouraged by providing some clue about the answer. ARMONI provides a positive feedback on accomplishing the right answer.
4. Coordination: ARMONI works as a learning tool for the participant. A table of two columns, first column having different color paints and second empty, is displayed on the screen. Images of same object in different colors are also shown on the bottom of screen. Participants are required to put the matching color image in empty column of table. The activity helps participants to learn about different colors and ordering.

2.2 Participants

The experiments for this case study were carried out with same set of six individuals who participated before in robot interaction study. These individuals are of different age, gender and intellectual disability levels. The assessment of these individuals was done by Assessment and Guidance Services for People with Disabilities (CAD Badal) organization of Government of Catalunya [2]. Details of all the individuals are presented in table 2.

2.3 Procedure

Participants performed the trials in four sessions, one session for each activity. All the participants in the study were familiar with ARMONI. So no major training about

[2] Generalitat de Catalunya, http://web.gencat.cat/ca/inici/

Table 2 Participants

ID	Gender	Age	Condition	Disability (%)
ID01	F	65 y, 4 m	Moderate Intellectual Disability, Affective Disorder, Right Hemiparesis, Mixed Cerebral Palsy	85
ID02	F	42 y	Autism, Severe Intellectual Disability	86
ID03	F	48 y, 8 m	Severe Intellectual Disability, Affective Disorder, encephalopathy	87
ID04	F	33 y, 2 m	Autism, Moderate Intellectual Disability	79
ID05	F	67 y, 10 m	Moderate Intellectual Disability, Tetraparesis	86
ID06	M	44 y, 6 m	Severe Intellectual Disability, Down Syndrome	75

Fig. 2 A participant working with ARMONI

use of ARMONI was given to them. However, before undergoing the actual trial, participants were administered with an initial short training about the execution of particular activity. Trials were performed only after ensuring that participants have understood the basic activity requirements. Only one session was conducted with one participant in a day to facilitate participants and care-taker. Figure 2 shows a general position of the participant in front of ARMONI workstation. During execution of the trials, a caretaker was accompanying the participants but her involvement was restricted to minimal in order to let the participants observe and respond without any restrictions. All the sessions were recorded to further analyze the response of participants.

2.4 Measurement and Evaluation

All the trials were evaluated using similar evaluation tools as were used in previous robot interaction study to enable a fair comparison. A brief description of measurement techniques are as follows:

1. Engagement rate by calculating the percentage of time, participants were engaged with the ARMONI. Engagement of the participants was decided by analyzing the video recordings of the trials with help of an expert psychiatrist.
2. A questionnaire adapted from GARS-2 [17], WHODAS 2.0 [18] and ABS-RC: 2 [19]. Sample questions from questionnaire are as follows:
   - Does individual stare or look unhappy or unexcited when praised, humored, or entertained?
   - Does individual generally understand what people/robot say?

   Detailed description of questionnaire can be found in our previous work [16].

In previous study of robot interactions, authors also used a performance measure of the participants against completion of activities in terms of time duration and correct/incorrect responses. However, in present work this measure could not be used as participants were not able to use touch screen very effectively due to several levels of limitations caused by their disabilities.

Evaluations of these trials are compared with the recorded responses from robot interaction study.

3 Results and Discussion

Table 3 shows the observed engagement rate for all participants during all the trials.

Important observations from table 3 can be summarized as follows:

1. Activity 1 does not comprise any interaction with participants. Hence, four out of six participants were not able to engage completely while watching the video of the dancing robot. Participant ID05 observed 100% engagement rate in both

Table 3 Engagement rate (% duration) of participants observed in each activity calculated from the interactive sessions with the ARMONI (AM) and the robot interaction (RI).

	Activity 1		Activity 2		Activity 3		Activity 4	
ID	**AM**	**RI**	**AM**	**RI**	**AM**	**RI**	**AM**	**RI**
ID01	75.76	96.60	100.00	100.00	100.00	100.00	100.00	100.00
ID02	100.00	64.56	100.00	100.00	100.00	100.00	100.00	100.00
ID03	78.67	93.20	100.00	100.00	100.00	100.00	100.00	100.00
ID04	71.43	100.00	100.00	100.00	100.00	100.00	100.00	100.00
ID05	100.00	100.00	100.00	100.00	100.00	100.00	100.00	100.00
ID06	82.00	98.06	100.00	100.00	100.00	100.00	100.00	100.00

categories, dancing video and with physical robot. Four other participants (ID01, ID03, ID04, ID06) observed lower engagement rates with the dancing video of the robot than the physical robot, while average difference in engagement rates for these participants was 20.00% . *It strongly indicates that physical robots have more engaging capability than video stimulations caused by the same type of robot, when no interaction is involved with participants.* Only one participant (ID02) observed higher engagement rate with dancing video in comparison to the physical robot. Overall, average engagement rates for all the participants with ARMONI was 84.64% and with that of robot was 92.07%, hinting to expect a slightly higher engagement rate with physical robot in this category. Small difference (7.43%) in these values also suggests strongly that *when no interaction is involved between the robot and participants, video stimulations caused by the robot videos can also be helpful to a big extent.* This conclusion is highly beneficial, especially in scenarios where physical robots can not be made available due to financial or several other limitations.

2. All participants observed a fascinating 100% engagement rate in all other activities with ARMONI. It shows strong engagement capability of ARMONI among these participants. It should be noted that same participants also observed 100% engagement rate in all other activities with robot. These observations indicate that participants are able to show complete engagement with robots as well as with tactile gaming consoles, when any sort of interaction is involved with them. Hence, *considering engagement ability point of view, tactile gaming consoles (such as ARMONI) and humanoid robots (such as NAO) are equally beneficial for these type of participants when any sort of interaction is involved.*

Results in figure 3, show disability behavior for all participants as observed during ARMONI stimulations, as observed during robot interactions and as observed during normal situations (before the trials) which serve as a reference value; all evaluated using the proposed questionnaire as explained before. The scores are calculated using above questionnaire, converted to percentages for each of the three methods (GARS2, WHODAS 2.0, ABS-RC-2) and are represented as average percentage values in the plot, where average was taken over three methods. Higher percentage value indicates more disability behavior observed while lower percentage value indicates less disability behavior observed.

It is evident from figure 3 that except for two occasions, all participants showed lower disability behaviors during all trials with ARMONI and robot. Although reduction in disability behavior in activity 1 was smaller than other three categories of activities. Given that no interaction was involved in first activity between participants and ARMONI or between participants and robot, this can be easily understood that *an interactive trial with tactile gaming console and robot is more effective than non-interactive session.* Other important observations can be summarized as follows:

1. In a total of 24 observations (6 *participants* $\times$ 4 *activities*), most reduced disability behavior was observed 13 times during robot interactions, 9 times during ARMONI trials and the same percentage of reduction in disability was observed

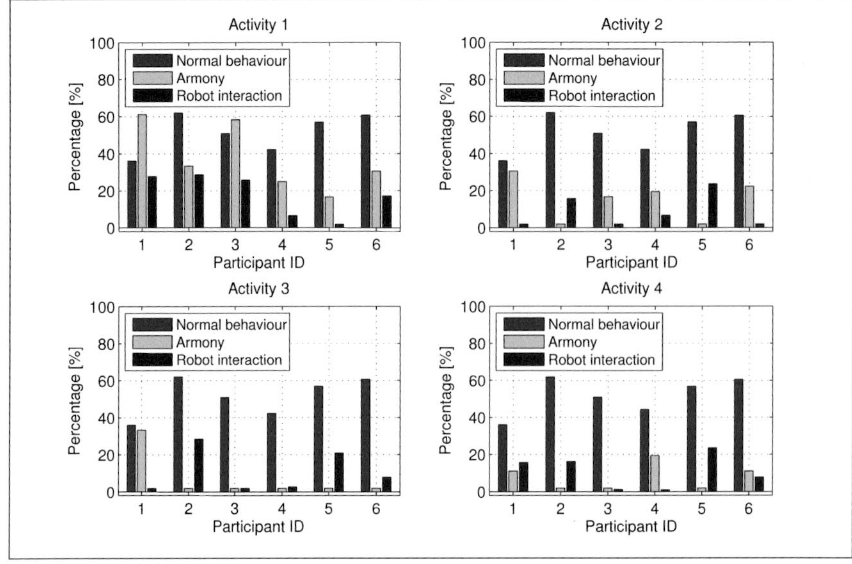

Fig. 3 Average disability percentage observed for each patient in the different scenarios.

twice among ARMONI and robot trials. This strongly indicates that *robot interactions among these participants are more effective than tactile gaming consoles in reduction of their disability behavior.*

2. Table 4 represents distribution of the lowest disability observed among ARMONI and robot interaction trials for all participants over all 4 categories of activities. * : *Same scores obtained during ARMONI and Robot interactions.* As can be observed from table 4, only twice ARMONI and robot interactions had same effect on participants else for all other twenty-two occasions the lowest disability behavior was observed 13 out of 22 times (59.09%) during robot interactions and 9 out of 22 times (40.91%) during ARMONI trials. It strongly

Table 4 Lowest Observed Disability Distribution of Participants Over All Categories

ID	Robot Interactions	ARMONI Trails
ID01	3	1
ID02	1	3
ID03	4	2*
ID04	3	1
ID05	1	3
ID06	3	1
Total	**13**	**11**

indicates that *robot interactions are more effective but at the same time ARMONI is serving as a complementary tool in therapeutic benefit of participants.*

3. The highest observed average reduction in disability behavior with ARMONI trials was 53.65% for Participant ID02.
4. The lowest observed average reduction in disability behavior with ARMONI trials was 2.08% for Participant ID01.
5. Participants ID01 and ID03 showed an increased disability behavior than normal situations during the first activity in ARMONI trials. Both patients suffer from Affective disorders, so it may indicate that *robot videos can cause adverse effect with patients suffering from Affective Disorders.*
6. An average difference of disability behavior between normal situations and ARMONI trials over all activities was 35.27%, which suggests about effectiveness of ARMONI trials among these types of patients.
7. An average reduction of 35.27% in the disability behavior with ARMONI trials vs. an average reduction of 39.94% in the disability behavior with robot trials, itself indicates that robot interactions among these type of participants are more effective.

Table 5 shows the average percentage of disability for all participants among normal situations, during ARMONI trials and during robot interactions over all four activities. All the values are calculated using proposed questionnaire method. Higher value represents higher observed disability.

Table 5 Average Percentage of Disability of Participants in Different Categories

ID	Disability (%): Normal Situation	Disability (%): ARMONI Trails	Disability (%): Robot Interactions
ID01	36.11	34.03	10.88
ID02	61.98	8.33	22.29
ID03	50.88	18.75	6.75
ID04	42.22	15.97	4.03
ID05	56.94	4.17	17.01
ID06	60.65	15.97	8.24

As can be noticed from table 5, four out of six participants performed better during robot interactions as compared to ARMONI trials. Participants who performed better with ARMONI trials are ID02 and ID05. Both of these participants have higher percentages (both 86%) of disability as can be seen from table 2. This might suggest that *Participants with higher disability level interact better with tactile gaming consoles than robot*. One important point to be noted is that individuals with higher disability levels have more difficulty in making an interaction. Hence, *robots need improved and advanced interaction abilities to exercise functional capabilities of the individuals with higher level of disabilities.*

4 Conclusion and Future Work

In order to effectively analyze the therapeutic effects of robot interactions among individuals with PMLD, a program with same participants and categories was created. Participants were exposed to robot interaction and tactile gaming console sessions in the same category of activities. Performance of the participants was evaluated using same methods for both type of trials and their disability behavior during normal situations was used as a ground truth for comparison purposes. A questionnaire using GARS-2, WHODAS 2.0 and ABS-RC2 was proposed solely to enhance the decision making about such trials. Detailed analysis of results indicated that by using robot interactions and tactile gaming console, a significant reduction in disability behavior among individuals with PMLD can be achieved. While robot interactions seem to be more effective than tactile gaming consoles but they serve better as a complementary tool than an alternative tool.

Robotic interactions and non-robotic stimulations, both offer unique advantages. It will be interesting to explore whether participants are able to carry forward advantages earned from one method to another. More importantly, how rehabilitation achieved among these individuals from these tools can be generalized to normal life with common people. Effectiveness determination of these therapies in improvement of specific cognitive abilities can help to provide customized treatment.

Acknowledgements This research work has been supported by the Industrial Doctorate program (Ref. ID.: 2014-DI-022) of AGAUR, Government of Catalonia. The authors gratefully acknowledge cooperation of participants and their guardians in this research.

References

1. Bellamy, G., Croot, L., Bush, A., et al.: A study to define: profound and multiple learning disabilities (PMLD). Journal of Intellectual Disabilities **14**(3), 221–235 (2010)
2. World Health Organization (2001). The World Health Report 2001 - Mental Health: New Understanding, New Hope. Geneva (2001)
3. Olness, K.: Effects on Brain Development Leading to Cognitive Impairment: A Worldwide Epidemic. Journal of Developmental & Behavioral Pediatrics **24**(2), 120–130 (2003)
4. Diehl, J.J., Schmitt, L.M., Villano, M., et al.: The Clinical Use of Robots for Individuals with Autism Spectrum Disorders: A Critical Review. Research in Autism Spectrum Disorders **6**(1), 249–262 (2012)
5. Davis, M., Robins, B., Dautenhahn., et al.: A Comparison of Interactive and Robotic Systems in Therapy and Education for Children with Autism (2013)
6. Warren, Z., Zheng, Z., Das, S., et al.: Brief Report: Development of a Robotic Intervention Platform for Young Children with ASD. Journal of Autism and Developmental Disorders, pp. 1–7 (2014). ISSN 0162–3257
7. Wainer, J., Robins, B., Amirabdollahian, F., et al.: Using the Humanoid Robot KASPAR to Autonomously Play Triadic Games and Facilitate Collaborative Play Among Children With Autism. IEEE Trans. on Autonomous Mental Development **6**(3), 183–199 (2014)

8. Robins, B., Dautenhahn, K.: Tactile Interactions with a Humanoid Robot: Novel Play Scenario Implementations with Children with Autism. International Journal of Social Robotics **6**(3), 397–415 (2014)
9. Miesenberger, K., Klaus, J., Zagler, W., et al.: User interface evaluation of serious games for students with intellectual disability. In: Lecture Notes in Computer Science, vol. 6179, pp. 227–234 (2010)
10. Lopez, A., Méndez, A., García, B., et al.: Serious games to promote independent living for intellectually disabled people: starting with shopping. In: Computer Games: AI, Animation, Mobile, Multimedia, Educational and Serious Games, CGAMES, pp. 1–4 (2014)
11. Rodríguez, A., Lopez, A., Méndez, A., et al.: Helping children with intellectual disability to understand healthy eating habits with an ipad based serious game. In: 2013 18th International Conference on Computer Games: AI, Animation, Mobile, Interactive Multimedia, Educational and Serious Games (CGAMES), pp. 169–173 (2013)
12. Salazar, C.P., Maldonado, J.G., García, M.M.F., et al.: Cognitive mechanisms underlying ARMONI: A computer-assisted cognitive training program for individuals with intellectual disabilities. Anales de Psicología (2015, to appear)
13. Schelhowe, H., Zare, S.: Intelligent mobile interaction: a learning system for mentally disabled people (IMLIS). In: Universal Access in Human-Computer Interaction. Addressing Diversity. Lecture Notes in Computer Science, vol. 5614, pp. 412–421 (2009)
14. Standen, P., Brown, D., Roscoe, J., et al.: Engaging students with profound and multiple disabilities using humanoid robots. In: Stephanidis, C., Antona, M. (eds.) Universal Access in Human-Computer Interaction. Universal Access to Information and Knowledge. LNCS, vol. 8514, pp. 419–430. Springer (2014)
15. Standen, P., Brown, D.J., Hedgecock, J., et al.: Adapting a humanoid robot for use with children with profound and multiple disabilities. In: 10th Int. Conf. Disability, Virtual Reality and Associated Technologies, pp. 205–211 (2014)
16. Shukla, J., Cristiano, J., Amela, D., et al.: A case study of robot interaction among individuals with profound and multiple learning disabilities. In: Tapus, A., Vincze, M., Martin, J.-C., André, E. (eds.) ICSR 2015. LNAI, vol. 9388, pp. 1–10. Springer, Heidelberg (2015)
17. Gilliam, J.E.: GARS-2: Gilliam Autism Rating Scale. Jour. of Psychoeducational Assessment **26**(4), 395–401 (2006). Second Edition. PRO-ED, Austin
18. Gold, L.H.: DSM-5 and the Assessment of Functioning: The World Health Organization Disability Assessment Schedule 2.0 (WHODAS 2.0). The Journal of the American Academy of Psychiatry and the Law **42**(2), 173–181 (2014)
19. Nihira, K., Leland, H., Lambert, N.M., et al.: ABS-RC:2: AAMR Adaptive Behavior Scale: residential and community, 2nd edn. Pro-Ed, Austin (1993)

Part X
Urban Robotics

Real-time Application for Monitoring Human Daily Activity and Risk Situations in Robot-Assisted Living

Mário Vieira, Diego R. Faria and Urbano Nunes

Abstract In this work, we present a real-time application in the scope of human daily activity recognition for robot-assisted living as an extension of our previous work [1]. We implemented our approach using Robot Operating System (ROS) environment, combining different modules to enable a robot to perceive the environment using different sensor modalities. Thus, the robot can move around, detect, track and follow a person to monitor daily activities wherever the person is. We focus our attention mainly on the robotic application by integrating several ROS modules for navigation, activity recognition and decision making. Reported results show that our framework accurately recognizes human activities in a real time application, triggering proper robot (re)actions, including spoken feedback for warnings and/or appropriate robot navigation tasks. Results evidence the potential of our approach for robot-assisted living applications.

1 Introduction

Mobile robots endowed with cognitive skills are able to help and support humans in an indoor environment, providing increased availability, awareness and access, as compared to static systems. Thus, a robot can act not only as assistant in the context of robot-assisted living, but also offer social and entertaining interaction experiences between humans and robots. For that, the robot needs to be able to understand human behaviours, distinguishing human daily routine from potential risk situations in order to react in accordance. In this work, we focus our attention on the domain

M. Vieira(✉) · D.R. Faria · U. Nunes
Department of Electrical and Computer Engineering, Institute of Systems and Robotics, University of Coimbra, Polo II, 3030-290 Coimbra, Portugal
e-mail: {mvieira,diego,urbano}@isr.uc.pt

U. Nunes—This work was supported by the Portuguese Foundation for Science and Technology (FCT) under the Grant AMS-HMI12: RECI/EEI-AUT/0181/2012.

L.P. Reis et al. (eds.), *Robot 2015: Second Iberian Robotics Conference*,
Advances in Intelligent Systems and Computing 418,
DOI: 10.1007/978-3-319-27149-1_35

of human-centered robot application, more precisely, for monitoring tasks, where a robot can recognize daily activities and unusual behaviours to react according to the situation. In this context, a robot that can recognize human activities will be useful for assisted care, such as human-robot or child-robot interaction and also monitoring elderly and disabled people regarding strange or unusual behaviours. We use a robot with an RGB-D sensor (Microsoft Kinect) on-board to detect and track the human skeleton in order to extract motion patterns for activity recognition. We present an application that combines different modules, allowing the robot localization and navigation in an indoor environment, and also to detect obstacles and human skeleton for motion tracking. In addition, we use modules for voice synthesizer and recognition, that will be triggered by our activity recognition module. The activity recognition module uses a Dynamic Bayesian Mixture Model (DBMM) [2] [1] for inference, in order to classify each activity, enabling the mobile robot to make a decision to react accordingly. The main contributions of this work are:

- Combining different ROS modules (navigation, classification and reaction module), towards a real time robot-assisted living application.
- Extending the use of DBMM to real-time applications using proposed discriminative 3D skeleton-based features, which can successfully characterize different daily activities.
- Assessment and validation: (i) leave-one-out cross validation of the activity recognition using our training dataset; (ii) comparison of different classification models using our proposed features; (iii) online validation of the integrated artificial cognitive system.

The remainder of this paper is organized as follows. Section 2 covers selected related work. Section 3 introduces our approach, detailing the proposed 3D skeleton-based features as well as the classification method. Section 4 describes how the approach is implemented in ROS. In section 5, the performance of the proposed application is presented. Finally, Section 6 brings the conclusion of this research pointing future directions.

2 Related Work

In order to have a fully operational robot-assisted living application, it is essential that the robot can recognize daily activities in real scenarios, in real-time. In spite of some proposed works that use inertial sensors for human activity recognition [3] [4], the most common approaches use vision-based depth sensors, even more nowadays, with low cost vision sensors (e.g. RGB-D sensors [5] [6]) that can track the entire human body accurately. In [7], a Microsoft Kinect sensor is used to track the skeleton and posteriorly extract the features. The action recognition is done using first order Hidden Markov Models (HMMs) and for every hidden state, the observations were modelled as a mixture of Gaussians. The work presented in [8] uses depth motion maps as features for activity recognition. Other works on the recognition of human activities focus their research on how to extract the right features in order to obtain better classification performance [9] [10] [11]. In the context of robot assisted living, [12] describes a behaviour-based navigation system in assisted living environments, using the mobile robot ARTOS. In [13] a PR2 robot is used to assist a person. The robot detects the activity being performed as well as the object affordances, enabling the robot to figure out how to interact with objects and plan actions. In [14], a mobile robot is used in a home environment to recognize activities in real-time by continuously tracking the pose and motion of the user and combining them with structural knowledge like the current room or objects in proximity. In our work, we use a Nomad Scout with a laser Hokuyo to assist the localization and navigation module, and an RGB-D sensor on-board to detect and track a person. It is a small mobile robot that monitors a person in an indoor environment, recognizing daily and risky activities and reacts with defined actions, assisting the person if needed. Our activity recognition module is based on the framework proposed in [1], where the features are also skeleton-based, however, herein we model different skeleton-based features, and in addition, we use a new collected dataset.

3 Activity Recognition Framework

3.1 Extraction of 3D Skeleton-Based Features

We have used a Microsoft Kinect sensor and the OpenNi's tracker package for ROS to detect and track the human skeleton. This package allows the skeleton tracking at 30 frames per second, providing the three-dimensional Euclidean coordinates of fifteen joints of the human body with respect to the sensor. Using this information, we compute a set of features as follows:

- Euclidean distances among the joints, all relative to the torso centroid, obtaining a 15×15 symmetric matrix with a null diagonal. Let *(x,y,z)* be the 3D coordinates of two body joints b_j with $j = 1, 2, ..., 15$ and b_i with $i = 1, 2, ..., 15$, then $\forall \{b_i, b_j\}$, the distances were computed as follows:

$$\delta(b_j, b_i) = \sqrt{(b_j^x - b_i^x)^2 + (b_j^y - b_i^y)^2 + (b_j^z - b_i^z)^2} \tag{1}$$

Subsequently, we removed the null diagonal, obtaining a 14×15 matrix $\mathbf{M}$ to compute its *log-covariance* as follows:

$$\mathbf{M}_{lc} = \mathbf{U}(\log(\text{cov}(\mathbf{M}))), \tag{2}$$

where $\text{cov}(\mathbf{M}_{i,j}) = (M_i - \mu_i)(M_j - \mu_j)$; $\log(\cdot)$ is the matrix logarithm function (logm) and $\text{U}(\cdot)$ returns the upper triangle matrix composed by 120 feature elements. The rational behind of log-covariance is the mapping of the convex cone of a covariance matrix to the vector space by using the matrix logarithm as proposed in [15]. A covariance matrix form a convex cone, so that it does not lie in Euclidean space, e.g., the covariance matrix space is not closed under multiplication with negative scalers. The idea of log-covariance is based on [16], where examples of manifold Riemannian metrics and log-covariance applied in 2D image features for activity recognition were used.

- The global skeleton velocities, assuming the 3D coordinates of 14 joints in the case of having the torso centroid as origin; and 15 joints in the case of having the sensor frame as origin were computed as follows:

$$v_j = \frac{\sqrt{(b^t_{j_x} - b^{t-t_w}_{j_x})^2 + (b^t_{j_y} - b^{t-t_w}_{j_y})^2 + (b^t_{j_z} - b^{t-t_w}_{j_z})^2}}{f_{rate} \times t_w}, \tag{3}$$

  where v_j is the velocity of a specific skeleton joint j; b_{j_d} represents the position $d = \{x, y, z\}$ of a skeleton body joint j in the current time t, and $t - t_w$ represents some preceding frames, herein $t_w = 10$; the frame rate is set to $f_{rate} = 1/30$.
- Differently of the aforementioned velocities in the torso frame of reference, herein, relative to the sensor frame, for all joints, for each dimension individually, we computed the difference $\delta(b^t_{j_d}, b^{t-t_w}_{j_d})$ between the position at a given frame and the preceding 10th frame. Using these values, we computed the velocities of the same joints for each dimension individually, $v_j = \frac{b^t_{j_d} - b^{t-t_w}_{j_d}}{f_{rate} \times t_w}$, obtaining additional 45 features.
- The angles variation of certain joints play a crucial role in carrying out many activities. We are interested in knowing whether a person is sitting or standing, so we compute the angles of both right and left elbows in the triangle formed by the hands, elbows and shoulders. We also compute the angles of the hip joints in the triangle formed by the shoulders, hips and knees and the angles of the knees in the triangles formed by the feet, knees and hips. The angle θ_i is given by:

$$\theta_i = \arccos\left(\frac{(\delta_{j_{12}})^2 + (\delta_{j_{23}})^2 - (\delta_{j_{13}})^2}{2 \times \delta_{j_{12}} \times \delta_{j_{23}}}\right), \tag{4}$$

  where $\delta_{j_{12}}$ is the distance between two joints, e.g. j_1 and j_2, that are forming a triangle in the skeleton. We have 2+2+2=6 features for angles, since we are considering the left and right side for the body joints. In addition, we compute the difference between these angles at a current frame and the preceding 10th frame, $\theta_{v_i} = \theta^t_i - \theta^{t-10}_i$, obtaining additional 2+2+2=6 features.

Thus, in total, we attained a set with 206 spatio-temporal skeleton-based features, useful to discriminate different classes of activities.

Features Pre-processing: Before using the features set in the classification module, we perform a pre-processing step. Normalization, standardization or filtering may be a requirement for many machine learning estimators, as they can behave badly if no pre-processing is applied to the features set. So, in the dataset case, we apply a moving average filter with 5 neighbours data points to filter the noise, smoothing the data. Subsequently, a normalization step is applied in such a way that the values of minimum and maximum obtained during the training stage were applied on the testing set as follows:

$$\mathbf{F}_{tr_i} = \frac{\mathbf{F}_{tr_i} - \min(\mathbf{F}_{tr})}{\max(\mathbf{F}_{tr}) - \min(\mathbf{F}_{tr})}, \text{ and } \mathbf{F}_{te_i} = \frac{\mathbf{F}_{te_i} - \min(\mathbf{F}_{tr})}{\max(\mathbf{F}_{tr}) - \min(\mathbf{F}_{tr})}, \tag{5}$$

where $\mathbf{F}_{tr}$ is the set of features for training and $\mathbf{F}_{te}$ is the set of features for test; i is an index to describe a set of features in a specific frame; $\max(\cdot)$ and $\min(\cdot)$ are functions to get the global maximum and minimum value of a feature set. In the real-time case, we did not apply the moving average filter because it returns worse results. The normalization step is done in the same way as in the offline tests because we keep the maximum and minimum values of the training set.

3.2 *Probabilistic Classification Model*

In this work, we adopt an ensemble of classifiers called Dynamic Bayesian Mixture Model (DBMM) proposed in [2] [1]. DBMM uses the concept of mixture models in a dynamic form in order to combine conditional probability outputs (likelihoods) from different single classifiers, either generative or discriminative models. A weight is assigned to each classifier, according to previous knowledge (learning process), using an uncertainty measure as a confidence level, and can be updated locally during the online classification. The local weight update assigns priority to the base classifier with more confidence along the temporal classification, since they can vary along the different frames. The key motivation of using a fusion model is because we are taking into consideration that an ensemble of classifiers is designed to obtain better performance than any of their individual classifiers, once there is diversity of the single components. Beyond of employing this classification model in an on-the-fly robot-assisted living application, we also compare the activity classification results with different well-known state-of-the-art classification models, such as Naive Bayes Classifier (NBC), Support Vector Machines (SVM) and k-Nearest Neighbours (k-NN). The DBMM general model for each class C is given by:

$$P(C|A) = \beta \times \underbrace{P(C^t|C^{t-1})}_{\text{dynamic transitions}} \times \underbrace{\sum_{i=1}^{n} w_i^t \times P_i(A|C^t)}_{\text{mixture model with dynamic w}}, \tag{6}$$

$$\text{with} \begin{cases} P(C^t|C^{t-1}) \equiv \frac{1}{C} \text{ (uniform)}, \; t = 1 \\ P(C^t|C^{t-1}) = P(C^{t-1}|A), \;\; t > 1 \end{cases},$$

where $P(C^t \mid C^{t-1})$ is the transition probability distribution among class variables over time, which a class C^t is conditioned to C^{t-1}. This means a non-stationary behavior applied recursively, then reinforcing the classification at time t; $P_i(A|C^t)$ is the posterior result of each ith base classifier at time t, becoming the likelihood in the DBMM model. The weight w_i^t in the model for each base classifier is initially estimated using an entropy-based confidence on the training set (offline), and afterwards ($t > 5$) it is updated as explained in our previous work [1]; $\beta = \frac{1}{\sum_j \left(P(C_j^t|C_j^{t-1}) \times \sum_{i=1}^{n} w_i \times P_i(A|C_j^t)\right)}$ is a normalization factor, ensuring numerical stability once continuous update of belief is done.

Base Classifiers for DBMM: In this work, we have used the NBC, SVM and k-NN as base classifiers for the DBMM fusion. The NBC assumes the features are independent from each other given a class, $P(C_i|A) = \alpha P(C_i) \prod_{j=1}^{m} P(A_j|C_i)$. For the linear-kernel multiclass SVM implementation, we adopted the LibSVM package [17], trained according to the 'one-against-one' strategy, and classification outputs were given in terms of probability estimates. A k-NN was also combined into the DBMM fusion. An object is classified by a majority vote of its neighbours, with the object being assigned to the class most common among its k nearest neighbours. The classification outputs of the adopted k-NN were given in terms of probability estimates as well.

4 Robot-Assisted Living Architecture in ROS

The proposed artificial cognitive system was implemented in ROS and comprises three main modules, as shown in Figure 1: classification, navigation and reaction modules.

In order to properly test the system in real scenarios, a mobile robot is used. Therefore, a personal robot endowed with cognitive skills, capable of monitoring the behaviours of a person should be able to autonomously navigate in an indoor environment. The navigation module uses odometry and laser scans from the robot to map the environment and self-localization, randomly navigating, avoiding obstacles. We use the navigation stack available in ROS distributions, more specifically, the *move_base* package to generate an appropriate collision free trajectory. For simultaneous localization and mapping (SLAM) the *hector_slam* package is used. While the robot is navigating, the MS-Kinect sensor is sending RGB-D data to the classification module. Once a skeleton is detected, the robot stops and the feature extraction process starts. Then, classification is done using the DBMM and an activity is recognized. Once the system knows the human activity being performed, the reaction module is in charge to select what the robot should do next. For each human activity, a predefined reaction in a lookup-table was associated, including warnings, questions or changes in navigation (Figure 2). In the event of a person telling the robot to follow him/her, a safe distance of 2.5 meters is maintained. A Kalman filter

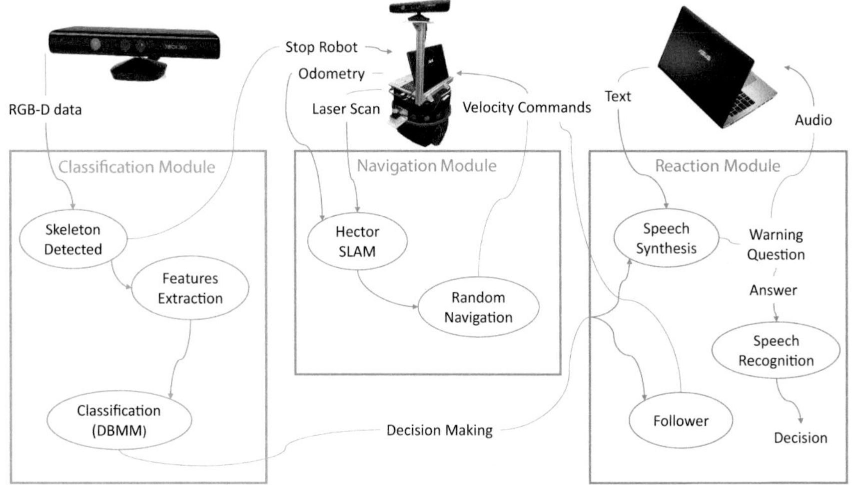

Fig. 1 System overview.

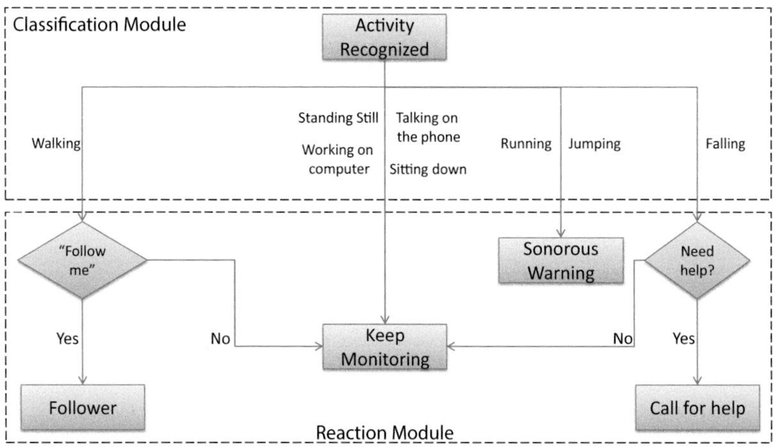

Fig. 2 Decision tree in reaction module.

is used to estimate the trajectory of the person one second ahead in order to avoid collision between the robot and the human. If a collision trajectory is estimated, the robot will step away, in order to the person walk through safely. For the prediction of the human motion, a position model was adopted, where the state includes position $(x(k); y(k))$ of the human target:

$$\begin{cases} x(k) = x(k-1) + v_x(k-1) \times \Delta t \\ y(k) = y(k-1) + v_y(k-1) \times \Delta t \end{cases} \tag{7}$$

with $\Delta t = t(k) - t(k-1)$. Using the torso coordinates as measures, it is possible to compute the x velocity v_x and y velocity v_y. For speech synthesis, we use the $sound\_play$ package that given a text input, it will be synthesized into sound output. For speech recognition, we use the $pocketsphinx$ package. This package recognizes a single word or a stream of words from a vocabulary file previously created. In our work, the vocabulary comprises the following words: "no", "yes", "please", "help", "follow", "me". The package can recognize combinations of these words, such as "please help me".

5 Experimental Results

5.1 Performance on Collected dataset

A new dataset of daily activities and risk situations, more complete and challenging than the one used in our previous work [1], was collected to train the activity recognition module. This dataset (Figure 3) comprises video sequences of two male subjects and two female subjects performing eight different activities in a living room. The daily activities are: *1-walking, 2-standing still, 3-working on computer, 4-talking on the phone, 5 sitting down*; and the unusual or risk situations are: *6-jumping, 7-falling down, 8- running*. This dataset is a challenging one, once there is significant

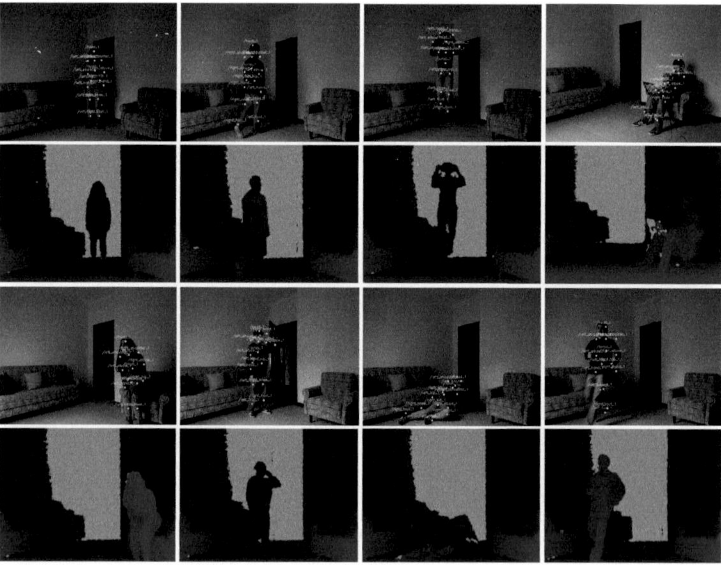

Fig. 3 Few examples of the dataset (RGB with skeleton joints and depth images) which was created to learn some daily and risk situations.

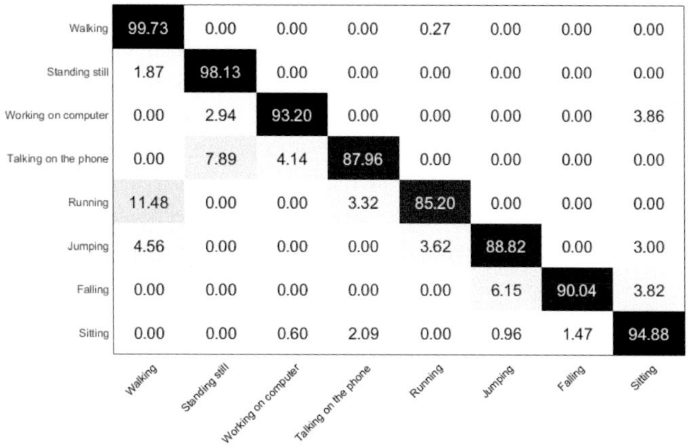

	Walking	Standing still	Working on computer	Talking on the phone	Running	Jumping	Falling	Sitting
Walking	99.73	0.00	0.00	0.00	0.27	0.00	0.00	0.00
Standing still	1.87	98.13	0.00	0.00	0.00	0.00	0.00	0.00
Working on computer	0.00	2.94	93.20	0.00	0.00	0.00	0.00	3.86
Talking on the phone	0.00	7.89	4.14	87.96	0.00	0.00	0.00	0.00
Running	11.48	0.00	0.00	3.32	85.20	0.00	0.00	0.00
Jumping	4.56	0.00	0.00	0.00	3.62	88.82	0.00	3.00
Falling	0.00	0.00	0.00	0.00	0.00	6.15	90.04	3.82
Sitting	0.00	0.00	0.60	2.09	0.00	0.96	1.47	94.88

Fig. 4 Confusion matrix obtained from the DBMM classification applied on the dataset

Table 1 Performance on the dataset ("new person"). Results are reported in terms of Precision (Prec) and Recall (Rec).

Activity	DBMM Prec	DBMM Rec
walking	89.63%	99.73%
standing still	94.86%	98.13%
working on computer	95.93%	93.20%
talking on the phone	93.64%	87.96%
running	92.81%	85.20%
jumping	92.52%	88.83%
falling down	97.24%	90.04%
sitting down	92.27%	94.88%
Average	**93.61%**	**92.25%**

Table 2 Global results using single classifiers, a simple average ensemble (AV) and the DBMM.

Method	Acc.	Prec.	Rec.
NBC	82.90%	85.79%	82.67%
SVM	88.47%	89.02%	87.62%
k-NN	87.98%	90.09%	87.06%
AV	85.29%	87.74%	84.68%
DBMM	**93.41%**	**93.61%**	**92.25%**

intra-class variation among different realizations of the same activity. For example, the phone is held with the left or right hand. Another challenging feature is that the activity sequences are registered from different views, i.e., from the front, back, left side, and so on. The classification results are presented in a confusion matrix and with the measures of Accuracy, Precision, Recall of the four tests. The idea is to verify the capacity of generalization of the classifier by using the strategy of "new person", i.e., learning from different persons and testing with an unseen person. Figure 4 shows the results in a single confusion matrix. Table 1 shows the performance in terms of Precision (Prec) and Recall (Rec) of this approach for each activity. The results show that using DBMM, improvements in the classification were obtained in comparison with using the base classifiers alone. The overall results attained were: accuracy 93.41%, precision 93.61% and recall 92.25%. For comparison purposes, Table 2 summarizes the results from single classifiers and an average ensemble compared with DBMM, showing the improvement achieved using the described skeleton-based features. The SVM was trained with *soft margin* (or Cost) parameter set to 1.0, and the k-NN was trained using 20 neighbours.

5.2 Performance On-the-Fly Using a Mobile Robot

The experimental tests using the proposed approach for a real time application is a little bit different than the experimental tests on the dataset. In this case, the robot will acquire 5 seconds of RGB-D sensor data for features extraction and classification. Only the NBC and SVM were used as base classifiers for the DBMM fusion, because they are enough for obtaining good results, thus, avoiding spending more processing time using other base classifiers. After 5 seconds of frames classification, a final decision is made for activity recognition to trigger a proper robot reaction. The proposed framework is capable of recognizing different activities transitions that happens sequentially in case of a person transit from one activity to another one, e.g., a person that is standing and sequentially pass to a sitting down position and consequentially working on the computer. Figure 5 shows some examples of tests of daily activities and unusual or risk situations that the mobile robot correctly recognized. Three tests were carried out for each activity with three different subjects. One of the subjects was already "seen" in the training, while the rest are "unseen" subjects.

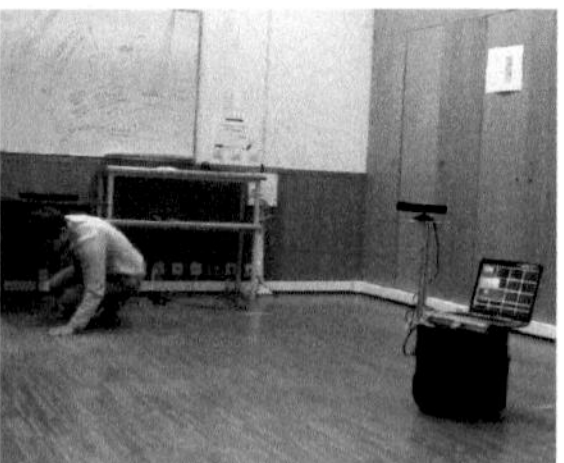

Fig. 5 Shots of tests of activity recognition ('unseen" person) using a mobile robot.

All activities were correctly classified, so that the overall performance of classification is shown in Figure 6. The overall (average) results attained in real-time experiments were: accuracy 90.55%, precision 90.84% and recall 90.55%. Table 3 shows the results in terms of recall of each test for each subject. Looking at the results

	Walking	Standing still	Working on computer	Talking on the phone	Running	Jumping	Falling	Sitting
Walking	85.77	3.25	0.00	0.82	4.48	3.29	1.19	1.19
Standing still	0.41	96.71	0.00	1.64	0.00	1.23	0.00	0.00
Working on computer	0.41	0.00	97.12	0.00	0.82	0.00	1.65	0.00
Talking on the phone	0.82	2.47	0.00	94.65	0.41	1.65	0.00	0.00
Running	5.27	2.88	2.22	1.98	83.21	1.48	2.96	0.00
Jumping	0.00	3.70	0.00	4.11	0.00	90.54	0.00	1.65
Falling	2.47	5.52	0.41	1.23	1.23	1.23	87.09	0.81
Sitting	0.00	4.12	5.35	0.00	0.00	1.23	0.00	89.30

Fig. 6 DBMM on-the-fly classification confidence (average) presented in a confusion matrix

attained, it is possible to conclude, as expected, that the best performance is achieved for the "seen" person (subject 1). However, the difference of results between subjects is not very significant, which indicates that the fact of being or not a "seen" person is not a key factor for the performance of the classification. The most important factor in a real-time application is that in the end, the activity being performed is correctly recognized. Since the robot correctly classified the activity performed, it also successfully reacted accordingly to the situation. Figure 7 shows a sequence of events from an activity that is being recognized (in this case falling) to react according to

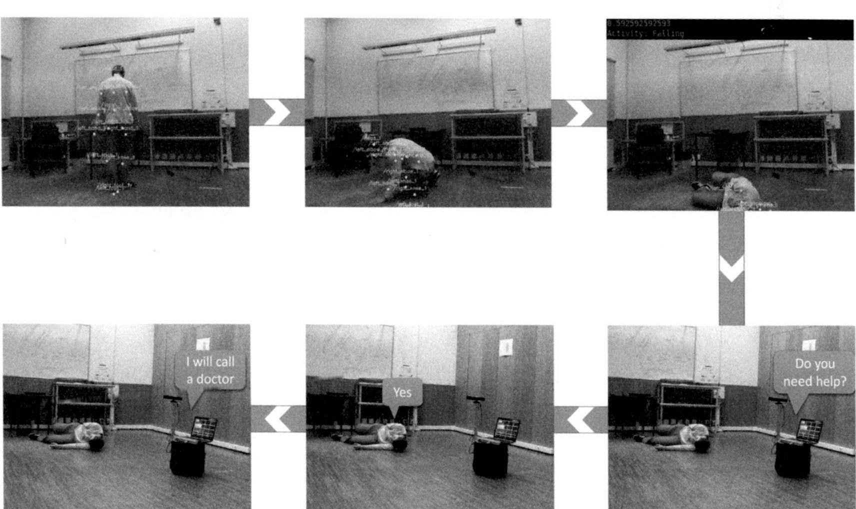

Fig. 7 Sequence of events on detecting a person falling and reacting

this activity. First, the skeleton of a person is detected and tracked, initiating the monitoring stage. Then, the person falls on the floor and the robot correctly recognizes the risk situation "falling". Detecting such a behaviour, the robot asks if the person needs help. The robot receives an affirmative answer from the person, recognizes the command and immediately calls for help.

Table 3 On-the-fly results in terms of recall for 3 different subjects. One subject seen and two unseen.

		Activity								Overall
	Test	walking	standing still	working on computer	talking on the phone	running	jumping	falling down	sitting down	
Subject 1 (seen)	1	96.30	100	100	59.26	85.19	85.19	85.19	96.30	88.43
	2	96.30	100	100	100	85.19	88.89	95.45	96.30	95.27
	3	92.59	100	92.59	100	85.19	88.89	92.86	96.30	93.55
	Average	95.06	100	97.53	86.42	85.19	87.65	91.17	96.30	**92.42**
Subject 2 (unseen)	1	66.67	100	96.30	100	96.30	81.48	74.07	70.37	85.65
	2	81.48	85.19	96.30	92.59	85.19	92.59	74.07	92.59	87.50
	3	81.48	100	88.89	100	85.19	92.59	95.45	92.59	92-02
	Average	76.54	95.06	93.83	97.53	88.89	88.89	81.20	85.18	**88.39**
Subject 3 (unseen)	1	82.14	96.30	100	100	73.33	96.30	85.19	88.89	90.27
	2	92.86	96.30	100	100	80.00	92.60	100	85.19	93.37
	3	82.14	92.59	100	100	73.33	96.30	81.48	85.19	88.88
	Average	85.71	95.06	100	100	75.55	95.07	88.89	86.42	**90.84**
Overall Average		**85.77**	**96.71**	**97.12**	**94.65**	**83.21**	**90.54**	**87.09**	**89.30**	**90.55**

6 Conclusions and Future Work

The main contribution of this work is a robotic application for real-time monitoring of daily activities and risk situations in indoor environments. A dynamic probabilistic ensemble of classifiers (DBMM) was used for daily activity recognition using a proposed spatio-temporal 3D skeleton-based features. We collected a dataset to endow a robot to recognize daily activities, and we used this dataset to compare our approach with other state-of-the-art classifiers. Using our proposed skeleton-based features, we attained relevant results using the DBMM classification, outperforming other single classifiers in terms of overall accuracy, precision and recall measures. More importantly, the experimental tests using a mobile robot presented good performance on the activity classification, allowing the robot to take appropriate actions to assist the human in case of risk situations, showing our framework has good potential for robot-assisted living. Future work will address addition of contextual information, such as "who", "where", "when" in order to fully understand human behaviours, as well as exploitation of our approach with more daily activities, risk situations and robot reactions.

References

1. Faria, D.R., Vieira, M., Premebida, C., Nunes, U.: Probabilistic human daily activity recognition towards robot-assisted living. In: IEEE RO-MAN 2015 (2015)
2. Faria, D.R., Premebida, C., Nunes, U.: A probalistic approach for human everyday activities recognition using body motion from RGB-D images. In: IEEE RO-MAN 2014, * Kazuo Tanie Award Finalist (2014)
3. Zhu, C., Sheng, W.: Realtime human daily activity recognition through fusion of motion and location data. In: IEEE International Conference on Information and Automation (2010)
4. Zhu, C., Sheng, W.: Human daily activity recognition in robot-assisted living using multi-sensor fusion. In: IEEE ICRA 2009 (2009)
5. Microsoft kinect. https://www.microsoft.com/en-us/kinectforwindows/ (accessed on June 2015)
6. Asus xtion. http://www.asus.com/multimedia/xtion_pro_live/ (accessed on June 2015)
7. Papadopoulos, G.T., Axenopoulos, A., Daras, P.: Real-time skeleton-tracking-based human action recognition using kinect data. In: 3D and Augmented Reality, pp. 473–483. Springer International Publishing (2014)
8. Chen, C., Liu, K., Kehtarnavaz, N.: Real-time human action recognition based on depth motion maps. Journal of Real-Time Image Processing (2013)
9. Sung, J., Ponce, C., Selman, B., Saxena, A.: Unstructured human activity detection from RGBD images. In: ICRA 2012 (2012)
10. Xia, L., Aggarwal, J.: Spatio-temporal depth cuboid similarity feature for activity recognition using depth camera. In: CVPR (2013)
11. Zhu, Y., Chen, W., Guo, G.: Evaluating spatiotemporal interest point features for depth-based action recognition. Image and Vision Computing (2014)
12. Mehdi, S.A., Armbrust, C., Koch, J., Berns, K.: Methodology for robot mapping and navigation in assisted living environments. In: 2nd International Conference on Pervasive Technologies Related to Assistive Environments (2009)
13. Koppula, H.S., Gupta, R., Saxena, A.: Learning human activities and object affordances from RGB-D videos. IJRR Journal (2012)
14. Volkhardt, M., Müller, S., Schröter, C., Gross, H.-M.: Real-time activity recognition on a mobile companion robot. In: 55th Int. Scientific Colloquium (2010)
15. Arsigny, V., Fillard, P., Pennec, X., Ayache, N.: Log-euclidean metrics for fast and simple calculus on diffusion tensors. Magnetic Resonance in Medicine **56**(2), 411–421 (2006)
16. Guo, K.: Action recognition using log-covariance matrices of silhouette and optical-flow features, Ph.D. dissertation, Boston University, College of Engineering (2012)
17. Chang, C.-C., Lin, C.-J.: LIBSVM: A library for support vector machines. ACM TIST (2011). http://www.csie.ntu.edu.tw/~cjlin/libsvm

Challenges in the Design of Laparoscopic Tools

J. Amat, A. Casals, E. Bergés and A. Avilés

Abstract The need to minimize trauma in surgical interventions has led to a continuous evolution of surgical techniques. The robotization of minimally invasive surgeries (MIS) through robotized instruments, provided with 2 or 3 degrees of freedom, aims to increase dexterity, accuracy… and thus, assist the surgeons. This work presents the challenges faced during the development of a surgical instrument, from the work carried out in the design and implementation of a complete surgical robotic system. After an overview of the surgical instruments associated to the alternative techniques in MIS, the process of designing laparoscopic instruments for the developed robotic system is described. Our approach focusses on the technological challenges of achieving a user-friendly laparoscopy, affordable for hospitals. These include the complexity of designing small-sized tools, which match the surgical requirements and introduce additional features, as haptic feedback. In addition, we explain the non-technological obstacles overcome to satisfy the commercialization requirements. The huge number of patents in this field acts as a spider web, which led us to seek for novelty. Although specific parts of the robotic system were not the core of our project, we needed to fit their design and obtain our own patents to grant the complete robotic system was free of patent infringement. On the other hand, complex regulatory procedures turn the whole commercialization process dilated and tedious. Finally, we present some of our ongoing research to improve performance of this kind of robot assisted surgery, as well as to other surgical fields.

Keywords Minimally invasive surgery · Laparoscopic instruments · Teleoperated surgery

1 Introduction

The continuous attempts to minimize the trauma associated to surgical interventions gave place in 1981 to a significant achievement, when the German gynecologist Kurt Semm performed the first intervention through tiny incisions into the

J. Amat(✉) · A. Casals · E. Bergés · A. Avilés
Universitat Politècnica de Catalunya, Barcelona-Tech, Barcelona, Spain
e-mail: josep.amat@upc.es

L.P. Reis et al. (eds.), *Robot 2015: Second Iberian Robotics Conference*,
Advances in Intelligent Systems and Computing 418,
DOI: 10.1007/978-3-319-27149-1_36

abdominal cavity [1]. This first intervention was an appendectomy, which led to the classical procedures in conventional laparoscopy based on the application of three incisions, as an alternative to the large incisions in open surgery. This intervention relied on previous experiences on the use of an optical pod, a cystoscope, which is inserted into the abdominal cavity inflated with CO2, and on trocars, for the insertion of the instruments. However, it was not until 1986 that the first cholecystectomy took place, a surgical technique that deserved the name of "magic surgery". These techniques have expanded more and more due to the development of a wide range of instruments. Micro cameras and screens for the visualization of the abdominal cavity are an example. This new scenario allows the assistant staff and the main surgeon to see the working scene.

Technological progress has been twofold. On the one side, towards the reduction of the diameter of the instruments, from 5 mm to even less than 2 mm. On the other side, the efforts have focused on providing the tools with degrees of freedom additional to the simple axial rotation.

It was with the launch of robotized laparoscopic surgery, in the early 2000s, that it became possible to use instruments with turning movements, pitch and yaw, which are too difficult to control in manual surgery. The higher complexity of these tools involves an increase of their diameter to achieve the same force performance. Being 8 mm the standard diameter, 5 mm diameters are also common, but with restrictions in the rotation ranges and reaching lower forces.

In the search of new performances, the development, from 1997, of trocars with three or up to four entries (access points, ports) to be able to operate through a unique incision offer the possibility of minimizing scares. Since these new trocars require larger diameters, from 12 to 20 mm, the incision is done through the belly button. This surgery has led to the development of instruments that can fold down, thus requiring thinner trocars. When they unfold, once inside the body, a better triangulation (concept used to indicate the three basic aim points, within the operation field) is achieved, which mitigates the difficulties of operating from a unique projection point. Nevertheless, the performance is worse compared to laparoscopic surgery, which allows reaching the target point from three or more directions. These operative limitations for the surgeon, in some determined interventions are compensated by the lesser esthetic affectation to the patient.

Looking for less invasive techniques, in 2005 new laparoscopic techniques, NOTES (Natural Orifice Transluminal Endoscopic Surgery) were introduced. Scare less surgery, or better, surgery without cutaneous incisions, uses as entry via, natural orifices [2].

In 2008, our research team assumed the challenge of developing a robotic system for laparoscopic surgery with the goal of providing a less complex and less invasive robot than the unique industrial system available on the market. This challenge not only implies the development of a robotic system, but also instruments with the three degrees of freedom necessary to provide high accessibility inside the abdominal cavity. The latter has represented in practice a major difficulty level than developing the robot itself. This is due, in part, to the mechanical complexity brought by size restrictions. The instrument pod is limited to a maximum diameter of 8 mm, needed to optimize the size-force tradeoff. An aggravation to the normal technical challenges is the need to skip solutions that are restricted by intellectual property barriers.

This paper presents the challenges we have achieved to design and develop a set of surgical instruments for a robotized system to be commercialized. The work also describes the future challenges our laboratory has raised in the field of robotized microsurgery instrumentation.

2 New Tools for Laparoscopic Surgery

Initially, laparoscopic surgery used an incision and a trocar for each instrument, so that the surgeon can select the entry points in function of the accessibility required in each intervention. This enables the surgeon to operate with different orientations, thus alleviating the limitations produced by the lack of rotation of conventional tools. The claims for a surgery with fewer incisions has led to the development of tools designed to operate from a unique access port. In order to assure a side access to the working area, even for frontal insertions, it has been necessary to design tools with a certain curvature, as shown in fig. 1.

The development of this unique incision technology, named Single Port (SILS, Single Incision Laparoscopic Surgery) or also LESS (Laparoendoscopic Single-site Surgery) has followed two well differentiated paths: a unique port, multi-access, with an access for each instrument, e.g. ASC TriPort® (Advanced Surgical Concepts, Wicklow, Ireland), or a port, also with a unique access for a tool [3]. In this case, once the tool reaches the abdominal cavity, it unfolds into two or more arms, and into the optics that protrudes over the rest to obtain the images, as The SPIDER surgical system by TransEnterix (Durham, NC, USA) [4], fig. 2.

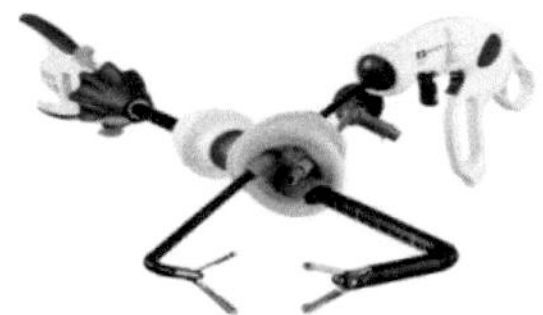

Fig. 1 Curved surgical instruments to facilitate a single-port access.

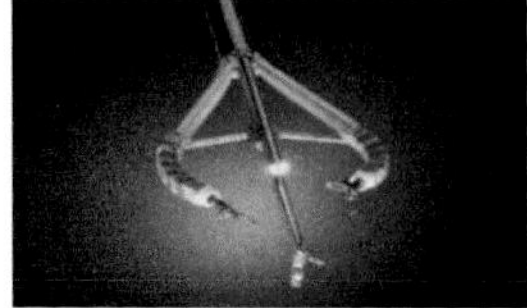

Fig. 2 Folding multiple surgical instruments.

Both in conventional laparoscopy and in SILS, with multi access port, the tool can be approached and retrieved, then generates two displacements, vertical and horizontal. They are executed from angular movements produced by pivoting the tool over the trocar, fig. 3.

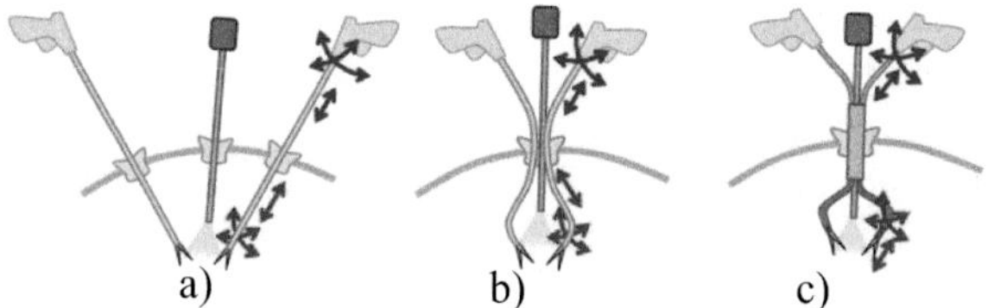

Fig. 3 Operability of surgical tools in a) Conventional laparoscopic surgery, b) Single port with multiple trocars and c) Single port and folding surgical instruments.

To achieve similar operability using SILS/LESS, with one access port, the tools that unfold should also be able to perform a similar vertical and horizontal deflection. In this case, they act over the tools that necessarily have a curvature. This curvature provides two different operating forms, with mirror effect (as conventional laparoscopic surgery) or without this effect (as in robotized laparoscopic surgery), fig. 4.

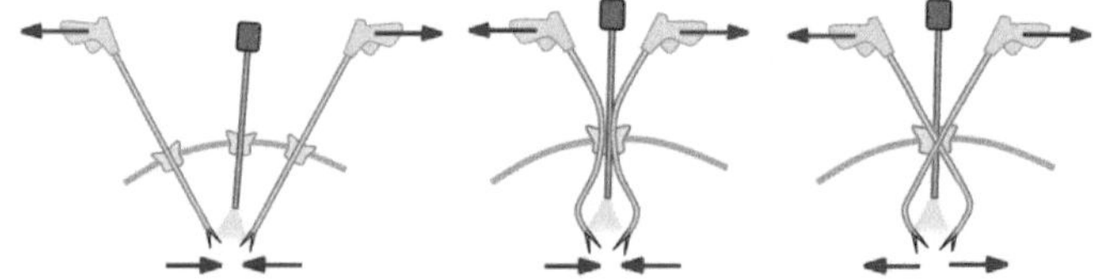

Fig. 4 Mirror effect due to surgical instruments working with trocars.

These instruments, in turn, have also two actuation typologies, rigid instruments, or instruments with variable curvature and/or longitude, which require a higher complexity technology.

The Single Port technology, though, has certain constrains regarding maneuverability. The fact that many instruments share a unique access point makes their positioning and orientation an uncomfortable task. To avoid these constrains, other tools have been developed to freely move along the abdominal cavity, achieving a widen accessibility margin. Usually, these tools are magnetically subjected to the upper wall of the abdominal cavity, in a way that they can be guided, so as to optimize the orientation of the working tool, fig. 5.

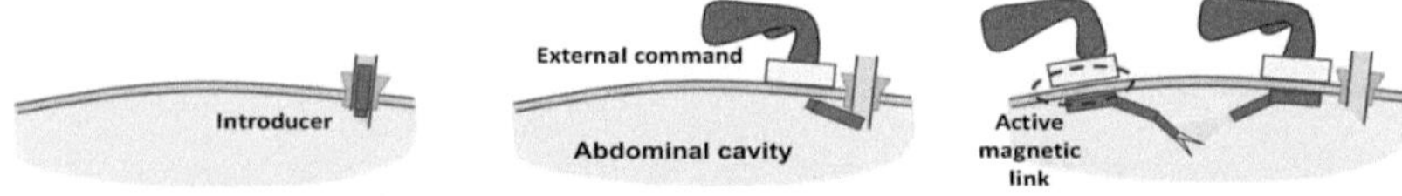

Fig. 5 Remote active magnetic actuation system.

Placing an introducer in an adequate place, dealing with both, the posterior visibility of the scar and the navigation of the instruments inserted into the abdominal cavity, it is possible to reach any point without needing additional holes. The first tools fabricated with the technology of magnetic support were the wireless cameras. The design of instruments for this technology is more complex, since it does not only require the positioning of the tool, but also its actuation, and in some cases, a certain orientation. In this case, the turn movements can be transmitted through the generation of magnetic fields, with magnet cylinders, as indicated in fig. 6.

Fig. 6 Remote magnetic4 DoF transmission.

3 Design of the Prototype. Technical Problems and Barriers

The challenge of our research group consisted on the development of a robotic system for laparoscopic surgery constituted by a set of two robots endowed with two arms each, and controlled by teleoperation. The system also requires the development of the working tools endowed with three degrees of freedom at the distal element. These should have a working force similar to that achieved by manual tools, and with the condition of not surpassing an 8 mm diameter.

The development of these instruments has been the heaviest part of the project. The main difficulties came from the technological design; constrains due to materials' biocompatibility requirements; the conditions imposed to obtain the regulatory approval, EC and FDA; and constrains derived of others' IP (Individual Property).

3.1 Technical Design and Operative Prototype

In what refers to technological challenges, the surgical instruments have to deal with four factors: efficient design, achievable force, cost and durability. They are characteristics opposed in many aspects and it has been necessary to look for a compromise to achieve their homologation, compatible with achieving its commercial applicability.

The need to actuate on grippers endowed with three rotations, requiring a certain working force, conditions the election of the way of transmitting movements, since the actuators cannot be at the distal part of the instrument, but outside the abdominal cavity. Looking for minimal cost solutions, it was decided to move the actuators outside the body, dividing the instrument into two parts. The first part is a motorized element integrated to the whole robotic equipment, and the second, the working tool that constitutes a reusable part after its sterilization or disposable after one use.

The technological alternatives to transmit movements can be rods, cables or gears. Belts are not considered, not only because of its size, but also due to biocompatibility problems. The mechanical solution using rods is the most frequently used and the one that transmits higher forces, as indicated in fig. 7. This type of transmission can either use rigid pods for straight instruments, or flexible tendons useful for curved instruments, fig. 8, specially used in SILS.

Fig. 7 Transmission of movements through rods.

Fig. 8 Tendon actuated bendable instruments.

This solution makes the transmission of the movement compatible with the rotation of the tool, but does not solve the problem of transmission of movements through a turn in its orientation. In this case, it is necessary to use gears, fig. 9 or cables, fig. 10. The solution with cables allows implementing more easily the pass through the two joints, necessary to perform the orientation turns, being thus the most frequently used, fig. 11.

Fig. 9 Gear based transmission.

Fig. 10 Instrument driven by cables and pulleys.

Fig. 11 Transmission by cables and pulleys.

Unfortunately, the election of a technology for the transmission does not depend only on technical considerations. With even more weight, it is necessary to obtain the "free to operate" with respect to existing patents. Solutions based on rods or gears admit then only few variants, hence we were forced to skip solutions based on these technologies.

The technology chosen has thus been that based on cables, as that of the currently hegemonic manufacturer. Both designs, though, are clearly differentiated because ours do not use pulleys to guide the cable in each joint interposed, but flexible cable-sheath, fig. 12. Sheathed cables are useful to transmit the movement through a joint, without varying the longitude of the cable, fig. 13. Hence, the movement becomes decoupled, contrarily to the case of pulleys shown in fig. 11.

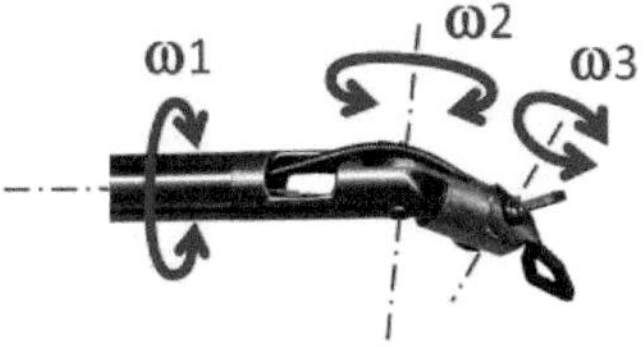

Fig. 12 The designed instrument, with cable and without pulley.

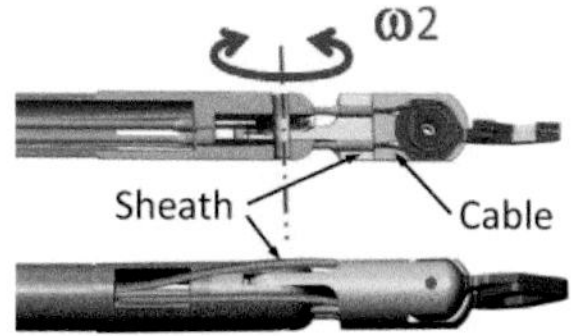

Fig. 13 Sheathed cables with decoupled movement.

This solution has been used for decades and thus it is not subject to IP claims. Although conceptually simple, this solution is difficult to be implemented at reduced scale. For this reason, it has been necessary to solve the problem of confronting the cable when leaving the sheath towards the actuation drum of the joint or the gripper. Although not optimal, this solution relies on the design of friction-based deflection nozzles, fig. 14. Therefore, it produces a certain wearing of the cable as it works under tension, thus limiting its durability. The profile studies of these nozzles have allowed optimizing performance and durability, solution that has resulted in a patent [5]. This design is not as efficient as a transmission with pulleys, presents fragments much higher, and thus, less performance. It also presents hysteresis, although it has been compensated by software.

The need to use trocar inputs of 8 mm limits the radius of the drum, which rolls the transmission cable, down to 2,5mm, fig. 14. To reach a torque on the gripper of approximately 0,2N m, it is necessary that the cable can apply a force of 0,2/0,003 = 66N, corresponding to cables of Ø0,45 mm. Due to the fragments of the transmission, it has been necessary to motorize the actuator in a way that it can transmit up to 100N force to the header .

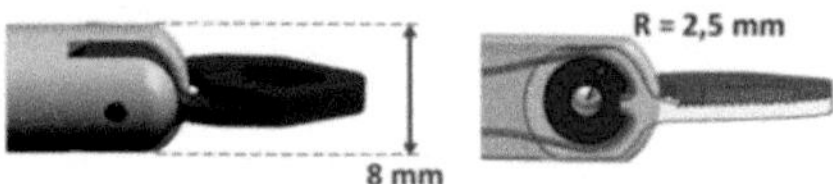

Fig. 14 Small working radius of the clamp, required for the 8mm trocar constrain.

The growing interest to endow these teleoperated instruments with haptic feedback makes necessary to sense the real forces exerted by the grippers on the tissue, either grasping, or cutting. The inherent problems of placing a sensor on the tool tip have led us to the development of a system for the indirect perception of forces, based on the tissues deformation model.

3.2 Non-technical Barriers

Up to this point, we have described the main challenges that we faced to achieve a suitable and reliable product. Nevertheless, there are additional requirements out of the scientific spotlight, which can also mean the product's success or failure. In this section we expose the difficulties and risks related to the management of intellectual property, as well as the legal requirements to put this product into the market.

It is obvious that the knowledge of the patents status regarding any technology is fundamental, whether to protect an own intellectual property or to avoid the infringement of others. In our case, robotic surgical instruments belong to the technological field with the most increasing number of patents applied in Europe for the last ten years. As fig. 15 shows, medical technology has reached the top of patent filing in Europe along the year 2014 [6] (we note the spike in 2010 due to a change in patent filing rules).

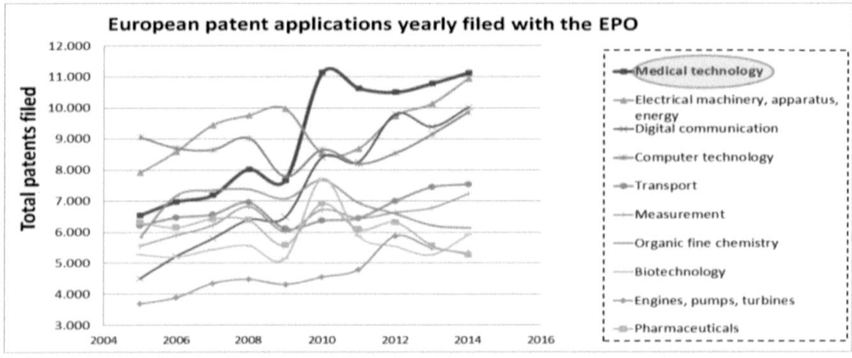

Fig. 15 European patents filed by the top ten technical fields from 2005 to 2014.

According to our experience, the patent filing has been a long time consuming and costly process, mainly due to the large number of documents, relevant and non-relevant, to be reviewed. Only the company Intuitive Surgical keeps exclusive rights to 1800 patents plus 1500 pending to be approved protecting the DaVinci surgical system [7]. Their patents strategy is clearly focused to control the industry of laparoscopic surgical robots as a monopoly. The development of these new instruments has forced us to patent them. Not necessarily to innovate, but just to avoid any patent infringement and be able to manufacture them.

There also exist non-reasonable patents, which lead to absurd situations. Patents claiming the color of an alarm as an invention, or a company claiming the concept of teleoperation as its own intellectual property are real examples, which would indeed block any development if they were taken into account. Their approval apparently denotes political reasons, rather than fairly protect genuine inventions.

In order to achieve the rights of commercialization of our surgical instruments, we have followed the *Council Directive 93/42/EEC on medical devices* for regulatory requirements. On the one hand, it has the advantage that the validation tests requested by the competent entity have to be designed by ourselves. The freedom to evaluate our technology by our own criteria, though, has also drawbacks. The lack of standard procedures involves a risk of rejection when presenting the results of the tests. The time needed for the regulatory approval becomes then uncertain.

Another obstacle is the fact that laws are differently applied from one country to another. Regulatory requirements may vary, depending on the territory the product is to be commercialized. Any additional requirement is inevitably translated into additional time and costs. It is then common to follow the so called *soft laws*. These are non-binding regulations, e.g. the ISO standards. Although they do not have legal effects, they provide efficient guidelines to grant the regulatory approval for products commercialization.

4 Results and Evaluation

During the design process of the surgical instruments, we considered two alternatives. One solution consisted of assuring long durability, with the cost of designing expensive tools, while the other was to design a low cost product, which allowed at least one complete surgical intervention. Given the limitations imposed by existing patents, which prevent logical solutions like wired transmission through drive pulleys in every joint, we opted for the design of single use instruments. Although they have a lower mechanical yield, we have reached a satisfactory trade-off between efficiency-force and durability in a tool endowed with three degrees of freedom. The instrument life is over 5 working hours, enough for a complete surgical intervention.

A test bed has been implemented to perform an objective and rigorous quality control process. The test monitors a repetitive task, measures the deflection angles of each degree of freedom, analyses their deterioration and quantifies the force exerted by the tool tips, fig. 16. The setup consists of a black box, with a camera for the monitoring of the roll, pitch and yaw measured angles, and with three elastic graspable elements with an embedded force sensitive resistor (FSR) in the two extreme positions of the yaw range for torque and force measurement, fig. 17.

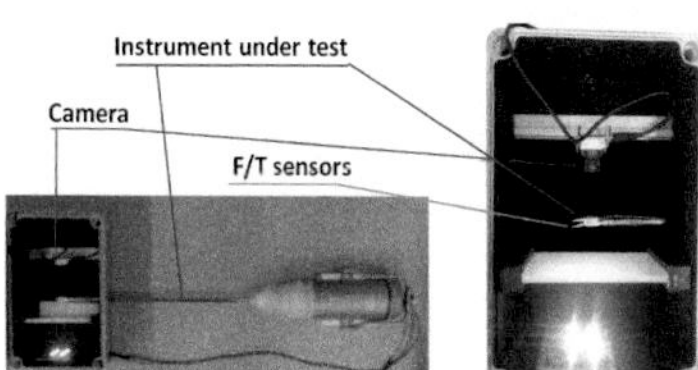

Fig. 16 Calibration and evaluation platform.

In order to measure the deflection angles, a vision system tracks the tree axis of each finger of the clamps. To validate this test, the measures are compared with the values given by the encoders at the outlet of the actuator. The discrepancy between both lectures quantifies the level of deterioration of the transmission system comprising the set wire-casing-deflection nozzle, fig. 18. Series 1 represents the encoder and real position relation with increasing angle, and series 2 with decreasing angle, indicating the level of hysteresis. Series 3 and 4 represent the same data after some operation time. The hysteresis, which indicates the level of degradation, increases with operation time.

Regarding the low costs requirements needed to reach a reasonable level as a fungible component, the followed strategy has been the reusability of the valuable parts of the instrument after each use, like its fingers and other steel pieces.

This strategy allowed us to verify the operative life of our design, which exceeds the 5 hours term. The working operability reached is ± 90° for each degree of freedom and the torques produced by the clamping forces is higher than 200 mN m in each actuator. Technically, it is proven that our design satisfies the needs of a laparoscopic instrument.

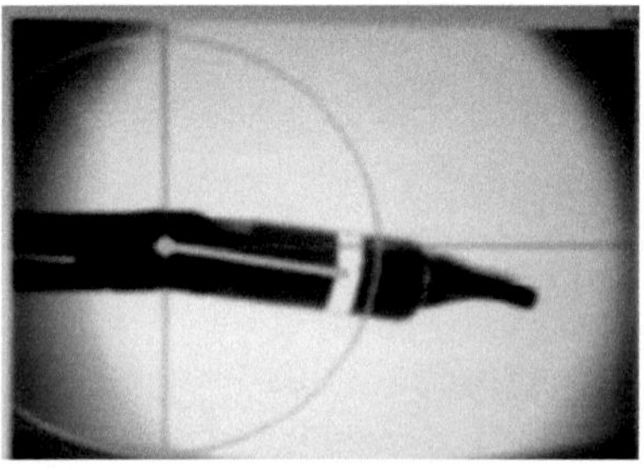

Fig. 17 Real deflection angle seen from the camera.

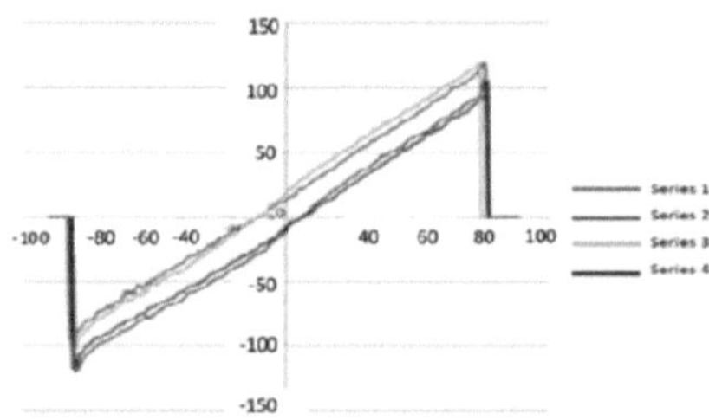

Fig. 18 Encoders' data vs real positions during the deterioration of the transmission.

5 Future Challenges

Medical robotics is still in its infancy and looking ahead there are still many challenges to be tackled, and many technological and cognitive limitations need to be overcome. Apart from robotic laparoscopy, our laboratory is also working in other robotic features and medical fields, i.e. haptic feedback and microsurgery. The following sections describe the ongoing projects.

5.1 Haptic Challenges

One of the major limitations in current robotic surgical systems is the lack of force feedback [8] which increases intraoperative injury and has a direct link to the completion time and on the accuracy achieved while decreasing transparency [9]. Although several attempts to develop force sensors able to provide data for haptic feedback have been done, the results still show limitations in size, cost, long-term stability, and biocompatibility and sterilization problems [10]. The complexity of direct measurements demands new alternatives. Our approach [11] is based on visual information of the deformation produced on the tissues by the applied forces [12], the so called: Vision-Based Force Measurement (VBFM) approach, fig. 19.

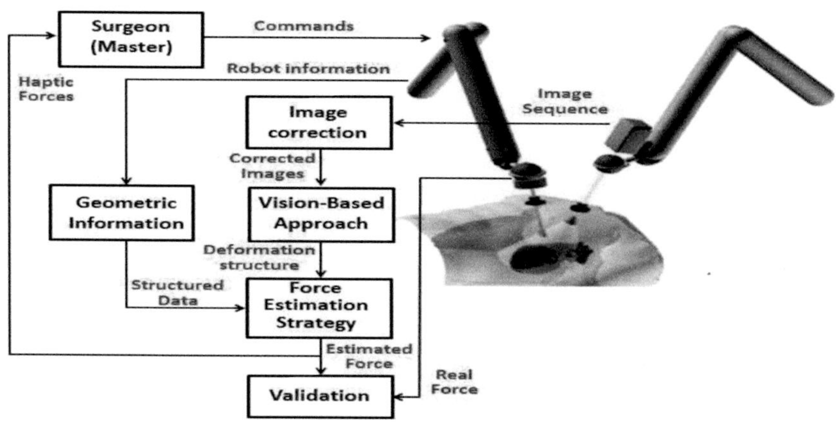

Fig. 19 Schema of a VBFM System.

Our work consists of a force estimation methodology for Minimally Invasive Surgery based on Recurrent Neural Networks and Long Short-Term Memory model (RNN-LSTM) to analyze the tissue deformation and associate this measurements to the equivalent force applied. The computation of the 3D deformation structure is based on the minimization of energy functional.

This information, together with the geometric data (i.e. robot information), constitutes the input to the RNN-LSTM architecture. The results, fig. 20, show that this method works with any tissue model.

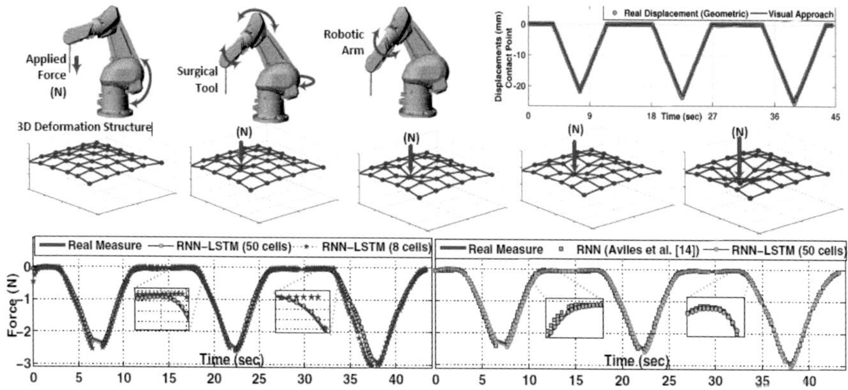

Fig. 20 Top row: Observed and measured displacements at contact point. Middle row: the 3D deformation structure of the tissue, obtained by our vision approach, plotted at different time instants. Bottom row: comparison between the real force measure and the estimated by the RNN-LSTM architecture and at right by our previous work.

5.2 The Microsurgery Worm

Our department closely collaborates with the prestigious medical center Parc Taulí, specialized in the revascularization of amputee limbs. One of their demands is the development of a robotized station for microsurgery; a surgical technique which requires operating with high accuracy in small workspaces. Hard training is needed to reach the desired dexterity, although the workspace limitations should not be a handicap for a teleoperated system. Our solution is based on designing small robots, working individually and cooperatively with their respective tools, compatible with the small working space. The teleoperated system is expected to contribute in a similar way as the microscope used for visualization does. While the microscope enlarges the visualized area, a change of scale in teleoperation reduces the magnitude of movements, increasing manual dexterity and accuracy.

In order to overcome these constrains, we have designed the M*icrosurgery worm*, a small robot with 6 degrees of freedom at the end effector. The robot architecture allows high precision movements at the tool tip. Due to the nature of this surgical technique, the forces applied by the tools are low; hence small motors with encoders are suitable for the intervention. Another characteristic of this robot is the end effector, which has been designed to hold standard tools currently used in conventional surgery. They are substantially cheaper than those articulated, or even those motorized robotic tools. Moreover, these tools can be easily replaced by a fast plug-and-play system. The purpose is to reduce the intervention times, whether for the setup preparation or in tool changes during the operation.

The general design of the microsurgery robot is shown in fig. 21. In working conditions, its dimensions are 30 cm height and the extended arm reaches a distance of 60 cm, approximately.

The M*icrosurgery worm* has been validated through simulation. The prototype is pending for construction to verify its viability and suitability for this kind of surgery.

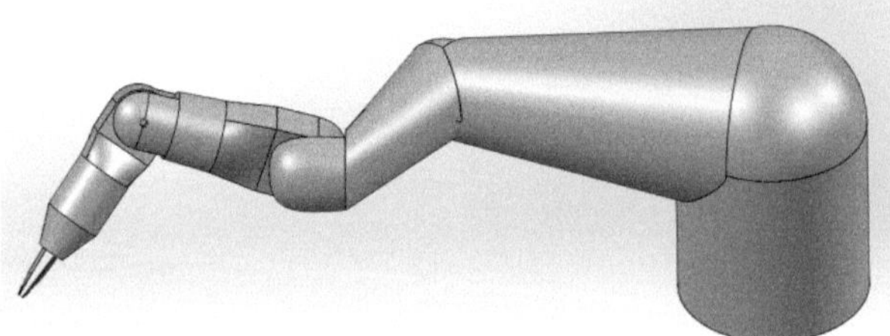

Fig. 21 The Microsurgery worm.

References

1. Mettlere, L.: Historical Profile of Kurt Karl Stephan Semm, Born March 23, 1927 in Munich, Germany, Resident of Tucson, Arizona, USA Since 1996. JSLS **7**(3), 185–188 (2003)
2. Rattner, D., Kalloo, A.: SAGES/ASGE Working Group on Natural Orifice Translumenal Endoscopic Surgery, "White Paper", Surg. Endosc. **20**, 329–333 (2006)
3. Nickles, A., et al.: Laparoendoscopic Single-Site Surgery in Gynaecology: A New Frontier in Minimally Invasive Surgery. Journal of Minimal Access Surgery **7**(1), 71–77 (2011)
4. Kim, S., Landman, J., Sung, G.: Laparoendoscopic Single-Site Surgery With the Second-Generation Single Port Instrument Delivery Extended Reach Surgical System in a Porcine Model. Korean J. Urol. **54**(5), 327–332 (2013)
5. Amat, J.: Robotized instruments for Laparoscopic Surgery, Patent number ES2388867
6. European Patent Office: Annual Report 2014. http://www.epo.org/about-us/annual-reports-statistics/annual-report/2014.html
7. Sparapani, T.: Surgical Robotics and the Attack of the Patent Trolls, FORBES/TECH. http://ieeexplore.ieee.org/xpls/abs_all.jsp?arnumber=1549967&tag=1
8. Bayle, B., Joinie-Maurin, M., Barbe, L., Gangloff, J., de Mathelin, M.: Robot interaction control in medicin and surgery: original results and open problems. In: Computational Surgery and Dual Training, pp 169–161 (2014)
9. Pacchierotti, C., Tirmizi, A., Prattichizzo, D.: Improving transparency in teleoperation by means of cutaneous tactile force feedback. ACM Transactions on Applied Perception **11**, 1 (2014)
10. Puangmali, P., Liu, H., Seneviratne, L.D., Dasgupta, P., Althoefer, K.: Miniature 3-Axis Distal Force Sensor for Minimally Invasive Surgical Palpation. IEEE Trans. on Mechatronics **17**(4), 646–656 (2012)
11. Aviles, A., Alsaleh, S.M., Sobrevilla, P., Casals, A.: Force-Feedback Sensory Substitution using Supervised Recurrent Learning for Robotic-Assisted Surgery. IEEE Int Conf. Engineering in Medicine and Biology (2015)
12. Greminger, M.A., Nelson, B.J.: Vision-based force measurement. IEEE Trans on Pattern Analysis and Machine Intelligence **26**(3) (2004)

Part XI
Visual Maps in Robotics

Ontologies Applied to Surgical Robotics

P.J.S. Gonçalves

Abstract The paper presents current efforts and methods presented by the research community to represent knowledge to be used, in a machine readable format, for surgical robotics. Ontologies from the medical field are surveyed, to be aligned with robotic ontologies to obtain proper surgical robotic ontologies. The later, are valuable tools that combine surgical protocols, machine protocols, anatomical ontologies, and medical image data. An orthopaedic robot surgical ontology, for knowledge representation, is presented and briefly discussed. The system based on ontologies uses dedicated algorithms, devices and merges existing medical and robotic ontologies to obtain a common ontology framework.

Keywords Surgical robotics · Ontologies · Knowledge representation

1 Introduction

Ontologies are considered to be a winning modelling approach in human–machine collaboration, regarding cognition and knowledge-sharing. A ontology can serve as a communication middle-layer between humans and robots. Classically, core-level and sub-level ontologies are distinguished, and this latter can be built based on certain tasks or applications [1]. One of the basic requirements for any type of robot communication (whether with other robots or humans) is the need for a common vocabulary along with clear and concise definitions. With the growing complexity of behaviours that robots are expected to perform as well as the need for multi-robot and human-robot collaboration, the need for a standard and well defined knowledge

P.J.S. Gonçalves(✉)
School of Technology, Polytechnic Institute of Castelo Branco, Castelo Branco, Portugal
e-mail: paulo.goncalves@ipcb.pt

P.J.S. Gonçalves
IDMEC / LAETA, Instituto Superior Técnico, Universidade de Lisboa, Lisbon, Portugal

L.P. Reis et al. (eds.), *Robot 2015: Second Iberian Robotics Conference*,
Advances in Intelligent Systems and Computing 418,
DOI: 10.1007/978-3-319-27149-1_37

representation is becoming more evident. The existence of such a standard knowledge representation will:

- more precisely define the concepts in the robots knowledge representation;
- ensure common understanding among members of the community;
- facilitate more efficient data integration and transfer of information among robotic systems.

Most recently, the IEEE RAS Ontologies for Robotics and Automation Working Group (ORA WG) presented a notable effort to provide a consensus-based core ontology in the domain. Their aim was to link existing ISO, IEC, etc. standards and current research efforts and new regulatory frameworks to a generic Robotics and Automation Ontology [2, 3]. From the perspective of this working group, an ontology can be thought of as a knowledge representation approach that represents key concepts with their properties, relationships, rules and constraints. Whereas taxonomies usually provide only a set of vocabulary and a single type of relationship between terms (usually a parent/child type of relationship), an ontology provides a much richer set of relationships and also allows for constraints and rules to govern those relationships. In general, ontologies make all pertinent knowledge about a domain explicit and are represented in a computer-interpretable format that allows software to reason over that knowledge to infer additional information. Having further defined the Core ontology [4], the next step is to develop sub-ontologies, e.g., for industrial robotics, surgical robotics, and so on. It is essential during the particular process affecting the medical domain to:

- ensure common understanding both among members of the engineering and clinical community;
- facilitate efficient data integration from medical ontologies;
- facilitate efficient component integration;
- facilitate more efficient information transfer among medical electrical equipment and robotic systems.

There have been some examples of medical robotic ontologies, including the REHABROBO-ONTO [5], the Surgical Workflow Ontology (SWOnt) [6], the Neurosurgery Robotic Ontology (NRO) [7], and (OROSU) [8, 9], Ontology for Robotic Orthopaedic Surgery. Recently the use of ontologies increased, e.g., in laparoscopic surgery (LapOntoSPM) [10] and for human–computer interaction in surgery [11].

Amongst other benefits, ontologies allow a perfect combination of surgical protocols, machine protocols, anatomical ontologies, and medical image data. With these four factors in the control loop of the robot, the surgical procedures will have an increase in the quality of monitoring and surgical outcomes assessment. Moreover the surgeon can perform surgical navigation with anatomical orientation, using state-of-the-art control architectures of robots, defined in the ontology.

The next sections of the paper will focus on generic medical ontologies, used in medicine, at section 2. Section 3 will present the existing ontologies for surgical robotics and interactions between humans and computer in the operating room.

Section four present further developments on the ontology for orthopaedic surgery. The paper ends with conclusions and presentation of possible future work.

2 Medical Knowledge Representation Using Ontologies

The medical community have a long history modelling knowledge, making the terminology used in clinical practice, clear and explicit to the community. As such, there exist a large number of databases of terminology. These databases have smoothly changed to ontologies over the years, making the modelled knowledge much more complete and rigorous. In the remainder of this section, will be presented the ontologies categorized by its purpose, e.g., data management, clinical diagnosis. In section 3, will be presented the ontologies that exist for the operating room, i.e., applied for surgery.

Since the mid 1990's, several projects have started this path: GALEN [12]; MENELAS [13]; SNOMED-RT [14]; UMLS©[15]; SNOMED-CT [16].

The results of the GALEN project are now open to the community, through Open-GALEN, although it is no longer actively maintained. It describes the anatomy, surgical deeds, diseases, and their modifiers used in the definitions of surgical procedures [12].

The MENELAS project delivered an access system for medical record using natural language, where a knowledge management tool was developed to browse the domain ontology and knowledge gathered via clinical data [13].

SNOMED-RT was initially developed as a reference terminology to enable user interfaces, electronic messaging, or natural language processing, to the medical community. This system evolved to SNOMED-CT ©, when its prior was merged to the United Kingdom National Health Service Clinical Terms. It is now a US standard for electronic health information exchange in Interoperability Specifications.

The Unified Medical Language System, UMLS©©[15], is a set of files and software that brings together many health and biomedical vocabularies and standards to enable interoperability between computer systems.

The presented projects, and the large amount of knowledge therein, led to large number of ontologies to serve the biomedical community, namely for: biomedical data management, clinical diagnosis and surgery. All of those are important to gather the knowledge needed to develop a robotic surgery ontology. In fact, we need to collect knowledge on how the biomedical data flows in the hospital ICT infrastructure, on how clinical diagnosis is performed (e.g., how to interpret medical images). In the following are presented state-of-the-art developments in these two topics.

Knowledge based systems are quickly gaining its position in healthcare institutions, giving in most cases a first diagnosis screening, based on the gathered clinical data of the patient and the knowledge models. Data mining or ontology based systems are two possible applications of artificial intelligence in this scope. In [17] was created a clinical recommender system based on a data mining system, using conditional probabilities to infer a recommendation. This system makes use of a medical information system, nowadays an essential part of healthcare institutions.

Ontologies can be used to model the knowledge system, as presented in the following cases. In [18] is developed an ontology driven system to obtain clinical guidelines to deliver a standardized care to patients. This system is a multi-agent system, composed by medical doctors, specific field ontologies, services, medical records, and so on, to obtain a medico-organisational ontology. In [19] the authors have designed a pre-operative assessment decision support system based in ontologies. The system is capable, based on the ontology and the patient data, to deliver personalised reports, perform risk assessment and also clinical recommendations. In [20] is presented and evaluated an ontology for guiding appropriate antibiotic prescribing, using them as an efficient decision support system. The paper focus on the development issues for representing and maintaining antimicrobial treatment knowledge rules, that generate alerts to provide feedback to clinicians during antibiotic prescribing. Also in [21] is presented and ontology for a healthcare network, based on a terminology database, to obtain a tool called "virtual staff" that enables cooperative diagnosis. Taking in mind the current section and the goal of the paper, the robot must be placed in the equation. In the next section will be presented ontologies currently developed for Medical/Surgical Robotics.

3 Ontologies for Medical/Surgical Robotics

Several works were performed to develop a first robot ontology, which are depicted in [2], and the references therein. Within these efforts, the current section presents an overview of the state-of-the-art ontologies for robotics, with special focus on the suited ones for Medical/Surgical Robotics, and also on how the surgical process management is performed in the operating room.

From the knowledge models presented in the previous ontologies, the next logic step is to enter the operating room, to obtain surgical models. In this section are presented ontologies that can model the surgical workflow.

For obtaining ontologies, information can be obtained from text-books, medical reports, or even by tracking surgical instruments during surgery. In [22] were developed strategies for neurosurgery, based on medical text-books. Actually the ontologies are used to obtain the requirements for a neurosurgery simulator. Also in [23] were developed ontologies from medical text reports, here applied to surgical intensive care. In [24] was developed a system based on a sensor based surgical device and an ontology to obtain Surgical Process Models. Surgical instruments were tracked during the surgical workflow, and the data gathered in an ontology server.

Further, surgical conceptual knowledge was modelled using ontologies, in [6], with special focus on Computer Aided Surgery. There, the authors developed Surgical Ontologies for Computer Assisted Surgery (SOCAS), based on the General Formal Ontology (GFO) [25] and the Surgical Workflow Ontology (SWOnt) [26].

The works related to the surgical workflow, recently ended in the development of an ontology for assessment studies in human–computer interaction [11]. This achievement is a major breakthrough since it firstly proposes a methodology to facilitate the planning, implementation and documentation of studies for human–computer interaction in surgery. It helps to navigate through the whole surgery (and its surgical procedures) in the form of a kind of standard or good clinical practice, based on the ontological frameworks. In other words, the framework is able to model the surgical workflow for clinical assessment and also for interaction with machines, e.g., a surgical robot.

At this stage medical ontologies that can be used in the operating room are defined in the literature, that allows the integration of robotic ontologies during surgery. In the following are depicted robot ontologies and the efforts to introduce them in the operating room.

Robotics development is pushing robots to closely interact with humans and real world unconstrained scenarios, that surely complicate robot tasks. In this context, complex control systems should be developed. These systems will have to interact with similar complex systems, like humans interact with each other and machines. For that, standardization and a common understanding of concepts in the domains involved with robots and its workplaces, should be pursued. Taking this in mind, IEEE started to gather knowledge from experts in academia and industry to develop ontologies for robotics and automation, [2]. The working workgroup developed ontologies for the Core Robotics domain [3], with possible applications in industrial, and/or surgical robots. The later with special interest to robotic surgery. As presented above, in [27] is presented the interconnections between ontologies and standards to obtain useful standardized systems to speed-up robotic development.

Related to the application to robotic ontologies, other ontologies arrive to the community. Existing ontologies comprise: rehabilitation robotics, [5], field robots [28], industrial robotics [29], amongst others surveyed in [3]. The fundamental part of this type of systems is its knowledge base that contains the necessary relationships and representations to allow for application in its specific domains. Moreover, its structure is agile, allowing the use of the knowledge model to ease the programming burden of new activities/tasks [29]. More related to the medical field, RehabRobo-Onto, [5], the rehabilitation robotics ontology provides a representation of the available information about rehabilitation robots in a structured form. From this, it provides the underlying mechanisms for translational physical medicine, from bench-to-bed and back, and personalized rehabilitation robotics [5].

There exist examples of surgical robotic ontologies, that try to model the knowledge related to the surgical procedures needed to perform specific surgeries. One of the first results, and more generic one, was the Surgical Workflow Ontology (SWOnt) [6]. This ontology was used as background knowledge basis for applications such as a workflow editor that records the flow of the events happening in the operating room.

The next steps were dedicated to the development of sub-domain ontologies for specific surgery fields, e.g., neurosurgery, orthopaedic, and laparoscopic.

The Neurosurgery Robotic Ontology (NRO) [7], was developed with the following objective: to specify the tools and procedures used for a particular operational scenario, using a standard tool to design each component. In other words, to have a final set-up which is safe and possibly interchangeable if the surgical scenario is changed, without re-designing the whole architecture, based on the ontological description of a surgical task to be performed by the surgical robot.

The Ontology for Robotic Orthopaedic Surgery (OROSU) [8, 9], was developed merging existing medical ontologies, i.e., the base ontologies, to represent the knowledge to be used, in a machine readable format, during surgeries. It was applied to Hip Surgery surgical procedures.

Recently the use of ontologies increased, e.g., in laparoscopic surgery (LapOntoSPM) [10]. This work, was developed to facilitate knowledge and data sharing. Using the modelled knowledge for laparoscopic surgeries. The authors applied it to situation interpretation, i.e., the recognition of surgical phases based on surgical activities. With this work, steps were done towards unified benchmark data sets for robotic surgical procedures.

The next section will present current efforts in orthopaedic surgery ontology development.

4 Example Application

In [9] are presented the foundations to obtain a solid ontology for robotic orthopaedic surgery, OROSU, the following parts are to be represented in such a model:

- a Biomedical / Human Anatomical Ontology;
- a Robot Ontology;
- how to represent and manage clinical data, e.g., image, case, patient data.

For the Biomedical / Human Anatomical Ontology, several ontologies have been proposed in the literature. The main sources are in the *NCBO BioPortal* [30], and the *Open Biological and Biomedical Ontologies* [31]. SNOMED-CT [16] within the *NCBO BioPortal* provided the basic definitions for the human anatomy, and the clinical concepts, used in OROSU.

Although other ontologies have been developed in recent years for the robotics domains, the orthopaedic ontology, is based in the recently develop IEEE ontology for the Core Robotics domain, *CORA*, [3], that is the standard ontology for robotics. This is the first ontological standard available for the robotics community.

KnowRob [32, 33] as defined by its authors is a "knowledge processing system that combines knowledge representation and reasoning methods with techniques for acquiring knowledge and for grounding the knowledge in a physical system. Also can serve as a common semantic framework for integrating information from different sources". In this paper, *KnowRob*

OROSU used *KnowRob* [32, 33] for task (surgical procedures) definition, and also as an engine to process the ontology, i.e., for reasoning. This framework is also capable to include knowledge provided: from the robot; from observation of humans (both trough perception capabilities); and also from the internet.

In the following are presented the main classes defined in the ontology and its relations. The CORA ontology takes basic definitions from the IEEE Suggested Upper Merged Ontology (SUMO) ontology (http://www.ontologyportal.org/), as depicted in the following figures.

Figure 1 depicts the *SUMO:Device* classes that exists in OROSU. The class *CORA:RobotPart*, the *SurgicalDevice* class, and the robot class *CORA:Robot*, all derive from a device *SUMO:Device*.

The OROSU ontology, applied to Robotic Hip Surgery, also contains a module related to the actions to be taken during orthopaedic surgery, e.g., to process the sensor data (point clouds) obtained during surgery. Figure 2 presents some algorithms performed during intra-operative or pre-operative scenarios for orthopaedics, related to Hip Resurfacing and developed in [34].

The sensor data is obtained from the devices depicted in figure 3, like *CTimaging* or *USimaging* to gather for example point clouds, and then obtain the 3D model of the bone pre- or intra-operatively.

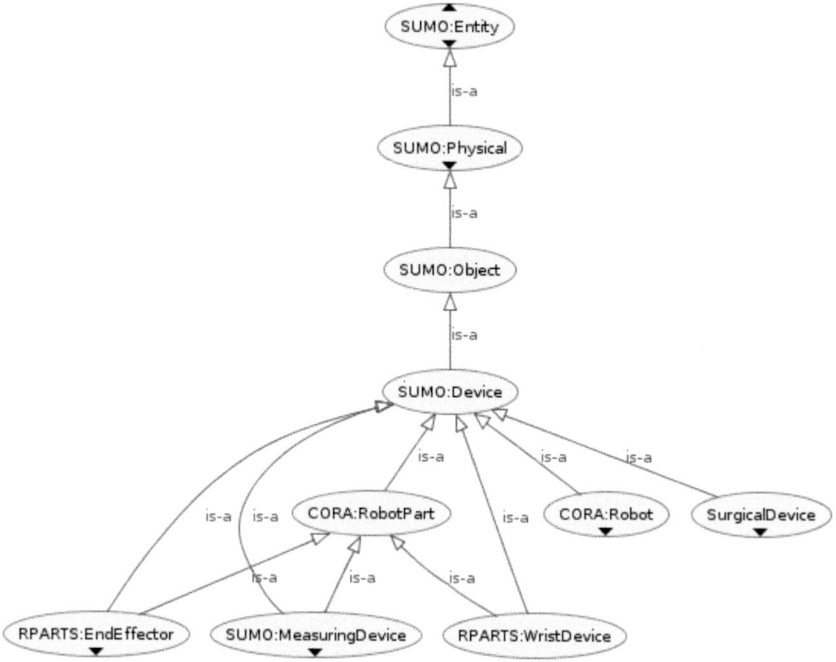

Fig. 1 An example of some devices defined in the ontology.

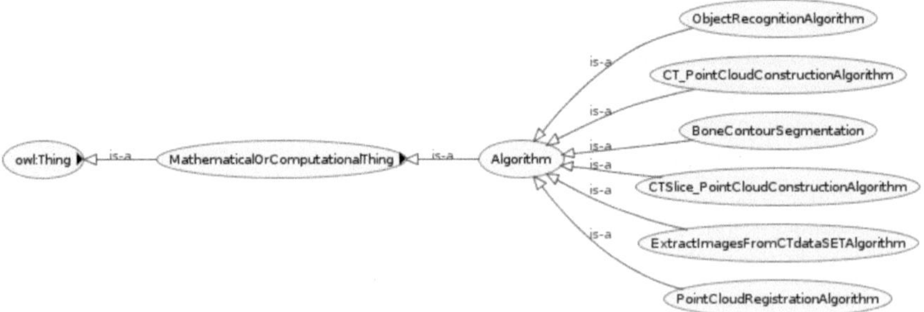

Fig. 2 An example of the algorithms needed to work with bone point clouds.

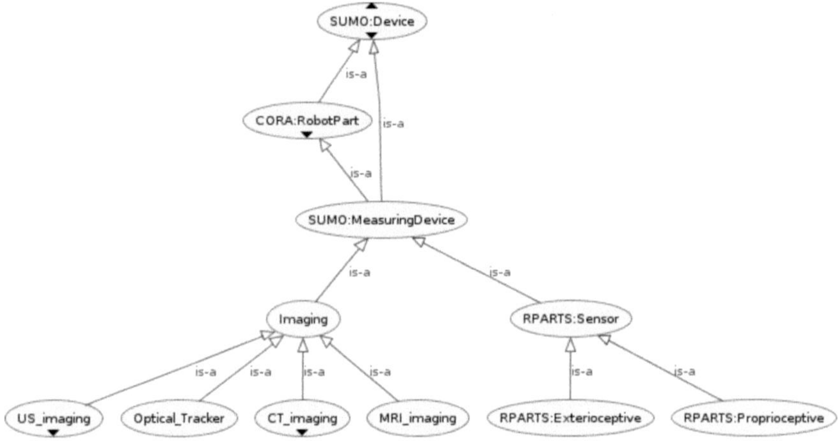

Fig. 3 An example of the measuring devices defined in the ontology.

5 Conclusions and Future Work

Ontologies allow a perfect combination of surgical protocols, machine protocols, anatomical ontologies, and medical image data. With this four factors in the control loop of the robot, the surgical procedures will have an increase in the quality of monitoring and surgical outcomes assessment. Moreover the surgeon can perform surgical navigation with anatomical orientation, using state-of-the-art control architectures of robots, defined in the ontology. The system based on ontologies can also be used to verify surgical protocols, by tracking surgical devices during surgery or simply by reasoning on the data gathered during surgery.

Although the previously defined aims to knowledge modelling using ontologies in surgical robotics, are not yet fulfilled, a large effort has been done by the research community. This fact is clear from the state-of-the-art presented. The developed frameworks however need further development, mainly during clinical trials in a realistic operating room. In the short term, the definition of benchmark tests to surgical workflow analysis is mandatory, in terms of standardization and also to obtain a robust knowledge representation to be used in a robotic operating room.

For the presented application, further work will be focused on the continuous development of the orthopaedic ontology, adding new tools and actions, e.g., a workflow analysis system and/or other surgical applications in orthopaedic surgery.

Acknowledgments This work was partly supported by the Project, UID/EMS/50022/2013, through FCT, under LAETA/IDMEC/CSI.

References

1. Guarino, N.: Semantic matching: formal ontological distinctions for information organization, extraction, and integration. In: Pazienza, M. (ed.) Information Extraction a Multidisciplinary Approach to an Emerging Information Technology. Lecture Notes in Computer Science, vol. 1299, pp. 139–170. Springer, Heidelberg (1997)
2. Schlenoff, C., Prestes, E., Madhavan, R., Gonçalves, P., Li, H., Balakirsky, S., Kramer, T., Miguelanez, E.: An IEEE standard ontology for robotics and automation. In: Proc. of the IEEE/RSJ Intl. Conf. on Intelligent Robots and Systems (IROS), pp. 1337–1342 (2012)
3. Prestes, E., Carbonera, J.L., Fiorini, S.R., Jorge, V.A., Abel, M., Madhavan, R., Locoro, A., Gonçalves, P., Barreto, M.E., Habib, M., Chibani, A., Gérard, S., Amirat, Y., Schlenoff, C.: Towards a core ontology for robotics and automation. Robotics and Autonomous Systems **61**(11), 1193–1204 (2013)
4. Fiorini, S.R., Carbonera, J.L., Gonçalves, P., Jorge, V.A., Rey, V.F., Haidegger, T., Abel, M., Redfield, S.A., Balakirsky, S., Ragavan, V., Li, H., Schlenoff, C., Prestes, E.: Extensions to the core ontology for robotics and automation. Robotics and Computer-Integrated Manufacturing **33**, 3–11 (2015). Special Issue on Knowledge Driven Robotics and Manufacturing
5. Dogmus, Z., Erdem, E., Patoglu, V.: RehabRobo-Onto: Design, development and maintenance of a rehabilitation robotics ontology on the cloud. Robotics and Computer-Integrated Manufacturing **33**, 100–109 (2015). Special Issue on Knowledge Driven Robotics and Manufacturing
6. Mudunuri, R., Burgert, O., Neumuth, T.: Ontological modelling of surgical knowledge. In: Fischer, S., Maehle, E., Reischuk, R. (eds.) GI Jahrestagung. LNI., GI, vol. 154, pp. 1044–1054 (2009)
7. Perrone, R., Nessi, F., De Momi, E., Boriero, F., Capiluppi, M., Fiorini, P., Ferrigno, G.: Ontology-based modular architecture for surgical autonomous robots. In: The Hamlyn Symposium on Medical Robotics, p. 85 (2014)
8. Gonçalves, P.: Towards an ontology for orthopaedic surgery, application to hip resurfacing. In: Proceedings of the Hamlyn Symposium on Medical Robotics, London, UK, pp. 61–62, June 2013
9. Gonçalves, P.J., Torres, P.M.: Knowledge representation applied to robotic orthopedic surgery. Robotics and Computer-Integrated Manufacturing **33**, 90–99 (2015). Special Issue on Knowledge Driven Robotics and Manufacturing
10. Kati, D., Julliard, C., Wekerle, A.L., Kenngott, H., Mller-Stich, B., Dillmann, R., Speidel, S., Jannin, P., Gibaud, B.: Lapontospm: an ontology for laparoscopic surgeries and its application to surgical phase recognition. International Journal of Computer Assisted Radiology and Surgery, 1–8 (2015)
11. Machno, A., Jannin, P., Dameron, O., Korb, W., Scheuermann, G., Meixensberger, J.: Ontology for assessment studies of humancomputer-interaction in surgery. Artificial Intelligence in Medicine **63**(2), 73–84 (2015)

12. Rector, A., Nowlan, W.: The GALEN project. Comput. Methods Programs Biomed. **45**, 75–78 (1994)
13. Zweigenbaum, P.: Menelas: Coding and information retrieval from natural language patient discharge summaries. In: Advances in Health Telematics, 82–89. IOS Press (1995)
14. Spackman, K.A., Campbell, K.E., Cote, R.A.: SNOMED RT: A reference terminology for health care. J. of the American Medical Informatics Association, 640–644 (1997)
15. Pisanelli, D.M., Gangemi, A., Steve, G.: An ontological analysis of the UMLS metathesaurus. Proceedings AMIA 1998, vol. 5, pp. 810–814 (1998)
16. Wang, A.Y., Sable, J.H., Spackman, K.A.: The snomed clinical terms development process: refinement and analysis of content. In: Proc AMIA Symp., pp. 845–849 (2002)
17. Duan, L., Street, W.N., Xu, E.: Healthcare information systems: data mining methods in the creation of a clinical recommender system. Enterprise Information Systems **5**(2), 169–181 (2011)
18. Isern, D., Snchez, D., Moreno, A.: Ontology-driven execution of clinical guidelines. Computer Methods and Programs in Biomedicine **107**(2), 122–139 (2012)
19. Bouamrane, M.M., Rector, A., Hurrell, M.: Experience of using owl ontologies for automated inference of routine pre-operative screening tests. In: Patel-Schneider, P., Pan, Y., Hitzler, P., Mika, P., Zhang, L., Pan, J., Horrocks, I., Glimm, B. (eds.) The Semantic Web ISWC 2010. Lecture Notes in Computer Science, vol. 6497, pp. 50–65. Springer, Heidelberg (2010)
20. Bright, T.J., Furuya, E.Y., Kuperman, G.J., Cimino, J.J., Bakken, S.: Development and evaluation of an ontology for guiding appropriate antibiotic prescribing. Journal of Biomedical Informatics **45**(1), 120–128 (2012)
21. Dieng-Kuntz, R., Minier, D., Ruzicka, M., Corby, F., Corby, O., Alamarguy, L.: Building and using a medical ontology for knowledge management and cooperative work in a health care network. Computers in Biology and Medicine **36**, 871–892 (2006)
22. Audette, M., Yang, H., Enquobahrie, A., Finet, J., Barre, S., Jannin, P., Ewend, M.: The application of textbook-based surgical ontologies to neurosurgery simulation requirements. International Journal of Computer Assisted Radiology and Surgery **6**(1), 138–143 (2011)
23. Moigno, S.L., Charlet, J., Bourigault, D., Degoulet, P., Jaulent, M.: Terminology extraction from text to build an ontology in surgical intensive care. In: Proceedings of the AMIA 2002 Annual Symposium, pp. 430–434 (2002)
24. Neumuth, T., Czygan, M., Goldstein, D., Strauss, G., Meixensberger, J., Burgert, O.: Computer assisted acquisition of surgical process models with a sensors-driven ontology. In: M2CAI workshop, MICCAI, London (2009)
25. Herre, H.: General formal ontology (GFO): a foundational ontology for conceptual modelling. In: Poli, R., Healy, M., Kameas, A. (eds.) Theory and Applications of Ontology: Computer Applications, pp. 297–345. Springer, Netherlands (2010)
26. Neumuth, T., Jannin, P., Strauß, G., Meixensberger, J., Burgert, O.: Validation of knowledge acquisition for surgical process models. Journal of the American Medical Informatics Association **16**, 72–80 (2009)
27. Haidegger, T., Barreto, M., Gonçalves, P., Habib, M.K., Ragavan, V., Li, H., Vaccarella, A., Perrone, R., Prestes, E.: Applied ontologies and standards for service robots. Robotics and Autonomous Systems **61**(11), 1215–1223 (2013)
28. Dhouib, S., Du Lac, N., Farges, J.L., Gerard, S., Hemaissia-Jeannin, M., Lahera-Perez, J., Millet, S., Patin, B., Stinckwich, S.: Control architecture concepts and properties of an ontology devoted to exchanges in mobile robotics. In: 6th National Conference on Control Architectures of Robots, Grenoble, France, INRIA Grenoble Rhône-Alpes, 24 p., May 2011
29. Balakirsky, S.: Ontology based action planning and verification for agile manufacturing. Robotics and Computer-Integrated Manufacturing 33, 21–28 (2015). Special Issue on Knowledge Driven Robotics and Manufacturing
30. Noy, N.F., Shah, N.H., Whetzel, P.L., Dai, B., et al.: Bioportal: Ontologies and integrated data resources at the click of a mouse. Nucleic Acids Research **37**(s2), 170–173 (2009)

31. Smith, B., Ashburner, M., Rosse, C., Bard, J., Bug, W., Ceusters, W., et al.: The OBO Foundry: Coordinated evolution of ontologies to support biomedical data integration. National Biotechnology **37**, 1251–1255 (2007)
32. Tenorth, M., Beetz, M.: Knowrob–knowledge processing for autonomous personal robots. In: Proc. of the IEEE/RSJ Intl. Conf. on Intelligent Robots and Systems (IROS), pp. 4261–4266 (2009)
33. Tenorth, M., Beetz, M.: KnowRob - A Knowledge Processing Infrastructure for Cognition-enabled Robots. International Journal of Robotics Research (IJRR) **32**(5), 566–590 (2013)
34. Gonçalves, P., Torres, P., Santos, F., António, R., Catarino, N., Martins, J.: A vision system for robotic ultrasound guided orthopaedic surgery. Journal of Intelligent & Robotic Systems, 1–13 (2014)

Low Cost, Robust and Real Time System for Detecting and Tracking Moving Objects to Automate Cargo Handling in Port Terminals

Victor Vaquero, Ely Repiso, Alberto Sanfeliu, John Vissers and Maurice Kwakkernaat

Abstract The presented paper addresses the problem of detecting and tracking moving objects for autonomous cargo handling in port terminals using a perception system which input data is a single layer laser scanner. A computationally low cost and robust Detection and Tracking Moving Objects (DATMO) algorithm is presented to be used in autonomous guided vehicles and autonomous trucks for efficient transportation of cargo in ports. The method first detects moving objects and then tracks them, taking into account that in port terminals the structure of the environment is formed by containers and that the moving objects can be trucks, AGV, cars, straddle carriers and people among others. Two approaches of the DATMO system have been tested, the first one is oriented to detect moving obstacles and focused on tracking and filtering those detections; and the second one is focused on keepking targets when no detections are provided. The system has been evaluated with real data obtained in the CTT port terminal in Hengelo, the Netherlands. Both methods have been tested in the dataset with good results in tracking moving objects.

Keywords Object detection · Object tracking · DATMO · Multi-hypothesis tracking · Autonomous driving · Autonomous transportation of cargo

1 Introduction and State of the Art

In the actual globalized world the volume of international trade, of which up to 90% is fully containerized, keeps rising up. Cargo handling at port terminals is a key step

V. Vaquero(✉) · E. Repiso · A. Sanfeliu
Institut de Robotica i Informatica Industrial, CSIC-UPC, Barcelona, Spain
e-mail: {vvaquero,erepiso,sanfeliu}@iri.upc.edu

J. Vissers · M. Kwakkernaat
TNO, Helmond, The Netherlands
e-mail: {john.vissers,maurice.kwakkernaat}@tno.nl

L.P. Reis et al. (eds.), *Robot 2015: Second Iberian Robotics Conference*,
Advances in Intelligent Systems and Computing 418,
DOI: 10.1007/978-3-319-27149-1_38

across the international supply chain network [2], and there is an increase demand of automatizing systems.

In order to improve efficiency, increase capacity and speed up operations to meet future demands, port terminals requires the use of advanced technologies and automation. Changes in port operations have been gradually introduced, and this trend seems to continue, as it is shown that one of the least efficient and most costly processes in ports come from internal transportation in non-automated terminals [13]. Therefore, Automated Container Terminals (ACTs) with Automated Storage and Retrieval Systems - ASRS [11] or Automated Guided Vehicles - AGVs [13] are nowadays getting more presence around the world. In Europe, the European Combined Terminal (ECT) in Rotterdam, the Netherlands, is one of the most automated container terminal in the world.

However, most of the automatic cargo handling systems existing nowadays in ports and terminals have a common drawback: they use predefined fixed travel paths based on different guidance systems (such as rails, markers, wires or magnetic paths), which do not exploit the whole port open area and make vehicles to suffer from problems as collisions or deadlocks [6].

The presented work, which is inside the European Project *Cargo ANTs*, aims to take a step further in the safe freight transportation by creating free-travelling smart vehicles - such as Automated Guided Vehicles (AGVs) and Automated Trucks (ATs) - that can co-operate in highly dynamic shared workspaces as cargo port terminals. Specifically, this paper presents a real time, low cost and simple - yet robust - Detector and Tracking of Moving Objects (DATMO) algorithm based on single layer laser-scanner data. Although the amount of information of a single layer scanner is low, our previous experience in using it for localization, navigation [12] and tracking people in urban sites allows us to explore its use in port terminals. In this way, one of the motivations of this work was to extend the limits of actual 2D rangefinders, squeezing to the maximum the little information obtained, but still providing a robust and real time performance in challenging environments.

DATMO algorithms have presence in research of autonomous vehicles and urban robotics since long time. Typical algorithms takes the general approach (extended from Computer Vision algorithms) of dividing between object detection in static frames (fixed time) and tracking the detected objects along the time [9], [8].

Differently, the presented moving objects detector not only uses single frames, but generates an inter-frame object association to filter static objects, so that reducing the tracking load. This technique is of special interest in places as the port container terminals where there is not a predefined environment structure. Container terminals have no walls, buildings or signs, and containers are stacked in different ways in specific areas and these stacks can change in the same day. Moreover, the containers are transported by straddle carriers, AGVs or trucks and they can be found in any place in the port terminal which make even more complicate to differentiate what is moving and what is static.

Existing tracker algorithms perform associations between objects on different frames by means of closeness and propagate the tracked object positions. In [7], this propagation is done by using a linear and curvilinear constant trajectory, in addition with an unscented Kalman filter. Other approaches use hypotheses to choose the best movement trajectory of the object such as [4], which also uses an extended Kalman filter to estimate the object motion using a different model for pedestrians and vehicles.

In contrast, the presented tracking approach uses a multi-hypothesis algorithm for objects association. And a Kalman filter with linear constant velocity propagation is also used for simplicity to predict positions, as it has been checked that performs as well as curvilinear trajectory models due to the closeness of the detections in time, and reduce considerably the algorithm complexity. Furthermore, the presented tracking system improves the estimation of object motions by analysing them through a window of the associated previous tracks.

The rest of this paper is organized as follows. Section 2 details the components of the presented DATMO system. Section 3 shows the dataset obtained from a real port environment, as well as the experiments that support the system performance. Finally, conclusions are presented in Section 4.

2 Detection and Tracking Moving Objects in Ports

Common DATMO systems divide the process in two separate phases: firstly an object detector extracts objects from static frames of data (frame by frame), and secondly, a tracker algorithm takes the detected object and assigns a coherent label thorough the time to each object, creating the moving object targets.

The contributions of the DATMO approach presented in this paper are:

- it uses only a single-layer rangefinders sensors to detect and track objects, extracting valuable information from small amount of data. It also speeds up the tracking phase by pre-selecting moving obstacles in the detection phase.
- it is robust to highly dynamic shared workspaces as cargo port terminals, where the location of the containers can vary greatly, obtaining different port configurations: without containers, medium occupied with containers or fully occupied.
- the method is able to detect and track moving objects (straddle carriers without and with containers, automated trucks, AGVs empty or carrying containers, cars and people) that can have the same shape of the static objects (containers).
- it manages to keep targets of objects even when crossing paths, get occluded or are temporally not detected, by means of the multi-hypothesis tracking and its probabilities.
- it filters real static objects from moving ones by comparing its velocities with the vehicle velocity and by taking into account the constancy of the moving detected objects. Also groups separated detections from big moving objects in a singular one obtaining a better estimation of the real shape.
- it has been created and tested for free-travelling autonomous vehicles in port terminal, which are known to be well structured but also highly dynamic environments.

The next sections will explain in more detail both the moving objects detector and the tracking steps of the developed DATMO system to be used for AGVs in the port terminals.

2.1 Moving Objects Detector Description

The different modules of the moving object detector explained next, are shown in the general schema of Figure 1.

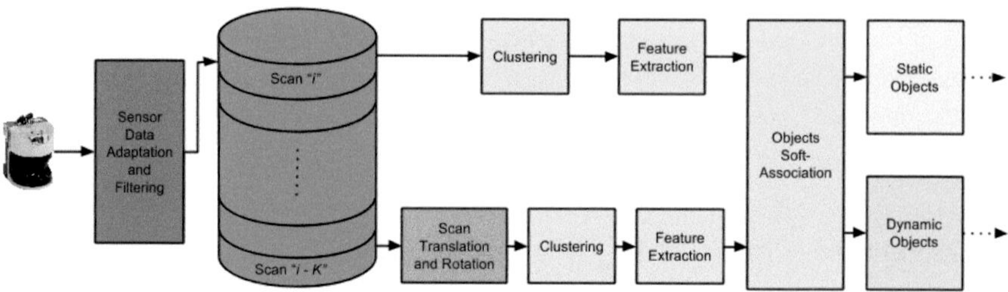

Fig. 1 General schema of the modules in the moving object detector.

Input Data Adaptation and Filtering. Every time i, a 2D laser scan is obtained, and points are filtered eliminating outlier samples (such as those coming from echoes of the laser-scan) as well as points beyond the interaction limit of the vehicle.

Scans Buffering. In contrast to other DATMO systems, this object detector is able to detect moving objects by comparing and matching different frames of the obtained data. These frames are hold in a buffer which size will depend on the applications of the detector. For the tested cargo terminal environment where vehicles speed is up to $6m/s$ (relatively slow) and the sensor provides data at $10Hz$, the buffer has been set to 10 frames so that, the actual objects will be compared with the ones detected 10 frames before (1 second).

Scan Translation and Rotation. Because the vehicle is moving, when the detector compares scan i with scan $i-k$, this last must be translated and rotated to be in the same actual reference frame. For doing this operation the detector use the information from the vehicle odometry. Note that odometry suffers from wheels drifting and other errors, but as in this case only the differential odometry between both scans is used, errors will not affect the algorithm performance.

Data Clustering. Groups of laser hits (clusters) are created by neighbours proximity. In single layer laser sensors, this can be done easily taking advantage of the provided ordered data, so that it is only needed to check the distance to the next point. However, when the detector use another sensor that provides unordered data, a k-nearest neighbours search must be performed in the euclidean space, keeping the simplicity and robustness of the algorithm.

After filtering clusters by a minimum number of points and size, the resulting ones are considered as *objects*. The general object model created is a rectangular box as shown red in Figure 2a, represented by the following state vector:

$$obj_n(frame_i) = [ref_{act}(x, y), w, l, \varphi, np]^T$$

which contains respectively the reference point of the object, its width, length and orientation with respect the longitudinal axis, as well as the number of laser points contained.

Geometric Primitives Extraction. One of the problems of having only a single slice of data from the environment, is the absence of reference points in the partially observed objects.

Objects in port environments, as containers, buildings or vehicles are very internally structured and in most of the cases have well defined straight lines and clear corners. Therefore, a fast way to find common characteristics is to extract its inherent geometric primitives (such as lines or corners) as shown in Figure 2b. At this step, the model of the found objects will be completed with the lists of its detected lines and corners.

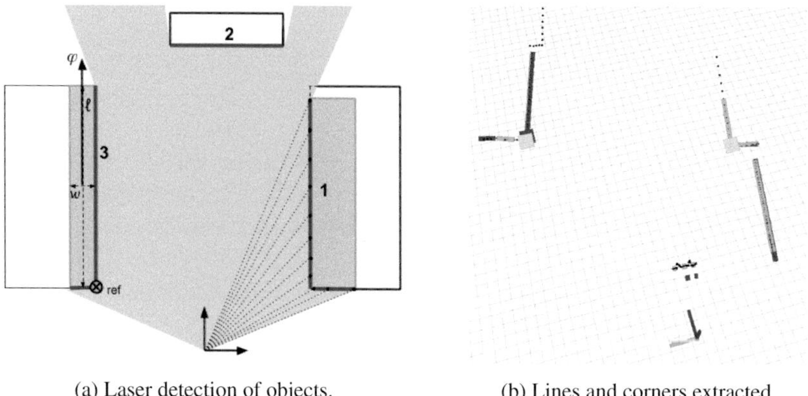

(a) Laser detection of objects. (b) Lines and corners extracted.

Fig. 2 Examples on laser detections of objects and the geometric primitives extracted from real data after the odometry propagation.

Lines are found by means of linear regression over the groups of consecutive points. Then a validation procedure is performed to check if the error is kept low. In a second step, consecutive lines are fused attending to the angle between them. The identified lines are parametrized in a state vector by its initial and final points, its linear regression error, and its director vector as follows:

$$line_j(obj_n) = [init_p(x, y), end_p(x, y), error, vector]$$

The corners are created when two lines cross at an angle larger than $\pi/6$, and are defined in a state vector by its position, orientation and aperture:

$$corner_j(obj_n) = [point(x, y), orientation, aperture]$$

Soft Object Association and Logical Reference Propagation. In this step, objects from the scan i and the scan $i - k$ under a common reference frame are compared to get whether or not an object has moved in time.

Each full object state vector (including lines and corners) of scan i is compared with every full object state vector of scan $i - k$ and its distances are measured. If the distance between an object in the actual scan with another in scan $i - k$ is less than an association threshold, it means that the object has not moved in time so that is tagged as static. Figure 3a shows an example of objects associated (in yellow) and a dynamic one (in red at the older scan and green the new, both in the common reference).

The soft association of objects allows also to propagate logical reference points for partially seen objects with no reference point, as well as information about size of the objects. The associations made by means of corners are considered very reliable and the corner position will be kept as the actual reference point of the object, even though it could not be seen in future frames. Figure 3b shows an example of how the reference of a corner is propagated over partial observations of an object (in T_1 and T_2).

2.2 *Tracking and Filtering Moving Objects*

The developed tracking algorithm for moving objects in port environments is based in the tracking approach developed by Reid [10], although the most probable hypothesis has been selected. Also, the tracker uses a similar approach for confirmation and elimination of the targets as the developed by [1], but in this case it has been modified to take into account the specificity of the port terminals. Moreover, it has been used the object tracking prediction of [5] and the local coordinates system described in [3].

Although this paper is focused in port environments, the tracker here explained has been also tested for people following in urban environments. A general schema of

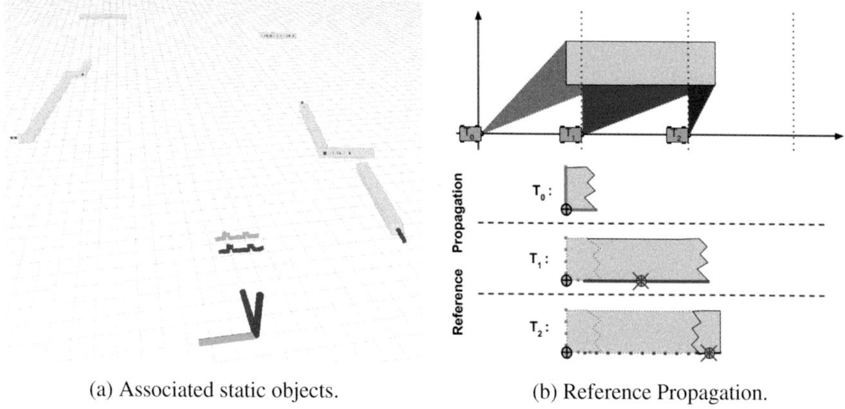

(a) Associated static objects. (b) Reference Propagation.

Fig. 3 Left, associated objects (yellow) tagged as static, and a dynamic detected object (in red the old detection and in green the current). Right, example of the actual reference propagation of a corner over associated objects with partial visibility.

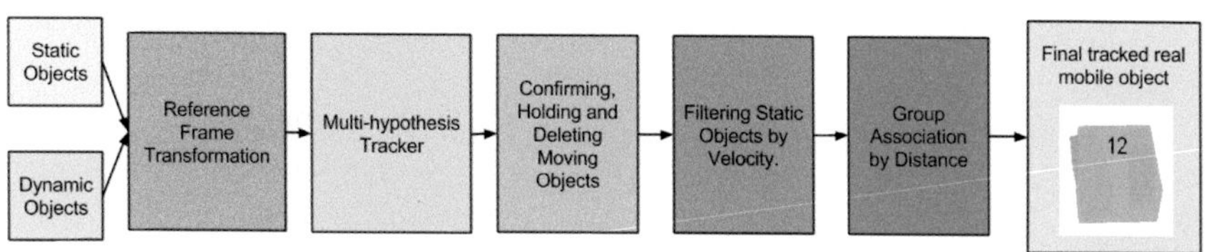

Fig. 4 Modules included in the Tracker system.

the tracker processes is shown in Figure 4. The presented implementation has several new functions and characteristics which will be explained in the next paragraphs.

Confirming, Holding and Deleting Moving Objects. The proposed tracker modifies the probability formula described by Reid in [10], as it can be seen in the equation 1. The first modification is to deal with the false positive detections (not real moving objects) and the second one to take into account the false negatives (non detected moving objects). These improvements allow to have a better control of the confirmation, hold and deletion of the moving targets.

The general formula of the probabilities taking into account the new two terms is (1).

$$P_i^k = \eta'' \cdot P_{\text{det}}^{N_{\text{det}}} \cdot (1 - P_{\text{det}})^{N_{\text{TGT}} - N_{\text{det}}} \cdot \beta_{\text{fal}}^{N_{\text{fal}}} \cdot \beta_{\text{new}}^{N_{\text{new}}} \cdot \\ \cdot \mathcal{N}_{detector}(Z_m - H\bar{x}, B) \cdot P_i^{k-1} \cdot f_{\text{c}}(t) \cdot f_{\text{d}}(t) \quad (1)$$

Where, $P_i{}^k$ is the probability of the actual hypothesis; η'' is the normalization factor; N_{det} is the number of detections; N_{TGT} is the number of existent targets;

N_{fal} is the number of false positives; N_{new} is the number of new targets; P_{det} is the probability value of the detection for the detector; $(1 - P_{\text{det}})$ is the probability of no detection for the detector; β_{fal} is the Poisson distribution that corresponds to the false alarm; β_{new} is the Poisson distribution that corresponds to the new target; $\mathcal{N}_{detector}(Z_m - H\bar{x}, B)$ is the Gaussian probability distribution of the detections for the detector; P_i^{k-1} is the probability of the previous hypothesis. All these terms are explained in detail in [10], where the reader is referred for a deeply explanation. On top of it, the presented tracker introduce two new functions. The first one, $f_{\text{c}}(t)$, is a *confirmation function* created to confirm the tracks as mobile objects, and the second, $f_{\text{d}}(t)$ is a *deletion function* which aim is to slowly erase tracks that have no detection associated. The tracker behaviour for confirming or erasing tracks can be balanced by modifying the parameters of both functions, so that the algorithm can be generalized and used in different environments.

In order to reduce the number of false positives (not real dynamic objects), the *confirmation function* (2) allows to control whether a target is confirmed or not by growing the probability in a slower way than other tracker approaches.

$$f_{\text{c}}(t) = \begin{cases} \beta_{\text{c}} & if\ l = 1 \text{ and target no confirmed} \\ \beta_{\text{c}} + (\triangle\beta_{\text{c}}) * (l - 1) & if\ l > 1 \text{ and target no confirmed} \\ 1 & if \quad \text{target confirmed} \end{cases} \tag{2}$$

The *confirmation function* (2) only applies when the target is still not confirmed and allows a slow rising of the target probability until it is confirmed. In it, the parameter $\beta_{\text{c}} = 0.02$ is the *confirmation constant* and is obtained experimentally for providing continuity to the target probability, which have to continue from the new target probability value of 0.5. The parameter $\triangle\beta_{\text{c}} = 0.01$ (can be between [0, 1]) is the *confirmation increment* that grows up each time the target has been associated to one detection (counted by l, being $l = 1$ for the first association after the new target generation). A low value of $\triangle\beta_{\text{c}}$ implies a slow probability growth.

In the case of false negatives (no detections of moving objects), the presented tracker holds the existent moving targets during short time by means of the new *deletion function* (3). The *deletion function* decreases slowly the probability so its targets take longer time to be deleted and therefore the IDs are kept longer. Figure 5b shows a real experiment where the tracker is able to hold a track when no detection exists. If the detection reappears and is associated to the target, it keeps its ID.

$$f_{\text{d}}(t) = \begin{cases} (\beta_{\text{d}})^{\Delta t} & if \quad \text{target no associated} \\ 1 & if \quad \text{target associated} \end{cases} \tag{3}$$

In the *deletion function* (3), Δt is the increment of time in seconds without detection associated to the target and $\beta_d = 0.99$ is the parameter that allows the slow decreasing of the target probability. The behaviour of this function is like a negative exponential that decrease very slowly.

Filtering Static Objects by Velocity. Apart from the information about the movement of the objects provided by the detector, the tracker algorithm filters static objects that have global velocity 0. This is equivalent to filtering objects which linear velocity, as seen locally by the vehicle, is equal to $\overrightarrow{V}_o^{\,v}$, obtained by means of translation and rotation: $\overrightarrow{V}_o^{\,v} = -(\overrightarrow{V}_R^{\,w} + \overrightarrow{W}_R^{\,w} \times \overrightarrow{r})$; where $\overrightarrow{V}_R^{\,w}$ is the linear velocity of the vehicle in the world coordinates, $\overrightarrow{W}_R^{\,w}$ is the angular velocity of the vehicle in the world coordinates and $\overrightarrow{r}$ is the distance between the vehicle and the tracked object.

Group Association by Distance. The targets of the same type are included in the same group using the following grouping distance (4), this equation is the same that uses [10], but changing the detection by a target.

$$(H\bar{x_1} - H\bar{x_2})^T (H\bar{P_1}H^T + H\bar{P_2}H^T)^{-1}(H\bar{x_1} - H\bar{x_2}) \leq \eta^2 \tag{4}$$

Where, $\bar{x_1}$ and $\bar{x_2}$ are the target states and P_1 and P_2 are its corresponding covariances; H is a measurement matrix defined in the same way as in [10]. Here, η^2 have the same meaning that in [10] but with biggest value, because is associating different targets of the same moving object.

When the tracker groups targets of the same moving object that are detected in a separate way, it takes into account the shape of all of the grouped targets to get the real shape of the moving object. In this case, a new ID for the final group is created, as it can be seen in Figure 5a.

3 Experiments

This work has been perfomed under the Cargo ANTs project, and a dataset has been created for testing the developed algorithms. Real data from the CTT port terminal in Hengelo, the Netherlands (Fig. 6b), has been captured using the TNO test vehicle that included specific sensors for this project (Fig. 6a). The dataset includes relevant data for *vehicle following* inside the port. For this dataset, a truck has been chosen as target (Fig 6c), because Cargo ANTs is about a mixed situations of AGV's and automated driving trucks. The sensor setup of the TNO test vehicle is shown below:

- A TNO test vehicle from where the host tracking signals were logged (e.g. longitudinal velocity and acceleration).

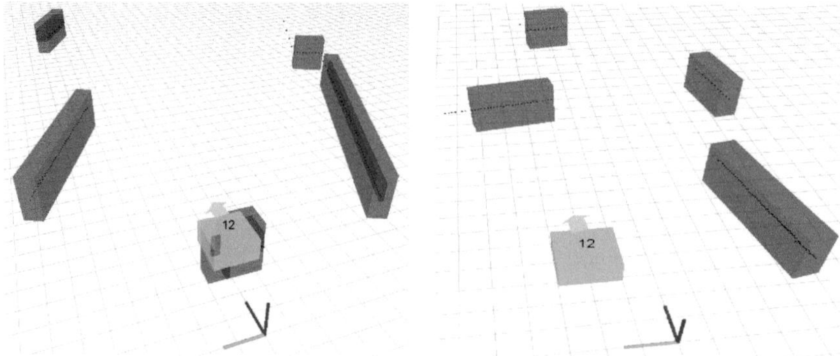

(a) Tracker filtering static objects and grouping dynamic objects.

(b) Tracker holding the real moving object in false negative situations.

Fig. 5 Tracking grouping and holding track examples over real data from a port terminal. In green the real mobile target with its identifier (12). Also, the red objects are the detections, the blue ones are the detection considered like a moving objects, and the reference frame represents the vehicle.

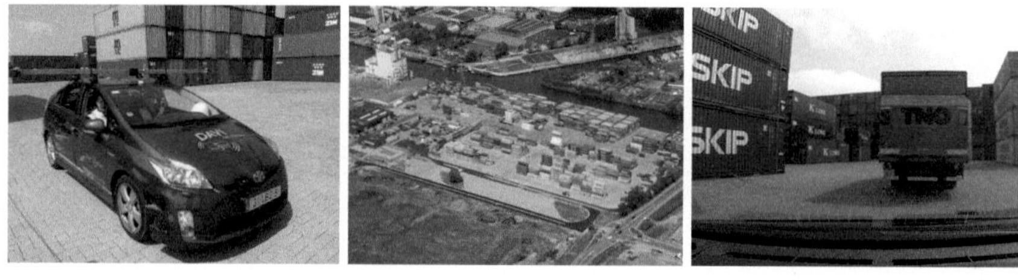

(a) TNO test vehicle setup. (b) Hengelo CTT terminal. (c) Vehicle following case.

Fig. 6 TNO test vehicle containing all the sensors used, CTT environment and image of the front view with the corresponding output of the algorithm.

- 3 Automotive radars (on top of the vehicle) which outputs are both object clusters and trackers.
- 6 Ibeo LUX laser scanners, mounted on the TNO test vehicle, capturing a 360 deg field of view as a pointcloud data.
- Accurate GPS measurement system of OxTS.
- Video camera which captures the front view of the TNO test vehicle.

For the purpose of the research described in this article, only the odometry measurements and a single layer laser scanner from the front view of the TNO test vehicle. Our results are extracted comparing directly to the ground-truth of the provided dataset, which contains 1065 frames of the vehicle following a truck. Two different DATMO system configurations, have been tested in the same dataset, a permissive case, and a restrictive one, which results can be seen in Fig. 7.

The first DATMO approach employs a more permissive detector with lower association threshold that introduce more false positives (static object considered as

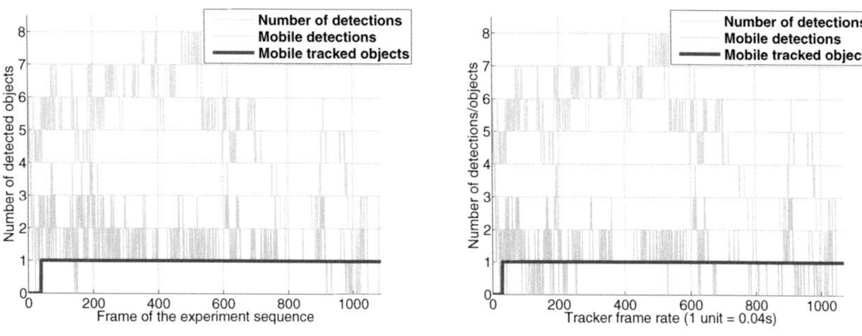

(a) Permissive detector and Tracker object filtering configuration.

(b) Restrictive detector and Tracker holding configuration.

Fig. 7 Results for two approaches of the DATMO system on a vehicle following scenario. Global number of detections are presented in red, dynamic ones in blue and final dynamic objects tracked in green. It can be seen how the restrictive case is more stringent when tagging objects as dynamic,and how the tracker is able to keep track of the vehicle even when no detection is provided.

moving ones), as can be seen in blue in Figure 7a. Over the whole number of 4266 detections, 1539 are considered as dynamic which translates on a 10.95% of false positives and 6.91% of false negatives on moving objects over all the detections. This version uses a tracker configuration focused on filtering those false positives to get the real dynamic detections, which output accomplished the task of following the moving vehicle as shown with the green line.

The second version has a bigger association threshold on the detector phase, and therefore is more restrictive on detecting moving vehicles. It generates less false positives (reduced to 5.09% of all the detections), but introduces more false negatives (moving objects not detected - 13.09%), which are compensated by a tracker focused on keeping track of the true positives. The results of this version can be seen on Figure 7b.

In both cases there exist a configurable initialization period for initializing targets of the objects, imposing a trade-off between performance and initialization time - the longer initialization time, the most real static objects will be filtered and better performance will be obtained. In the permissive approach, an initialization time of 1.5 seconds shows very good results being able to filter correctly all the static objects and keeping track of the moving vehicle. For the restrictive case, as the tracker trust more in the detections and does not have to filter them so much, an initialization time of less than 1 second is enough.

4 Conclusions

This work describes a detection and tracking system for moving objects for AGVs in port container terminals using single layer laser scanner.

The detection algorithm is able to recognize the objects of the port container scenario including its real size. It has been defined two types of detector configurations, one more restrictive than the other, to study the performance of reduction of false positive and false negative.

The developed tracker allows to track only the correct dynamic objects, by eliminating all the false positives. It is also able to keep the target when there are false negatives (no dynamic objects detections). The tracker is able to group the different detections of objects of the same type, thus creating a target of the whole real object with its real shape. Finally, there have been verified two types of tracker configurations, corresponding with the two detector configurations.

The system has been evaluated with real data obtained in the CTT port terminal in Hengelo, the Netherlands. The results shows that the DATMO system accomplish the vehicle following task although little information provided by one single layer laser scanner has been used.

Acknowledgments This work has been partially funded by the EU project CargoANTs FP7-SST-2013-605598 and the Spanish CICYT project DPI2013-42458-P.

References

1. Arras, K.O., Grzonka, S., Luber, M., Burgard, W.: Efficient people tracking in laser range data using a multi-hypothesis leg-tracker with adaptive occlusion probabilities. In: IEEE International Conference on Robotics and Automation (2008)
2. Carlo, H.J., Vis, I.F., Roodbergen, K.J.: Transport operations in container terminals: Literature overview, trends, research directions and classification scheme. European Journal of Operational Research **236**, 1–13 (2014)
3. Corominas-Murtra, A., Pages, J., Pfeiffer, S.: Multi-target & multi-detector people tracker for mobile robots. In: IEEE European Conference on Mobile Robotics (2015)
4. Ess, A., Schindler, K., Leibe, B., Van Gool, L.: Object Detection and Tracking for Autonomous Navigation in Dynamic Environments. The International Journal of Robotics Research **29**(14), 1707–1725 (2010)
5. Ferrer, G., Sanfeliu, A.: Bayesian Human Motion Intentionality Prediction in urban environments. Pattern Recognition Letters **44**, 134–140 (2014)
6. Kim, K.H., Jeon, S.M., Ryu, K.R.: Deadlock prevention for automated guided vehicles in automated container terminals. In: Container Terminals and Cargo Systems: Design, Operations Management, and Logistics Control Issues (2007)
7. Kim, Y.S., Hong, K.S.: A tracking algorithm for autonomous navigation of agvs in an automated container terminal. Journal of Mechanical Science and Technology **19**(1), 72–86 (2005)
8. Mendes, A., Bento, L., Nunes, U.: Multi-target detection and tracking with a laserscanner. In: IEEE Intelligent Vehicles Symposium, pp. 796–801 (2004)
9. Mertz, C., Navarro-Serment, L.E., Maclachlan, R., Rybski, P., Steinfeld, A., Urmson, C., Vandapel, N., Hebert, M., Thorpe, C., Duggins, D., Gowdy, J.: Moving Object Detection with Laser Scanners. Journal of Field Robotics, 1–27 (2012)
10. Reid, D.: An algorithm for tracking multiple targets. IEEE Transactions on Automatic Control **24**(6), 843–854 (1979)
11. Roodbergen, K.J., Vis, I.F.: A survey of literature on automated storage and retrieval systems (2009)
12. Trulls, E., Corominas Murtra, A., Pérez-Ibarz, J., Ferrer, G., Vasquez, D., Mirats-Tur, J.M., Sanfeliu, A.: Autonomous navigation for mobile service robots in urban pedestrian environments. Journal of Field Robotics **28**(3), 329–354 (2011)
13. Vis, I.F.A.: Survey of research in the design and control of automated guided vehicle systems. European Journal of Operational Research **170**(3), 677–709 (2006)

Observation Functions in an Information Theoretic Approach for Scheduling Pan-Tilt-Zoom Cameras in Multi-target Tracking Applications

Tiago Marques, Luka Lukic and José Gaspar

Abstract The vast streams of data created by camera networks render unfeasible browsing all data, relying only on human resources. Automation is required for detecting and tracking multiple targets by using multiple cooperating cameras. In order to effectively track multiple targets, autonomous active camera networks require adequate scheduling and control methodologies. Scheduling algorithms assign visual targets to cameras. Control methodologies set precise orientation and zoom references of the cameras. We take an approach based on information theory to solve the scheduling and control problems. Each observable target in the environment corresponds to a source of information for which an observation corresponds to a reduction of the uncertainty and, as such, a gain in the information. In this work we focus on the effect of observation functions within the information gain. Observation functions are shown to help avoiding extreme zoom levels while keeping smooth information gains.

Keywords Active cameras scheduling · Multi-target tracking · Information gain

1 Introduction

The increased need for surveillance in public places and recent technological advances in embedded video compression and communications have made camera networks ubiquitous. However, at the moment, there are still missing suitable algorithms that are capable of automatically processing so much data captured by so many cameras having few staff members.

One of the issues associated with this problem is the decision on which control action to send to the pan-tilt-zoom cameras (PTZ) in order to successfully carry on the

T. Marques · L. Lukic · J. Gaspar(✉)
Institute for Systems and Robotics (ISR/IST), LARSyS, University of Lisbon, Lisbon, Portugal
e-mail: tiago.d.oliveira.marques@tecnico.ulisboa.pt, luka.lukic@epfl.ch, jag@isr.ist.utl.pt

L.P. Reis et al. (eds.), *Robot 2015: Second Iberian Robotics Conference*,
Advances in Intelligent Systems and Computing 418,
DOI: 10.1007/978-3-319-27149-1_39

desired surveillance tasks. In simple words, the cooperative problem of controlling a network of PTZ cameras for the purpose of active surveillance in a dynamic scenario is that one wants to maintain high zoom levels without losing track on the targets in the scene.

The first autonomous surveillance systems were composed of multiple static cameras, working together to solve some practical tasks of interest such as tracking moving objects. The need for considering overlapping and wide fields (resulting in low-resolution images) of views of deployed cameras has led to the deployment of pan-tilt-zoom cameras in modern surveillance systems.

A number of new architectures appeared, such as master-slave camera configurations, and cooperative smart networks [11]. In the master-slave configuration, static cameras are used for event detection in order to direct the PTZ camera to the target of interest. In contrast, with more complex architecture, both static and PTZ camera streams are used for event analysis. This way, the global state of these systems is composed of both the individual state of each target and the camera state. In both architectures, there is a need for efficient and reliable target tracking methodologies and, in more complex cooperative architectures, there is also a need for camera management methodologies, in order to compute the optimal configuration of the network in order to coordinately maximize the coverage of tracked visual targets in the scene.

1.1 Related Work

The surveillance problem in active camera networks can be divided into two main components. The detection and tracking of visual targets of interest in the scene and the computation of a scheduling policy to control each of the cameras' degree of freedom, in order to take into account the dynamics of the scene.

There are many generic methodologies used for target tracking. The commonly taken approach, based on the Bayes filter [1], [11], [10], is characterized at each iteration by the update of the state estimate based on the predicted state given by the motion model of the target and the observations given by the sensor. Another approach is taken in [7], [6], [3] and [2], where, contrary to the recursive approach, the trajectory is estimated in batches, making available both past and future observations for the estimation of the trajectory at a given time instance.

Regarding camera scheduling, Starzyk *et al.* [12] proposed a complete system for tracking multiple targets using cooperative cameras. The conflicts in behaviors are resolved using a central processor, which combines the individual desired behaviors in a single behavior, which reflects the best compromise between all of them.

The Multi-armed bandit algorithm is introduced in [13], [9] and [8], primarily not being a camera management system, but a decision methodology for coordinately allocating *resources* to *projects*, e.g. robots that can travel to certain locations in order to discover events or network packets that can be routed to various channels in order to maximize the throughput. This framework can be used to model the problem of camera management, as well. Each camera is considered as a resource and each target as a project. The objective is to allocate cameras (resources) to targets (projects) in order to maximize some measure of reward over time.

Another approach, based on the information theory framework, is applied to the problem of camera management in [11] and [10]. This approach is based on the previous work on automatic zoom selection in [5], in which the camera parameters (i.e. the target-camera assignment) are chosen based on the mutual information gain between the state estimate and the estimate given an observation. In this approach, the Extended Kalman Filter is used as the selected tracking algorithm.

The aforementioned approach is the one adopted in this work. However, it is noteworthy that the design choice of the observation functions is still an open and ongoing challenge with the family of the information theory approaches. In the following sections, we will present our particular approach.

1.2 Problem Formulation

A set of pan-tilt-zoom cameras is supposed to track and maintain trajectories of as many targets as possible. The targets are circulating in the environment.

We use an Information Theory framework in which the optimal control policy a^* for each camera is defined by the following maximization problem

$$a^* = \arg\max_a I(x; o) = \arg\max_a H_a(x) - H_a(x|o), \tag{1}$$

where $I(x; o)$ denotes the mutual information gain between the state estimate and the observation, and $H_a(x|o)$ is the conditional entropy of the state estimate given the observation, given that some control a was sent to the camera.

Each target has an assigned Extended Kalman Filter, with the motion model dependent on the targets being tracked and the pinhole camera model. When a target is in the field of view of a camera, a new observation is available and its value is used to update the EKF. In this scenario, there is a reduction of the uncertainty in the target's state and, as such, a positive information gain. On the other hand, when a target is not in the camera's field of view, the observation will not contribute to the reduction of the state uncertainty. In other words, the entropy $H(x) = H(x|o)$ and the information gain will be zero. This way, by maximizing the mutual information gain, the cameras will effectively be in configurations in which more targets are present in the field of view.

The aforementioned problem boils down to how to compute $H(x|o)$ before making an actual observation. This entropy can be computed by taking into account that the tracking is being made by using an EKF. In the Kalman Filter (and so in the Extended Kalman Filter), the assumption is that the state distribution follows a Gaussian distribution with the mean x (the state estimate) and covariance P.

Under this property, $H(x)$ and $H(x|o)$ are both differential entropies of Gaussian distributed variables, given by

$$H(x) = \frac{k}{2}(1 + log(2\pi)) + \frac{1}{2}log(|\Sigma|), \tag{2}$$

where k is the dimension of x and Σ is the covariance matrix of $p(x)$.

The first result consists in that the entropy depends only on the covariance matrix and thus the problem of computing $H(x)$ and $H(x|o)$ can be reduced to computing the covariance of the state estimate P_k and the state estimate after the observation P_{k+1}.

The equations of the Extended Kalman Filter show that the observation z_k is only incorporated in the innovation equation and later used in the state update equation. Therefore, the update covariance matrix $P_{k|k}$ can be computed prior to any observation. The same applies to the conditional entropy $H(x|o)$, as well.

In the most general case, the optimization can be achieved by an exhaustive search procedure over the parameter space of the camera. However, the structure of the sensor can be considered in order to find better optimization strategies.

2 System Overview

The surveillance system consists of a set of active pan-tilt-zoom cameras which acquire images to be processed by a central controller. The controller is responsible for image processing tasks and for keeping track of each of the targets moving in the environment.

Each camera feeds the controller with new image frames and its corresponding state. The controller is responsible for processing the image and extracting target observations by fusing the available information and updating the targets state estimates (see Fig. 1).

The result of the update is then fed to the decision controller, which is responsible for computing a new control policy for each camera.

The detection of new targets can be made both by carefully placed static cameras, or by the same active cameras used in the tracking process. The latter approach is the one followed in [10], however, this topic will not be addressed in this work.

3 Scheduling Cameras

Using the Information Theoretic approach proposed in [10], the camera scheduling problem is modeled as an information gathering problem. For each camera, a choice on the pan-tilt-zoom parameters is made based on the information gain in the state distribution of all targets given possible observations. The optimal pan-tilt-zoom configuration is the one which maximizes the information gain $I(x; o)$ between the

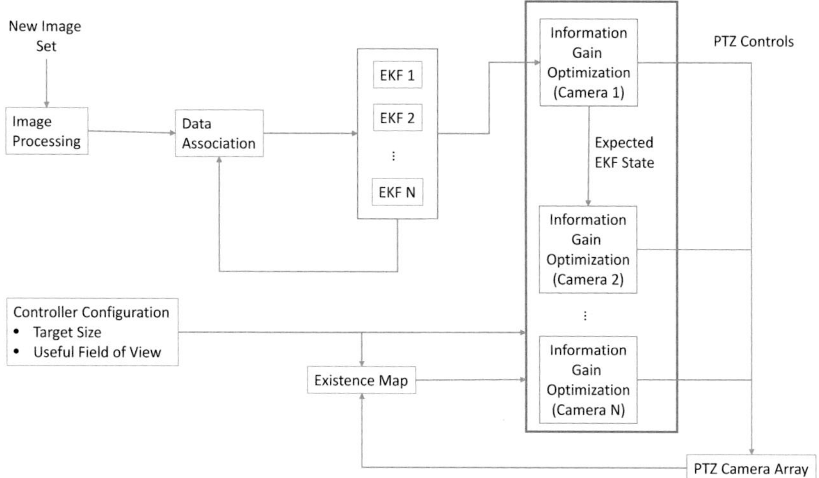

Fig. 1 System architecture

state estimate and the observation. In other words, it is the one that leads to a greater increase in the certainty of the state estimate.

3.1 Single Camera Scenario

In a single camera scenario, a single camera is responsible for tracking and maintaining the trajectories of various targets which move freely across the environment.

Each target is tracked using an Extended Kalman Filter (EKF) with the motion model depending on the kind of target being tracked (e.g. pedestrian or vehicle) and the pinhole camera model as the observation model. Let $h(x)$ denote the sensor model and H_k denote its Jacobian around the predicted state. Converting the pinhole camera model in projective coordinates

$$\begin{pmatrix} u' \\ v' \\ w \end{pmatrix} = P_{3\times4} \begin{pmatrix} x' \\ y' \\ z' \\ h \end{pmatrix} \Rightarrow \begin{cases} u = \frac{f_1(x,y,z)}{f_3(x,y,z)} = \frac{p_{11}(\frac{x}{h})+p_{12}(\frac{y}{h})+p_{13}(\frac{z}{h})+p_{14}h}{p_{31}(\frac{x}{h})+p_{32}(\frac{y}{h})+p_{33}(\frac{z}{h})+p_{14}h} \\ v = \frac{f_2(x,y,z)}{f_3(x,y,z)} = \frac{p_{21}(\frac{x}{h})+p_{22}(\frac{y}{h})+p_{23}(\frac{z}{h})+p_{24}h}{p_{31}(\frac{x}{h})+p_{32}(\frac{y}{h})+p_{33}(\frac{z}{h})+p_{14}h}, \end{cases} \tag{3}$$

where $f_k(x, y, z)$ is the internal product between the kth row of the projective matrix and the real world position in projective coordinates and thus p_{ij} is the value of the projective matrix in the ith row and j^{th} column.

By taking into account the structure of the sensor model, the Jacobian $H(x, y, z) \in \mathbb{R}^{2\times3}$ can be easily computed. Let x_i be the ith state variable of the set $\{x, y, z\}$, and

g_i the ith observation variable from the set u, v, then the partial derivative in the position (i, j) of the matrix is given by

$$\frac{\partial g_i}{\partial x_j}(x, y, z) = \frac{\frac{\partial f_i}{\partial x_j}(x, y, z) f_3(x, y, z) - \frac{\partial f_3}{\partial x_j}(x, y, z) f_i(x, y, z)}{f_3(x, y, z)^2}. \quad (4)$$

The decision on which pan-tilt-zoom command to send to each camera is done by maximizing the mutual information gain $I(x_t;\ o_t)$ for all targets, with x_t being the target state estimate at time t and o_t the observation made.

Using the definitions of mutual information gain $I(x;\ o)$ and conditional entropy $H(x|o)$ [4], the cost function can be expanded

$$a^* = \arg\max_a I(x;\ o) = \arg\max_a H_a(x) - H_a(x|o) \quad (5)$$

$$= \arg\max_a H_a(x) + \int p(o) \int p(x|o) \log p(x|o) \partial x \partial o \quad (6)$$

$$= \arg\max_a H_a(x) + \int_v p(o) \partial o \int p(x|o) \log p(x|o) \partial x + \int_{\bar{v}} p(o) \partial o \int p(x|o) \log p(x|o) \partial x, \quad (7)$$

where distribution $p(o)$ denotes the probability of making an observation of the target. Its domain of integration can be divided into two subdomains: v, which denotes all the camera configurations in which the target is visible and $\bar{v}$, which represents all the configurations in which the target is not visible.

Without further assumptions, this problem is hard to solve. However, recalling the EKF structure, predict and update steps, one obtains the following properties: (i) The probability distribution $p(x|o)$ corresponds to the state distribution after the update step. In other words, $p(x|o) \sim \mathcal{N}(x_{k|k}, P_{k|k})$. (ii) All state variables are gaussian distributed. This means all differential entropies can be computed in closed form according to equation (2). (iii) When there is no observation of the target, the update step is skipped. This means that the state is independent from the observation and $p(x|o) = p(x) \sim \mathcal{N}(x_{k|k-1}, P_{k|k-1})$. Applying these properties into equation (7), the objective function is simplified:

$$a^* = \arg\max_a \left(H_a(x) + \omega(a) H(x^+) + (1 - \omega(a)) H(x^-)\right), \quad (8)$$

where

$$\omega(a) = \int_v p(o) do \quad (9)$$

represents the observation function, which depends on the action a.

Note that all the state variables are Gaussian distributed and, by definition, the differential entropy for Gaussian distributed variables depends only on the variable covariance matrix. By observing the EKF equations, it can be seen that the observation o_k is only used in the innovation equation and in the state update equation. This enables the computation of P^+ without having an observation. Using the definition of differential entropy for a multivariate gaussian distribution (2) with the covariance update equation in the Extended Kalman filter, the cost functional can be simplified

$$a^* = \arg\min_a \omega(a)(\log|P^+| - \log|P^-|) \tag{10}$$

$$= \arg\min_a \omega(a)\log(I - K_k H_k). \tag{11}$$

The choice of the optimal configuration is made by optimizing the sum of the mutual information gains $I(x;\ o)$ for all targets. Despite the elegance of this cost function, in the general case its evaluation is intractable because it requires the exhaustive search over the configuration space of the camera. For each pan-tilt-zoom configuration of the camera, the observation model must be linearized anew and an EKF update must be performed to obtain the new Jacobian matrix H_k and K_k. In order to overcome that problem, in this work each target is modeled as a circle projected into the ground plane. The visible region v of each camera, in the camera coordinates, is also modeled as an ellipse around the center of the image. The term $w(a)$ in equation (11) is computed by projecting the target ellipse onto the image plane and computing its intersection with the visible region v of the camera.

3.2 Multiple Camera Scenario

In a multiple camera scenario, multiple cameras can have observations of the same target. The incorporation of these observations is done using a sequential Kalman Filter. In this variation, a single prediction step is made and each observation is used to make an update so that the update from the observation k uses the estimated position and covariance of the update from the observation $k - 1$.

Each observation contributes to the covariance matrix with a factor of $(I - K_c H_c)$, where K_c is the Kalman gain from the observation of camera c and H_c is the Jacobian of the observation model for camera c.

The mutual information gain of a target for multiple observations

$$I(x;\ o_1, ..., o_C) \propto \sum_{c \in C} log|I - \omega(a) K_c H_c|, \tag{12}$$

is then obtained by combining equation (10) with the new covariance matrix update

$$P^+ = \left(\prod_{c \in C}(I - K_c H_c)\right) P^- . \tag{13}$$

3.3 Observation Function

The observation function $\omega(a)$, equation (11), is a central component of the target tracking methodology as it effects on the convexity of the information gain.

The observation function can be just a flag indicating whether a point representing the target location, in world coordinates, is visible or not in the image:

$$\omega(a) = \begin{cases} 1, & \mathcal{P}(X_{gnd};\ a) \in E_{img} \\ 0, & \text{otherwise} \end{cases} \tag{14}$$

where X_{gnd} is a point in the ground plane representing the target position, $\mathcal{P}(\cdot;\ a)$ is the projection operator on a set of points, which is defined by action a, i.e. in (3) the projection matrix $P_{3\times 4}$ is modified by action a and E_{img} is the ellipse in the image plane concentric with the image rectangle. However, this observation function does not perform a smooth regularization of the cost function, making difficult to design iterative optimization algorithms.

By changing the model of the target from a single point in ground plane into a surface in the ground plane, one obtains a smoother optimization function. Modeling the shape of the target as a rectangle, taking into account the visible part normalized by the area of the imaged shape not truncated by the field of view of the camera, the observation function takes the form

$$\omega(a) = \frac{A(\mathcal{P}(R_{gnd};\ a) \cap E_{img})}{A(\mathcal{P}(R_{gnd};\ a))} \tag{15}$$

where $A(\dot{})$ indicates area of a convex hull of points and R_{gnd} is a rectangle in the ground plane. Alternatively, modeling the target as an ellipse in the ground plane, the function becomes

$$\omega(a) = \frac{A(\mathcal{P}(E_{gnd};\ a) \cap E_{img})}{A(\mathcal{P}(E_{gnd};\ a))} \tag{16}$$

where E_{gnd} is an ellipsis in the ground plane.

4 Experimental Results

The experiments described in this section are based in a virtual reality environment simulating a parking lot (see Fig. 2(a-f)). A number of buses, with the dimension of 2x10x2 m, cross the scene according to predefined trajectories shown in Fig. 2(g-i). These trajectories are generated using the car model $\dot{x} = cos(\theta)sin(\theta)V$, $\dot{y} = sin(\theta)cos(\phi)V$, $\dot{\theta} = \frac{sin(\phi)}{L}V$ and $\dot{\phi} = \omega_s$, where V is the linear velocity of the bus and ϕ is the steering angle, both set using a joystick interface. Tracking is performed

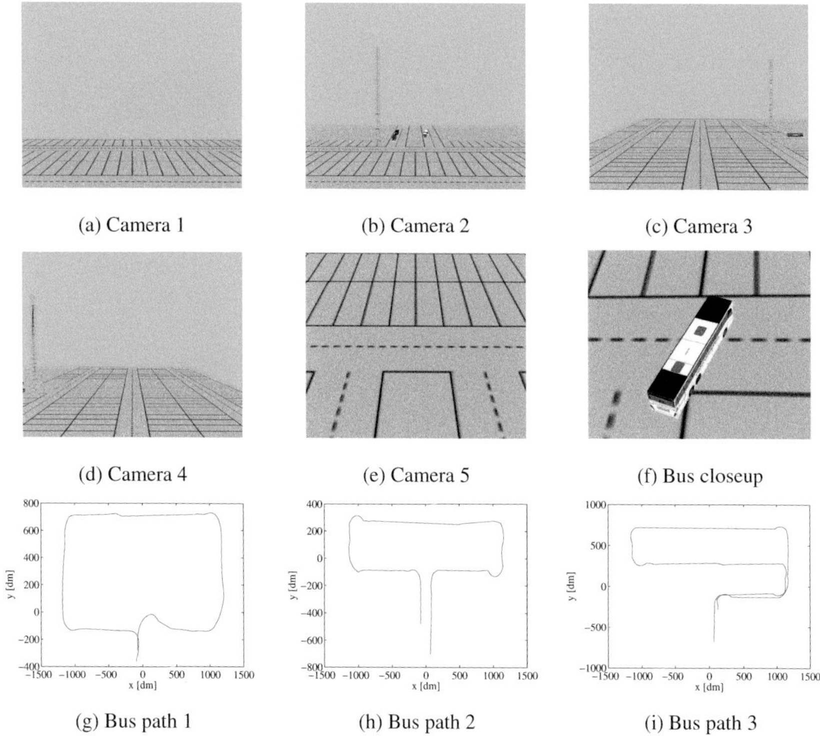

(a) Camera 1 (b) Camera 2 (c) Camera 3

(d) Camera 4 (e) Camera 5 (f) Bus closeup

(g) Bus path 1 (h) Bus path 2 (i) Bus path 3

Fig. 2 Field of view of the cameras at rest orientation and typical bus trajectories.

using five pan-tilt-zoom cameras located at fixed positions on the sides of the parking lot and in the entrance.

In order to evaluate the geometry of the cost function using different observation functions, the cost function was evaluated from a set of values from the pan-tilt-zoom space for a camera. This was achieved by placing a target in a fixed location in front of the camera and making an observation. The camera model and its Jacobian were computed for the new pan-tilt-zoom configuration and, along with the new observation, used to update an existing EKF. The resulting Kalman gain and the linearized observation model were obtained and used in the cost function (11).

Figures 3(a) show the cost function evolution with both the zoom level and the pan-tilt values using the observation function (14). Figures 3(b) show the same evolution as in the previous case, but now using the observation function (16). The term $w(a)$ defined as in (11) changes the cost minimum from extreme values of zoom, pan and/or tilt to within-range, central, values.

The second set of figures was generated in a similar way from the former ones, however sampling the whole camera parameters space. Figure 4 shows slices of the cost function in the camera parameter space (pan-tilt-zoom) using different observation functions as described in Section 3.3. Each row represents a fixed zoom

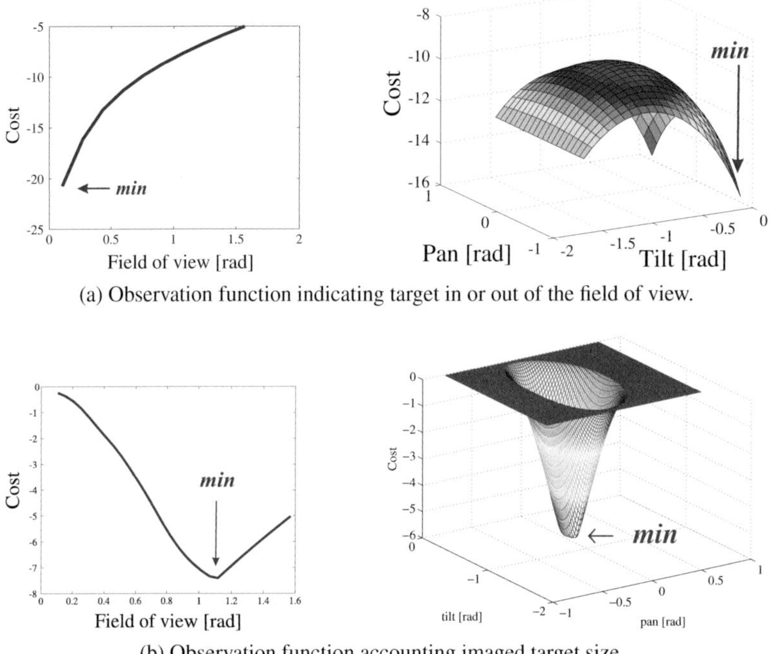

(a) Observation function indicating target in or out of the field of view.

(b) Observation function accounting imaged target size.

Fig. 3 Effect of the observation function into the cost function. In case (a) the observation function $\omega(a)$ is defined by (14) and therefore just indicates the target in or out the field of view. In case (b) $\omega(a)$ is defined by (16) and thus takes in account the imaged size of the target and the field of view. Two plots for each one of the cases, cost vs zoom (left) and cost vs pan and tilt (right).

configuration, with lower rows representing configurations with higher field of view (less zoom).

In column (a) of Fig. 4, is used $w(a)$ defined in (11). In column (b) is considered the same $w(a)$ however, in this experiment, there are two targets in the scene in different locations and with different state estimate covariances. In columns (c) and (d) the term $w(a)$ is the one defined by (15) and (16), respectively.

Figure 4(a) shows that the cost function is generally lower in configurations which make the target appear in the image plane with higher resolution, that is, when the camera is at its highest zoom or when the camera has the target of interest in the corner of the image. This result comes from the influence of the observation model (in particular, its Jacobian) used in the EKF in the cost function. While this is a good result in the sense resolution is maximized, uncontrolled zoom onto the target is not the desired behavior, since a minimum movement can make the target disappear from the camera field of view. Term $\omega(a)$, as defined in (15) or (16), acts a regularizing factor, managing the optimal zoom level to observe the target at constant zoom level and preventing observations just using the "‘corner of the eye"’.

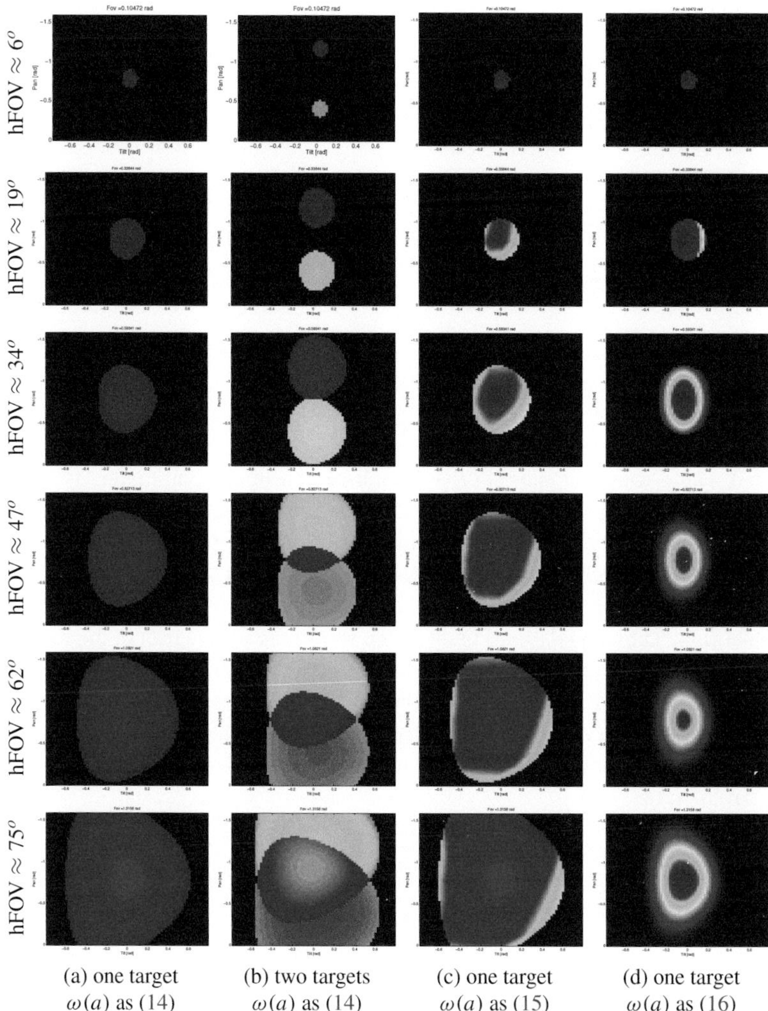

Fig. 4 Cost function slices at fixed field-of-view (zoom). Each row represents a different zoom level (field of view, top row 6[deg] $\approx$ 0.1 [rad], bottom row 75[deg] $\approx$ 1.3 [rad]). Colder colors represent lower cost functions. Configurations in which the target was not visible were assigned high cost. In cases (a,c,d) one camera observes one target. In case (b) one camera observes two targets. The observation function $\omega(a)$ is defined by (14) in cases (a) and (b), and is defined by (15) or (16) in cases (c) or (d), respectively.

In the last experiment four cameras are active and four targets (buses) enter the scene sequentially, separated by approximately 10sec, following different trajectories. The proposed scheduling methodology is tested with two alternative observation functions, (15) or (16). The pan, tilt and zoom parameters are estimated for all cam-

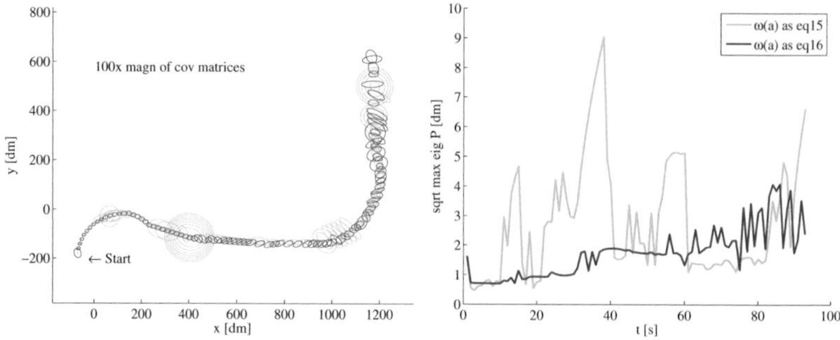

Fig. 5 Scheduling experiment. Uncertainty at the ground plane (left) and uncertainty along time (right), for observation functions defined in (15) or (16).

eras in a round-robin manner (Fig. 1). Each search of the parameters is based in Nelder-Mead simplex direct search and is limited in the number of iterations.

Figure 5 shows the tracking of the first bus along the first 100sec. The plot in the left shows the uncertainty of the EKF of the first bus, on the ground plane, as ellipses corresponding to a 50% confidence level. Covariances are magnified 100× for readability. The second plot, Fig. 5(right), shows the square root of the maximum eigenvalue of the covariance matrix, along time, for both observation functions. Results show that (16) allows for lower and smoother in time localization uncertainty in presence of distractors, as the other buses entering the fields of view of the cameras, due to its smoother nature allowing for more effective searches of the pan, tilt and zoom parameters.

5 Conclusions and Future Work

The results show the influence of the target modeling in the cost function proposed by [10]. By modeling the buses as ellipses projected onto the ground plane and the visible region of the camera as an ellipse in the image plane, the cost functions gain a defined structure, which simplifies the design of fast algorithms to search for the optimal policy.

When a single target is present inside the range of the active camera, the optimal command is to zoom on the target, according to the cost function. The regulator term controls the optimal zoom on the target, independently of the target location.

In a multi-target scenario, the overlapping cost functions can make the optimal command to keep both targets inside the camera's field of view.

The challenges to be addressed in the future include the integration with policies for exploration of unobserved regions, making the system independent of the need to have carefully placed static cameras.

Acknowledgments This work has been partially supported by the FCT project [UID / EEA / 50009 / 2013], and by the EU Project POETICON++ EU-FP7-ICT-288382.

References

1. Arulampalam, M.S., Maskell, S., Gordon, N., Clapp, T., Arulampalam, S.: A tutorial on particle filters for online nonlinear/non-Gaussian Bayesian tracking. IEEE Trans. on Signal Processing **50**(2), 174–188 (2002)
2. Ben Shitrit, H., Berclaz, J., Fleuret, F., Fua, P.: Tracking multiple people under global appearance constraints. In: IEEE Int. Conf. on Computer Vision (ICCV), pp. 137–144. IEEE (2011)
3. Berclaz, J., Fleuret, F., Turetken, E., Fua, P.: Multiple object tracking using k-shortest paths optimization. IEEE Trans. on Pattern Analysis and Machine Intelligence **33**(9), 1806–1819 (2011)
4. Cover, T.M., Thomas, J.A.: Elements of information theory. John Wiley & Sons (2012)
5. Denzler, J., Zobel, M., Niemann, H.: Information theoretic focal length selection for real-time active 3D object tracking. In: IEEE Int. Conf. on Computer Vision (ICCV), pp. 400–407. IEEE (2003)
6. Fleuret, F., Berclaz, J., Lengagne, R., Fua, P.: Multicamera people tracking with a probabilistic occupancy map. IEEE Trans. on Pattern Analysis and Machine Intelligence **30**(2), 267–282 (2008)
7. Jorge, P.M., Abrantes, A.J., Marques, J.S.: Automatic tracking of multiple pedestrians with group formation and occlusions. In: VIIP, pp. 613–618 (2001)
8. Liu, K., Zhao, Q.: Indexability of restless bandit problems and optimality of whittle index for dynamic multichannel access. IEEE Trans. on Information Theory **56**(11), 5547–5567 (2010)
9. Ny, J.L., Dahleh, M., Feron, E.: Multi-UAV dynamic routing with partial observations using restless bandit allocation indices. In: American Control Conf., pp. 4220–4225. IEEE (2008)
10. Sommerlade, E., Reid, I.: Probabilistic surveillance with multiple active cameras. In: IEEE Int. Conf. on Robotics and Automation (ICRA), pp. 440–445. IEEE (2010)
11. Sommerlade, E., Reid, I., et al.: Cooperative surveillance of multiple targets using mutual information. In: Workshop on Multi-camera and Multi-modal Sensor Fusion Algorithms and Applications-M2SFA2 2008 (2008)
12. Starzyk, W., Qureshi, F.Z.: Multi-tasking smart cameras for intelligent video surveillance systems. In: 8th IEEE Int. Conf. on Advanced Video and Signal Based Surveillance (AVSS), pp. 154–159, August 2011
13. Whittle, P.: Restless bandits: Activity allocation in a changing world. Journal of applied probability, 287–298 (1988)

Nearest Position Estimation Using Omnidirectional Images and Global Appearance Descriptors

Yerai Berenguer, Luis Payá, Adrián Peidró, Arturo Gil and Oscar Reinoso

Abstract This work presents an algorithm to estimate the position and orientation of a mobile robot using only the visual information provided by a catadioptric system mounted on the robot. Each omnidirectional scene is described with a single global appearance descriptor. We have developed a description method which is based on the Radon transform. Our localization method compares the visual information captured by the robot from an unknown position with the visual information stored in a previously built map. As a result it estimates the nearest position of this map and the orientation of the robot. We have tested all the algorithms with a virtual database we have built. This database is composed of a set of omnidirectional images captured from different points of an indoor virtual environment. The experiments have allowed us to tune the main parameters and the results show the effectiveness and the robustness of our method.

Keywords Grid map · Omnidirectional images · Global appearance · Radon transform · Computer vision

1 Introduction

Nowadays there are countless kinds of robots with many configurations. Among them, mobile robots have extended due to their flexibility, as they are able to change their position during operation. Usually, these robots have to solve a task autonomously in an unknown environment, so the robot must estimate its position and orientation to be able to arrive to the target point avoiding obstacles. There are

Y. Berenguer(✉) · L. Payá · A. Peidró · A. Gil · O. Reinoso
Departamento de Ingeniería de Sistemas Y Automática, Miguel Hernández University, Elche, Spain
e-mail: {yberenguer,lpaya,apeidro,arturo.gil,o.reinoso}@umh.es
http://arvc.umh.es

L.P. Reis et al. (eds.), *Robot 2015: Second Iberian Robotics Conference*, Advances in Intelligent Systems and Computing 418,
DOI: 10.1007/978-3-319-27149-1_40

two main approaches to solve the localization problem. First, the robot may have a previously created map of the environment. In this case the robot has to estimate its position within this map. Second, the robot may not have any a priori knowledge of the environment thus it has to create the map and calculate its position simultaneously. This problem is named SLAM (Simultaneous Localization And Mapping).

These mapping and localization processes are possible thanks to the robot sensors, such as lasers, encoders, cameras, etc. They provide environmental information to the robot in different ways (e.g. lasers measure the distance to the nearest objects around the robot). This information is processed by the robot to build a map and to estimate its position and orientation. With these data, the robot must be able to carry out its work autonomously.

Along the last years much research has been developed about robot mapping and localization using different kinds of sensors and many algorithms have been proposed to solve these problems. A lot of these works use visual sensors to carry out the localization because this kind of sensors has many possible configurations, relative low cost and they provide the robot with very rich information from the environment. This images permit carrying out other high-level tasks. In this work we use the omnidirectional configuration [15]. We can find many previous works that use omnidirectional images in mapping and localization tasks, such as [3, 9, 14]. Valiente et al. [14] present a comparison between two different visual SLAM methods using omnidirectional images. Mohai et al. [9] propose a topological navigation system using omnidirectional vision. At last, Garcia et al. [3] make a survey of vision-based topological mapping and localization methods.

Traditionally, the developments in mobile robotics using visual sensors are based on the extraction and description of some landmarks from the scenes, such as SIFT (Scale-Invariant Feature Transform) [8] and SURF (Speeded-Up Robust Features) [1] descriptors. This approach presents some disadvantages: the computational time to calculate and compare the descriptors is usually very high thus these descriptors may not be used in real time, and it leads to relatively complex mapping and localization algorithms.

More recently some works propose using the global information of the images to create the descriptors. These techniques have demonstrated to be a good option to solve the localization and navigation problems on the ground plane. Chang et al. [2], Payá et al. [12] and Wu et al. [16] propose three examples of this. In [11], several methods to obtain global descriptors from panoramic scenes are analyzed and compared to prove their validity in map building and localization. The majority of these global appearance descriptors can be used in real time because the computational time to calculate and handle them is low, and they usually lead to more straightforward mapping and localization algorithms.

In this work we propose a solution to the localization problem using only the visual information captured by an omnidirectional system mounted on the robot. This system is composed of a camera pointing to a hyperbolic mirror and it provides the robot with omnidirectional scenes from the environment. We describe each scene using one global-appearance descriptor. Our starting point is a database of omnidirectional images captured on a grid of points in the environment where the robot has to navigate. We face the localization problem as an image retrieval problem.

Comparing to previous works, the contribution of this paper is twofold. First we define a new method to describe the global appearance of omnidirectional images. This method is based on the Radon transform. We have not found any previous work that uses this mathematical transformation in the field of robotics localization. Second we optimize the localization process in a previously built map.

The experiments have been carried out with our own images database that has been created from a synthetic indoors environment.

The remainder of this paper is structured as follows. Section 2 introduces the concept of global appearance and the description method we propose. Section 3 describes our localization method. In section 4 the experiments and results are presented. At last, section 5 outlines the conclusions.

2 Global Appearance of Omnidirectional Images: Radon Transform

Methods based on the global appearance of the scenes constitute a robust alternative compared with methods based on landmarks extraction. The key is that the global appearance descriptors represent the environment through high-level features that can be interpreted and handled easily, and with a reasonably low computational cost.

This section presents the transform we have employed to describe the scenes (omnidirectional images). Each scene is represented through a single descriptor that contains information of the whole appearance without any segmentation or local landmark extraction. We also present the distance measure we use to compare descriptors. Any novel global appearance description method should satisfy some properties: (a) it should make a compression effect in the image information, (b) there should be a correspondence between the distance between two descriptors and the metric distance between the two positions where the images were captured, (c) the computational cost to calculate and compare them should be low, so that the approach can be used in real time, (d) it should provide robustness against noise, changes in lighting conditions, occlusions and changes in the position of some objects in the environment, (e) at last, it should contain information of the orientation the robot had when it captured the image.

The description method we have developed is mainly based on the Radon transform.

2.1 Radon Transform

The Radon transform was initially described in [13]. It has been used in some computer vision tasks, such as shape description and segmentation, such as [5] and [4].

The Radon transform in 2D consists of the integral of a 2D function over straight lines (line-integral projections). This transform is invertible. The inverse Radon transform reconstructs an image from its line-integral projections. By this reason it was initially used in medical imaging (such as CAT scan and Magnetic Resonance Imaging (MRI)).

The Radon transform of a 2D function $f(i, j)$ can be defined mathematically as:

$$\mathcal{R}\{f(i, j)\} = \lambda_f(p, \phi) = \iint_{-\infty}^{+\infty} f(i, j)\delta(p - \overrightarrow{r}\,\widehat{\overrightarrow{p}})didj \tag{1}$$

Where δ is the Dirac delta function ($\delta(x) = 1$ when $x = 0$, and $\delta(x) = 0$ elsewhere). The integration line is specified by the radial vector $\overrightarrow{p}$ that is defined by $\overrightarrow{p} = \widehat{\overrightarrow{p}} \cdot p$ where $\widehat{\overrightarrow{p}}$ is a unitary vector in the direction of $\overrightarrow{p}$. p is the $\overrightarrow{p}$ magnitude:

$$p = |\overrightarrow{p}| \tag{2}$$

The line-integral projections evaluated for each azimuth angle, ϕ, produce a 2D polar function, λ_f, that depends on the radial distance p and the azimuth angle ϕ. $\overrightarrow{r}$ is a cluster of points which is perpendicular to $\overrightarrow{p}$.

The Radon transform of an image $im(i, j)$ along the line $c_1(d, \phi)$ (Figure 1) can be expressed more clearly by the following equivalent expression:

$$\mathcal{R}\{im(i, j)\} = \int_{\mathbb{R}} im(i' \cos\phi - j' \sin\phi, i' \sin\phi + j' \cos\phi)\, ds \tag{3}$$

where

$$\begin{bmatrix} i' \\ j' \end{bmatrix} = \begin{bmatrix} \cos\phi & \sin\phi \\ -\sin\phi & \cos\phi \end{bmatrix} \cdot \begin{bmatrix} i \\ j \end{bmatrix} \tag{4}$$

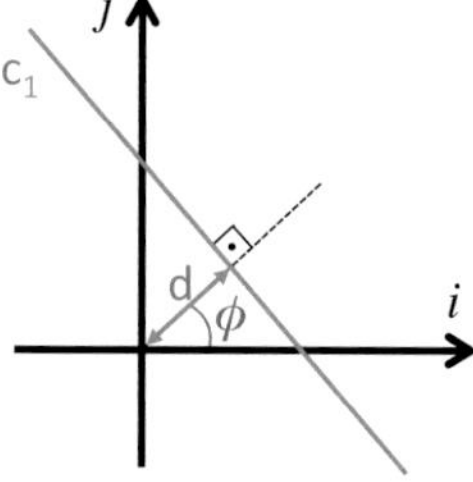

Fig. 1 Line parametrization through the distance to the origin d and the angle between the normal line and the i axis, ϕ.

When the Radon transform is applied to images, it calculates the image projections along the specified directions through a cluster of line integrals along parallel lines in this direction. The distance between the parallel lines is usually one pixel. The Figure 2(a) shows the integration paths to calculate the Radon transform of an image in the ϕ direction, and the Figure 2(b) shows the value of each component of the Radon transform in a simplified notation.

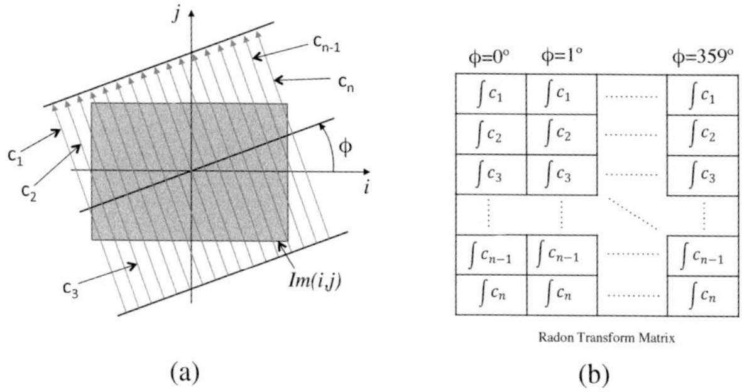

Fig. 2 (a) Integration paths to calculate the Radon transform of the image $im(i, j)$ in the ϕ direction. (b) Radon transform matrix of the image $im(i, j)$.

The Figure 3 shows a sample black and white image, on the left, and its Radon transform, on the right. Furthermore it shows graphically the process to calculate the Radon transform.

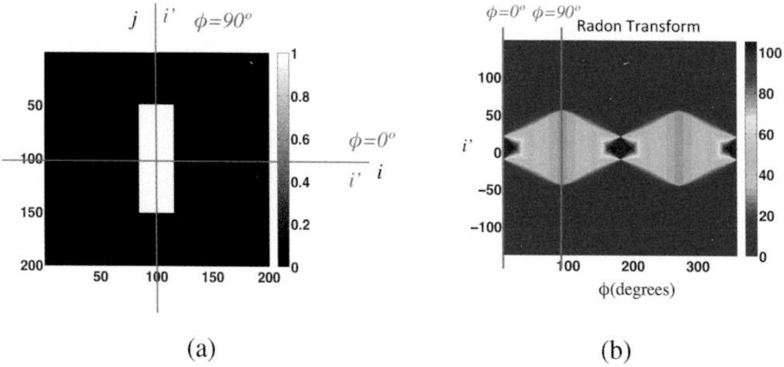

Fig. 3 (a) Sample image. (b) Radon transform of the sample image.

Radon Transform Properties. The Radon transform has several properties that make it useful in localization tasks using images. These properties are the following:

- Linearity: The Radon transform has the linearity property as the integration operation is a linear function of the integrand:

$$\mathcal{R}\{\alpha f + \beta g\} = \alpha \mathcal{R}\{f\} + \beta \mathcal{R}\{g\} \tag{5}$$

- Shift: The Radon transform is a variant operation to translation. A translation of the two-dimensional function by a vector $\overrightarrow{r_0} = (i_0, j_0)$ has a translation effect on each projection. This translation is given by a distance $\overrightarrow{r} \cdot (\cos\phi, \sin\phi)$.
- Rotation: If the image is rotated an angle ϕ_0 it implies a shift ϕ_0 of the Radon transform along the variable ϕ (columns shift).
- Scaling: A scaling of f by a factor b implies a scaling of the d coordinate and amplitude of the Radon transform by a factor b:

$$\mathcal{R}\left\{f\left(\frac{i}{b}, \frac{j}{b}\right)\right\} = |b|\lambda_f\left(\frac{d}{b}, \phi\right) \tag{6}$$

2.2 POC (Phase Only Correlation)

In this subsection we present the method we use to compare the Radon transform of two images.

In general, a function in the frequency domain is defined by its magnitude and its phase. Usually, only the magnitude is taken into account and the phase information is discarded. However, when the magnitude and the phase features are examined in the Fourier domain, it follows that the phase features contain also important information because they reflect the characteristics of patterns in the images.

Oppenheim and Lim [10] have demonstrated this by reconstructing images using the full information from the phase with unit magnitude. This shows that the images resemble the originals, in contrast to reconstructing images using the full information from the magnitude with uniform phase.

POC (Phase Only Correlation), proposed in [7], is an operation made in the frequency domain that provides a correlation coefficient between two images [6]. In our case we compare two Radon transforms but this does not affect the POC performance because the Radon transform can be interpreted as an image.

The correspondence between two images $im_1(i, j)$ and $im_2(i, j)$ calculated by POC is given by the following equation:

$$C(i, j) = \mathcal{F}^{-1}\left\{\frac{\mathbf{IM}_1(u, v) \cdot \mathbf{IM}_2^*(u, v)}{\left|\mathbf{IM}_1(u, v) \cdot \mathbf{IM}_2^*(u, v)\right|}\right\} \tag{7}$$

Where $\mathbf{IM}_1$ is the Fourier transform of the image 1 and $\mathbf{IM}_2^*$ is the conjugate of the Fourier transform of the image 2. $\mathcal{F}^{-1}$ is the inverse Fourier transform operator.

To estimate the distance between two images we have used the following expression:

$$dist(im_1, im_2) = 1 - max\{C(i, j)\} \tag{8}$$

$max\{(C(i, j)\}$ is a coefficient that takes values in the interval $[0, 1]$ and it measures the similitude between the two images.

This operation is invariant against shifts in the i and j axes of the images. Furthermore, it is possible to estimate these shifts Δ_i and Δ_j along both axes by:

$$(\Delta_i, \Delta_j) = argmax_{(i,j)}\{C(i, j)\} \quad (9)$$

If we use Radon transforms of omnidirectional images, the value Δ_i allows us to estimate the change of the robot orientation when capturing the two images.

This way, POC is able to compare two images independently on the orientation and it is also able to estimate this change in orientation.

3 Localization Method

In this section we address the localization problem. Initially, the robot has a map of the environment. It is composed of a set of omnidirectional images along with the position of the capture points (coordinates with respect to a reference system). Then, the robot captures an image from an unknown position (test image). Comparing this image with the visual information stored in the map, the robot must be able to estimate the nearest position of the map.

Since the positions where the map images were captured are known the method we develop is a pure localization method. Also, the robot has no information about its previous position neither its path, so we face the problem as an absolute localization.

This method allows us to know the nearest image of the map (the most similar to the test image). Thanks to this information we know that the robot is located in the surrounding of the point where the corresponding image was captured. The method is detailed in the following subsection.

3.1 Nearest Position of the Map (Image Retrieval Problem)

The operation consists of the following steps:

1. The robot takes an omnidirectional image from its current unknown position (test image). The objective is to estimate this position and the orientation of the robot on the ground plane.
2. This image is transformed using the Radon transform.
3. It is compared with all Radon transforms of the map using the POC comparison. As a result, we know which is the most similar image.
4. The position where this omnidirectional image was captured is the nearest neighbor.
5. Once the nearest position is known, we estimate the orientation of the robot. We compare the Radon transform of the test image with the radon transform of the image extracted at step 3. (Eq. (9)).

After this process we assume the robot is located around this position. The accuracy depends mainly on the distance between the capture points of the map images.

4 Experiments and Results

In this section we present the virtual database created to test our method, we use it to test the method and we show the results obtained.

4.1 *Virtual Database*

In order to check the performance of the proposed technique, we have created a virtual environment that represents an indoor room. In this environment it is possible to create omnidirectional images from any position. This is an advantage since it allows us to create a versatile database to test the localization algorithm. The Figure 5(a) shows a bird's eye view of the environment.

The omnidirectional images have 250x250 pixels and they have been created using a catadioptric system composed of a camera and a hyperbolic mirror whose geometry is described in Figure 4. The parameters used in the mirror equation are $a = 40$ and $b = 160$.

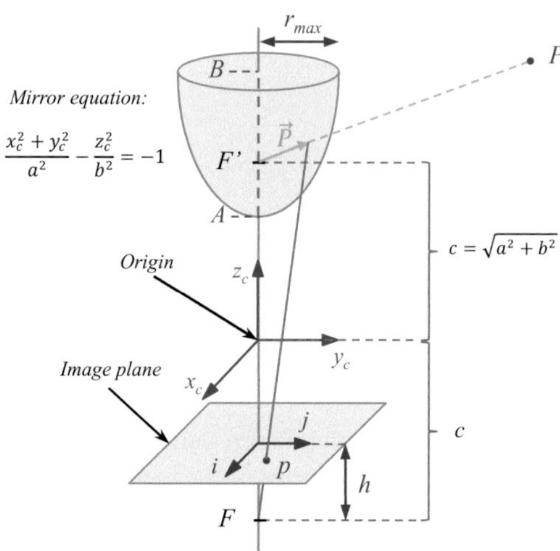

Fig. 4 Catadioptric system used to capture the synthetic omnidirectional images.

Several images have been captured in the environment to create the map from several positions on the floor. The map is composed of 4800 images captured on a 8x6 meters grid with a step of 10 centimeters between positions. To carry out the experiments we can change the number of map images to test the performance of the map when the distance between capture points increases. The Figure 5(b) shows one sample omnidirectional image of the environment created with our program.

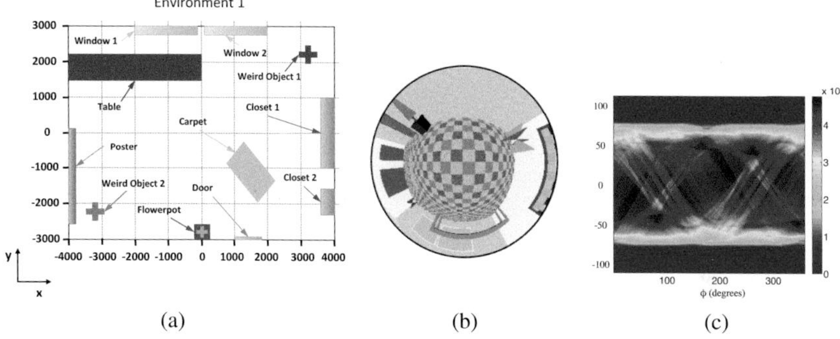

Fig. 5 (a) Bird's eye view of the virtual environment. (b) Example of an omnidirectional image captured from the point x=1500 mm and y=1500 mm. (c) Radon transform of (b).

4.2 *Results Obtained with the Virtual Database*

The performance of our method depends on some parameters, such as the size of the map and the distance between images. In this subsection we compare the results of different tests changing the values of these parameters to test our method. We will study the influence of the distance between map positions in the same area of the environment (size of the grid), so the number of map images is different in each case.

Nearest Position. In this experiment we analyze 4 different step sizes between consecutive map positions. The distances that we will use are 100, 200, 300 and 400 mm. The size of the grid is 8m x 6m in all cases, so the number of map images depends on the step size.

The Figure 6 shows the distance (Eq. (8)) between the Radon transform of a test image and each image of the map (200x200). The position of the test image is x=-2239 mm and y=-1653 mm. And the corresponding position according to this figure (the minimum of the 2D function) is x=-2200 and y=-1600. As we can see, the distance decreases sharply around this position.

The average computation time per iteration in each case is shown in Figure 7. The 100 mm distance between map positions is not advisable because the computational time is much higher than the others.

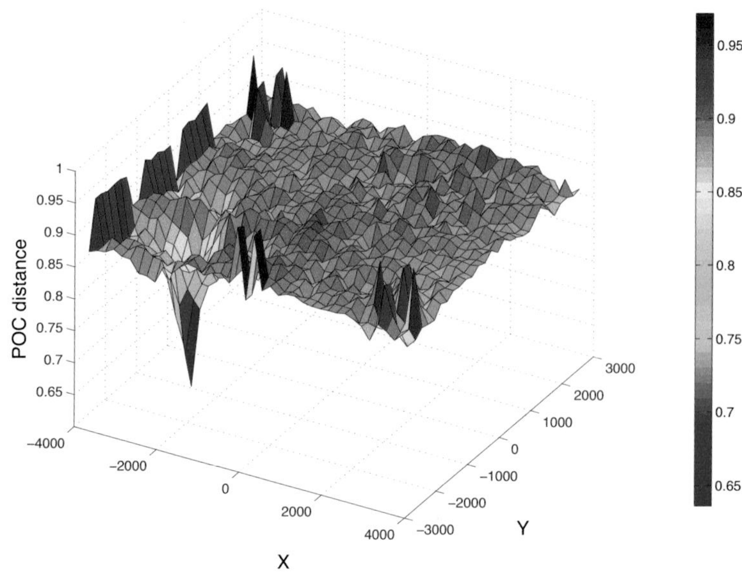

Fig. 6 Example of an iteration of our method in a random position. The position of the test image is x=-2239 mm and y=-1653 mm.

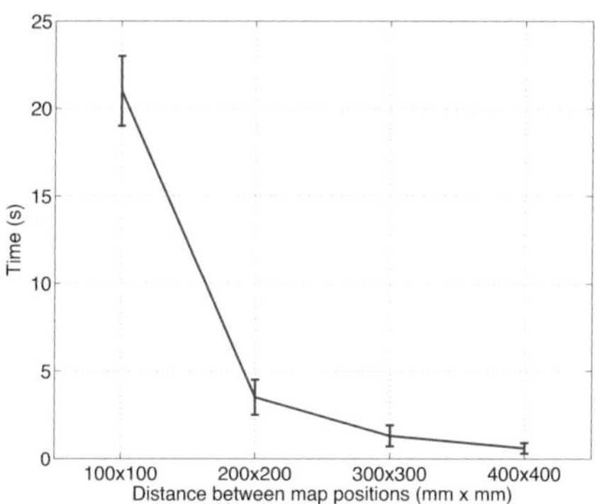

Fig. 7 Computational time for each iteration.

The Figure 8 shows the global results of this test. We have used 3500 test images captured from different random positions of the environment. In each bar, the blue part represents the proportion of correct localizations, the green part represents that the method localizes the robot around the second nearest position, i.e. the position calculated is not the nearest position but rather the second nearest position of the map, and the red part is the proportion of errors for each map size.

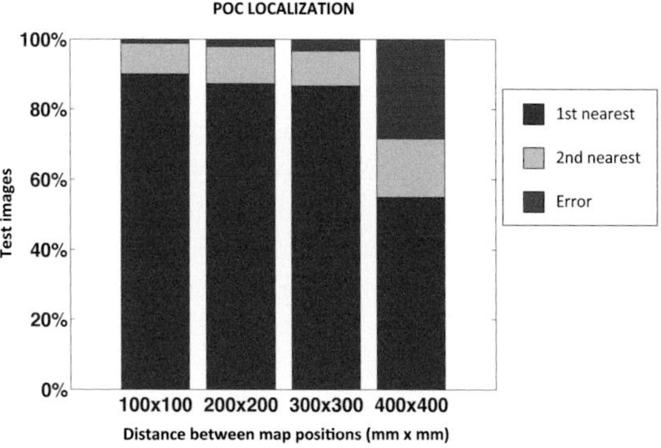

Fig. 8 POC localization experiment.

The average orientation versus the distance between map positions (Eq. (9)) is shown in Figure 9. This error increases when the distance between map positions is higher because the test image and the corresponding map image are less similar.

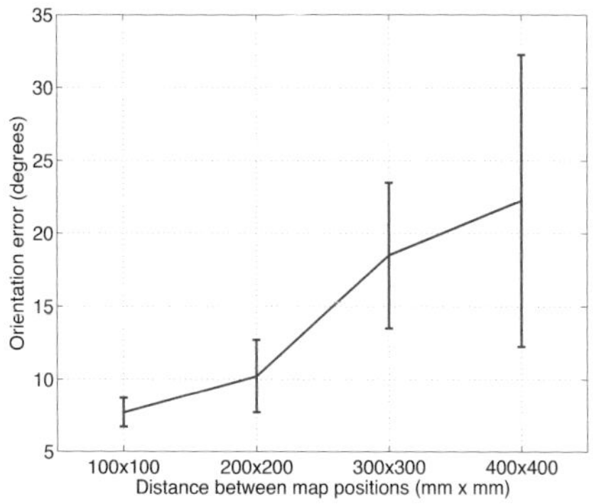

Fig. 9 Average orientation error of the experiment.

We can observe that the method is working properly with a distance between map positions not exceeding 300x300 millimeters.

5 Conclusions

In this paper we have presented a method to estimate the position and orientation of a mobile robot in an environment that has been previously mapped. We propose a solution to the absolute localization problem and we face it as an image-retrieval problem. We have also developed a novel description method for omnidirectional scenes based on the Radon transform. At last, all the algorithms have been tested with our own virtual database.

The results have demonstrated that it is a very reliable method to estimate the nearest position of the map. As for the values of the parameters, the distance between map positions is the main parameter to tune in this method. The best choice is 200x200 mm as it offers a good balance between accuracy and computational time.

The results presented in this paper show the effectiveness of the global appearance descriptors of omnidirectional images to locate the robot. We are now working to improve this method and we are trying to carry out the localization in a map with more degrees of freedom. We are also working to include the color information of the images as it can also provide useful information.

Acknowledgments This work has been supported by the Spanish government through the project DPI2013-41557-P and by the Generalitat Valenciana (GVa) through the project GV/2015/031.

References

1. Bay, H., Tuytelaars, T., Gool, L.: Surf: Speeded up robust features. Computer Vision at ECCV **3951**, 404–417 (2006)
2. Chang, C., Siagian, C., Itti, L.: Mobile robot vision navigation and localization using gist and saliency. In: IROS 2010, International Conference on Intelligent Robots and Systems, pp. 4147–4154 (2010)
3. Garcia-Fidalgo, E., Ortiz, A.: Vision-based topological mapping and localization methods: A survey. Robotics and Autonomous Systems **64**, 1–20 (2015)
4. Hasegawa, M., Tabbone, S.: A shape descriptor combining logarithmic-scale histogram of radon transform and phase-only correlation function. In: 2011 International Conference on Document Analysis and Recognition (ICDAR), pp. 182–186, September 2011
5. Hoang, T., Tabbone, S.: A geometric invariant shape descriptor based on the radon, fourier, and mellin transforms. In: 20th International Conference on Pattern Recognition (ICPR), pp. 2085–2088, August 2010
6. Kobayashi, K., Aoki, T., Ito, K., Nakajima, H., Higuchi, T.: A fingerprint matching algorithm using phase-only correlation. IEICE Transactions on Fundamentals of Electronics, Communications and Computer Sciences, 682–691 (2004)
7. Kuglin, C., Hines, D.: The phase correlation image alignment method. In: Proceedings of the IEEE, International Conference on Cybernetics and Society, pp. 163–165 (1975)
8. Lowe, D.: Object recognition from local scale-invariant features. In: ICCV 1999, International Conference on Computer Vision, vol. 2, pp. 1150–1157 (1999)
9. Maohai, L., Han, W., Lining, S., Zesu, C.: Robust omnidirectional mobile robot topological navigation system using omnidirectional vision. Engineering Applications of Artificial Intelligence **26**(8), 1942–1952 (2013)
10. Oppenheim, A., Lim, J.: The importance of phase in signals. Proceedings of the IEEE **69**(5), 529–541 (1981)
11. Payá, L., Amorós, F., Fernández, L., Reinoso, O.: Performance of global-appearance descriptors in map building and localization using omnidirectional vision. Sensors **14**(2), 3033–3064 (2014)
12. Payá, L., Fernández, L., Gil, L., Reinoso, O.: Map building and monte carlo localization using global appearance of omnidirectional images. Sensors **10**(12), 11468–11497 (2010)

13. Radon, J.: Uber die bestimmung von funktionen durch ihre integralwerte langs gewisser mannigfaltigkeiten. Berichte Sachsische Akademie der Wissenschaften **69**(1), 262–277 (1917)
14. Valiente, D., Gil, A., Fernández, L., Reinoso, O.: A comparison of EKF and SGD applied to a view-based SLAM approach with omnidirectional images. Robotics and Autonomous Systems **62**(2), 108–119 (2014)
15. Winters, N., Gaspar, J., Lacey, G., Santos-Victor, J.: Omni-directional vision for robot navigation. IEEE Workshop on Omnidirectional Vision, pp. 21–28 (2000)
16. Wu, J., Zhang, H., Guan, Y.: An efficient visual loop closure detection method in a map of 20 million key locations. In: 2014 IEEE International Conference on Robotics and Automation (ICRA), pp. 861–866, May 2014

Part XII
Visual Perception for Autonomous Robots

Accurate Map-Based RGB-D SLAM for Mobile Robots

Dominik Belter, Michał Nowicki and Piotr Skrzypczyński

Abstract In this paper we present and evaluate a map-based RGB-D SLAM (Simultaneous Localization and Mapping) system employing a novel idea of combining efficient visual odometry and a persistent map of 3D point features used to jointly optimize the sensor (robot) poses and the feature positions. The optimization problem is represented as a factor graph. The SLAM system consists of a front-end that tracks the sensor frame-by-frame, extracts point features, and associates them with the map, and a back-end that manages and optimizes the map. We propose a robust approach to data association, which combines efficient selection of candidate features from the map, matching of visual descriptors guided by the sensor pose prediction from visual odometry, and verification of the associations in both the image plane and 3D space. The improved accuracy and robustness is demonstrated on publicly available data sets.

Keywords SLAM · Point features · Tracking · Factor graph · RGB-D data

1 Introduction

Some types of robots, like quadrotors [19] and legged machines [2, 17] require accurate pose estimates in 3D for reliable navigation. Recently, solutions to the SLAM problem that employ the compact RGB-D sensors, such as Kinect or Xtion, became popular on those robotic platforms. Approaches that use dense depth data, like Kintinuous [25], demonstrate high accuracy, but they require hardware acceleration on high-end GPGPU cards, which cannot be packed on-board in small mobile robots. A viable alternative is the pose-based approach to SLAM with RGB-D data, which computes the sensor motion between consecutive frames in order to estimate the

D. Belter · M. Nowicki · P. Skrzypczyński(✉)
Institute of Control and Information Engineering, Poznań University of Technology,
ul. Piotrowo 3A, 60-965 Poznań, Poland
e-mail: {dominik.belter,michal.nowicki,piotr.skrzypczynski}@put.poznan.pl

L.P. Reis et al. (eds.), *Robot 2015: Second Iberian Robotics Conference*,
Advances in Intelligent Systems and Computing 418,
DOI: 10.1007/978-3-319-27149-1_41

trajectory. The obtained sensor poses and the motion-related constraints are treated respectively as the vertices and edges of a factor graph, which is then optimized [14]. Loop closures detected whenever the robot comes back to an already visited location introduce pose-to-pose constraints that enable the graph optimization in SLAM to correct the trajectory drift. The sensor motion estimate between two frames may be obtained in several ways, e.g. applying dense optical flow [12] or sparse optical flow [16], but the most common approach is to match sparse features by using local visual descriptors [6, 11]. As we have demonstrated in [3] the pose-based approach to RGB-D SLAM results in accurate real-time trajectory estimation without hardware acceleration. However, this approach neglects a large part of the feature-to-pose constraints resulting from frequent re-observations of the features. Because the loop closure detection has linear complexity in the number of locations, in the pose-based SLAM the data associations are usually established only between a relatively small fraction of the previous keyframes memorized along the trajectory.

In contrast, in the bundle adjustment (BA) approach [23], widely employed by modern visual SLAM [13, 21], the feature-to-pose constraints are directly used in optimization. Henry *et al.* [11] already applied the two-view BA to improve the frame-to-frame motion estimation in RGB-D visual odometry (VO), whereas Scherer and Zell [19] presented a solution to the RGB-D SLAM based on the structure of Parallel Tracking and Mapping (PTAM) [13]. The use of local BA defined on submaps for 3D object reconstruction from RGB-D data was presented in [15]. Very recently, the SlamDunk system has been presented [7], which uses a pool of point features (map) to track the sensor motion and then includes these features in optimization.

In this paper we investigate how to efficiently implement a RGB-D SLAM based on factor graph optimization including 3D features. The crucial components of such a solution are fast and accurate sensor tracking, map maintenance, and robust data association between the map and the perceived features. This work contributes a novel architecture of the map-based RGB-D SLAM system, which differs from other similar approaches [7, 15, 19] by employing the VO to track the sensor pose, by defining a flexible structure of the factor graph with a buffered part that enables real-time optimization in the background, and by exploring new techniques for outlier rejection in data association. The experimental evaluation focuses on the influence of the VO pipeline implementation on the accuracy of the trajectory estimation, and on the behavior of features in the map. Moreover, a comparison to selected state-of-the-art approaches to the RGB-D SLAM problem is included.

2 System Architecture

2.1 RGB-D Data Processing

The implementation of our RGB-D SLAM system[1] is divided into the front-end, which implements the VO pipeline and data association procedures, and the

[1] Source code is available at https://github.com/LRMPUT/PUTSLAM/tree/release

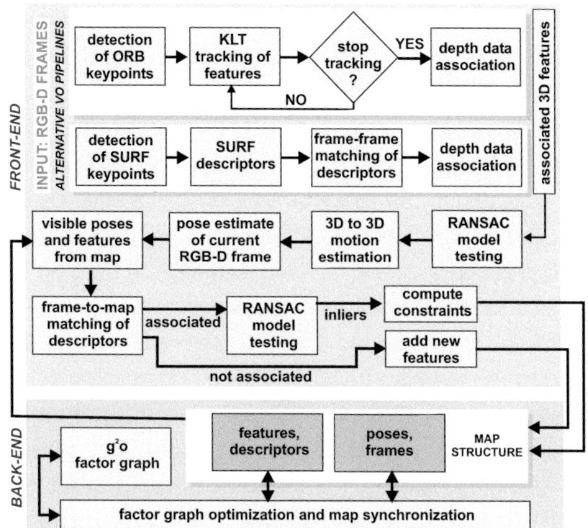

Fig. 1 Block scheme of the feature-based RGB-D SLAM system

back-end implementing the map structure and factor graph optimization. This architecture is presented in Fig. 1.

The main tasks of the front-end are to estimate the local sensor motion and to associate the observed point features to the existing map. In the front-end the sensor displacement guess is obtained from the VO pipeline, which considers 3D-to-3D feature correspondences for frame-to-frame motion estimation, like in stereo vision [17]. The 3D features are obtained by augmenting the 2D features (keypoints) extracted from the RGB images by the readily available depth data. The correspondences between the 2D keypoints can be accomplished either by matching of visual descriptors, or by optical flow tracking. The matching-based approach with local descriptors is popular in SLAM research [8]. The alternative approach we demonstrate here detects the keypoints only in the first RGB image of a sequence of n frames (n is small), and then tracks these features through the n images using the Lucas-Kanade algorithm [20]. This variant provides a very fast VO pipeline, which is also quite accurate [16].

Once the correspondences between the two RGB-D frames are established, the **SE(3)** rigid transformation between these two frames is estimated using the Umeyama algorithm [24]. The estimation procedure is embedded in the RANSAC scheme to make the computed transformation robust to outliers resulting from imperfect tracking or wrong descriptor associations. The RANSAC returns a transformation consistent with the largest number of feature pairs, which are considered inliers. The final transformation is computed using the g^{2}o optimization library to perform the two-view BA on the set of inlier features.

The resulting sensor motion estimate is added head-to-tail to the last known sensor pose in the factor graph in order to compute the pose estimate of the last RGB-D frame. Using this initial pose guess from the VO, features selected from the map become candidates for matching with the current set of local features. The current set of features is established by the points that were tracked over the last n frames, or have been detected in the current RGB-D data frame, if the matching-based VO is used. In the implementation demonstrated in this paper we employ ORB features [18] in the tracking-based pipeline, and SURF [1] in the matching-based version. For the sake of computation efficiency the descriptors of the features employed in the VO are then re-used to obtain associations between the local features and the map. The features used by the Lucas-Kanade tracker have to be corner-like, thus ORB is selected. The matching-based version employs SURF, which has a blob-like detector, but provides a more discriminative descriptor.

Regardless of the feature type being used it is necessary to place the keypoints in all parts of the image frame in order to achieve good estimation of the sensor motion. Thus, the RGB images are divided into 80×80 pixels square subimages, and the feature detection is performed individually in each subimage with adaptation of the detector parameters ensuring that roughly the same number of features in each square is obtained. After feature detection the DBScan unsupervised clustering algorithm is used to detect groups of keypoints in the image, and then each group is represented only by a single, strongest keypoint [16].

Having the sensor displacement guess, the front-end attempts to associate the local features and the features already stored in the map by using the visual descriptors (ORB or SURF). Having these associations the front-end computes the **SE(3)** transformation between the current pose and the map (in the global frame), by applying the Umeyama algorithm to the set of corresponding feature pairs. Again, the inliers are determined by RANSAC. For the set of inliers $\mathbb{R}^3$ constraints (translations) are added between the pose of the current frame, and the re-observed features in the map.

2.2 Map Optimization

The back-end holds the map and creates the factor graph, which is then optimized using the g^{2}o library [14]. The map contains a set of 3D point features augmented by visual descriptors, and the sensor poses related to the keyframes. The features are connected to the poses by constraints in $\mathbb{R}^3$. The poses are expressed in the global reference frame, whereas the features are anchored in the coordinate systems of the poses, from which they have been observed for the first time. The sensor poses are represented by their Cartesian positions (x,y,z) and quaternions (q_w,q_x,q_y,q_z) for the sensor orientation. The features $\mathbf{p}_j^f$ $(j = 1 \dots m)$ are represented as 3D points. The scalability of the system that incorporates a high number of features in the factor graph is limited, because the computational complexity of graph optimization is cubic in the number of variables. Therefore, our architecture adopts the Double Window optimization framework [21] by defining an inner window of keyframes around the current frame in the factor graph, and an outer window of peripheral keyframes. The parts of the feature map outside of the inner window are marginalized out, which renders the factor graph optimization efficient.

However, the graph optimization procedure is still much slower than the front-end. Thus, the front-end and back-end are implemented in separate threads, similarly to the PTAM architecture [13]. The front-end and the back-end work asynchronously, and they get synchronized on specified events, such as a query for visible features and insertion of a new set of constraints. New poses, features, and measurements are buffered in a smaller temporary graph, which is not used by g^{2}o until the back-end finishes the on-going graph optimization session. At this moment the temporary structure is merged with the main graph.

Unlike typical visual SLAM systems, our algorithm does not track the camera directly against the map, but employs an efficient VO pipeline to predict the sensor location. This allows our SLAM system to handle situations when there is a very small overlapping between the current view and the map, and we cannot re-observe the mapped features. In such cases the direct pose-to-pose constraints are introduced to the factor graph to stabilize the optimization in g^{2}o.

The optimal sensor poses $\mathbf{p}_1^c, \ldots, \mathbf{p}_n^c$ and feature positions $\mathbf{p}_1^f, \ldots \mathbf{p}_m^f$ are found by minimizing the function:

$$\underset{\mathbf{p}}{\operatorname{argmin}}\, F = \sum_{i=1}^{n} \sum_{j=1}^{m} e(\mathbf{p}_i^c, \mathbf{p}_j^f, \mathbf{m}_{ij})^T \Omega_{ij} e(\mathbf{p}_i^c, \mathbf{p}_j^f, \mathbf{m}_{ij}), \tag{1}$$

where $e(\mathbf{p}_i^c, \mathbf{p}_j^f, \mathbf{m}_{ij})$ is the error function between the estimated and the measured pose of the vertex related to the measurement $\mathbf{m}_{ij}$. In our formulation $\mathbf{m}_{ij}$ is $\mathbf{t}_{ij} \in \mathbb{R}^3$ for the feature-to-pose constraints, or $\mathbf{T}_{ij} \in \mathbf{SE(3)}$ for the pose-to-pose constraints. The information matrix $\boldsymbol{\Omega}$ represents the accuracy of each measurement. For the feature-to-pose constraints this matrix is obtained by inverting the covariance matrix of the feature uncertainty. In this work we assume isotropic spatial uncertainty of the features, so we set $\boldsymbol{\Omega}$ to an identity matrix, but as demonstrated by our recent research [4] computing $\boldsymbol{\Omega}$ as inversion of a realistic feature covariance matrix improves the accuracy of trajectory estimation. The $\boldsymbol{\Omega}$ of a pose-to-pose constraint is also set to identity, but it can be estimated from the uncertainty of the measurements that are marginalized when the transformation between two poses is computed [9].

3 Data Association in the Map

Our system handles local, metric loop closures implicitly, by establishing feature-to-pose constraints in the map. The map contains features that were discovered at various time instances along the sensor trajectory. Only features anchored to the sensor poses that are in the Euclidean neighborhood of the current sensor pose are considered for matching. In the current implementation the system does not account for global loop closures that require purely appearance-based place recognition [5]. However, integration of such a method is considered as future work.

The front-end queries the back-end (which holds the map data structure) for the set of candidate features that could be re-observed from the current pose of the sensor.

The set of features resulting from this query is projected into the current frame. If the projection results in an acceptable position of the feature on the image, this feature is considered for matching. According to the guided matching principle the candidate matches are considered only in a small neighborhood of the predicted feature in the image plane. As a feature from the map is often observed from various viewpoints, we store for a single point feature in the map multiple ORB/SURF descriptors obtained at various observation angles. Then, when matching the features we choose the descriptor with the smallest difference between its stored observation angle and the observation angle from the current frame. The observation angles are represented in the global coordinate system, and the angle between them is computed quickly using the dot product. We reject matches altogether whenever the difference in the observation angle between the most similar viewpoint of the mapped feature and the current sensor viewpoint is larger than 30°.

When we attempt to associate the local features with the map it is important to limit the number of map features that are considered for matching. Thus, we eliminate features that are occluded when observed from the current pose. This is accomplished using the dense depth measurements readily available in the current frame. We compare the distance between each feature projected from the map to the current coordinate frame and the image plane (i.e. "predicted depth" of the feature) with the average depth measured around the projected image coordinates of the map feature. If the measured depth is shorter by more than 0.1 m than the predicted depth, we consider the feature as occluded (Fig. 2a). This approach is more precise than the one based on considering only features visible from the local neighborhood [21] (Fig. 2b).

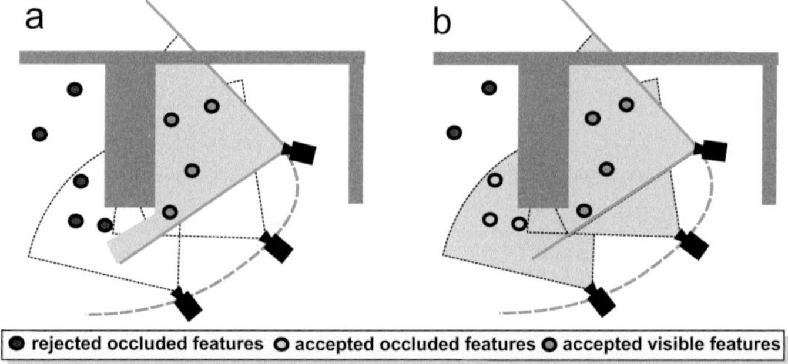

Fig. 2 Discrimination of the occluded features: features within the depth sensor field of view (a), and all features visible from the neighboring keyframes (b)

From the set of visual correspondences determined by descriptor matching, the correct matches (inliers) are estimated using the preemptive RANSAC framework. The outlier rejection test in RANSAC can use either the feature re-projection error in the image plane, or the Euclidean distance between the observed and the predicted feature in 3D. The re-projection error is given as:

$$E_{\text{Rep.}} = \max\left(d_{2D}(\Pi_a(\mathbf{p}_i^f), \Pi_a(\mathbf{p}_j^f)), d_{2D}(\Pi_b(\mathbf{p}_i^f), \Pi_b(\mathbf{p}_j^f))\right), \tag{2}$$

where Π_a represents the operator of projection onto the image plane a, $d_{2D}(x, y)$ represents computation of image distance between two image points x and y, and $\mathbf{p}_i^f$ is the 3D position of the i-th feature. The alternative Euclidean norm between the previously mentioned features is computed as:

$$E_{\text{Euc.}} = d_{\text{3D}}(\mathbf{p}_i^f, \mathbf{p}_j^f), \tag{3}$$

where $d_{\text{3D}}(x, y)$ is the computation of the Euclidean distance between two 3D points x and y. We have implemented both outlier rejection approaches in order to test which one is more efficient in practice.

4 Map Maintenance

If the point features discovered in the current keyframe cannot be matched to the already mapped features, they are added to the map. When adding features we try to avoid the possible aliasing of nearby features, which may further result in false matches. Thus, we add features to the map only from the keyframes, which contribute enough to the environment exploration. A keyframe is selected from the incoming RGB-D data stream if the number of features re-projecting from the map into the current image is below a given threshold (usually set to 80) or the map matching procedure resulted in less feature-to-pose constraints than a preset threshold (usually set to 25). From each keyframe, no more than 40 point features can be added to the map, and a preference is given to the keypoints having strongest detector response. Features that are located too close or too far from the sensor are not added to the map, as their locations may be highly uncertain due to the range measurement errors.

When a new feature is added to the map, we also verify that this new point is distant enough to the features already existing in the map. There is a minimal distance threshold in the Euclidean space, and a threshold on the minimal distance in the image plane. However, these parameters often need to be adjusted for scenes or sensor motions of different characteristics in order to achieve best accuracy.

5 Evaluation and Results

5.1 Investigation of the Map Evolution

We performed a series of experiments on the ICL-NUIM data set [10] in order to understand how the accuracy of estimated trajectories is influenced by such factors as VO-based motion estimation, the choice of error metrics in RANSAC, and the distribution of feature measurements in the map. The choice of ICL-NUIM, which is rendered in a synthetic environment is motivated by the perfect ground-truth sensor trajectories provided in this data set, and the availability of two variants of the depth data: one assuming no noise, and another one with simulated noise. We investigated

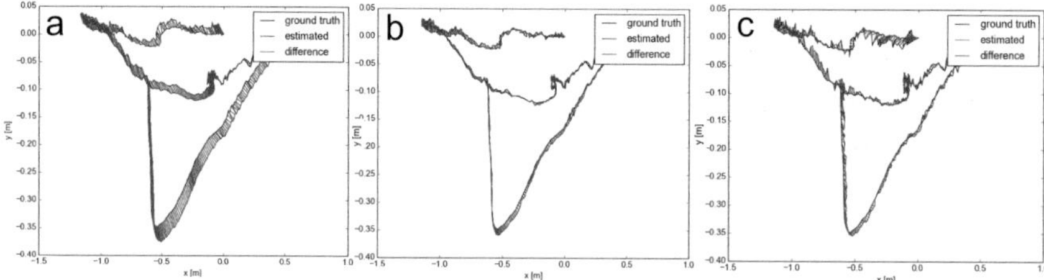

Fig. 3 ICL-NUIM data set `office_room/kt2` trajectory recovered from the noiseless data with two versions of our RGB-D SLAM: the tracking-based version (a), and the matching-based version (b), and the same trajectory recovered from noisy data by using the matching-based SLAM (c)

the evolution of features in the map, and evaluated quantitative results using the well-established ATE (Absolute Trajectory Error) and RPE (Relative Pose Error) metrics [22].

Figure 3 shows trajectories recovered from the exemplary `office_room/kt2` sequence. The trajectory estimated by the tracking-based version (Fig. 3a) is slightly less accurate than the one estimated by the matching-based variant (Fig. 3b). These example trajectories have been obtained from the data assuming no noise, however, the robust feature matching and outlier rejection mechanisms allow our SLAM to obtain similar results on the noisy version of this sequence (Fig. 3c).

Table 1 Performance for the representative configurations of our SLAM system measured on the ICL-NUIM `office_room/kt2` sequence without and with depth noise

RANSAC variant and noise in ICL-NUIM data	Tracking-VO SLAM					Matching-VO SLAM				
	ATE [m]	ATE std dev. [m]	RPE [m]	No. of features	FPS [Hz]	ATE [m]	ATE std dev. [m]	RPE [m]	No. of features	FPS [Hz]
$E_{\mathrm{Euc.}}$ no noise	0.028	0.010	0.006	1566	28.8	0.010	0.006	0.007	2479	2.9
$E_{\mathrm{Rep.}}$ no noise	0.022	0.011	0.007	2279	25.9	0.009	0.005	0.007	3328	2.8
$E_{\mathrm{Euc.}}$ with noise	0.050	0.024	0.016	2027	29.6	0.021	0.009	0.013	2480	2.4
$E_{\mathrm{Rep.}}$ with noise	0.045	0.021	0.015	2032	28.6	0.019	0.009	0.011	2541	2.2

Table 1 summarizes the quantitative results obtained on the `office_room/kt2` sequence by the tracking-based and matching-based versions of our SLAM demonstrating also how the outlier rejection metrics in RANSAC influences the accuracy of the obtained trajectory. All frame rates (FPS) were measured on a PC with Intel Core i7-2600 3.4GHz CPU and 16GB RAM. Figure 4a provides a visualization of the feature-based map for the same sequence[2]. The matching-based version achieves better accuracy (smaller ATE), but at the cost of being much slower. The influence of the error metric in RANSAC is small, but the re-projection error metric constantly provides slightly better accuracy. However, the more complicated computations in eq. (2) slow down the front-end.

[2] A short video clip is available at http://lrm.cie.put.poznan.pl/robot15.mp4

Comparing the ATE RMSE results in Tab. 1 to the figures given in [10] for five SLAM or VO systems, we can see that our matching-based SLAM is more accurate than all of the feature-based approaches evaluated on the `office_room/kt2` sequence in [10]. Only the dense ICP-based Kintinuous achieved smaller ATE on this sequence, but it requires hardware acceleration to run. Even the tracking-based version, which runs at the high frame rate allows us to localize the sensor with a positional error of about 4.5 cm (with noise).

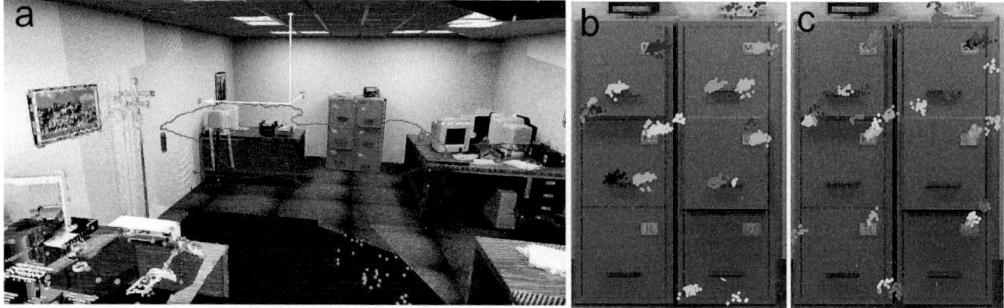

Fig. 4 Exemplary view of the map features (green points) and sensor trajectory (red) recovered from the ICL-NUIM `office_room/kt2` sequence (a), and distribution of the keypoints extracted by the front-end in the tracking-based (b), and matching-based (c) version. Points belonging to different features are shown in different colors

In order to investigate in more detail the behaviour of the features in the map, we conducted an experiment, which shows the distribution of all measured features in the common coordinate system. We run our SLAM algorithm on the same `office_room/kt2` sequence. Whenever the algorithm computes a transformation between two frames we substitute the computed value by the perfectly known transformation from the ground truth. Thus, the error of feature positions is caused only by the imperfect feature detection, and tracking or matching in the front-end. Then, we transform all measured positions (u, v) of features into the first RGB frame using the ground truth data. A close-up of a selected, feature-rich region of this RGB image is presented in Fig. 4b and 4c. The tracking-based front-end yields more measurements per feature, which is advantageous for the factor graph optimization. On the other hand, when the front-end tracks a single feature for a long sequence of frames, the measurements start to drift. The features form an oval shape on the image (Fig. 4b). The features produced by the matching-based front-end are more scattered on the image (Fig. 4c), and in the Euclidean space. The matching-based version re-establishes the features at each frame, and thus is more discriminative in accepting the feature-to-map associations on the basis of their descriptors. Thus, whenever a feature starts to drift, the matching-based front-end creates a new feature. This approach gives better precision in SLAM (smaller drift of features), but finally, the features are close to each other, and the map expands quickly. The larger number of features, together with the relatively slow SURF detector/descriptor, contribute to the low frame rate (cf. Tab. 1) achieved by the matching-based version.

5.2 Accuracy Assessment on Benchmark Data

We evaluated the accuracy of our map-based RGB-D SLAM on the TUM RGB-D benchmark [22], which has been already used to demonstrate performance of many RGB-D SLAM and VO systems [3, 6, 25]. We have tested our SLAM on five sequences, that are representative for indoor navigation: `fr1_desk` (571 frames), `fr1_desk2` (611 frames), `fr1_room` (1352 frames), `fr2_desk` (2217 frames), and `fr3_long_office_household` (2486 frames).

Table 2 Positional ATE and RPE for the two main configurations of our SLAM system measured on five TUM RGB-D benchmark sequences

TUM RGB-D benchmark sequence	Tracking-VO SLAM				Matching-VO SLAM				Known accuracy ATE RMSE [m]
	ATE [m]	ATE std dev. [m]	RPE [m]	FPS [Hz]	ATE [m]	ATE std dev. [m]	RPE [m]	FPS [Hz]	
`fr1_desk`	0.044	0.021	0.020	24.9	0.027	0.013	0.011	3.5	0.026 [6]
`fr1_desk2`	0.084	0.034	0.028	32.7	0.040	0.016	0.015	2.9	0.043 [6]
`fr1_room`	0.155	0.051	0.014	38.6	0.133	0.049	0.010	3.3	0.084 [6]
`fr2_desk`	0.095	0.034	0.012	19.7	0.067	0.016	0.008	2.1	0.076 [15]
`fr3_office`	0.057	0.018	0.013	26.1	0.023	0.012	0.009	3.0	0.035 [15]

The positional ATE RMSE and positional RPE RMSE results achieved on these sequences by our SLAM with two variants of the front-end are summarized in Tab. 2 (note that `fr3_office` stands for `fr3_long_office_household`). In these tests parameters that considerably depend on the characteristics of the data, like the distance thresholds in the map maintenance procedure have been adjusted individually for each sequence. For all five sequences the matching-based version was more accurate than the tracking-based one. In the TUM sequences there are many RGB frames with a considerable amount of motion blur and the tracking-based front-end using the corner-like ORB detector had problems with the drift and perhaps aliasing of keypoints in such frames. The matching-based version managed to produce more repeatable point features using the multi-scale SURF, which resulted in very accurate pose estimation. However, this was at the cost of low frame rate, an order of magnitude smaller than for the tracking-based version. The tracking-based version achieved the trajectory estimation accuracy that is satisfactory for most tasks of a mobile robot, being able to work in real-time without any hardware acceleration. For comparison we provide in Tab. 2 also the positional ATE RMSE values achieved on the respective sequences by other state-of-the-art systems, namely the RGB-D SLAM [6] and the Submap-based BA described in [15]. As far as we can tell from the literature analysis, these were the best ATE results published till now for the respective TUM benchmark sequences. Our approach achieves similar or even smaller positional ATE values. Trajectories recovered for three exemplary sequences are visualized in Fig. 5.

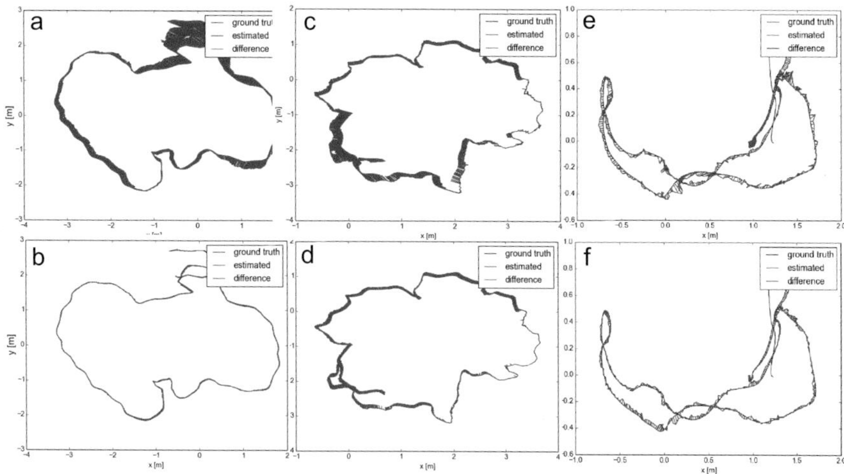

Fig. 5 Estimated trajectories with the ATE error for TUM RGB-D benchmark sequences: `fr3_long_office_household` (a,b), `fr2_desk` (c,d), and `fr1_desk` (e,f) recovered by our SLAM with tracking-based (a,c,e), and matching-based (b,d,f) front-end

6 Conclusions

We have presented a new feature-based RGB-D SLAM system, which employs the factor graph optimization approach, but unlike the more popular pose-based SLAM approaches builds a persistent map of 3D features. The new concept in this system is the use of a VO pipeline to compute the sensor motion guess, instead of direct tracking of the sensor against the map of features. This concept implemented by using the very fast sparse optical flow tracker results in a map-based RGB-D SLAM that can run at the Kinect frame rate without any hardware acceleration on a commodity PC. The high number of feature-to-pose constraints established and maintained by our algorithm enables more precise trajectory reconstruction than in the pose-based SLAM systems. The robustness was improved by new feature management techniques: the detection of occluded features by directly using the dense depth image, and multiple-view-angle descriptors associated with the mapped features. The tracking-based front-end enables real-time operation (up to 38 Hz), but it achieves lower accuracy of the trajectory estimation than the matching-based variant implemented for comparison. This is attributed to the choice of feature type: ORB keypoints are less repeatable, and more prone to errors due to motion blur. Thus, the tracked features drift more than the points generated by SURF. The SURF keypoints are more evenly distributed and stable on the image, but SURF is too slow to compute and match for real-time operation.

Acknowledgement This work was financed by the National Science Centre under decision DEC-2013/09/B/ST7/01583.

References

1. Bay, H., Ess, A., Tuytelaars, T., Van Gool, L.: Speeded-up robust features (SURF). Computer Vision and Image Understanding **110**(3), 346–359 (2008)
2. Belter, D., Skrzypczyński, P.: Precise self-localization of a walking robot on rough terrain using parallel tracking and mapping. Industrial Robot: An International Journal **40**(3), 229–237 (2013)
3. Belter, D., Nowicki, M., Skrzypczyński, P.: On the performance of pose-based RGB-D visual navigation systems. In: Cremers, D., et al. (eds.) Computer Vision – ACCV 2014. LNCS, vol. 9004, pp. 407–423. Springer (2015)
4. Belter, D., Skrzypczyński, P.: The importance of measurement uncertainty modeling in the feature-based RGB-D SLAM. In: Proc. Int. Workshop on Robot Motion and Control, Poznań, pp. 308–313 (2015)
5. Cummins, M., Newman, P.: Accelerating FAB-MAP with Concentration Inequalities. IEEE Trans. on Robotics **26**(6), 1042–1050 (2010)
6. Endres, F., Hess, J., Sturm, J., Cremers, D., Burgard, W.: 3-D Mapping with an RGB-D Camera. IEEE Trans. on Robotics **30**(1), 177–187 (2014)
7. Fioraio, N., Di Stefano, L.: SlamDunk: affordable real-time RGB-D SLAM. In: Computer Vision – ECCV 2014 Workshops. LNCS, vol. 8925, pp. 401-414. Springer (2015)
8. Gil, A., Martinez Mozos, O., Ballesta, M., Reinoso, O.: A comparative evaluation of interest point detectors and local descriptors for visual SLAM. Machine Vision and Applications **21**(6), 905–920 (2010)
9. Grisetti, G., Kümmerle, R., Ni, K.: Robust optimization of factor graphs by using condensed measurements. In: Proc. IEEE/RSJ Int. Conf. on Intelligent Robots & Systems, Vilamoura, pp. 581–588 (2012)
10. Handa, A., Whelan, T., McDonald, J. B., Davison, A. J.: A benchmark for RGB-D visual odometry, 3D reconstruction and SLAM. In: IEEE Int. Conf. on Robotics & Automation, Hong Kong, pp. 1524–1531 (2014)
11. Henry, P., Krainin, M., Herbst, E., Ren, X., Fox, D.: RGB-D mapping: Using Kinect-style depth cameras for dense 3D modeling of indoor environments. Int. Journal of Robot. Res. **31**(5), 647–663 (2012)
12. Kerl, C., Sturm, J., Cremers, D.: Robust odometry estimation for RGB-D cameras. In: Proc. IEEE Int. Conf. on Robotics & Automation, Karlsruhe, pp. 3748–3754 (2013)
13. Klein, G., Murray, D.: Parallel tracking and mapping for small AR workspaces. In: Proc. Int. Symp. on Mixed and Augmented Reality, Nara, pp. 225–234 (2007)
14. Kümmerle, R., Grisetti, G., Strasdat, H., Konolige, K., Burgard, W.: g2o: A general framework for graph optimization. In: IEEE Int. Conf. on Robotics & Automation, Shanghai, pp. 3607–3613 (2011)
15. Maier, R., Sturm, J., Cremers, D.: Submap-based bundle adjustment for 3D reconstruction from RGB-D Data. In: Pattern Recognition. LNCS, vol. 8753, pp. 54–65. Springer (2014)
16. Nowicki, M., Skrzypczyński, P.: Combining photometric and depth data for lightweight and robust visual odometry. In: European Conf. on Mobile Robots, Barcelona, pp. 125–130 (2013)
17. Ozawa, R., Takaoka, Y., Kida, Y., Nishiwaki, K., Chestnutt, J., Kuffner, J., Inoue, H.: Using visual odometry to create 3D maps for online footstep planning. In: IEEE Int. Conf. on Systems, Man and Cybernetics, Hawaii, pp. 2643–2648 (2005)
18. Rublee, E., Rabaud, V., Konolige, K., Bradski, G.: ORB: an efficient alternative to SIFT or SURF. In: IEEE Int. Conf. on Computer Vision, pp. 2564–2571 (2011)
19. Scherer, S., Zell, A.: Efficient onboard RGBD-SLAM for autonomous MAVs. In: Proc. IEEE/RSJ Int. Conf. on Intelligent Robots & Systems, Tokyo, pp. 1062–1068 (2013)
20. Shi, J., Tomasi, C.: Good features to track. In: IEEE Conf. on Comp. Vis. and Pattern Recog., Seattle, pp. 593–600 (1994)
21. Strasdat, H., Davison, A. J. Montiel, J., Konolige, K.: Double window optimisation for constant time visual SLAM. In: Proc. Int. Conf. on Computer Vision, Los Alamitos, pp. 2352–2359 (2011)

22. Sturm, J., Engelhard, N., Endres, F., Burgard, W., Cremers, D.: A benchmark for the evaluation of RGB-D SLAM systems. In: IEEE/RSJ Int. Conf. on Intelligent Robots & Systems, Vilamoura, pp. 573–580 (2012)
23. Triggs, B., McLauchlan, P.F., Hartley, R.I., Fitzgibbon, A.W.: Bundle adjustment – a modern synthesis. In: Vision Algorithms: Theory and Practice. LNCS, vol. 1883, pp. 298–372. Springer (2000)
24. Umeyama, S.: Least-squares estimation of transformation parameters between two point patterns. IEEE Trans. on Pattern Analysis & Machine Intelligence **13**(4), 376–380 (1991)
25. Whelan, T., Johannsson, H., Kaess, M., Leonard, J., McDonald, J.: Robust real-time visual odometry for dense RGB-D mapping. In: IEEE Int. Conf. on Robotics & Automation, Karlsruhe, pp. 5704–5711 (2013)

Onboard Robust Person Detection and Tracking for Domestic Service Robots

David Sanz, Aamir Ahmad and Pedro Lima

Abstract Domestic assistance for the elderly and impaired people is one of the biggest upcoming challenges of our society. Consequently, in-home care through domestic service robots is identified as one of the most important application area of robotics research. Assistive tasks may range from visitor reception at the door to catering for owner's small daily necessities within a house. Since most of these tasks require the robot to interact directly with humans, a predominant robot functionality is to detect and track humans in real time: either the owner of the robot or visitors at home or both. In this article we present a robust method for such a functionality that combines depth-based segmentation and visual detection. The robustness of our method lies in its capability to not only identify partially occluded humans (e.g., with only torso visible) but also to do so in varying lighting conditions. We thoroughly validate our method through extensive experiments on real robot datasets and comparisons with the ground truth. The datasets were collected on a home-like environment set up within the context of RoboCup@Home and RoCKIn@Home competitions.

Keywords Person detection · Tracking · Domestic environment · RoboCup · Service robotics · Benchmarking

D. Sanz
Robotics and Cybernetics Research Group, Center for Automation and Robotics,
Jose Gutierrez Abascal, 2, 28006 Madrid, Spain
e-mail: d.sanz@upm.es

A. Ahmad(✉)
Max Planck Institute for Biological Cybernetics, Spemannstrae 44, 72076 Tübingen, Germany
e-mail: aamir.ahmad@tuebingen.mpg.de

P. Lima
Institute for Systems and Robotics, Instituto Superior Tcnico,
Universidade de Lisboa, Lisboa, Portugal
e-mail: pal@isr.ist.utl.pt

L.P. Reis et al. (eds.), *Robot 2015: Second Iberian Robotics Conference*,
Advances in Intelligent Systems and Computing 418,
DOI: 10.1007/978-3-319-27149-1_42

1 Introduction

The demographics of human society is changing rapidly. World population is living longer and, therefore, getting older. Diseases that were once incurable are either mostly eradicated or their fatal effects significantly reduced. At the same time, modern family structures and societal roles encourage nuclear/singular families, eventually resulting in an increasing number of older people living alone. All these aspects together entail an exponential increment in the number of people requiring regular and detailed domestic assistance. Consequently, domestic service robotics has been recently identified as a potential solution to the societal challenge of population aging.

Fig. 1 RoCKIn@Home testbed at the Institute for Systems and Robotics (ISR), Lisbon. The colored hat on the robot and the color coded jacket on the person are for obtaining ground truth poses of the robot and the person, respectively.

Assistive tasks can vary widely, including domestic tasks, personal assistance and health care. Since any domestic service robot is expected to perform most of these tasks, it requires a smart combination of carefully designed robotic functionalities. The functionalities include, among others, robot localization, mapping, object detection and tracking, mobile manipulation, decision making and so forth. Among these, person detection and tracking is not only an indispensable functionality to perform all human-robot interaction-related (HRI) tasks but often also necessary for the success of other basic functionalities, e.g., human-aware safe navigation within a domestic environment. At the same time visitors at home also need to be detected and tracked in order to perform adequate behavior given some specific situation (e.g., leading a nurse to the owner's room, a deliveryman to the kitchen and friends to the living space) and guarantee personal security (e.g., reporting potential intrusions at home).

In this article we present a robust approach to the robotic functionality of person detection and tracking. The novelty of our approach is twofold. Primarily, we achieve a very high degree of detection success rate and tracking precision through onboard sensors only. Real robot experiments and comparisons with the ground truth support this. Secondly, our method is robust to partial occlusions (e.g., if only the person's torso is visible) as well as to varying lighting conditions. Moreover, due to its modular implementation on the Robot Operating System (ROS) middleware, it was flexibly integrated with other functionalities, e.g., localization, navigation, path planning, object manipulation and speech-based human-robot interaction to perform complete tasks. While in this article we focus solely on the person detection and tracking functionality of our robot, all functionalities run integrated when it performs the full task [2]. The experiments presented in this article were performed on real robot datasets, including ground truth (GT) collected specifically for this purpose on a home-like environment setup (see Figure 1). We identify these datasets, which we make publicly available, as an additional contribution of our work in this article.

The rest of the article is organized as follows. In Section 2 we situate our work within the state-of-the-art, followed by the details of our approach in Section 3. Experiments, results and discussions are provided in Section 4. We conclude with a comment on ongoing and future work in Section 5.

2 Challenges and State of the Art

Person detection and tracking (PDT) has been addressed in several different ways in the literature. Some of the most popular approaches are based on static sensors using either RFID tags [10], intelligent floors [19] or, quite frequently, fixed-(multi)camera systems [6]. These methods provide very accurate and reliable person detection and tracking estimates. However, their major drawback is that they require application-specific environmental adaptation, which is not only quite expensive but also rigid. Other solutions have attempted to provide affordable and flexible solutions, e.g. using wearable devices [11]. However, they proved to be awkward and inefficient in various situations, e.g., when dealing with visitors or dementia patients. Consequently, it is expected that the next generation PDT functionality needs to be fully onboard the robots without any sort of modifications to the environment or to the person's attributes. At the same time, PDT functionality should be modular enough to take advantage of the networked sensors in the environment, if they are present.

Various approaches have been proposed to provide robots with PDT functionality. Initially, laser range finders were used due to their success in other fields, e.g., robot localization. However, the applicability of these techniques, mostly based on legs or torso detection, were quite limited, mainly due obstacles in domestic environments like furniture [14]. Similarly, human motion characteristics-based methods were also considered for PDT. These required, e.g, pyro-electric infrared (PIR) detectors and ultrasonic sensors [9] but their downside was their poor accuracy.

Within the context of PDT, vision-based methods have emerged as the most successful ones so far. Such methods include face detectors [14], body recognizers [7] and blob classifiers [20]. However, despite their successes in many laboratory environments, it has been shown that none of these techniques are actually robust enough for uninterrupted use in real environments. For instance, face detectors work well only when faces/heads are fully visible with enough quality. This is not usual in domestic environments where people are often moving and turning. Likewise, visual body

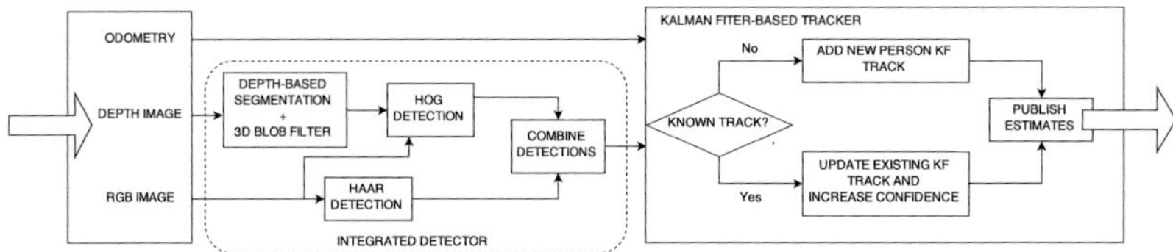

Fig. 2 Person Detection and Tracking (PDT) pipeline.

recognizers, most of them based on color, silhouette, luminance or pattern identification, work smoothly only under controlled situations [16]. To cope with changes in the illumination, clothing or human postures, remains a difficult challenge. As it is made clear in further sections, our integrated method for PDT systematically overcomes these limitations.

Among blob segmentation-based approaches, background subtraction methodologies provide good results only if using static cameras [8] but not with moving robots consisting of onboard cameras. On the other hand, motion estimation-based techniques provide successful results when monitoring moving people [18] but are useless when people are engaged in standing activities (i.e. conversations, talking on phone). To the rescue of both of the aforementioned extreme situations, point-cloud and distance-based segmentation provide good results in both static and dynamic situations [20]. However, their drawback lies in their poor accuracy that does not guarantee robust detection.

Considering the limitations of all the methods described above, a potential solution lies in developing an integrated method that combines the strengths of several individual methods in a realistic manner. Consequently, very recent methods result from the combination of vision and distance-based approaches [5]. Moreover, since the availability of low-cost RGB-D sensors (e.g., Microsoft Kinect or Asus Xtions), many authors have integrated contour recognition methodologies (specially based on head templates) with the traditional visual approaches [4]. Although this integration has provided good results, contour-based approaches have not been effective when taking into account partial occlusions. As it is demonstrated in the further sections, our method overcomes this issue. In summary, as most promising state-of-the-art approaches for PDT rely on the integration of complementary information, such as semantic context [6] or behavioral tracking data [13], in this article we present a robust PDT approach that stitches depth-based blob segmentation and human statistical feature-based blob analysis with traditional visual inspection techniques. Furthermore, using measurements from our integrated detector coupled with the robot's odometry-based ego-motion compensation we construct a robust and reliable person tracker.

3 Integrated Approach to Person Detection and Tracking

In this section we present our integrated approach to PDT. The overall pipeline of our proposed method can be visualized in Figure 2. The core novelty lies in

- a depth-based segmentation approach that is robust to lighting and color changes,
- integrating segmentation, statistical 3D blob filtering, histogram of oriented gradients (HOG) and Haar detection methods in a way such that the the benefits of each individual method is secured without significantly increasing the overall computational load.

Measurements from this integrated detector updates a multi-track Kalman-filter that uses robot odometry to perform ego-motion compensation. We further describe the most significant blocks of our PDT pipeline.

3.1 Robust Depth-Based Segmentation

The first step of our detection method comprises of a depth-based segmentation scheme to identify the presence of potential persons in the scene.

Regular thresholding-based depth segmentation[1] usually provides poor results because it is highly dependent on a heuristic: the threshold value. The left-most images in Fig. 3 neatly illustrate this. The person in the images is standing close to a wall behind, causing him to be clustered almost with the walls. On the other hand, adaptive depth-thresholding methods based on, e.g., Otsu's method [12], are capable of only distinguishing borders of similar-depth regions but not the regions themselves (see second column of Fig. 3). Our approach combines these two methods to achieve the best of both. We use Otsu's method-based multi-level adaptive depth thresholding that employs entropy search and results in extracting the borders of different depth regions in an image. On this we apply regular thresholding that eventually provides significant, distinguishable and meaningful regions (see third column of Fig. 3). As this method is only depth-based, the color of person's clothing or changes in the ambient light do not have any effect on the segmentation. Our 3D depth-segmentation is expressed as follows. Let $d_{x,y}$ denote the depth of a pixel at the image coordinate (x, y). We compute

$$s_{x,y} = t_s(d_{x,y})(1 - t_o(d_{x,y})), \tag{1}$$

where $s_{x,y}$ is the resulting binary value of the image pixel at (x, y) in our 3D depth-based segmentation. $t_s(.)$ and $t_o(.)$ are the static and Otsu-based method's binary results, respectively. These are computed as follows.

[1] Note that the depth-segmentation is performed on a gray-scale image where the gray levels denote the depth value of each pixel, not the color.

$$t_s(d_{x,y}) = \begin{cases} 1 & \text{if } d_{x,y} < T_{\text{static}} \\ 0 & \text{otherwise} \end{cases} \tag{2}$$

$$t_o(d_{x,y}) = \begin{cases} 1 & \text{if } d_{x,y} < \frac{t_s[min(\sigma_{wd}^2)]+t_s[max(\sigma_{bd}^2)]}{2} \\ 0 & \text{otherwise} \end{cases} \tag{3}$$

where T_{static} is the static threshold. $t_s[min(\sigma_{wd}^2)]$ is the value that minimizes the within-class distance variance, σ_{wd}^2; and $t_s[max(\sigma_{bd}^2)]$ is the value that maximizes the between-class distance variance, σ_{bd}^2, in the Otsu-based thresholding process [15].

Fig. 3 Comparison of depth-based segmentation methods. First column: regular thresholding. Second column: Otsu's method. Third column: combination of Otsu's method-based multi-level adaptive depth thresholding and regular thresholding. Fourth column: final detection result overlaid on the grayscale image. Note that the final detection is the result of the full integrated detector (combining 3D blob filtering, HOG and HAAR) as depicted in Figure 2.

3.2 3D Blob Filter and Combined Detection

Depth-based segmentation produces a number of 3D blobs that are either potential candidates for being detected as a person or spurious. To this end, we apply a human statistical characteristics-based filter that eliminates blobs not adhering to the standard [1] human size features, e.g., human-body aspect ratio and typical person width. Figure 4 illustrates this filtering. Image regions corresponding to 3D blobs not adhering to these statistical characteristics (e.g., the 3D blobs corresponding to the couch and wall on the left side of the person as shown in Fig. 4) are discarded. This allows us to drastically reduce the image space on which the rest of our integrated detector's processing takes place, thereby significantly increasing the computational speed.

The next step in our PDT pipeline is to classify a filtered 3D blob as either a person or not. Here we apply the histogram of oriented gradients (HOG) because of its very high accuracy. Although it has a high computational cost [7] in general, the significantly reduced image space allows HOG to function in real-time.

The 3D depth-based segmentation might still result in a few false negatives, e.g., when a person stands touching a large obstacle (wall/furniture) such that the distance between the person and the obstacle becomes negligible. In such situations the 3D blobs will be discarded by the human statistical characteristics-based filter and there might be no candidate blobs at all to apply the HOG detection. To overcome this issue and to make the overall detection even more reliable, we stitch another layer of detection based on Haar cascades [17]. It rapidly performs a multi-scale search based on an upper body training set. In itself, Haar cascades are quite model dependent and perform poorly in changing environments. Therefore, in our detection scheme, we invoke it only in case of failure of the HOG-based detection layer. As it will be evident with experimental results in the next section, it indeed provides an additional boost to the overall detection capability of our system.

Finally, few false positives might originate from uncontrollable factors, such as, human shape-like elements (e.g., wardrobes and decorative plants) as well as there exists considerable positional uncertainty in each positive detection measurement. Therefore, we construct a Kalman filter-based (KF) multi-track estimator that uses our integrated detector's measurements to update the tracks and odometry measurements to compensate for the robot's ego motion.

3.3 Person Tracking

The state components of our KF-based estimator are the 2D position (assuming that the person is always on the same ground plane as the robot) and velocity of the person in the robot's reference frame. For the prediction step of the KF, we first compensate for the ego-motion of the robot using the odometry measurements and then apply a constant velocity motion model with zero mean Gaussian acceleration noise to the person's position. The update step then uses our integrated detector's measurements. However, before the update, we perform a blob matching step to determine whether a detected 3D blob actually belongs to an existing estimator track. This is done by comparing the blobs' distance (from the robot) and size w.r.t. to each track's last known state under the assumption that a moving person does not displace beyond a certain threshold within 2 successive frames. Note that since our detections are in 3D, moving persons that might overlap (cross each other while moving) in a 2D image

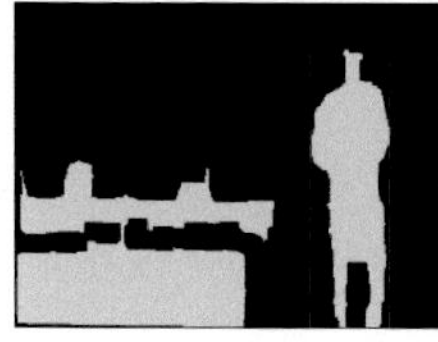

Fig. 4 (Left) Illustration of the result of human statistical characteristics-based filter on the 3D depth-segmented image. (Right) Final detection result overlaid on top of the grayscale image

for a certain time-window would still be distinguishable as their corresponding 3D blobs would not overlap in at least one dimension. On positive blob matching, the KF track is updated based on the measurements of the corresponding 3D blob. If the blob matching steps results negatively, we assume a new person has been detected and a separate new estimator track is initiated.

4 Experiments and Results

4.1 Testbed and Dataset

A vital aspect of our work in this article concerns experiments not only with real robots but also in fully real-world setting. In order to do so, we performed all the experiments in a home-like environment (Figure 1), set up within the context of RoCKIn@Home competitions. The set up consists of a living room, kitchen with dining space, a bedroom, two hallways and various furniture[2]. Furthermore, the test bed also consists of stereo vision-based ground truth acquisition system [3] that uses colored markers on the robot and the person to obtain their GT poses during the time intervals in which they are within the field of view of the stereo cameras. We use these GT poses to compute our estimation errors.

The robot used in our work is a fully custom made 4-wheeled omni-directional platform. In addition to the other sensors and actuators, it is equipped with two laser range finders which were used for mapping, navigation, and obstacle avoidance. For PDT we used the onboard Kinect RGBD sensor and the platform's odometer.

In order to perform rigorous experiments that are not only repeatable but also allow us to have the exact same experimental situation to compare various different approaches, we made 2 datasets. As we used the ROS middleware, the datasets were stored in the *rosbag* format. They consist of timestamped data that includes the robot's localization poses (running augmented-MCL), laser scans, odometry, RGB, gray-scale and depth images (all at 640×480 pixel resolutions) from the Kinect, as well as images from both the cameras of the GT system. The video attached with this article (http://youtu.be/7-WONyiszj4) uses the images from one of the GT cameras. During the data collection we remotely moved the robot in the environment. Simultaneously, a person, whom the robot is expected to detect and track in our experiments, arbitrarily moved near the robot. It must be noted that the ambient light in the environment was not uniform over the whole space (visible in the video as well as in Figure 6 (right) where the left part is well lighted compared to the right) and the colored jacket worn by the person was only for the GT acquisition. The dataset is versatile not only because it consists of real world setting and non-uniform ambient light but because it can also be used for testing other robot functionalities than PDT, e.g., localization, mapping and object tracking. The dataset is available online[2].

[2] http://datasets.isr.ist.utl.pt/lrmdataset/PersonTrackingDataset/ROSbags/

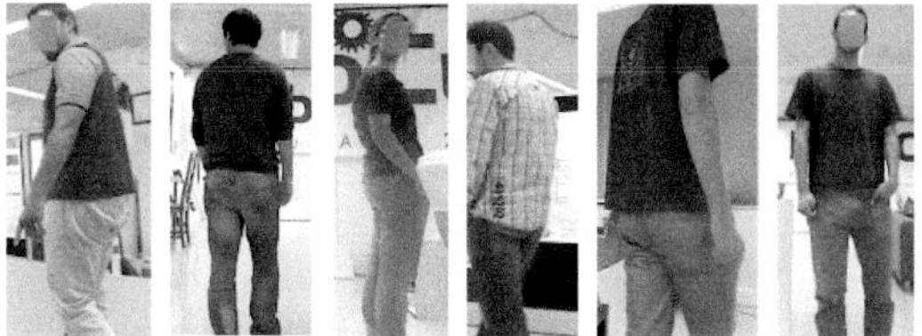

Fig. 5 Example images from the training set

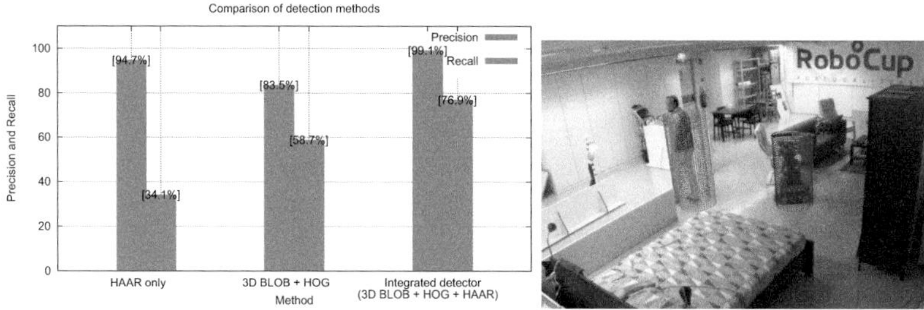

Fig. 6 (Left)Detection method comparison. (Right) Tracking results overlaid on a GT image. Blue cylinder centered at robot GT position and red line on top denotes GT orientation. Red cylinder centered at the person's GT position while green centered at the tracked person position estimate.

4.2 Implementation

The HOG descriptor that we used was trained using a support vector machine (SVM) and a set of 350 samples (positives and negatives) corresponding to different persons in the testbed. Figure 5 illustrates that different poses, clothing and distances to the camera were considered. Furthermore, special attention was paid to occlusions, including multitude of samples where some parts of the person were missed (e.g., head, arms, legs or portions of the torso). For Haar detection, the upper body training set was used.

Implementation of the integrated-detector and Kalman filter-based tracker modules (as shown in the PDT pipeline in Fig. 2), were done on the ROS platform, in C++, on a system with the following configuration: Intel(R) Pentium(R) Dual CPU T3400 @ 2.16GHz and 3GB of RAM (without GPU support).

4.3 Experimental Results

Integrated Detector's Precision and Recall. In Figure 6 (left) we present a precision and recall comparison of the individual detection methods and our integrated detector. Both datasets used in the experiments together consist of $\sim$ 12000 depth images and the same number of corresponding rgb images from the Kinect sensor. To calculate the precision/recall of the detection methods we followed the following procedure. We picked a sequence of 5161 RGB images from the Kinect (from our real robot dataset described earlier in this section) and overlaid on each of those images, the results of i) HAAR-only detection; and ii) 3D segmentation, blob-filtering and HOG detection. We then manually (by visual inspection) counted the relevant images (those in which the person was actually present), true positives (if a method correctly detected a person) and false positives (if a method incorrectly detected a person). In a total of 4496 relevant images (RI), HAAR-only method resulted in 1533 true positives (TP) and 85 false positives (FP). 3D segmentation, blob-filtering and HOG detection together resulted in 2639 TPs and 520 FPs. The integrated detector (combination of all individual detection methods as described previously and depicted in the PDT pipeline in Fig. 2) resulted in 3459 TPs and only 32 FPs. From Fig. 6 (left), it can be seen[3] that not only the combination of our novel 3D segmentation, blob-filtering and HOG detection substantially outperforms the HAAR-only method but through our integrated detector we are also able to achieve a near-perfect precision ($\sim$ 99%) and a high degree of recall ($\sim$ 77%). It must be noted that we considered an RGB image from the Kinect as RI even if a small fraction of the person's body was visible.

Computation Time Comparison. On an average, Haar detection method took $\sim$ 170 milliseconds (ms) to perform detection on one image. On the other hand the combination of depth-based segmentation, 3D blob filter and HOG detection took only $\sim$ 59ms. However, since each detection method runs asynchronously, the combined detection took an average time of only $\sim$ 120ms on one image. This is because the combined detector waits for the Haar detection's thread only if the rest (depth-based segmentation, 3D blob filter and HOG) of the combined detector's thread results in no detection. Given that our system had quite low computational capacity compared to most contemporary standard configurations, an average detection achievement of $\sim$ 8.3 frames per second should be considered high.

Tracking Accuracy and Precision. In the video accompanying this article we present the results of our KF-based tracker by re-projecting and overlaying the filter estimates over a concatenation of the image stream from one of the GT cameras. The tracker uses the combined detector, as described previously, to update the filter. Parts of the video footage where the GT poses of the robot or the person is not available (corresponding to $\sim$ 30% of the datasets), due to their occlusions from the GT cameras or other factors, have been removed from the footage as well as from the error estimation w.r.t. the GT.

Figure 6 (right) is one of the images from the video footage that illustrates the overlaid attributes. These attributes are defined as follows. Let the GT pose of the robot at time instant t be denoted by $\mathbf{x}_t^G = \{x_t^G, y_t^G, \theta_t^G\}$ in a world reference frame G, that we can simply assume, without loss of generality, coincides with the reference frame of our fixed GT system. A blue cylinder is overlaid centered at $\{x_t^G, y_t^G\}$ and

[3] We used the standard definitions of precision ($\frac{\text{TP}}{\text{TP+FP}}$) and recall ($\frac{\text{TP}}{\text{RI}}$).

Table 1 Error statistics of the person's tracked position

	Mean error (m)	RMSE (m)	MSE (m^2)
Dataset 1	0.34	0.20	0.04
Dataset 2	0.37	0.21	0.05

a dark-red line on top of the cylinder denotes the robot's GT orientation θ_t^G. Let the 2D GT position of the person be denoted by $\mathbf{p}_t^G = \{p_{x_t}^G, p_{y_t}^G\}$. A red cylinder is overlaid centered at $\{p_{x_t}^G, p_{y_t}^G\}$. Recall that the detection and tracking is performed in the robot's reference frame. Therefore, we use the robot's GT pose to transform the tracked local position of the person from the robot frame to the world frame in order to evaluate the error in the tracked position with respect to its GT value. Using the robot's self-localization to do this transformation would incorporate error that is not due to the PDT functionality but because of the self-localization itself. Hence, using the robot's GT pose isolates that source of error. Let the 2D tracked position of the person be denoted by $\mathbf{q}_t^R = \{q_{x_t}^R, q_{y_t}^R\}$ in the robot reference frame R. The 2D tracked position of the person $\mathbf{q}_t^G = \{q_{x_t}^G, q_{y_t}^G\}$ in the world frame is then given as

$$\mathbf{q}_t^G = \begin{bmatrix} \cos\theta_t^G & -\sin\theta_t^G \\ \sin\theta_t^G & \cos\theta_t^G \end{bmatrix} \mathbf{q}_t^R + \left[x_t^G \; y_t^G\right]^\top . \quad (4)$$

A green cylinder is overlaid centered at $\{q_{x_t}^G, q_{y_t}^G\}$. At every time step t if the GT is present we calculate the tracking error as $||\mathbf{q}_t^G - \mathbf{p}_t^G||$ where $||.||$ denotes the 2D Euclidean norm. Table 1 presents the statistics of this tracking error. It can be observed that our tracker performed consistently similar over both the datasets and achieved a root mean square error (RMSE) of $\sim$ 20cm. A person cannot be considered as a rigid body and is constantly turning and changing postures w.r.t. the robot during the experiment. The width of a bounding box around the person (from the onboard RGBD camera) will, therefore, vary with the person's posture. Since the position estimate $\mathbf{q}_t^G$ of the tracker refers to the center of the bounding box of the person's body in any posture, it is obvious that even in the ideal case (most accurate detection) $\mathbf{q}_t^G$ will not match (or be at a constant shift) with the center of the person's color-coded jacket $\mathbf{p}_t^G$ which we consider to be the GT of the person's pose. Thus, the RMSE of our tracker can be considered as significantly small.

5 Conclusions and Future Work

In this article we presented a new integrated method for onboard person detection and tracking functionality in domestic service robots. We showed how using a novel depth-based segmentation method and 3D blob filter combined with HOG and Haar detectors, we were able to achieve a robust person detector that overcame occlusions and non-uniform ambient lighting issues. Using this detector we constructed a Kalman-filter based tracker. Real robot experiments not only validate our proposed method but comparisons with the ground truth also show the success and precision of our person detector and tracker functionality. Furthermore, as the experiments were done in a real home-like environment in the context of RoCKIn@Home, it is important to note that the validation of our onboard PDT functionality goes beyond typical laboratory settings where most state-of-the-art methods are usually implemented and tested. Finally, we also make the experimental datasets of the real environment and robots publicly available.

In our approach the tracker loses the estimates once the person is out of the field of view of the RGB-D sensor for a certain chunk of time, although a new tracker is initialized the moment the person reappears. The immediate next step in our work is to integrate the laser scans (also available in the datasets) from the front and rear laser range finders in PDT to facilitate tracking even when the robot is not facing the person. Furthermore, some of our ongoing work is to integrate the information from the available network sensors with onboard PDT in order for the robot to have uninterrupted position estimates of all the people within a home-like environment.

References

1. Human figures average measurements. http://www.fas.harvard.edu/~loebinfo/loebinfo/Proportions/humanfigure.html
2. Rockin@home 2014 rulebook. http://rockinrobotchallenge.eu/rockin_home_rulebook.pdf
3. Ahmad, A., Xavier, J., Santos-Victor, J., Lima, P.: 3D to 2D bijection for spherical objects under equidistant fisheye projection. Computer Vision and Image Understanding **125**, 172–183 (2014)
4. Camplani, M., Salgado, L.: Background foreground segmentation with RGB-D kinect data: An efficient combination of classifiers. Journal of Visual Communication and Image Representation **25**(1), 122–136 (2014)
5. Cruz, L., Lucio, D., Velho, L.: Kinect and rgbd images: Challenges and applications. In: 25th SIBGRAPI Conference on Graphics, Patterns and Images Tutorials (SIBGRAPI-T), pp. 36–49. IEEE (2012)
6. Cucchiara, R., Prati, A., Vezzani, R.: A multi-camera vision system for fall detection and alarm generation. Expert Systems **24**(5), 334–345 (2007)
7. Dalal, N., Triggs, B.: Histograms of oriented gradients for human detection. In: IEEE Computer Society Conference on Computer Vision and Pattern Recognition, CVPR 2005, vol. 1, pp. 886–893. IEEE (2005)
8. Horprasert, T., Harwood, D., Davis, L.: A robust background subtraction and shadow detection. In: Proc. ACCV, pp. 983–988 (2000)
9. Kaushik, A., Celler, B.: Characterization of pir detector for monitoring occupancy patterns and functional health status of elderly people living alone at home. Technology and Health Care **15**(4), 273–288 (2007)
10. Kulyukin, V., Gharpure, C., Nicholson, J., Pavithran, S.: RFID in robot-assisted indoor navigation for the visually impaired. In: 2004 IEEE/RSJ International Conference on Proceedings of the Intelligent Robots and Systems, (IROS 2004), vol. 2, pp. 1979–1984, September 2004
11. Noury, N., Herve, T., Rialle, V., Virone, G., Mercier, E., Morey, G., Moro, A., Porcheron, T.: Monitoring behavior in home using a smart fall sensor and position sensors. In: 1st Annual International, Conference on Microtechnologies in Medicine and Biology, pp. 607–610 (2000)
12. Otsu, N.: A threshold selection method from gray-level histograms. Automatica **11**(285–296), 23–27 (1975)
13. Satake, J., Miura, J.: Robust stereo-based person detection and tracking for a person following robot. In: ICRA Workshop on People Detection and Tracking (2009)

14. Scheutz, M., McRaven, J., Cserey, G.: Fast, reliable, adaptive, bimodal people tracking for indoor environments. In: 2004 IEEE/RSJ International Conference on Proceedings of the Intelligent Robots and Systems, (IROS 2004), vol. 2, pp. 1347–1352, September 2004
15. Srichumroenrattana, N., Lursinsap, C., Lipikorn, R.: 2D face image depth ordering using adaptive hillcrest-valley classification and Otsu. In: 2010 IEEE 10th International Conference on Signal Processing (ICSP), pp. 645–648, October 2010
16. Vezzani, R., Grana, C., Cucchiara, R.: Probabilistic people tracking with appearance models and occlusion classification: The ad-hoc system. Pattern Recognition Letters **32**(6), 867–877 (2011)
17. Viola, P., Jones, M.: Rapid object detection using a boosted cascade of simple features. In: Proceedings of IEEE Computer Society Conference on Computer Vision and Pattern Recognition, CVPR 2001, vol. 1, pp. I-511 (2001)
18. Viola, P., Jones, M., Snow, D.: Detecting pedestrians using patterns of motion and appearance. In: Ninth IEEE International Conference on Proceedings of the Computer Vision, pp. 734–741. IEEE (2003)
19. Wen-Hau, L., Wu, C., Fu, L.: Inhabitants tracking system in a cluttered home environment via floor load sensors. IEEE Transactions on Automation Science and Engineering **5**(1), 10–20 (2008)
20. Xia, L., Chen, C., Aggarwal, J.: Human detection using depth information by kinect. In: 2011 IEEE Computer Society Conference on Computer Vision and Pattern Recognition Workshops (CVPRW), pp. 15–22. IEEE (2011)

Visual-Inertial Based Autonomous Navigation

Francisco de Babo Martins, Luis F. Teixeira and Rui Nóbrega

Abstract This paper presents an autonomous navigation and position estimation framework which enables an Unmanned Aerial Vehicle (UAV) to possess the ability to safely navigate in indoor environments. This system uses both the on-board Inertial Measurement Unit (IMU) and the front camera of a AR.Drone platform and a laptop computer were all the data is processed. The system is composed of the following modules: navigation, door detection and position estimation. For the navigation part, the system relies on the detection of the vanishing point using the Hough transform for wall detection and avoidance. The door detection part relies not only on the detection of the contours but also on the recesses of each door using the latter as the main detector and the former as an additional validation for a higher precision. For the position estimation part, the system relies on pre-coded information of the floor in which the drone is navigating, and the velocity of the drone provided by its IMU. Several flight experiments show that the drone is able to safely navigate in corridors while detecting evident doors and estimate its position. The developed navigation and door detection methods are reliable and enable an UAV to fly without the need of human intervention.

Keywords UAV · AR. Drone · Indoor autonomous navigation · Vanishing point · Door detection · Location estimation

---

F. de Babo Martins(✉)
DEEC, FEUP, Faculdade de Engenharia, Universidade do Porto, Porto, Portugal
e-mail: ee09153@fe.up.pt
http://www.fe.up.pt

L.F. Teixeira
DEI, FEUP, Faculdade de Engenharia, Universidade do Porto, Porto, Portugal
e-mail: luisft@fe.up.pt
http://www.fe.up.pt

R. Nóbrega
INESC TEC, Instituto de Engenharia de Sistemas e Computadores - Tecnologia e Ciência, Rua Dr. Roberto Frias, s/n, 4200-465 Porto, Portugal
e-mail: ruinobrega@fe.up.pt
http://www.fe.up.pt

L.P. Reis et al. (eds.), *Robot 2015: Second Iberian Robotics Conference*,
Advances in Intelligent Systems and Computing 418,
DOI: 10.1007/978-3-319-27149-1_43

1 Introduction

In recent years, UAVs have become affordable and relevant in several research areas such as military applications and surveillance systems. These reliable and low-cost devices are normally equipped with high definition cameras and can be used as an autonomous image gathering device. The captured images can be processed in order to extract and obtain useful information that may be used for a variety of tasks and applications.

Being able to hover and fly laterally at low speeds, makes UAVs an ideal platform to accomplish different military and civilian tasks such as reconnaissance support in hazardous zones, visual surveillance and inspection. In addition, some relevant industries are starting to use drones for other tasks beyond surveillance (e.g Amazon's "Prime Air"). Moreover, the most important task in order to achieve UAV autonomy is autonomous navigation. This may prove useful in a near future for tasks in indoor environments such as indoor transportation, object retrieval (e.g a missing part in an assembly line), monitoring misplaced books in a library and autonomously reporting sports events[1].

Although years of research of GPS position and data tracking have improved outdoor navigation and localization, in environments such as indoors or dense urban areas where maps are unavailable and the GPS signal is weak, an UAV will operate in high hazardous regions, running the risk of becoming lost and colliding with obstacles.

Since the scope of this project consists of enabling an UAV to autonomously navigate in an GPS impaired environment, the main challenge is using visual odometry and on-board IMU to develop navigation and position estimation algorithms to achieve an autonomous and robust navigation.

The various contributions that resulted from all the work developed on the scope of this paper are the development of a computer vision framework for autonomous navigation and position estimation in corridors and the implementation of a vanishing point and door detection methods.

2 Related Work

There has been a significant amount of work in regards to making robotic systems understand the environment they are in and reacting accordingly towards reaching a goal.

A team from the University of Texas [14] developed an assistive guide-robot that provides vision-based navigation as well as laser-based obstacle and collision avoidance to a visually-impaired person. In [13] the authors conducted experiments with a visual system able to detect obstacles as their relative size changes when the drone is approaching them.

Moreover, [10] presents a method able to autonomously steer a Micro Aerial Vehicles (MAV) towards distant features called vistas while building maps of unexplored regions. Also, in Bills et al. [9] is presented a method that classifies the types of indoor environments, such as corridors and stairs, while the MAV is traversing them, by using the perspective cues vision algorithm for estimating the desired steering direction.

Other approaches involved using stereo cameras[15] and the Microsoft Kinect RGB-D sensors [16] to solve the SLAM problem and find the best solution for a robust state estimation and control methods. These approaches perform local estimation of the vehicle position and build a 3D model of the environment (SLAM) to plan trajectories through an environment.

Finally, a team from University of Texas [8] developed a method aimed at navigating autonomously in indoor environments (corridors) and industrial environments (production lines) and detecting and avoiding obstacles (people). The navigation was accomplished by using the vanishing point algorithm, the Hough transform for wall detection and avoidance and HOG (histogram of oriented gradients) descriptors using SVM (support vector machine) classifiers for detecting pedestrians.

However, when there is no GPS coverage, some visual navigation autonomous vehicle approaches struggle to estimate with high confidence the location of the drone indoors while enabling the drone to maintaining a stable wall-avoidance trajectory. The approach described in this paper focuses on the autonomous aerial navigation using the frontal camera of a drone to detect the vanishing point of each frame and thus keeping the drone aligned with it and the door detection for estimating the position of the drone.

3 System Architecture

The proposed method consists of a modular algorithm designed using image processing methods. The result of the algorithm is a generated map in which the position of the drone and the doors of a corridor are estimated and displayed.

As most indoor environments satisfy the Manhattan World assumption [2], i.e., most planes lie in one of three mutually orthogonal orientations, all the work developed made use of this assumption.

3.1 Framework

A detailed framework was created to tackle the challenge of autonomous navigation and position estimation using a monocular camera. A graphical representation of the algorithm is shown in Figure 1.

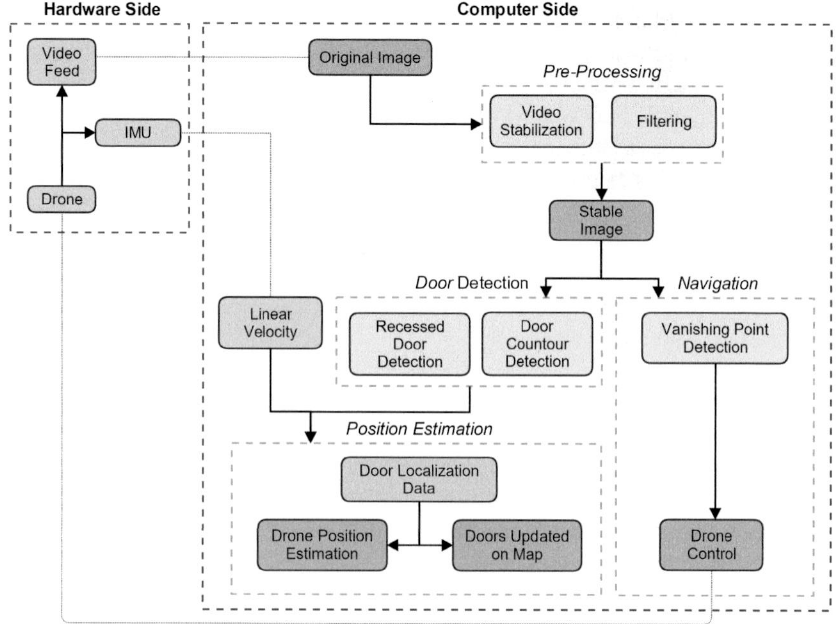

Fig. 1 The developed framework which bridges the gap between the hardware of the drone and the external processing unit using Wifi.

This framework is devised into two parts:

- **Hardware Side:** Image and sensory data from the vehicle.
- **Software Side:** All the image processing methods.

In order to reduce the on-board payload data processing, the drone sends the acquired data to a laptop running a linux distribution with an i7-4790K (4.40 GHz) processor.

The hardware side is comprised of the drone mechanism and stability, the camera input and the medium through which the video feed is transmitted. The computer side processes a series of algorithms and methods that process the video feed and output the drone commands.

3.2 Setup

The live video feed comes from a 720p camera on-board the AR. Drone. The video is recorded at 30 frames per second while the drone navigates in a corridor.

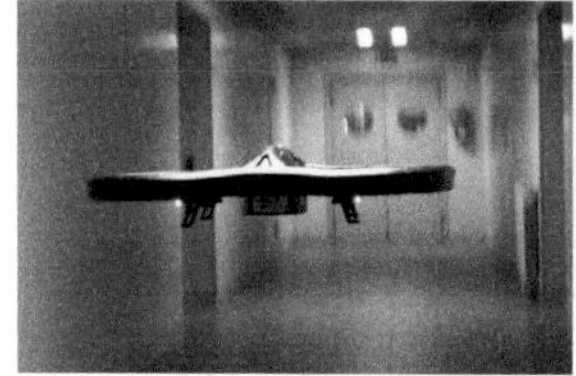

Fig. 2 Drone navigating inside a corridor.

The AR. Drone has limited computational power which may prove a problem as image processing using computer vision algorithms require a lot of resources. Therefore, the analysis of the video feed of the drone is performed remotely in real-time using the OpenCV library for C++ in a computer that is connected to the AR. Drone via wireless. The communication is established using the Robot Operation System (ROS) software framework running on the computer and using a custom made third-party driver. The reason for using this driver rather than a built-in driver is because the development team behind ROS has yet to build one. Therefore, the Ardrone_autonomy[1] library was used.

3.3 Pre-processing

In order to extract useful information from the live video feed, some methods were implemented having in mind that the AR. Drone is susceptible to oscillations caused by either external and uncontrollable factors, like airflows, as well as internal factors such as its own dynamic.

a) Video Stabilization

The aim of the image stabilization method developed for this project was providing useful stabilization for a forward-moving and panning video stream.

The first step is to find the transformation from the previous to the current frame consisting in a rigid Euclidean transform, achieved by using optical flow for all frames. In order to detect corner feature points for optical flow tracking, the OpenCV implementation of *goodFeaturesToTrack* was used which is based on the Shi-Tomasi corner detector algorithm[3]. Retrieved feature points consist of 2D pixel locations that are matched in consecutive frames and afterwords are tracked using the OpenCV Lucas-Kanade Optical Flow implementation[2].

The second step is getting the trajectory for x,y and the angle at each frame by accumulating the frame-to-frame transformations and the third step consists of smoothing the trajectory by using a sliding average window.

---

[1] https://github.com/AutonomyLab/ardrone_autonomy

[2] http://docs.opencv.org/modules/video/doc/motion_analysis_and_object_tracking.html

The final step consists of creating a new transformation and applying it to the current frame from the video.

As seen in Figure 3, the developed stabilization method sets out to perform as intended as it compensates undesirable motion when a strong oscillation occurs. The first picture represents the original frame and the second one the stabilized frame.

Fig. 3 Result of the Stabilization method when a strong oscillation occurs.

b) Filtering

After video stabilization and before releasing the frame to the other modules, some relevant image operation needs to take place. In order to reduce some noise in each frame, some image smoothing (blurring) was applied. This was done by applying the *opening* morphology operation and a Gaussian filter to remove noise and unimportant edges.

3.4 Vanishing Point Detection

Any set of parallel lines on the plane define a vanishing point and when traversing a corridor or a hallway, one can easily observe four lines drawn to the ends of that corridor or hall. Therefore, an image from a corridor may have more than one vanishing point. Following the vanishing point will enable the drone to avoid collisions with the walls.

The process of detecting the vanishing point in a corridor consists of the following steps:

- **Step 1:** Detecting the lines in the frame using the Hough Transform[5].
- **Step 2:** Filtering out horizontal and vertical lines: only lines which respect the condition $170° < \theta < 10° \cap 180° < \theta < 100°$ are diagonal;
- **Step 3:** Calculating the intersection of all the diagonal lines.
- **Step 4:** Obtaining the vanishing point which is the point with the highest number of intersections.
- **Step 5:** Removing outliers and using a Kalman filter to maintain the vanishing point stable throughout the whole video stream.

Figure 4 shows the result of this module. The drone will adjust its yaw to the left or to the right depending on the location of the vanishing point.

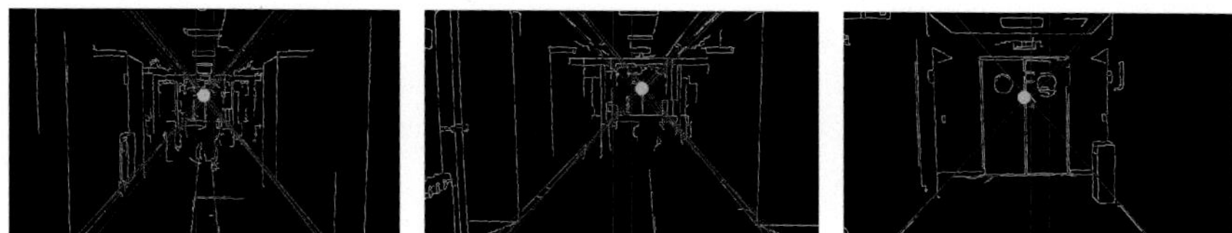

Fig. 4 Result of the vanishing point detection method.

3.5 *Door Detection*

As a means to achieve localization estimation, the approach taken was that of detecting structural elements of the corridors (like doors) in which the drone would be navigating. A single detection method was perceived as not being enough due to the reason that while traversing the corridor, the front facing camera of the drone does not see the doors, but instead sees recessed doors.

a) Recessed Door Detection

This method aims at detecting the recesses in the corridor which would mean that a door was detected. This method was divided into two parts:

a).1 Floor Segmentation using the watershed segmentation algorithm which performs a non-parametric marker-based image segmentation. This is an interactive image segmentation in which different labels are assigned to the known objects in the frame. Since the floor is the foreground and the rest of the frame is background, they are labelled accordingly.

a).2 Vertex Detection which is achieved by applying a mask to isolate an extract the ground in each frame and using the Hough transform[5] to detect horizontal and vertical lines. The vertexes in the far left and far right side of the frame, which correspond to a recess in the left and the right side of the corridor, are then acquired. Each new vertex detection is cross-validated with the previous one and if successive vertex detections occur during a certain period of time, a door is detected. If the coordinates of the newly detected vertex are within a certain range of the previously detect vertex, it is considered a valid door detection.

As seen in Figure 5, which shows the results of the developed method,the vertexes of the doors are successfully detected.

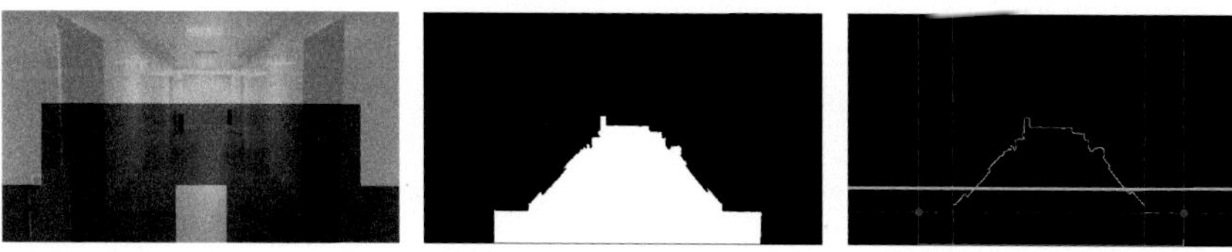

Fig. 5 The various steps of the recessed door detection method.

b) Door Contour Detection

This method aims at detecting the contours of doors in corridors, filtering out the unnecessary contours in order to detect those who correspond to doors. To achieve this, the following sequence of operations is used:

- **Step 1:** Apply a threshold to the current frame in order to obtain a binary image.
- **Step 2:** Detect the edges in the current frame using the Canny edge detector.
- **Step 3:** Retrieve the contours from the binary image using [6].
- **Step 4:** Approximate the contours to a rectangle, which will make the contour sides be a lot more regular, using the Douglas-Peucker algorithm.
- **Step 5:** Segment the rectangular contours using geometry. All the contours whose height is not bigger than its width are discarded.

Resembling the recessed door detection method, this method also uses temporal information in order to avoid unreliable detections. This is achieved by simply keeping a counter for both left and right door contour detections. When that counter reaches a specific value, a door is detected and the corresponding counter is reset.

In the end, the contours of the doors in the video feed are detected properly as seen in figure 6.

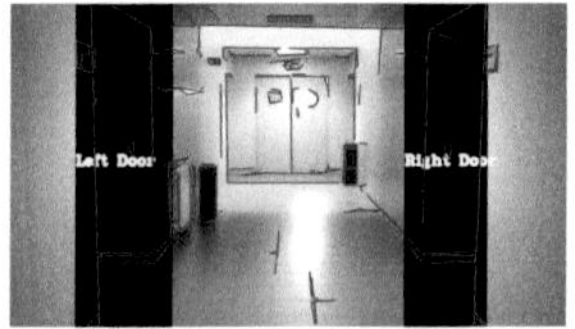

Fig. 6 Result of the door contour detection method.

3.6 Position Estimation

The final module of the developed framework is responsible for the position estimation of the vehicle traversing the corridor. In order to estimate this information, the following data is used:

- **The velocity of the drone:** either a real-time value provided by the drone, or, assuming that the vehicle is moving at a steady pace, a constant value.
- **Precoded Map Information:** a text file containing relevant metric data from the corridor in which the drone will be traversing such as the distance from the starting point of each door. Before each flight the framework loads the corresponding floor map data containing the information shown in Table 1 (all the measurements are in centimetres).
- **The door detection information:** obtainable from the door detection module.

After reading the floor information provided by the data file, a visual representation of the floor map is created. The map itself never changes, however, two types of information are updated/drawn on the map:

Table 1 Example of corridor data: details, door numbers (I00X) and distances.

storey	length	start	rooms	left	right	left	right	left	right
0	2300	north	6	I006 500	I011 500	I007 1700	I010 1700	I008 2300	I009 2300

- **The position of the drone:** updated depending on its current velocity and the initial point of departure.
- **The location of the detected doors:** drawn on the map when the current position of the drone is near the location of a door and the door detection module recognizes it.

The result of the this module, which uses all the previously developed modules, can be seen in Figure 7.

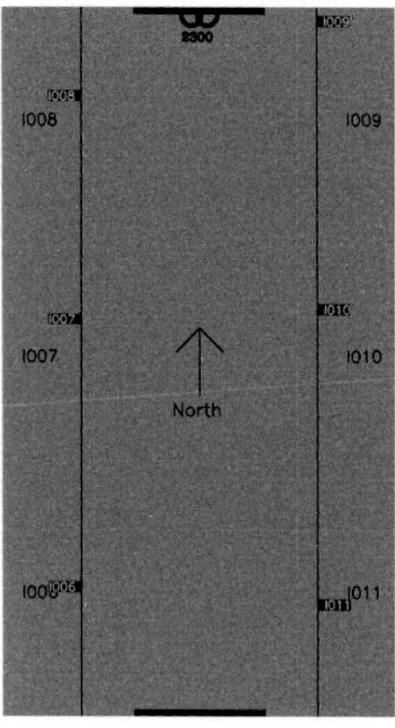

Fig. 7 Example of a successful door detection and position estimation.

4 Experiments

After testing simple take-off, hover and navigation commands on the Parrot AR. Drone, it was noticeable that it could not maintain a stable trajectory on narrower corridors (width lower than 1,5 meters). The reason for the loss of stability was due to different airflows:

- Originating from the various rooms behind the doors in the corridor.
- Caused by the rotors of the drone. Since the corridor is so narrow, the airflow caused by the drone does not disperse uniformly causing the drone to destabilize.

Therefore, the narrower corridor was used to test the behaviour of the developed framework with a previously recorded video instead of a live video from the Parrot AR. Drone.

The corridor used for the flight tests with the drone has a length of 23 meters and 6 rooms: 3 on the left and 3 on the right (all with recessed doors).

Two different trajectories were used as test scenarios:

- **Experiment #1:** Northern entrance as the starting point.
- **Experiment #2:** Southern entrance as the starting point.

The narrow corridor that was used has a length of 25 meters and 12 rooms: 8 on the left and 4 on the right (only the latter were recessed doors).

All the experiments were conducted in optimal conditions, meaning that:

- There were no people walking along the corridor.
- All the doors were completely shut in order to reduce airflow interference.

Figure 8 shows the typical output of the framework during an experiment.

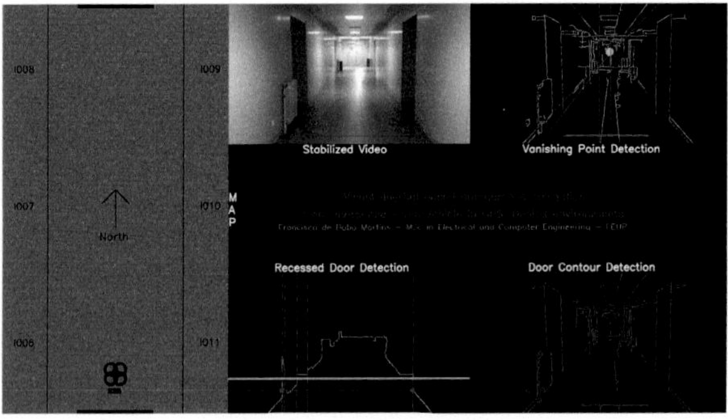

Fig. 8 Example of the control window provided by the framework.

The developed modules proved to work accordingly and as desired. Table 2 provides a summary of the 3 experiments conducted.

Table 2 Summary of the experiments.

	Experiment #1	Experiment #2	Experiment #3
Vanishing Point Standard Deviation	[11.47 ; 10.88]	[25.23 ; 10.85]	[18.60 ; 16.91]
Door Detection Success Rate	83.33%	75%	53.33%
Number of Doors	6	4	12
Arrival Success Rate	80%	80%	N/A

The majority of the doors were successfully detected in all the experiments. Each experiment had 10 trials, the average frame rate was around 26 frames per second, and the mean time required by the drone to traverse the corridors was 1 minute and 20 seconds (for experiments #1 and #2). However, the door contour detection proved to be not so robust, leading to a door detection that was almost totally dependent on the recess detection module. The *arrival success rate* corresponds to the number of trials in which the drone was able to travel from one end of the corridor to the other without colliding with the walls. Since experiment #3 was conducted without a flying drone, the *arrival success rate* is not relevant.

The experiment with the best overall results was the first one due to its lower *vanishing point standard deviation*, and higher *door detection success rate*. This results demonstrate that the drone was able to keep aligned with the center of the corridor without colliding with the walls, reach the end of such corridor and detect most of the its doors. Some simulations and flight experiments can be seen on youtube: https://youtu.be/xjIpChc9Oio.

5 Conclusions

This project proved that it is possible to enable an UAV to possess a system capable of providing a collision-free navigation, a door detection system and a robust position estimation of both a vehicle and doors in corridors.

Using only the vanishing point, the drone was able to fly with a wall collision-free navigation and doors with different formats were successfully identified thanks to the use of two different detection methods: *recessed door* and *door contour detection*. Also the position of the drone and the position of the doors in the floor map were estimated with minimal errors and were quite satisfactory.

The developed framework is constructed in such a way that the inclusion of additional modules would be a simple and accessible task. Since each module is independent of one another, sharing just one video feed among each other, it would be as simple as switching a specific module on or off. This feature enables the framework to have a considerable degree of flexibility as it is not bound to a specific type of robot, being versatile to the point of working properly on either aerial or ground vehicles. A ground vehicle would provide a higher stability and lower external interferences like airflows.

Concluding, the proposed method has proven to be flexible and versatile enough in order to be considered a positive asset in applications such as: auto-pilot systems; mobile surveillance; traffic management; domestic, industrial and military applications; and search and rescue missions.

5.1 Future Work

By refining the door and vanishing point detection modules, the door detection success rate would increase for corridors without recessed doors and the vanishing point detection would not be susceptible to errors in corridors with poster and boards hanged on the walls. Also, using a drone with a faster data exchange rate would speed up the video input stream processing and therefore make and provide quicker navigation decisions.

Furthermore, adding a collision avoidance and person detector module would provide a safer navigation for coexisting drones and people. Furthermore, processing frames and performing image operations with the GPU instead of the CPU would lead to an improved performance.

Using additional hardware (external or internal to the drone), such as range-finders, would provide additional data regarding possible obstacles such as pillars, windows and foreign objects.

References

1. Ferreira, F.O.R.T.: Video Analysis in Indoor Soccer with a Quadcopter. Master Thesis, Faculdade de Engenharia da Universidade do Porto (2014)
2. Coughlan, J.M., Yuille, A.L.: The Manhattan world assumption: regularities in scene statistics which enable Bayesian inference. In: NIPS, pp. 845–851 (2000)
3. Stavens, D.: The OpenCV library: computing optical flow (2007)
4. Van Den Heuvel, F.A.: Vanishing point detection for architectural photogrammetry. International Archives of Photogrammetry and Remote Sensing **32**, 652–659 (1998)
5. Duda, R.O., Hart, P.E.: Use of the Hough transformation to detect lines and curves in pictures. Communications of the ACM **15**(1), 11–15 (1972)
6. Suzuki, S.: Topological structural analysis of digitized binary images by border following. Computer Vision, Graphics, and Image Processing **30**(1), 32–46 (1985)
7. Welch, G., Bishop, G.: An introduction to the kalman filter 2006. University of North Carolina: Chapel Hill, North Carolina, US (2006)
8. Lioulemes, A., Galatas, G., Metsis, V., Mariottini, G.L., Makedon, F.: Safety challenges in using AR. Drone to collaborate with humans in indoor environments. In: Proceedings of the 7th International Conference on PErvasive Technologies Related to Assistive Environments, p. 33. ACM (2014)
9. Bills, C., Chen, J., Saxena, A.: Autonomous MAV flight in indoor environments using single image perspective cues. In: IEEE international conference on Robotics and automation (ICRA 2011), pp. 5776–5783. IEEE (2011)
10. Saska, M., Krajník, T., Faigl, J., Vonásek, V., Přeučil, L.: Low cost MAV platform AR-drone in experimental verifications of methods for vision based autonomous navigation. In: 2012 IEEE/RSJ International Conference on Intelligent Robots and Systems (IROS), pp. 4808–4809. IEEE (2012)
11. Engel, J., Sturm, J., Cremers, D.: Camera-based navigation of a low-cost quadrocopter. In: 2012 IEEE/RSJ International Conference on Intelligent Robots and Systems (IROS), pp. 2815–2821. IEEE (2012)
12. Bills, C., Chen, J., Saxena, A.: Autonomous MAV flight in indoor environments using single image perspective cues. In: 2011 IEEE international conference on Robotics and automation (ICRA), pp. 5776–5783. IEEE (2011)
13. Mori, T., Scherer, S.: First results in detecting and avoiding frontal obstacles from a monocular camera for micro unmanned aerial vehicles. In: 2013 IEEE International Conference on Robotics and Automation (ICRA), pp. 1750–1757. IEEE (2013)
14. Galatas, G., McMurrough, C., Mariottini, G.L., Makedon, F.: eyeDog: an assistive-guide robot for the visually impaired. In: Proceedings of the 4th International Conference on PErvasive Technologies Related to Assistive Environments, p. 58. ACM (2011)
15. Achtelik, M., et al.: Stereo vision and laser odometry for autonomous helicopters in GPS-denied indoor environments. In: SPIE Defense, Security, and Sensing. International Society for Optics and Photonics (2009)
16. Huang, A.S., et al.: Visual odometry and mapping for autonomous flight using an RGB-D camera. In: International Symposium on Robotics Research (ISRR) (2011)

Ball Detection for Robotic Soccer: A Real-Time RGB-D Approach

André Morais, Pedro Costa and José Lima

Abstract The robotic football competition has encouraged the participants to develop new ways of solving different problems in order to succeed in the competition. This article shows a different approach to the ball detection and recognition by the robot using a Kinect System. It has enhanced the capabilities of the depth camera in detecting and recognizing the ball during the football match. This is important because it is possible to avoid the noise that the RGB cameras are subject to for example lighting issues.

Keywords Image processing · Kalman filter · Kinect · Robotic football

1 Introduction

The evolution of technology in recent years has allowed the emergence of sensors that enable new and improved methods of interpreting the environment. This is relevant for many areas especially for robotics, as it allowed the creation and development of algorithms that perceive the world around us.

However, these algorithms are usually computationally demanding, making them unviable for robotic systems based on artificial vision that require real-time processes.

A. Morais · P. Costa
INESC TEC (formerly INESC PORTO) - Robotics and Intelligent Systems,
Faculty of Engineering of University of Porto, Rua Dr. Roberto Frias, 4200-465 Porto, Portugal
e-mail: {ee10276,pedrogc}@fe.up.pt

J. Lima
INESC TEC (formerly INESC PORTO) - Robotics and Intelligent Systems,
Rua Dr. Roberto Frias, 4200-465 Porto, Portugal

J. Lima(✉)
Polytechnic Institute of Bragança, Campus Sta Apolónia, 5301-857 Bragança, Portugal
e-mail: jllima@ipb.pt

L.P. Reis et al. (eds.), *Robot 2015: Second Iberian Robotics Conference*,
Advances in Intelligent Systems and Computing 418,
DOI: 10.1007/978-3-319-27149-1_44

Thus the development and creation of increasingly fast and efficient algorithms has been a challenge for the scientific community.

The aim of this work involves the development of a detection and object recognition system using the RGB-D sensors to perform the recognition of the ball in a robotic football game, that is computationally efficient and robust from outside interference such as lighting problems [9].

The motivation of this project is mainly to integrate all knowledge acquired in robotic football team FEUP (RoboSoccer-5DPO Team) and to get a good performance in a future robotic football competition, mainly because competitions have become more competitive and challenging which encourages students to seek new alternatives that have better results in order to improve their systems.

This article is divided into six sections, excluding this introduction section. The section 2 details the related work that has been found in the available literature. It is approached in the section 3 the detection system which aims to image segmentation and the identification of different image objects to be subsequently recognized. The football ball recognition system is specified in the section 4 where is specified the algorithm developed to recognize the different objects as a football ball. It is described in the section 5 the implementation and incorporation of the Kalman filter in the developed system. The conclusions and future developments are presented in the section 6.

2 Related Work

The detection and recognition of the objects using vision systems is a big part of the RoboCup competition. Through the years some rules have changed to improve the competition. This work comes in response to the latest change to the ball rules which is that the ball can be of any color. In the previous years the ball has always the same color and the teams use mainly the color information to localize the ball position. This new improvement to the competition presents a challenge to the participants because they cannot rely only in the color information.

The techniques used to detect and recognize objects using RGB-D sensors were analyzed because this systems are widely used for this type of applications [1], [2], [10].

In [3] a new perspective to the Hough Transformation is presented as a technique used in image processing in the detection of circular/lines features of the image. This article shows that this technique can be optimized using a set of descriptions and feature matching techniques following by a comparison with the data training that was done.

In [4] is detailed the SIFT method that enables the recognition of an object that is invariant to the object rotation, scale or projections in the 3D space. Also in [6] an improvement of the SIFT method techniques increasing the algorithm processing time with similar results.

The techniques presented in [5] allow to measure the similarity of the object that has to be identified in the image from any other using histograms and descriptors acquired in a training phase.

In [7] the RANSAC algorithm is presented which finds in a set of data the presence of a mathematical model for example a plane.

3 Object Detection

This section presents the techniques used to make the detection of the ball. It is observed that during a football game, the ball may be present on the ground or in the air and these two cases have different properties. Thus, it was decided to follow this philosophy and detect the ball in two different cases. The following sections presents the procedures used to achieve effectively the detection of the football ball.

3.1 Aerial Ball

Clustering. In order to separate the different objects in the depth image (see figure 1) it was chosen to separate the image into different clusters by applying the K-Means Clustering Technique based on OpenCV Libraries.

From figure 1 is possible to get the different clusters that are shown in figure 2 and also it shows that in the 2d sub-figure the ball is well defined and disconnected from other objects. The process time of the clustering and the recognition system is $\sim$30ms that is quite fast, on an Intel i5 processor and using five clusters.

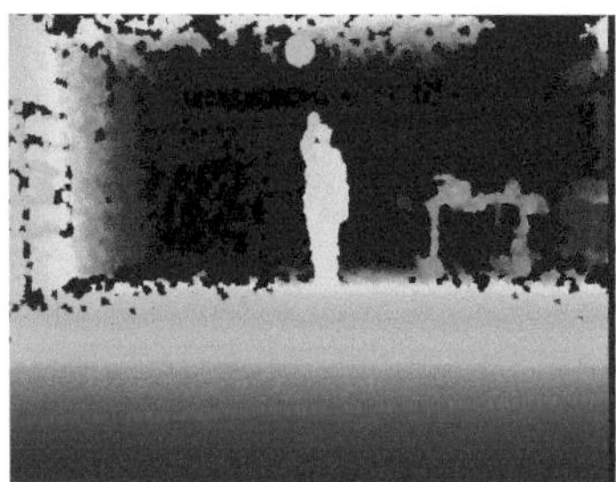

Fig. 1 Depth Image with an Aerial Ball present.

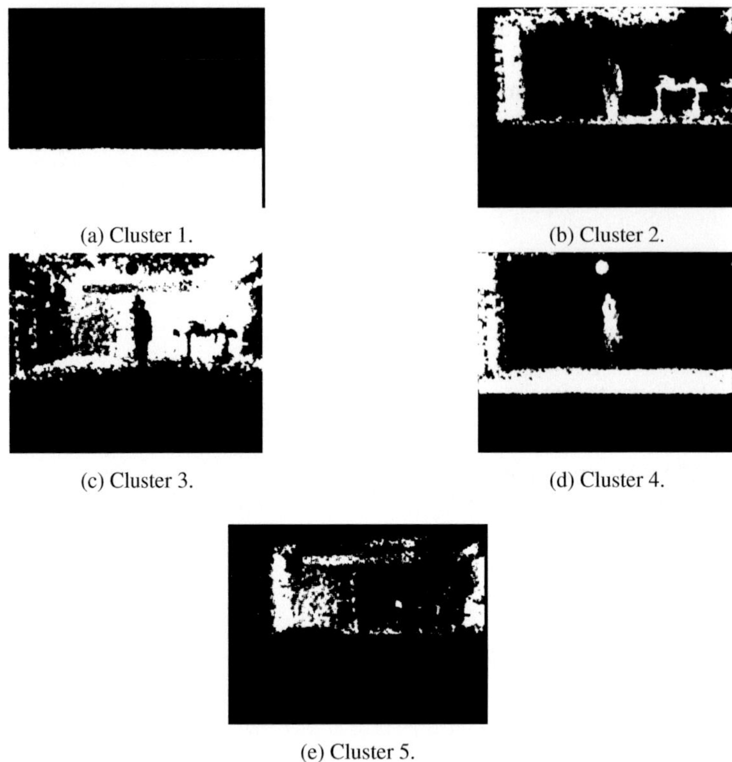

(a) Cluster 1. (b) Cluster 2.

(c) Cluster 3. (d) Cluster 4.

(e) Cluster 5.

Fig. 2 *Clustering* of the Depth Image in the figure 1.

3.2 Ground Ball

For the detection of the ball when it is in the ground it is not advantageous to use the technique used in the aerial ball detection without any pre-processing, because a large number of cluster was needed to archive similar results as in the aerial ball detection.

The increasing of the number of clusters will affect the computational time. This has to be avoided because it will reduce the reaction time by the goalkeeper, increasing the chances of suffer goal.

The technique used for this case will be detailed in the next sections.

Field Segmentation. It is known that the ground field is green, so this fact is used in order to find a set of points for calculate the ground plane equation.

The use of a RGB image for the threshold of color is not the best option because of the high correlation between the different components, the image is highly sensitive to the scene illumination, etc [8]. To counter act this problem it is transformed the RGB image to the YUV color space. The YUV color space allows the separation of the luminance and the color in different components allowing a better threshold than the RGB image.

Ground Plane. After the threshold is selected a set of points acquiring the correspondent depth for each point based on a $3 * 3$ mask.

These points are used to calculate the ground plane parameters using the RANSAC algorithm on the beginning of the match. In order to remove the ground floor thus allowing to cluster the image and get similar results as in the aerial ball clustering.

Results. The next procedure is the same as used in the aerial ball which is the clustering of the example figure 3. In the figure 4 and 5 it is possible to see that after the football field segmentation the ball appears disconnected from other objects allowing to perform the recognition of the different image objects.

Fig. 3 Depth Image with Ground Ball.

It was possible to see that the clustering of the depth image allows the detection of the ball in the image as an object disconnected from others objects with defined characteristics that are analyzed by the recognition system detailed in the following section. The process time of this task (including the recognition process) is slower than the aerial ball case taking a total time of ~80ms, on an Intel i5 processor, due the elimination of the ground planes takes a long time.

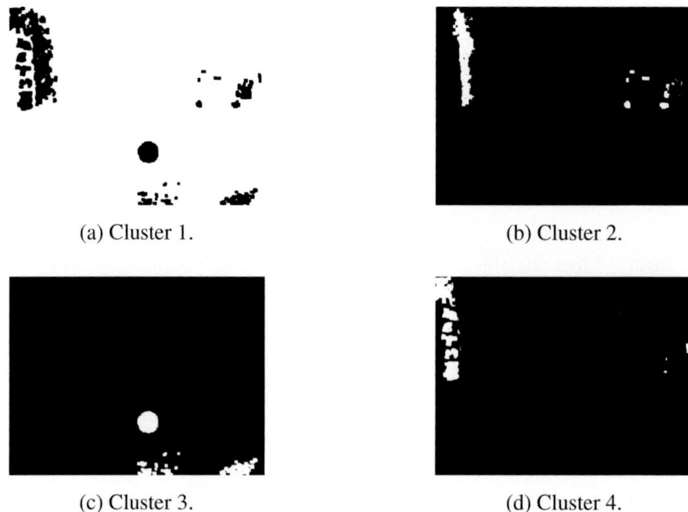

(a) Cluster 1. (b) Cluster 2.

(c) Cluster 3. (d) Cluster 4.

Fig. 4 *Clustering* of the Depth Image in the figure 3.

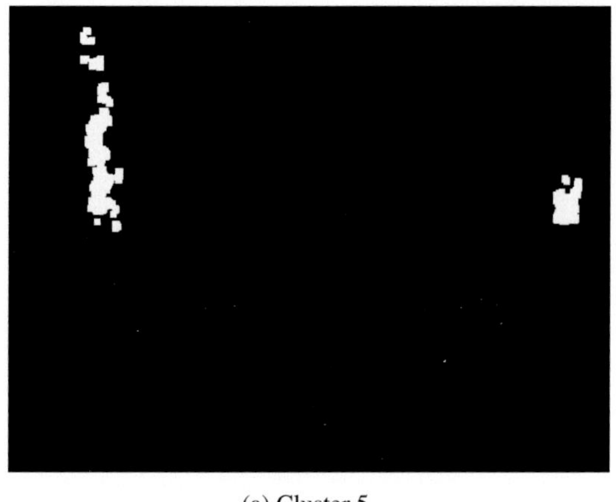

(a) Cluster 5.

Fig. 5 *Clustering* of the Depth Image in the figure 3.

4 Object Recognition

This section addresses the recognition process that is performed after the segmentation previously presented.

To develop an algorithm capable of identifying the presence of football ball is used the knowledge of the ball geometrical properties.

The features that yields better results to the conditions that are imposed to the system are: area, perimeter, circularity and the radius of the ball.

To create a dynamic recognition model was analyzed the variations of the considered object features through a series of test yielding the results present in the figure 6.

This results are approximated by a polynomial function that models the value of that feature.

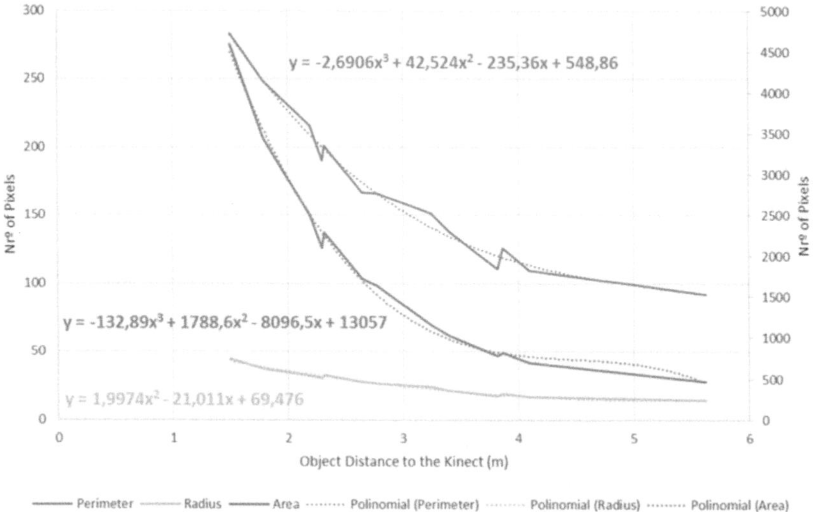

Fig. 6 Fitting curves to the different measured ball features, based on the distance tests to the Kinect.

The developed ball recognition system initially removes the objects that exceed the maximum and minimum geometric characteristics considerated. These limits were experimentally found taking into account the limitations of the equipment used, that is, the boundaries of the distance sensor that is able to return consistent results.

After this first step it is obtained, for the remaining objects, their distance to the camera using a 3x3 mask in order to avoid noise that may arise in the center of mass pixels of the object. Using the functions that were previously calculated that indicate the expected value for each characteristics. The objects that are out of the defined interval error around the expected value are discarded.

At the end it is considered that the object that passes all of these steps is considered to be a football ball.

4.1 Results

The results of this recognition system is presented in figures 7 and 8 where the football ball is shown with a red point.

(a) Aerial Ball Result 1.

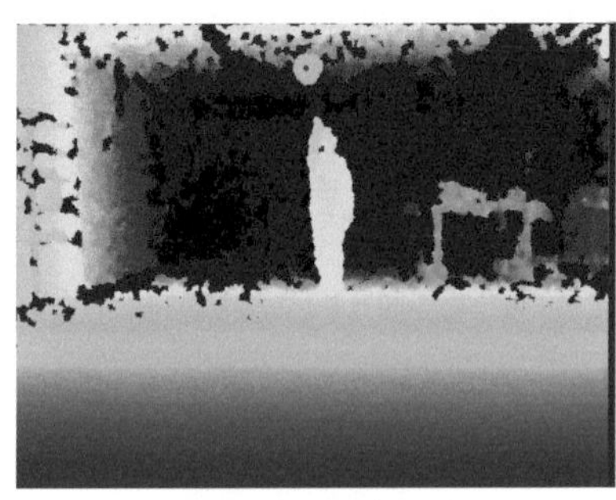

(b) Aerial Ball Result 2.

Fig. 7 System Recognition Results.

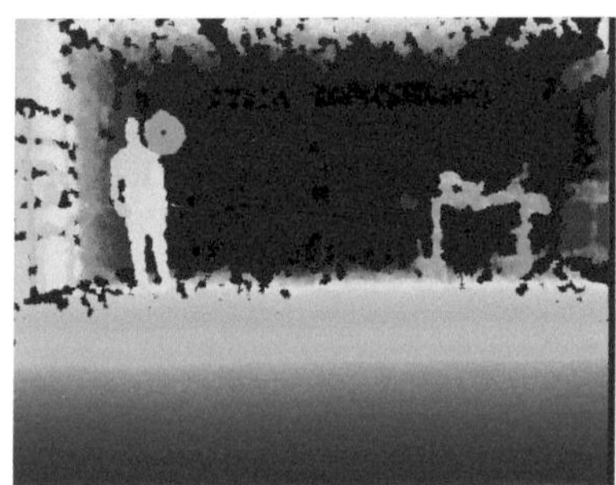

(a) Aerial Ball Result 3.

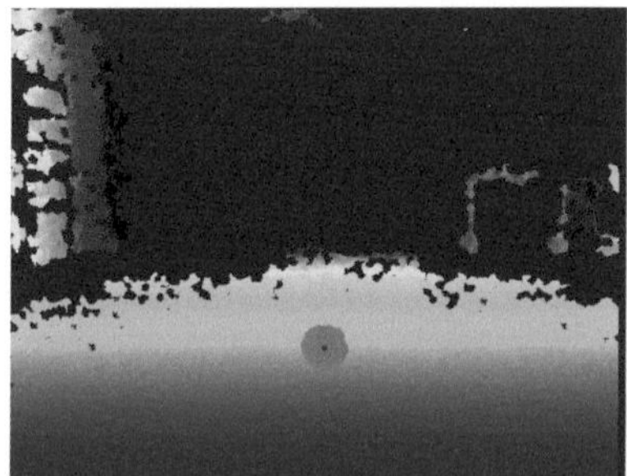

(b) Ground Ball Result 1.

(c) Ground Ball Result 2.

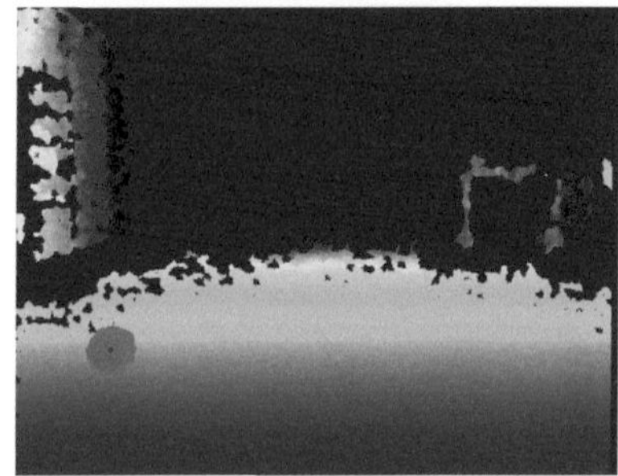

(d) Ground Ball Result 3.

Fig. 8 System Recognition Results.

Finally it is possible to see that the recognition system yields good results. However it is not always possible to recognize the ball in every image frame because of the image acquisition rate or the natural occlusion that can occur, so to counter act this problem a Kalman Filter is implemented to predict the ball position when it is not possible to detect the presence of the ball as present in next section.

5 Kalman Filter

5.1 Aerial Ball Model

The model used for the case of the aerial ball is the model with constant acceleration. This model is used because being the ball in the air is knew that the gravity force is always present in the ball movement.

Equation 1 details the state of the ball used by the Kalman Filter. Where x, y and z are the position of the ball in the world.

$$X = \begin{bmatrix} x\ y\ z\ \dot{x}\ \dot{y}\ \dot{z}\ \ddot{x}\ \ddot{y}\ \ddot{z} \end{bmatrix}^T \tag{1}$$

The calculation of the next state of the ball is done by using the equation 2 taking into account that there is no control system, the value B matrix is zero.

$$X_{k+1} = A \cdot X_k + B \cdot u_k \tag{2}$$

$$X_{k+1} = \begin{bmatrix} 1 & 0 & 0 & \Delta t & 0 & 0 & \frac{1}{2}\Delta t^2 & 0 & 0 \\ 0 & 1 & 0 & 0 & \Delta t & 0 & 0 & \frac{1}{2}\Delta t^2 & 0 \\ 0 & 0 & 1 & 0 & 0 & \Delta t & 0 & 0 & \frac{1}{2}\Delta t^2 \\ 0 & 0 & 0 & 1 & 0 & 0 & \Delta t & 0 & 0 \\ 0 & 0 & 0 & 0 & 1 & 0 & 0 & \Delta t & 0 \\ 0 & 0 & 0 & 0 & 0 & 1 & 0 & 0 & \Delta t \\ 0 & 0 & 0 & 0 & 0 & 0 & 1 & 0 & 0 \\ 0 & 0 & 0 & 0 & 0 & 0 & 0 & 1 & 0 \\ 0 & 0 & 0 & 0 & 0 & 0 & 0 & 0 & 1 \end{bmatrix} \cdot X_k \tag{3}$$

The calculation of the object's position is possible using the equation 4.

$$Y = \begin{bmatrix} 1 & 0 & 0 & 0 & 0 & 0 & 0 & 0 & 0 \\ 0 & 1 & 0 & 0 & 0 & 0 & 0 & 0 & 0 \\ 0 & 0 & 1 & 0 & 0 & 0 & 0 & 0 & 0 \end{bmatrix} \cdot X \tag{4}$$

5.2 Ground Ball Model

For the next case the gravity force is present however, it doesn't make much interference in the movement. In this case the ball movement is detailed by the force applied by the kick executed by other robot.

It was considered that the ball velocity is equal through its movement, so it was used the constant velocity model to filter the ball position, as present in equation 5.

$$X = \begin{bmatrix} x & y & z & \dot{x} & \dot{y} & \dot{z} \end{bmatrix}^T \tag{5}$$

The calculation of the next state of the ball is done by using the equation 6 taking into account that there is no control system, the value B matrix is zero and A is shown in 6.

$$X_{k+1} = A \cdot X_k + B \cdot u_k \tag{6}$$

$$X_{k+1} = \begin{bmatrix} 1 & 0 & 0 & \Delta t & 0 & 0 \\ 0 & 1 & 0 & 0 & \Delta t & 0 \\ 0 & 0 & 1 & 0 & 0 & \Delta t \\ 0 & 0 & 0 & 1 & 0 & 0 \\ 0 & 0 & 0 & 0 & 1 & 0 \\ 0 & 0 & 0 & 0 & 0 & 1 \end{bmatrix} \cdot \begin{bmatrix} x \\ y \\ z \\ \dot{x} \\ \dot{y} \\ \dot{z} \end{bmatrix}_k \tag{7}$$

The calculation of the object's position is possible using the equation 8.

$$Y = \begin{bmatrix} 1 & 0 & 0 & 0 & 0 & 0 \\ 0 & 1 & 0 & 0 & 0 & 0 \\ 0 & 0 & 1 & 0 & 0 & 0 \end{bmatrix} \cdot X \tag{8}$$

5.3 Results

The results of the Kalman Filter are showed in the figure 9, where the result of the recognition system are showed using a red point and the result of the Kalman Filter is showed using a green line.

Table 1 shows the mean error of the actual position of the ball in comparison with the position returned by the Kalman filter. It is possible to see that the Kalman Filter gives a larger error in the aerial ball case than in the ground ball case. This fact is because the process and measure noise matrix are very challenging to determine. Where the number of pixels is the difference between the calculated center of the ball and the true center of the ball.

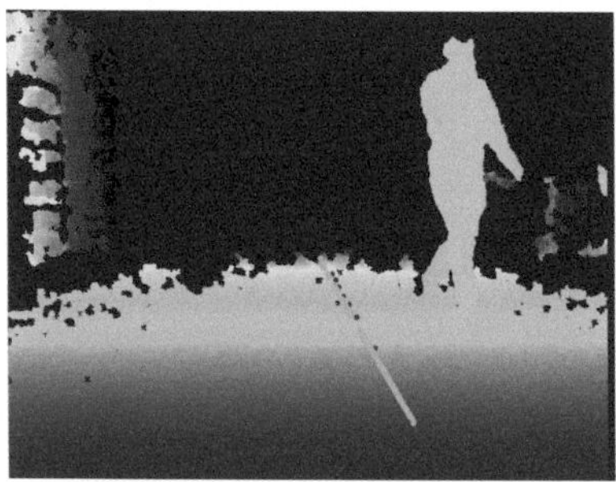
(a) Kalman Ground Ball Result 1.

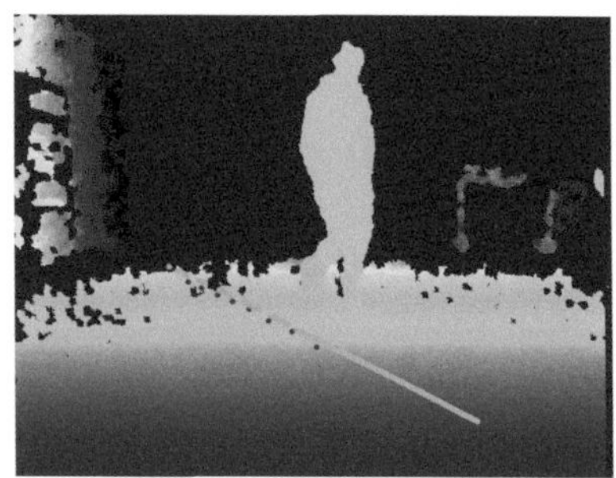
(b) Kalman Ground Ball Result 2.

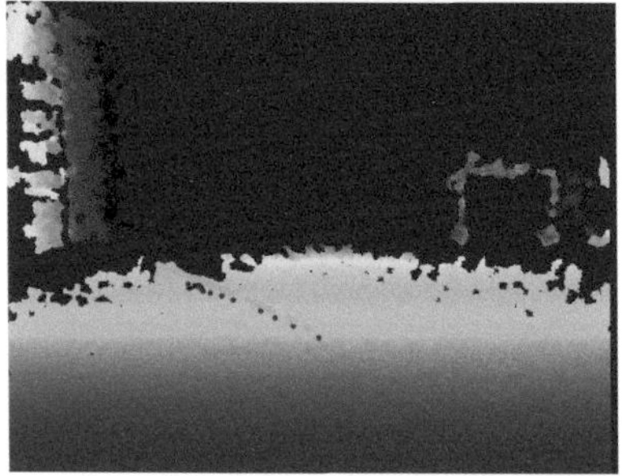
(c) Kalman Ground Ball Result 3.

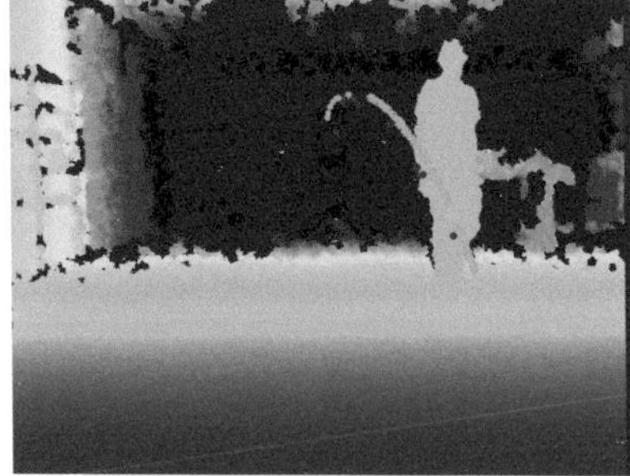
(d) Kalman Aerial Ball Result 1.

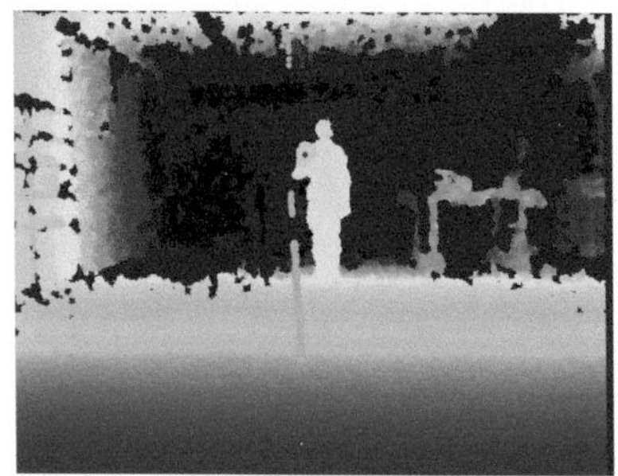
(e) Kalman Aerial Ball Result 2.

(f) Kalman Aerial Ball Result 3.

Fig. 9 Kalman Filter Results.

Table 1 Mean error of the actual position of the running ball in comparison with the position returned by the Kalman filter.

	Number of Pixels
Kalman Aerial Ball 1	12.3241
Kalman Aerial Ball 2	15.0333
Kalman Aerial Ball 3	13.684
Kalman Ground Ball 1	3.2083
Kalman Ground Ball 2	3.6056
Kalman Ground Ball 3	6.2111

6 Conclusions

Summarily, it was observed that the separation of the detection and recognition of the football ball problem in two cases, aerial ball and ground ball, was beneficial because they present different problems which allowed the increasing of the algorithm speed.

The aerial ball case is quite interesting because it appears that while it is in the air it is disconnected from any image element which allows to quickly cluster the image based in the depth and make the ball recognition three times faster than for the ground ball case.

In the ground ball case it appears that the object is connected to the ground which difficults the cluster processing because it needs a lot of more cluster to provide the same results as in the aerial ball case.

So, it was necessary to eliminate the points that correspond to the ground plane. This process slows the algorithm ~40ms, on an Intel i5 processor, but this strategy does not become a bad choice.

The recognition system returns good results using that four features of the ball but it does not guarantee the recognition of the object in every frame due to the ball quick movement or occlusions.

To counter act this problem a Kalman filter was introduced to fill the gaps of the recognition algorithm. This filter can predict the position of the ball in an effective manner according to the inserted system model. It was observed that the biggest challenge is the modeling of the process noise and measurement matrix that can interfere significantly in its predictions.

This article illustrates one more time the utility of RGB-D sensors in robotic football. In the future it is expected that this algorithm is enhanced to detect and recognize the ball faster and more effectively.

For future developments it is proposed to improve the recognition algorithm that aims to circumvent the effects of the image depth by rapid movement of the ball. Also the method of choosing the aerial ball method or the ground ball method in a given time can be more studied.

The use of a depth sensor with better quality would be a plus because it was possible to get more accurate results avoiding some noise that comes in the picture.

Acknowledgment This work is financed by the ERDF – European Regional Development Fund through the COMPETE Programme (operational programme for competitiveness) and by National Funds through the FCT – Fundação para a Ciência e a Tecnologia (Portuguese Foundation for Science and Technology) within project FCOMP-01-0124-FEDER-037281.

References

1. Marcel, J.: Object Detection and Recognition with Microsoft Kinect (2012)
2. Pheatt, C., Ballester, J.: Using the Xbox Kinect Sensor for Positional Data Acquisition, 1–14 (2011)
3. Tombari, F., Stefano, L.D.: Hough Voting for 3D Object Recognition under Occlusion and Clutter. IPSJ Transactions on Computer Vision and Applications
4. Lowe, D.G.: Object recognition fromlocal scale-invariant features. In: IEEE International Conference on Computer Vision (1999)
5. Rusu, R.B.: Semantic 3D Object Maps for Everyday Manipulation in Human Living Environments (2010)
6. Bay, H., Tuytelaars, T., Van Gool, L.: SURF: speeded up robust features. In: Computer Vision – ECCV 2006. Proceedings of the 9th European Conference on Computer Vision, Graz, Austria, May 7–13, 2006. Lecture Notes in Computer Science, vol. 3951, pp. 404–417 (2006)
7. Fischler, M.A., Bolles, R.C.: Communications of the ACM **24**, 381–395 (1981)
8. Qu, Z., Wang, J.: A color YUV image edge detection method based on histogram equalization transformation. In: Natural Computation (ICNC) (2010)
9. Budden, D., Fenn, S., Walker, J., Mendes, A.: A novel approach to ball detection for humanoid robot soccer. In: AI 2012: Advances in Artificial Intelligence. LNCS, vol. 7691, pp. 827–838 (2012)
10. Dias, P., Silva, J., Castro, R., Neves, A.J.R.: Detection of aerial balls using a kinect sensor. In: RoboCup 2014: Robot World Cup XVIII. LNCS, vol. 8992, pp. 537–548 (2015)

Real Time People Detection Combining Appearance and Depth Image Spaces Using Boosted Random Ferns

Victor Vaquero, Michael Villamizar and Alberto Sanfeliu

Abstract This paper presents a robust and real-time method for people detection in urban and crowed environments. Unlike other conventional methods which either focus on single features or compute multiple and independent classifiers specialized in a particular feature space, the proposed approach creates a synergic combination of appearance and depth cues in a unique classifier. The core of our method is a Boosted Random Ferns classifier that selects automatically the most discriminative local binary features for both the appearance and depth image spaces. Based on this classifier, a fast and robust people detector which maintains high detection rates in spite of environmental changes is created.

The proposed method has been validated in a challenging RGB-D database of people in urban scenarios and has shown that outperforms state-of-the-art approaches in spite of the difficult environment conditions. As a result, this method is of special interest for real-time robotic applications where people detection is a key matter, such as human-robot interaction or safe navigation of mobile robots for example.

Keywords People detection · RGBD · Learning · Boosted Random Ferns

1 Introduction

From social robotics aiming to help people in different ways, to autonomous vehicles that needs to detect pedestrians and obstacles in order to avoid them and provide a safe navigation, robots these days are designed to share spaces with humans. Hardware and Software have evolved rapidly incorporating better sensors and algorithms on perception systems. However, robust algorithms are required for robots

V. Vaquero(✉) · M. Villamizar · A. Sanfeliu
Institut de Robotica i Informatica Industrial - CSIC-UPC, Barcelona, Spain
e-mail: {vvaquero,mvillami,sanfeliu}@iri.upc.edu
http://www.iri.upc.edu

L.P. Reis et al. (eds.), *Robot 2015: Second Iberian Robotics Conference*,
Advances in Intelligent Systems and Computing 418,
DOI: 10.1007/978-3-319-27149-1_45

coexisting with humans in populated environments and people detection algorithms are therefore fundamental.

Typical computer vision approaches on people detection use monocular vision systems and analyse individual appearance (RGB) images looking for a set of pre-learned features, as in [3]. However, other imaging spaces exist, as for example Histograms of Oriented Gradients (HOG) [1]. It has been proved that image HOG space is more robust to illumination and object appearance changes, obtaining substantial gains over features based on the appearance RGB domain. For a deeply analysis of the state-of-the-art in vison-based pedestrian detectors, the reader is referred to recent surveys and benchmarks, i.e [2], [6], [11], [5], [7].

We here present a real-time, robust and reliable method for detecting people in RGB-D images (Figure 6). This is done by computing a classifier that is able to learn how to combine in an effective way cues from both the RGB and Depth image spaces. In our approach we use Random Ferns (RFs), to compute simple and fast Local Binary Features (LBFs). Other works as [15], make use of RFs over the image appearance space (RGB intensities or HOG computed from greyscale images). However, our novelty resides in creating a synergistic combination of RGB and Depth spaces that allows the classifier to keep detecting people when one space is badly degraded, as it will be backed up by the other.

To summarize, the main contributions of the proposed approach are:

- A robust people detector able to learn by means of boosting algorithms the best combination of cues from different image spaces, creating a synergic environment where one space is still able to maintain high detection rates in case that others are spoiled. Our method outperforms the state of the art detectors achieving around an 89% of EER in the people database [9] which has challenging variability of poses, shapes and illumination.
- A fast and simple classifier that uses Random Ferns to compute and evaluate features over the defined spaces. This allows real-time applications because are based on Local Binary Comparisons and therefore does not require computationally expensive calculus. We have obtained performance results of 15-20 fps with a C++ version of the algorithm running in a standard computer (64 bits Intel Core-i7-3770 with 8Gb RAM, running Ubuntu 14.04). There is no need for using GPUs computation as in [13], [8].
- An open approach for a single classifier independently of the sources of information. Unlike other approaches, we create one unique classifier that seeks the best discriminative features combining any input spaces. This means that other imaging data (i.e, thermal information), could be easily used.

2 Related Work

This section presents the related work on people detection over RGB-D images as well as some other works where Random Ferns where used for similar purposes.

2.1 People Detection on RGB-D Images

Recent technologies that allows the easy capture of RGB-D images of a scene have upraised the limits of standard vision-based detectors. Affordable technologies such as stereo-vision systems (*Point Grey's - Bumblebee*), structured-light (*Microsof's - Kinect*) and time-of-flight cameras (*Asus' Xtion*), are nowadays being used for building new approaches on people detection.

In this way, Spinello *et al* used in [13] RGB-D data from a *Kinect* camera and trained two different classifiers separately, one for the RGB (appearance space) while the other for the Depth image. In a second step both classifiers were combined in their *Combo* classifier. A double effort is done in this work, firstly creating two separate classifiers, and secondly merging their outputs.

In contrast, we have created a single unique classifier devoted to both sources of information at the same time, which is able to learn and take the most discriminative features from each of the input image spaces (Appearance and Depth).

We have tested our method over the same database than [13] and [8] and under similar conditions, obtaining better and remarkable results with an average Equal Error Rate around 89%. Furthermore, as our Boosted Random Ferns performs simple binary comparisons, lower computational effort is required allowing real time performance without the need of GPUs.

Further works have also studied the way to leverage the RGB-D data, adapting the information that the classifier used from each of the sources depending on its arriving quality [4], [14]. On the contrary our Shared RFs and Boosting approach allows to developed a single strong classifier that learn and take the best discriminative RFs over each of the sources of information, which allows to keep detecting people when one of the input sources is distorted, as the other source is still able to produce good results.

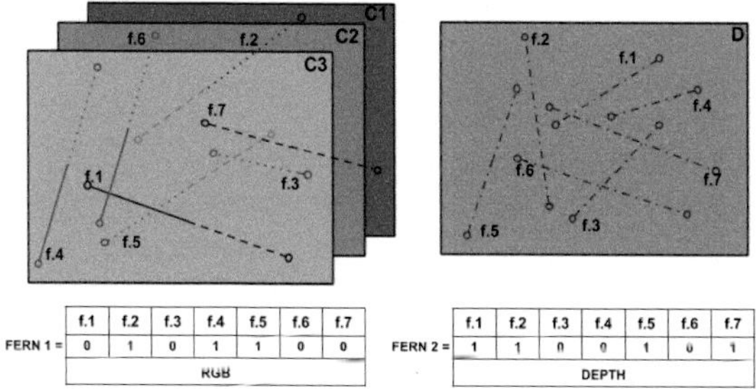

Fig. 1 Example of Ferns devoted to different spaces. Images presents three and one images spaces (i.e: RGB and Depth). Ferns compute Local Binary Features (LBFs, named as f_i) from random positions and channels over the image space creating weak classifiers as presented at the bottom of the image.

2.2 *Boosted Random Ferns (BRFs)*

Random Ferns (RFs), are presented in [10] and use hundreds of simple binary comparisons as features for modelling a class posterior probability. Extensions on this work, have been done in [17], where a *pool* of shared RFs was created obtaining a faster and efficient method for detecting even multiple classes. On top of that, in [16], a boosting algorithm was used for training the classifier with the best samples over the image domain.

In our method, we keep the pool of shared ferns, but we distinct ferns depending on the input space, being able to work in different spaces simultaneously. Ferns for RGB access to three different channels of information (Figure 1 - left), whereas others exist in only one dimension (Figure 1 - right). The boosting algorithm decides which ferns and where should be used to build the strongest possible classifier, as explained in Section 3. With this strategy, we keep the simplicity, speed and efficiency of the algorithm allowing its real time execution.

3 Developed Approach

3.1 *Random Ferns on RGB-D Domains*

In contrast to the original formulation of the Random Ferns of [10], proposed for keypoint classification, we write the Ferns expression in terms of likelihood ratios between classes. The aim is then to find by means of the boosting algorithm the combinations of features and its positions that maximize this ratio.

A posterior object class probability given a set of n Local Binary Features (LBFs) can be expressed by means of the Bayes rule as,

$$P(C|f_1, f_2, ..f_n) = \frac{P(f_1, f_2, ..f_n|C)P(C)}{P(f_1, f_2, ..f_n)}, \tag{1}$$

where C refers to class *People* (C^p) or *Background* (C^b), and f_i are the LBFs computed by the Ferns as shown in Figure 1. Therefore, we aim to maximize the posterior probability ratio of class *People*, with respect to the *Background* class.

Prior probabilities, $P(f_1, f_2, ..f_n)$, are common for all the classes, so it can be removed. Moreover, assuming uniform prior probabilities for both classes, $P(C_p) = P(C_b)$, the posterior probability is written by the likelihood ratio as,

$$\log \frac{P(C^p|f_1, f_2, ..f_n)}{P(C^b|f_1, f_2, ..f_n)} = \log \frac{P(f_1, f_2, ..f_n|C^p)}{P(f_1, f_2, ..f_n|C^b)} \tag{2}$$

Computing the complete joint probability for a large feature set is not feasible. A solution is to split the previous equation into m subsets ($F_i = \{f_1, f_2, ..f_r\}$), with $r = n/m$. These feature subsets will be our Ferns, and assuming they are independent, their joint log-probability can be computed as,

$$\log \frac{\prod_{i=1}^{m} P(F_i|C^p, g_i, d_i)}{\prod_{i=1}^{m} P(F_i|C^b, g_i, d_i)} = \sum_{i=1}^{m} \log \frac{P(F_i|C^p, g_i, d_i)}{P(F_i|C^b, g_i, d_i)} , \tag{3}$$

where the parameter g_i ($g \in \mathbb{R}^2$) corresponds to the image coordinates location where the Fern F_i is evaluated, and d_i belongs to any of the image spaces evaluated (RGB, HOG, Depth or HOD, as will be explained in Section 4.1).

In this way, each Fern captures the co-occurrence of r binary features computed locally on the working spaces of the image, and therefore encodes people local features. Its response is represented by a combination of boolean outputs as seen at the bottom of Figure 1, where for instance the Fern F_1 (which applies to the RGB space - left of the Figure) is made of $r = 7$ features, results in 0, 1, 0, 1, 1, 0, 0, that outputs a value of $(0101100)_2 = 44$.

The Fern probability may then be written using the class conditional probability (for people, C^p and for background C^b), the Fern location g, the image space where the Fern works d, and the feature set of observations z_i as:

$$\sum_{i=1}^{m} \log \frac{P(F_i|C^p, g_i, d_i, z_i = k)}{P(F_i|C^b, g_i, z_i = k)} , \quad k = 1, 2, ...K , \tag{4}$$

with k, the observation index.

3.2 Combining RGB-D Spaces in a Single People Classifier

As have been seen, a weak classifier is created when any of the Shared RFs of our initial pool is taken and computed over its corresponding image domain at a certain position, having a score which corresponds to the Fern observation. Our aim is then to learn which are the most discriminative weak classifiers and build a single and robust classifier, which is done by means of boosting. Algorithm 1 summarizes the following described steps in order to build our final classifier.

More formally, we want to build a single people classifier $H(x)$, using the most discriminative Shared RFs F_i from our pool while maximizing Eq. 4. In this way, a Real Adaboost algorithm [12] is used for learning the best combination of these RFs over the image space locations g_i by iteratively assembling weak classifiers and adapting their weighting values.

The final people classifier is therefore defined as a sum of T weak classifiers,

$$E(x) = \sum_{t=1}^{T} h_t(x) > \beta_e , \tag{5}$$

where β_e is a threshold with a zero default value and $h_t(x)$ is the value of weak classifier, defined by

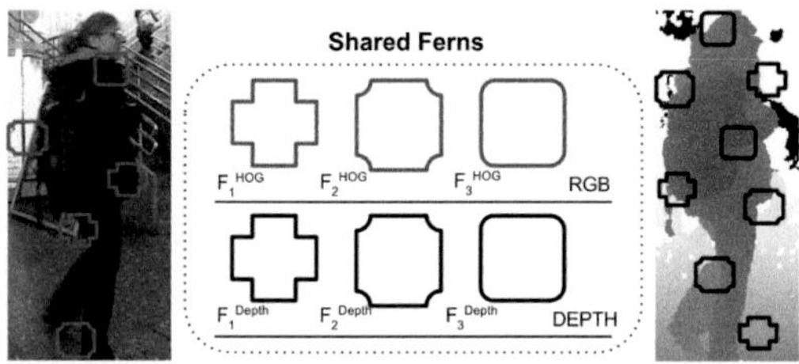

Fig. 2 Example of a pool of Shared Random Ferns (center) applied to an image of the database. In this case, three ferns are devoted to RGB space and other 3 to depth. The boosting algorithm applies the RFs over different positions generating weak classifiers, and choose the most discriminative ones. Here, five weak classifiers are chosen over the RGB space and eight over depth.

$$h_t(x) = \frac{1}{2} \log \frac{P(F_t|C^p, g_t, d_t, z_t = k) + \epsilon}{P(F_t|C^b, g_t, d_t, z_t = k) + \epsilon} , \quad k = 1, .., K , \tag{6}$$

being ϵ a smoothing factor.

Figure 2 shows a naive example of our people classifier, in which a pool of six shared random Ferns is created for a $d = 2$ image space, three RFs for the RGB space (d_1) and other three for Depth one (d_2).

At each iteration t of the boosting step, the classifier tries to find the most discriminative weak classifiers according to a sample weight distribution $D(i)$ by calling a weak learner. At iteration t, the probability $P(F_t|C^p, g_t, d_t, z_t)$ is computed under the $D(i)$ distributions as,

$$P(F_t|C^p, g_t, d_t, z_t = k) = \sum_{i:z_t(x_i)=k} D_t(i) , \quad k = 1, .., K \tag{7}$$

The classification power of each weak classifier is measured by means of the Bhattachryya distance between people and background distributions as,

$$Q_t = 2 \sum_{k=1}^{K} \sqrt{P(F_t|C^p, g_t, d_t, z_t = k) P(F_t|C^b, g_t, d_t, z_t = k)} \tag{8}$$

Figure 3 shows a density map of weak classifiers applied over the HOG and Depth image spaces, resulting after identifying in the boosting step, which areas of each space are the most discriminative in the training classifier set.

The final classifier will therefore be composed by the combination of the selected weak classifiers from the image spaces. Figure 4 shows an example of a resulting distribution from a pool of twelve shared RFs and 600 weak classifiers. In this example, around 350 weak classifiers are applied over the HOG space, whereas 250 are in the Depth space. Although the Depth space have less weak classifiers, its distribution is more focused on certain areas (as seen in Figure 3), so it still manages to have a high accuracy so that being able to correctly driving the detector even if the HOG space is corrupted at certain point.

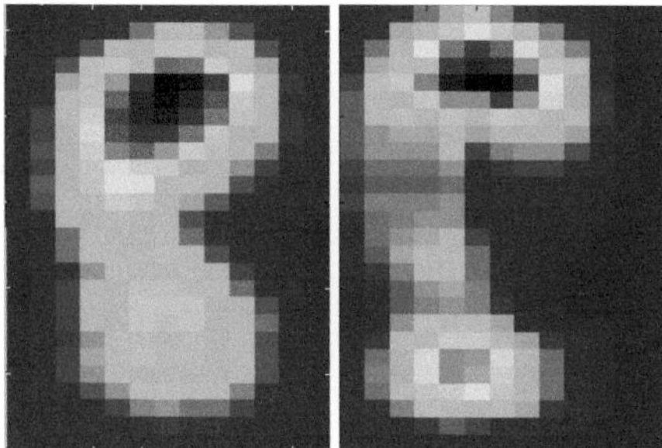

Fig. 3 Resulting weak classifiers density map over the HOG space (left) and the Depth space (right) of a real database image. The boosting algorithm find the best position to apply the Random Ferns from the pool which would produce the most discriminative results. The weak classifiers are mainly focused in the head area and uniformly distributed over the rest of the body for the HOG space whereas for the Depth space main discriminative areas are the head and feet zones. High density areas are represented in red colors.

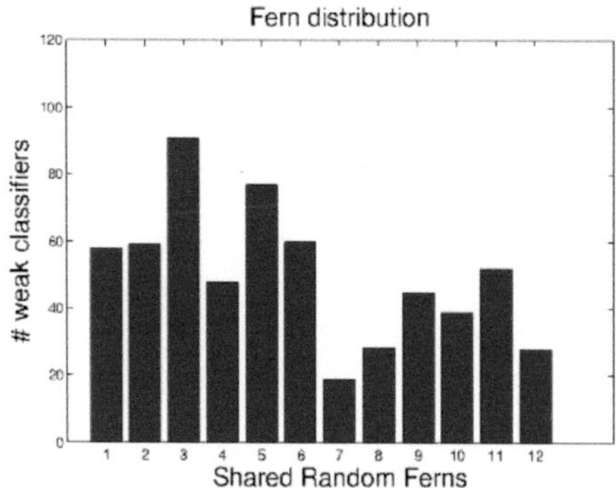

Fig. 4 Example of the resulting distribution of the most discriminative selected Ferns. In this case, a pool of 12 Shared Random Ferns where created, being the 6 first devoted to the HOG domain, whereas the rest to the Depth space. As it can be seen, the boosting algorithm has given more weight to the HOG image.

4 Experiments

In this Section, the experiments that validate our people detector will be presented. In order to show how our contributions outperforms other similar algorithms for people detection in RGB-D, we compare the results with the ones obtained in [13] using the same RGB-D database and under the same circumstances.

This public database for people detection [9] was collected indoor by three different *Kinect* sensors in the lobby of a university canteen. We have noticed some missing tracks of annotated people in the public website, and therefore no full groundtruth

Algorithm 1. Detector computation.

Given a number of weak classifiers T, and N RGB-D image samples labelled $(x_1, y_1)...(x_n, y_n)$, where $y_i \in \{+1, -1\}$ is the label for people category (C^p) and background classes (C^b), respectively:

Construct a shared feature pool ϑ consisting of M Random Ferns divided uniformly to be devoted to appearance or depth spaces $d_i, i \in 1, 2$.

Initialize sample weights $D_1(i) = \frac{1}{N}$.

for $t = 1$ to T **do**

  for $m = 1$ to M **do**

    Under current distribution D_t, calculate h_m and its Bhattachryya distance Q_m.

  end for

  Select the h_t that minimizes Q_m.

  Update sample weights.

  $D_{t+1}(i) = \frac{D_t(i)\exp[-y_i h_t(x_i)]}{\sum_{i=1}^{N} D_t(i)\exp[-y_i h_t(x_i)]}$

end for

Final strong classifier.

$H(x) = sign\left(\sum_{t=1}^{T} h_t(x) - \beta_e\right)$

information in available. This make the database a challenging one, but even under these circumstances our people detector outperforms previous cited works obtaining remarkable results of almost 89% of EER. In the fourth column of Figure 6, some of the missing annotations that our detector has positively find are shown.

The dataset in its actual version contains 3399 images (for each RGB and depth source) from which, up to 2351 are annotated providing 4498 2-D boxes of people. In total, 1748 images include full people, and from all the annotations, up to 2248 of the boxes corresponds to people that is fully visible, whereas the rest is considered as occluded.

In each implementation of our experiments, 1096 random crops from the 2248 of fully visible people annotations have been extracted to be used as positive training samples. The remaining 1152 annotated people on each implementation where used for testing purposes. Unlike [13] did using 5000 negative samples of background for training, we just also use 1096, in order to keep the equity between people and background classes on training.

Alike the authors did and aiming to reproduce the same circumstances of their experiments for a fair comparison, a *no-reward-no-penalty* policy was used as also introduced in [3]. Therefore, partially occluded annotated people do not count for true positives nor false positives. Also, in the same way a detection is considered as true positive if its intersections with the corresponding annotation is at least the 40%.

4.1 Approaches for Combining RGB-D Data

In our people detecton approach, we have mainly work with four different image spaces. Apart from standard RGB and Depth spaces, Histograms of Oriented Gradients (HOG) and Histograms of Oriented Depths (HOD) has been used for creating

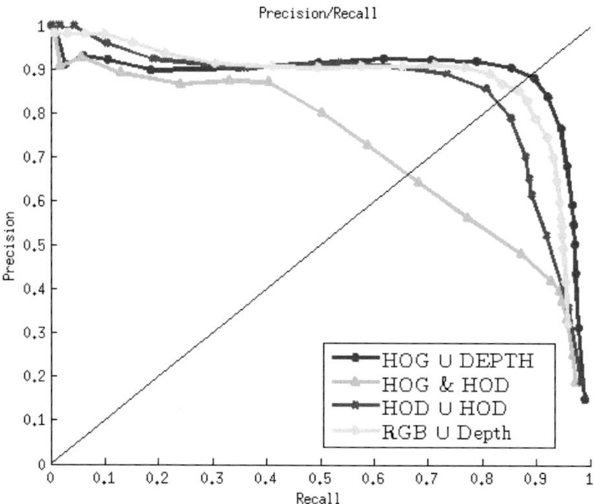

Fig. 5 Detection performance evaluation in terms of the Precision-Recall plots for different depth- and appearance-based feature configurations.

the combined classifier. HOG image domain was introduced in [1] and its features showed to obtain substantial gains over standard RGB features, even though it is calculated over the greyscale transformed image. As the Depth of a scene can be observed as a kind of a greyscale image, the same process of computing HOG can be extrapolated to it, obtaining in this way the areas where there exists variations in depth. In [13], authors called this technique Histogram of Oriented Depths (HOD), so in order to keep notation simply we have adopted the same name.

At this point, to validate our people detector method for combining Appearance and Depth information in any of the described spaces, we have implemented 4 different combination approaches. Each combination have been tested up to 10 times,

Table 1 Qualitative results on the performance of the different classifier combinations approaches tested. Columns present the Equal Error Value (EER), along with the recall and precision values achieved at this point. In addition the mean number of True Positives (tp), False Positives (fp) and False Negatives (fn) detections over the set of 1152 test images are included. However, some of these fp, are existing people correctly detected, but without background associated.

Method	EER Value	Recall	Precis	tp	fp	fn
HOG & HOD	0.6534	0.6516	0.6552	747.333	394.000	399.667
HOG ∪ HOD	0.8199	0.7995	0.8403	917.000	174.333	230.000
RGB ∪ DEPTH	0.8758	0.8767	0.873	1017.667	148.333	130.333
HOG ∪ DEPTH	0.8861	0.8881	0.8841	1018.667	133.667	128.333

Fig. 6 Sample images showing the output of the proposed approach for people detection in urban and crowded scenarios. Full people annotations (bounding boxes) are indicated by blue boxes in images, whereas magenta rectangles correspond to occluded people. Correctly detections (true positives) provided by our classifier are represented in green, whereas false positives are shown by red boxes. Columns 1 and 2 depicts some true positive detections, while column 3 shows some false positives. Column 4 shows some examples of not detected people, as well as a *false* false positive in the middle (a correct detection but not annotated in the dataset). Finally, column 5 presents some other *false* false positives, which really penalises our detector. However, despite the fact of the presence of this *false* false positives, our experiments on people detection combining appearance and depth images spaces achieves very good results around a 89% of EER, which surpass other methods.

calculating afterwards the mean of values. In each experiment, a shared pool of 12 RFs was created (6 for each of the 2 dimensions), with 9 LBFs each. These values were chosen after previous experiments showed to be the ones with best computational efficiency vs robustness ratio. Final qualitative results of the experiments are shown in Table 1, and Precision-Recall curves for the image spaces combinations can be consulted in Figure 5.

HOG ∪ Depth
This combination has obtained the best results for people detection over the database. The union of an already proved accurate image feature domain as HOG, along with the depth information of the scene, is very robust and even under the presence of *false* false positives due to the lack of annotations. In our experiments

it has achieved remarkable results outperforming other methods such as the one presented in [13], with an EER of almost 89%.

- HOG ∪ HOD
  In the same way as the authors of the used database did in [13], we have build a classifier based also in the union of both HOG and HOD image domains in order to do a full and fair comparison of the results. The results for this approach can been observed in magenta color at Figure 5, obtaining a EER of around the 82%, which is 7 points less than our best combination.
- HOG & HOD
  In this approach both the HOG and HOD information is fused, instead of combining it. For this, two options have been evaluated. In one hand we added the results of both histograms, which would favor the areas where both gradients exists. On the other hand the multiplication has been done, which would penalize more the areas where there are no gradients while at the same time would exalts the places where both domains have gradients. However, both of these approaches have thrown a really bad performance of around an 65% of EER when applied in the current database.
- RGB ∪ Depth
  A final combination of both RGB and depth raw information has been done. Results for this approach over the database can be observed in yellow in Figure 5. Quite good results are appreciated, of around an EER of 87%, and as here only raw data is used, less computational time is employed so better speeds of around 15 frames per second are obtained.

5 Conclusions

We have presented a robust, fast and accurate method for people detection in RGB-D data that obtains remarkable results compared to other classical people and pedestrian detectors. The presented approach is based on the combination of RGB and Depth image spaces, as well as its derived spaces (Histograms of Oriented Gradients and Depths) by means of extracting features using Shared Random Ferns. A learning approach making use of a boosting algorithm selects in the training phase the most discriminative weak classifiers between all the possible permutations of Random Ferns over the combined image domains. This combination of Random Ferns and boosting, allows to create a single classifier where its components act synergistically providing accurate detections even when one of the spaces are spoiled or distorted.

The proposed method has been validated in a recent and challenging RGB-D database of people with four different ways of combining the image spaces, and shows remarkable results of around an 89% of EER in spite of the difficult environment conditions.

Acknowledgements This work has been partially funded by the EU project CargoANTs FP7-SST-2013-605598 and the Spanish CICYT project DPI2013-42458-P.

References

1. Dalal, N., Triggs, B.: Histograms of oriented gradients for human detection. In: Proceedings - 2005 IEEE Computer Society Conference on Computer Vision and Pattern Recognition, CVPR 2005, vol. I, pp. 886–893 (2005)
2. Dollár, P., Wojek, C., Schiele, B., Perona, P.: Pedestrian detection: an evaluation of the state of the art. IEEE Transactions on Pattern Analysis and Machine Intelligence **34**(4), 743–761 (2012)
3. Enzweiler, M., Gavrila, D.M.: Monocular pedestrian detection: Survey and experiments. IEEE Transactions on Pattern Analysis and Machine Intelligence **31**, 2179–2195 (2009)
4. Enzweiler, M., Gavrila, D.M.: A multilevel mixture-of-experts framework for pedestrian classification. IEEE Transactions on Image Processing **20**, 2967–2979 (2011)
5. Gandhi, T., Trivedi, M.M.: Pedestrian protection systems: Issues, survey, and challenges. IEEE Transactions on Intelligent Transportation Systems **8**, 413–430 (2007)
6. Gerónimo, D., López, A.M., Sappa, A.D., Graf, T.: Survey of pedestrian detection for advanced driver assistance systems. IEEE Transactions on Pattern Analysis and Machine Intelligence **32**, 1239–1258 (2010)
7. Guo, L., Wang, R.B., Jin, L.S., Li, L.H., Yang, L.: Algorithm study for pedestrian detection based on monocular vision. In: 2006 IEEE International Conference on Vehicular Electronics and Safety, ICVES, pp. 83–87 (2006)
8. Luber, M., Spinello, L., Arras, K.O.: People tracking in RGB-D data with on-line boosted target models. In: IEEE International Conference on Intelligent Robots and Systems, pp. 3844–3849 (2011)
9. Luber, M., Spinello, L., Arras, K.O.: RGB-D People Dataset. annotated people and tracks in rgb-d kinect data (2011). http://www2.informatik.uni-freiburg.de/spinello/RGBD-dataset.html (accessed September 30, 2014)
10. Özuysal, M., Fua, P., Lepetit, V.: Fast keypoint recognition in ten lines of code. In: IEEE Conference on Computer Vision and Pattern Recognition (2007)
11. Porikli, F., Davis, L., Hussein, M.: A Comprehensive Evaluation Framework and a Comparative Study for Human Detectors. IEEE Transactions on Intelligent Transportation Systems **10**, 417–427 (2009)
12. Schapire, R.E., Singer, Y.: Improved boosting algorithms using confidence-rated predictions. Machine Learning **37**, 297–336 (1999)
13. Spinello, L., Arras, K.O.: People detection in RGB-D data. In: IEEE International Conference on Intelligent Robots and Systems, pp. 3838–3843 (2011)
14. Spinello, L., Arras, K.O.: Leveraging RGB-D data: adaptive fusion and domain adaptation for object detection. In: IEEE International Conference on Robotics & Automation, pp. 4469–4474, May 2012
15. Villamizar, M., Andrade-Cetto, J., Sanfeliu, A., Moreno-Noguer, F.: Bootstrapping Boosted Random Ferns for discriminative and efficient object classification. Pattern Recognition **45**(9), 3141–3153 (2012)
16. Villamizar, M., Moreno-Noguer, F., Andrade-Cetto, J., Sanfeliu, A.: Efficient rotation invariant object detection using boosted random ferns. In: Proceedings of the IEEE Computer Society Conference on Computer Vision and Pattern Recognition, pp. 1038–1045 (2010)
17. Villamizar, M., Moreno-Noguer, F., Andrade-Cetto, J., Sanfeliu, A.: Shared random ferns for efficient detection of multiple categories. In: International Conference on Pattern Recognition (ICPR), pp. 2–5 (2010)

Visual Localization Based on Quadtrees

Francisco Martín

Abstract Autonomous mobile robots moving through their environment to perform the tasks for which they were programmed. The robot proper operation largely depends on the quality of the self localization information used when globally navigating in its environment. This paper describes a method of maintaining a self-location probability distribution of a set of states, which represents the robot position. The novel feature of this approach is to represent the state space as a Quadtree that dynamically evolves to use the minimum set of statements without loss of accuracy. We demonstrate the benefits of this approach in localizing a robot in the RoboCup SPL environment using the information provided by its camera.

Keywords RoboCup · Localization · Computer vision

1 Introduction

Robot self-localization is a classic problem in Robotics, which has been addressed since the first mobile robot wanted to navigate somewhere. It is necessary to know what is the starting pose to calculate a route, or evaluate the progress of this route. Since these early works to our times, the state of the art of robot self-localization has advanced a lot. Although many roboticist believe that this problem is solved, there is still much work to be done in this field. An evidence of this is the intense work done through competitions like Robocup@home, or RoboCup Rockin [1] where self-localization is considered a challenge that not all teams are able to overcome robustly and efficiently.

There are several applications where an autonomous robot needs to know its position. The Grand Challenge is a competition in which autonomous vehicles have to

F. Martín(✉)
Robotics Group, Rey Juan Carlos University, C/ Tulipán s/n, 28933 Móstoles, Madrid, Spain
e-mail: francisco.rico@urjc.es

L.P. Reis et al. (eds.), *Robot 2015: Second Iberian Robotics Conference*,
Advances in Intelligent Systems and Computing 418,
DOI: 10.1007/978-3-319-27149-1_46

cross the desert. To achieve this goal, they need to know their position at all times to find their way to their destination. This competition addressed as a challenge the problems of developing autonomous vehicles. In this case, the calculation of the position is solved by GPS devices present int these vehicles. This sensor solves this problem in part, because if the requirements are accuracy, fast movements with a high refresh rate, it is necessary to combine GPS with other methods. In indoor environments, where GPS can not be used, and there are other alternatives. Applications such as a robotic messenger environments offices, a museum guide, or guard of buildings, the robot has to use other methods that integrate local information from sensors (laser, go, sonar, cameras, ..) with information a priori (usually a map) to know its position and reach the destination.

Probabilistic approaches are far the most used. It seems a fact that the most effective way to model the uncertainty in the real world is using probabilistic models. You can statistically analyze the accuracy of a sensor and its error, the possibility of detecting an obstacle or even the probability that if we move a robot one meter, it really does. Applied to self-localization, we denominate *state* a probable position of the robot in its environment, which would usually associate other features like uncertainty of this state, or quality indicators. A self-localization method can be analyzed taking into account these aspects.

- **Number of states.** The Gaussian, like Kalman Filters [3], methods define the position as a single state and its associated uncertainty. The status is updated with the odometry of the robot and sensory information. Topological [8] and grid based [4] methods typically have a set of states whose position is fixed, and which is updated is the probability of each. Sampled methods, Monte Carlo Localization (MCL) [2], define a set of states sampled from a probability distribution. After updating each sample position and probability, a new probability distribution is recalculated. In grid based methods, the accuracy and efficiency largely depends on the number of states. The more states we have, the more computation is required to update each of the states. Also, the more states we have, a greater accuracy in determining the actual position of the robot. This commitment is critical, and it may cause you cannot apply this method in very large environments or requiring great precision.
- **Global vs local localization.** Some approaches require to initialize the actual state of the robot and, thereafter, performing a robot tracking. Global methods are able to correctly estimate the state of the robot from full uncertainty, or even erroneous estimation. Gaussian methods generally fall within the methods of tracking, while those based on grid or sampled are global methods.
- **Multi-hypothesis.** The multi-hypothesis approaches are those that are able to maintain multiple hypotheses about the position of the robot. This property is often useful when applied to symmetric environments or similar perceptions in different positions. Topological methods or grid are multi-hypothesis.

Currently, sampled approaches are considered the state of the art of self-localization in indoor environments. They are considered the most efficient and flexible. Grid approaches are typically not used by the large number of states required for acceptable accuracy. Therefore, we believe that the latter approaches provide a number of benefits that justify its use. For this reason, we have developed a method that uses a grid approach, but whose cells do not have a fixed size and provide a set of fixed orientations. This size varies, requiring precision only in those areas where it is most likely that the robot is, while large areas with low probability are covered with few states. Thus, we get the benefits of grid methods (multi-hypothesis, global positioning), while we limit the number of states, making possible to use it in large environments or in applications requiring high accuracy. A key element in our approach is to use a Quadtree to represent the states of the robot, and define a split / merge dynamically applied based on the probability of the leaves of this tree.

This paper is structured as follows. After discussing related work in the following section, we will present our Quadtree-based approach to self-localization in Section 3. In Section 4 we will present experimental results that demonstrates the advantages of our algorithm.

2 Related Work

Markov methods have traditionally been successfully used in self-localization of mobile robots. These methods are based on discretizing the environment in a set of states where the robot can find, and keep the probability of being in each of these states. States may have a regular size, building a grid [4], or have topological meaning [8][9]. Typically one location contains several states to represent the orientation of the robot. This method is simple and very flexible to the requirements of many applications, integrating robot self-localization and actuation using partially observable markov decision process models (POMDPS) [10]. This is a global method, so it can start from complete unknown. It also can maintain several hypothesis about the robot position. However, the major problem of this approach arises when the set of states is high because the environment is too large or because the application requires high precision. some researchers have developed methods to overcome this performance problem trying to reduce the space of states. In the environment of the robot soccer, in [11] is proposed that the orientation is not encoded as a set of states, but as a fuzzy variable, reducing computing needs. Although these requirements are reduced, it still remains a compromise between performance and accuracy. Our approach overcomes this problem by maintaining a grid of cells of different sizes with different orientations. This allows unlikely areas are represented with few cells, while the most likely represent areas with smaller cells. In this way, the cell number is drastically lower, allowing its use in large environments or in applications requiring high accuracy.

More recent work in self-localization of mobile robots are based on particle filters [12], also called Monte Carlo methods [13][14][15]. Particle filters metric determine the robot location by sampling the state space where the robot can be. Each particle represents a possible robot location with an associated weight. The weight of each particle varies depending on the robot's sensory information. Those particles with low weights are replaced by others close to those with more weight. The accumulation of particles in a certain location determines where the robot is. This approach is attractive because it does not depends on the the size of the environment and it is able to recover from initial unknowledge or kidnappings, and therefore it is used for many applications [16][17][18]. In particular, some works in soccer robotic bring new features to this method. In [19], it takes advantage of the existing mark locations in the environment to better suit kidnappings and restarts. Lines are also used in [20], in which this algorithm is tuned for an excellent solution, highly adapted to this environment. This method, however, has some problems when there are symmetries in the environment can often decant for the wrong choice prematurely. Furthermore, although able to recover from kidnappings, the dynamics of the particles can make recovery take longer than desired. In contrast, our approach has the advantages of grid-based approaches, which can maintain multiple hypotheses. Recoveries are much faster, especially in larger environments.

Extended kalman filter is a widely used method of estimating the state of a process. The system state is represented as a Gaussian distribution, with an associated uncertainty. This system has been applied in the location of a mobile robot in several works [6][7]. All of them seeks to address the problem of initialization from the total ignorance, but the solutions are not definitive. The biggest problem with this method is that it is a local localization method, also called tracking methods. Once initialized Kalman filter is very difficult to recover from erroneous estimates. Furthermore, in symmetric environments suffer the consequences of not being a multi-hypotheses method. There are previous works using populations extended Kalman filter to solve these problems through dynamic creation / destruction of estimates and techniques for evaluating combined estimates.

3 Q-Tree Based Localization

The main feature of the proposed work is the representation of the position of the robot in its environment through a Quadtree [23]. The environment is represented using this structure. Each region of the 2D space is represented using a leaf. This leaf contains a set of states that are the different orientations of the robot. Initially, the environment is represented as a large region. At this time, this region can be divided into four regions. Each of these regions can be further divided in 4 other regions, and thereby recursively, as shown in Figure 1. In this section we describe how we used this structure to represent the position and orientation of the robot, how to update the probability of each of the states and how the structure of Quadtree is modified over time.

3.1 Quadtree to Represent the States Space

The state of the robot is defined as $s_i = (x, y, \theta)$. Each state s_i has associated a probability, that means the probability of a robot of being in that position. Each leaf

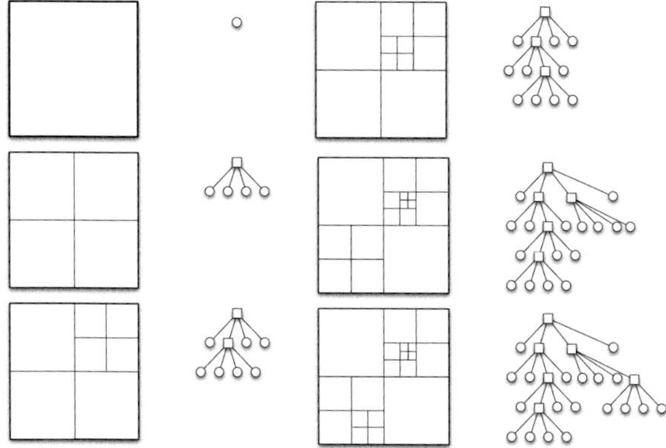

Fig. 1 Recursive division of the space in regions, and the Quadtree that represents this division. In the Quadtree, nodes are represented by squares and leafs are represented by circles.

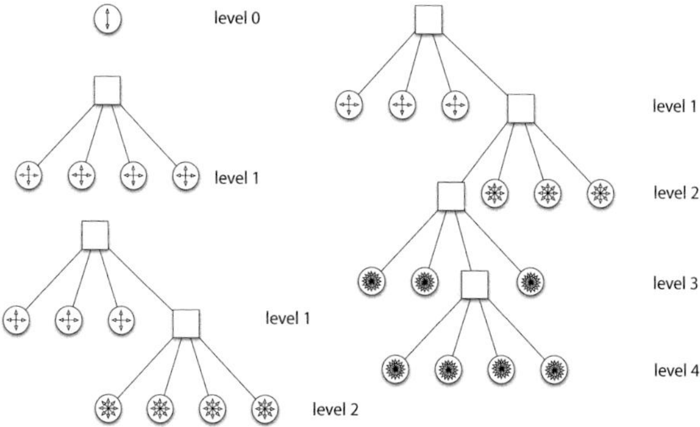

Fig. 2 Number of orientations depending on the level.

of the tree contains several states, corresponding to the same position (x, y), with different orientations θ. In our approach, the number n of different states on the same leaf depends on its depth, being $n = (level + 1)^2$, as shown in Figure 2. We limit the number of states per leaf to 16 having the same states depth levels above 3. We believe that this angular resolution is enough for most navigation applications.

It's simple to perform certain recursive operations on any node or leaves of the tree.

```
int getSum(Node *node)
{
   return getSum(lnode->child[0]) + getSum(lnode->child[1])
   +  getSum(lnode->child[2]) + getSum(lnode->child[3]);
}

int getSum(Leaf *leaf)
{
```

```
    float ret=0.0;

    for(state s in leaf)
     ret = ret + probability(s);

    return ret;
}

bool isFinal(Node *node)
{
    return isLeaf(node->child[0]) and isLeaf(node->child[1])
    and isLeaf(node->child[2]) and isLeaf(node->child[3]);
}
```

3.2 Updating the Probabilities Using Observations

During operation of the robot, it perceives observations $o_t = (\rho, \theta)$, as shown in Figure 3. Besides of this, this observation has an associated uncertainty $(\sigma_\rho, \sigma_\theta)$ which depends on sensor accuracy.

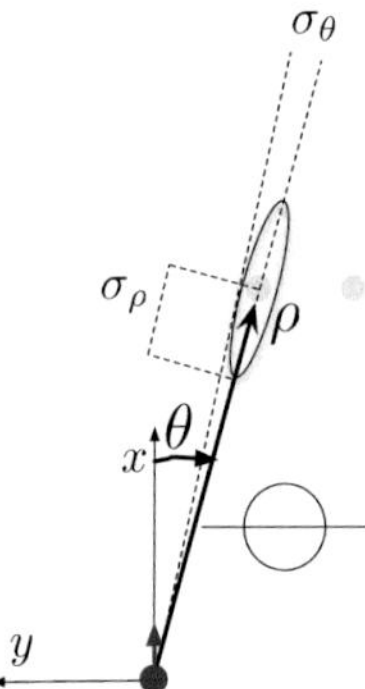

Fig. 3 Observation model based on $o_t = (\rho, \theta)$ and the uncertainty $(\sigma_\rho, \sigma_\theta)$.

$$p(s_i|o_t) = \frac{p(o_t|s_i) * (p(s_i)}{p(o_t)} \tag{1}$$

$$p(o_t|s_i) = N(\rho, \sigma_\rho + \frac{size}{2}) * N(\theta, \sigma_\theta + \frac{2\pi}{resolution}) \tag{2}$$

where $resolution$ depends on the number n of states in the same leaf, being $\frac{2\pi}{n}$, and $size$ is the half of the size of the region containing the evaluated state s_i.

3.3 Quadtree Dynamic

The Quadtree varies in a split and merge dynamics. When the sum of the probabilities of the states of a leaf exceeds a threshold, this sheet becomes a tree node with 4 new leaves resulting from the previous leaf. The condition of this split is that the probability of the sum of the states of the resulting leaf must be equal to the sum of the states of the original leaf. If the number of states of the new leaves is similar to the original leaf, a copy of the original leaf is performed, dividing the probability by 4. If the number of states of new leaves is higher than the original leaf, a new distribution is made from the original probabilities.

When the sum of the probabilities of the states of all the leaves of a node is less than a threshold, this all becomes a single leaf from the union of the original leaves. Again, the condition for this union is that the probability of all states in the resulting sheet must be equal to the sum of the original sheets.

3.4 Algorithm Description

The self-localization method is shown in Algorithm 1. Once thresholds to split and merge are set, for each observation o_t, all states update its probability. Once performed a normalization so that $\sum(s_i) = 1$, leaves with high probability states are split, and leaves with low probability states are merged.

Algorithm 1. Quadtree based approach algorithm

```
Initialize Qtree with the initial leaf;
Merge_threshold = 0.00001 ;
Split_threshold = 0.01 ;
while true do
    wait for a new observation o_t
    foreach leaf_n do
        foreach s_i in leaf_n do
            Update p(s_i) = p(s_i|o_t);
        end
    end
    normalize();
    foreach node_i do
        if isFinal(node_i) and sum(node_i) < Merge_threshold then
            merge(node_i);
        end
    end
    foreach leaf_i do
        if (sum(leaf_i))> Split_threshold then
            split(leaf_i)
        end
    end
end
```

You may have noticed that our algorithm does not use the robot motion information to update the states of the tree. We are still working on a method that allows us to get the neighbors of each region. It is complex, especially when neighbors can be any size.

4 Experiments

To validate the algorithm presented in this paper, we developed an implementation in the environment of the RoboCup SPL. This environment is composed of a field of 7x5 meters, with a set of visually detectable elements, whose position is known a priori. To measure the error in this experiment we have a ground truth system. This system is composed by two zenithal cameras that detect a pattern placed on the robot.

The experiment consists on moving the robot in a straight line from one goal to the other, doing a 180 turn and return to midfield.

In this experiment we compare the results of our algorithms implementations of a grid-based methodand a sampled method based on Monte Carlo (Figure 4). To compare the results of these algorithms, we used a system composed of two ground truth zenithal cameras that detect a pattern positioned above the robot. The variables to be measured are the errors in the estimation of the position and the number of states required to determine the position of the robot. To consider the self-location information useful spatial accuracy should be less than 20 cm, and the angular resolution of 20. For the MCL method we use a set of 100 particles. None of these algorithms performs prediction phase to incorporate the robot motion. This allows us to compare the algorithms under equal conditions.

Figure 5 shows how the regions dynamically evolve to represent the robot position. The position of the red arrows correspond to the center of the regions (a leaf in the Quadtree), and its orientation correspond to the most probable state in that region.

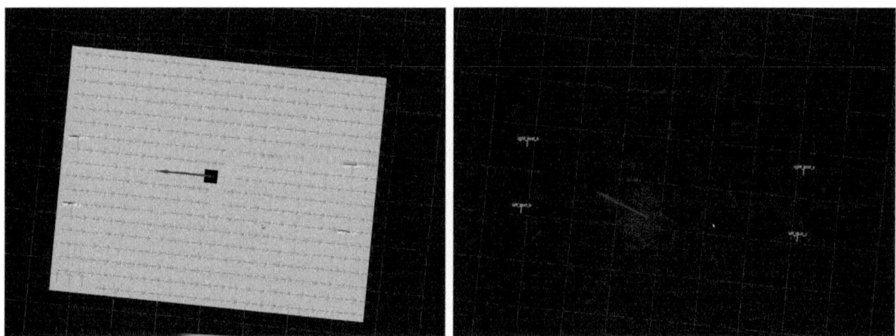

Fig. 4 Grid based self-localization approach (left) and Monte Carlo Approach (right).

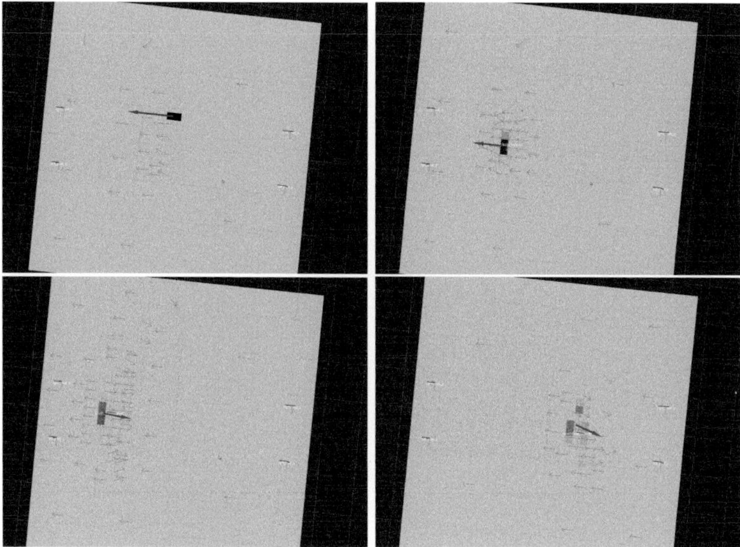

Fig. 5 Quadtree based approach.

The result of the error in position and orientation (Figure 6) show that this our approach maintains a good estimation in both cases, while requiring less states than the grid based approach, as shown in Figure 7. During this experiment we noticed that the error was high when splitting a region was needed. Sometimes it tooks more time than expected. We believe that incorporating the odometry information to our method would fix this problem.

5 Future Work and Conclusions

The main weakness of our system is that we have not provided the Quadtree leaves a simple way to know all their neighbors leaves. This is necessary to properly incorporate the odometry information. Even in the absence of information on landmarks, the probability should *move* to other states when the robot moves. In addition to completing this phase, another future work is to test this algorithm in large environments. We believe that this approach has much to contribute to these environments.

The main contribution of this work is that we have achieved to successfully model the set of states as a Quadtree. This allows us to use Markov methods without using a regular grid, avoiding the problem of having an intractable number of states. The Markov methods provide global positioning multi-hypothesis and a quick recovery from errors or kidnappings.

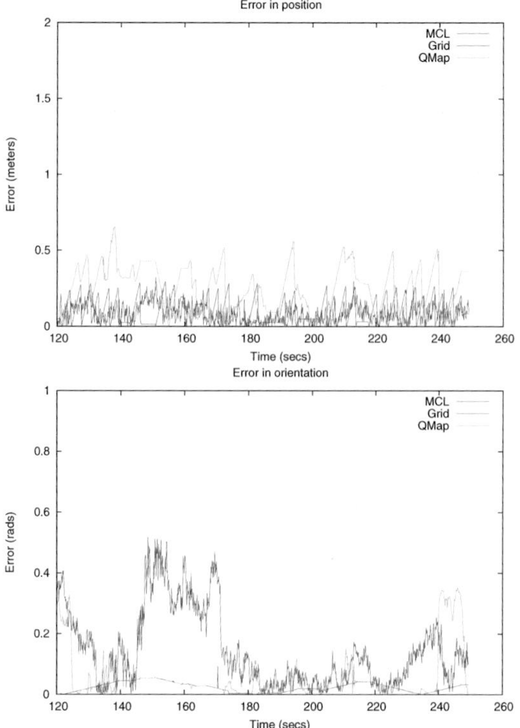

Fig. 6 Error in estimating (x, y) position (left). Error in estimating orientation(right)

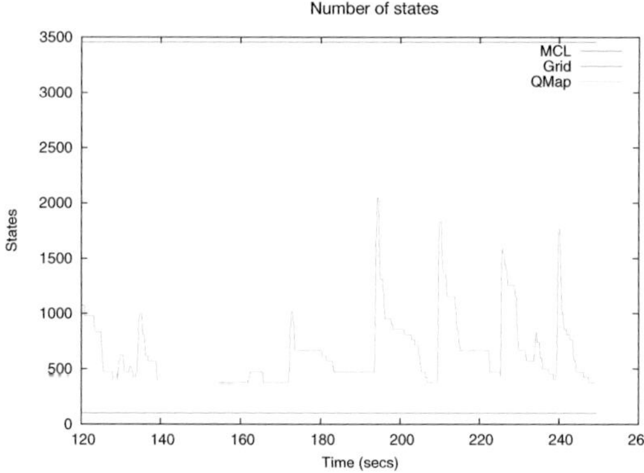

Fig. 7 Evolution of number of states.

We have implemented this method in a humanoid robot Nao, in the environment of the RoboCup SPL, appearing as effective as the reference methods in the state of the art. Regarding to the grid, the number of states is much less, consuming less computation. In relation to Monte carlo, recovery is faster, and have multi-hypothesis.

Acknowledgements This research has been sponsored by grant No. DPI2013-40534-R by Spanish Ministry of Economy and Competitiveness corresponding to SIRMAVED project[24].

References

1. Kitano, H.: RoboCup-97: robot soccer world cup I. In: LNCS, vol. 1395. Springer (1998)
2. Dellaert, F., Fox, D., Burgard, W., Thrun, S.: Monte carlo localization for mobile robots. In: IEEE International Conference on Robotics and Automation (ICRA 1999), pp. 1322–1328 (1999)
3. Kalman, R.E.: A New Approach to Linear Filtering and Prediction Problems. Transactions of the ASME, Journal of Basic Engineering **82**(Series D), 34–45 (1960)
4. Fox, D., Burgard, W., Thrun, S.: Markov Localization for Mobile Robots in Dynamic Environments. Journal of Artificial Intelligence Research **11**, 391–427 (1999)
5. Aguero, C., Cañas, J.M., Martín, F., Perdices, E.: Behavior-based Iterative component architecture for soccer applications with the nao humanoid. In: Proceedings of the 5th Workshop on Humanoid Soccer Robots, inside 10th IEEE-RAS Int. Conf. Humanoid Robots. Nashville, TN (2010)
6. Kyriy, E., Buehler, M.: Three-state Extended Kalman Filter for Mobile Robot Localization, Carnegie Mellon Technical Report, Pittsburgh, PA (2002)
7. Lastra, R., Vallejos, P., Ruiz-del-Solar, J.: Self-Localization and ball tracking for the robocup 4-legged league. In: Proceeding of the 2nd IEEE Latin American Robotics Symposium LARS 2005, Santiago de Chile, Chile (2005)
8. Thrun, S., Bücken, A.: Integrating grid-based and topological maps for mobile robot navigation. In: Proceedings of the AAAI Thirteenth National Conference on Artificial Intelligence, vol. 2, Portland, OG, pp. 944–950 (1996)
9. Martín, F., Matellán, V., Cañas, J.M., Barrera, P.: Visual based localization for a legged robot. In: LNCS (LNAI), vol. 4020, pp. 708–715 (2006)
10. Koenig, S., Simmons, R.: Xavier: A Robot Navigation Architecture Based on Partially Observable Markov Decision Process Models. Artificial Intelligence Based Mobile Robotics: Case Studies of Successful Robot Systems, 91–122 (1998) (MIT Press)
11. Buschka, P., Saffiotti, A., Wasik, Z.: Fuzzy landmark-based localization for a legged robot. In: Proceedings of the International Conference on Intelligent Robots and Systems 2000, Takamatsu, Japan, pp. 1205–1210 (2000)
12. Thrun, S.: Particle filters in robotics. In: Proceedings of the 18th Annual Conference on Uncertainty in Artificial Intelligence (UAI 2002), San Francisco, CA, pp. 511–518 (2002)
13. Fox, D., Burgard, W., Dellaert, F., Thrun, S.: Monte carlo localization: efficient position estimation for mobile robots. In: Proceedings of the Sixteenth National Conference on Artificial Intelligence (AAAI 1999), Orlando, FL, pp. 343–349 (1999)
14. Burchardt, A., Laue, T., Röfer, T.: Optimizing particle filter parameters for self-localization. In: RoboCup 2010: Robot Soccer World Cup XIV. LNCS, vol. 6556, pp. 145–156 (2010)
15. Hornung, A., et al.: Monte Carlo Localization for Humanoid Robot Navigation in Complex Indoor Environments Int. J. Human. Robot. **11**, 1441002 (2014)

16. Thrun, S., Beetz, M., Bennewitz, M., Burgard, W., Cremers, A.B., Dellaert, F., Fox, D., Hähnel, D., Rosenberg, C., Roy, N., Schultea, J., Schulz, D.: Probabilistic Algorithms and the Interactive Museum Tour-Guide Robot Minerva. International Journal of Robotics Research **19**(11), 972–999 (2000)
17. Conde, R., Ollero, A., Cobano, J.A.: Method based on a particle filter for UAV trajectory prediction under uncertainties. In: 40th International Symposium of Robotics, Barcelona, Spain (2009)
18. Ko, N.Y., Kim, T.G., Noh, S.W.: Monte carlo localization of underwater robot using internal and external information. In: 2011 IEEE Asia-Pacific Services Computing Conference, APSCC 2011, Jeju, South Korea, pp. 410–415 (2011)
19. Lenser, S., Veloso, M.: Sensor resetting localization for poorly modelled mobile robots. In: Proceedings of ICRA-2000, the International Conference on Robotics and Automation, San Francisco, CA, pp. 1225–1232 (2000)
20. Röfer, T., Laue, T., Thomas, D.: Particle-filter-based self-localization using landmarks and directed lines. In: RoboCup 2005: Robot Soccer World Cup IX. LNAI, vol. 4020, pp. 608–615. Springer (2006)
21. Martín, F., Matellán, V., Barrera, P., Cañas, J.M.: Localization of legged robots combining a fuzzy-Markov method and a population of extended Kalman filters. Robotics and Autonomous Systems **55**, 870–880 (2007)
22. Daronkolaei, A.G., Shiry, S., Menhaja, M.B.: Multiple target tracking for mobile robots using the JPDAF algorithm. In: Proceedings of the 19th IEEE International Conference on Tools with Artificial Intelligence, vol. 01, Washington, DC, pp. 137–145 (2007)
23. Finkel, R., Bentley, J.L.: Quad Trees: A Data Structure for Retrieval on Composite Keys. Acta Informatica **4**(1), 1–9 (1974)
24. Cazorla, M., García-Rodríguez, J., Cañas Plaza, J.M., García Varea, I., Matellán, V., Martín Rico, F., Martínez-Gómez, J., Rodríguez Lera, F.J., Suarez Mejias, C., Martínez Sahuquillo, M.E.: SIRVAMED: development of a comprehensive robotic system for monitoring and interaction for people with acquired brain damage and dependent people. In: XVI Conferencia de la Asociación Española para la Inteligencia Artificial (CAEPIA) (2015)

A Simple, Efficient, and Scalable Behavior-Based Architecture for Robotic Applications

Francisco Martín, Carlos E. Aguero and José M. Cañas

Abstract In the robotics field, behavior-based architectures are software systems that define how complex robot behaviors are decomposed into single units, how they access sensors and motors, and the mechanisms for communication, monitoring, and setup. This paper describes the main ideas of a simple, efficient, and scalable software architecture for robotic applications. Using a convenient design of the basic building blocks and their interaction, developers can face complex applications without any limitations. This architecture has proven to be convenient for different applications like robot soccer and therapy for Alzheimer patients.

Keywords Behavior architectures · Autonomous humanoid robots · Real-time processing

1 Introduction

In recent years, many developments have been made in mobile robotics. Robots are equipped with more complex actuators (even for diving or flying), richer sensors (as cameras RGB-D), batteries that increment robot autonomy, and more powerful processors that let the robot process huge data onboard. Moreover, many commercial robotic platforms are available now at a low cost. This enables us to focus on the development of software without addressing the development of a complete robot from scratch. Despite this advantage, the development of software for robots is a complex task.

F. Martín(✉) · J.M. Cañas
Robotics Group, Rey Juan Carlos University, C/ Tulipán s/n, 28933 Móstoles, Madrid, Spain
e-mail: {francisco.rico,josemaria.plaza}@urjc.es

C.E. Aguero
Open Source Robotic Foundation, 419 N Shoreline Blvd, Mountain View, CA 94043, USA
e-mail: caguero@osrfoundation.org

L.P. Reis et al. (eds.), *Robot 2015: Second Iberian Robotics Conference*,
Advances in Intelligent Systems and Computing 418,
DOI: 10.1007/978-3-319-27149-1_47

Software for robots defines robot operation. Roodney Brooks [1] presented the basis of behavior-based robotics. This paradigm describes how the complex behaviors are built up to decompose into simpler behavior modules, which can be organized as modules. How these simple modules work, how they organize, and how they interact among them are open questions that continue to receive attention from the robotics community. The active ROS [11] community and its impact on the industry denotes this importance during the last few years.

We have developed a behavior-based architecture, which includes simple and novel ideas for effective development of robotic applications. We have defined a basic building block and an effective and simple mechanism for making them cooperate. Each behavior is implemented as a basic building block that decomposes its complexity, explicitly executing another building block. We will describe mechanisms to warranty that the information produced by perceptual building blocks does not expire before being used, avoiding race conditions using a single thread scheme, and recovering from high-load situation using a graceful degradation approach. These concepts are related to the scheduling of real-time systems, which is a very convenient approach when developing software for a robot that interacts with the real world.

An important part of this paper is the focus on a complete, deep technical description. We think that this approach is valid for making this work really useful, letting robotic software developers include some of these ideas in their software architectures.

We develop complex mechanisms using the following building blocks: visual attention, perception, or debugging mechanisms. These capabilities can be considered basic to the software for robots, but they are crucial during the development of high-level behaviors. An effective, clean, and scalable design lets developers face complex behaviors without any limitations. Monitoring and debugging tools are essential when developing robotic software, in which the usual techniques from computers (messages to stdout or debuggers) are not effective on robots.

An important aspect of this work is that all these ideas have been implemented in a real robot, and used to implement real applications. Because of the experience and feedback along these years, some ideas have been incorporated, redefined, or discarded. Nowadays, we can affirm that this architecture is mature and effective. Using this architecture, we have developed a complete software system for a team of robots that participates in the Standard Platform League of the RoboCup. This competition presents a dynamic scenario, a soccer match, where we have implemented fast response behaviors [2], self-localization [3], navigation, coordination, attention, and perception [4] algorithms using this architecture. We also used this architecture to develop a complete therapy system [5] for Alzheimer patients by using robots as a cognitive activator actor.

In section 2 we will present the existing works on software architectures. Section 3 describes the core ideas of the architecture presented in this work. In section 4 we extend these general principles with some optional communication or debugging mechanism. Section 5 presents mechanisms, such as perception and visual attention, particular to robotic humanoids equipped with cameras. A brief description of two successful uses of this architecture is presented in section 6.

2 Related Work

Robotic frameworks can be grouped into two main paradigms: those tightly coupled with a cognitive model in their designs and those designed just from pure engineering criteria. The former *forces* the user to follow a set of rules to program certain robotic behaviors, while the latter are just a collection of tools that can flexibly be put together in several ways to accomplish the task.

Cognitive robotic frameworks were popular in the 1990s and strongly influenced by the AI, where planning was one of the main keys. Indeed, one of the strengths of such frameworks was their planning modules built around a sensed reality. A good example of cognitive frameworks was Saphira [6], based on a behavior-based cognitive model. Some of its low-level functionality was rewritten as a C++ library called ARIA [7] that it is still supplied with the popular robotic platforms from MobileRobots/ActivMedia. Even though the underlying cognitive model is usually a good practice guide for programming robots, this hardwired coupling often leads the user to problems that are difficult to solve while trying to do something the framework is not designed to do.

Current robotic frameworks focus their design on the requirements that robotics applications need and let the user (the programmer) choose the organization that better fits with the specific application. The main requirements driving the designs are: multitasking, distribution, ease of use, and code reusability. Another requirement, that we believe is the main key, is the open source code, which creates a synergy between the user and the developer.

The key achievements of modern frameworks are the hardware abstraction, hiding the complexity of accessing heterogeneous hardware (sensors and actuators) under standard interfaces, the distributed capabilities that allow running complex systems spread over a network of computers, the multiplatform and multilanguage capabilities that enable the user to run the software in multiple architectures, and the existence of big communities of software that share codes and ideas.

As mentioned before, we believe that open source plays a major role in the development of modern robotic frameworks. Proof of this is the two most popular robotic frameworks: Player/Stage [8, 9, 10], which has been the *standard de facto* in the last decade and ROS [11], which is taking its place currently. As seen in other major software projects as GNU/Linux kernel or the Apache web server, to name but a few, the creation of communities that interact and share codes and ideas could be greatly beneficial to the robotic community. The main examples of open source modern frameworks are the aforementioned Player/Stage and ROS. Another important example is ORCA [12, 13]. In this work, we describe them briefly.

There are other open source frameworks that have had some impact on the current state of the art, such as RoboComp [14] by Universidad de Extremadura,

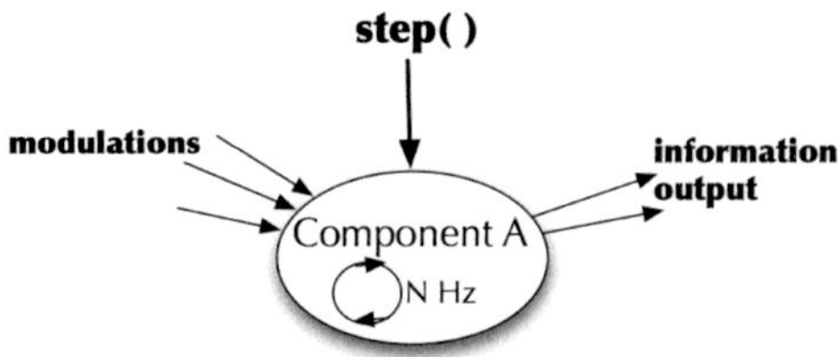

Fig. 1 BICA component.

CARMEN [15] by Carnegie Mellon, and Miro [16] by University of Ulm. All these use some component-based approach to organize robotic software using ICE, IPC, and CORBA, respectively, to communicate their modules.

We can find non-open-source solutions as well, such as Microsoft Robotics Studio or ERSP by Evolution Robotics.

3 Basic Principles

The basic building block in BICA is the **component** (Figure 1), which is the basic unit of functionality. The main idea is to build a component that does only one thing, but efficiently. A component is composed of three main parts:

- **Modulations:** The modulation methods set operation modes or set up the next component iterations.
- **Execution:** All the components inherit from the virtual class `component`, which defines the mandatory methods to be implemented. The most important method is `step()`. This method performs an iteration of this component. This is the entry point for a **component-explicit execution**.
- **Output:** The results method is used to get the information produced in the last iteration.

It is noteworthy to highlight that modulations and results methods only write or read internal variables; so the execution time of these methods is in the range of a few microseconds. The computation time of a component can be assumed to be the `step()` method execution time.

Class `component` also determines that each component is a Singleton. This makes it easy to have only one instance of each component, and obtain a reference to it from any other component. The next code describes the initialization and `step()` method of the component A, that uses the component B to get some information. This information is used to modulate the execution of component C. Note the order of the calls. If the component needs information, it executes the `step()` method first. If the component modulates another component, it calls the modulation method before calling the `step()` method.

```
//Two components used by A
B A::b;
C A::c;

A::A()
{
    b = B::GetInstance();
    c = C::GetInstance();
}

A::step()
{
    b->step();

    int info = b->getInfo();

    //Do component A work

    c->setInfo(info);

    c->step();
}
```

The scheduler implemented in this architecture calls the `step()` method of a list of components. Usually, this list only contains a reference to the top-level component, which defines a behavior that can be specified in a setup file. Additionally, we can add more references to components from the monitoring tools in order to execute more components. The implementation is simple:

```
while(true)
    for(i=CompList.begin();i!=CompList.end();++i)
        i->step();
```

Each component is set up to a different frequency, depending on the particular function it does. As a simple orientation, the frequency set to any perceptive component depends on the time for which this information can be considered valid for the current actuation. As an example, image processing is usually made at the camera frame rate (30 Hz) and the locomotion controller sends to the walking engine speed commands at 2 Hz. To carry out this property, the class `component` implements methods `setFreq()` (called during component initialization) and `isTime2Run()`, that returns true if the elapsed time since the last time it returned true is longer than the one specified by the frequency (500 ms if 2 Hz, for example). Using this method, a usual `step()` implementation is

```
A::step()
{
    b->step();

    if(isTime2Run())
    {
        int info = b->getInfo();
        c->setInfo(info);
    }

    c->step();
}
```

The ideas presented above add two important characteristics to this architecture:

- The behavior architecture is **thread safe**. The scheduler does not create multiple threads to execute components. Only one thread calls sequentially to the scheduler list of components. This thread executes in cascade the components, in the order defined depending on the relation of the components (modulation or results).

- If any of the components sporadically spends more time than that desired, the systems suffers from what in real-time literature is called **graceful degradation**. The execution of the other components is delayed, but no executions are cancelled or overlapped. In the development phase of the components, offender components are detected because `istime2Run()` methods of each component periodically test if the frequency is achieved, generating a warning if not.
- The set of of components that a component can activate varies dynamically. As there is no explicit deactivation method, its step function is simply not called anymore; we have to design the component having in mind that a component does not know when it is going to be called again. This is called **quiet shutdown**.

Components can be very simple or very complex. Simple components communicate with the underlaying system methods to communicate with sensors or motors, or use a fixed number of components. Complex components can be implemented as finite state machine, changing the set of components it activates dynamically depending on the state. We have developed a useful tool for designing these complex components. This tool generates the code of the graphically represented behavior.

4 Extensions

The previous section described the basic principles of our software architecture. These principles are the core of the implementation of an architecture that follows the principles described in this paper. This section describes optional extension to this architecture that we have developed to have more functionality when developing any of the possible applications.

4.1 Communications

By the communication mechanism, components can be acceded from remote applications or even components running in other robots. This mechanism is implemented using the ZroC ICE communications framework. This framework hides all the complexity of programming over sockets and provides a convenient RPC paradigm for our communications. Each robot runs an ICE broker with a predefined set of interfaces that connects directly to the component's methods. This is allowed only to the modulation or retrieving information method of components. This is a convenient way to implement teleoperation applications or cooperating behaviors in groups of robots.

Communication mechanism uses a separate thread for managing the remote calls. The components that can be remotely called must implement a mutual exclusion mechanism to have one thread active executing in a component (in the `step()` method or in any modulation/retrieving information method).

The next section presents a debugging mechanism using the communication approach described in this section. The data exchanged, a list of structures that defines graphical elements (circles, boxes, etc.), are defined as an ICE data type. When running, each part of the communication (computer GUI and BICA) runs a broker, and the call to the ICE interface`getDebugData()` is directly mapped to the `Debug::getDebugData()` method.

4.2 *Debug*

Debug mechanisms are critical while developing new components. The low-level mechanisms (stout, stderr, and gdb) are assumed to be available all the time. Actually, class `component` already contains two functions (`startDebugInfofo()` and `endDebugInfo()`) used to periodically (each 10 s) print to stdout info about the real frequency and CPU time.

This mechanism is a convenient way of detecting if a component is consuming too much time, or if the set frequency is not reached because the system has a high load, probably produced by a high consumption component. Despite this, the development of robotic software also needs higher level of debugging mechanisms. We need to know if the sensor information (image, ultrasound, etc.) is correct, or if a self-localization algorithm is working correctly, for example.

We have developed a mechanism, which lets us graphically debug the internal information of the components. An external application can connect to this architecture to retrieve a list of graphical primitives (points, lines, ellipses, boxes, text, and images) produced by the components. A component can be debugged in the image space, in robot relative coordinates and global coordinates. To illustrate the importance of this concept, we can think of a self-localization component based on Extended Kaman Filter that could be implemented to return a list of ellipses in global coordinates, representing the state and uncertainty representing the robot position, while a self-localization component based on Particle Filters could return a list of point with an arrow, representing the position and orientation of each particle. Only the component developer knows which information is crucial for debugging, and this model affords us this functionality.

We have implemented an external graphic application that communicates with the scheduler to add components to the execution list and with a `Debug` component, also in the execution list, for which the set of components are marked for debugging. Any component able to be debugged, has to inherit from the abstract class `debuggeable` and implement a `getDebugInfo()` method:

This debugging mechanism works as follows:

- The GUI activates a set of components and marks them for debugging.
- Once selected, the debugging space (image, relatives, or absolutes), it periodically asks to the `debug` component for a list of primitives in that space.

- The `debug` component calls the `getDebugInfo()` of every component marked for debugging and returns to the GUI a list with the information retrieved.
- The GUI draws the graphical primitives retrieved.

5 Basic Capabilities Applied to Humanoid Robot

The BICA Architecture, in which we have shown the key ideas in the last section, has been implemented in the humanoid robot Nao using NaoiQi, a programming framework provided by the manufacturer. It is important to note that these ideas do not depend on the robot, and they are suitable to be implemented in any other platform. Actually, it would be very easy to be implemented inside a ROS node for using all the services it provides, and to be available for a great variety of robots. The design of the contributions described in the next sections are focused on robots with legs, whose main sensor is a camera with a limited field of view, that can be oriented to cover all the surroundings.

Nao is a fully programmable humanoid robot. It is equipped with a x86 AMD Geode 500 Mhz CPU, 1 GB flash memory, 256 MB SDRAM, two speakers, two cameras (nonstereo), wi-fi connectivity and ethernet port. It has 25 degrees of freedom. The operating system is Linux 2.6 with some real-time patches. The robot is equipped with a microcontroller ARM 7 allocated in its chest to control the robot motors and sensors, called DCM. NaoQi is a distributed object framework which allows several distributed binaries (called brokers), each containing several software modules to communicate together. Robot functionality is encapsulated in software modules, so we can communicate to specific modules in order to access sensors and actuators.

BICA is implemented inside one of these modules. Only a limited set of BICA components, those that provide access to motors and sensors, are NaoQi-dependent, making nonblocking calls to the services that the robot provides. The rest of the components are independent of this platform.

5.1 *Perception and Visual Memory*

The main source of information about the robot environment is mainly provided by cameras. There are components that provide information from ultrasound sensors or buttons to the behaviors, but the camera provides the richest information, and is also a more complex sensor to manage.

The processing of the image has a set of visual stimuli relevant to the robot behaviors as output: obstacles, human faces, other robots, landmarks, balls, for example. This detection is based on the color, shape, size, and position. This processing starts labeling the color of each pixel in the image. This step is common to the detection of every visual stimuli. Next steps are particular for each stimuli.

We have taken advantage of the component-based architecture that BICA provides to avoid unnecessary processing, when some of these visual stimuli are not needed by the active behaviors during the robot operation. First of all, we have developed a component called `Camera` which labels each pixel with its color using a fast lookup table, and makes this information available for other components. Particular stimuli detection is made by particular components, using the labeled image as input, and performing the rest of the processing.

Detectors are also responsible for updating a local estimation of the local stimuli with the detection made in every cycle. Detectors use a list of extended kalman filters to maintain the detected stimuli. Behaviors retrieve this filtered information to make decisions.

5.2 Attention

The robot's camera has a limited field of view. Because of this, the robot has to move the neck to cover all the surroundings and perceive with the camera, fitted in its head, all the relevant visual stimuli. This is carried out by a visual attention system. This system is in charge of searching, tracking, and revisiting the visual stimuli.

The visual attention system developed inside the BICA architecture has the `Attention` component as the central element. This component has three functions:

1. It sends to the `Head` component the 3D points where the cameras have to be orientated.
2. It receives the perceptive requirements from the behavior components with perceptive needs. Using this information, it decides which visual stimulus governs attention in every moment.
3. It asks to the detector in charge of the visual stimulus that governs attention for a 3D point. This point is sent to the `Head` component.

The most relevant benefit of this system is that the searching and tracking is specialized for each visual stimulus. Some stimulus can be searched on the floor whereas others in the skyline; it can only have one instance of each stimulus, or more. This is decided by each detectors developer. In [4], we have presented three different implementations of attention mechanism using the flexibility and modularity that this design allows.

6 Robotic Applications

In the previous sections we have described the key ideas of the BICA architecture. In this section, we will present two different scenarios where we have applied this architecture.

6.1 Robot Soccer

Using the BICA architecture we have developed a complete set of behaviors for a team of robotic players [2] of the Standard Platform League of the RoboCup. This competition presents a challenging and dynamic scenario where teams of robots have to play in a soccer match. In this league, in particular, all the robots are similar so all the efforts are focused on developing software capable of dealing with this problem.

This application needs reactive behaviors, where collisions, error in perception, and kidnappings are common. We have successfully developed a set of behaviors to deal with this problem, including cooperation among robots, navigation and self-localization algorithms, perception, and reactive visual attention. The details of this implementation can be found in[5].

6.2 Therapy for Alzheimer Patients

This architecture has also shown to be adequate for a completely different application, such as using this robot for therapies in patients having Alzheimer's disease. For this scenario, we have developed a component that plays therapy scripts using new components capable of playing music and speech and managing the robots leads. In addition, the reproduction of the script is controlled by applications running in a tablet, or receiving commands from a wiimote.

7 Conclusion

This paper has presented novel ideas in the designing of behavior-based architecture. This architecture decomposes complex behaviors into building blocks that cooperate among them, called components. These concepts provide an effective, clean, and scalable way of designing complex behaviors in limited resources robots, with real-time requirements. Our architecture has no separation into layers, but the organization is made explicit by the relation between component activation. This contrasts with the classical approaches, such as Xavier [19] or the one proposed by Arkin [20]. In Xavier, the work is made out of four layers: obstacle avoidance, navigation, path planning, and task planning. The behavior arises from the combination of these separate layers, each with a specific task and priority. Arkin designed a hybrid architecture, in which the behavior is divided into three components: deliberative planning, reactive control, and motivation drives. Deliberative planning made the navigation tasks. Reactive control provided with the necessary sensorimotor control integration for response reactively to the events in its surroundings. Motivation drives were responsible for monitoring the robot behavior.

ROS [11] is nowadays the reference in software architectures. It decomposed applications into nodes that can communicate among them using direct messages or notifications to published data. It is thread safe and provides many graphical and test-based tools for debugging, monitoring, and developing. ROS community is very active, developing many software libraries and drivers that can be easily reused. Our approach is a more compact design. Instead of making each component a separate process, all the BICA components share the same memory space and the calls are local, making our approach more efficient for embedded applications, but limited in distributed applications. The design of each component as a singleton makes it another very easy to use component without any risk.

We have described the details of the execution of these components. These details give this architecture some important characteristics when developing real-time robotic applications: thread safe approach, graceful degradation to high-load events, and efficient perceptive pipeline. We have also described how we have used these concepts to develop visual attention or debugging mechanisms. One of the most important characteristic of this paper is that these concepts are described from a technical point of view, being useful to incorporate these ideas to any robotic software.

This architecture has been successfully used to deal with two complex applications such as the robot soccer or the therapy for Alzheimer patients. The robot soccer presents a challenging environment where the robot perceives the relevant visual stimulus, processes this information, and generates a fast response in terms of actuation and active perception commands. We have demonstrated how our approach is efficient enough to deal with perception, navigation, self-localization, team coordination, action generation, and active vision when all the processing is performed onboard. The perception requirements and component activations dynamically change during the robot operation, and our approach adapts to these changes, saving computing resources. The other application, focused on using new interfaces to communicate with the robot during therapies, shows how the approach is also valid in such a different environment.

We have also shown that this architecture is convenient for developing robotic applications. Our approach is scalable and provides a simple way to decompose the functionality in components that can be run isolated during development phase and being debugged with the efficient debugging mechanism described in this paper. At this time, this architecture is mature enough to be used in any application. It has been developed for the humanoid robot Nao, but we are currently working on an implementation inside an ROS node, obtaining the benefits of ROS, and the properties of our approach.

Acknowledgments This research has been sponsored by grant No. DPI2013-40534-R by Spanish Ministry of Economy and Competitiveness corresponding to SIRMAVED project[21].

References

1. Brooks, R.A.: Intelligence Without Representation. Artificial Intelligence **47**, 139–159 (1991)
2. Martín, F., Aguero, C., Cañas, J.M., Perdices, E.: Humanoid soccer player design. In: Papic, V. (ed.) Robot Soccer, pp. 67–100. IN-TECH (2010)

3. Martín, F., Aguero, C., Cañas, J.M.: Localization of legged robots combining a fuzzy-Markov method and a population of extended Kalman filters. Robotics and Autonomous Systems **55**, 870–880 (2007)
4. Martín, F., Aguero, C., Rubio, L., Cañas, J.M.: Comparison of Smart Visual Attention Mechanisms for Humanoid Robots. International Journal of Advanced Robotic Systems: Smart Sensors for Smart Robots **9**, 1–10 (2012)
5. Martín, F., Aguero, C., Cañas, J.M., Martínez, P., Valenti, M.: RoboTherapy with Alzheimer Patients. International Journal of Advanced Robotic Systems: Humanoid **9**, 1–7 (2012)
6. Konolige, K., Myers, K., Ruspini, E., Saffiotti, A.: The Saphira architecture: A design for autonomy. Journal of Experimental & Theoretical Artificial Intelligence **9**(2–3), 215–235 (1998)
7. Konolige, K.: Saphira robot control architecture, Technical Report, SRI International, Menlo Park, Calif, USA (2002)
8. Gerkey, B., Vaughan, R.T., Howard, A.: The player/stage project: tools for multi-robot and distributed sensor systems. In: Proceedings of the 11th International Conference on Advanced Robotics (ICAR 2003), Coimbra, Portugal, pp. 317–323, June 2003
9. Collett, H.J., MacDonald, B.A., Gerkey, B.: Player 2.0: toward a practical robot programming framework. In: Proceedings of the Australasian Conference on Robotics and Automation (ACRA 2005), Sydney, Australia, December 2005
10. Vaughan, R.T.: Massively multi-robot simulations in Stage. Swarm Intelligence **2**(2–4), 189–208 (2008)
11. Quigley, M., Ken, C., Gerkey, B., Faust, J., Foote, T., Leibs, J., Wheeler, R., Andrew, Y.: ROS: an open-source robot operating system. In: Proc. of the IEEE Intl. Conf. on Robotics and Automation (ICRA). Workshop on Open Source Robotics (2009)
12. Brooks, A., Kaupp, T., Makarenko, A., Orebck, A., Williams, S.: Towards component-based robotics. In: IEEE/RSJ International Conference on Intelligent Robots and Systems (IROS 2005), pp. 163–168 (2005)
13. Makarenko, A., Brooks, A., Kaupp, T.: On the benefits of making robotic software frameworks thin. In: IEEE/RSJ International Conference on Intelligent Robots and Systems (IROS 2007). Workshop on Evaluation of Middleware and Architectures (2007)
14. Cintas, R., Manso, L.J., Pinero, L., Bachiller, P., Bustos, P.: Robust Behavior and Perception using Hierarchical State Machines: A Pallet Manipulation Experiment. Proceedings, Journal of Physical Agents **5**(1), 35–44 (2011). ISSN 1888–0258
15. Montemerlo, M., Roy, N., Thrun, S.: Perspectives on standardization in mobile robot programming: the carnegie mellon navigation (CARMEN) toolkit. In: IROS 2003, pp. 2436–2441 (2003)
16. Kraetzschmar, G.K., Utz, H., Sablatnög, S., Enderle, S., Palm, G.: Miro - middleware for cooperative robotics. In: Proceedings of RoboCup-2001 Symposium. LNAI, vol. 2377, pp. 411–416. Springer-Verlag, Heidelberg (2002)
17. Henning, M.: The Rise and Fall of CORBA. ACM Queue Magazine **4**(5), June 2006
18. Canas, J.M., Matellán, V.: From bioinspired vs psychoinspired to ethoinspired robots. Robotics and Autonomous Systems **55**, 841–850 (2007)
19. Simmons, R., Goodwin, R., Haigh, K., Koenig, S., O'Sullivan, J., Veloso, M.: Xavier: Experience with a Layered Robot Architecture. SIGART Bull. **8**(1–4), 22–33 (1997)
20. Stoytchev, A., Arkin, R.C.: Combining deliberation, reactivity, and motivation in the context of a behavior-based robot architecture. In: Proceedings of the 2001 IEEE International Symposium on Computational Intelligence in Robotics and Automation, pp. 290–295 (2001)
21. Cazorla, M., García-Rodríguez, J., Cañas Plaza, J.M., García Varea, I., Matellán, V., Martín Rico, F., Martínez-Gómez, J., Rodríguez Lera, F.J., Suarez Mejias, C., Martínez Sahuquillo, M.E.: SIRVAMED: development of a comprehensive robotic system for monitoring and interaction for people with acquired brain damage and dependent people. In: XVI Conferencia de la Asociacin Española para la Inteligencia Artificial (CAEPIA) (2015)

Analysis and Evaluation of a Low-Cost Robotic Arm for @Home Competitions

Francisco J. Rodríguez Lera, Fernando Casado, Vicente Matellán Olivera and Francisco Martín Rico

Abstract This paper reviews the design design, construction and performance of an affordable robotic arm of four degrees of freedom based on an Arduino controller in a home-like environment. This paper describes the kinematic design of our 4 DOF arm and the physical restrictions that this design imposes. We have also proposed two types of end-effectors to address two types of manipulation tasks: to grasp objects and to push different light switches. The arm was on board of the MYRABot platform and both were evaluated in the RoCKIn competition. This competition involves grasping and manipulation tasks that are described in the paper as well. Comments on the results of the competition and their implication in further improvement of the robot are also described in the paper.

Keywords Manipulators · Low-Cost · Design · Four Degrees of Freedom · Manufacturing · End Effector · @home Competitions

1 Introduction

In these days, robot manipulation appears as a mature technology, particularly in manufacturing domain. However, robots in industrial environments do not deal with unpredictable external factors as human interference, dynamic environments, or variable lighting conditions. These are key points on open environments, particularly in

F.J.R. Lera(✉) · F. Casado · V.M. Olivera
School of Industrial and Computer Engineering, University of León, León, Spain
e-mail: {fjrodl,fcasag00,vicente.matellan}@unileon.es
http://robotica.unileon.es/

F.M. Rico
School of Telecommunication Engineering, Rey Juan Carlos University, Fuenlabrada, Madrid, Spain
e-mail: fmartin@gsyc.es

L.P. Reis et al. (eds.), *Robot 2015: Second Iberian Robotics Conference*, Advances in Intelligent Systems and Computing 418,
DOI: 10.1007/978-3-319-27149-1_48

inhabited indoor environments. Consequently robot manipulation remains as a scientific and technological challenge in home-like environments.

Many competitions are run every year to foster research in these domains, as for instance RoboCup@Home or RoCKIn challenge. In these events the organizers prepare a home-like environment where it is possible to test the robots. They are considered as both scientifically and technologically valid tool [9] to assess mobile manipulators.

However the arenas created to test them is partially adapted to robot characteristics and they do not always respect more realistic situations. For instance a robot grasping a picture frame in cluttered shelf, instead of an empty one.

Moreover, the robots that take part of these competitions present different drawbacks: the robots are heavy, they generate loud noises, their footprints cover a big area or the arm workspace is not feasible for real crowd life conditions. In addition, the robots employed in these competitions are really expensive. With this drawbacks in mind, we can perceive that it is not easy to move these robots from these scientific events to a real home environment.

Under those circumstances, we propose the use of a low-cost mobile manipulator for robotics competitions. We think that once it has been tested in these competitions, it could also be moved and tested to real environments. Our proposal aims to solve some the tasks performed in the @home challenges from a simplified point of view. We want to address these tasks modeling the problem and proposing an affordable solution.

We faced some issues observed in these competitions: the robot weight, an acceptable robot footprint, and an affordable arm with a suitably workspace for an average home. The two first problems have already been described in [6]. The third issue has been faced in this paper.

In our proposal we decided to face the problem using human arm simplification from 7DOF to 4DOF as other researchers had previously done [4],[8]. It is possible to emulate the seven degrees of freedom of a human arm (shoulder abduction, shoulder flexion, shoulder flexion, rotation and elbow flexion) using only a 4DOF. Nevertheless, the designs reviewed in this research present a four DOF arms which only provide: should pitch, shoulder yaw, elbow pitch and wrist pitch. We lose two degrees of freedom in the wrist and one DOF on the shoulder.

We are going to present in this paper a review of commercial-off-the-shelf solutions and discuss our election. Then we describe the design of our proposal, and its integration in our mobile platform, analyzing the main characteristics of our decisions from a kinematic point of view. We test the arm performance in laboratory conditions and finally we show the possibilities of our arm in a naive experience performed during RoCKIn competition.

In this way the rest of the paper is organized as follows. First, a brief analysis of available commercial arms is made to determine the common features that we want to include in our robot. Then, the mechanical design of the arm is made, both the arm itself, the effectors as well as the kinematics calculations for its control. Then the experiments to validate it are described. Finally some conclusions are discussed and future work is presented.

2 Analysis of Commercial Arms

In this section we present the review of three robotic arms available from two point of view, commercial (two models) and scientific (one model).

The main design restriction (besides its limited cost) came from the previously designed robot frame. The robot imposed two physical limitations: 1) arm weigh (taking into account the lifting capacity); and 2) arm workspace. The arm has to be able to perform simple manipulation and grasping tasks, such as handling a small plastic bottle or soda can. The last restriction was that the arm has to be compatible with Atmega controllers like Arduino.

Due to our previous experience with TurtleBot, we started with Wise solution [12], [13]. It was designed to TurtleBot robot and it was based on three elements: Bioloid components for the links, Dinamixel servomotors, and an ArbotiX-M Robocontroller (ATMEGA644p).

We also reviewed a commercially known platform called OWI 535 Robotic Arm [10]. It offers a really low cost solution using simple motors.

The physical characteristics according to the specifications are: it has a wrist with a radial motion range of 120°, the elbow joint has a working range of 300°, and finally the base has two joints with a rotation of 270° in base plane and 180° of vertical motion. The maximum working space is about 40cm vertically and 34 cm horizontally with a lifting capacity of 100g. Attending the end effector it has jaws. Its open size is only 4.5 cm, which suposses a big handicap to grasp daily objects. Finally, attending to software side, there are two interfaces available for controlling this arm from a computer: one using OWI which only has Windows support; and one using Arduino Uno that uses open-source Arduino Software (IDE) and runs in Windows and Linux .

We also analyzed ad-hoc solutions from the scientific literature, as for instance, Elfasakhany work [3] who presents a 4DOF platform made of acrylic material. It uses Hextronik HX12K servomotors and includes a commercial gripper as a final effector (BCM Gripper with Servo). The problem of this approach lies in this gripper, because the stall torque (0.225 N-m) it is not enough for competition requirements. Regarding to the software side, this arm is managed by an Arduino Atmega, it means that can be controlled using open source solutions.

Table 1 outlines the main characteristics of each arm. The arms have similar workspace characteristics. The first main difference is the customized possibilities. OWI arm 535 is a commercial platform with closed and defined hardware limits; it does not appear to be easily customized. Second difference is the total weight of the platform taking into account the payload. The willow garage approach appeared as the lighter solution because in the Elfasakhany work it is not clear the total weight of the proposed solution, they do not present the arm support weight. Consequently we chose the Willow Garage design to adapt it to our purposes.

Table 1 Main characteristics of reviewed arms.

Arm	Vertical Workspace	Horizontal Workspace	Weight	Payload	DOF	Custom	Controller
Wise-TurtleBot	0.375m	0.345m	[325g, 400g]	100g	4+1	Yes	Arbotix Atmega 644p
OWI 535	0.40m	0.34m	658g	100g	4+1	No	OWI Interface Arduino Uno
Elfasakhany (Estimated)	0.47m	0.30m	[340g, 500g]	100g	4+1	Yes	Arduino Atmega 368

3 Mechanical Design

The arm design had to face simple tasks in domestic environments. In particular, we needed to adapt Willow Garage design attending the RoCKIn competition regulations. With this intention, the mobile manipulator had to be able to grasp small objects (less than 100gr of weight) and also to turn on-off different types of switches (pressing, turning, moving..).

The arm base has two degrees of freedom for simulating the human shoulder, the first motor is able to turn horizontally 280° in parallel to the ground; the second one is mounted on top of the first one and has two steps, the first one runs with maximum work space, one from parallel to base position to 90° up, perpendicular to the ground, and the second step is around 40° down. The second step runs 100° and maintains the first arm join fixed. Another DOF simulates robot elbow changing the end effector position. The last DOF is used for the wrist. It allows us to change the end effector vertical orientation from parallel to perpendicular to the ground.

This configuration should let us perform the proposed tasks, grasping objects from a table or switching the lights on/off. Of course, the path planning is restricted to these 4 DOFs, so we are not able to reach all poses in the working space. Adding two more motors will allow us to define the end effector to whatever position but this will also add extra weight to the arm.

We made the calculations of the joints that will have the largest loads to determine which type of servomotors should be used in the robot's arm. Every other joints will use the same motor to simplify maintenance and design. The force diagram used is shown in Fig. 1 where the red arrows states the forces on link BC, whose links the (B and C) carry the heavies load. The weights of links are $W_{l_{BC}} = W_{l_{CD}} = 0.025$ kg, $W_{l_{DE}} = 0.009$ kg and the length of the links are $l_{BC} = l_{CD} = 0.12$ m and $l_{DE} = 0.031$ m.

The weights of the servomotors[1] are:

- $W_A = W_B = W_C = W_D = W_E = 55$g

The sum of forces are calculated through the Y axis (Fig. 1) using the RoCKin specified load of 1kg . We solve for C_Y and C_B, see equations 1-2. The sum of moments around the joint C ($\sum M_C$, Eq. 1), and joint B ($\sum M_B$, Eq. 2), are used to obtain the torque in joint C (M_C, Eq. 3) and joint B (M_B, Eq. 4).

[1] Motor Dinamixel AX12A: http://www.generationrobots.com/media/Dynamixel-AX-12-user-manual.pdf

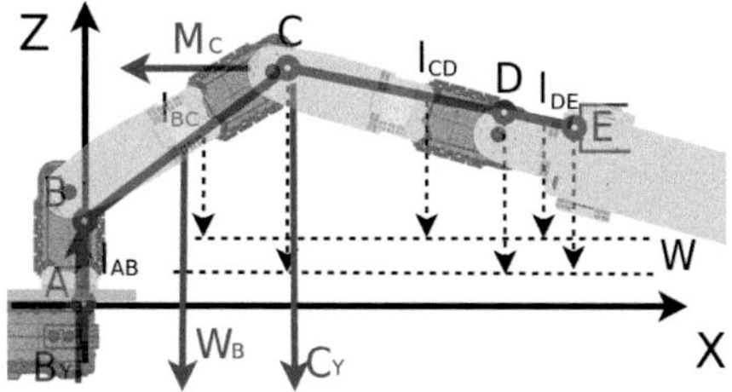

Fig. 1 Arm forces, links, diagram

$$\sum M_C = (-\left(\frac{W_{l_{CD}} * l_{CD}}{2}\right) - W_{l_{DE}}\left(l_{CD} + \frac{l_{DE}}{2}\right) - L(L_{CD} + L_{DE}) - W_D(L_{CD})) * g + M_C = 0 \quad (1)$$

$$\sum M_B = (-L(l_{BC} + l_{CD} + l_{DE}) - W_{l_{DE}}\left(l_{BC} + l_{CD} + \frac{l_{DE}}{2}\right) - W_D(L_{BC} + L_{CD}) - W_C\left(L_{BC}) + \frac{L_{CD}}{2}\right) - W_C\left(\frac{L_{BC}}{2}\right)) * g + M_B = 0 \quad (2)$$

$$M_C = (\left(\frac{W_{l_{CD}} * l_{CD}}{2}\right) + W_{l_{DE}}\left(l_{CD} + \frac{l_{DE}}{2}\right) + L(L_{CD} + L_{DE}) + W_D(L_{CD})) * g = 1.85024Nm \quad (3)$$

$$M_B = (L(l_{BC} + l_{CD} + l_{DE}) + W_{l_{DE}}\left(l_{BC} + l_{CD} + \frac{l_{DE}}{2}\right) + W_D(L_{BC} + L_{CD}) - +W_C\left(L_{BC}) + \frac{L_{CD}}{2}\right) + W_C\left(\frac{L_{BC}}{2}\right)) * g = 3.167556Nm \quad (4)$$

The motors that we considered are shown in Table 2. We chose them according the torque requirements and a budget restriction of 400 euros.

Table 2 Motor Selection

	Theoretical	AX-12A	AX-18F	DX-117
Stall Torque	Nm	Nm	Nm	Nm
M_C	1.85	1.50	1.8	
M_B	3.16			3.77
Price - Euros x unit		43,00	85,00	180,00

Attending to joint C, we need a torque of 1.85 Nm so AX-12A and AX-18F motors could barely afford it. Even the AX-18F motor is near the theoretical Nm needed for joint C, we consider that the extra 0.3 Nm does not justify the double price versus the AX-12A. For this reason we use the AX-12A.

Under those circumstances, we also have other problem, the joint B needs more torque than the other joints, and we are not able of reach the total torque even if we use two motors. This issue degrades the overall performance of the arm.

Once we have chosen the servos, we calculate the maximum charge that the robots will be able to lift according to our design restriction (it has a length of 0.345 m). Attending the RoCKIn competition rulebook it is 1 kg. The theoretical torque for a weight of a 1 kg should be 3.98 Nm. Attending our torque we are able to lift around 0.1 kg.

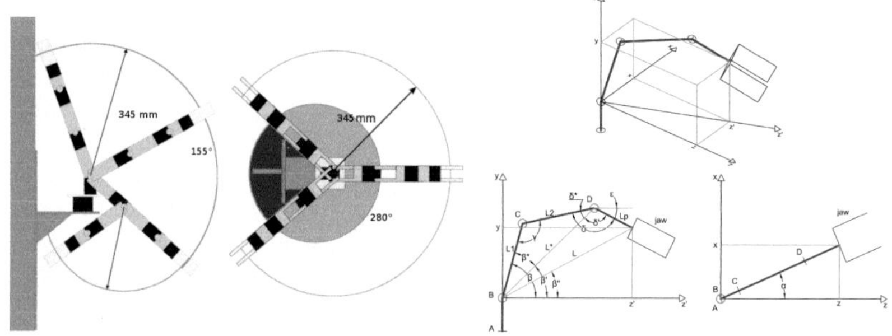

Fig. 2 Arm workspace and kinematics calculation

3.1 *Kinematics*

Due to the 4 DOF design, the rotational motion of the joints is made around two axes in the arm base (the shoulder) and around only one in the other joints (elbow and wrist). Once assembled, the arm work space is shown in Fig.2. Using the inverse kinematics calculations that follows (equations 5 to 16), we can get the angle of each motor from any given position according the the coordinate system of the robot.

$$\alpha = tan^{-1}\left(\frac{x}{z}\right) \quad (5)$$

$$z' = \sqrt{z^2 + x^2} \quad (6)$$

$$L = \sqrt{z'^2 + y^2} \quad (7)$$

$$L* = \sqrt{(y + Lp.sin\varepsilon)^2 + (z' - Lp.cos\varepsilon)^2} \quad (8)$$

$$\beta' = tan^{-1}\left(\frac{y + Lp.sin\varepsilon}{z' - Lp.cos\varepsilon}\right) \quad (9)$$

$$\beta'' = tan^{-1}\left(\frac{y}{z'}\right) \quad (10)$$

$$\beta* - cos^{-1}\left(\frac{L1^2 + L*^2 - L2^2}{2.L1.L*}\right) \quad (11)$$

$$\beta = \beta' + \beta* \quad (12)$$

$$\gamma = cos^{-1}\left(\frac{L1^2 + L2^2 - L*^2}{2.L1.L2}\right) \quad (13)$$

$$\delta * = 180 - (\beta * + \gamma) \quad (14)$$

$$\delta' = cos^{-1}\left(\frac{L *^2 + Lp^2 - L^2}{2.L * .Lp}\right) \quad (15)$$

$$\delta = \begin{cases} \delta' + \delta * & \text{if } \beta' \geq \beta'' \\ 360 - (\delta' - \delta *) & \text{if } \beta' < \beta'' \end{cases} \quad (16)$$

When the end effector is out of radial axis (equations from 17 to 23), Dx and Dy are the deviation with respect to radial axis, and Dz is the length increase of the end effector position .

$$z'_0 = \sqrt{z^2 + x^2} \quad (17)$$

$$\alpha_0 = sin^{-1}\left(\frac{Dx}{z'_0}\right) \quad (18)$$

$$\alpha_1 = tan^{-1}\left(\frac{x}{z}\right) \quad (19)$$

$$\alpha = \alpha_1 - \alpha_0 \quad (20)$$

$$x = x + Dx|cos(\alpha)| \quad (21)$$

$$z = \begin{cases} z + |Dx.sin(\alpha)| & \text{if } \alpha \geq 0 \\ z - |Dx.sin(\alpha)| & \text{if } \alpha < 0 \end{cases} \quad (22)$$

$$y = y + Dy \quad (23)$$

4 Final Effector

Attending RoCKIn requirements, we needed to integrate an end-effector to perform manipulation or grasping tasks. Its design has to maintain the arm features. With this intention, we have designed two different effectors for our platform. At present, it is possible the find versatile and affordable end effectors as Dollar adaptative SDM hand [2]. However, as a result of our arm design it is not possible to use it. Dollar design increments the length of the arm making impossible to maintain a minimal payload of 100g. For this reason we decide to propose a multiple effector solution.

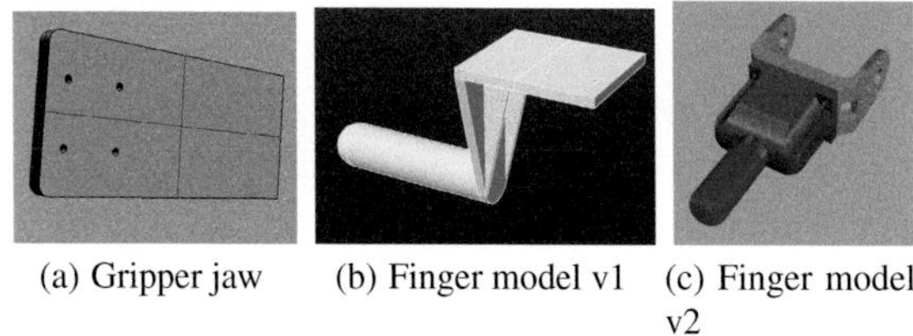

(a) Gripper jaw (b) Finger model v1 (c) Finger model v2

Fig. 3 End effector Designs

First effector uses a pair of jaws (Fig. 3(a)) to build a basic gripper. The proposed design attaches one of the jaws to the motor base, so only one of the jaws has movement. The maximum opening angle of this gripper is 60 degrees and the mobile jaw can reach the fixed jaw. Using a gripper the robot can grasp a small objects, for instance empty bricks of 1l or plastic bottles of 0.5 l.

The second effector is a finger for switching on and off the light switches defined in RoCKIn challenge. We developed two finger models. First model (v1) shown in Fig. 3(b) was designed for working with the proposed arm workspace. In this case we used the feedback of end effector for knowing the push status. The second model (v2) presented in Fig. 3(c) was designed with the same purpose but with different pattern We have to remove the motor of the end effector and attach the finger directly to the wrist motor along with a piezoelectric sensor.

These components were manufactured in two different ways, for the jaw, we used a residual methacrylate sheet that we cut with a laser machine. Both fingers models were manufactured using a rapid prototyping technique named addition, through a Prusa i3-3D printer using ABS as a base material. The characteristics of the real effectors and their interactions with the arm are presented in Table 3 and the CAD models in figure 3.

In real competition we finally used the model v1 of the effector because it increases the workspace in 1 cm versus the model 2 that decreases the workspace in 5 cm. Weights of the effectors are similar in both solutions, finger model (v2) saves us 5 gr. but the increase in the workspace is worth it.

Table 3 Dimensions of final effectors

End effector	Weight (gr)	Dimensions (length)	Arm workspace (length)
Gripper(Jaws)	22 (11x2)	9 cm	34.5 cm
Finger (v1)	21	10 cm	35.5 cm
Finger (v2)	16	7 cm	29.5 cm

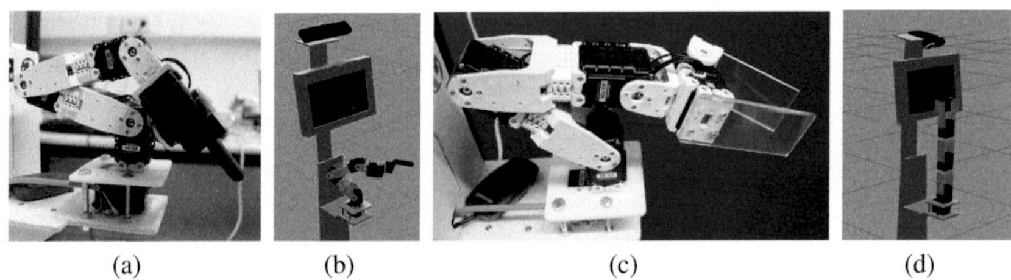

(a) (b) (c) (d)

Fig. 4 End effectors in real and Gazebo environments. Pictures (a,b) present the finger Solution and pictures (c,d) shows the jaws solution.

5 Software Design

ROS[11](Robot Operating System) provides a set of libraries and tools for developing robotic applications. We have made extensive use of this software in our robot. We used the following ROS components:

- *ULEKinematics*: We developed our own component using the arm geometry and kinematics calculations summarized previously.
- *MoveIt!*: It is consider the standard software for manipulation. This software provides motion planning, manipulation, 3D perception, kinematics, control and navigation.

We also use standard libraries like: *TF* for managing the transformation (rotation and translation) among coordinate frames; and *Find_object* [5] to detect objects in an image planes, and calculate their position relative to the camera.

These software applications are integrated using *BICA*. It is an architecture of behaviors described in [6], that has also been implemented as a ROS node. Within this architecture can implement high-level behaviors, using a scheme of subsumption [1]. Each component can be implemented as a reactive controller, a fuzzy controller, or even a finite state machine (FSM). BICA software coordinates the behaviors that solve the various problems that the robot has to solve.

The procedure to grasp an object involves two component of BICA: one related with the vision and one related with the arm. When the vision component estimates the bottle position a command signal is sent to the manipulation component (MC). Then the MC activates the grasp or manipulation behavior which has a connection with the kinematics component (which is composed by MoveIt! or ULEKinematics). In this state it is calculated the position of the end-effector and it is sent to the kinematics component. When the position is reached the behavior returns to the inactive state. Fig. 5 shows the FSM associated to the manipulation component using the MoveIt! kinematics component.

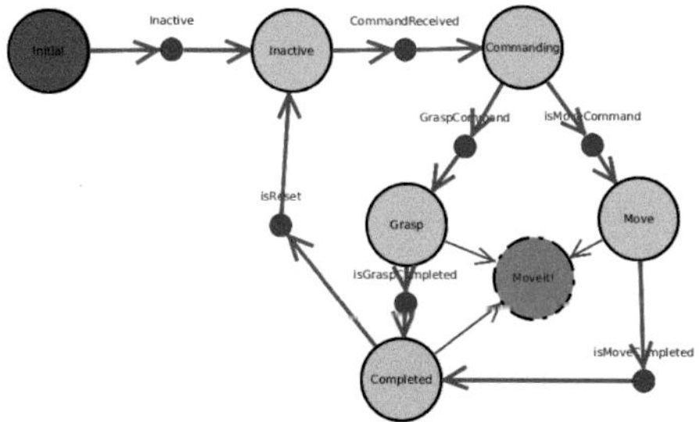

Fig. 5 FSM of the Manipulation component designed in BICA.

6 Experimentation

We performed three experiences for analyzing the arm performance of the MYRABot platform. They are divided in two groups, synthetic (benchmarking) and real experiences.

6.1 Benchmarking Experience

This test was made for evaluating the motor performance and the overall behaviour of the arm. Particularly we want to measure arm accuracy [7], it means the closeness of the final effector to the expected position. We made two test, in each test the arm end position has to reach a fixed (x,y,z) from robot reference axis. We made ten trials in each test and all the measurements are in millimetres.

The first trial was to reach the (0,100,150) position. We obtained (-1.61, 88.73, 150.56) in average with a standard deviation of (0.61, 4.43, 1.77). This means poor results in Y axis and acceptable in X and Z.

The second test goal was to reach the point in (-80,0,230) obtaining (-83.03,-18.45,230.64) in average with standard deviation of (3.22, 1.60, 2.13). Therefore, the arm presents poor reliability in Y axis and acceptable in X and Z axis.

6.2 Real Experience 1 - Grasping

This experience is linked with the task benchmark 1 of RoCKIn challenge. The robot has to help Grannie Annie in the RoCKIn arena. The robot would need to grasp an object in order to achieve Grannie commands.

In this experiment we have launched an open loop program using our kinematics approach. The robot arm tries to grasp a small bottle of water (98gr) located in its workspace. We provide the cartesian coordinates of the bottle in the reference system of the arm.

We performed this experiment ten times in our lab. In two cases the end-effector finished in a position where it is only possible to take the bottle with the jaws end, so the bottle fell down after the grasping approach.

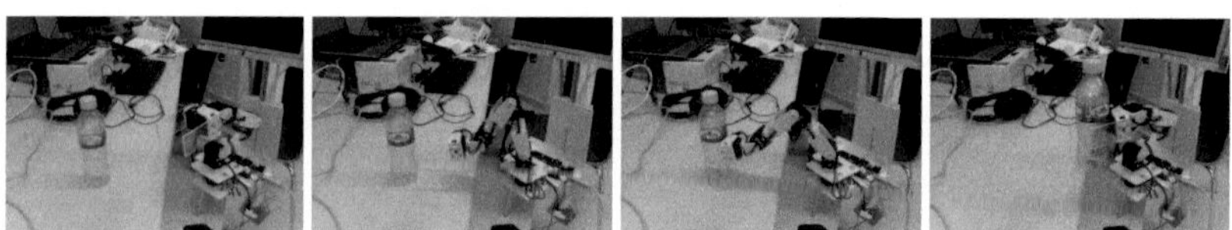

Fig. 6 Real experience test 1: Grasping

6.3 Real Experience 2 - Manipulation

This experience corresponds to the functionality benchmark 2 proposed by RoCKIn challenge. In this task the robot has to switch on/off a set of switches attending the order specified by the organization. Each switch has a state given by a light bulb located next to the button.

In this experiment we launched a close loop program that tries to switch the light on/off using the finger effector. This task started with robot situated in front of the switches panel. When the vision component perceives the first switch and their state, it is passed to the manipulation component and the robot starts to work.

In our approach we have a big handicap versus other robots. Our robot, as other platforms do, needs to optimize the distance to the panel to perceive the maximum number of buttons. Under these conditions, the workspace of our robot arm is not enough to reach the target button. Consequently, we need to perform a previous navigation step to approximate the robot to the switch panel and then, recalculate the switch position to push it.

Despite the prevailing circumstances, we pushed and changed the button position during training tests[2] as is presented in Fig. 7.

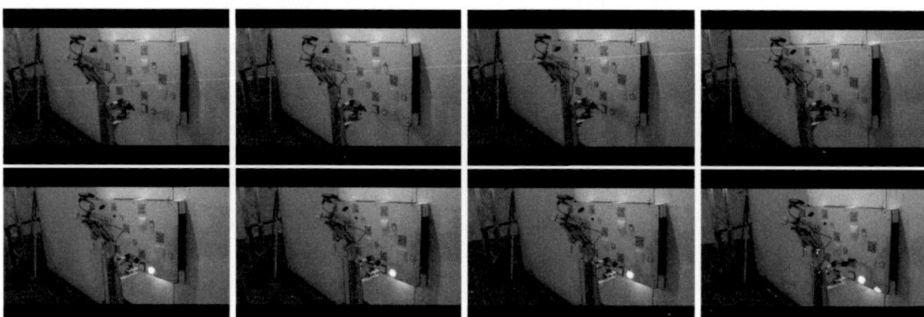

Fig. 7 Real experience: RoCKIn functionality benchmark 2.

7 Conclusion

We have presented in this paper the design, development, implementation and test of a low-cost robot arm with four degrees of freedom. It is able of accomplish basic grasping and manipulation tasks in a home-like environment. This proposal is based on Wise [12] design and using COTS solutions: Bioloid joints, Dinamixels servomotors and Arduino Mega. We have built also two final effectors, one for grasping tasks, using methacrylate material and a finger for a manipulation task using a 3D printer and ABS material.

---

[2] https://vimeo.com/127464373

The arm is controlled using BICA architecture running on ROS. The Manipulation component of BICA depends on a kinematic component that can be the ULEKinematics ROS node, or the MoveIt! one.

Attending benchmarking results, the arm presents good precision with a linear systematic error in Y axis. The arm was tested in the last RoCKIn competition (Toulouse, November 2014) our results were not fully successful.

However, we can say that our solution is good for modeling and evaluating the tasks involved in real robotic competition environment. Although we may not be able to perform the complete task, our research models can be extrapolated to robots with better performance.

Acknowledgments This work has been partially funded by Spanish Ministerio de Economía y Competitividad under grant DPI2013-40534-R

References

1. Brooks, R.: Intelligence without reason. In: Proceedings of the 1991 International Joint Conference on Artificial Intelligence, pp. 569–595 (1991)
2. Dollar, A.M., Howe, R.D.: The highly adaptive SDM hand: Design and performance evaluation. The International Journal of Robotics Research **29**(5), 585–597 (2010)
3. Elfasakhany, A., Yanez, E., Baylon, K., Salgado, R., et al.: Design and development of a competitive low-cost robot arm with four degrees of freedom. Modern Mechanical Engineering **1**(02), 47–55 (2011)
4. Hersch, M., Billard, A.G.: A model for imitating human reaching movements. In: Proceedings of the 1st ACM SIGCHI/SIGART conference on Human-robot interaction, pp. 341–342. ACM (2006)
5. Labb, M.: Find-object: Object detection using surf/sift like features (2015). http://introlab.github.io/find-object/
6. Martin Rico, F., Rodriguez Lera, F.J., Matellan Olivera, V.: Myrabot+: a feasible robotic system for interaction challenges. In: 2014 IEEE International Conference on Autonomous Robot Systems and Competitions (ICARSC), pp. 273–278. IEEE (2014)
7. Menditto, A., Patriarca, M., Magnusson, B.: Understanding the meaning of accuracy, trueness and precision. Accreditation and Quality Assurance **12**(1), 45–47 (2007)
8. Moubarak, S., Pham, M.T., Pajdla, T., Redarce, T.: Design and modeling of an upper extremity exoskeleton. In: World Congress on Medical Physics and Biomedical Engineering, September 7–12, 2009, Munich, Germany, pp. 476–479. Springer (2009)
9. Negrijn, S., van Schaik, S., Haber, J., Visser, A.: UvA@Work Team Description Paper - RoCKIn@Work Camps 2014 - Rome, Italy (2013)
10. OWI Inc: Robotic Arm OWI 535 - Assembly and Instruction Manual (2006)
11. Quigley, M., Conley, K., Gerkey, B., Faust, J., Foote, T., Leibs, J., Wheeler, R., Ng, A.Y.: Ros: an open-source robot operating system. In: ICRA workshop on open source software, vol. 3, p. 5 (2009)
12. Wise, M.: Make projects: Build an arm for your turtlebot. Make Website (2012). http://makezine.com/projects/build-an-arm-for-your-turtlebot/
13. Wise, M.: Make projects: Wiring and attaching an arm to your turtlebot. Make Website (2012). http://makezine.com/projects/wiring-and-attaching-an-arm-to-your-turtlebot/

Object Categorization from RGB-D Local Features and Bag of Words

Jesus Martínez-Gómez, Miguel Cazorla, Ismael García-Varea and Cristina Romero-González

Abstract Object categorization from robot perceptions has become one of the most well-known problems in robotics. How to select proper representations for these perceptions, specially when using RGB-D images, has received a significant attention in the last years. We present in this paper an object categorization approach from RGB-D images. This approach is based on the BoW representation, and it allows to integrate any type of 3D local feature implemented in the Point Cloud Library. The experimentation performed over the challenging RGB-D Object dataset shows how competitive object categorization systems can be developed using this procedure.

Keywords Object categorization · 3D features · Classification · Robotics

1 Introduction

Nowadays robots are requested to interact with humans and other robots in daily life environments [17]. Human robot interaction relies on the use of several sensors and actuators, taking advantage of multimodal information [3]. The human robot interaction process may also involve object manipulation capabilities. Manipulation tasks require to previously identify objects in the environment that would be suitable for manipulation [10].

The use of RGB-D sensors (e.g. Microsoft Kinect and Asus Xtion) has become very popular during last years. RGB-D images allow to cope with challenging lighting

J. Martínez-Gómez(✉) · I. García-Varea · C. Romero-González
Computer System Department, University of Castilla-La Manchas, Ciudad Real, Spain
e-mail: {jesus.martinez,ismael.garcia,cristina.rgonzalez}@uclm.es

J. Martínez-Gómez · M. Cazorla
Department of Computer Science and Artificial Intelligence,
University of Alicante, P.O. Box 99, 03080 Alicante, Spain
e-mail: miguel.cazorla@ua.es

L.P. Reis et al. (eds.), *Robot 2015: Second Iberian Robotics Conference*,
Advances in Intelligent Systems and Computing 418,
DOI: 10.1007/978-3-319-27149-1_49

conditions. This is done by encoding depth in conjunction to the color information. We can find a large set of mobile robots fitted with this type of cameras. Moreover, novel RGB-D datasets with object annotations have been released like NYU Depth V2 [20], ViDRILO [15], or RGB-D Object [11].

In this article, we present an experimentation where several object categorization systems are evaluated. Using the RGB-D Object dataset as benchmark, the systems are generated by: a) extract 3D local features from input RGB-D images, b) generate image descriptors following a Bag-of-Words approach [4], and c) train a classification model from the descriptors and object ground truth annotations. The experimentation includes several types of 3D local features, dictionary sizes and classification models. All these internal parameters are analyzed, and the internal details of the most promising object categorization systems are widely discussed. This work is intended to serve as baseline procedure for future developments, where we expect to generate solutions outperforming the use of depth kernel [2] or convolutional k-means descriptors [1].

The rest of the article is organized as follows: the object categorization problem is presented and formulated in Section 2. Section 3 defines the methodology proposed and presents the 3D local features in conjunction with the BoW approach. The experimentation is carried out in Section 4. Finally, Section 5 outlines the main conclusions drawn from the experimentation as well as the future work.

2 Object Categorization

Object categorization refers to the problem of predicting the category of a perception based on the object imaged in its content. This problem has been widely studied [5, 8, 16, 26], as there are several applications where object categorization is mandatory.

Object categorization can be formulated as a supervised classification problem as follows. Let I be an input RGB-D image representing a 3D object, $g(I) = \mathbf{x}$ a function that generates a descriptor $\mathbf{x}$ from I, and M a classification model that provides the class posterior probability $P_M(o|\mathbf{x})$, where o is an object label from a set of predefined objects $\mathcal{O}$. Then, the object categorization problem can be stated, without loss of generality, as:

$$\hat{o} = \arg\max_{o \in \mathcal{O}} P_M(o|g(I)) \tag{1}$$

that is, to find the optimal object label $\hat{o}$ from the set $\mathcal{O}$ of predefined label that provides the maximum posterior probability $P_M(\hat{o}|g(I))$, given a specific classification model M.

In this work, we will focus on defining $g(I)$. For that, we assume the use of RGB-D images. That is, I consists of a set of N points $p_1 \ldots p_N$. Each point p_i represents its own coordinate $< x_i, y_i, z_i >$, as well as color information $< R_i, G_i, B_i >$. Using a

Bag-of-Words approach, the process of generating an appropriate image descriptor ($g(I)$) is carried out from any set of 3D local features extracted from I, once the dictionary has been generated. With regard to the classification model (M), we opt for evaluating some state-of-the-art alternatives.

3 RGB-D Image Descriptor Generation

The image descriptor generation process proposed in this work includes three main steps. First, given an input RGB-D image, a selection of keypoints over the whole image (single region of interest) is carried out. Then, from these keypoins, a set of 3D local features is extracted. Finally, 3D local features are used to generate image descriptors following the BoW approach. This process is show in Fig. 1. It is worth noting that the generated descriptors will serve as inputs for subsequent classification stages. In the following subsections a detailed description of this three steps is provided.

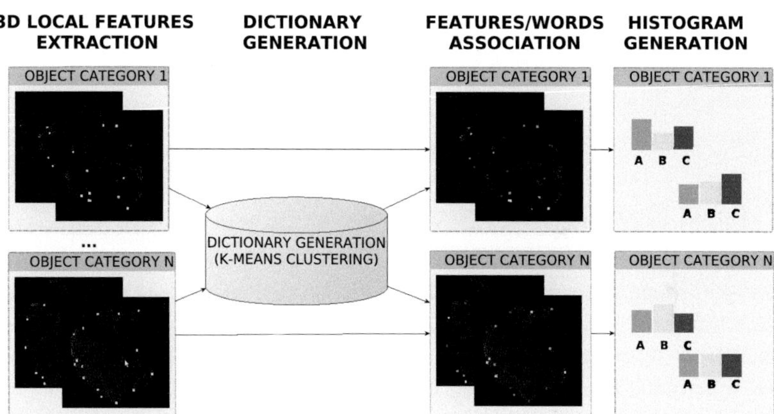

Fig. 1 Bag-of-Words from 3D local features scheme.

3.1 Keypoint Selection

As mentioned above, we cope with segmented RGB-D images as those presented in the RGB-D Object dataset [11]. The segmentation process is out of the scope of this work, but it can be carried out by removing the predominant plane in the scene, and then using region growing algorithms [7], as it is done for tabletop objects in [12]. In addition to this, hand-held objects can also be successfully segmented by exploiting the skeleton information [14].

To reduce the amount of information to work with, we adopted a uniform sampling approach for keypoint selection. This approach uniformly selects keypoint candidates from the whole segmented 3D point cloud. An example is shown in Fig. 2. Other keypoint detection techniques (as Harris3D [21] or NARF [22]) can also be applied, but the detection criteria would affect to the diversity of local features extracted (in subsequent stages) from such candidates.

Fig. 2 Uniform Sampling over a segmented RGB-D image.

3.2 Local Features Extraction

The keypoint candidates detected in the previous stage are used as reference points for the extraction of 3D local features. Such features are computed from the neighborhood of the reference point, and the extraction process can present some requirements. When such requirements are not met (e.g. a minimum number of 3D in the neighborhood) for a candidate 3D point, no features are extracted. This implies a number of extracted featured less or equal to the number of keypoint candidates detected in the sampling approach.

In this work, we opted for five different 3D local features implemented in the Point Cloud Library (PCL [19]):

- Fast Point Feature Histogram (FPFH [18])
- PFHRGB, which is the same as FPFH but integrating color information.
- Signature of Histograms of OrienTations (SHOT) [23]
- CSHOT [24], which is SHOT with and additional color histogram.
- Spin Images [9].

The dimensionality of all these features, as well as their main characteristics are summarized in Table 1.

Table 1 3D Local features summary.

	FPFH	FPHRGB	SHOT	CSHOT	SPIN IMAGES
Dimension	33	250	352	1344	153
Computed From	Geometry	Geometry	Orientation	Orientation	Surface
Color Information	No	Yes	No	Yes	No

3.3 Descriptor Generation

The use of the well-known Bag-of-Words approach [4] allows for the generation of fixed-dimensionality image descriptors from local features. This approach has been extensively used and studied [25], and several improvements and modifications like the use of a spatial pyramid [13] have been proposed. Taking into account the large set of 3D local features already implemented in the PCL, we can generate a great variety of BoW descriptors.

Given a set of 3D local features already extracted, the BoW approach would include the following steps:

1. Dictionary generation
2. Features/words association
3. Construction of the histogram of word frequencies

The dictionary generation step relies on a set of images $\mathcal{I} = \{I^1, ..., I^N\}$ and their local features $\{\mathcal{V}^1, ..., \mathcal{V}^N\}$ (where $\mathcal{V}^h = \{v_1^h, ..., v_m^h\}$ represents the local descriptors extracted from the m keypoint of the image I^h). The dictionary is generated by performing a k-means clustering (with $k = |D|$) on those local descriptors, where the set of computed centroids represents the dictionary of words or codebook $\{w_1, ..., w_k\}$.

$$D(w_1, ..., w_k) = \text{k-means}(k, \{\mathcal{V}^1, ..., \mathcal{V}^N\}) \tag{2}$$

Next, each feature has to be assigned to the closest word in the dictionary, according to the distance to the corresponding local descriptor:

$$word_{f_j^h} \leftarrow \arg\min_{w \in D} |v_j^h - w| \tag{3}$$

Finally, each image I^h is represented using a vector $\{n(w_{1,h}), ..., n(w_{k,h})\}$ where $n(w_{i,h})$ denotes the number of occurrences of word w_i in the image I^h.

Taking into account the use of RBG-D images (encoded as point clouds), each descriptor v_i^h corresponds to the 3D local feature extracted from the coordinate of the i-th keypoint (x_i, y_i, z_i) from the image I^h. The set of m coordinates is given by the keypoint detection algorithm, while the type of descriptor depends of the 3D local feature selected. The size of the dictionary $k = |D|$ should be previously established.

4 Experimental Results

We evaluate our proposal on the RGB-D Object dataset [11]. This dataset contains 300 object instances organized in 51 different categories. Eight of these categories are shown in Fig. 3. RGB-D images were captured individually on a turntable and segmented from the background. For each object instance, we can find 3 sequences which differ in the distance between the camera and the object (camera height). We used a BoW implementation over the PCL, which allows to be used in conjunction with any keypoint detectors and feature extractors already included in the library.

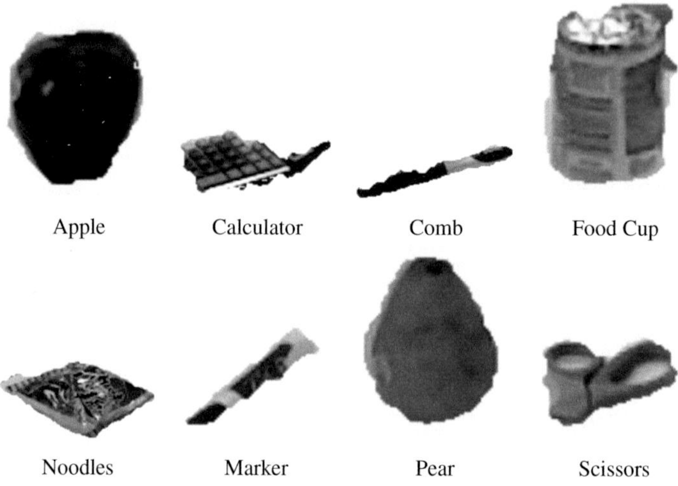

Fig. 3 Exemplar visual images for some of the object categories included in the RGB-D Object dataset.

As proposed in [11], the category recognition has been carried out by: a) selecting every 5th image to obtain a dataset with around 45000 RGB-D images, b) leaving one object instance out from each category for testing, and c) generating a classification model from the remaining object instances. This procedure and training/test partitions are those proposed by the authors of the dataset (we refer the readers to [11] for more details).

The challenging of this experimentation comes from the requested generalization capabilities, as well as from the large number of categories (51). Namely, there exist strong differences between object instances from the same category. This is exposed in Fig. 4, where it can be observed the color and geometric variations between instances belonging to the same object category.

Regarding the internal parameters used in the experimentation, we evaluated the following configurations:

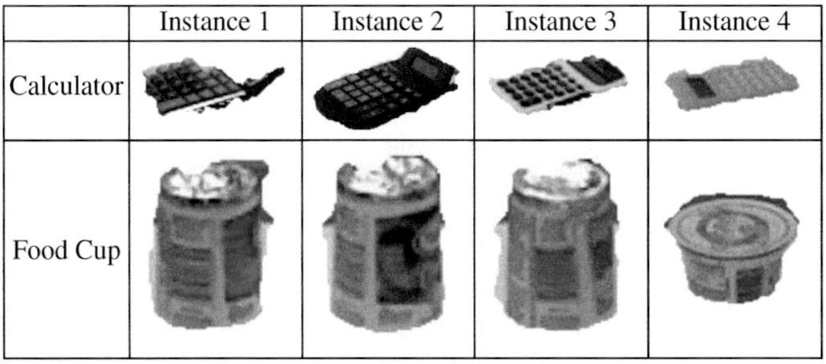

	Instance 1	Instance 2	Instance 3	Instance 4
Calculator				
Food Cup				

Fig. 4 Differences between four different instances of two objects in the RGB-D Object dataset: calculator and food cup.

- Keypoint Detector (1): Uniform Sampling
- Feature Extractors (5): FPFH, PFHRGB, SHOT, CSHOT, and Spin Images
- Dictionary sizes (7): 4, 8, 16, 32, 64, 128, 256
- Classifiers (4): SVMs (linear and $\chi^2 kernel$), Random Forest (50 trees), and k-NN ($k = 7$).

From this set of parameters, we computed the average accuracy using the category recognition procedure. The obtained results are shown in Fig. 5, where lowest accuracy values are obtained with k-NN, and most promising with χ^2 SVM. Regarding the 3D features, Spin Images and the two variations of Point Feature Histogram (FPFH and PFHRGB) outperform the use of SHOT and CSHOT. The low performance of both SHOT and CSHOT comes from the nature of the RGB-D Object dataset. Namely, this two features are computed from the orientation information of the input RGB-D image. As all the objects are imaged from a wide range of orientations in the dataset, no discrimination capabilities can be extracted from the orientation information. In the problem we are facing, the color information is not as relevant as it could be expected. In fact, PFHRGB obtains worse results than its pure depth alternative FPFH. This can be explained due to the large color variations between instances of the same object in the dataset (e.g. red and green apples).

With respect to the different dictionary sizes, χ^2 SVM is the classifier that more properly manages the increase of this parameter. Concretely, larger dictionary sizes are tipically translated into higher accuracy values, and this tendency is more clearly exposed with small values ($n \leq 64$). Increasing the dictionary size has a negative effect for k-NN, while the impact for linear SVM and Random Forest depends on the 3D feature.

While larger dictionary sizes can lead into higher accuracy, they present two main drawbacks. Firstly, larger descriptors involve larger training and classification stages due to their dimensionality. However, this effect can be mitigated using dimension reduction techniques like PCA or SVD [6]. The second drawback refers to the generalization capabilities. That is, the use of large dictionaries increases the over-fitting risk, which prevents the classifier from presenting generalization capabilities.

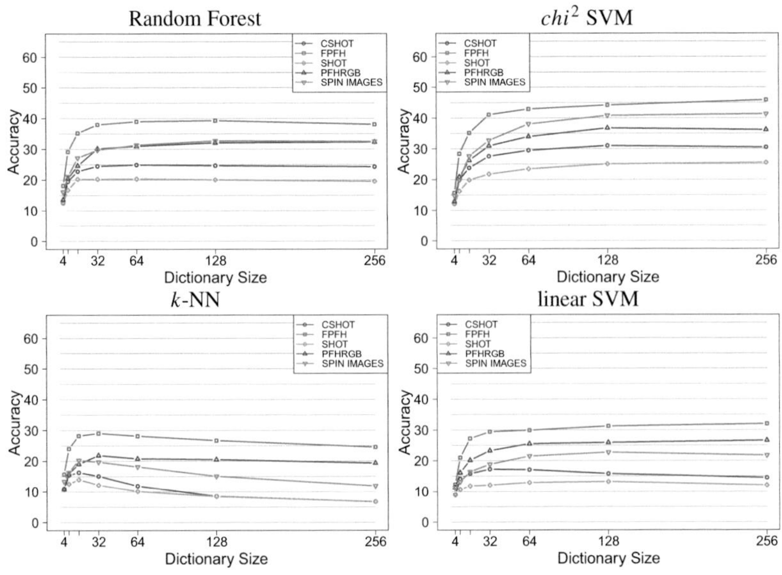

Fig. 5 Overall Results. Accuracy values over the RGB-D Object dataset.

The highest accuracy, 45.85%, was obtained using a χ^2 SVM classifier in conjunction with FPFH and a 256 sized-dictionary. This value is still far away from the highest accuracy values presented in the original RGB-D Object dataset (66.8 % using depth information, and 74.7 % from visual one). However, there is a point to be taken into account: our proposal relies on a pure point cloud representation, whereas other approaches use depth and perspective images as input. This encoding, in conjunction with the Point Cloud Library, allows for several capabilities that can be required before classifying an input RGB-D image, such as segmentation, planar removal or skeleton identification.

5 Conclusions and Future Work

We have presented a throughout methodology for the generation of 3D object categorization systems using the Bag-of-Words approach. This proposal relies on the use of the PCL and allows for the integration of current and future 3D features included in it. The methodology has been evaluated with the challenging RGB-D Object dataset. The experimentation carried out has tested several dictionary sizes, classification models and different 3D local features. This experimentation has served us to reduce the range of parameter combination in further developments, which would allow us to focus on the most promising features, classifiers, and dictionary sizes. Based on the results obtained, the χ^2 kernel SVM is the best classification model as it properly manages larger dictionary sizes. Moreover, overall higher accuracy values are obtained with such classifier. FPFH and Spin Images are the most favorable local features.

As future work, we have in mind the use of larger dictionary sizes and a wider range of 3D local features. We will also exploit the combination of two or more descriptors generated from the BoW approach, the introduction of contextual information, and we will test our approach against different datasets as the one used here.

Acknowledgements This work was supported by grant DPI2013-40534-R of the Ministerio de Economia y Competitividad of the Spanish Government, supported with FEDER funds, and by Consejería de Educación, Cultura y Deportes of the JCCM regional government through project PPII-2014-015-P. Cristina Romero-González is funded by the MECD grant FPU12/04387, and Jesus Martínez-Gómez is also funded by the JCCM grant POST2014/8171.

References

1. Blum, M., Springenberg, J.T., Wülfing, J., Riedmiller, M.: A learned feature descriptor for object recognition in RGB-D data. In: 2012 IEEE International Conference on Robotics and Automation (ICRA), pp. 1298–1303. IEEE (2012)
2. Bo, L., Ren, X., Fox, D.: Depth kernel descriptors for object recognition. In: 2011 IEEE/RSJ International Conference on Intelligent Robots and Systems (IROS), pp. 821–826. IEEE (2011)
3. Bustos, P., Martínez-Gómez, J., García-Varea, I., Rodríguez-Ruiz, L., Bachiller, P., Calderita, L., Manso, L., Sánchez, A., Bandera, A., Bandera, J.P.: Multimodal interaction with loki. In: Workshop de Agentes Físicos, Madrid-Spain (2013)
4. Csurka, G., Dance, C.R., Fan, L., Willamowski, J., Bray, C.: Visual categorization with bags of keypoints. In: Workshop on Statistical Learning in Computer Vision, ECCV, pp. 1–22 (2004)
5. Duchenne, O., Joulin, A., Ponce, J.: A graph-matching kernel for object categorization. In: 2011 IEEE International Conference on Computer Vision (ICCV), pp. 1792–1799. IEEE (2011)
6. Fodor, I.K.: A survey of dimension reduction techniques (2002)
7. Holz, D., Behnke, S.: Fast range image segmentation and smoothing using approximate surface reconstruction and region growing. In: Intelligent Autonomous Systems 12, pp. 61–73. Springer (2013)
8. Jie, L., Tommasi, T., Caputo, B.: Multiclass transfer learning from unconstrained priors. In: 2011 IEEE International Conference on Computer Vision (ICCV), pp. 1863–1870. IEEE (2011)
9. Johnson, A.E., Hebert, M.: Using spin images for efficient object recognition in cluttered 3d scenes. IEEE Trans. Pattern Anal. Mach. Intell. **21**(5), 433–449 (1999)
10. Koppula, H.S., Gupta, R., Saxena, A.: Learning human activities and object affordances from RGB-D videos. The International Journal of Robotics Research **32**(8), 951–970 (2013)
11. Lai, K., Bo, L., Ren, X., Fox, D.: A Large-Scale Hierarchical Multi-View RGB-D Object Dataset. In: IEEE International Conference on Robotics and Automation (ICRA), pp. 1817–1824, May 2011
12. Lai, K., Bo, L., Ren, X., Fox, D.: Detection-based object labeling in 3d scenes. In: 2012 IEEE International Conference on Robotics and Automation (ICRA), pp. 1330–1337. IEEE (2012)
13. Lazebnik, S., Schmid, C., Ponce, J.: Beyond bags of features: spatial pyramid matching for recognizing natural scene categories. In: 2006 IEEE Computer Society Conference on Computer Vision and Pattern Recognition, vol. 2, pp. 2169–2178. IEEE (2006)
14. Lv, X., Jiang, S.Q., Herranz, L., Wang, S.: RGB-D hand-held object recognition based on heterogeneous feature fusion. Journal of Computer Science and Technology **30**(2), 340–352 (2015)

15. Martinez-Gomez, J., Cazorla, M., Garcia-Varea, I., Morell, V.: Vidrilo: The visual and depth robot indoor localization with objects information dataset. The International Journal of Robotics Research (2015)
16. Martínez-Gómez, J., Gámez, J., García-Varea, I., Matellán, V.: Using genetic algorithms for real-time object detection. In: RoboCup 2009: Robot Soccer World Cup XIII. Lecture Notes in Computer Science, vol. 5949, pp. 215–227. Springer, Heidelberg (2010)
17. Romero-Garcés, A., Calderita, L.V., Martínez-Gómez, J., Bandera, J.P., Marfil, R., Manso, L.J., Bandera, A., Bustos, P.: Testing a fully autonomous robotic salesman in real scenarios. In: 2015 IEEE International Conference on Autonomous Robot Systems and Competitions (ICARSC), pp. 124–130. IEEE (2015)
18. Rusu, R., Blodow, N., Beetz, M.: Fast point feature histograms (FPFH) for 3d registration. In: IEEE International Conference on Robotics and Automation, ICRA 2009, pp. 3212–3217, May 2009
19. Rusu, R., Cousins, S.: 3D is here: Point Cloud Library (PCL). In: IEEE International Conference on Robotics and Automation (ICRA), Shanghai, China, May 9–13 2011 (2011)
20. Silberman, N., Hoiem, D., Kohli, P., Fergus, R.: Indoor segmentation and supportinference from rgbd images. In: ECCV (2012)
21. Sipiran, I., Bustos, B.: Harris 3d: a robust extension of the harris operator for interest point detection on 3d meshes. The Visual Computer **27**(11), 963–976 (2011)
22. Steder, B., Rusu, R.B., Konolige, K., Burgard, W.: Narf: 3d range image features for object recognition. In: Workshop on Defining and Solving Realistic Perception Problems in Personal Robotics at the IEEE/RSJ Int. Conf. on Intelligent Robots and Systems (IROS), vol. 44 (2010)
23. Tombari, F., Salti, S., Di Stefano, L.: Unique signatures of histograms for local surface description. In: Computer Vision-ECCV 2010, pp. 356–369. Springer (2010)
24. Tombari, F., Salti, S., Di Stefano, L.: A combined texture-shape descriptor for enhanced 3d feature matching. In: 2011 18th IEEE International Conference on Image Processing (ICIP), pp. 809–812, september 2011
25. Yang, J., Jiang, Y.G., Hauptmann, A.G., Ngo, C.: Evaluating bag-of-visual-words representations in scene classification. In: Proceedings of the International Workshop on Workshop on Multimedia Information Retrieval, MIR 2007, pp. 197–206. ACM, New York (2007)
26. Zhang, C., Cheng, J., Liu, J., Pang, J., Liang, C., Huang, Q., Tian, Q.: Object categorization in sub-semantic space. Neurocomputing **142**, 248–255 (2014)

A Multisensor Based Approach Using Supervised Learning and Particle Filtering for People Detection and Tracking

Eugenio Aguirre , Miguel García-Silvente and Daniel Pascual

Abstract People detection and tracking is an interesting skill for interactive social robots. Laser range finder (LRF) and vision based approaches are the most common although both present strengths and weaknesses. In this paper, a multisensor system to detect and track people in the proximity of a mobile robot is proposed. First, a supervised learning approach is used to recognize patterns of legs in the proximity of the robot using a LRF. After this, a tracking algorithm is developed using particle filter and the observation model of legs. Second, a Kinect sensor is used to carry out people detection and tracking. This second method uses a face detector in the color image, the color of the clothes and the depth information. The strengths and weaknesses of the second proposal are also commented. In order to put together the strengths of both sensors, a third algorithm is proposed. In this third approach both laser and Kinect data are fused to detect and track people. Finally, the multisensory approach is experimentally evaluated in a real indoor environment. The multisensor system outperforms the single sensor based approaches.

Keywords People detection and tracking · Multisensor based tracking · Social robot · Human-robot interaction

1 Introduction

In order to focus its attention on humans, a social robot needs to be aware of their presence. Therefore, people detection and tracking is an interesting skill. This is a challenging task because people move freely by the environment around the robot,

E. Aguirre(✉) · M. García-Silvente · D. Pascual
Department of Computer Science and A.I., CITIC-UGR., E.T.S. Ingenierías en Informítica y en Telecomunicaciones, University of Granada, 18071 Granada, Spain
e-mail: {eaguirre,M.Garcia-Silvente}@decsai.ugr.es, dpascual@correo.ugr.es
http://decsai.ugr.es

E. Aguirre—This work have been partially supported by the Spanish Government project TIN2012-38969.

L.P. Reis et al. (eds.), *Robot 2015: Second Iberian Robotics Conference*,
Advances in Intelligent Systems and Computing 418,
DOI: 10.1007/978-3-319-27149-1_50

moreover the sensory systems could suffer from the presence of false positives, noise and vagueness in the sensorial data.

Several approaches have been proposed to carry out people detection and tracking using different kinds of sensors, being the most popular sensors for those tasks laser sensors and cameras. Regarding laser based approaches, for instance, in [7] authors propose detection and tracking schemes for human legs by the use of a single LRF. In [15] a systematic comparative analysis of laser based tracking methods, at feet and upper-body height, is performed. In [16] a system for tracking a variable number of pedestrians in crowded scenes by exploiting laser range scanners is proposed. On these works some conclusions arise. Compared with vision approaches, the use of laser sensors is advantageous since they are robust against illumination changes in the environment and the tracking algorithms are faster and more efficient. However, laser sensors have some limitations because the robot only can obtain distance information from a 2d-plane located at a certain height. A 3D-laser could solve this limitation but other problems arise as the cost of this device or the time of the data acquisition. In regards to the vision based approaches, mono, stereo cameras and RGB-Depth, as the Kinect sensor [12], have been used to detect and track people. Stereo and RGB-Depth cameras provide color and depth information. In [13] a fuzzy algorithm for detection and tracking of people in the proximity of a robot by using stereo vision is proposed. In [19] the depth information obtained from a Kinect sensor is used to detect humans. A 2-D head contour model and a 3-D head surface model is shown. Then a tracking algorithm is proposed based on their detection results. Vision based approaches also have some limitations. Illumination conditions can affect the performance of these methods, depth information is not always reliable and false positives in the detection methods are possible. In order to improve the results of laser and vision based approaches, multisensor solutions propose to use several kinds of sensors to achieve a more robust solution. In [3] a people following behavior is developed fusing information provided by a laser sensor and a stereo camera. In [5] authors propose multisensor data fusion using a laser scanner and a monocular camera. In [17] a multiple sensor fusion approach is proposed using three kinds of devices, Kinect, laser and a thermal sensor mounted on a mobile platform. It is shown that combination of different sensory cues increases the reliability of their people following system.

In this paper we propose a multisensor system to detect and track people in the proximity of a mobile robot. First, a supervised learning approach is used to identify patterns of legs in the proximity of the robot. This method analyses certain geometric features present in the laser data in order to detect possible legs of people. A classifier is trained using Support Vector Machines (SVM) [6] to classify the data obtained from laser using instances of patterns of legs. After this, a tracking algorithm is developed using particle filter and the observation model of the legs. The tracking algorithm is experimentally evaluated and some strengths and weaknesses are commented. Second, a Kinect sensor is used to carry out the people detection and tracking. This second method uses a face detector in the color image in order to perform people detection and the color of the clothes and the depth information is used to track people. Again a particle filter is developed using only the Kinect sensor to compare the results against the first proposal. The strengths and weaknesses of the second method are commented as well. In order to put together the strengths of both sensors, a third algorithm is proposed. In this third approach both laser and Kinect are used to detect and track people in the proximity of the robot. In the same way, the multisensory approach is experimentally evaluated. In the current work the robot is standing still and looking the motion of people but in next works this limitation will be removed taking into account the required changes.

The rest of this paper is organized as follows. Section 2 describes the hardware that has been used in our system. Section 3 briefly describes the human legs detection algorithm, including the supervised learning algorithm, and the laser based tracking proposal. Section 4 shows the people detection and tracking algorithm based on Kinect. In Section 5 the multisensor approach is shown and experimental results are compared with both previous approaches. Finally some concluding remarks and future works are commented in Section 6.

2 System Description

Our hardware system comprises a PeopleBot mobile robot equipped with a LRF SICK LMS200 and a Kinect sensor. Laser sensor scans 180° with a 1° resolution at 75 Hz. Its maximum range of distance in the current operation mode is 8 meters. It is mounted at a height of 30 cm above the ground. The Kinect features a RGB camera, a depth sensor and a multi-array microphone. Kinect uses an infrared projector and an infrared camera which are able to compute depth [14]. The Kinect depth sensor range is: minimum 800 mm and maximum 4000 mm. The resolution of both color and depth images is 640x480 at 30 fps. Because it uses IR, Kinect will not work under direct sunlight, e.g. outdoors. Since our system is intended to allow human robot interaction in indoor environments, these features of Kinect are suitable for our specifications. More information on Kinect is available in [20]. A laptop has been used to run the software due to the onboard computer is not powerful enough to perform video processing. The laptop has an Intel Core i5 with 4 GB DDR3 RAM and it is wired connected to the onboard computer. The laptop receives the laser data from the onboard computer while the Kinect sensor is directly connected to the laptop.

3 People Detection and Tracking Using Laser Sensor (LRF Based Method)

The objective of this work is the design of a system capable to be used in Human Robot Interaction (HRI) applications. People interested in establishing interaction with the robot should be close to the robot; thus, an operation range of 1 to 3 m is defined. The first approach to detect and track people is based on a previous work by the authors [1] so that only a brief description is given below.

Table 1 Contingency table for the SVM classifier

		Observation class	
		Positive	Negative
Predicted class	Positive	88.71 %	10.62 %
	Negative	11.29 %	89.38 %

3.1 Leg Detection Method

The idea is to detect the legs when people are both moving or static. It is a challenging task because legs patterns are different in both situations. To do so, the laser measurements are clusterized and their geometrical properties are then analyzed. The considered properties comprise width, depth and size and all of them have been used successfully by others authors [7]. In our approach, a SVM classifier is trained by using the properties of the detected clusters and a large data set that contains positive and negative instances of patterns of legs. Positive instances were registered with people walking and standing in the proximity of the robot. Negative instances include objects such as table legs, bins, boxes and various kinds of fire extinguishers. Note that some of this object could have geometrical properties similar to those of human legs. A balanced dataset containing 7802 instances of both, positive and negative samples, was used to train the SVM classifier.

In order to apply the SVM classifier, LibSVM was used [6]. Different kernels have been considered and the best precision is obtained with radial basis function (RBF). A wide grid-search using cross-validation has been performed in order to find the optimal value for these parameters, obtaining a precision of 89% which is suitable for this kind of application. Table 1 shows the results of a 10-fold cross validation. Results are acceptable since the rates of true positives and true negatives are high.

3.2 Particle Filter Based Tracking

Particle filter is well known for its many applications in tracking. Target tracking problem is expressed by recursive Bayesian estimation. Essentially, two steps are given in each iteration: prediction and estimation. Both steps take into account the information of an observation model. Equations of particle filter are well known [4]. The vector of state, the definition of state transition and the model of noise is described in [1].

The LRF based method uses the leg detection algorithm as observation model so that each laser reading set is analyzed and the positions of possible legs are obtained. The probability for each particle is computed taking into consideration the distance between the position of the nearest detected legs to the evaluated particle and the last known position of legs of the tracked person. Details are also described in [1].

3.3 Experimental Results of LRF Based Method

In order to test the accuracy of the LRF based method several experiments have been carried out in a real indoor environment. A set of five paths on the floor were defined taking into account different trajectories. Two trajectories are straight, one is a circle and the last two are curves. The experiments consist of tracking people whom are following those trajectories. The trajectories have been manually mapped to serve as ground truth on people motion. Five persons participated in the experiments. It is important to acquire data from different people since each person has a particular gait. Every person walked on each trajectory three times. Thus, 75 different samples can be analyzed to measure the accuracy of the proposal. Notice that laser and images were collected at the same time to build a dataset which is used to evaluate the three approaches shown in this work.

The performance of the LRF based tracking on a trajectory J, is measured taking into account for each time t, the euclidian distance from the hypothesis computed by the tracker h_t, to the real position $p_t \in J$. The correspondence between h_t and p_t should not be made if its distance d_t exceeds a certain threshold H. If $d_t > H$ then the tracker has missed the person in the time t. The tracking error $TE_1(J)$, given a trajectory J, is computed by $TE_1(J) = \frac{\sum_{t=1}^{m_j} d_t}{m_j}$ where m_j is the total number of matches made in the trajectory J.

The algorithm has been evaluated using different numbers of particles: 50, 100, 150, 200 and 250. Table 2 shows the results obtained for each trajectory T and each number of particles. Tracking error $TE_1(J)$ and standard deviation Std_1, in mm, are indicated. The error of tracking decreases when the number of particles increases. For a number of particles higher than 200 the error decreases at lower rate, hence, the final algorithm uses 200 particles. Using 200 particles, the average tracking error for all the trajectories is computed obtaining a value of 34,33 mm and its standard deviation is 9,93 mm. The processing framerate obtained using 200 particles has been 40,02 Hz. Therefore the LRF based approach has a good precision to track a person and it can be used in real time.

Strengths of LRF based approach are the precision, performance rate and wide field of view of the sensor. However if the model of observation is not sufficiently discriminatory then it is not possible to distinguish between two people when their trajectories intersect. In such situations, the tracker can confuse the targets.

Table 2 Results in mm for $TE_1(J)$

	Number of particles									
	50		100		150		200		250	
J	TE_1	Std_1	TE_1	Std_1	TE_1	Std_1	TE_1	Std_1	TE_1	Std_1
1	28.82	47.60	28.90	51.03	27.26	50.71	27.11	47.90	25.32	39.52
2	57.04	53.53	55.60	56.40	49.43	51.75	49.38	63.16	47.83	28.83
3	28.95	30.75	31.18	31.61	27.47	28.56	24.14	31.08	23.87	39.82
4	43.11	39.80	38.65	39.12	32.10	38.88	33.32	44.75	33.07	34.12
5	49.62	40.24	46.67	41.39	42.36	42.27	37.68	40.14	36.84	44.97

This problem is illustrated by Fig. 1. In this figure, two persons are tracked by the system. Red and green points represent two different persons and blue points are the laser readings. From up left to down right, the two first scenes show the system properly tracking both people. The two scenes situated below show the situation when two persons are intersecting their trajectories and the system is confused as both trackers end up following the same person. Some proposals try to overcome this problem by including a model of human walking motion [7] or by using a more complex state and observation models and then applying data association techniques [5]. However false positives and tracking errors are still possible. In this work, a multisensor based tracking is proposed to overcome this problem, and therefore, to achieve multiple people tracking.

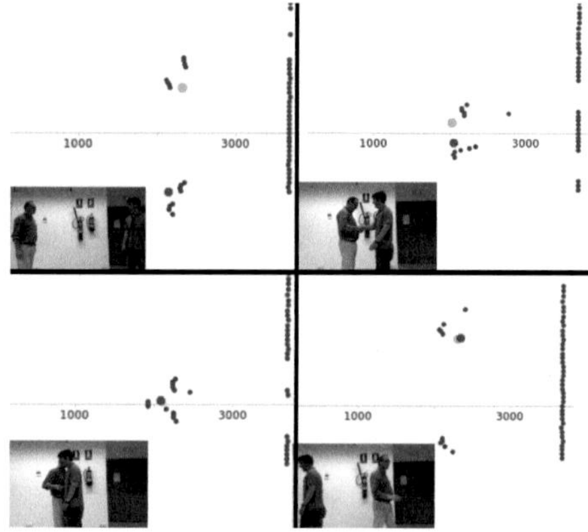

Fig. 1 LRF based approach fails to discriminate between people that get close enough.

4 People Detection and Tracking Using Kinect Sensor

Kinect sensor provides both 640x480 distance map and color image. Therefore both color and depth information can be used for people detection and tracking. The Kinect based approach used in this work has been adapted from a previous work by the authors [13]. In the previous work, a traditional stereo camera was used and the color and depth information was fused using fuzzy logic. Now, Kinect is the sensor used and color and depth information is fused using a particle filter. Below Kinect based approach is described.

Notice that in this work people detection and tracking comprise separate processes. When new people are detected then independent trackers are created for every one. People detection is made by using the frontal face detector of OpenCV [10] based on the Viola and Jones' method [18]. Once a person is detected, a model of that person is built by our method. This model is an elliptical region placed at the height of the person chest. Standard anthropomorphic measurements have been taken into account to build this model. The model is resized depending on the distance of people to the Kinect sensor. The center of this elliptical model will be the target to track on the color image. This is done by analyzing its color and depth information. To do so, a color histogram $\hat{q}$ of the elliptical region is calculated using the HSV color space [8]. HSV color space is relatively invariable to illumination changes. A color histogram $\hat{q}$ comprises $n_h n_s$ bins for the hue and saturation. However, chromatic information cannot be considered reliable when the value component is too small or too big. Therefore, pixels on this situation are not used to describe the chromaticity. Because of these pixels might have important information, the histogram includes also n_v bins to capture its luminance information. Thus, the resulting histogram is composed by $m = n_h n_s + n_v$ bins.

As stated above, we consider an elliptical region of the image to create the color model whose horizontal and vertical axes are h_x and h_y respectively. Let p_c be the ellipse center and $\{p_j\}_{j=1,\ldots,n}$ the locations of the interior pixels. Let's also define a function $b : \Re^2 \rightarrow 1, \ldots, m$ which associates to the pixel at location p_j the index $b(p_j)$ of the histogram bin corresponding to the color u of that pixel. It is now possible to compute the color density distribution for each bin $\hat{q}(u)$ of the elliptical region with:

$$\hat{q}(u) = \frac{1}{n} \sum_{j=1}^{n} k[b(p_j) - u], \tag{1}$$

where the parameter k is the Kronecker delta function. Please notice that the resulting histogram is normalized, i.e., $\sum_{u=1}^{m} \hat{q}(u) = 1$. After calculating the color model, the Bhattacharyya coefficient as described in [2] can be computed. In the case of a discrete distribution it can be expressed as indicated in Eq. 2. The result expresses the similarity between two color models in the range of [0, 1] where 1 means that they are identical and 0 means that they are completely different.

$$\rho(\hat{q}, \hat{q}') = \sum_{u=1}^{m} \sqrt{\hat{q}(u)\hat{q}'(u)}. \tag{2}$$

Once the Bhattacharyya coefficient is computed, two models of color $\hat{q}$, $\hat{q}'$ can be compared through the Bhattacharyya distance [2]:

$$BD(\hat{q}, \hat{q}') = \sqrt{1 - \rho(\hat{q}, \hat{q}')} \tag{3}$$

It provides values near 0 when two color models are similar and tends to 1 as they differ. An important feature of ρ is that both color models, $\hat{q}$ and $\hat{q}'$, can be compared even if they have been created using regions of different sizes.

4.1 Particle Filter Based Tracking

A particle filter is again used to achieve a robust tracking of detected people. In this case the state at time t is defined as a pair of coordinates (x, y) on the image plus the information of depth Z of that pixel. These coordinates correspond to the pixel centered on the elliptical region that is used to model the detected person. That is, the people position S_t is represented by the state model $S_t = [x_t, y_t, Z_t]$. The prediction is carried out by the model of the state transition. The state transition is defined as $S_t = S_{t-1} + R_{t-1}$ where S_{t-1} is the previous state vector and R_{t-1} is the process noise. A model of people velocity is not explicitly considered in order to manage the unpredictability of human behaviors. The noise is modeled using a Gaussian with average μ_R and standard deviation σ_R. Experimental data have been taken into account to establish the values of μ_R and σ_R in order to model the conditions of the real world.

Condensation algorithm [11] is used to generate a weighted set of particles $(s_i(t), \Pi_i(t))$ where $s_i(t)$ represents an hypothesis of the position of the person being tracked, and $\Pi_i(t)$ is a factor called *importance weight* which provides an estimation of the observation. At the beginning the algorithm is provided by an initial sample $(s_i(0), \Pi_i(0))$ of N equally weighted particles. At each iteration, the algorithm uses the sample set $(s_i(t-1), \Pi_i(t-1))$ to create a new one. A resample mechanism is used to solve the divergence problem by eliminating particles having low importance weights. Afterwards, the model of state transition is used to predict the motion of the person obtaining the prediction of the state S'_t. The weight $\Pi_i(t)$ of each particle is computed based on the new observation $O(t)$. Then the weights are normalized so that $\sum_{i=1}^{N} \Pi_i(t) = 1$.

The observation model is required to carry out the update. As model of observation, position (x, y) on the image, depth Z and color information are used. On one hand, let $f_{x,y}$ be the euclidian distance in pixels between the position of the particle $s_i(t)$ on the image and the last known state S_t and f_Z the difference of depth between both positions. On the other hand, let $BD(\hat{q}, \hat{q}')$ be the Bhattacharyya distance (Eq. 3) between the corresponding elliptical regions centered on the particle $s_i(t)$ and the last known position S_t. Then, the importance weight of each particle is computed by:

$$\Pi_i(t) = e^{-\frac{1}{2}\left(\frac{f_{x,y}}{\sigma_1}\right)^2} \cdot e^{-\frac{1}{2}\left(\frac{f_Z}{\sigma_2}\right)^2} \cdot (1 - BD(\hat{q}, \hat{q}')). \tag{4}$$

Parameters σ_1, σ_2 correspond to the standard deviations of two zero centered normal distributions, respectively. σ_1, σ_2 have been experimentally tuned. The final person position corresponds to the mean of the state $\mathcal{E}[S(t)]$, calculated as $\mathcal{E}[S(t)] = \sum_{i=1}^{N} \Pi_i(t) s_i(t)$. Please, note that face detection is only used in the detection phase but it is not used in the tracking phase. Therefore, once a person is detected, this person can be tracked using its people model although his or her face is not again detected.

4.2 Experimental Results of Kinect Based Method

The goal is to compare the results of the Kinect based approach with the LRF based approach. Therefore the same dataset collected to test the LRF based approach is used. However both systems use different coordinates systems. As Kinect sensor provides the information of depth, the estimated position on the image x, y can be projected to the LRF coordinates system. Also it is required to have into account that the ground truth was built by measuring the positions of the middle point between the legs and now the target is located at the height of the chest. All these details have been taken into consideration in order to achieve comparable results. The algorithm has been evaluated using different numbers of particles: 50, 100, 150, 200 and 250. Table 3 shows the results obtained for each trajectory T and each number of particles. Tracking error $TE_2(J)$ and standard deviation Std_2 in mm are computed in a similar way to that of the first approach. Once again, the tracking error decreases as the number of particles increases. Using 200 particles, the average tracking error for all the trajectories is computed obtaining a value of 66,17 mm and its standard deviation is 29,29 mm. The processing framerate obtained using 200 particles has been 2,46 Hz. These results point out that the precision of this approach is lower than LRF based approach although it is enough to develop Human-Robot Interaction applications. The main problem is the processing framerate since which is low due to the computing time required to process the color image and the usage of face detector on each frame in order to detect a new people in the frame.

However the Kinect based approach has a main advantage over the LRF based approach. When two or more people are tracked and the color of their vests are different, the Kinect based approach can still track people without confusing them and it can cope as well with certain level of occlusion. This situation is shown by Fig. 2.

Fig. 2 Tracking two people using the Kinect based approach.

Table 3 Results in mm for $TE_2(J)$

	Number of particles									
	50		100		150		200		250	
J	TE_2	Std_2	TE_2	Std_2	TE_2	Std_2	TE_2	Std_2	TE_2	Std_2
1	73.63	37.90	58.02	67.69	58.61	72.12	52.01	40.46	50.33	49.72
2	127.27	135.03	111.41	130.71	100.18	126.65	97.80	123.60	97.18	124.62
3	114.44	99.18	98.24	93.33	92.87	81.71	96.45	100.35	86.96	89.32
4	44.04	67.36	39.35	32.86	38.50	32.62	32.92	25.47	33.15	20.15
5	89.38	90.98	67.90	68.50	51.03	45.87	51.68	56.88	51.04	34.21

5 People Detection and Tracking Using a Multisensor Approach

On one hand, LRF based approach is precise and fast but it can confuse the targets when two or more people are being tracked. On the other hand, the Kinect based approach has a lower precision and is slower but it can distinguish people by using color and depth information. Both approaches can suffer from false positives detection. That is, the SVM classifier can recognize laser data as legs in a false way and the OpenCV face detector can recognize faces in the color image erroneously as well. The multisensor approach fuses information from both sensors in order to achieve a more robust people detection and tracking system. In the detection phase, first the leg detector is used to recognize possible pairs of legs in the proximity of the robot. Second, the possible detected faces are matched to the possible legs and both observations have to be coherent to consider that a new person has been detected. Notice that the fields of view of both devices are different. That is, there can be legs detected but if the person is out of the field of view of Kinect then it is not possible to find the corresponding face. Only when both, legs and face, are detected the system creates a new tracker if the person was not already being tracked.

5.1 *Particle Filter Based Tracking*

The state definition is similar to the Kinect based approach, but it now includes the information on the position $h_t = \{hx_t, hy_t\}$ of the people legs. Thus, people position S_t is represented by $S_t = [x_t, y_t, Z_t, h_t]$. The state transition and noise models are similar to those explained in Sect. 3 and Sect. 4. The observation model includes both previous kinds of information $f_{x,y}$ and f_Z, so that the importance weight of the particle is computed by:

$$\Pi_i(t) = e^{-\frac{1}{2}\left(\frac{f_{x,y}}{\sigma_1}\right)^2} \cdot e^{-\frac{1}{2}\left(\frac{f_Z}{\sigma_2}\right)^2} \cdot (1 - BD(\hat{q}, \hat{q}')) \cdot e^{-\frac{1}{2}\left(\frac{f_h}{\sigma_3}\right)^2} \quad (5)$$

being f_h the euclidian distance between the position of the nearest detected legs to $s_i(t)$ and the last known position of legs of the tracked person. Parameters σ_1, σ_2 are the same as of those in Eq. 4 and σ_3 correspond to the standard deviation of a zero centered normal distribution. σ_3 has been experimentally tuned. The final person position corresponds to the mean of the state $\mathcal{E}[S(t)]$, calculated as $\mathcal{E}[S(t)] = \sum_{i=1}^{N} \Pi_i(t)s_i(t)$.

5.2 Experimental Results of Multisensor Based Method

The idea is to compare the results of the multisensor based approach with the two previous approaches. Therefore the same dataset is used. The transformation of the coordinates is also done in this case. The algorithm has been evaluated using different numbers of particles: 50, 100, 150, 200 and 250. Table 4 shows the results obtained for each trajectory T and each number of particles. Tracking error $TE_3(J)$ and standard deviation Std_3, in mm, are indicated. Again the error of tracking decreases as the number of particles increases. Using 200 particles, the average tracking error for all the trajectories is computed, obtaining an average value of 40,88 mm and its standard deviation is 10,18 mm. The processing framerate obtained using 200 particles has been 3,24 Hz. The results point that the precision of this approach is lower than LRF based approach but higher than the Kinect based approach. Nevertheless, we think that it is enough to develop Human-Robot Interaction applications. Also, the multisensor approach can distinguish several people depending on the color of their vest and avoid some false positives of both face and legs detectors. It is the most robust approach and certain level of occlusion can been managed by the system. The processing framerate has improved regarding to the Kinect based approach due to that the face detector is not used for each frame but only when an additional legs are detected. However it is still slow due to the computing time required to process the color image. Although the frame rate is low, some applications on mobile robots have been developed using similar processing framerates taking into account certain limits. For instance, in [5] a multisensor human detection and tracking system at 4 Hz is used to follow people. Also, in [9] a multisensor system running a face detector at a rate of 3 Hz provides good results to follow people moving in a standard office domain.

Table 4 Results in mm for $TE_3(J)$

	Number of particles									
	50		100		150		200		250	
J	TE_3	Std_3	TE_3	Std_3	TE_3	Std_3	TE_3	Std_3	TE_3	Std_3
1	44.16	40.61	42.08	44.77	38.96	43.23	32.54	42.39	30.87	42.95
2	78.69	52.55	66.15	55.45	61.55	54.95	58.63	48.67	58.11	63.11
3	50.12	52.48	49.86	58.52	41.74	55.37	37.64	51.54	34.63	55.44
4	56.76	50.45	53.24	58.44	39.50	60.11	37.42	57.35	35.97	55.27
5	45.17	28.55	42.15	27.49	38.97	39.20	38.15	28.64	37.86	41.64

6 Conclusions and Future Work

In this paper a new multisensor system to detect and track people in the proximity of a mobile robot has been proposed. The multisensor approach tries to put together the strengths of both LRF and Kinect sensors. To explain the develop of the multisensor system and its advantages, first the LRF based approach is shown and experimentally evaluated. This method analyses certain geometric features present in the laser data in order to detect possible legs of people. A classifier is trained using SVM to classify the data obtained from laser using instances of patterns of legs. The LRF based approach is briefly described because is based on a previous work by the authors. Second, a new Kinect based approach has been developed for this work. The second approach has been also experimentally evaluated and results show less precision and more computation time than the first one but it can distinguish people using the color of their vests. The best results have been obtained by the multisensor system. The multisensor based approach is able to detect and track people in a real indoor environment obtaining average tracking error of approximately 4 cm. The main contributions of this work are the development of the multisensor based approach and the method to fuse color and depth information of the Kinect sensor with distance information of a LRF sensor using a particle filter. As future work the goal will be to improve the processing framerate. To do so, for instance, one possibility is to reduce the resolution of the images so that less data have to be processed. Another idea is to execute the face detector only on certain parts of the images instead of on the whole frame. Also, parallel computing can be used to improve the speed. Finally, depending on the required precision, the number of particles used can be lowered in order to reduce the processing time.

References

1. Aguirre, E., Garcia-Silvente, M., Plata, J.: Leg detection and tracking for amobile robot and based on a laser device, supervised learning and particle filtering. In: ROBOT2013: First Iberian Robotics Conference, volume 252 of Advances in Intelligent Systems and Computing, pages 434–440. Springer Int. Publis. (2014)
2. Aherne, F., Thacker, N., Rockett, P.: The bhattacharyya metric as an absolute similarity measure for frequency coded data. Kybernetica **32**, 1–7 (1997)
3. Ansuategui, A., Ibarguren, A., Martíonez-Otzeta, J.M., Tubío, C., Lazkano, E.: Particle filtering for people following behavior using laser scans and stereo vision. International Journal on Artificial Intelligence Tools **20**(02), 313–326 (2011)
4. Arulampalam, M.S., Maskell, S., Gordon, N., Clapp, T.: A tutorial on particle filters for online nonlinear/non-gaussian bayesian tracking. IEEE Transactions on Signal Processing **50**(2), 174–188 (2002)
5. Bellotto, N., Hu, H.: Multisensor-based human detection and tracking for mobile service robots. IEEE Transactions on Systems, Man, and Cybernetics, Part B: Cybernetics **39**(1), 167–181 (2009)
6. Chang, C.-C., Lin, C.-J.: LIBSVM: A library for support vector machines. ACM Transactions on Intelligent Systems and Technology, 2:27:1–27:27, 2011. Software available at http://www.csie.ntu.edu.tw/cjlin/libsvm

7. Chung, W., Kim, H., Yoo, Y., Moon, C.-B., Park, J.: The detection and following of human legs through inductive approaches for a mobile robot with a single laser range finder. IEEE Transactions on Industrial Electronics **59**(8), 3156–3166 (2012)
8. Foley, J.D., van Dam, A.: Fundamentals of Interactive Computer Graphics. Addison Wesley (1982)
9. Fritsch, J., Kleinehagenbrock, M., Lang, S., Plótz, T., Fink, G.A., Sagerer, G.: Multi-modal anchoring for human-robot interaction. Robotics and Autonomous Systems **43**(2–3), 133–147 (2003)
10. Intel-Corporation. Opencv (2015)
11. Isard, M., Blake, A.: Condensation-conditional density propagation for visual trackings. International Journal of Computer Vision **29**, 5–28 (1998)
12. Microsoft. Kinect official webpage (2010)
13. Paúl, R., Aguirre, E., García-Silvente, M., Muñoz-Salinas, R.: A new fuzzy based algorithm for solving stereo vagueness in detecting and tracking people. International Journal of Approximate Reasoning **53**, 693–708 (2012)
14. Primesense. Primesense official webpage (2005)
15. Schenk, K., Eisenbach, M., Kolarow, A., Gross, H.: Comparison of laser-based person tracking at feet and upper-body height. In: Bach, J., Edelkamp, S. (eds.) KI 2011: Advances in Artificial Intelligence. Lecture Notes in Computer Science, vol. 7006, pp. 277–288. Springer, 2011
16. Shao, X., Katabira, K., Shibasaki, R., Zhao, H., Nakagawa, Y.: Tracking a variable number of pedestrians in crowded scenes by using laser range scanners. In: SIEEE International Conference on ystems, Man and Cybernetics, SMC 2008, pp. 1545–1551, October 2008
17. Susperregi, L., Martínez-Otzeta, J.M., Ansuategui, A., Ibarguren, A., Sierra, B.: RGB-D laser and thermal sensor fusion for people following in a mobile robot. International Journal of Advanced Robotic Systems **10**, 271 (2013)
18. Viola, P., Jones, M.: Rapid object detection using a boosted cascade of simple features. In: IEEE Conf. on Computer Vision and Pattern Recognition, pp. 511–51 (2001)
19. Xia, L., Chen, C.-C., Aggarwal, J.K.: Human detection using depth information by kinect. In: 2011 IEEE Computer Society Conference on Computer Vision and Pattern Recognition Workshops (CVPRW), pp. 15–22, June 2011
20. Zhang, Z.: Microsoft kinect sensor and its effect. IEEE MultiMedia **19**(2), 4–10 (2012)

Incremental Compact 3D Maps of Planar Patches from RGBD Points

Juan Navarro and José M. Cañas

Abstract The RGBD sensors have opened the door to low cost perception capabilities for robots and to new approaches on the classic problems of self localization and environment mapping. The raw data coming from these sensors are typically huge clouds of 3D colored points, which are heavy to manage. This paper describes a premilinary work on an algorithm that incrementally builds compact and dense 3D maps of planar patches from the raw data of a mobile RGBD sensor. The algorithm runs iteratively and classifies the 3D points in the current sensor reading into three categories: close to an existing patch, already contained in one patch, and far from any. The first points update the corresponding patch definition, the last ones are clustered in new patches using RANSAC and SVD. A fusion step also merges 3D patches when needed. The algorithm has been experimentally validated in the Gazebo-5 simulator.

Keywords 3D point cloud · Mapping · Computer vision

1 Introduction

In late 2010 Microsoft introduced the Kinect sensor device with its Xbox 360 game console. Since them many Kinects have been sold, and other low cost RGBD sensors have appeared like Asus Xtion and Kinect-2. Opening the SDK to manage these devices has brought applications for them in other areas like robotics [11] , health, security and industrial applications. Now they are about to be included in mobile phones and tablets for general public applications (Apple bought PrimeSense in 2013, Google is developing its Tango project[1] and both have interesting prototypes).

J. Navarro · J.M. Cañas(✉)
Universidad Rey Juan Carlos, Madrid, Spain
e-mail: josemariaplaza@urjc.es

[1] https://www.google.com/atap/project-tango/

L.P. Reis et al. (eds.), *Robot 2015: Second Iberian Robotics Conference*,
Advances in Intelligent Systems and Computing 418,
DOI: 10.1007/978-3-319-27149-1_51

In robotics, maps are useful for robot navigation and even self localization. Taking into account the information contained in maps a robot can plan its paths and make better movement decisions. Many map building techniques have been developed to create maps from robot sensors . They gather noisy sensor data and merge them into a more abstract, reliable and compact description of the 2D or 3D environment. In the last years many SLAM techniques ([9][4]) have been developed that simultaneously cope with the Localization and Mapping problems for mobile robots, using different sensors as input. In particular many recent SLAM works focus on the use of RGBD sensors ([10][3][5]).

Many 3D spatial primitives have been proposed in the literature to represent the 3D maps of scene or objects: 3D Points (dense point clouds), regular 3D voxels, octree maps [6], Surfels [2], Signed Distance Fusion (SDF) [8], patch volumes [7] and planar patches [1] among others.

This paper presents a preliminary work on an algorithm that builds incremental 3D maps from 3D point clouds using planar patches as compact spatial primitive. The continuous 3D pose estimation of the sensor is outside the scope of this paper, and we assume that a localization algorithm is working in the background. We focus on the incremental map construction and update.

2 Incremental 3D Planar Patches Map Building

We have designed and developed an algorithm for building a 3D map of the environment surrounding a mobile RGBD sensor using its readings. We define the map as a collection of 3D planar patches. A *planar patch* is defined as an area of a plane bounded by a contour (Fig. 1). Each planar patch (patch with hourglass shape, in green) is described by the general equation of the corresponding plane (square plane, in red), its *contour* (yellow line of the hourglass shape) and associated *cell image* (black and white binary image).

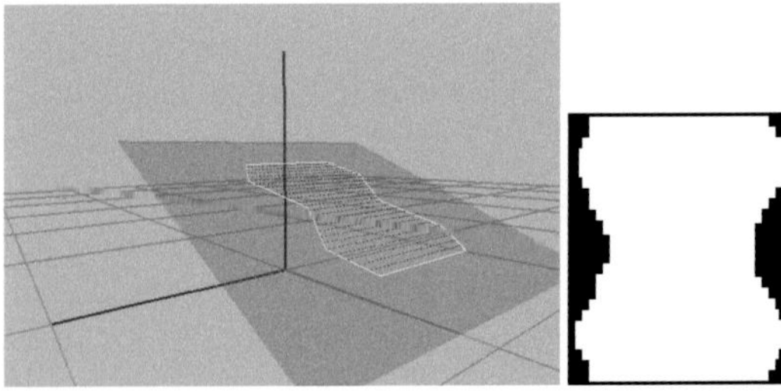

Fig. 1 Example of 3D points, a planar patch in 3D and its cell image.

A *cell image* gives us information on the distribution of points within the planar patch. It representes an image of the occupied and empty zones after dividing the area of each planar patch into fixed-size cells (pixels).

2.1 Design

The map building algorithm runs iteratively following the next steps:

1. Data acquisition. At the beginning of each iteration the current 3D position of the RGBD sensor and the current depth image are collected. The depth image is transformed into 3D point cloud applying a depth sampling and translation to absolute coordinates.
2. Point Classification. Each 3D point from the current cloud is classified into one of the following three categories according their relationship with the existing planar patches: (a) Belonging to one planar patch; (b) Close to one planar patch, and (c) Far from any existing patch, and so, not explained.
3. With the points belonging to one of existing planar patch we do nothing.
4. Patch redefinition. Each set of points close to an existing planar patch are used to redefine its contour.
5. The mapping algorithm generates (several) new planar patches that group unexplained points. This stage involves several steps:

   (a) RANSAC algorithm is used to group the unexplained points in one or more tentative planar patches.
   (b) Each tentative patch is evaluated with a *coherence test* to ensure that holds four criteria to be considered valid. If the patch meets all criteria it will be considered valid and added to the map. If the patch fails any of the first three criteria, then its points will be returned to RANSAC algorithm to continue considering them. If the patch only fails in the connexity criterion then it will be refined into several smaller refined patches. This situation is typical of very large patches with holes inside and in the case of several noncontiguous but coplanar areas.

   The RANSAC algorithm is iterative itself and ends when the number of unexplained points is less than a certain amount, or if it exceeds a certain number of iterations without extracting any valid new planar patch.
6. Fusion step. Finally the algorithm explores all possible pairs of existing patches to check whether some of them can be joined together. The generated map is a vector of planar patches and a vector of three-dimensional unexplained points.

The algorithm has several thresholds, which were finally selected after testing in the experiments. It has been programmed as a C++ component in the open-source *JdeRobot* framework[2].

[2] http://jderobot.org

2.2 Data Acquisition

In each iteration of the algorithm, the position and orientation of the device is captured first, and then the depth image from the RGBD sensor. A sampling is performed when translating the Depth image into a 3D point cloud. Without sampling the density of 3D points near the sensor is greater than the density of distant points. With our sampling all the Depth pixels are separated into layers according its distance to the sensor, and a different sampling rate is used on each layer to preserve an homogeneous 3D density of points between close and far areas. Fig. 2 shows a raw point cloud (left) and the sampled one (right).

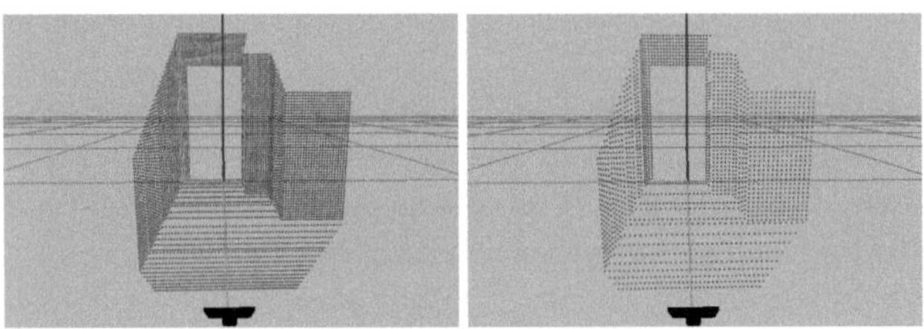

Fig. 2 Raw point cloud from the RGBD sensor and the sampled one.

So far the points are computed in coordinates relative to the sensor, but now the algorithm computes their absolute position incorporating the current 3D position and orientation of the RGBD sensor. Due to this, the algorithm can segment the planes from various RGBD sensors (or a mobile one) as all the points lie in the same absolute frame of reference.

2.3 Point Classification

Second step of each iteration classifies the 3D points into three categories according to their relation with the existing planar patches until the previous iteration (Fig. 3):

1. *Belonging to a planar patch.* These points do not give us any new information, only confirm some existing patch.
2. *Close to a planar patch.* These points will be useful to extend and redefine the planar patches which they are associated to.
3. *Unexplained by any existing planar patch.* The algorithm will try to group them in new planar patches.

For each point of the cloud we will calculate its normal distance to all planar patches. For those points whose normal distance is less than a configurable threshold, we also calculated the minimum lateral distance to the contour of the corresponding

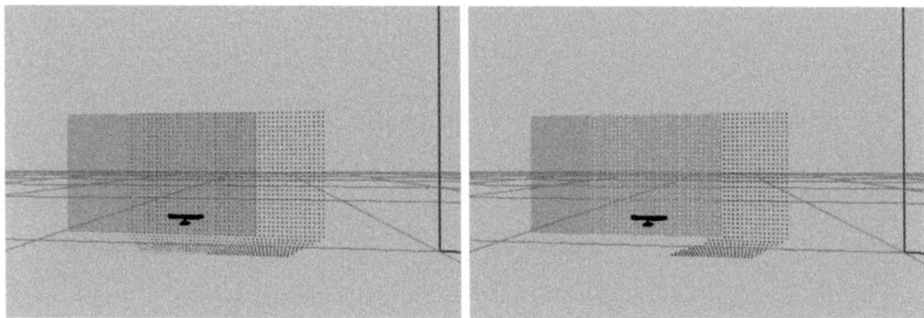

Fig. 3 Existing planar patches, point cloud (left) and their classification (right): close (yellow), nearby (blue) and unexplained (red).

planar patches. If this lateral distance is less than another threshold, we consider that point *belonging* to the plane. If not, if the lateral distance is less than a second threshold, we will consider *close* to the planar patch. One point may fall into different categories with respect to several planar patches and then, it is classified according to the planar patch whose distance is smallest. The points that do not fit into any of the previous categories are classified as *unexplained by any planar patch*.

2.4 Redefinition of Patches

With the set of points that are close to an existing planar patch, we extend this patch by redefining its cell image and its contour, but without affecting the equation of its plane (Fig. 4).

For the redefinition of the cell image, first we add the new points to the corresponding cells, adding new rows and/or columns of cells at the edges of the image where necessary. Then we employ *dilate* and *erode* functions of OpenCV to apply dilation a predefined number of iterations, erosion the same number of iterations plus one, and another iteration of expansion (more information about this in the subsection 2.5). This preprocessing is necessary to join slightly separated areas, and to eliminate the isolated areas.

Once the cell image is preprocessed, we obtain the new contour of the image using the *findContours* function of OpenCV. The row and column coordinates of the cells that form the contour of the image are translated into absolute 3D coordinates, and a Principal Component Analysis of the new cell image is performed. Initially, we did use PCL to obtain the contour by calculating the Concave Hull of the points that support a patch, but for some concave patches the contours were inadequate.

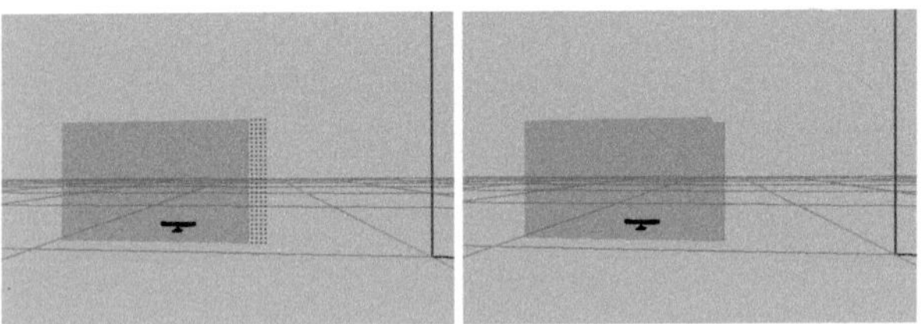

Fig. 4 (Left) Wall patch and close points. (Right) Redefined, wider, patch.

2.5 Generation of New Patches

With the point cloud of *unexplained by any existing patch* we seek for new patches that may explain these points using the RANSAC *(RANdom SAmple Consensus)* iterative algorithm. At each iteration it randomly selects three points of the cloud as candidates to form a plane. It classifies the remaining points into *inliers* (low distance to that candidate plane) or *outliers* (big distance), according to normal distance of each point relative to the candidate plane. Also, it gets the average distance associated with the inlier points. It stores the plane with the highest number of inliers, and in case of a tie, the plane that has the lower average distance. When a plane remains as the best for a number of iterations (which is customizable), RANSAC presents it as tentative patch.

Some of the tentative patches obtained by RANSAC are unsatisfactory. For instace, it extracts only one tentative patch several patches that are coplanar but not connected. Therefore we implemented a *coherence test* to study the validity of these tentative patches before incorporating them into the map. If they include separate coplanar clusters, the algorithm refines the tentative patch into smaller patches which must also pass the coherence test themselves.The coherence test consists of four criteria:

1. Have a sufficient number of points that support it.
2. Exceeds a nonlinearity threshold, to ensure that the distribution of the points does not correspond to a line.
3. The size of its surface is large enough.
4. The occupied cells in the corresponding cell image are connected.

After verifying that meets the first criterion, the algorithm builds the *cell image* associated to the tentative patch to evaluate the remaining criteria. For generating the cell image, a Principal Component Analysis (PCA) is performed on the cloud of points of this patch , using a *Singular Value Decomposition (SVD)*. The PCA extracts the three perpendicular directions of greater variability in the distribution of points . With this analysis we obtain the three principal singular vectors with their associated eigenvalues, although we are only interested the first two. We also get the centroid of the distribution.

Using the centroid as the origin and the two principal singular vectors (which are unitary) as basis vectors, we pass the points to this new two-dimensional base. Looking for the minimum and maximum values in both directions, we save the minimum value and the geometric distance in each direction. Using the geometric distance, along with the size chosen for the side of the cells we get the number of cells in each direction, and we round it upwards.

With the points in two-dimensional format we generate a raw cell image with the occupied cells (Fig. 5(a)). Then we perform a preprocessing using the *dilate* and *erode* functions of *OpenCV*, consisting in a number of iterations of dilation (Fig. 5(b)), the same number plus one of erosion (Fig. 5(c)), and other of dilation (Fig. 5(d)), to join groups of cells near each other but eliminate isolated cells.

To prevent that dilation damaged the real perimeter estimation, during preprocessing we added an edge of pixels that simulate empty cells. At the end we will cut the edges leaving only one pixel around (Fig. 5(e)), necessary when the calculating of the contour with *findContours* function of OpenCV is performed.

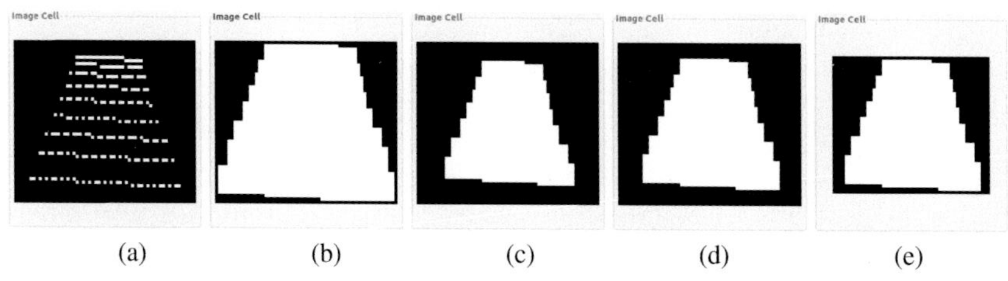

Fig. 5 Preprocessing of cell image: (a) Initial image. (b) First dilation. (c) Erosion. (d) Second dilation. (e) Cut of the edge.

To check the linearity criterion, we compare if the geometric distance corresponding to the second singular vector exceeds a certain threshold. For the minimum size criterion, we compare the product of the two geometric distances with a threshold which represents the minimum area to be considered.

And for the criterion of connectivity, we perform a growing region algorithm on the cell image using *findContours* and *floodFill* functions of *OpenCV* (Fig. 6). We use as seed of region growth the first vertex of each contour. If only one region is obtained, the patch meets the criterion. If there are more regions then the tentative patch will be refined in a number of patches equal to the number of obtained regions, which are formed by the points in the cells of each region.

If a tentative patch holds the four criteria then is considered valid, and its contour is calculated using the *findContours* function of *OpenCV* on its cell image (Fig. 7). The two-dimensional coordinates of the cells that form the vertices are converted to absolute three-dimensional coordinates. Then we store the planar patch on the map, defined by its equation of the plane, its contour and its cell image.

If a tentative patch fails during any of the first three criteria, it is discarded directly and their points are returned to the RANSAC algorithm to continue taking them

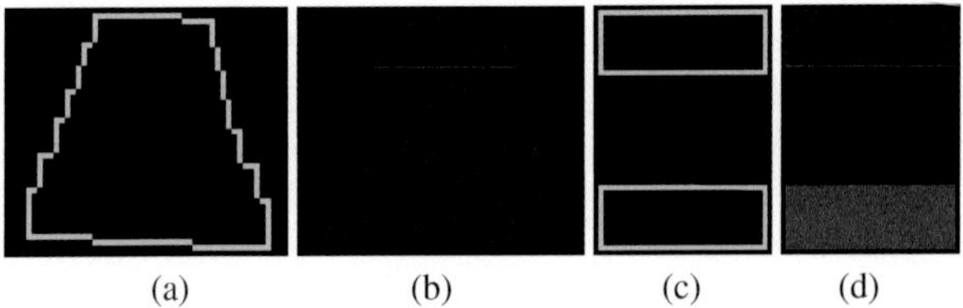

Fig. 6 Connectivity criterion: (a) *findContours* and (b) *floodFill*, for a connected patch. (c) and (d) for a non-connected patch.

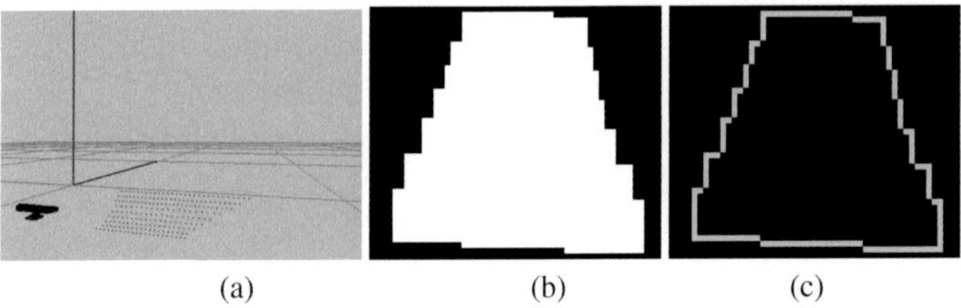

Fig. 7 (a) Points of a tentative patch (b) Cell image (c) Contour.

into account in the search for new tentative patches. If it only fails the ultimate criterion, because it is formed by non-connected groups (Fig. 8), it will be refined into several smaller tentative patches. The failed coherence test delivers the sets of points corresponding to the tentative refined smaller patches. These refined patches are passed to a new coherence test for themselves. This time, those which meet the four criteria will be incorporated into the map, those failing any of the criteria will be discarded and their points returned to the RANSAC algorithm.

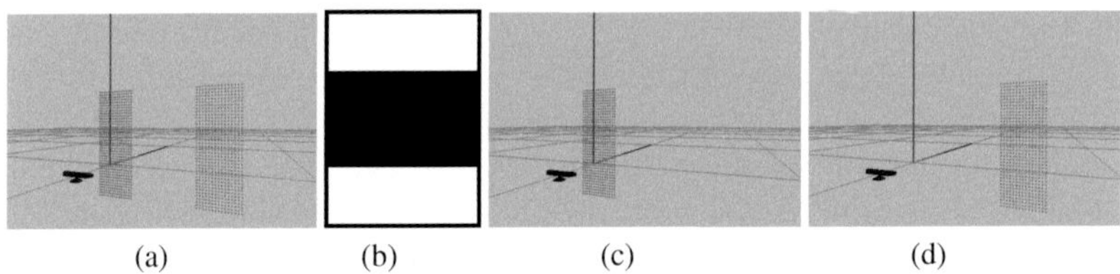

Fig. 8 (a) Points of non-connected patch. (b) Cell image of non-connected patch. (c) Points of first refined patch. (d) Points of the second refined patch.

Since RANSAC is an iterative algorithm we need a stop condition. We used any of these two: first, if at the beginning of an iteration the number of remaining unexplained points is less than a threshold; second, if the number of iterations exceedes a certain maximum without extracting any valid patch. Fig. 9 shows the result of applying this algorithm.

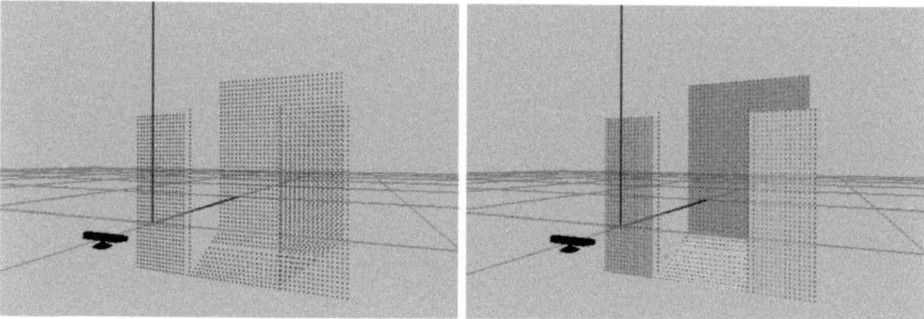

Fig. 9 Generation of new planar patches: (left) Initial points, (right) Obtained patches.

2.6 Fusion of Patches

The final step of the iterative mapping algorithm is the patch fusion. We explore the planar patches inside the map to check whether some of them correspond to different areas of a single real larger patch and can be joined together. We compare each planar patch with all others. For each pair of patches we first check their parallelism calculating the cross product of their normals to plane equations. If they can be considered parallel, we studied their coplanarity through its normal distance. To calculate the normal distance between two planar patches we get the minimum distance of the contour points of each planar patch to the plane equation of the other patch. If one of the two minimum distances is less than a threshold we consider the two patches coplanar.

For a pair of coplanar patches, we also calculate the minimum lateral distance between them. If any vertices of the contour of a planar patch is contained within the contour of the other, the lateral distance is zero. If the lateral distance is not zero, we calculate the minimum distance between all of the vertices of the contour of each patch to all the sides the contour of the other patch. If this lateral distance is less than a threshold we consider this pair of patches amenable of fusion.

When two patches are amenable of fusion we redefine the first one using the second . To redefine a planar patch by fusion with another patch, we first obtain a point cloud formed by the vertices of the contours of both patches and two sets of synthetic points within their cell images. Generating synthetic points from the cell image consists in generating one point in three-dimensional coordinates per each occupied cell With that point cloud we obtain the equation of the plane that best fits it using SVD (Singular Value Decomposition). We also define a new cell image using that point cloud, and with such cell image we get the contour of the planar patch resulting from the fusion. This new patch is compared with successive patches of the map for the remaining of the fusion step. This allows to join several patches that correspond to the same one in a single pass of the fusion algorithm. The Fig. 10 shows a fusion between planar patches.

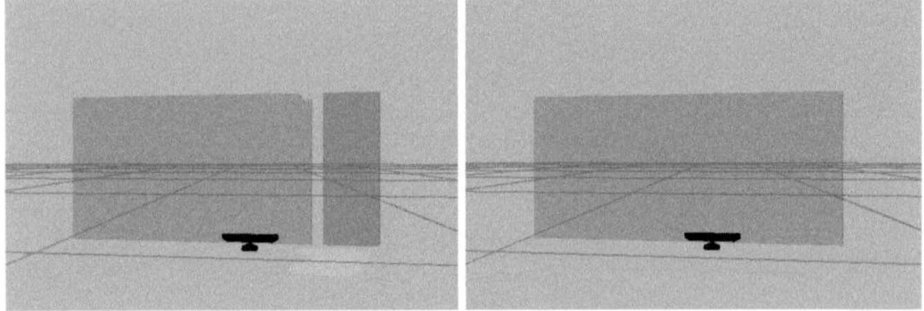

Fig. 10 Fusion of patches: (left) Scene before, (right) scene after.

3 Experiments

This section describes some experiments made with the software implementation of the developed algorithm in a simulated environment. To perform these experiments we used a scenario consisting of an apartment in Gazebo simulator. In particular, the world `GrannyAnnie` of the Rockin@Home[3] project, shown in Fig. 11. The experiments test the algorithm output positioning the simulated sensor at different locations of the apartment. For this experiments, before run the algorithm in each location, we wait until the 3D sensor location reading is stabilized.

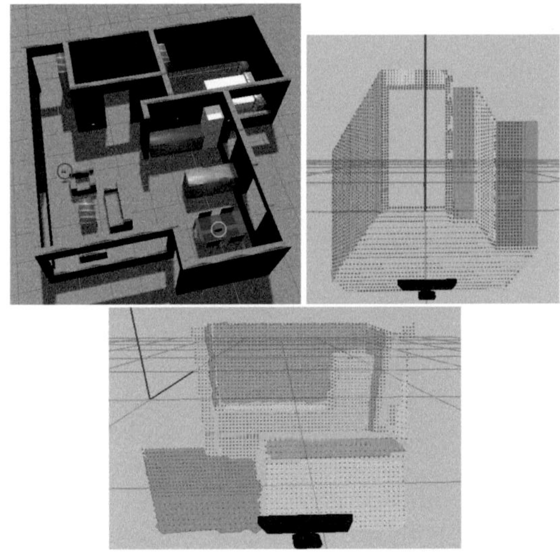

Fig. 11 Apartment GrannyAnnie in Gazebo. Patches and point cloud for the hall and for the kitchen.

[3] http://rockinrobotchallenge.eu/home.php

The used hardware is a desktop PC with a processor Intel Core 2 Quad CPU Q6600 @ 2.40GHz(x4) and 2x2GB(4GB) DDR2 @ 800MHz of RAM memory. The Gazebo simulator and our component simultaneosly run in this computer. The minimum, average and maximum times for each iteration of our algorithm are 27 ms, 77 ms and 304 ms, respectively. These numbers are obtained after multiple runs of the algorithm working with a cloud of 1500 $\sim$ 3500 points and up to 25 patches. We expect similar costs for the real scenario but we have not tried yet because not have solved the reliable 3D location.

3.1 Typical Execution from Static Position

In the first experiment the sensor is located oriented to the hall with a shelf (red circle, at the left in Fig. 11). The center of the same figure shows the point cloud. The variable depth sampling rate allows a good distribution of points in all patches, resulting in 3361 points. It also shows the 6 patches resulting from the algorithm. The right side of the Fig. 11 shows the point cloud (in this case of 3046 points) and the obtained planar patches (10) from the mapping algorithm when the sensor is located at the kitchen. In both situations the obtained patches are very satisfactory and correspond pretty well to the scenario.

3.2 Incremental Map from Several Positions

In this experiment the ability of the algorithm to incrementally modify its planar patch collection has been tested. The resulting map coming from incorporating two readings from two different sensor locations was compared to real scenario. We placed the sensor in two positions inside of the bedroom, keeping overlapping areas between the two observations (Fig 12). This scenario has furniture that includes patches of different sizes.

First, we placed the sensor in front of the bed and left (sensor, with the red circle, in the left of Fig. 12), facing slightly clockwise and towards the ground. The center of the same figure shows the cloud with 3453 points and the 6 patches obtained by the mapping algorithm.

Second, we placed the sensor in front of the bed, but this time right (sensor, with the green circle, in the left of Fig. 12), facing slightly counterclockwise and towards the ground. In this position, several iterations of the algorithm were executed, allowing the redefinition of existing patches from the previous position. In the right image of the same figure we have the cloud with 3117 points, and the 9 patches resulting of this process. We can see how the patches of the background wall (red), the top of the bed (yellow), and the part of the foot of the bed (magenta) have extended their surfaces absorbing points.

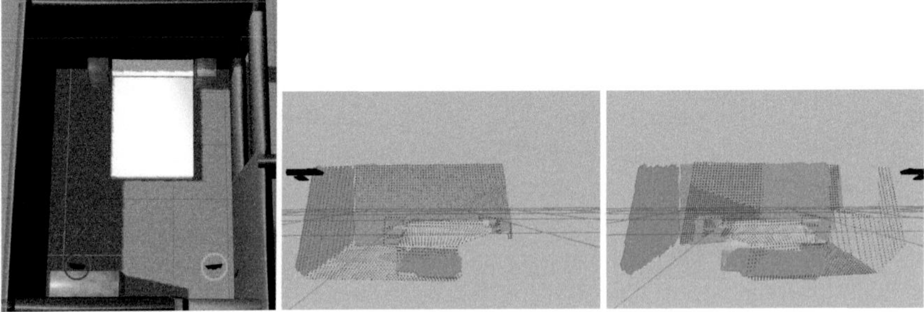

Fig. 12 Bedroom of the apartment in Gazebo and planar patches and point cloud for the bedroom in positions 1 and 2.

4 Conclusions

In this paper a preliminary work with an incremental 3D map building algorithm from RGBD sensor data has been presented. It uses planar 3D patches as spatial primitive, and faces typical problems when working with noisy low cost RGBD devices like gap filling, extension with close points and segment fusion. The algorithm computes point-plane associations and so classifies new coming points as belonging to an existing patch, close to one of them or far from any. Close points redefine patch spatial equation and its contour. Far points are clustered together using RANSAC an a coherence planar test.

The generated 3D maps are compact, dense and simple, with tens of patches instead of millions of points. They can be used in 3D path planning and robot navigation. The algorithm is incremental. It has been experimentally validated in a cutting edge simulator, Gazebo-5, showing promising results.

We have assumed that the 3D localization was already solved by a different algorithm, so all the 3D raw data are located in absolute coordinates. We want to test the algorithm with noise and false readings in both the real depth data and the received 3D position from a real 3D localization algorithm.

Acknowledgements This work has been supported by the Spanish Government DPI2013-40534-R Grant, supported with Feder funds. The research leading to these results has also received funding from the RoboCity2030 III-CM project (S2013/MIT-2748), funded by Programas de Actividades I+D en la Comunidad de Madrid and cofunded by Structural Funds of the EU.

References

1. Erdogan, C., Paluri, M., Dellaert, F.: Planar segmentation of RGBD images using fast linear fitting and markov chain monte carlo. In: 2012 Ninth Conference on Computer and Robot Vision (CRV), pp. 32–39, (2012)
2. Henry, P., Krainin, M., Herbst, E., Ren, X., Fox, D.: RGB-D Mapping: Using Kinect-style Depth Cameras for Dense 3D Modeling of Indoor Environments. International Journal of Robotics Research (IJRR) **31**, 5 (2012)
3. Endres, F., Hess, J., Sturm, J., Cremers, D., Burgard, W.: 3D Mapping with an RGB-D Camera. IEEE Transactions on Robotics **30**(1), January 2012
4. Endres, F., Hess, J., Engelhard, N., Sturm, J., Cremers, D., Burgard, W.: An evaluation of the RGB-D SLAM system. In: Proc. of the IEEE Int. Conf. on Robotics and Automation, (ICRA) (2012)
5. Endres, F., Hess, J., Sturm, J., Cremers, D., Burgard, W.: 3D Mapping with an RGB-D Camera. IEEE Transactions on Robotics (2014)
6. Sückler, J., Behnke, S.: Multi-Resolution Surfel Maps for Efficient Dense 3D Modeling and Tracking. Journal of Visual Communication and Image Representation **25**(1), 137–147 (2014). Springer
7. Henry, P., Fox, D.: Patch Volumes: Segmentation-based consistent mapping with RGB-D cameras. In: 2013 International Conference on 3D Vision (2013)
8. Newcombe, R.A., Izadi, S., Hilliges, O., Molyneaux, D., Kim, D., Davison, A.J., Kohli, P., Shotton, J., Hodges, S., Fitzgibbon, A.: KinectFusion: real-time dense surface mapping and tracking. In: 10th IEEE International Symposium on Mixed and Augmented Reality (ISMAR), October 2011
9. Newcombe, R.A., Lovegrove, S., Davison, A.J.: DTAM: dense tracking and mapping in real-time. In: IEEE International Conference on Computer Vision, ICCV 2011, Barcelona, pp. 2320–2327 (2011)
10. Audras, C., Comport, A.I., Meilland, M., Rives, P.: Real-time dense appearance-based SLAM for RGB-D sensors. In: Australian Conference on Robotics and Automation (2011)
11. Calderita, L.V., Manso, L.J., Nuñez, P.: Assessment of primesense RGB-D camera for using in robotics: application to human body modeling. In: Proc. of XIII Workshop of Physical Agents, WAF 2012, September 2012

Computing Image Descriptors from Annotations Acquired from External Tools

Jose Carlos Rangel, Miguel Cazorla, Ismael García-Varea, Jesús Martínez-Gómez, Élisa Fromont and Marc Sebban

Abstract Visual descriptors are widely used in several recognition and classification tasks in robotics. The main challenge for these tasks is to find a descriptor that could represent the image content without losing representative information of the image. Nowadays, there exists a wide range of visual descriptors computed with computer vision techniques and different pooling strategies. This paper proposes a novel way for building image descriptors using an external tool, namely: Clarifai. This is a remote web tool that allows to automatically describe an input image using semantic tags, and these tags are used to generate our descriptor. The descriptor generation procedure has been tested in the ViDRILO dataset, where it has been compared and merged with some well-known descriptors. Moreover, subset variable selection techniques have been evaluated. The experimental results show that our descriptor is competitive in classification tasks with the results obtained with other kind of descriptors.

Keywords Descriptor generation · Computer vision · Semantic localization · Robotics

1 Introduction

Representing images in an appropriate way is essential for tasks like image reconstruction, image search or place recognition [17]. Comparisons between image

J.C. Rangel(✉) · M. Cazorla
Computer Science Research Institute, University of Alicante, P.O. Box 99, 03080 Alicante, Spain
e-mail: jcrangel@dccia.ua.es

I. García-Varea · J. Martínez-Gómez
University of Castilla-La Mancha, Albacete, Spain

É. Fromont · M. Sebban
Jean Monnet University, Saint Etienne, France

L.P. Reis et al. (eds.), *Robot 2015: Second Iberian Robotics Conference*,
Advances in Intelligent Systems and Computing 418,
DOI: 10.1007/978-3-319-27149-1_52

descriptors can be used to determine the similarity between pairs of images [16]. Moreover, they can also be used for generalization capabilities. This is usually done by learning classification models, where the class corresponds to a desired image category. We can find binary categorization [14], and also multi-nominal proposals [9].

The main goal of an image descriptor is to find a proper representation minimizing the loss of information. Besides, some well-known approaches like histograms of gradients (HoG [4]) or Centrist [19], we can find some novel and interesting approaches. Among these alternative representations, Fei et al. [10] propose the use of an Object Filter Bank (OFB) for scene recognition. OFB contains the responses produced for objects detectors formerly trained. The work presented in [15] includes the generation of a model that simultaneously classifies and obtains a list of annotations from images. Zhou et al. [20] suggest the application of the Super Vector (SV) coding, a non linear method to compute image descriptors. Lampert et al. [8] employ an attribute classification based object recognizer. The proposal relies on semantic attributes like shape or color of an object to perform a high-level description. Banerji et al. [1] construct a descriptor based on the color, shape and texture, through the fusion of two different descriptors using an feature representation technique.

Nowadays, it is very common the use of an external and/or remote tools that provides some functionalities or information. We can find traffic or meteorology information systems [13], but also some technologies offering processing capabilities, like grid and cloud computing [5]. These systems allow the access to a wide range of services and novel capabilities. In this sense, the Clarifai system [1] provides the technology to analyze images and identify descriptive annotations (i.e. tags) related to them. Clarifai offers an Application Programming Interface (API) that obtains the 20 most descriptive annotations from a submitted image.

This article proposes (and analyzes) the use of Clarifai to build an image descriptor based on the labels got by means of this system. To carry out the experimentation, we extract Clarifai descriptors from the visual images included in the ViDRILO dataset [11]. These descriptors are then evaluated and compared with two well-known visual and depth descriptors (GIST [12], and the Ensemble of Shape Functions (ESF [18])) in the scene classification problem, using Support Vector Machines(SVMs [2]) as classifier. Clarifai descriptors are also tested when combined with ESF and GIST. Furthermore, a subset variable selection algorithm is applied to the Clarifai descriptors.

The rest of the paper is organized as follows: Section 2 exposes how the Clarifai annotation system works. Section 3 describes our proposal to build the Clarifai descriptor. The computed descriptors used in this paper are explained in Section 4. Section 5 shows the experimentation and the obtained results. Finally, in Section 6 the conclusions obtained for this work and the future work are presented.

[1] http://www.clarifai.com/

2 Remote Image Annotation: Clarifai Technology

Clarifai technology relies on the use of Convolutional Neural Networks(CNN [6] to process an image, and then generates a list of tags describing the image. CNNs are defined as hierarchical machine learning models, which learn complex images representations from large volumes of annotated data. They use multiple layers of basic transformations that finally generate a highly sophisticated representation of the image [3]. The Clarifai approach was firstly proposed to the ImageNet classification challenge [7] in 2013, where the system produced the top 5 results.

Clarifai works through the analysis of images to produce a list of descriptive tags representative of a given image. For each tag in this list, the system also provides a probability value. This probability represents the likelihood of describing the image using the specific tag. The Clarifai API can be accessed as a remote web service. The working scheme of the Clarifai technology is shown in Fig. 1. In this scheme, Clarifai uses a ViDRILO image as input and uses the CNNs to analyse it and produce the list of labels and probabilities.

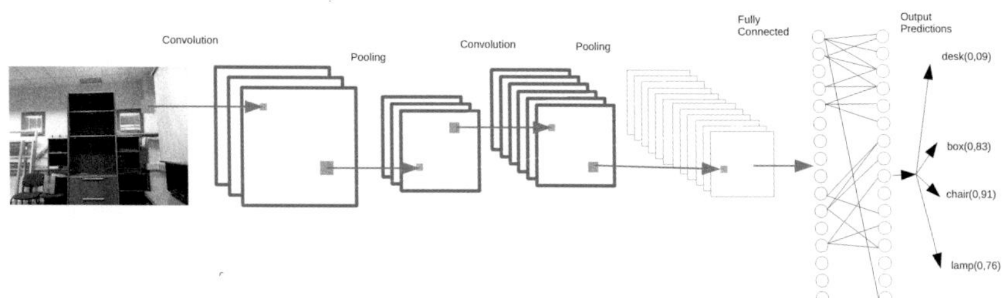

Fig. 1 Processing Scheme of the Clarifai API.

The Clarifai API can be accessed using a trial option or used under a payment mode. In this work, the trial mode has been used to obtain the tags of the images. This mode allows to get the tags for a restricted number of images (10000), and it limits the number of calls per hour to 1000. Using the trial mode, we can only obtain 20 tags per image with their associated probability. In Fig. 2, we show the image sent to the API and the returned labels/probabilities. From these labels, Fig. 3 shows some examples of representative images produced by the API. These images are generated using the annotations from ImageNet.

Regarding processing time, Clarifai has a latency time of almost 1 minute. Clearly it is not a real-time option, but new approaches for deep learning (using Caffe with Googlenct, for example) will provide a local solution (speeding-up with GPU architectures) which good time responses.

Image	Label	Probability
	Indoors	0.9936
	Seat	0.9892
	Contemporary	0.9787
	Chair	0.9779
	Furniture	0.9744
	Room	0.9634
	Interior design	0.9627
	Window	0.9505
	Table	0.9428
	Computer Technology	0.9417

Fig. 2 Labels and probabilities obtained with the Clarifai API for a ViDRILO image.

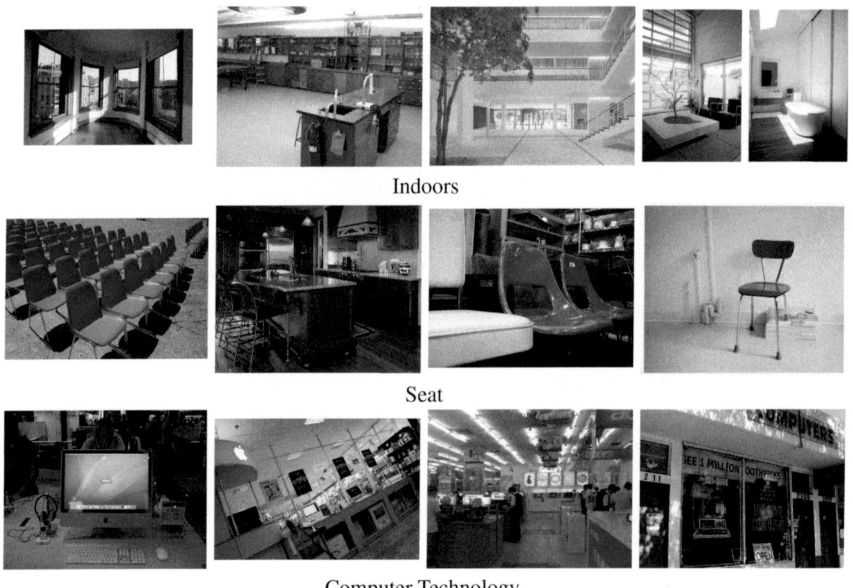

Fig. 3 Representative images for the labels extracted in Fig. 2.

3 Image Descriptors from Clarifai Annotations

The process to generate Clarifai descriptors starts with an early step where we discover all the labels included in our problem domain. This is done by processing all the available images with the Clarifai API, and then removing the duplicated label values. This step can be seen as a codebook or dictionary generation process. Each input image is then encoded as a list of probability values whose length is determined by the size of the dictionary. The i-th value in this descriptor will correspond to the probability returned by Clarifai for the i-th label, or zero otherwise. This process is shown in Table 1.

Table 1 Generation of Clarifai descriptors

	$label_1$	$label_2$	$label_3$	$label_4$	$label_5$	$label_6$	...	$label_N$
$Image_1$	0.99	0.98	0.97	0	0	0	...	0
$Image_2$	0	0.94	0	0.93	0.94	0	...	0
...	...	...	...	...	...	...	...	...
$Image_M$	0	0	0	0	0	0	...	0.91

4 Computing ViDRILO Descriptors

To make a comparison among existing descriptors and Clarifai ones, the first step is to generate all these descriptors for the images from the chosen ViDRILO dataset. We select GIST [12], and the Ensemble of Shape Functions (ESF [18]) as they obtain the highest baseline results using visual and depth information respectively. Moreover, they are released in conjunction with the ViDRILO dataset and its processing toolbox[2]. We briefly describe these two descriptors and the proposed one.

4.1 GIST

The purpose of the GIST descriptor is to represent the shape of the image using a holistic representation of the space envelope. To generate the descriptor, an image is split into $N \times N$ patches. A low-dimensional vector is then generated from each of these patches. The size of these vectors depends on the number of orientations and scale variations initially selected. Using 16 patches ($N = 4$), 4 scales and 8 orientations (default values) we generate GIST descriptors whose dimensionality is 512.

4.2 Ensemble of Shape Functions(ESF)

This depth descriptor produces 10 different histograms as a result of three shape functions. It codifies the relation existing among the 3D points of a depth image. It uses 64 bins for histogram, which result in a descriptor with size 640.

[2] http://www.rovit.ua.es/dataset/vidrilo/

4.3 Clarifai Descriptor

The Clarifai descriptor is generated as explained in Section 3. To compute the descriptor for the ViDRILO dataset, we firstly tagged all the images of the dataset. The next step was to count the unique produced labels, and we obtained 793 different values. This is the final dimensionality of the Clarifai descriptors. Hence, the Clarifai descriptors are built for every image placing the probability of the label under the respective label in the descriptor. Table 2 contains the 10 most frequent labels in the ViDRILO dataset and their appearance ratio by sequence. This table shows that the most common identified labels of the dataset are the most common in each sequence too with small variations. It is worthy of note the fact that most common labels neither show nor describe a specific scene or object name.

Table 2 Most frequent labels in the ViDRILO sequences and their appearance ratio by sequence.

Labels	Sequence 1	Sequence 2	Sequence 3	Sequence 4	Sequence 5
Indoors	91%	94%	81%	79%	94%
Room	90%	89%	80%	79%	90%
Window	85%	86%	91%	74%	82%
House	83%	80%	88%	73%	79%
Door	76%	77%	74%	68%	70%
Dwelling	69%	71%	60%	64%	70%
Nobody	62%	75%	67%	83%	54%
Architecture	62%	65%	63%	63%	49%
Contemporary	58%	61%	40%	44%	61%
Floor	59%	50%	38%	42%	62%

5 Experiments and Results

All the experiments included in this paper have been performed using ViDRILO [11] as benchmark. The main characteristics of this dataset, as well as the distribution of its sequences, are shown in Table 3. The RGB-D images from this dataset have been acquired with a mobile robot in an indoor office environment.

In ViDRILO , each image is annotated with the category of the scene it was acquired, from a set of 10 room categories. Fig. 4 shows some representative images for the categories.

Table 3 Overall ViDRILO characteristics and sequences distribution.

Sequence	Number of Frames	Floors imaged	Dark Rooms
Sequence 1	2389	1st,2nd	0/18
Sequence 2	4579	1st,2nd	0/18
Sequence 3	2248	2nd	4/13
Sequence 4	4826	1st,2nd	6/18
Sequence 5	8412	1st,2nd	0/20

Fig. 4 Exemplar visual images for the 10 room categories in ViDRILO .

5.1 Experimentation

In order to carry out the evaluation of the proposed descriptor, the experiments consist of training a classifier using a sequence of the ViDRILO dataset and then use another sequence to test the trained model. From the 5 available sequences, we trained five different classifiers and all of them were evaluated tested against five different sequences. This resulted into 25 possible scenarios where each descriptor (GIST, ESF and Clarifai) was evaluated. To determine the effectiveness of the Clarifai descriptor, we measure the accuracy of the results produced by the classifier. This is defined as the percentage of well classified images in the test sequence of the dataset. As the experimentation produces 75 accuracy values (3 descriptor and 25 scenarios), we averaged the results by training and test sequence to better visualize these results. All the experiments were performed using a χ^2 SVM classifier.

5.2 Baseline Results

The baseline results are obtained by comparing the accuracy values obtained with Clarifai, ESF and GIST descriptors. The Table 4 contains the average results obtained by the descriptors grouped by the train sequence. Fig. 5 shows the results obtained by descriptor and averaged over the sequence used to train (left) and test (right) the classifier. Here we see that our proposal obtains results similar to those obtained with the rest of descriptors. Moreover, Clarifai descriptor outperforms ESF and GIST when using Sequence 5 for training. This indicates a capability of generalization, as Sequence 5 was taken in another building where structure and color was different.

Table 4 Accuracy values obtained with the 3 evaluated descriptors and averaged by training sequence.

Sequence	GIST	ESF	Clarifai Tags
Sequence 1	65,38	56,59	58,32
Sequence 2	66,80	60,32	56,10
Sequence 3	60,55	60,34	59,43
Sequence 4	66,01	62,24	58,69
Sequence 5	57,94	54,82	63,17

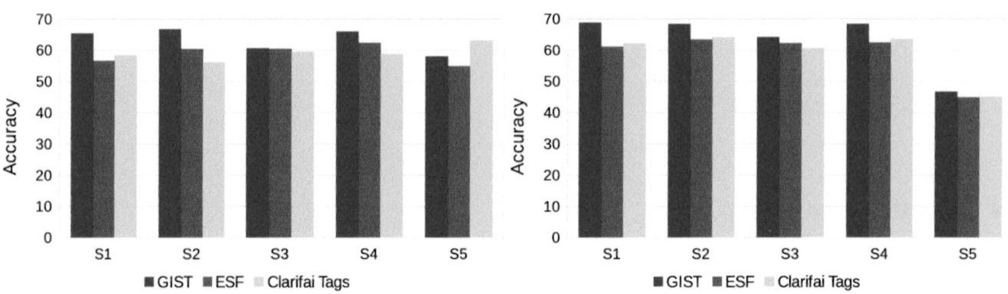

Fig. 5 Clarifai Descriptor averaged by Train (left) and Test Sequence(right).

5.3 *Descriptor Combination*

Here we present the results training the classifier using a combination of 2 descriptors. The Table 5 contains the average results obtained using a merge of descriptors grouped by the train sequence. Fig. 6 shows the results obtained by the merge of descriptors and averaged over the sequence used to train (left) and test (right) the classifier. Fig. 7 shows how the accuracy of the Clarifai descriptors improves by using a merge of descriptors. These results are grouped by the training sequence.

Table 5 Merge of Descriptors averaged by Train Sequence

Sequence	ESF+GIST	ESF+Clarifai	GIST+Clarifai
Sequence 1	65,77	58,38	58,36
Sequence 2	66,82	58,60	58,51
Sequence 3	61,76	58,98	59,34
Sequence 4	66,38	58,67	58,65
Sequence 5	53,04	63,18	63,21

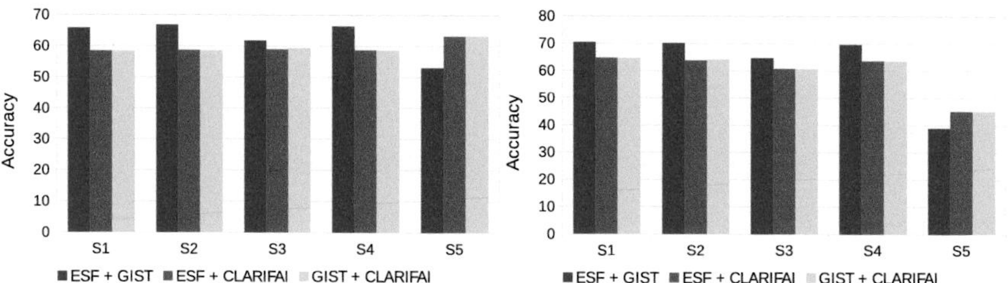

Fig. 6 Results with merge of descriptors averaged by Train Sequence (left) and by Test Sequence (right).

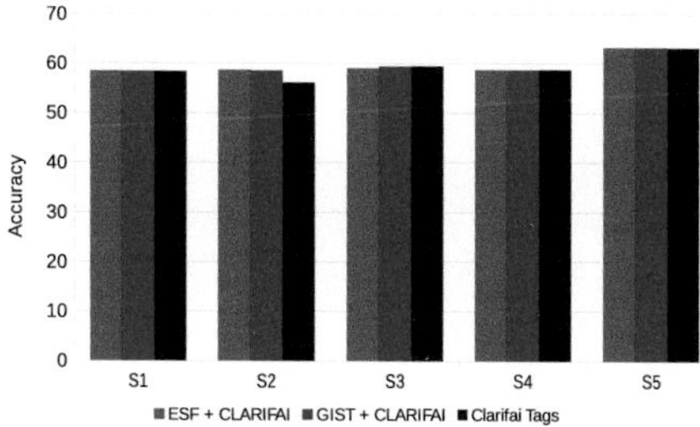

Fig. 7 Comparison of baseline Clarifai against merge of descriptors.

5.4 *Label Subset Selection*

The experiments for this section included an additional step before the classification. This step consists in the application of a subset variable selection process in the Clarifai descriptors to reduce the number of attributes used to describe the image. The descriptor still gets good results with this process, obtaining even better results using a reduced amount of labels to train the classifiers. Fig. 8 shows the result of selecting 10, 20, 50, 75, 100, 150, 200 variables and then training the SVM classifier. It shows that the use of a few labels reduces the accuracy results, but when the number increase the accuracy outperforms the one obtained using the 793 labels of the descriptor. Hence the use of 75 labels gets higher accuracy than the use of all the labels.

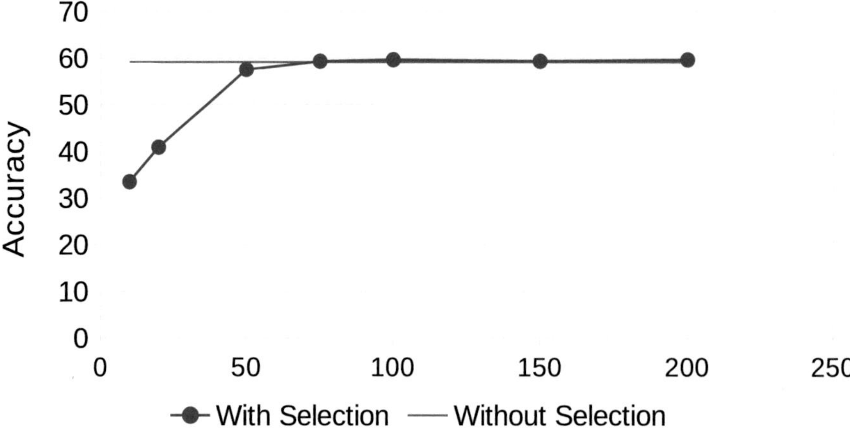

Fig. 8 Accuracy of the Clarifai Descriptors using a Subset Variable Selection Process

6 Conclusions and Future Work

We have presented a new way to build an image descriptor using Clarifai, an external image labeling tool. We have compared this descriptor with other two descriptors, achieving competitive results in the different scenarios that we have tested. The Clarifai descriptor shows that even with a reduced dimensionality, outperforms its own results in classifications tasks. As future work, we want to test this system using all the possible tags for a given image.

Acknowledgements This work was supported by the Ministerio de Economia y Competitividad of the Spanish Government, supported with Feder funds under grant DPI2013-40534-R; Consejería de Educación, Cultura y Deportes of the JCCM regional government under project PPII-2014-015-P. Jesus Martínez-Gómez is also funded by the JCCM grant POST2014/8171.

References

1. Banerji, S., Sinha, A., Liu, C.: Novel color, shape and texture-based scene image descriptors. In: 2012 IEEE International Conference on Intelligent Computer Communication and Processing (ICCP), pp. 245–248, August 2012
2. Chang, C.C., Lin, C.J.: LIBSVM: A library for support vector machines. ACM Transactions on Intelligent Systems and Technology **2**, 271–2727 (2011)
3. Clarifai: Clarifai: Amplifying Intelligence (2015). http://www.clarifai.com/
4. Dalal, N., Triggs, B.: Histograms of oriented gradients for human detection. In: Int. Conf. on CVPR, vol. 1, pp. 886–893. IEEE (2005)
5. Foster, I., Zhao, Y., Raicu, I., Lu, S.: Cloud computing and grid computing 360-degree compared. In: Grid Computing Environments Workshop, GCE 2008, pp. 1–10. Ieee (2008)

6. Krizhevsky, A., Sutskever, I., Hinton, G.E.: Imagenet classification with deep convolutional neural networks. In: Advances in Neural Information Processing Systems, pp. 1097–1105 (2012)
7. Krizhevsky, A., Sutskever, I., Hinton, G.E.: Imagenet classification with deep convolutional neural networks. In: Pereira, F., Burges, C., Bottou, L., Weinberger, K. (eds.) Advances in Neural Information Processing Systems, vol. 25, pp. 1097–1105. Curran Associates, Inc. (2012)
8. Lampert, C., Nickisch, H., Harmeling, S.: Attribute-based classification for zero-shot visual object categorization. IEEE Transactions on Pattern Analysis and Machine Intelligence **36**(3), 453–465 (2014)
9. Li, L.J., Fei-Fei, L.: What, where and who? Classifying events by scene and object recognition. In: IEEE 11th International Conference on Computer Vision, ICCV 2007, pp. 1–8. IEEE (2007)
10. Li, L.J., Su, H., Lim, Y., Fei-Fei, L.: Objects as attributes for scene classification. In: Kutulakos, K. (ed.) Trends and Topics in Computer Vision. Lecture Notes in Computer Science, vol. 6553, pp. 57–69. Springer, Heidelberg (2012)
11. Martinez-Gomez, J., Cazorla, M., Garcia-Varea, I., Morell, V.: Vidrilo: The visual and depth robot indoor localization with objects information dataset. International Journal of Robotics Research (2015)
12. Oliva, A., Torralba, A.: Modeling the shape of the scene: A holistic representation of the spatial envelope. International Journal of Computer Vision **42**(3), 145–175 (2001)
13. Petty, K.F., Moylan, A.J., Kwon, J., Mewes, J.J.: Traffic state estimation with integration of traffic, weather, incident, pavement condition, and roadway operations data (February 5, 2014), uS Patent App. 14/173,611
14. Szummer, M., Picard, R.W.: Indoor-outdoor image classification. In: IEEE International Workshop on Proceedings of the Content-Based Access of Image and Video Database, pp. 42–51. IEEE (1998)
15. Wang, C., Blei, D., Li, F.F.: Simultaneous image classification and annotation. In: IEEE Conference on Computer Vision and Pattern Recognition, CVPR 2009, pp. 1903–1910, June 2009
16. Wang, G., Hoiem, D., Forsyth, D.: Learning image similarity from flickr groups using fast kernel machines. IEEE Transactions on Pattern Analysis and Machine Intelligence **34**(11), 2177–2188 (2012)
17. Winder, S., Brown, M.: Learning local image descriptors. In: IEEE Conference on Computer Vision and Pattern Recognition, CVPR 2007, pp. 1–8, June 2007
18. Wohlkinger, W., Vincze, M.: Ensemble of shape functions for 3D object classification. In: 2011 IEEE International Conference on Robotics and Biomimetics (ROBIO), pp. 2987–2992. IEEE (2011)
19. Wu, J., Rehg, J.M.: Centrist: A visual descriptor for scene categorization. IEEE Transactions on Pattern Analysis and Machine Intelligence **33**(8), 1489–1501 (2011)
20. Zhou, X., Yu, K., Zhang, T., Huang, T.: Image classification using super-vector coding of local image descriptors. In: Daniilidis, K., Maragos, P., Paragios, N. (eds.) Computer Vision ECCV 2010. Lecture Notes in Computer Science, vol. 6315, pp. 141–154. Springer, Heidelberg (2010)

Keypoint Detection in RGB-D Images Using Binary Patterns

Cristina Romero-González, Jesus Martínez-Gómez, Ismael García-Varea and Luis Rodríguez-Ruiz

Abstract Detection of keypoints in an image is a crucial step in most registration and recognition tasks. The information encoded in RGB-D images can be redundant and, usually, only specific areas in the image are useful for the classification process. The process of identifying those relevant areas is known as keypoint detection. The use of keypoints can facilitate the following stages in the image processing process by reducing the search space. To properly represent an image by means of a set of keypoints, properties like repeatability and distinctiveness have to be fullfilled. In this work, we propose a keypoint detection technique based on the Shape Binary Pattern (SBP) descriptor that can be computed from RGB-D images. Next, we rely on this method to identify the most discriminative patterns that are used to detect the most relevant keypoint. Experiments on a well-know benchmark for 3D keypoint detection have been performed to assess our proposal.

Keywords Keypoint detection · RGB-D images · Binary patterns · Local descriptors · Shape binary pattern

1 Introduction

The preprocessing step of most robot vision tasks usually includes the detection of relevant parts in an image. In general, a set of points in the image that encodes the most distinctive information is selected. This selection is based on different characteristics

C. Romero-González(✉) · J. Martínez-Gómez · I. García-Varea · L. Rodríguez-Ruiz
Computer Systems Department, University of Castilla-La Mancha,
Campus Univ. s/n, 02071 Albacete, Spain
e-mail: {Cristina.RGonzalez,Jesus.Martinez,Ismael.Garcia,Luis.RRuiz}@uclm.es

J. Martínez-Gómez
Computer Science and Artificial Intelligence Department, University of Alicante,
P.O. Box 99, 03080 Alicante, Spain

L.P. Reis et al. (eds.), *Robot 2015: Second Iberian Robotics Conference*,
Advances in Intelligent Systems and Computing 418,
DOI: 10.1007/978-3-319-27149-1_53

of the local neighborhood of such points. These points are called keypoints or interest points, and they allow to reduce the search space for later steps in typical machine vision applications. In 3D, these keypoints are expected to represent distinctive shapes in an image.

Repeatability is the most important property in any keypoint detector. In sort, a good detection method should be able to find the same set of points in different images representing the same object or scene. Usually, these images may differ due to changes on the point of view [5], occlusion [11] or noise [25].

In this paper, we propose the use of Shape Binary Patterns (SBP) from RGB-D images [14] to detect keypoints. The proposal consists in identifying the patterns corresponding to basic forms in the image, like lines or corners, and use the less frequent of them as keypoints. Intuitively, the less frequent patterns should correspond to the parts of the image with more variation and, thus, with more relevant information. For example, it is expected to find more patterns corresponding to planes than corners, but corners are more representative of the shapes found in the image. Also, to validate this proposal we present experimental results on the 3D Keypoint detection benchmark proposed in [19], obtaining a performance similar to other state-of-the-art keypoint detectors.

The remaining of the paper is organized as follows. In Section 2 we make a brief introduction to the state of the art in keypoint detection methods. Section 3 includes a description of the generation process on binary patterns in RGB-D images. Section 4 describes our proposal for keypoint detection. In Section 5 experimental results of the proposed keypoint detector against a well known evaluation benchmark are presented. Finally, Section 6 presents the conclusions that can be drawn from this work.

2 Related Work

The keypoint detection problem is a widely studied topic in the computer vision community [20]. Traditionally, SIFT [9] and SURF [2] have been the most used methods of keypoint detection despite their high computational requirements. Recently, some new methods based on binary descriptors have been proposed to decrease the those computational requirements, while maintaining similar performance. In 2010, the FAST [15] method was proposed. FAST is a keypoint detector based on the corners found in an image and, it was designed specifically for a high processing speed. It determines that a point corresponds to a corner if n contiguous pixels in a circle around a point have a brighter intensity than the center point. To achieve its high speed they only perform the minimum comparisons necessary to determine whether a point is a corner or not. Similar approaches based on the same idea are: AGAST [10], which increases its performance by providing an adaptive and generic accelerated segment test; ORB [16] which adds an orientation component to FAST for its keypoint detection; and BRISK [8] which is an extension aimed to achieve invariance to scale. Another not so related proposal, is FREAK [1], which computes a cascade of binary strings by comparing intensities over a retinal sampling pattern.

All those proposals have achieved remarkable results in keypoint detection for 2D images. In the meantime, in 3D, and in RGB-D specifically, keypoint detectors is also a widely studied problem [6, 19]. In this case, detectors are focused on finding distinctive shapes within the image based on the 3D surface.

Some of those detectors are the MeshDoG [24], that is designed for uniformly triangulated meshes, invariant to changes in rotation, translation, and scale. The Laplace-Beltrami Scale-Space (LBSS [23]), which pursues multi-scale operators on point clouds that allow detection of interest regions. the KeyPoint Quality (KPQ [21]) or the Salient Points (SP [4]). Other detectors require a scale value to determine the search radius for the local neighborhood of the keypoint. Some of those detectors are the Intrinsic Shape Signature (ISS [25]), that relies in a Eigen Value Decomposition (EVD) of the points in the support radius, the Local Surface Patches (LSP [5]), based on point-wise quality measurements, or the Shape Index (SI [7]), that uses the maximum and minimum principal curvatures at the vertex.

3 Binary Patterns in RGB-D Images

The Shape Binary Pattern (SBP [14]) is a binary descriptor inspired in LBP [13]. SBP generates a unique pattern based on the shape of the local neighborhood of a point in a RGB-D image. This descriptor overlaps a 3D grid over those points in the local neighborhood of a given point p_c, and assigns a binary value to each bin depending on the presence of points inside the bin. This grid is oriented according to a local Reference Frame (RF) to achieve rotation invariance. Also, the bin size is determined dynamically, so the pattern can be adapted to the scale of the image. In general, SBP is a simple but efficient descriptor, that effectively represents the shape surrounding an interest point with invariance to rotation and scale. Additionally, it can be computed fast and with low computational requirements. For completeness, in this section we include a description of its generation process.

First, a set of N points $\mathcal{P}_c = \{p_1, ..., p_N\}$ corresponding to the nearest points to p_c in a search radius R is taken. The neighbors covariance matrix M for p_c is computed as:

$$M = \frac{1}{N}\sum_{i=1}^{N}(p_i - \mu)(p_i \quad \mu)^T \qquad (1)$$

where N is the number of points in the local neighborhood of p_c and μ is the centroid of $\mathcal{P}_c$. Then, an eigenvalue decomposition of M that results in three orthogonal eigenvectors is performed. This procedure is usually used to approximate the normal of p_c [17], as the eigenvector corresponding to the smallest eigenvalue. In this case, the SBP generation requires two vectors as RF1, so the eigenvector with the highest eigenvalue is selected along with the approximated normal. Their orientation is determined to be coherent with the vectors they represent [3, 22].

Once the two vectors of the RF have been computed, their orientations can be used to overlap a cubic 3D grid over $\mathcal{P}_c$ of size $k = 64$ ($4 \times 4 \times 4$). The bin size l is computed from the cloud resolution. Namely,

$$l = 4 \times \min_{p_i, p_j \in \mathcal{I}} |p_i - p_j| \tag{2}$$

where $\mathcal{I}$ is the set of points in the RGB-D image, and $\min_{p_i, p_j \in \mathcal{I}} |p_i - p_j|$ represents the minimum distance between two any points in the cloud. This formula allows SBP for adapting the size of the 3D grid to the dispersion/concentration of the points in the cloud. It provides the descriptor with invariance to scale transformations. As l is obtained prior to the SBP descriptor computation, the search radius R for the local nearest neighbors can also be dynamically determined. This is carried out by assuming that the 3D grid is fitted into the sphere formed by the search radius R of $\mathcal{P}_c$:

$$R = \frac{l \cdot \sqrt[3]{k} \cdot \sqrt{3}}{2} \tag{3}$$

Finally, each point in $\mathcal{P}_c$ is assigned to the corresponding bin based on its position in the local RF. Then, the binary pattern T that encodes the presence (or absence) of at least one point inside each bin is created. That is,

$$T = \langle f(b_1), ..., f(b_k) \rangle, \quad f(x) = \begin{cases} 1 & \text{if } |x| > 0 \\ 0 & \text{if } |x| \leq 0 \end{cases} \tag{4}$$

where T represents the final SBP descriptor, b_i the set of points assigned to the i-th bin, and $f(x)$ is a function that returns a binary value depending on the number of points in the bin. In Figure 1 we can see an illustration of the process for generating a SBP descriptor.

4 Keypoint Detection Based on Uniform Patterns

Uniform patterns have been widely studied in LBP. They are a subset of all the possible patterns of the LBP descriptor. Specifically, uniform patterns refers to those patterns that have all their ones together, in other words, there are no more than two changes between ones and zeros along the pattern. These uniform patterns are the most frequent and represent basic shapes found in an image [13].

Given the low computational requirements of the SBP pattern, it can be used to efficiently detect the keypoints in RGB-D images. Basically, we can distinguish two

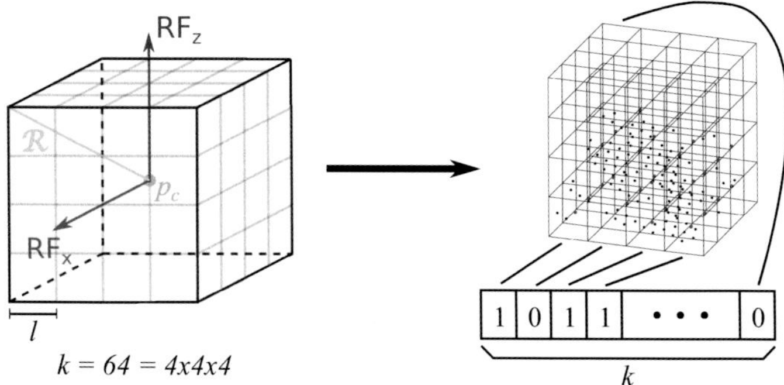

Fig. 1 SBP generation. First, a 3D grid is overlapped over p_c based on its calculated RF, and then a pattern is generated based on the apparition of points in the corresponding bin.

basic steps in this keypoint detection process: a) identify the points corresponding to uniform patterns, and b) select the most representative patterns in the image.

To address the first step, we define a uniform pattern by the disposition of ones in the 3D grid. Specifically, we consider that a pattern is uniform if all the ones are in contiguous positions in the 3D grid. In the second step, we will select the less frequent patterns, as they represent the basic forms that are more representative of a given scene or object.

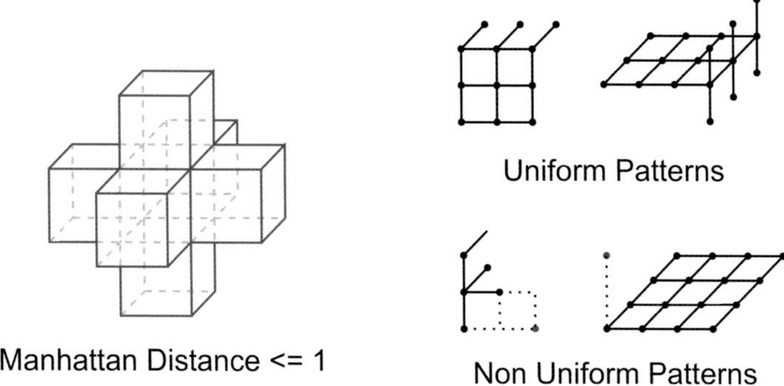

Fig. 2 Support area in the 3D grid of the SBP pattern for Manhattan distance ≤ 1, and examples of uniform and non uniform patterns.

Formally, given the 3D grid disposition of the binary values used to generate a SBP pattern T, we consider that pattern as uniform if the highest Manhattan distance between the ones in the grid is one. In other words, given the subset $\mathcal{B}_1$ of bins with value 1, the closest bin to any bin $b_i \in \mathcal{B}_1$ must be inside a radius of Manhattan distance of 1 (see Figure 2). In summary, the condition for a pattern T to be considered uniform is:

$$\forall b_i \in \mathcal{B}_1, \min_{b \in \mathcal{B}_1} |b - b_i| \leq 1 \tag{5}$$

where $|b - b_i|$ is the Manhattan distance between two bins. Then, to identify the possible uniform patterns that fulfill this criteria, we assign to each pattern T an index value U_T based on the number of ones that it contains:

$$U_T = \begin{cases} \sum_{i=1}^{k} f(b_i) & \text{if } T \text{ is uniform} \\ k+1 & \text{if } T \text{ is not uniform} \end{cases} \tag{6}$$

where k is the size of the pattern (in this case $k = 64$, see Section 3) and $f(x)$ is the function defined in Eq. (4) to calculate the binary value of the bins. With this index value, we cluster the patterns according to their number of ones, and independently of its distribution in the final pattern.

Finally, to select the most representative points, we generate a histogram of the index assigned to the patterns. Then we sort them by increasing frequency and select the index values with less frequency than a threshold. This threshold establishes the number of pattern index to select, for example a threshold of 10% would result in the selection of $65 \times 0.1 = 7$ pattern indexes. Finally, we select as keypoints on the image the points corresponding to the selected U. In Figure 3 we can see the selected keypoints for different frequency values in a example scene with 5 objects.

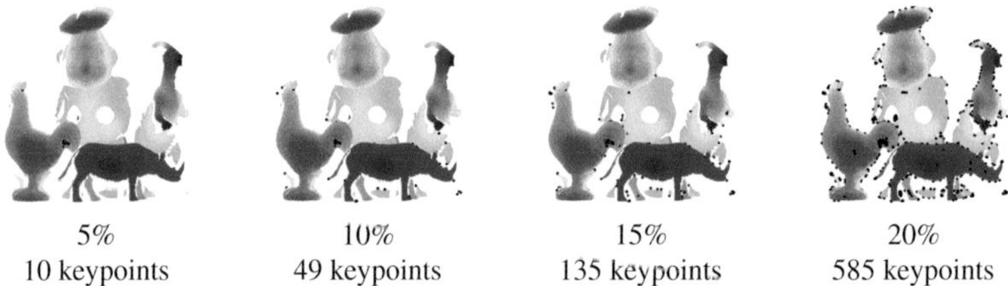

5%	10%	15%	20%
10 keypoints	49 keypoints	135 keypoints	585 keypoints

Fig. 3 Selected keypoints for different frequency values.

5 Experiments

To evaluate the performance of a keypoint detector, it is necessary to measure its repeteability [20]. In our experiments we have used the evaluation benchmark and repeteability measures proposed in [19]. They define the repeatability of a keypoint if, after its transformation according to the ground-truth information, the distance from its nearest neighbor is less than a threshold. Among the proposed evaluation databases, we will use the Laser Scanner dataset [11, 12][1], that contains 3D models

[1] http://vision.deis.unibo.it/keypoints3d/

of 5 objects and 50 scenes with the objects placed in it. The scenes were scanned with the Minolta Vivid 910 scanner to get a 2.5D view (see Figure 4). We will compare the SBP keypoint detector against MeshDoG [24], LBSS [23], KPQ [21] and SP [4]. The source code for these experiments has been implemented using the Point Cloud Library (PCL [18])[2].

Fig. 4 Example images from the laser scanner dataset. The top row shows the 5 object models: cheff, chicken, parasaurolophus, rhino and T-rex. The bottom row contains 5 sample scenes containing the previous models.

First, we evaluate the absolute and relative repeteability of our proposal. In Figure 5 we show the absolute and relative repeatability for different values of selection frequency. Ideally, the detector should find a high number of repeteable keypoints with small frequencies. The figures show that the SBP detector starts showing these desirable results from a frequency of 30%. This seems to be ideally the best value, as it gives a good repeatability without selecting an excesive number of points. This is the frequency that we will use to compare with other descriptors.

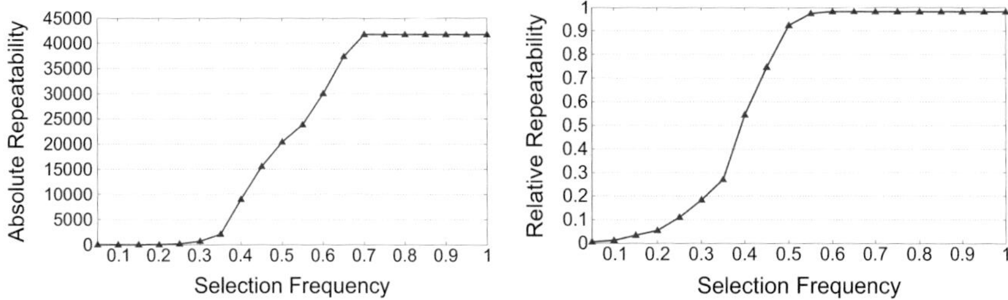

Fig. 5 Absolute and relative repeatability by selection frequency.

In Figure 6 we compare the SBP keypoint detector with other scale-invariant detectors. In this case, our detector has a similar behavior to previously proposed detectors. It outperforms LBSS, and has a slightly lower repeteability than KPQ, MeshDoG and SP.

---

[2] http://simd.albacete.org/supplements/sbp-keypoints/

In addition, in Figure 7, we present the results for scale repeatability. In this case, the SBP keypoint detector clearly outperforms the other descriptors. So, considering that this proposal is in an early stage and that it outperforms state-of-the-art descriptors in some of the tests, we consider that the SBP keypoint detector presents a promising idea that should be tested more extensively in future works to truly assess its remarkable capabilities shown in these experiments.

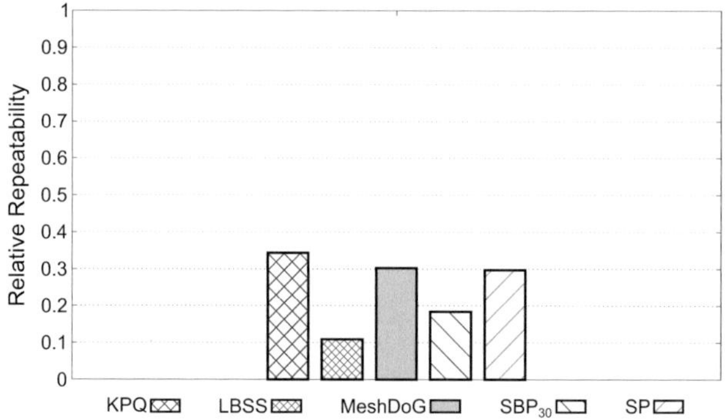

Fig. 6 Relative repeatability for scale-invariant keypoint detectors.

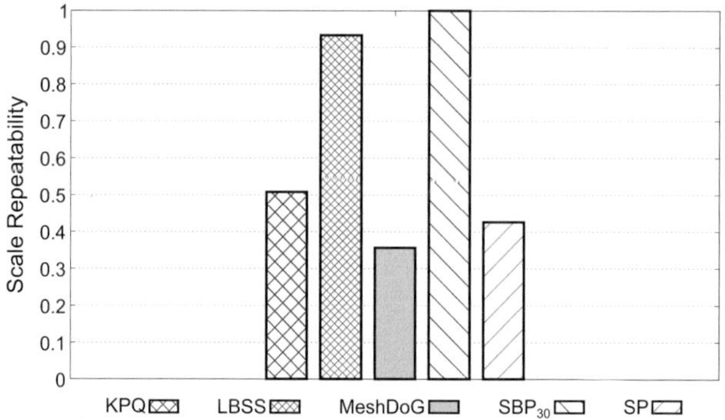

Fig. 7 Scale repeatability for scale-invariant keypoint detectors.

6 Conclusions

In this paper we propose the use of the binary description of the local neighborhood of a point as a decision criteria for a keypoint detector. As this approach is based on a binary discretization can be computed efficiently. We introduce the definition of uniform pattern for this case and test several selection criteria.

In general, the performance of the detector is similar to existing ones, as shown by the experiments carried out over a well known 3D keypoint detection benchmark. However, the results also show that the detector still needs to be improved in future works by, for example, providing a more accurate definition of uniform patterns or creating a more adaptive base descriptor.

Acknowledgements This work has been partially funded by FEDER funds and the Spanish Government (MICINN) through projects TIN2013-46638-C3-3-P and DPI2013-40534-R and by Consejería de Educación, Cultura y Deportes of the JCCM regional government through project PPII-2014-015-P. Cristina Romero-González is funded by the MECD grant FPU12/04387, and Jesus Martínez-Gómez is also funded by the JCCM grant POST2014/8171.

References

1. Alahi, A., Ortiz, R., Vandergheynst, P.: FREAK: fast retina keypoint. In: 2012 IEEE Conference on Computer Vision and Pattern Recognition (CVPR), pp. 510–517, June 2012
2. Bay, H., Tuytelaars, T., Van Gool, L.: SURF: speeded up robust features. In: Computer Vision ECCV 2006. Lecture Notes in Computer Science, vol. 3951, pp. 404–417 (2006)
3. Bro, R., Acar, E., Kolda, T.G.: Resolving the sign ambiguity in the singular value decomposition. Journal of Chemometrics **22**(2), 135–140 (2008)
4. Castellani, U., Cristani, M., Fantoni, S., Murino, V.: Sparse points matching by combining 3D mesh saliency with statistical descriptors. Computer Graphics Forum **27**(2), 643–652 (2008)
5. Chen, H., Bhanu, B.: 3D free-form object recognition in range images using local surface patches. Pattern Recognition Letters **28**(10), 1252–1262 (2007)
6. Filipe, S., Alexandre, L.A.: A comparative evaluation of 3D keypoint detectors in a RGB-D object dataset. In: 9th International Conference on Computer Vision Theory and Applications. Citeseer, Lisbon (2014)
7. Koenderink, J.J., van Doorn, A.J.: Surface shape and curvature scales. Image and Vision Computing **10**(8), 557–564 (1992)
8. Leutenegger, S., Chli, M., Siegwart, R.: BRISK: binary robust invariant scalable keypoints. In: 2011 IEEE International Conference on Computer Vision (ICCV), pp. 2548–2555, November 2011
9. Lowe, D.G.: Distinctive Image Features from Scale-Invariant Keypoints. International Journal of Computer Vision **60**(2), 91–110 (2004)
10. Mair, E., Hager, G., Burschka, D., Suppa, M., Hirzinger, G.: Adaptive and generic corner detection based on the accelerated segment test. In: Computer Vision ECCV 2010. Lecture Notes in Computer Science, vol. 6312, pp. 183–196 (2010)
11. Mian, A., Bennamoun, M., Owens, R.: On the Repeatability and Quality of Keypoints for Local Feature-based 3D Object Retrieval from Cluttered Scenes. International Journal of Computer Vision **89**(2–3), 348–361 (2010)
12. Mian, A., Bennamoun, M., Owens, R.: 3D Model-Based Object Recognition and Segmentation in Cluttered Scenes. IEEE Transactions on Pattern Analysis and Machine Intelligence **28**(10), 1584–1601 (2006)
13. Ojala, T., Pietikainen, M., Maenpaa, T.: Multiresolution Gray-Scale and Rotation Invariant Texture Classification with Local Binary Patterns. IEEE Transactions on Pattern Analysis and Machine Intelligence **24**(7), 971–987 (2002)
14. Romero-González, C., Martínez-Gómez, J., García-Varea, I., Rodríguez-Ruiz, L.: Binary patterns for shape description in RGB-D object registration. In: 2016 IEEE Winter Conference on Applications of Computer Vision (WACV) (2016). Submission ID 135
15. Rosten, E., Porter, R., Drummond, T.: Faster and Better: A Machine Learning Approach to Corner Detection. IEEE Transactions on Pattern Analysis and Machine Intelligence **32**(1), 105–119 (2010)

16. Rublee, E., Rabaud, V., Konolige, K., Bradski, G.: ORB: an efficient alternative to SIFT or SURF. In: 2011 IEEE International Conference on Computer Vision (ICCV), pp. 2564–2571, November 2011
17. Rusu, R.B.: Semantic 3D Object Maps for Everyday Manipulation in Human Living Environments. Ph.D. thesis, Computer Science department, Technische Universitaet Muenchen, Germany, October 2009
18. Rusu, R.B., Cousins, S.: 3D is here: Point Cloud Library (PCL). In: IEEE International Conference on Robotics and Automation (ICRA), Shanghai, China, May 9–13 2011
19. Salti, S., Tombari, F., Di Stefano, L.: A performance evaluation of 3D keypoint detectors. In: 2011 International Conference on 3D Imaging, Modeling, Processing, Visualization and Transmission (3DIMPVT), pp. 236–243, May 2011
20. Schmid, C., Mohr, R., Bauckhage, C.: Evaluation of Interest Point Detectors. International Journal of Computer Vision **37**(2), 151–172 (2000)
21. Somanath, G., Kambhamettu, C.: Abstraction and generalization of 3D structure for recognition in large intra-class variation. In: Computer Vision ACCV 2010. Lecture Notes in Computer Science, vol. 6494, pp. 483–496 (2011)
22. Tombari, F., Salti, S., Di Stefano, L.: Unique signatures of histograms for local surface description. In: Computer Vision ECCV 2010. Lecture Notes in Computer Science, vol. 6313, pp. 356–369 (2010)
23. Unnikrishnan, R., Hebert, M.: Multi-scale interest regions from unorganized point clouds. In: IEEE Computer Society Conference on Computer Vision and Pattern Recognition Workshops, CVPRW 2008, pp. 1–8, June 2008
24. Zaharescu, A., Boyer, E., Varanasi, K., Horaud, R.: Surface feature detection and description with applications to mesh matching. In: IEEE Conference on Computer Vision and Pattern Recognition, CVPR 2009, pp. 373–380, June 2009
25. Zhong, Y.: Intrinsic shape signatures: a shape descriptor for 3D object recognition. In: 2009 IEEE 12th International Conference on Computer Vision Workshops (ICCV Workshops), pp. 689–696, September 2009

Unsupervised Method to Remove Noisy and Redundant Images in Scene Recognition

David Santos-Saavedra, Roberto Iglesias and Xose M. Pardo

Abstract Mobile robotics has achieved important progress and level of maturity. Nevertheless, to increase the complexity of the tasks that mobile robots can perform in indoor environments, we need to provide them with a scene understanding of their surrounding. Scene recognition usually involves building image classifiers using training data. These classifiers work with features extracted from the images to recognize different categories. Later on, these classifiers can be used to label any image taken by the robot. The problem is that the training data used to recognize the scene might be redundant and noisy, thus reducing significantly the performance of the classifiers. To avoid this, we propose an unsupervised algorithm able to recognize when an image is unrepresentative, redundant or outlier. We have tested our algorithm in real and difficult environments achieving very promising results which take us a step closer to a complete unsupervised scene recognition with high accuracy.

Keywords Scene recognition · Canonical views · Unsupervised solution · Mobile robotics

1 Introduction

Nowadays, one of the limitations of robots is scene understanding. Robots are not able to understand high level concepts such as "kitchen", "corridor" or "laboratory". Nevertheless, knowing 'where i am', i.e., the categorization of the scene, is very useful in robotics. Thus, robot localization and robot learning are some examples, to mention but a few, where scene recognition would be very useful. In the case of robot localization, scene recognition can help the robots to distinguish the different places

D. Santos-Saavedra(✉) · R. Iglesias · X.M. Pardo
CITIUS (Centro Singular de Investigación En Tecnoloxías da Información),
Universidade de Santiago de Compostela, Santiago de Compostela, Spain
e-mail: {david.santos,roberto.iglesias.rodriguez,xose.pardo}@usc.es

L.P. Reis et al. (eds.), *Robot 2015: Second Iberian Robotics Conference*,
Advances in Intelligent Systems and Computing 418,
DOI: 10.1007/978-3-319-27149-1_54

where they are moving. Once we are able to recognize different places, we can apply this ability, for example, to improve the estimate of the pose of the robot. Despite it is not an accurate source of information, scene recognition can provide useful data to help to obtain a robust pose estimation of the robot [1] in real conditions. There are many approaches that utilize the information from different sensors [2][3][4] to determine the topological localization of the robot. However, this localization is based in coordinates, their comprehension of the space is based usually in a map and the different positions. Regarding robot learning, scene recognition would allow the robot associating what it learns to where it is learning. This can be applied, for example, to exploration tasks: when the robot enters in a new room, if it has the same category than a previously known place, the robot will be able to launch the correct controllers or learned knowledge for that place. However, the use of scene recognition opens other challenges that must be dealt with. An important problem that needs to be solved and that appears in real experiments is the acquisition frequency of data. In scene recognition, this frequency affects directly to the size and quality of the captured dataset: low frequencies can cause the loss of relevant information and high frequencies will collect many identical or irrelevant data. In this line, we have carried out a preliminary study for the detection of representative views of a scene (canonical views [10][11]). Our idea is based on allowing high frequencies of images acquisition and then doing a noise filtering. When we talk about noise, we are referring to two types of images that we want to avoid: First, we are not interested on unrepresentative images, i.e., images that not contain enough information to recognize the place (Figures 1c and 1d) or images that are very similar to each other (Figures 1a and 1b).

In this paper, we present an algorithm inspired in [12] to reduce the size of datasets detecting these unrepresentative and redundant images.

2 Description of the Unsupervised Algorithm

This work follows the objective of reducing the size of dataset. We want to contribute to remove unnecessary data from datasets without information loss. Our sets of images are usually composed by raw data ranging from perfect samples, to images that only disturb the experiments. For this reason, we believe that is important to reduce the size of datasets removing these bad samples. In our research, we consider that there are three conditions that the images must fulfill to be part of the final set of images: Coverage, Orthogonality and likelihood [10]. We think that the samples of a good set of images must provide enough information (Coverage), they must be highly probable (Likelihood) and different amongst them to avoid redundancy (Orthogonality).

In this work, we studied how to identify which images are highly probable and representative. We develop an unsupervised algorithm which has important similarities with the one described in [12], although in this work is used for dimension reduction. Our objective is to create dense clusters of images, i.e., groups of very

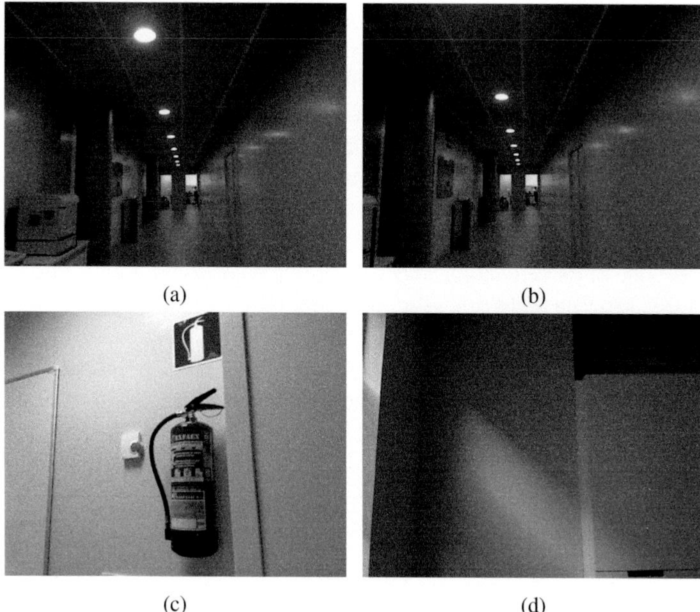

Fig. 1 *(a)(b) Redundant images.* Due to the high acquisition frequency, the position of the camera will not change significantly from one image to other. *(c)(d) Unrepresentative images.* Samples without useful information to identify the scene.

similar images. One of the images of the cluster can be selected as representative and the rest of then could be discarded as redundant. During this clustering, the images without cluster will be considered as outliers. In addition, we want to achieve the reversibility of the process, i.e., when a new image appears, if it was assigned in the past to one cluster or it is similar to the centroid of one of them, it should be associated to the same cluster for each appearance. However, if their similarity is low, the image must be categorized as outlier until more samples like that appear allowing us to create a new dense cluster. With the definition of dense clusters we want to achieve Voronoi regions [13] where each cluster must be within a region.

2.1 Image Representation and Measurement of Similarity

To carry out the comparison of the images, first it is necessary to decide how to represent the images. In our case, we have chosen *Local Difference Binary Pattern* [8] due to the excellent results we achieved with this descriptor in our previous works [5] [6]. We took into consideration LDBP and SURF [7] as candidates to be the descriptor but LDBP got the best results for a preliminary experiment carried out with the both options. In addition, the solution based in SURF and bag of words requires a previous training step to create the visual words. Therefore, LDBP is a faster option without intermediate steps. This representation captures properties, such as, rough geometry and generalizability by modeling the distribution of local structures. The LDBP has two components: Local Different Sign Binary Pattern (LSBP) and Local Different Magnitude Binary Pattern (LMBP). The LSBP, also referred to as the Census Transform (CT) [9], is a non-parametric local transform based on the comparison amongst the intensity value of each pixel of the image with its eight neighboring pixels, as illustrated in Figure 2. As we can see in this figure, if the center pixel is bigger than (or equal to) one of its neighbors, a bit 1 is set in the corresponding location. Otherwise a bit 0 is set. The eight bits generated after all the comparisons have to be put together following always the same order, and then they are converted to a base-10 number in the interval [0, 255]. This process maps a 3×3 image patch to one of 256 cases, each corresponding to a special type of local structure, and it is repeated for every pixel of the original image.

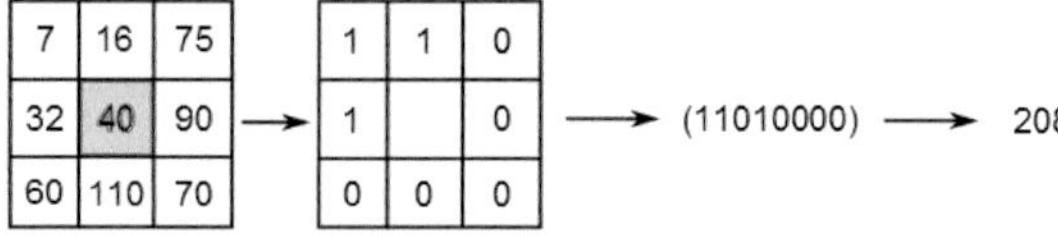

Fig. 2 Illustration of the Census Transform Process on a 3×3 image patch

The component LMBP serves to complete the representation of the image. In this case, the LMBP is computed as the intensity difference between the center pixel and its neighboring pixels. If the difference in intensity amongst the center pixel and one of its neighbors is higher than a threshold T, a bit 1 is set, otherwise a bit 0 is set. Like in the case of the LSBP, the eight bits generated after all the comparisons have to be put together following always the same order, and then they are converted to a base-10 number.

Thus, for every image we can compute the holistic representation given by the combination of the *LMBP* and the *LSBP*. Once this process is over, we can compute the histogram of LSBP and LMBP as a feature representation. Both the LSBP and the LMBP histograms are 256 dimensions (the bins of the histograms are each one of the values that the LMBP and LSBP can take), therefore, the new feature representation is 512 dimensional (256×2). It is a common practice to suppress the first and the last bins of these histograms, due to noise cancellation and the removal of not significant information, that is the reason why the dimension that appears in Figure 3 is 508. We will call *LDBP* (local difference binary pattern) to the combination of the LSBP and LMBP histograms.

Distance Metric. Once we have selected a representation for the images, it is necessary to decide the metric to evaluate how similar two images are. In our case, we

Fig. 3 The combination of LSBP and LMBP will be used as image descriptor.

will use the correlation of the descriptors. The image descriptors are histograms so we will use the Eq. 1 to estimate the correlation between them. Due to the correlation value of two histograms is higher while they are similar, we estimate it as $d(H_1, H_2) = 1 - correlation$. This change will allow us to use the correlation as a distance metric, where low values are related to close samples and high values to different images.

$$d(H_1, H_2) = 1 - \frac{\sum_I (H_1(I) - \bar{H}_1)(H_2(I) - \bar{H}_2)}{\sqrt{\sum_I (H_1(I) - \bar{H}_1)^2 \sum_I (H_2(I) - \bar{H}_2)^2}} \quad (1)$$

where $I = 1..N$, N is the number of histogram bins and $\bar{H}_k$ is defined by:

$$\bar{H}_k = \frac{1}{N} \sum_J H_k(J) \quad (2)$$

where $J = 1..N$.

2.2 *Description of the Algorithm*

The algorithm 1 describes the solution that we propose to identify relevant images and remove redundant data. It is important to remind that this algorithm will detect three types of images: references (centroids of the dense clusters), redundant images and outliers. Also, it is important to notice that at the beginning of the process, it is necessary to decide an initial k. This k defines the number of component of the first cluster to create. As our objective is to obtain dense clusters, we will search the cluster with k elements with lowest $r_{i'}^k$. This is the distance from the centroid of the cluster to the k-neighboring image. Lower values of $r_{i'}^k$ are related to clusters were the images are closer to each other and a higher value of $r_{i'}^k$ means that the k-th element of the cluster is far of the centroid and their similarity is lower. This parameter will be used as density reference for the search of the next clusters. Once the first cluster is defined, we will search more clusters with the same number of elements k and a density value lower or equal to $r_{i'}^k$. If there are not more clusters

Algorithm 1.

Let's define R as the whole set of images, $refs$ represents the set with the relevant images and O is the set of outlier images.
Step 1: Choose an initial value for k where $k \leq$num of images-1.
Step 2: For each image i of R, we estimate r_i^k where
$r_i^k = d(H_i, H_{k\_nearestNeighbor\_of\_i})$.
Step 3: Search the image i' with minimum $r_{i'}^k$.
Step 4: Build the set of redundant images of i':
Redundant(i') = $\{j \in R \mid d(H_j, H_{i'}) \leq r_{i'}^k\}$,

Remove the images included in Redundant(i') from R:
$R \Leftarrow R \setminus Redundant(i')$.
Step 5: Define a threshold $\epsilon = r_{i'}^k$.
Step 6: Include i' in the set of relevant images:
$refs = \{refs \cup i'\}$.
Keep the radius of this relevant image: $r(i') = r_{i'}^k$.
if cardinality($refs$) > 1 **then**
(For every redundant image, check which one of the relevant images in $refs$ is the closest. Move the redundant image to the correct set of redundant images if it is necessary).
For every ref_i included in $refs$ do:
$\forall j \in Redundant(ref_i)$:
$BMU = argmin_m d(H_j, H_{ref_m})$.

if $BMU \neq ref_i$ **then**
Reallocate j: $Redundant(BMU) = \{Redundant(BMU) \cup j\}$
end if
Update the radii of the relevant images:
$r(i') = max_m d(H_{i'}, H_m)$, $\forall i' \in refs$ and $\forall m \in Redundant(i')$.
end if
Update the threshold $\epsilon = max\ r(i), \forall i \in refs$.
$k = k_{initial}$.
if $k > cardinality(R) - 1$ **then**
$k = cardinality(R) - 1$.
end if
if $k = 1$ **then**
Go to step 8.
end if
while $r_i^k > \epsilon$ **do**
(a)
k = k - 1.
$r_i^k = inf_{i \in R}\{r_i^k\}$.
(b)
if $k = 1$ **then**
Go to step 8.
end if
end while
Paso 7: Go to step 2.
Paso 8: Return $refs$ and Redundancy(i'), $\forall i' \in refs$.

with k elements to satisfy the density condition, then the size of the cluster to search is reduced by one until a new candidate appear. The clustering stops when the size of the clusters to form drop to one. This methodology would allow us the definition of dense clusters but it does not ensure Voronoi regions. We need to be sure about the correct separation of the clusters, i.e., we want to achieve a partition of the space into a set of Voronoi regions so that each cluster is within the same region. We achieved this by checking if each image belongs to its cluster when a new group is created. If the process detects an image that is miss-assigned, i.e., that the distance between it and the centroid of another cluster is lower than with the centroid of its current cluster, the image is re-assigned. With this strategy we will get well-formed clusters with a clear separation amongst them resulting in Voronoi regions.

As result of the execution of our algorithm it is possible to detect three types of data. First, the algorithm will return a set of images, that we will refer as *relevant images*, which have high probability of appearance because they were selected as representation for different similarity groups. The rest of images in the same cluster or Voronoi region are discarded as redundant. Together with the relevant images, the algorithm provides a radius ϵ_i that defines the maximum distance to be associated to the relevant image i. We can remove or identify the redundant images with this information easily. Finally, the algorithm provides us an additional information as result to the previous group: the outliers, the remaining images which were not classified into a group. This means that they are different enough to the previous selected images but they are not frequent enough to be taken into consideration and therefore create a new group. For static dataset, these images can be removed, but for incremental sets where the images are been continuously included, this type of images must not be ignored at all. In real time applications, the appearance of new images can become some outlier into a relevant view. They could be classified as outlier because the camera still has not captured enough samples of the same type. For this reason, the appearance of new samples can become a set of outlier images in a new dense cluster. Nevertheless, since our research was oriented to datasets already created, we have not designed the steps to build new clusters from outliers and new images.

3 Experiments

3.1 Experimental Setup

We want to use the algorithm described in the previous section in the image acquisition routine of our robot. For this reason, we have collected samples from the research center where the robot is usually moving to analyze the performance of the algorithm. This set consists in 3078 images in gray scale. They were obtained using a Pioneer P3DX robot and a Kinect camera with an acquisition frequency of 1 image/sec in different places of the building. We have considered several places of the center but they can be summarized into eleven classes attending to the similarities of their spatial distribution:

- Office.
- Entrepreneurship Lab.
- Staircase.
- Common Staff areas (floor 1 and 2).
- Assembly Hall.
- Common Staff areas (S1 floor).
- Kitchen.
- Laboratories.
- Common Staff areas (Ground floor).
- Instrumentation Lab.
- Robotics Lab.

This is useful for image classification tasks like the experiments we did in the past. We have categorized the images in eleven classes because they correspond to different places. However, some categories are very similar despite not being the same. Before carrying out the experiment with the algorithm, we wanted to analyze the capability of LDBP as image descriptor and correlation as distance metric to match similar images. To do this, we carried out an 1-NN clustering to associate each image of the dataset with the closest sample amongst the 3078. The experiment showed that 2980 images, the 96.81% of the cases, were matched with another image of the same category and only the 3.19% of the cases the category was different. This allow us to confirm that the image descriptor and the distance metric selected are good options for the algorithm.

3.2 Experimental Results

We have carried out an experiment to analyze the performance of the algorithm with our dataset. In this case, we have not provided any information about the category of each image. The experiment was realized with a initial $k = 100$ and the complete dataset. The results show that the algorithm was able to identify 336 relevant images and 1852 (60.16%) redundant images, leaving the rest of the dataset as outliers.

The unsupervised creation of clusters of redundant images does not provide enough information about the quality of the results, i.e., we actually do not know whether these clusters of similar images group together images that represent the same scene/room, etc. Because of this, we decided to analyze the results obtained when our unsupervised algorithm is applied on an image set which we have already used in the past, [5][6], and which was manually labeled. The images included in this set, described in section 3.1, have been manually labeled, according to their belonging to 11 possible classes. The comparison amongst the clusters detected by the unsupervised algorithm described in this paper, and the classes described in section 3.1, and which have been created manually, will provide some clues about whether this clusters put together images that really represent the same environment or scene. We are interested on knowing whether the clusters have some kind of correspondence with the predefined categories that have been manually created, and whether these clusters cover all the categories or only a few. This would open the possibility for an unsupervised solution for on-line dynamic image categorization. Therefore, to be able to analyze this information, we will use the class of each image. Nevertheless, we want to emphasize that this class information if used only to analyze the results. The unsupervised algorithm described in Alg. 1 does not use any type of class information at all. Table 1 shows the relation between the original label of the images classified as redundant (rows) and the label of the centroid to which they were associated (columns). In the 94.81% of the cases, the images where assigned to a cluster where the centroid belong to the same scene. It is also important to realize that most cases where there is discrepancy between the places to where an image and its centroid were taken are related to the classes 5, 6 and 7 (common staff areas but at different floors). This is a clear example of the difficulty to decide the category to where some images belong due to the similarities amongst them. Despite of that concrete problem, the algorithm seems to have a good performance creating reliable clusters.

Table 1 Confusion matrix to summarize the results of the experiment for an initial $k = 100$. The diagonal shows the correct classified images and the other positions show the number of miss-classifications.

Class	0	1	2	3	4	5	6	7	8	9	10	Total
0	282				2							284
1	1	67										68
2			45									45
3				63								63
4				1	143		1					145
5						281	2	70				353
6						2	319	1				322
7				1		15		305				321
8									74			74
9										105		105
10											72	72

4 Conclusions

In this paper we proposed a solution to solve one of the problems that appear when scene recognition is used in real time application: high acquisition frequency of images. Due to the large number of samples captured, the resulting dataset usually contains noisy and redundant images. Our objective was to identify and remove this information to get a shortest dataset with only images that provide relevant information. For this reason, we proposed an algorithm inspired in [12] to create dense clusters and remove redundant images. With this algorithm we want to identify three types of information: relevant images, redundant images and outliers. Additionally to this clustering, we want to get the reversibility of the process, i.e., if a redundant image appears again, it will be assigned to the same cluster and discarded and for this reason, we built Voronoi regions. During this process, the appearance of new outliers can become a unrepresentative image in a relevant image. This would allow the dynamic creation of new dense clusters. We have carried out an experiment with a dataset obtained in our research center to analyze the performance of the algorithm. The results shown the capability of the unsupervised algorithm to create dense and homogeneous clusters, i.e., clusters where the images belong to the same place. This confers to the algorithm more value due to possibility of applying the proposed solution to scene recognition as an unsupervised solution for image classification and robot localization.

Acknowledgment This work was supported by the research grant TIN2012-32262 (FEDER) and by the Galician Government (Xunta de Galicia), Consolidation Program of Competitive Reference Groups: GRC2013/055 (FEDER cofunded), GRC2014/030 (FEDER cofunded).

References

1. Santos-Saavedra, D., Canedo-Rodriguez, A., Pardo, X.M., Iglesias, R., Regueiro, C.V.: Scene recognition for robot localization in difficult environments. LNCS, vol. 9108, pp. 193–202 (2015)
2. Canedo-Rodriguez, A., Alvarez-Santos, V., Santos-Saavedra, D., Gamallo, C., Fernandez-Delgado, M., Iglesias, R., Regueiro, C.V.: Robust multi-sensor system for mobile robot localization. In: Natural and Artificial Computation in Engineering and Medical Applications. 5th International Work-Conference on the Interplay between Natural and Artificial Computation. LNCS, vol. 7931, pp. 92–101 (2013)
3. Tardos, J.D., Neira, J., Newman, P.M., Leonard, J.J.: Robust mapping and localization in indoor environments using sonar data. IJRR **21**(4), 311–330 (2002)
4. Christian, F., et al.: RFID-based hybrid metric-topological SLAM for GPS-denied environments. In: 2013 IEEE International Conference on Robotics and Automation (ICRA). IEEE (2013)
5. Santos-Saavedra, D., Pardo, X.M., Iglesias, R., Álvarez-Santos, V., Canedo-Rodríguez, A., Regueiro, C.V.: Global image features for scene recognition invariant to symmetrical reflections in robotics. In: XV Workshop of Physical Agents, pp. 29–37 (2014)
6. Santos-Saavedra, D., Pardo, X.M., Iglesias, R., Canedo-Rodríguez, A., Álvarez-Santos, V.: Scene recognition invariant to symmetrical reflections and illumination conditions in robotics. LNCS, vol. 9117, pp. 130–137 (2015)
7. Bay, H., Ess, A., Tuytelaars, T., Van Gool, L.: Speeded-up robust features (SURF). Computer Vision and Image Understanding **110**(3), 346–359 (2008)
8. Meng, X., Wang, Z., Wu, L.: Building global image features for scene recognition. Pattern Recognition **45**(1), 373–380 (2012)
9. Wu, J., Rehg, J.M.: CENTRIST: A Visual Descriptor for Scene Categorization. IEEE Trans. Pattern Analysis and Machine Intelligence **33**(8), 1489–1501 (2011)
10. Simon, I., Snavely, N., Seitz, S.M.: Scene summarization for online image collections. In: ICCV, vol. 7, pp. 1–8 (2007)
11. Hall, P.M., Owen, M.: Simple canonical views. In: BMVC (2005)
12. Mitra, P., Murthy, C.A., Pal, S.K.: Unsupervised feature selection using feature similarity. IEEE Transactions on Pattern Analysis and Machine Intelligence **24**(3), 301–312 (2002)
13. Aurenhammer, F., Klein, R.: Voronoi diagrams. In: Sack, J.-R., Urrutia, J. (eds.) Handbook of Computational Geometry, pp. 201–290. North-Holland, Amsterdam (2000)

Part XIII
16th Workshop on Physical Agents

Procedural City Generation for Robotic Simulation

Daniel González-Medina, Luis Rodríguez-Ruiz and Ismael García-Varea

Abstract In robotics, simulation plays a fundamental role for testing the models and techniques in a controlled environment prior to conducting experiments on real physical agents. In addition, some kind of scenarios can be easily reproduced within a simulator which is not always possible with a real robot. Building simulation environments, however, can be a tiresome and complex task. For robots performing in an urban environment, manually designing a city for testing navigation or localization algorithms can be prohibitive. As an alternative, in this work, we propose the use of procedural graphic techniques aimed at producing synthetic cities that can be employed within a robotic simulator. Experiments with the generated environments have been performed on a real simulation tool to assess the viability of the approach here proposed.

Keywords Procedural generation · Robotic simulator · Automatic robotic benchmarks generation

1 Introduction

When it comes to developing a new model, method or technique in robotics, the use of simulators have become an invaluable tool that is widely used prior to carry out experiments on real robots. Robotic simulators are focused on mimicking the behaviour of robots and their surrounding environments, using 3D virtual scenarios

---

D. González-Medina(✉)
Albacete Research Institute of Informatics, University de Castilla-La Mancha,
Campus Univ. s/n, 02071 Albacete, Spain
e-mail: Daniel.Gonzalez@uclm.es

L. Rodríguez-Ruiz · I. García-Varea
Computer Systems Department, University of Castilla-La Mancha,
Campus Univ. s/n, 02071 Albacete, Spain
e-mail: {Luis.RRuiz,Ismael.Garcia}@uclm.es

L.P. Reis et al. (eds.), *Robot 2015: Second Iberian Robotics Conference*,
Advances in Intelligent Systems and Computing 418,
DOI: 10.1007/978-3-319-27149-1_55

without the need of using a real robot. Some of the main advantages of robotic simulation, in contrast to real agents are:

- Saving cost related to actual robotic platforms and its maintenance.
- No mechanical and electric problems.
- No need to charge batteries.
- Saving time on the tests.

A robotic simulator makes use of different real or synthetic environments to simulate the robot operation mode. We can classify the environments in two categories: indoor and outdoor. Indoor environments correspond to the inside of any building and outdoor environments refer to any outdoor place, like a park, a city, a beach, etc. From all these outdoor environments, cities are one of the most common places where a robot (or a swarm of robots) can be deployed. As a first example, we can cite the case of autonomous vehicles, where a robotic car navigates and drives entirely on its own with neither human driver nor remote control. In this sense, the *DARPA Grand Challenge* for autonomous robotic vehicles was proposed, where every participant robot had to complete a series of paths along different streets and avoiding obstacles to reach the finish line. The first edition was in 2004 but, it was in 2005 when a robot was capable of completing the challenge (*Stanley*) [5]. Afterwards, other challenges were proposed in urban scenarios, including traffic, performing complex manoeuvres such as merging, passing, parking and negotiating intersections. Also, in 2014, *Google* developed its own autonomous robotic car [7], which is able to self-drive in a city and to solve most of the problem that can arise. Besides, the company *Amazon* is developing a project named *Amazon Prime Air*, which is based on using drones and GPS systems to deliver packages to any place [2].

In all the previous examples, the availability of virtual cities as benchmark testbeds could boost the development of autonomous vehicles since we could provide an environment where all the models, techniques and algorithms can be tested in changing situations. Designing a virtual city, however, can be a tiresome and hard task because each element (building, street, etc.) has to be manually designed and appropriately placed. As an alternative, the use of procedural graphic techniques can be considered. Here, the city and its constituent elements are automatically (or semi-automatically) designed. In this work, we propose a framework to produce procedural cities that can be used as testbeds in outdoor robotic simulators.

The rest of the paper is organized as follows. In section 2 an overview of procedural graphics is provided. In section 2.1 the steps followed to produce virtual cities are described. Next, in section 3 we discuss how to integrate the generated cities into a robotic simulator. Finally, in section 5 some conclusions and future work are outlined.

2 Procedural Graphics

Procedural graphics is about the generation of any type of graphic content (geometry, textures, etc.) by using mathematical models and algorithms instead of manually designing the graphical elements [6]. The use of a procedural approach has several advantages such as:

- **Reducing cost and time** : Producing a detailed 3D graphic model requires a skilled designer (or designers) working for weeks, months or even years, depending on the complexity of the project.
- **Accurate and easy modeling of some type of objects**: Some objects that exist in nature, like plants, trees, terrains, etc. are hard to be designed by hand but they can be precisely approached by using mathematical models.
- **Saving memory and computational costs:** A scene designed by hand usually requires that all the elements needed to render the models are stored in memory, which result in high memory and computational requirements. On the contrary, the elements needed by a procedural model can be generated on-the-fly, only when they are actually needed and then freed from memory.
- **Control parameters:** The models used for procedural generation normally rely on parameters that can easily define the high and low level details of the model. For example, in the case of tree generation we could control the height and the width of the trunk or the density and size of leaves.
- **Generation of similar models:** The parameters described in the previous items, along with some other internal parameters can be used to produce hundred of thousands of models with a common structure but introducing slight changes that help produce different results.
- **Multiple levels of detail:** With a procedural model it is easy to generate objects with different level of detail (number of polygons, image resolution, etc.) This way, highly detailed models can be obtained when needed (for instance when an object is close to the point of view) and less detailed objects can be used when the object is, for instance, far away from the scene point of view, this way saving rendering time.

There are, however, some drawbacks in procedural modelling. Keeping a full control on the final result is really difficult for most of the techniques employed. Besides, the algorithms for producing the content are usually computationally expensive. In spite of this fact, procedural techniques are extensively used to produce a great variety of graphical content, from basic textures, like wood or grass, with techniques like Perlin Noise and Tiling [6], to complex models such as the recreation of organic structures like snowflakes or trees, using models like L-Systems [10].

2.1 City Generation

In order to construct a procedural city we have to initially generate its basic elements [8]. Specifically, these elements are: the road network, the blocks and the buildings. By adequately combining and distributing those elements we can recreate the structure of a typical city [1, 11].

The procedural city generation procedure employed in this work comprises six stages, as it can be seen in the pipeline shown in Figure 1, where this process is controlled by some generation parameters. Some of these parameters are: the global size of the city, the road width or the allotment areas for buildings.

In the following sections, we describe all the stages followed to achieve the final result.

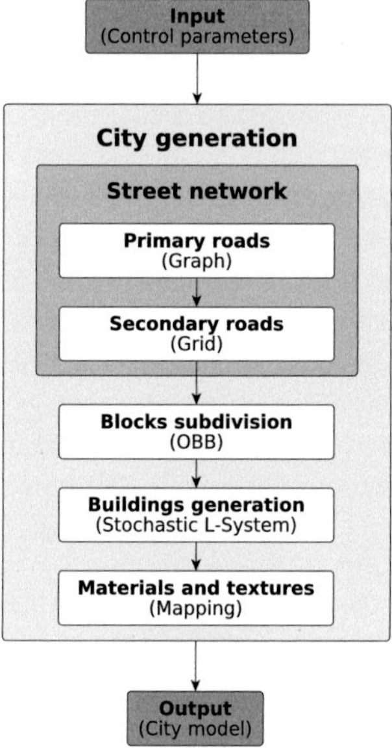

Fig. 1 Pipeline of the city creation. The element to be created (stage) is shown in bold and, below, the technique employed to this end is also shown.

2.1.1 Road Network

The road network is actually composed of a primary and a secondary network. According to several studies on the structure and distribution of the roads [1, 11], we can conclude that this is basically a random process, where no pattern can be clearly identified. From this, we can use the following approach to produce the primary network.

The primary roads will be represented by a graph, where each node is linked to a point in a 2D space. That means that each node stores a pair that determines the coordinates of the node. This way, a node represent a road intersection (which position is determined by the coordinates) and, an edge between two nodes, represents a road. The graph is constructed by initially generating the nodes (where random coordinates are associated to each node). Next we produce a set of edges that connect the nodes following a set of rules:

- Two nodes are connected if the distance between their coordinates is less than a predefined distance.
- Two nodes are joined together if they are too close.
- Two nodes are connected if the edge which connect them does not cut any other edge already created.

The final step is to transform the graph into a 2D geometry according to the nodes and edges previously defined.

Once the primary roads been created, we have a set of regions that are delimited by those roads. The next step consists in building the secondary roads (or secondary network) on these regions. To this end, a grid is overlapped on the geometry resulting from the primary network graph [4]. As a result of this, we have the primary roads constructed before and the secondary roads produced by the grid. The intersection of these two networks will form the final road map of the city. In Figures 2 and 3, a fragment of the New York city road map and an example of a road network built following the approach just described are shown. Here, we can see the similarities between them, where the primary roads follow a random distribution while the secondary network has a grid structure.

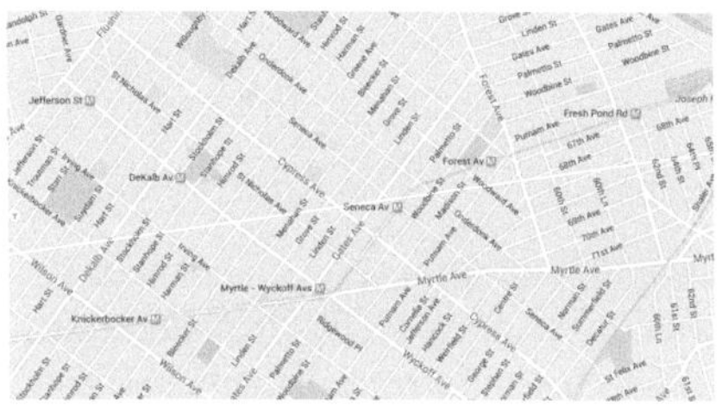

Fig. 2 Part of the urban map of New York city [7]

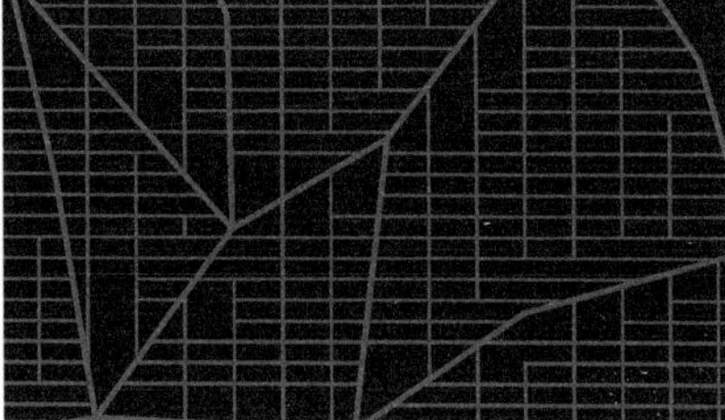

Fig. 3 Part of the road network generated.The areas surrendered by roads in black are the corresponding blocks

2.1.2 Blocks

Once the road network has been defined, the blocks delimited by roads have to be filled with buildings or other structures. The first task to perform is to subdivide each block in different smallholdings that will be assigned to each building. To this end, we have relied on the OBB (*Oriented Bounding Box* [15]) algorithm. OBB is a recursive algorithm that subdivides any polygon or geometric shape in smaller fragments. The process of subdivision consists of 3 steps:

1. **Oriented Bounding Box Creation:** a bounding box (4-side polygon), which delimits the current geometric shape, is computed.

2. **Subdivision:** the box is split in two parts along the longest sides of the oriented bounding box. Next the current geometric shape is subdivided according to the new box subdivision.
3. **Recursion:** the previous two steps are then applied to the new subdivisions of the geometric shape.

In Figure 4 the operation mode of the OBB algorithm is shown. In our block subdivision, the recursive procedure is applied until reaching an area size below a specific threshold. In addition, if any new subdivision creates an inner area (that is, a householding that does not have access to a road) the recursion is also stopped.

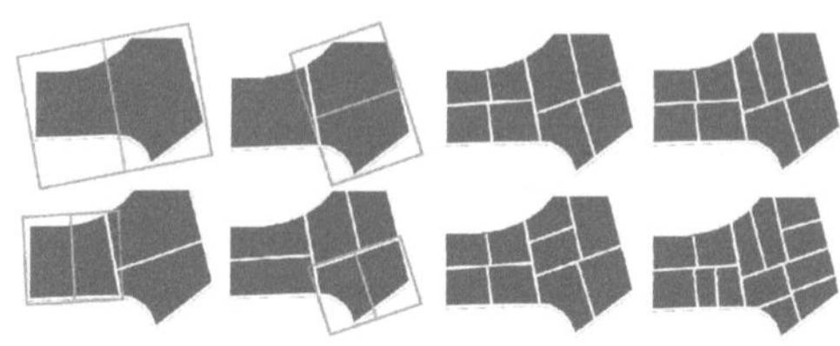

Fig. 4 Oriented bounding box subdivision. This algorithm recursively splits a geometry shape into two smaller areas [15].

2.1.3 Buildings

For every allotment generated by the OBB algorithm, a building is constructed. The design of the buildings is approached by using the L-system modeling technique [10]. L-systems are actually a variation of formal grammars and were originally proposed to model the growing of plants [13]. Formally, an L-system is a grammar defined by the tuple $G = (V, \omega, P)$: where V is an alphabet, ω is the grammar initial symbol and P is a set of productions. Starting from ω, the productions are iteratively applied to derive strings that belong to the language specified by the grammar.

L-systems are very similar to Context-Free grammars but there are some differences that are worth mentioning. The first one is that all the symbols in the alphabet are non-terminal symbols. Besides, for each symbol $X \in V$ we assume the existence of the implicit production $X \rightarrow X$. This way, the derivation process for a string never ends since every symbol in the current string can be derived (either using a production in P or using the implicit production $X \rightarrow X$). The second main difference is that at each derivation steps as many symbols as possible are derived in parallel.

In an L-system, the symbols in V are mapped to a set of modeling operators so that each string actually codes a set of geometric operations. For example, let's define an alphabet $V : \{E, S\}$ where E is mapped to the geometric operator *Extrude* and S is mapped to the operator *Scale*. In Figure 5 we can see the process of translation of the string "ESE" (which we can assume as produced by an L-system) into the corresponding geometric operators. Here, we assume that this process is performed on an initial object that is a plane.

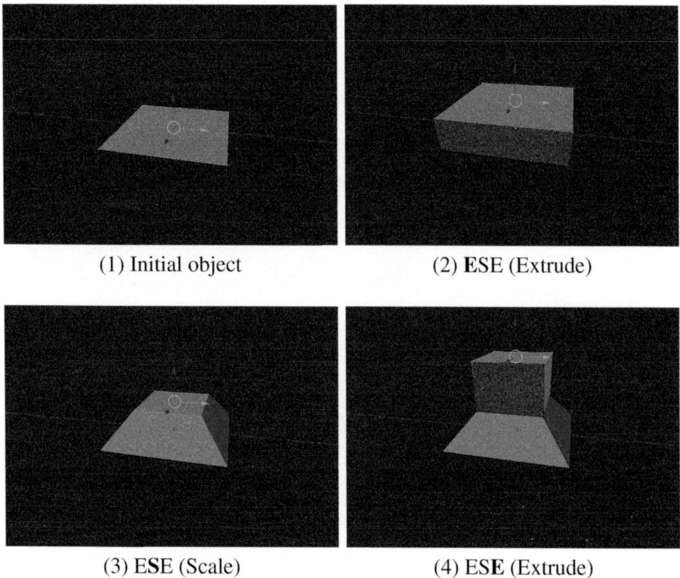

Fig. 5 Apllying the translation of the string "ESE" into an object (plane) sequentially.

To design an L-system for procedurally constructing buildings, we have to carry out the following steps:

1. **Alphabet generation and mapping:** a set of modeling operators have to be defined. Next the alphabet of the L-system is defined by mapping each operator to a symbol.
2. **L-system definition:** Along with the alphabet defined in the previous step, a set of production rules that will determine how the L-system will produce strings has to be built.
3. **String derivation:** Once the L-system has be specified, a string will be derived from the initial symbol and the set of productions. Here, we have to specify a stop condition to finish the derivation process. This condition usually consists in reaching a previously specified number of derivation steps.
4. **Translation:** Finally, each symbol in the string derived is translated into a modeling operators (as defined in the first step) and the corresponding operator is applied. As a result a geometric 3D model is obtained.

Depending on the set of productions, L-systems can be either deterministic or non-deterministic. Non-deterministic models are preferable for graphic generation because they can easily produce different variations from a single model. This way we can, for instance, build an L-system to produce trees in such a way that every time the L-system is used to derive a string, different strings are produced and, therefore different tree designs are then obtained.

Table 1 Some of the modeling operators mapped to the symbols of the stochastic L-system used in this work

Symbol	Modeling operator	Initial object	Resulting object
E	Extrude	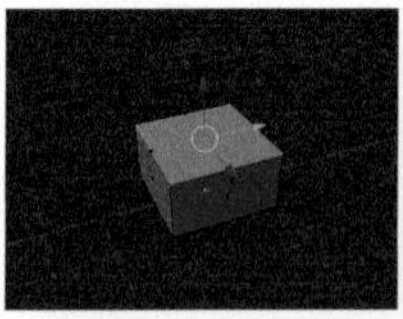	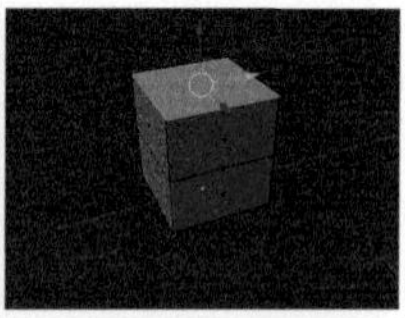
S	Scale	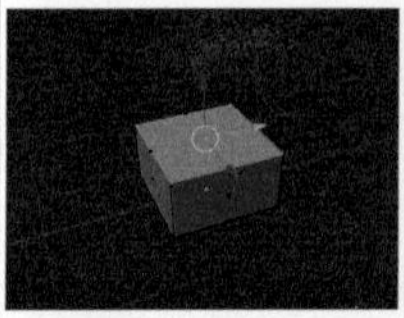	
F	Terraced roof shape	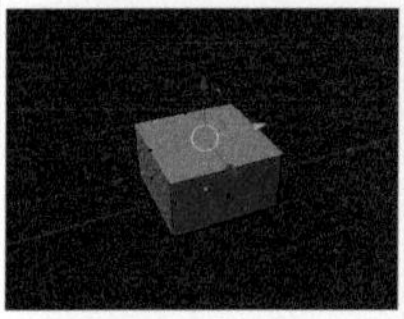	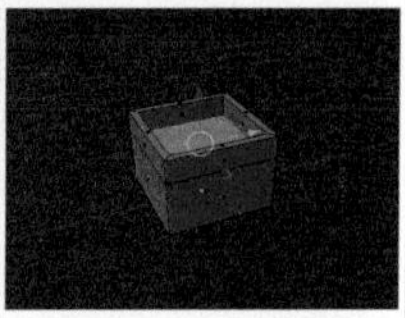
P	"Balcony" shape	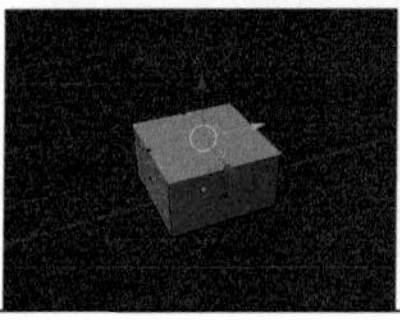	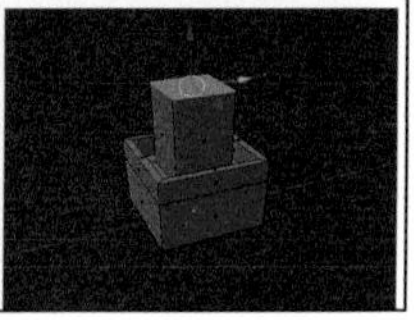

Here, we propose the use of a stochastic and, hence, non-deterministic L-systems to generate buildings. In a stochastic L-systems, each string is produced with certain probability or, in other words, a probability distribution is defined among all the strings in the language. A simple way to define a stochastic L-system is by using probabilistic productions. This way, each symbol in the alphabet V can be derived into different strings based on a probability distribution.

Below, the stochastic L-system $G = (V, \omega, P)$ used in this work is shown, where each production of P is labeled with a probability:

- $V : \{E, S, I, \#, F, P, +, -, >, <, X\}$
- $\omega : X$
- P :
  - $X \rightarrow (0.8)EE \mid (0.1)EECI \mid (0.1)EEF$
  - $E \rightarrow (0.7)EE \mid (0.1)E\#E \mid (0.1)EPE \mid (0.05) < -E \mid (0.05) > +E$

Some of the modeling operators mapped to the symbols in V are shown in Table 1, where we can see how each symbol is mapped to a modeling operator and the result of applying it to an initial object.

From this stochastic L-system, we can obtain buildings as those shown in Figure 6.

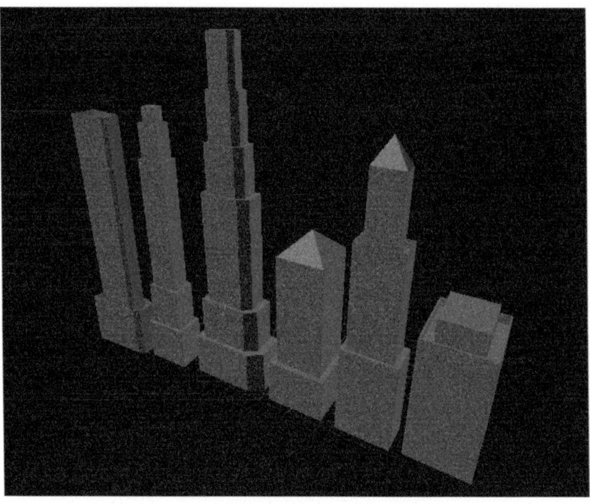

Fig. 6 Example of different buildings generated with a stochastic L-System

2.1.4 Materials and Textures

Once the 3D geometric models of the building have been constructed it is necessary to apply some properties to the model surfaces to achieve an appropriate look when they are rendered. Materials define a set of properties like color, brightness, diffuse and specular reflections, etc., that give an homogeneous appearance to the surfaces. Two kind of materials are used for the buildings in our procedural city. The first one is a granite-like material that will be applied on the roof of all the buildings and on the facade of some of them. The second one is a mirror-like material that will be used for skyscrappers.

In addition to materials, textures are also used to provide the surfaces with a higher level of detail. Textures are usually 2D images that are mapped onto the surfaces so that the color of a point on the surface is a combination of the material properties defined for that surface and the texture applied. Procedural textures have been used in this work to simulate windows and walls. These textures are dual color images where one of the colors is completely transparent and the other is a predefined RGB color (see Figures 7 and 8).

The rationale behind this kind of textures is to show the surface as defined by the material in those areas where the color is transparent and use the non-transparent color to provide some heterogeneity to the surfaces. For instance, skyscrappers are defined using a highly reflective material (mirror) in such a way that the texture uses a transparent color as background color and a black color to simulate the space between two windows as it is usually seen in a skyscrapper. The type of textures employed is shown in Figure 7.

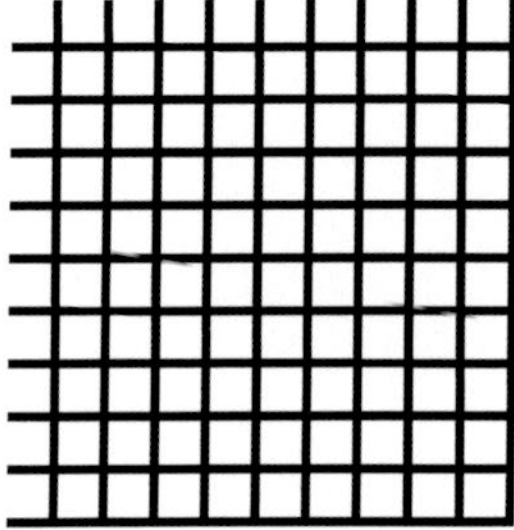

Fig. 7 Texture to simulate windows used to the surfaces with mirror-like material.

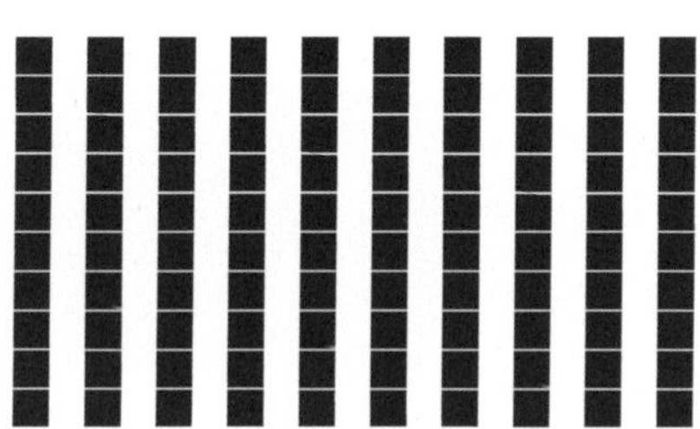

Fig. 8 Texture to simulate windows used to the surfaces with granite-like material.

The second texture class (Figure 8) uses a different approach. Here the transparent color is used to simulate the walls (the granite-like material is shown in this areas) and a second color is employed to show the windows.

An example of the resulting buildings after applied these textures is shown in Figure 9.

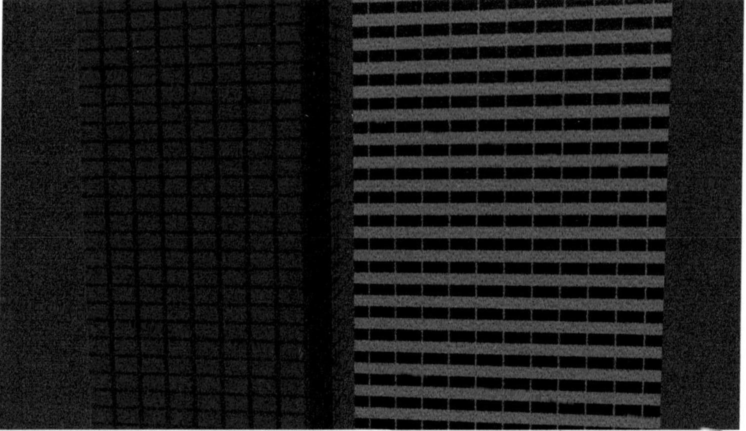

Fig. 9 Examples of buildings after applying materials and textures. The left building is using the mirror-like material and its corresponding texture (we can see reflections from the other building). The right building is using the granite-like material and its related texture.

2.2 *Implementation*

All the techniques and models described in previous sections have been implemented within Blender [3]. Blender is an open-source 3D modeling tool that incorporates a rendering engine among other tools.

The whole implementation process has been made in Python, since it is the Blender scripting language [14]. This has facilitated the process of coding and testing the algorithms although the Blender API is quite limited in terms of procedural design and, for that reason, a procedural graphics library was firstly designed and coded to implement the process of building generation. On the other hand, for texture mapping and rendering purposes, the Blender API has been employed.

3 Integration

In most physical agent simulators we can define an environment so that the agents are tested in a specific scenario. Environment definition is usually made by relying on a 3D world definition using to this end a file format to specify a 3D scene. Examples of such format are: Collada, DFX, 3DS Max, or VRML, among others.

To make the procedural cities usable within a robotic simulation tool, we took advantage of Blender exporting tools so that the 3D scene can be exported to one of the formats previously mentioned. This way, the process of producing a procedural city to be employed in a robot simulation tool can be described as the scheme shown in Figure 10. On account of this, the procedural cities are ready to be used in any simulation tool.

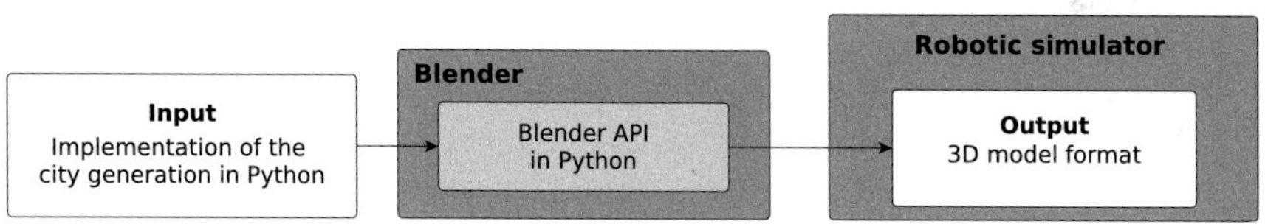

Fig. 10 Whole process from the city generation to the test in the robotic simulator. Using the Blender API, the city model is created with the implementation of all procedural techniques. Then, it exports the model into a robotic simulator with a specific 3D model format.

4 Evaluation

In order to assess the work presented here, several cities were created and then exported into a set of file formats in order to test them in a real robotic simulator. The simulator chosen was MORSE [9, 12]. In Figure 11 we can see a capture of a procedural city being used as an environment in MORSE . Here is showing the simulation of a autonomous vehicle navigating around the city provided. In addition, we can appreciate how the city generation parameters have been tuned so that the scale of the city is consistent with the scale of the vehicle.

Additionally, an example of a city rendered in Blender is showed in Figure 12. As we can see the MORSE capture lack of some graphic details due to real-time rendering constraints.

Fig. 11 Screenshot of a robotic simulator (MORSE) while is using a procedural city generation

Fig. 12 City generation on the platform Blender

5 Conclusion and Future Work

In this work we have presented an approach to automatically produce urban environments to be used in robotic simulation. Several procedural graphics techniques have been used to achieve this goal. In spite of the fact that this work constitutes only an initial approximation to this topic, the approach adopted seems to be suitable to easily design different environments for testing autonomous vehicles, drones, etc.

As future work, we propose to include more elements in the cities produced. For instance, items like trees, streetlights, traffic signs, etc., to create a more realistic environment. In addition the inclusion of dynamic and autonomous agents like cars or pedestrians could notably improve the simulation quality as it would allow the agent to be simulated to interact with other agents.

References

1. Alexander, C., Ishikawa, S., Silverstein, M.: A pattern language: towns, buildings, construction, vol. 2. Oxford University Press (1977)
2. Amazon. Amazon inc. (2015). http://www.amazon.com/b?node=8037720011 (last access July 7, 2015)
3. Blender. Blender wiki (2015). http://wiki.blender.org/ (last access July 10, 2015)

4. Chen, G., Esch, G., Wonka, P., Müller, P., Zhang, E.: Interactive procedural street modeling. ACM Transactions on Graphics (TOG) **27**, 103 (2008). ACM
5. DARPA. Darpa grand challenge (2015). http://archive.darpa.mil/grandchallenge/ (last access July 8, 2015)
6. Ebert, D.S.: Texturing & modeling: a procedural approach. Morgan Kaufmann (2003)
7. Google. Google inc. (2015). http://us.blizzard.com (last access July 7, 2015)
8. Kelly, G., McCabe, H.: An interactive system for procedural city generation. Institute of Technology Blanchardstown (2008)
9. LAAS-CNRS. Laboratory of analysis and architecture of systems (2015). https://www.laas.fr/public/ (last access July 15, 2015)
10. Lindenmayer, A.: Mathematical models for cellular interactions in development I. filaments with one-sided inputs. Journal of Theoretical Biology **18**(3), 280–299 (1968)
11. Lynch, K.: The image of the city, vol. 11. MIT press (1960)
12. MORSE. Morse simulator (2015). http://www.openrobots.org/morse/doc/stable/morse.html (last access July 15, 2015)
13. Prusinkiewicz, P., Lindenmayer, A.: The algorithmic beauty of plants. Springer Science & Business Media (2012)
14. Python, A.P.I.: Blender/python documentation, p. 3. Blender Index (2011)
15. Vanegas, C.A., Kelly, T., Weber, B., Halatsch, J., Aliaga, D.G., Müller, P.: Procedural generation of parcels in urban modeling. In: Computer Graphics Forum, vol. 31, pp. 681–690. Wiley Online Library (2012)

A New Cognitive Architecture for Bidirectional Loop Closing

Antonio Jesús Palomino, Rebeca Marfil, Juan Pedro Bandera and Antonio Bandera

Abstract This paper presents a novel attention-based cognitive architecture for a social robot. This architecture aims to join perception and reasoning considering a double interplay: the current task biases the perceptual process whereas perceived items determine the behaviours to be accomplished, considering the present context and role of the agent. Therefore, the proposed architecture represents a bidirectional solution to the perception-reasoning-action loop closing problem. The proposal is divided into two levels of performance, employing an Object-Based Visual Attention model as perception system and a general purpose Planning Framework at the top deliberative level. The architecture has been tested using a real and unrestricted environment that involves a real robot, time-varying tasks and daily life situations.

Keywords Cognitive architecture · Attention model · Social robot · Bidirectional loop closing

1 Introduction

An autonomous robot placed in a real world has to deal with a lot of visual information. At the same time, the agent has to address different actions, different tasks that vary over the time, reacting to unexpected situations. When developing a perception system for such a robot, some key questions come up: is it possible to modify the way a robotic agent perceives the world depending on its current responsibilities? And, vice versa, are new interesting objects able to modify the ongoing task? How can perception and reasoning interoperate simultaneously in an autonomous robot?

A.J. Palomino
Fundación Magtel, 14014 Córdoba, Spain

R. Marfil · J.P. Bandera · A. Bandera(✉)
University of Málaga, 29071 Málaga, Spain
e-mail: ajbandera@uma.es

L.P. Reis et al. (eds.), *Robot 2015: Second Iberian Robotics Conference*,
Advances in Intelligent Systems and Computing 418,
DOI: 10.1007/978-3-319-27149-1_56

Analysing the problem from a deliberative point of view, the behaviours to be accomplished depend on the perception of a specific set of objects. From that definition, the effects on deliberative planning can be deduced: partial observability and uncertainty, since the attention model constrains the information that the robot perceives. Simultaneously, accomplishing a certain behaviour is likely to require attention to be focused on a specific kind of objects. In other words, there exists a very close relationship between an attention-driven perception system and a deliberative planner, typically included in the reasoning phase of the classical perception-reasoning-action loop.

Physiological observations suggest that certain perceptual characteristics, such as location or shape, engage actions related to those characteristics[1, 2]. Thus, the processing of stimuli related to these actions should be primed. In consequence, a classical unidirectional assumption of the perception-reasoning loop is not enough. A complete solution must cover the double interplay between perception, reasoning and action.

Although some authors has tackled several parts of the loop-closing problem ([3–7]), they often propose "isolated" solutions. Therefore, there is a lack of proposals providing integration solutions, especially when introducing attention-based models of perception. This paper presents a **novel need-based cognitive architecture** for an attention-based and bidirectional perception-reasoning loop closing. The architecture is based on the interaction of a general purpose Planning Framework, that produces plans constrained by the information perceived from the vision system, and an object-based attention model, able to highlight elements that are suitable for the ongoing task. Through the exchange of *relevant elements* and *sets of perception parameters*, the cognitive system is able to decide what tasks are going to be executed, following a $need-based$ approach. At last, the predominant task selects what kind of elements are going to be searched next and the system performs the loop again in a cyclic way.

The remainder of this paper is organised as follows. The employed attention-based perception system is introduced in section 2. Section 3 describes the proposed cognitive architecture and the bidirectional solution given to the loop-closing problem. Finally, experimental results are presented in section 4 and discussed in section 5.

2 Perception Based on Attention: Object-Based Visual Attention System (OBVIAS)

The proposed architecture employs an object-based attention model, OBVIAS [8] as perception system. OBVIAS is a two-stages attention systems that integrates task-independent bottom-up processing and task-dependent top-down selection (see Fig. 1). The units of attention are the so-called proto-objects [9], which are defined as units of visual information that can be bounded into a coherent and stable object.

In the *pre-attentive stage*, the most relevant elements in the scene are obtained by computing a saliency measure based on a set of basic features (colour, shape…),

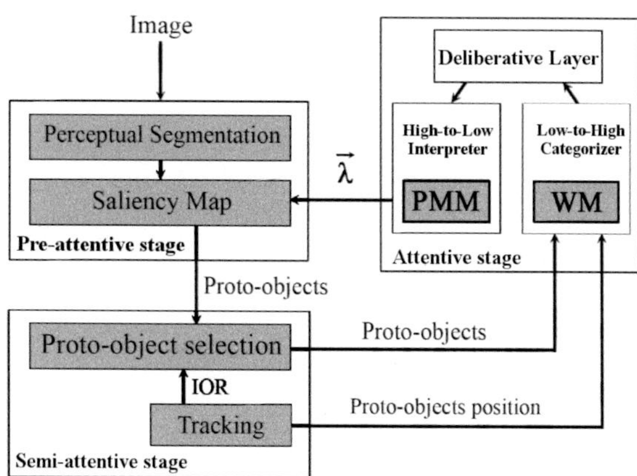

Fig. 1 Object-Based Visual Attention Model (OBVIAS).

biased by a set of perception parameters that determine the influence of the current task to accomplish. First, the different proto-objects in the image are extracted using a perceptual segmentation algorithm. Then, the relevance of each proto-object is computed taking into account different low-level features weighted by an *attentional set* ($\boldsymbol{\lambda}$) [10], defined as a collection of different perception parameters able to highlight proto-objects with specific features [10]. The attentional set is stored in a *Perception-Modulation Memory* (PMM), that implements the top-down component of attention. Depending on the specific values in the attentional set, the system is able to modify the influence of each low-level feature in the global saliency computation, so attention can be guided in a top-down way. That is, the saliency of the same element could be different depending on the ongoing task. As the result of this stage, a set of proto-objects ordered by their saliency is obtained and saved in a *Working Memory* (WM).

The next stage, the *semi-attentive* stage, deals with the management of the WM and the *Inhibition of Return (IOR)*. IOR avoids the system getting stuck in the most relevant element, ignoring the rest of the scene. Thus, the use of IOR becomes mandatory in the context of computational attention systems. In the case of OBVIAS, a tracker module updates the position of each element in the WM, allowing to manage not only moving objects but also camera and robot egocentric movements.

See [8] for further details about the pre-attentive and the IOR processes of OBVIAS.

Regarding WM management, following the work of Bundesen in his *Theory of Visual Attention* [11], the number of proto-objects stored at the same time in the WM in OBVIAS has a fixed value of 5. Every proto-object in the WM is characterised by a set of descriptors: saliency value, depth, orientation, colour, area, position, low-level features, time-to-live and a copy of the region of interest (ROI) occupied by the proto-object in the image. A new proto-object gets into the WM if and only if it is more salient than the currently stored elements. If the memory is full, the least salient element is dropped out. A proto-object can also be removed from WM if it is lost by the image tracker. Total saliency of the elements in the WM is recomputed in every perception cycle in terms of the new attentional set. Thereby, the saliency of the stored proto-objects is always kept up-to-date.

The saliency of a proto-object also depends on the time-to-live parameter [12]. A proto-object already stored in WM has, at the beginning of its life, an increment in saliency due to a high time-to-live. On the contrary, older proto-objects receive a decrement of saliency directly proportional to their time-to-live. In order to implement this behaviour in the WM, an exponential function is applied to time-to-live. In summary, the WM operates as a cache memory. While a proto-object is relevant for the ongoing task, it is kept in memory. The content of the WM is only fully swapped when a new task comes out and no proto-object in the WM is suitable to accomplish the new behaviour.

Both Working Memory (WM) and Perception Modulation Memory (PMM) are the interface between the early attention-based perception module and the rest of the system, including the deliberative level. On the one hand, the WM provides the most relevant proto-objects to other modules in the system. On the other hand, other modules are able to guide the attention system to specific and useful objects in the scene, just swapping the attentional set stored in the PMM.

3 Need-Based Cognitive Architecture

As it was pointed out in the Introduction section, most of the artificial attention systems show a lack of connection with abstract reasoning layers. In order to overcome that issue, this section introduces the proposed cognitive architecture that, using as starting point the attention-based perception system presented in the previous section, goes beyond, considering that attention-based perception and abstract reasoning present a close and symbiotic relationship: they mutually modulate each other simultaneously.

The proposed need-based cognitive architecture can be seen in fig. 2. It is divided into two levels. The higher level, or *Knowledge-based level*, determines the current context and the role of the agent. Hence, the aim of this abstract reasoning level is the selection of the tasks that will be active, by adding or removing them from the lower level. It also adjusts their priorities, depending on context data and the state of achievement of each action. The core of this level is a Planning Framework that could be aided by other modules (Machine Learning, Knowledge, Machine Learning, etc.) to figure out the context.

The different tasks that can be performed at each moment are placed at the *Rule-based level*. Depending on the perceived elements, a task can be executed or not. This level is concerned with quantitative models of execution and tasks are based on **needs**, defined as perceptual categories of visual elements. In other words, the accomplishment of a task is closely linked to the presence of specific elements in the scene. Finally, the architecture is linked with the previously described attention system using the aforementioned PMM and WM memories.

3.1 Knowledge-Based Level

The **knowledge-based level** is responsible for the coordination and management of the tasks located at the rule-based level. When the role of the agent or the environment conditions change, it is necessary to obtain a new set of rules that fit the scenario at hand. These rules are translated into a new set of tasks that can handle specific parts and expected situations of the environment.

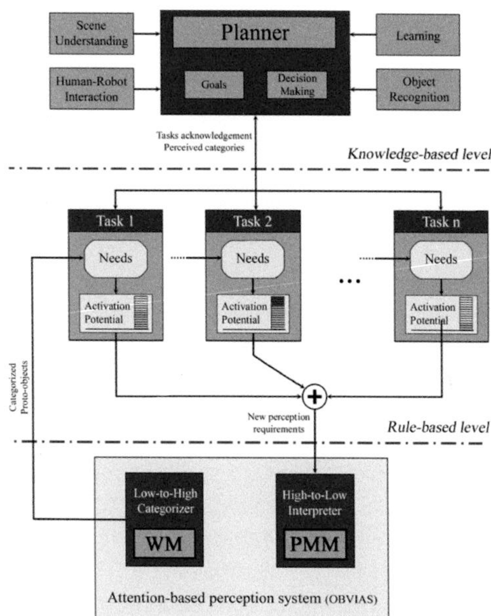

Fig. 2 Need-based cognitive architecture.

To carry out the overall planning, goals definition and decision making processes, a **General-purpose Planning Framework** is set as the core of this level. It receives information both from the sensory input and the tasks located at the rule-based level. The visual information is expressed in form of abstract predicates, derived from the perceived visual categories (e.g. "near red square", "far human face", "round green shape", etc.). Since the visual data is supplied by an attention system, the Planning Framework must be able to deal with partial information. Thereby, a continuous adaptation of the plan to the perceived elements should be allowed. Concretely, we have used a Planning, Monitoring and Learning Framework (PELEA) based on Oversubscription Planning (OSP) [13].

As a result, a set of actions (tasks) that solve different parts of the whole problem is obtained. For each task, a set of preconditions and needs (i.e. visual categories to be perceived) as well as its priority are defined. In addition, the effects of tasks over attentional guidance is also established, selecting a top-down template from the ones available in the domain definition.

Finally, other modules helping to figure out context information from visual data could be placed at this level (Machine Learning, Scene Understanding, Human-Robot Interaction …).

3.2 Rule-Based Level

As it was mentioned, the tasks that better fit the current context and role of the agent, detected by the Knowledge-based level, are placed in this level. Each task has associated two main parameters: i) **Needs:** visual categories that must be perceived in order to execute the corresponding action (e.g. "face", "green things", etc.). ii) **Activation potential:** this factor measures the number of satisfied needs and, consequently, the probability of executing a task in the future. Therefore, an action is executed if its visual requirements (needs) are fulfilled. On the contrary, an attentional guidance is applied based on a top-down template.The activation potential, for each task, is obtained as a weighted linear combination of the needs currently satisfied. These weights are defined in the domain definition. The influence of the attentional guidance of each task depends on the activation potential. Thereby, those elements in the scene that can fully cover the needs of the task with the highest probability to be executed can be highlighted. Note that the top-down template for visual guidance is selected by the knowledge-based level when obtaining the tasks to be located at the rule-based level. It is possible that some specific needs produce higher levels of activation. For example, a robotic arm employing the task "press the orange SOS button if a red danger light is turned on" among others. This task should have a high activation potential so that the "orange button" condition is executed as soon as the "red light" category is filled, regardless other conditions.

3.3 Closing the Loop

In order to finally connect the reasoning levels of architecture with the attention model used as perception system, the proto-objects stored in the WM must be translated into abstract predicates whereas the abstract templates for attentional guidance associated to the tasks in Rule-based level must be converted to Attentional Sets to be stored in the PMM. The former process is carried out in the **Categorizer** whereas the latter is performed in the **Interpreter**.

Regarding the Categorizer, proto-objects in WM are classified into semantic categories according to their basic features. However, the categorisation process must be fast and computationally efficient since it is part of the early-vision stage. That is, proto-objects do not need to be exactly recognised but classified into generic classes. An accurate object recognition is not necessary indeed. The category assigned to each proto-object is also stored in the WM to make it available to other systems. Finally, a proto-object may be classified as not belonging to any currently known category.

The Interpreter module translates the effects for visual biassing, expressed as semantic predicates (e.g. "look for green and rounded objects"), to a new Attentional Set stored in the PMM. Although some approaches has tried to deduce the biasing parameters from a specific target object using Bayesian inference [14], there is no general solution for this kind of interpreters at present. The procedure strongly depends on the number and type of features employed in the attention model and the guidance to be achieved. So, at the moment, the solution should be manually developed, or the interpreter should be previously trained using Machine Learning techniques, for each concrete action and problem.

4 Experimental Results

In this section the whole cognitive architecture is tested. The evaluation is mainly focused on analysing the reliability of the need-based architecture when addressing complex environments and problems, with time varying tasks. The experiment involves a real social robot (NOMADA), placed in a natural environment with no specific restrictions about illumination or background elements. In the experiment, the robot has to classify different objects (coloured balls), given by a human, into the correspondent container. In order to simplify the algorithms corresponding to the *categorizer* and the *interpreter*, some simplifications have been assumed: (i) people always give balls to the robot; (ii) the containers are represented by coloured rectangles in a wall; (iii) the robot has to bring each ball to the rectangle that matches its colour.On the other hand, there are no constraints about illumination, elements or number of people in scene. In order to validate the robustness of the system, some balls are placed as distractors in the middle of the environment.

By default, the robot is looking for humans to assist them. In this task, people are the most relevant elements in the scene. Concretely, the nearest person is always preferred. Therefore, these requirements are translated by the interpreter into increasing parameters related to skin colour, roundness and proximity in the correspondent attentional set. When the robot has found people, it moves towards them and detects if they have a ball. If a ball is perceived, the robot opens a compartment to receive the ball. In this case, the parameters to be increased in the attentional set are roundness and proximity, maintaining the presence of humans as a need. Once the robot has obtained the ball, the next step consists in looking for the nearest correspondent rectangle in the wall and moving towards it. People and balls are not relevant now, so the attentional set is configured to increase the parameters related to symmetry, proximity and the specific ball colour. Finally, the robot drops the ball and returns to the default task.

Fig. 3 summarises the tasks placed at the rule-based level, including their needs and their effects over attentional sets. Depending on the current behaviour of the robot, the knowledge-based level adds or removes certain tasks. Concretely, when the *default* behaviour is accomplished (a person is detected), the knowledge-based level removes the search-for-human task and the tasks corresponding to the *assisting human* activity (search for ball, move towards human and get ball) are added. In a similar way, these tasks are replaced by the ones belonging to the *recycling* behaviour when a ball is obtained. Thereby, the tasks that can be activated vary depending on the current context.

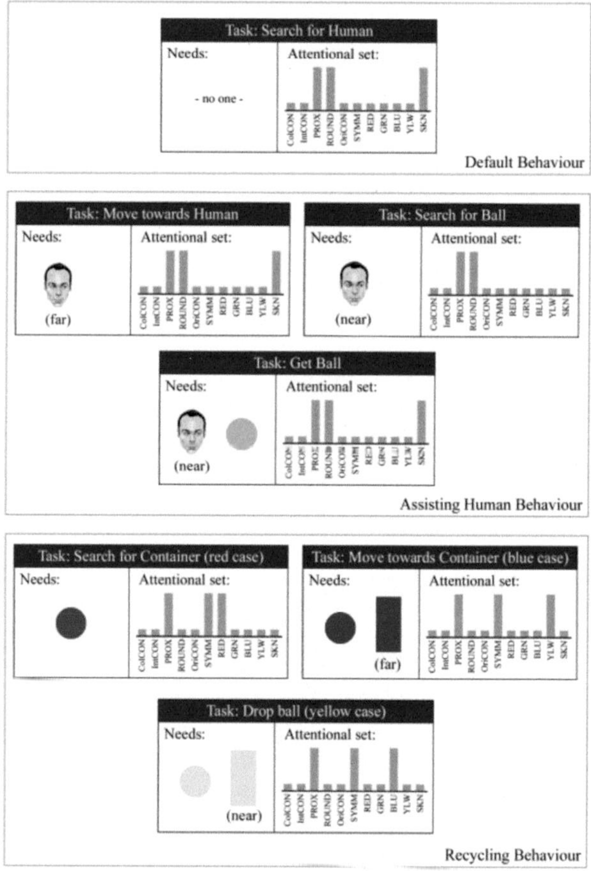

Fig. 3 Tasks placed at rule-based level for the experiment.

Fig. 4 shows one successful execution of the experiment 2, including a SMPTE Timecode. The categorised content of the Working Memory, the plan to be executed and the specific attentional set to be stored in the Perception Modulation Memory are also displayed. As it can be observed, the number of relevant proto-objects varies in each frame. In the first two frames (first row in figure 4), the robot is executing its default behaviour, looking for humans (note the high relevance given to skin colour in the attentional set). In the third frame (first in the second row), the knowledge-based level has swapped the tasks to those corresponding to the assisting behaviour. In this case, nothing but a person is visible so two tasks has all their needs covered now: moving towards human and search for ball. The moving task is already completed (the robot is near the person). Hence, the *search for ball* task is executed. Otherwise, the moving task would have been executed to approach the far person.

When the robot finds a ball and a person, the *get ball* task is executed (fourth displayed frame). Immediately, another change of behaviour is performed and the tasks corresponding to the recycling behaviour are loaded on the rule-based level. While the container is not visible, the *search container* task is executed (frames 5 and 6). Notice that the colour corresponding to the obtained ball (in this case, blue) is highlighted in the attentional set in order to find the correct marker. In frames 7 and 8 (fourth row), the *moving* task is activated by a visible far marker. Finally, once the robot is near the "container", the ball is dropped (frames in the last row) and the behaviour is changed again to the default one. The robot employs about 55 seconds in executing all the tasks.

This experiment has been repeated using different colours, with people located at different places. Due to the continuous adaptability of the cognitive architecture, the robot is not likely to fall into inaction or to accomplish a wrong behaviour. Nevertheless, a successful execution is not always achieved. This is mainly caused by a wrong categorisation of the perceived proto-objects. Another frequent errors are related to a bad colour assignation to the collected ball (often caused by people's clothes) or to a wrong identification of a goal marker with non useful objects in the scene (e.g., the "red container" category can be easily assigned to a red extinguisher). On the contrary, the inherent robustness of the architecture is remarkable. For instance, occlusions and lost items are handled in a natural manner. If a target (e.g. a "container" marker) is lost or occluded during navigation, the needs of the *move towards container* task can not be completely fulfilled. As a consequence, the *search for container* task becomes executable and the target is sought again. Besides, attention is not drawn to elements belonging to other behaviours. If the robot is in the *recycling* behaviour, people walking or standing near it and distracting balls are ignored by the perception system.

A total of **20 executions** of the experiment has been carried out in a similar scenario with a successful rate of **85%**. The initial position of the human and the colours involved vary throughout the different evaluations. Some statistics have been measured from these executions. The robot needs **130 iterations** on average of the perception-reasoning-action loop to completely solve the problem. The standard deviation of this measure has a high value (**60**) due to the diversity of people's initial location. The execution time has been also measured: the robot employs about

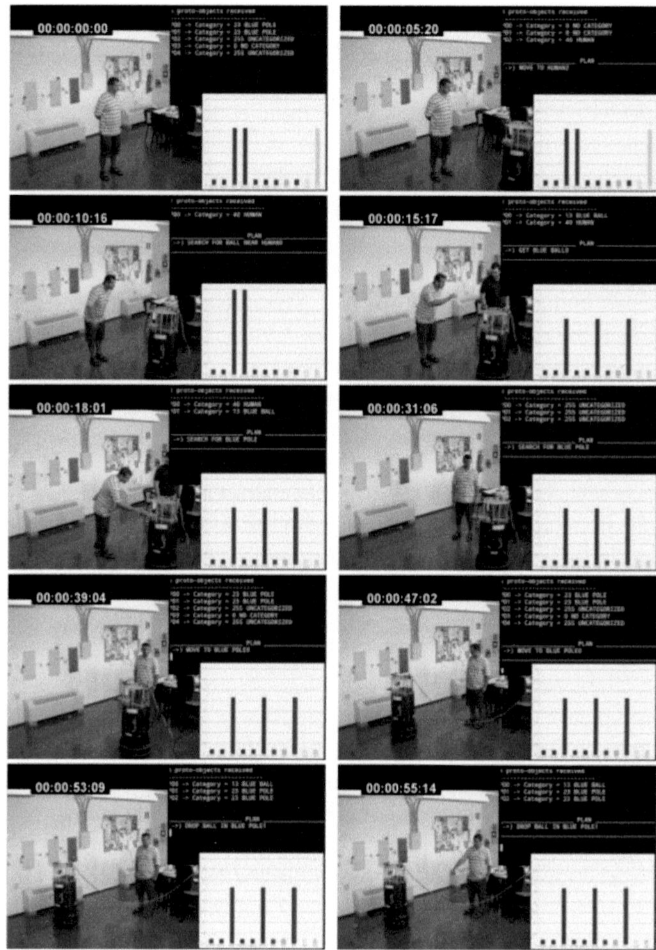

Fig. 4 Some significant frames from an execution of experiment 2. The output of the *Deliberative* component is also included, showing the categorised content of the WM, the plan provided by the Panning Framework and the attentional set to be stored in the PMM.

53 seconds on average. This produces a ratio of about **2.4** $\frac{iterations}{second}$. Hence, the architecture is fast enough to react to behavioural changes. The amount of activations of each task during a complete evaluation has been also measured (see Table 1). Tasks involving visual search are activated considerably more times than the others. This behaviour is completely reasonable since these tasks present less needs to be satisfied and they are performed as default tasks when no other possible action is available. The case of *search for human* task is specially significant due to the absence of people in the scene at the beginning of some evaluations. On the contrary, tasks concerning ball getting and dropping are the less overall executed because these tasks require very specific conditions to be executed.

Table 1 Activation distribution on average among the different tasks.

Search for Human	Search for Ball	Search for Container	Move towards Human	Move towards Container	Get Ball	Drop Ball
73%	15%	6%	2%	2%	1%	1%

5 Discussion

The cognitive architecture introduced in this paper has been evaluated following the criteria proposed by Langley et al. [15].

The experimental results show the ability of the proposal to discriminate different kind of information depending on time-varying tasks and behaviours. Thus, complex environments, where the computational cost associated with a full scene processing is unaffordable, can be addressed by highlighting simple, specific and suitable features, reducing the received amount of information. The **bidirectionality** of the proposal has also been demonstrated. Not only the perception system is guided in terms of the ongoing task, but also the presence of significant objects triggers awaiting tasks or even changes the current behaviour of the robot (e.g., the detection of a person moves the robot to assist her).

The modular configuration of the proposed need-based architecture provides a high degree of *versatility*. The proposal can be adapted to different uses by defining the involved tasks (including their perceptual needs and effects) as well as providing the new domain definition. Task modularisation makes easier the development of new behaviours because interdependencies should not be taken into account at the task level. They are handled as abstract predicates at the knowledge-based level.

Rationality is also inherent to the proposed architecture definition. Only tasks that are supposed to accomplish behavioural useful goals are placed at the rule-based level. Besides, only those tasks getting all their needs fulfilled can be activated. Hence, only actions that accomplish goals can be selected. The *persistence* of the architecture is also easily observable. For instance, when the robot is approaching a "container", people are only taken into account as obstacles to be avoided, but the robot does not try to assist them. That is, small variations in the scene, that are not expected to change the ongoing behaviour, do not alter the current activity to be accomplished.

Finally, *reactivity* is intimately related to the rule-based level. Unexpected categories may be handled by a high priority task with only one need to be satisfied. Although there is no such a task among the ones presented in the evaluation experiment, its functionality may be similar to any standard task. Thereby, the system could instantaneously react to any situation.

Acknowledgements This work has been partially granted by the Spanish Ministerio de Economía y Competitividad (MINECO) and FEDER funds, project no. TIN2012-38079-C03-03.

References

1. Collins, T., Heed, T., Rder, B.: Visual target selection and motor planning define attentional enhancement at perceptual processing stages. Frontiers on Human Neuroscience **4**, 1–10 (2010)
2. Fagioli, S., Hommel, B., Schubotz, R.I.: Intentional control of attention: action planning primes action-related stimulus dimensions. Psychological Research **71**, 22–29 (2007)
3. Langley, P., Cummings, K., Shapiro, D.: Hierarchical skills and cognitive architectures. In: Proceedings of the twenty-sixth Annual Conference of the Cognitive Science Society, Citeseer, pp. 779–784 (2004)
4. Bischoff, R., Graefe, V.: Hermes-a versatile personal robotic assistant. Proceedings of the IEEE **92**(11), 1759–1779 (2004)
5. Gilet, E., Diard, J., Bessière, P.: Bayesian action-perception computational model: Interaction of production and recognition of cursive letters. PloS One **6**(6), e20387 (2011)
6. Sridharan, M., Wyatt, J., Dearden, R.: Planning to see: A hierarchical approach to planning visual actions on a robot using pomdps. Artificial Intelligence **174**(11), 704–725 (2010)
7. Ferreira, J.F., Lobo, J., Bessière, P., Castelo-Branco, M., Dias, J.: A bayesian framework for active artificial perception. IEEE Transactions on Systems, Man and Cybernetics, Part B: Cybernetics **43**(2), 699–711 (2013)
8. Palomino, A.J., Marfil, R., Bandera, J.P., Bandera, A.: Multi-feature bottom-up processing and top-down selection for an object-based visual attention model. In: Procedings of the II Workshop on Recognition and Action for Scene Understanding (REACTS), pp. 29–43 (2013)
9. Rensink, R.: Seeing, sensing, and scrutinizing. Vision Research **40**(10–12), 1469–1487 (2000)
10. Corbetta, M., Shulman, G.L.: Control of goal-directed and stimulus-driven attention in the brain. Nature Reviews Neuroscience **3**(3), 201–215 (2002)
11. Bundesen, C., Habekost, T., Kyllingsbaek, S.: A neural theory of visual attention and short-term memory (ntva). Neuropsychologia **49**(6), 1446–1457 (2011)
12. Klein, R.M.: Inhibition of return. Trends in Cognitive Sciences **4**(4), 138–147 (2000)
13. García-Olaya, A., de la Rosa, T., Borrajo, D.: Using the relaxed plan heuristic to select goals in oversubscription planning problems. In: Lozano, J., Gámez, J., Moreno, J. (eds.) Advances in Artificial Intelligence. Lecture Notes in Computer Science, vol. 7023, pp. 183–192. Springer, Heidelberg (2011)
14. Yu, Y., Mann, G.K., Gosine, R.G.: A goal-directed visual perception system using object-based top-down attention. IEEE Transactions on Autonomous Mental Development **4**(1), 87–103 (2012)
15. Langley, P., Laird, J.E., Rogers, S.: Cognitive architectures: Research issues and challenges. Cognitive Systems Research **10**(2), 141–160 (2009)

A Unified Internal Representation of the Outer World for Social Robotics

Pablo Bustos, Luis J. Manso, Juan P. Bandera, Adrián Romero-Garcés, Luis V. Calderita, Rebeca Marfil and Antonio Bandera

Abstract Enabling autonomous mobile manipulators to collaborate with people is a challenging research field with a wide range of applications. Collaboration means working with a partner to reach a common goal and it involves performing both, individual and joint actions, with her. Human-robot collaboration requires, at least, two conditions to be efficient: a) a common plan, usually under-defined, for all involved partners; and b) for each partner, the capability to infer the intentions of the other in order to coordinate the common behavior. This is a hard problem for robotics since people can change their minds on their envisaged goal or interrupt a task without delivering legible reasons. Also, collaborative robots should select their actions taking into account human-aware factors such as safety, reliability and comfort. Current robotic cognitive systems are usually limited in this respect as they lack the rich dynamic representations and the flexible human-aware planning capabilities needed to succeed in these collaboration tasks. In this paper, we address this problem by proposing and discussing a deep hybrid representation, DSR, which will be geometrically ordered at several layers of abstraction (deep) and will merge symbolic and geometric information (hybrid). This representation is part of a new agents-based robotics cognitive architecture called CORTEX. The agents that form part of CORTEX are in charge of high-level functionalities, reactive and deliberative, and share this representation among them. They keep it synchronized with the real world through sensor readings, and coherent with the internal domain knowledge by validating each update.

Keywords Social robotics · World internalization · Deep representations

P. Bustos · L.J. Manso · L.V. Calderita
RoboLab Group, University of Extremadura, Cáceres, Spain
e-mail: {pbustos,ljmanso,lvcalderita}@unex.es
https://robolab.unex.es/

J.P. Bandera · A. Romero-Garcés · L.V. Calderita · R. Marfil · A. Bandera(✉)
Dept. Tecnología Electrónica, University of Malaga, Málaga, Spain
e-mail: {jpbandera,adrigtl,rebeca,ajbandera}@uma.es
http://www.grupoisis.uma.es/

L.P. Reis et al. (eds.), *Robot 2015: Second Iberian Robotics Conference*,
Advances in Intelligent Systems and Computing 418,
DOI: 10.1007/978-3-319-27149-1_57

1 Introduction

While the economic benefits of robotics in industry are already clear, it is expected that their inclusion in everyday life will have a tremendous impact. The EU's H2010 initiative states that, as human assistants, tomorrow's robots will have the capacity to resolve many of the future economic and social challenges faced by European society, such as aging and well-being. However, to access these new markets and to be competitive, robots have to be dependable, smarter and able to work in closer collaboration with humans. In these scenarios, human-robot interaction (HRI) is now envisioned more as a relationship among companions than a mere master-slave relationship. It can be considered that the complete design of a real co-robot is beyond the scope of what can be achieved technically today. Still, to make progress along this path there are several important issues that are currently being discussed as a way to facilitate the design of new cognitive robotic architectures that will, one day, show real human-robot collaboration (HRC).

If a collaborative robot has to cooperate with a human partner as a work companion, it should be endowed with the abilities to consider its environmental context and assess how external factors could affect its action, including the role and activity of the human interaction partner in the joint activity. Efficient collaboration not only implies a common plan for all involved partners, but also the coordination of the behavior of each agent with those of the other ones, i.e. to gain a joint intention. This coordination should be simultaneously addressed at different levels of abstraction -e.g. semantic, situational or motor, and the robot has to internalize a coherent representation about the motions, actions and intentions -including abilities and preferences- of the rest of partners. Additionally, a major difficulty in HRC scenarios is that people can exhibit a rather non-deterministic and unstable behavior, but they also tend to perceive current robots as slow and unintelligent. These factors difficult HRC. To overcome them the robot should continuously try to guess their partners' goals and intentions, trigger appropriate reactions and, ultimately, be socially proactive.

In this paper we will argue that to fully develop HRI, and to pave the way into HRC, a cognitive robotics architecture should use a deep, central representation shared among the agents composing it, which codes information at different levels of abstraction. We will explore here this issue, unfolding the arguments that support it and the design decisions taking during its development.

To our knowledge, the first works that proposed a graph as an internal representation for a robotics architecture focused only in geometric data. ROS' transform library, *tf* [1], BRICS Robot Scene Graph [2] and RoboCog's InnerModel [3] all appeared in 2013 as a response to the need for an structured, centralized representation of the robot and world kinematics. These constructions are important advances towards better robotic architectures, a richer, and deeper representation was needed to hold the complete set of beliefs of the robot. The concept of deep representations was first described by Beetz et al. [28] as *representations that combine various levels of abstraction, ranging, for example, from the continuous limb motions required to perform an activity to atomic high-level actions, subactivities, and activities*. This definition is however provided in a paper where the robot performs its activities alone. In a collaborative scenario, we should also consider representation and inference mechanisms for models including the persons bodies, actions, abilities and intentions.

Separately, symbolic and metric representations have been proposed in many different forms and uses. Symbolic knowledge representation have been at the core of AI since its beginnings [4] [5] and cover all forms of relational formalizations such as production rules, frames, schemes, cases, semantic nets, first order logic or situational calculus. At a high level of abstraction, the Robot Learning Language (RoLL) [27] could be used for learning models about human behaviour and reactions, joint plan performance or recognizing human activity. Also, human models have been employed by the Human-Aware Task Planner (HATP) [29]. A symbolic graph structure was proposed in [6] as part of our previous architecture RoboCog [7] and it will be described in later sections. Metric and kinematic representations are commonly used as part of 3D simulators and graphics engines[1,2].

However, the concept of deep representations implies an unified, hierarchical organization of the knowledge that ranges from the symbolic layer to the motor one, mapping abstract concepts to, or from, geometric environment models and sensor data structures of the robot. The presence of a detailed representation of the spatial state of the problem is also required in the work of S. Wintermute: *... actions can be simulated (imagined) in terms of this concrete representation, and the agent can derive abstract information by applying perceptual processes to the resulting concrete state* [30]. The use of a situational representation of the outer world to endow the robot with the ability to understand physical consequences of their actions can be extended, in a collaborative scenario, to support proactive robot behaviors. This possibility has been addressed in the LAAS Architecture for Autonomous Systems proposed by Ali et al. (2009).

The rest of the paper is organized as follows: Section 2 presents arguments and examples that support the former claims. Section 3 presents an application scenario where the world model is currently been tested. Conclusions and future work are drawn at Section 4.

2 The Deep State Representation

CORTEX is an agent-based new cognitive robotics architecture designed as an evolution of our former RoboCog architecture that provides the agents with a shared, hybrid representation of the robot's belief about itself and its environment. This graph-like structure is called DSR and can be accessed by all agents during their operations. DSR is the only means for the agents to communicate among them. Figure 1 shows a small DSR graph with multiple labeled edges representing heterogeneous attributes.

---

[1] http://wiki.ros.org/urdf

[2] https://www.khronos.org/collada/

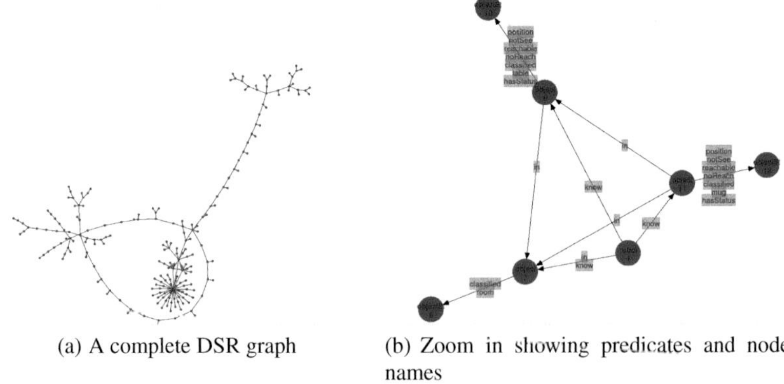

(a) A complete DSR graph

(b) Zoom in showing predicates and node names

Fig. 1 The DSR graph at different levels of resolution, showing metric and symbolic properties.

The idea of a shared representation among agents has its roots in several classical papers [8] [9] [10] that developed the concept of *blackboard architecture*. Later, Hayes-Roth [11] extended this idea into a complete control architecture. In the original blackboard systems, agents where conceived more as problem solvers, heterogeneous experts that contribute to the overall problem in a hybrid planned-opportunistic way. They communicate through a shared structure where goals, sub goals and problems state were incrementally updated. In CORTEX, agents solve not only deliberative tasks but also perceptual, motor and behavioral ones, so their communication needs are somewhat different. Nevertheless, we gather some ideas from these architectures [12] [13] and also others from graph theory and distributed databases[3]. We present now some arguments supporting the need of a graph representation if the robot's inner beliefs.

The first reason to use a graph in CORTEX is because all internal information defining the state of the robot and its beliefs about the environment can be stored according to a *generic structure*. As generic data structures, graphs can hold any relational knowledge composed of discrete elements and relations among them. In this broad category falls almost all symbolic knowledge representation methods including frames, schemes, production rules and cases, and also the geometric knowledge that the robot has to maintain about itself and the environment. This geometric knowledge includes instances of the types of objects recognizable in the world like i.e. chairs, tables, cups or generic obstacles of undefined form. Also human bodies and its parts like arms, heads, legs, etc. All these parts are kinematically related through 3D transformations forming a scene-tree.

A second reason is that the graph can be made to evolve under some *generative rules*. Assuming that the type of nodes and edges are predefined, the graph can evolve by inclusions or deletions of parts, causing structural changes. Also it can evolve by changing the value of the attributes stored in nodes and edges. The structural changes can be regulated by a generative grammar that defines how the initial model can

[3] http://rethinkdb.com

change. A typical example would be that of the robot entering a new room and, after exploring it, it would add the a new node to the graph. The grammar would impede the new node to be connected to something else but the corresponding door, and, maybe, it would have to be oriented parallel to one of the walls of the proceeding room. So graphs give us the capacity needed to store objects and their relationships and, combined with a grammar, a means to control its evolution to produce a growing model coherent with some initial domain knowledge. Figure 2 shows how the graph changes when a person enters the scene. In the left side only the robot and the rooms are represented. In the right side, a person enters the room and the graph incorporates her as sub graph, correctly related to the existing structure and with symbolic attributes denoting what is known about her. Thus, the graph is not only a means of storage but a way to articulate information coming from sensors and processed by agents. Once in the graph, information can be accessed and interpreted by other agents.

A third reason to use a graph structure is the possibility of translating it into a *PDDL* instance. There are certain restrictions that depend on what is stored in the graph and the PDDL version used, but it allows a direct use of start of the art planning algorithms that otherwise would have required an important effort. Further details on how this translation is done can be found in [6].

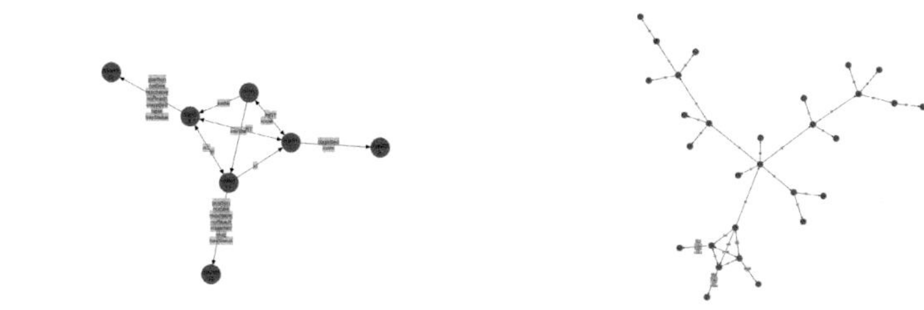

(a) Initial world model in DSR with the robot and the room.

(b) A person enters the room and is inserted in the DSR when detected by the *Person* agent.

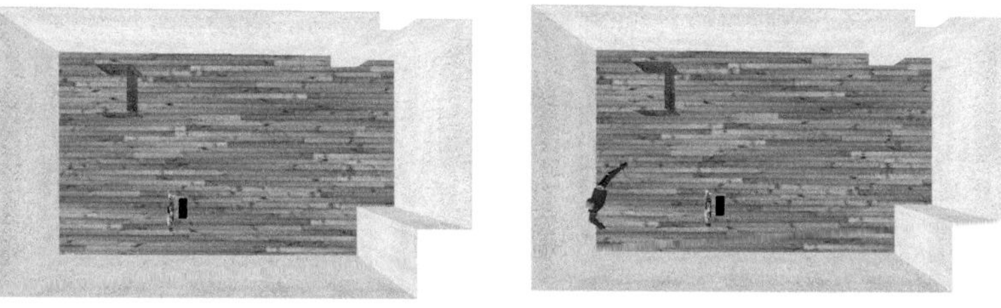

(c) Graphic representation of the geometric view.

(d) Graphic representation of the geometric view when a person is inserted in the DSR.

Fig. 2 Two states of the DSR graph, before a person enters the room (a,c), and after she is detected and inserted in the DSR, (b,d).

A forth reason to support the choice of graphs is the facility to *visualize* its contents. Graph's contents can be displayed in multiple ways using available 3D technology and this is a crucial feature to debug the code of the agents, specially when interacting among them. In CORTEX, visualization of the DSR is done using the open source 3D scene-graph OpenSceneGraph, OSG[4] and a class implementing the observer pattern that keeps DSR and OSG synchronized. The DSR graph can be drawn in different ways. The geometric nodes and edges are drawn as a normal 3D scene, using the meshes and 3D primitives that can be stored as attributes in DSR. The symbolic relations can be drawn as an independent graph or as a superimposed structure on its geometric counterpart.

An additional reason to use a graph is because it is possible to *share* it efficiently among the agents using different techniques. The goal is to provide the agents with a mechanism to modify the graph and propagate that modification to all others. As long as this is achieved, all agents will have access to the global represented state and will be able to use it as a broad context to select the best possible action. There are several options that can be analyzed:

- A first option is to use an existing graph database server running as an agent and use its API to modify and query the graph. The database should allow for multi-graphs with a variable number of attributes in nodes and edges and work with low latency and high throughput. Also, the model checking functionality that filters candidate updates would have to be coded outside the server. We have not made tests with currently available graph databases like Neo4j[5] or Sparksee[6] but we expect to be a reasonable option if some latency is allowed.
- A better solution in this line would be a database with a notification service, so changes were automatically propagated to a set of clients. There is at a least one open source database that we now of that provides this capability, RethinkDB, but it is a document oriented database and conversion between database types and agents language types will penalize the overall process.
- A second option is to use a communications middleware -Ice in RoboComp- that provides a publication-subscription service. Using a set of topics, all agents would publish their changes and all would receive the updates. In this distributed solution, the graph would not have a central store, but it would exist as a set of local copies. This solution needs a synchronization mechanism distributed in all the agents to guarantee the global coherence of the graph, similar to the ones used in collaborative editing [14] or BASE databases[15][16].
- A third option would be to let the agents push partial or global updates on the graph to a known server agent. This agent would process the updates to guarantee the global coherence of the graph and would publish the new versions back to the agents. A similar solution was proposed in [17] for a distributed scene-graph to be used in shared virtual reality scenarios. Also, this approach is similar to the

4 www.openscenegraph.com

5 http://neo4j.com/

6 http://sparsity-technologies.com/

one used in modern code repositories, such as Git[7]. Each agent works with a local copy of the graph while new updated versions are arriving by subscription to the server. The local management of this flow is responsibility of the local agent until it decides to push the changes up to the server. After that it receives a confirmation that the changes are valid or a denying response with the error. This is the solution currently implemented in CORTEX. Performance is more than enough for our current needs and comparative tests will be done when the other implementations be available.

In the next section we present a brief formalization of DSR in its current state.

2.1 DSR Formalization

DSR is a multi-label directed graph which holds symbolic information as logic attributes related by predicates. These are stored in nodes and edges respectively. Also, DSR holds geometric information as predefined object types linked by $4x4$ homogeneous matrices. Again, these are stored in nodes and edges respectively. With DSR, the hand of the robot can be at a 3D pose and, at the same time, it can be *close_to_the_door_knob*, being this a predicate computed by measuring the distance between the hand and the knob, in the graph representation. Note that this distance could also be measured with more precision by direct observation of both the knob and the hand once they are inside the frustum of the robot's camera but, at the end, that information would have to be stored in the graph and propagated to the other agents.

As a hybrid representation that stores information at both metric and symbolic level, the nodes store concepts that can be symbolic, geometric or a mix of them. Metric concepts describe numeric quantities of objects in the world that can be structures like a three-dimensional mesh, scalars like the mass of a link, or lists like revision dates. Edges represent relationships among symbols. Two symbols may have several kinds of relationships but only one of them can be geometric. The geometric relationship is expressed with a fixed label called "RT". This label stores the transformation matrix between them. A formal definition of DSR can be given as a multi-label directed graph $G = (N, E)$ where N represents the set of nodes $\{n_1, ...n_k\}$ and E the set of edges $\{e_1....e_r\}$. An edge e joining the nodes u and v will be expressed as $e = uv$.

$$G = (N, E) \text{ where } E \subseteq N \times N, uv \neq vu (\text{ without loops } vv) \tag{1}$$

According to its nature, the properties of symbolic edges are:

1. Given a symbolic edge $e = uv$, we cannot infer the inverse $e^{-1} = vu$
2. A symbolic edge $e = uv$ can store multiple values
3. The set of e is defined as $L = \{e_1, ...e_r, (l_1, l_2, ...l_s)\}$ where $l_i \neq l_j$

---

[7] https://git-scm.com/

On the other hand, according to its geometric nature and the properties of the transformation matrix RT, the characteristics of geometric edges are:

1. For each geometric edge $e = uv$, e is unique
2. For each geometric edge $e = uv = RT$, we can define the inverse of e as $e^{-1} = vu = RT^{-1}$

Therefore the kinematic chain $C(u, v)$ is defined as the path between the nodes u, v and an equivalent transformation $RT*$ can be computed by multiplying the equivalent transformations corresponding to the sub paths from each node to their closest common ancestor. Note that sub path from the common ancestor to v will be obtained multiplying the inverse transformations. These geometrical relations are showed in Figure 3.

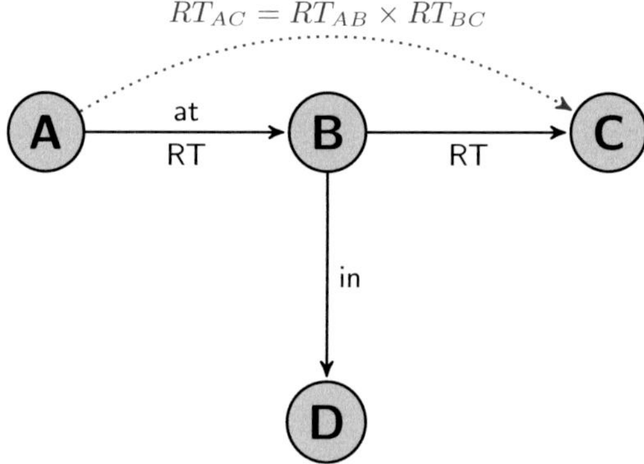

Fig. 3 Unified representation as a multi-labeled directed graph. Edges are labeled "at" and "in" denoting logic predicates between nodes. Also, edges between A,B and B,C have a geometric type of label, "RT" that codes a rigid transformation between them. Geometric transformations can be chained or inverted to compute changes in coordinate systems.

3 Experimental Results

As an initial validation of CORTEX and DSR in a real robot interacting with humans, we tested these ideas in Gualzru [23]. Gualzru is a salesman robot that works autonomously in crowded scenarios and has to step out when a potential client passes by. He will approach the customer and start a conversation trying to convince her to go to an interactive sales panel. If the robot succeeds, it will walk the person to the panel and then will start a new search.

In previous versions of the robot, we found that some synchronization problems were caused by having a fragmented internal representation. The robot used two separated graphs, one for the kinematic state and one for symbolic attributes and predicates, as many current robotic architectures [18]. Agents injecting data in both graphs at different rates and expecting changes to occur under timeout restrictions, caused unpredictable behavior very hard to debug. This not well understood complexity caused a steady decrease of productivity in the project, to a point where it was difficult to go on. The substitution of the graphs by the new integrated DSR considerably improved the working conditions again. Not all problems were gone and new debugging and monitoring tools are still needed, but communication among agent started to work flawlessly and more complex behaviours are now being explored.

A simple example shows how DSR enables the coordination of several agents in a primitive HR collaboration scenario. DSR provides the agents a common context with multi-modal, semantically distant information, to take the correct decisions. When a person entered the robot's field of view, the *Person* agent would inject a simplified skeleton in the kinematic graph at the right position relative to the floor. Perceptive updates on this representation were performed smoothly as long as the person remained in view and all agent could access that cognitive object. At the same time, other agent *Dialog* was trying to maintain a conversation with the person following steps of a plan hold in the symbolic graph. This agent will keep talking under the condition that the person is paying attention, which is computed as a simple function of some person-robot relational parameters such as presence, proper distance, and face and eyes correctly detected. Those parameters were also being computed by the first agent *Person* and injected as attributes in the symbolic graph.

The existence of an integrated representation also helps to the redesign of the software architecture. For instance, one important drawback of Gualzru was related to its limited conversational abilities. These limitations greatly affects its performance. Speech recognition is hard to solve in noisy, crowded scenarios in which even people find difficulties in understanding each other (see Figure 4). It is also difficult to understand what the robot is saying. To improve the ability of the robot to communicate within this scenario, we have added a tactile screen on the robot. The screen

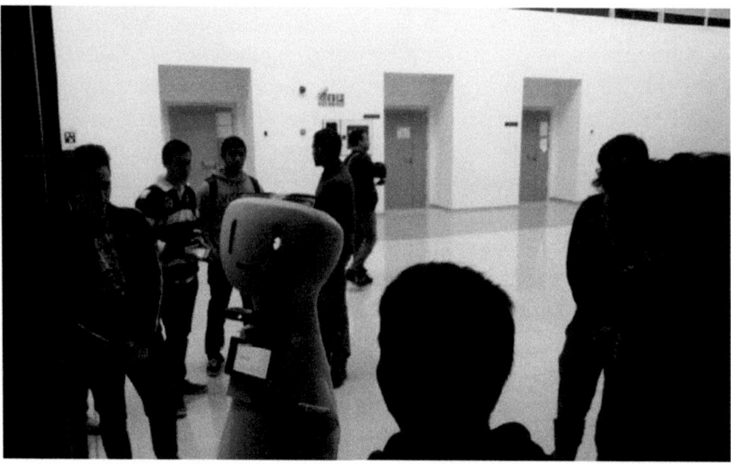

Fig. 4 The Gualzru robot interacting with people at the University of Malaga.

is controlled by a specific agent, but shares information with the rest of agents on the framework, such as the Dialog one. Thus, it was easy that this screen displayed what the robot is saying. The screen also allows the person to answer to the robot by touching the desired response on the screen. This information, although captured by the Screen agent, is injected in the graph and made available to the rest, so the agent in charge of the ASR/Dialog can use it to complete missing data. It is important to note that these concepts can be updated by the agents at interaction rates.

The ADAPTA project, which gave birth to Gualzru and the advertisement scenario, started in 2012 and many versions and options of the current DSR have been evaluated since then. The last demonstration tests will be held in October 2015 a will show us if the DSR graph is able to the sustain the whole architecture at human interaction rates.

4 Conclusions and Future Work

This paper has presented our proposal for internalizing a deep state representation of the outer world. After testing the previous approaches in very demanding scenarios, the unified representation arises as our final approach for

- solving the synchronization problem;
- endowing the full kinematic tree with symbolic information; and
- providing the geometric information to the high-level planner

The unified representation is currently interfaced by a set of task-related networks of agents, which will provide broad functionalities such as navigation, dialog or multi-modal person monitoring. The current implementation guarantees that the agents are able to feed the unified representation with new geometric models or symbolic concepts, and that the data stored in the representation is kept synchronized with the real world by updating actions performed by different agents. Also, the whole graph is kept synchronized among the agents by using an efficient publishing mechanism.

Future work will focus on injecting raw data directly in the graph and let the agents build on it more abstract representations. The processing schema that we propose admits the inclusion of active perception strategies by mixing top down -planned- and bottom up -reactive- trends through the agents interaction with the DSR. Also, we plan to exploit the hierarchical structure in the graph to optimize the communication mechanism by, for example, allowing temporal subscriptions to specific parts of the representation -e.g. the person or the robot arm. It is also needed to evaluate the computational effort associated to the management of graphs such as the one in Figure 1. Although initially the number of nodes/arcs may be relatively small, the inclusion of raw data in the leaves, of new spatial structures discovered during navigation or new predicates relating logical attributes, might introduce delay or throughput problems affecting the overall performance.

Acknowledgments This paper has been partially supported by the Spanish Ministerio de Economía y Competitividad TIN2012-TIN2012-38079 and FEDER funds, and by the Innterconecta Programme 2011 project ITC-20111030 ADAPTA.

References

1. Foote, T.: tf: The transform library. In: 2013 IEEE International Conference on Technologies for Practical Robot Applications (TePRA). Open-Source Software Workshop, pp. 1–6 (2013)
2. Blumenthal, S., Bruyninckx, H., Nowak, W., Prassler, E.: A scene graph based shared 3d world model for robotic applications. In: 2013 IEEE International Conference on Robotics and Automation, pp. 453–460, May 2013
3. Bustos, P., Martinez-Gomez, J., Garcia-Varea, I., Rodriguez-Ruiz, L., Bachiller, P., Calderita, L., Manso, L., Sanchez, A., Bandera, A., Bandera, J.: Multimodal interaction with loki. In: Workshop of Physical Agents, pp. 1–8 (2013)
4. Rusell, S., Norvig, P.: Artificial Intelligence: A Modern Approach, 3rd edn. Pearson (2009)
5. Poole, D., Mackworth, A.: Artificial Intelligence: Foundations of Computational Agents. Cambridge University Press (2010)
6. Manso, L.J.: Perception as stochastic sampling on dynamic graph spaces, Ph.D. dissertation (2013)
7. Calderita, L.V., Bustos, P., Suárez Mejías, C., Fernández, F., Bandera, A.: Therapist: towards an autonomous socially interactive robot for motor and neurorehabilitation therapies for children. In: 7th International Conference on Pervasive Computing Technologies for Healthcare and Workshops, vol. 1, pp. 374–377 (2013)
8. Selfridge, O.G.: Pandamonium: a paradigm for learning. In: Proceedings of the Symposium on the Mechanization of Thought Processes, pp. 511–529 (1959)
9. Newell, A.: Some problems of basic organization in problem-solving programs, Tech. Rep. (1962)
10. Erman, L.D., Hayes-Roth, F., Lesser, V.R., Reddy, D.R.: The hearsay-ii speech-understanding system: Integrating knowledge to resolve uncertainty. ACM Computing Surveys **12**(2), 213–253 (1980)
11. Hayes-Roth, B.: A blackboard architecture for control. Artificial Intelligence1 **26**(2), 251–321 (1985)
12. McManus, J.W.: Design and analysis of concurrent blackboard systems, Ph.D. dissertation (1992)
13. Corkill, D.D.: Blackboard systems. AI Expert **6**(9) (1991)
14. Shapiro, M., Preguiça, N., Baquero, C., Zawirski, M.: Convergent and commutative replicated data types. Bulletin of the European (104) (2011)
15. Balegas, V., Ferreira, C., Rodrigues, R., Shapiro, M., Preguic, N., Najafzadeh, M.: Putting consistency back into eventual consistency. In: EuroSys 2015 (2015)
16. Pritchett, D.: Base: An acid alternative. Queue, June 2008
17. Naef, M., Lamboray, E., Staadt, O., Gross, M.: The blue-c distributed scene graph. In: Proceedings - Virtual Reality Annual International Symposium, pp. 275–276 (2003)
18. Lemaignan, S., Ros, R., Mösenlechner, L., Alami, R., Beetz, M.: Oro, a knowledge management platform for cognitive architectures in robotics. In: 2010 IEEE/RSJ International Conference on Intelligent Robots and Systems, pp. 3548–3553, October 2010
19. Hofland, K., Jørgensen, A.M., Drange, E., Stenström, A.: A Spanish spoken corpus of youth language. Corpus Linguistics (2005)
20. Jelinek, F.: Statistical methods for speech recognition. MIT Press (1997)
21. Chamorro, D., Vazquez Martin, R.: R-ORM: relajación en el método de evitar colisiones basado en restricciones. In: X Workshop de Agentes Físicos, Cáceres, España (2009)
22. Calderita, L.V., Manso, L.J., Bustos, P., Suárez-Mejías, C., Fernández, F., Bandera, A.: THERAPIST: Towards an Autonomous Socially Interactive Robot for Motor and Neurorehabilitation Therapies for Children. JMIR Rehabil. Assist. Technol. (2014)

23. Romero-Garcés, A., Calderita, L.V., González, J., Bandera, J.P., Marfil, R., Manso, L.J., Bandera, A., Bustos, P.: Testing a fully autonomous robotic salesman in real scenarios. In: Conference: IEEE International Conference on Autonomous Robot Systems and Competitions (2015)
24. Manso, L.J.: Perception as stochastic sampling on dynamic graph spaces, Ph.D. dissertation, Univ. of Extremadura, Spain (2013)
25. Tomasello, M., Carpenter, M., Call, J., Behne, T., Moll, H.: Understanding and sharing intentions: The origins of cultural cognition. Behavioral and Brain Sciences **28**(5), 675–691 (2005)
26. Bauer, A., Wollherr, D., Buss, M.: Human-robot collaboration: a survey. Int. Journal of Humanoid Robotics (2007)
27. Kirsch, A., Kruse, T., Mösenlechner, L.: An integrated planning and learning framework for human-robot interaction. In: 4th Workshop on Planning and Plan Execution for Real-World Systems (held in conjunction with ICAPS 2009) (2009)
28. Beetz, M., Jain, D., Mösenlechner, L., Tenorth, M.: Towards performing everyday manipulation activities. Robotics and Autonomous Systems (2010)
29. Alami, R., Chatila, R., Clodic, A., Fleury, S., Herrb, M., Montreuil, V., Sisbot, E.A.: Towards human-aware cognitive robots. In: AAAI 2006, Stanford Spring Symposium (2006)
30. Wintermute, S.: Imagery in cognitive architecture: Representation and control at multiple levels of abstraction. Cognitive Systems Research **19–20**, 1–29 (2012)
31. Ali, M.: Contribution to decisional human-robot interaction: towards collaborative robot companions, PhD Thesis, Institut National de Sciences Appliquées de Toulouse, France (2012)
32. Clark, A.: An embodied cognitive science? Trends in Cognitive Sciences **3**(9) (1999)
33. Holland, O.: The future of embodied artificial intelligence: machine consciousness? In: Embodied Artificial Intelligence, pp. 37–53 (2004)
34. Manso, L.J.: Perception as Stochastic Sampling on Dynamic Graph Spaces, PhD Thesis, University of Extremadura, Spain (2013)

A Navigation Agent for Mobile Manipulators

Mario Haut, Luis Manso, Daniel Gallego, Mercedes Paoletti, Pablo Bustos, Antonio Bandera and Adrián Romero-Garcés

Abstract Robot navigation and manipulation in partially known indoor environments is usually organized as two complementary activities, local displacement control and global path planning. Both activities have to be connected across different space and time scales in order to obtain a smooth and responsive system that follows the path and adapts to the unforeseen situations imposed by the real world. There is not a clear consensus in how to do this and some important problems are still open. In this paper we present the first steps towards a new navigation agent controlling both the robot's base and the arm. We address several of theses problems in the design of this agent, including robust localization integrating several information sources, incremental learning of free navigation and manipulation space, hand visual servoing in camera space to reduce backslash and calibration errors, and internal path representation as an elastic band that is projected to the real world through measurements of the sensors. A set of experiments are presented with the robot Ursus in real and simulated scenarios showing some encouraging results.

Keywords Robotics · Navigation · Social robots

1 Introduction

Robot indoor navigation and manipulation are crucial components of current autonomous robot control architectures. As other complex modules that form part of these architectures, navigation and manipulation are usually decomposed in several

M. Haut · L. Manso · D. Gallego · M. Paoletti · P. Bustos(✉)
University of Extremadura, 10003 Cáceres, Spain
e-mail: pbustos@unex.es
http://robolab.unex.es

A. Bandera · A. Romero-Garcés
ISIS Group, University of Málaga, Málaga, Spain

L.P. Reis et al. (eds.), *Robot 2015: Second Iberian Robotics Conference*,
Advances in Intelligent Systems and Computing 418,
DOI: 10.1007/978-3-319-27149-1_58

loosely coupled elements that form a distributed system. Typical navigation elements are collision avoidance, environment perception, (re)planning of safe optimal paths and (re)localization. One challenge in the design of these architectures is the "gap" problem, that arises when two different elements have to share enough context as to take proper informed decisions. A typical example would be the gap between the local collision controller and the path planner. Another one is the gap between the (re)localization algorithm and the environment perception element, when a moving human crosses in front of the robot, or when a new structural element appears in the environment.

Different solutions have been proposed to this problem, mainly to specific versions of it. One of the best known ideas is the concept of elastic bands, introduced by [1]. An elastic band is a theoretical construction obtained by a path planner that gets grounded to the real world by means of the interaction with a range sensor. The band works as a glue filling the gap between the internal representation of the path and the constraints imposed by the world physics. The path could be "broken" by a human passing by and restored afterwards. The local controller only "sees" a small perturbation that might involve a change in speed. In this paper we present ongoing work on the design of a new navigation and manipulation agent for the RoboCog architecture, that is based on the idea of elastic bands. Figure 1 shows a schematic representation of the robot in a navigation task that ends with the manipulation of the cup on the table.

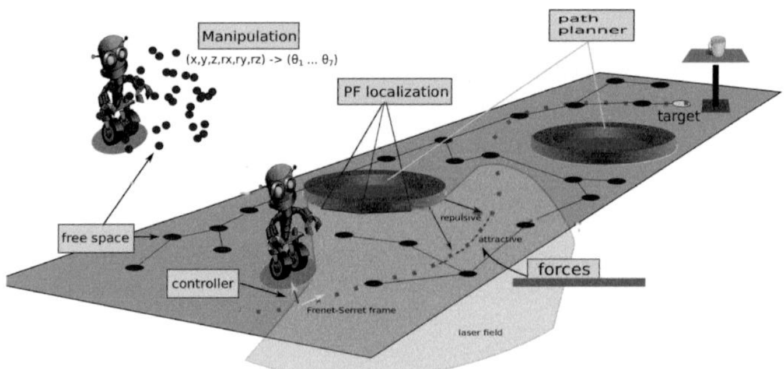

Fig. 1 Schematic view of a robot in a BringMe(x) task showing representing the free space for base navigation and manipulation. Dotted line is the planned path partially smoothed by laser projecting forces. Knowlegde of obstacles provides localization information.

Following these ideas, we will address several problems in the design of a new manipulation and navigation agent that will replace our current technology. We will describe improvements for indoor localization, lifelong free space learning and manipulation control with poorly calibrated and backslash mechanical arms. These three issues are crucial to achieve the mid-term goal of having a reliable BringMe(X) skill in the robot. There other elements necessary to accomplish this endeavor but they are discussed elsewhere [2] [3]. The algorithms described here are based on a shared representation of the robot and the environment. The different software components can access this structure to share and become aware of what other modules are doing. We review now work done in the elastic band concept and in visual servoing, since those are the main theoretical motivations in this paper.

Elastics bands were introduced by Quinlan and Kathib [1] as a method to close the gap between path planning and real-time collision avoidance. The technique has not received too much attention in the robotics community, maybe due to the independent interest on more specialized methods of local control and path planning. Later on an extension of the idea was introduced as elastic strips to cope with robots with many degrees of freedom[4]. All DOFs of the manipulator where included combining obstacle avoidance with desired posture behavior. A last generalization to the original idea was published as elastic roadmaps [5] to include planning in the loop.

When there are unmodeled misalignments and backslash in a low cost robotic arm, the place where the hand will get and the place that the camera is selecting, will not match. One elegant way to solve this problem was described by Hutchinson [6]. The idea is to bring the hand and the target to the camera space and perform a visual control loop there minimizing the observed visual error. The process can be shown to converge under mild conditions. The theory comes form visual servoing, a mature discipline originated in the control arena that has spread many new research lines and applications. The original work of [7] was followed by many others [8] [9] [23] [10] and also in different areas such as tracking of unknown objects [11] or binocular heads [12].

The rest of the paper describes the overall system and each of the problems and improvements introduced in the architecture. A final section discusses two experiments involving the robot Ursus that validate the choices made here and the way they have been implemented and integrated.

2 Overview of the System

The navigation agent proposed here is being built as part of the robotics cognitive architecture RoboCog [13] [2][3]. The overall goal is to integrate navigation and manipulation in a common framework so more complex tasks can be handled through body, head and arm coordinated movements. In this paper we will focus on the first steps of the design of the agent, describing localization, learning of free space, path planning, path execution and trajectory control of the arm. Each part is described as an improvement over their previous versions in RoboCog. We have identified important limitations in each one and introduced the necessary changes to solve the existing problems. From the point of view of the software, the complete agent is being built using the robotics framework RoboComp [14]. Each functionality is implemented as a set of interconnected components, many of which are shared among the others.

2.1 Internal Representation of Space

In RoboCog, the robot and its environment is represented as a graph $I = (N, E)$, also called *InnerModel*. The nodes in I correspond to parts of the robot and to elements in the world. They belong to one of the following types $N \rightarrow \{Robot, Joint, Laser, IMU, RGBD, Object, Mesh\}$. The edges in I are rigid euclidean transformations linking the nodes and represented as $4x4$ homogeneous matrices. The nodes can be extended with a list of $< attribute, value >$ pairs so they can be annotated with semantic information encountered during robot operation. In building the new Navigation agent, we have extended the initial set of types with a new one, named FreeSpace. This type defines a graph $G = (N, E)$ representing what the robot believes about its free configuration space. We now define separately the C-space of the robot base and the arm. Let's as $B \in \mathbf{R^2}$ be the base's C-space in which orientation is ignored, and $A \in \mathbf{R^7}$ the C-space of the arm. From them we can define two free configuration spaces, $B_f = B/C_{obs}$ for the displacement of the base, ignoring orientation, and $A_f = A/C_{obs}$ for the movement of the arm. Therefore, two graphs will be created, $G_B = (N_B, E_B)$ and $G_A = (N_A, E_A)$. In the next sections, the construction of these spaces and their use in computing safe paths will be described.

2.2 Localization

Localization is a crucial task that updates the estimated position of the robot in its internal model. Probabilistic algorithms based on partially known geometric information of the environment constitute the focus of this task. We use the recently introduced particle filter variant CGR [15] to obtain an an estimate of the pose of the robot with respect to an initial global reference system. The map of the environment is represented as a list of lines corresponding to the lower part of walls and known objects in the world. This algorithm is very efficient in cpu cycles and number of particles because the measurement function is analytic and derivable. A pre-ordering of the lines accounts for occlusions and an internal minimization loop improves the particles position using the Jacobian of the measurement. To obtain a reasonable estimate of the ground truth pose of the robot, a set of AprilTags [16] marks have been distributed in the apartment. Specialized components detect the marks and compute the error between the robot's current belief and its real position.

Figure 2 shows the current organization of components that implement localization. Arrows in the graph can be interpreted as one component sending information to another. *Localizer* receives poses from Base, AprilLoc and CGR along with their uncertainty, although only the base odometry and CGR estimation are fused. Localization based on AprilTags are used only for evaluation. Currently, an empirically obtained threshold over the variance is used to combine them and produce the current belief for the robot.

2.3 *Free Space Representation for the Robot*

If we want the robot to be perceived by the human as real collaborative object, it has to react and operate at human rates. One limitation to this requirement is the delay introduced by path planning algorithms searching the free space, such as RRT [17]. For our agent we use the probabilistic road map algorithm, PRM [18] to create a graph representing the free space, and RRT only to search for paths when unconnected islands remain in the PRM graph. The resulting path is inserted into the graph to connect the isolated islands. RoboComp currently includes a wrapper for OMPL [19] but, as of today, OMPL's PRM implementations does not allow to store the computed graph on disk. As that feature is crucial for long term robot operation we implemented a version of PRM using the Boost Graph Library, BGL.

A free space graph created with our PRM algorithm is shown in Figure 3. The upper-right rooms has been intentionally placed there to create an isolated region in the graph, giving a finite construction time. In (b) the algorithm calls RRT and obtains a feasible path. The robot traverses the path in (c) and in (d) it is added to the graph a only one connected component remains.

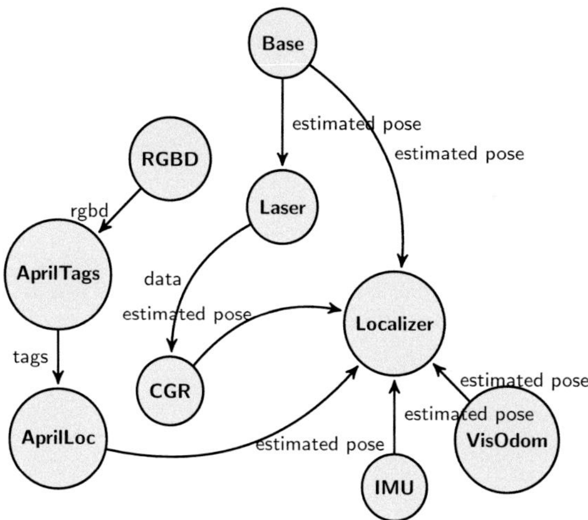

Fig. 2 Components used for localization in the Navigation Agent: Base controls the robot displacement, RGBD provides access to the camera, Laser provides access to the laser, AprilTags detects and estimates the pose of the tags, AprilLoc computes the localization error between the current estimate and what the tag provides, CGR implements the Corrective Gradient Refinement algorithm, IMU and VisOdom estimate indepent robot poses but their ouputs are not included here, and Localizer is the component that integrates all the sources to maintain the current belief.

2.4 Free Space Representation for the Robot's Arm

A similar argument concerning the planning time can be applied to arm manipulation, where seven DOFs expand the configuration space. Natural HRI demands that repetitive actions improve over time. If the robot grabs a cup from the same table several times, the accompanying person will expect that the robot reduces its execution time down to human standards. Doing otherwise, human confidence on a helpful interaction decreases. To avoid this situation we adopt a similar solution as in the previous section. In this case, instead of a randomly sampled graph of free C-space, a regular 3D grid sampling the working volume of the arm will be used. Each element in the grid codes an euclidean 6D pose for the hand tip and a set of configurations in free C-space that correspond to that pose. In this initial model, only the most common hand orientation for grasping a common small object is assigned to each point in space. As we will see later, this is not a limitation since the final manipulation plan includes a last refinement step, using inverse kinematics, to reach the target orientation and position.

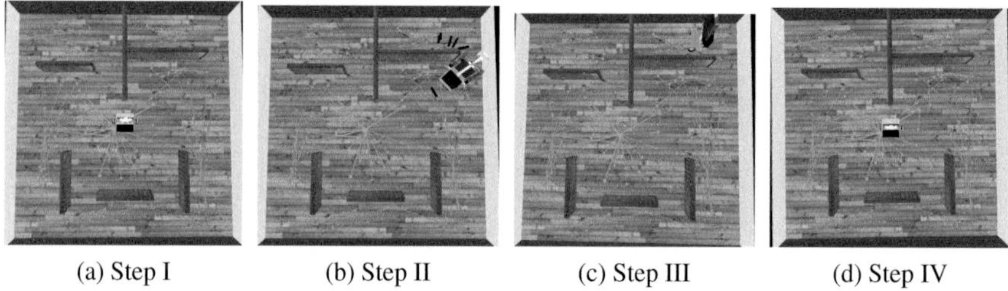

(a) Step I (b) Step II (c) Step III (d) Step IV

Fig. 3 Sequence showing graph learning with PRM in N_B created by PRM. (a) The upper right room has a very narrow entrance to force the creation of an island. (b) When the robot is sent there, RRT is activated and finds a route to the isolated spot. (c) The robot traverses the path and (d) the steps in the path are learned into the graph. Only one connected component remains

2.5 Path Planning in Free Space

Once these graphs have been created using inverse kinematics computation, the process to obtain a safe path is quite similar in both robot navigation and arm positioning. Note that both graphs are currently kept separated although they will be integrated in a near future. The path is created by first searching the closest point in the graph to the current position -robot or hand-, then the closest point in the graph to the target position and, finally, a path through the graph linking both points.

The plan constructed by the path planner can be seen as a *theoretical* construct based on the robot's beliefs about the world. As such, it will not be exact or even precise. Therefore, another components are needed to *ground* this construct into the real word using the information coming for the sensors. These new components update the path as it is traversed, adapting it to unexpected events and detect critical

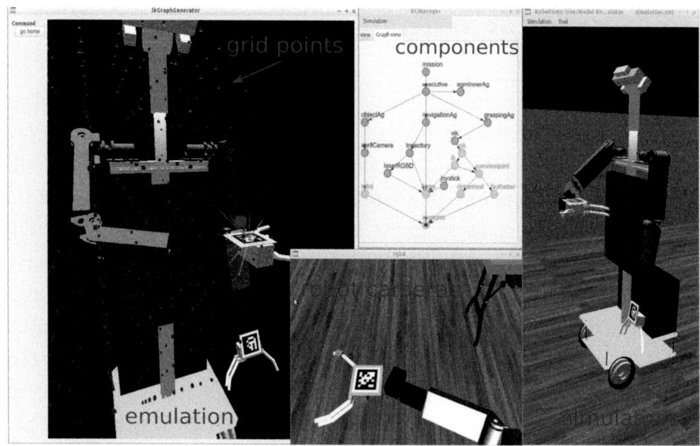

Fig. 4 Free space graph for the arm. The four captures express different aspects of the process. On the left, the robot is shown with grid during the learning process -emulation. Middle up, is the graph of components deployed for this activitiy. Middle down, is the arm as seen by the robot's camera. Left, is the robot in RoboComp simulator.

conditions blocking the robot's path. So, as the path is a shared structure, when a blocking situations occurs, the planning agent is aware and reacts computing a new path. This dynamic process will continue until the path is completed or no solution can be found in a certain amount of time. Note that the path connects the global planner with the robot controller providing the necessary persistence to avoid local minima.

As illustrated in Figure 1, the path is *analyzed* under the laser field and two virtual forces are created that push the path away from the obstacles while keeping it from bending too much [1]. Currently we use the following robot controller to compute the final forward speed V_a and rotation speed V_r [20]:

$$\begin{aligned} V_a &= V_M e^{\alpha C} G(\beta V_r) \\ V_r &= \phi + \arctan(d) + C \end{aligned} \tag{1}$$

where V_M is the maximum advance speed, C is the curvature of the path, G is the Gaussian function, V_r is the rotation speed and β is a gain. Note that V_a is computed as the inhibition of the initial V_M caused by world interactions. For V_r, ϕ is the angle formed by the robot and the tangent to the path at the closest point to the robot, d is the perpendicular distance to the road and C is again the curvature.

The dynamics created by the algorithm make the path adapt to narrow passages, moving objects and even can be broken by a person passing by and restored afterwards. As the robot traverses the path, the steps left behind are deleted and the process ends when the target is reached and the path is completely erased.

2.6 Path Execution by the Arm

Low-cost arms present an important limitation that must be solved. Backslash and inaccurate calibrations induce errors in the end effector's position. The errors can easily ruin any intent of grasping objects if the only feedback used is the one from the joints' encoders. An elegant way out of this situation is to use a two-step procedure. First, the hand is brought to a place close to the target using a path through N_A. This movement is based only on joints feedback and is executed in parallel with a movement of the eyes whose purpose is to bring the hand and the target inside the camera's frustum. The planning of that movement using the graph of free space N_A was described in Section 2.5. The second step is a closed-loop visual servo control that brings the hand to its final grasping positions. This second part is the one that can cancel out the errors caused by backslash and mechanical misalignments. To detect and track the position of the hand, an AprilTag [16] is used. We present now this algorithm.

To describe the scenario and the operations involved we follow the notation in [6]:

- ${}^{x}R_y$: expresses the rotation matrix of the coordinate frame y with respect the x frame
- ${}^{x}t_y$: the position of the frame y with respect x
- ${}^{x}t_y$: the position of the frame y with respect x

We also define the following poses[1]:

- v_i: The coordinate frame of the end effector, as provided by the visual feedback in the ith iteration of the algorithm.
- t: The target pose.

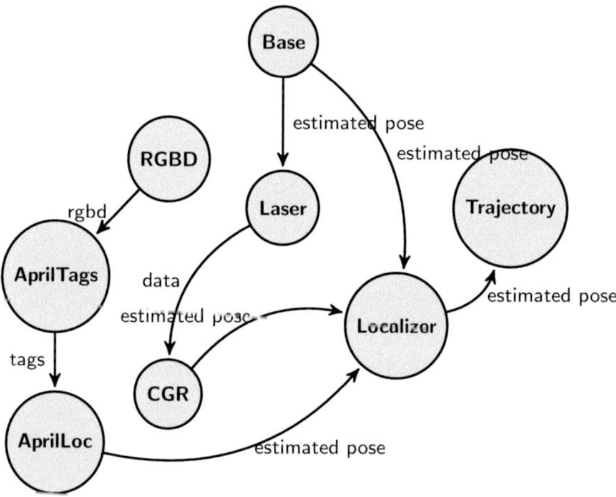

Fig. 5 Components involved in free space learning, path generation and navigation. Note that most components are also used in graphs shown in other sections.

[1] Here, as in [6], we use the terms *coordinate frame* and *pose* interchangeably.

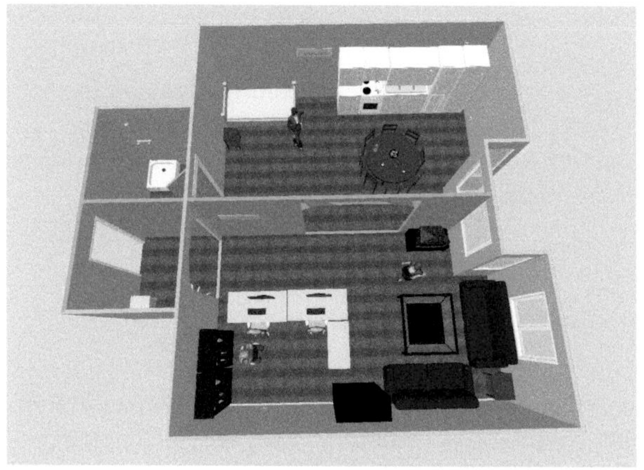

(a) RoboHome

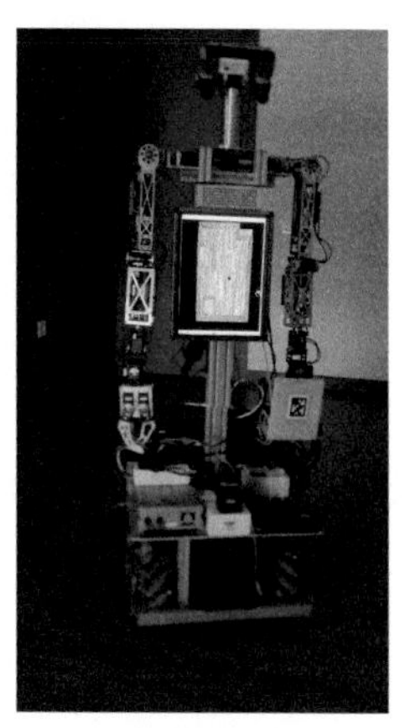

(b) Ursus 3

Fig. 6 (a) Autonomy Lab in RoboLab, UEx. A 70m apartment where robots can interact with people in HRI experiments. b)Ursus is the third generation of RoboLab's mobile manipulator [21]. It has a head, two 7 DOFs arms and an omnidirectional base custom built with Mecanum wheels.

- k_i: The pose that is actually sent to the IK module in the ith iteration of the algorithm. The existence of backslash and calibration inaccuracies make t and k differ.

Thus, the aim of the algorithm is to reduce to the maximum the difference between the effector's pose as given by visual feedback (v) and the target pose (t). Following the previously mentioned notation, the translation error is computed as ${}^{t}t_{v}$ and the rotational error as ${}^{t}R_{v}$. This way, the goal is to make ${}^{t}t_{v}$d as close to zero as possible and ${}^{t}R_{v}$ as close to an identity matrix as possible.

The position of the effector as seen (v) is updated in real time each time the camera sees the corresponding tag. The algorithm works by sending the inverse kinematics module a series of intermediate positions (k_n) of the effector as seen by the camera (v) and as close to the target (t) as possible.

The algorithm works as follows:

1. The IK target k is initialized as the actual target: $k = t$
2. The translation and rotational $errors$ between the visual position and the IK position are calculated: ${}^{t}t_{v}$ and ${}^{t}R_{v}$, respectively.
3. The algorithm successfully stops if the errors are small, or unsuccessfully if it has been running for a long period of time.
4. A corrected IK position (k_i) so that: ${}^{k_i}t_t = -{}^{v_i}t_t$ and ${}^{k_i}R_t = {}^{v_i}R_t{}^{-}1$ are calculated and sent to the IK module.
5. If $error > threshold$ go to step 2.

3 Experiments

Two experiments have been made to test and validate the current state of the new navigation agent within the robot Ursus, shown in Figure 6. Also in the same Figure it is depicted a 3D drawing of RoboLab's Autonomy Lab, a $70m^2$ apartment conditioned for social robotics research 6.

3.1 *Localization and Navigation Experiment*

The first experiment was designed to measure the robustness of the CGR localization algorithm in real world, highly perturbing conditions. A list with the 2D segments representing the walls and furniture on the floor was obtained from the construction blueprint and by manual measurement. The robot was sent to a set of random locations in the apartment and the ground truth position was recorded using a set of AprilTags [22]. Figure 7 shows the evolution of the accumulated error during 90 meters of unstopped navigation. The mean is 8cm and the standard deviation 3cm showing a very good localization performance for this kind of social tasks.

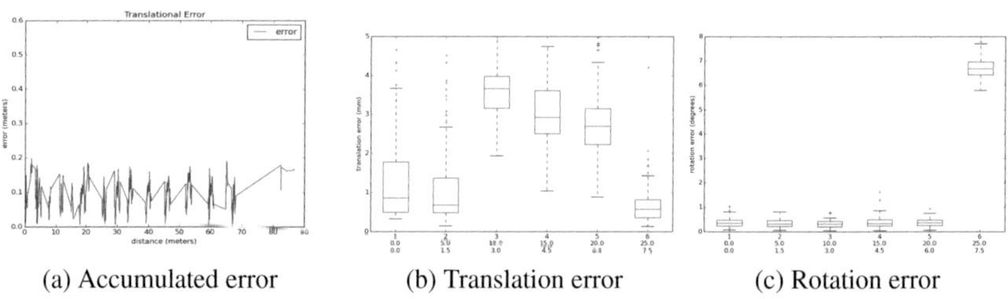

(a) Accumulated error (b) Translation error (c) Rotation error

Fig. 7 (a) Accumulated error of the robot after 90m of continuous navigation in a noisy environment, with unmodeled objects and people walking around. Mean 0.08 m. Variance 0.0014. Standard deviation 0.0367. b)The y axis shows the errors (euclidean norm) obtained in translation. The X axis shows the experiment number and the standard deviation of the errors. c) The Y axis shows the errors obtained in rotation. The X axis shows the experiment number and the standard deviation of the errors.

3.2 *Arm Control Experiment*

A simulated experiment with increasing synthetic calibration errors was conducted to show how the visual servoing algorithm is capable of moving the effector to target poses with good precision. Six experiments were performed sending the arm to 100 different positions in each one. An increasing level of error was injected for each experiment in the arm's kinematic configuration. Errors were assigned in increments of 1.5 degrees for joint angular position (encoder error) and of 5 mm for translational joint position (mechanical misalignment). The 100 targets of each experiment were randomly generated from a work space of dimensions $X \in 140 - 130$ $Y \in 780 - 800$ and $Z \in 300 - 900$ in front of the robot, where $+Z$ points outwards perpendicular to robot's chest. The robot's arm movement would stop whenever the rotational error was lower than 0.1 rad and the translational error was lower than 5 mm. Errors were recorded at the end of each robot's arm movement. As can be seen in Figure 7, even in the presence of very high calibration errors the end effector ended with translation errors under the 5 mm threshold, and rotational errors were widely bellow the 0.1 rad. threshold except for the most adverse situation in which the mean settled close to 8 degrees.

4 Conclusions

In this paper we have shown work in progress on the construction of a new navigation software agent for the RoboCog architecture. Five interrelated functionalities have been reassessed and new algorithms have replaced the existing ones. As a result we have now a very robust localization component that can take pose estimations of the robot from a number of other components and maintain a reliable pose for extended periods of time under real life aggressive conditions. In this group, CGR provides map based localization and its performance is excellent. Robot navigation was formerly based on the combination of RRT and VFH+ and their integration presented important problems. With the introduction of a persistent structure coding the current path, those problems no longer exist and a whole new set of new possibilities bounce into. The path allows for smoother paths, easier recovery, better controllers and a simpler and more rational software architecture. As a future line of work, we want to enrich the path with semantic annotations, leading also to the idea of semantic path planning and execution. Arm planning has also received an important improvement with this work. The early problems that we had with very low repeatability and precision during grasping operations, have been reduced to a point where we can grab a cup with much more reliability. The decomposition of arm grabbing gestures in two actions, one driven only by internal feedback and the other by visual feedback in the camera space, is a promising line of future research, markless visual servoing being the next challenge.

Acknowledgements This paper has been partially supported by the Spanish Ministerio de Economía y Competitividad TIN2012-TIN2012-38079-C01, ADAPTA project of the INTERCONNECTA program and FEDER funds.

References

1. Quinlan, S., Khatib, O.: Elastic bands: connecting path planning and control. In: Proceedings IEEE International Conference on Robotics and Automation (1993)
2. Marfil, R., Calderita, L.V., Bandera, J.P., Manso, L.J., Bandera, A.: Toward social cognition in robotics: extracting and internalizing meaning from perception. In: Workshop of Physical Agents WAF2014, pp. 1–12, June 2014

3. Calderita, L.V., Bustos, P., Mejías, C.S., Fernández, F., Bandera, A.: Therapist: towards an autonomous socially interactive robot for motor and neurorehabilitation therapies for children. In: 7th International Conference on Pervasive Computing Technologies for Healthcare and Workshops, vol. 1, pp. 374–377 (2013)
4. Brock, O., Khatib, O.: Elastic strips: A framework for motion generation in human environments. The International Journal of Robotics Research **21**(12), 1031–1052 (2002)
5. Yang, Y., Brock, O.: Elastic roadmaps - motion generation for autonomous mobile manipulation. Autonomous Robots **28**(1), 113–130 (2009)
6. Hutchinson, S., Hager, G.D., Corke, P.I.: A tutorial on visual servo control. IEEE Transactions on Robotics and Automation **12**(5), 651–670 (1996)
7. Espiau, B., Chaumette, F., Rives, P.: A new approach to visual servoing in robotics. IEEE Transactions on Robotics and Automation **8**(3), 313–326 (1992)
8. Malis, E., Chaumette, F.: 2 1/2 d visual servoing with respect to unknown objects through a new estimation scheme of camera displacement. International Journal of Computer Vision **37**(1), 79–97 (2000). Irisa, Inria Rennes, and France Received
9. Drummond, T., Cipolla, R.: Application of lie algebras to visual servoing. International Journal of Computer Vision **37**(1), 21–41 (2000)
10. Baumgartner, E.T., Seelinger, M.J., Fessler, M., Aldekamp, A., Gonzalez-Galvan, E., Yoder, J.-D., Skaar, S.B., Dame, N.: Accurate 3-d robotic point positioning using camera space manipulation, pp. 2–5
11. Gratal, X., Romero, J., Bohg, J., Kragic, D.: Visual servoing on unknown objects. Mechatronics **22**(4), 423–435 (2012)
12. Sapienza, M., Hansard, M., Horaud, R.: Real-time visuomotor update of an active binocular head. Autonomous Robots, September 2012. (July 2011)
13. Manso, L.J., Calderita, L.V., Bustos, P., Garc, J., Fern, F.: A general-purpose architecture to control mobile robots. In: Waf 2014, pp. 1–12, June 2014
14. Manso, L., Bachiller, P., Bustos, P., Calderita, L.: Robocomp: a tool-based robotics framework. In: SIMPAR, Second International COnference on Simulation, Modelling and Programming for Autonomous Robots (2010)
15. Biswas, J., Coltin, B., Veloso, M.: Corrective gradient refinement for mobile robot localization. In: 2011 IEEE/RSJ International Conference on Intelligent Robots and Systems, pp. 73–78, September 2011
16. Olson, E.: Apriltag: A robust and flexible visual fiducial system. In: Proceedings - IEEE International Conference on Robotics and Automation, pp. 3400–3407 (2011)
17. LaValle, S.M.: Planning algorithms. Methods **2006**(2), 842 (2006)
18. Kavraki, L.E., Svestka, P., Latombe, J.C., Overmars, M.: Probabilistic roadmaps for path planning in high-dimensional configuration spaces. IEEE Transactions on Robotics and Automation **12**(4), 566–580 (1996)
19. Şucan, I., Moll, M., Kavraki, L.: The open motion planning library. IEEE Robotics and Automation Magazine **19**(4), 72–82 (2012)
20. Thrun, S.: Probabilistic algorithms in robotics. AI Magazine **21**(4), 93–109 (2000)
21. Martín, F., Mateos, J., Lera, F.J., Bustos, P., Matellán, V.: A robotic platform for domestic applications. In: WAF 2014, pp. 1–8, June 2014
22. Olson, E., Leonard, J., Teller, S.: Fast iterative alignment of pose graphs with poor initial estimates. In: Proceedings 2006 IEEE International Conference on Robotics and Automation, ICRA 2006, pp. 2262–2269, May 2006
23. Chaumette, F., Hutchinson, S.: Visual Servo Control (II). IEEE Robotics and Automation Magazine, March 2007

Building a Warehouse Control System Using RIDE

Joaquín López, Diego Pérez, Iago Vaamonde, Enrique Paz, Alba Vaamonde and Jorge Cabaleiro

Abstract There is a growing interest in the use of Autonomous Guided Vehicles (AGVs) in the Warehouse Control Systems (WCS) in order to avoid installing fixed structures that complicate and reduce the flexibility to future changes. In this paper a highly flexible and hybrid operated WCS, developed using the Robotics Integrated Development Environment (RIDE), is presented. The prototype is a forklift with cognitive capabilities that can be operated manually or autonomously and it is now being tested in a warehouse located in the *Parque Tecnológico Logístico (PTL)* of Vigo. The main advantages and drawbacks on this kind of implementation are also discussed in the paper.

Keywords Warehouse control system · Autonomous guided vehicle · Robot control architecture · Navigation

1 Introduction

It is well documented that warehouse automation can increase throughput speed, accuracy, safety, and traceability [1]. The major activity in a warehouse is material handling, which is focused on the input and output of stored goods usually packed in pallets. The service that this kind of warehouses provides to its clients requires guaranteeing some time restrictions in delivering the products. The functions of the kind of warehouse we are dealing with can be divided into storage and repackaging of products.

J. López(✉) · E. Paz · A. Vaamonde · J. Cabaleiro
Systems Engineering and Automation Department, University of Vigo, Vigo, Spain
e-mail: joaquin@uvigo.es

D. Pérez · I. Vaamonde
Robotics and Control Unit. AIMEN Technology Centre, Vigo, Spain

L.P. Reis et al. (eds.), *Robot 2015: Second Iberian Robotics Conference*,
Advances in Intelligent Systems and Computing 418,
DOI: 10.1007/978-3-319-27149-1_59

Different factors are forcing changes in the logistics market, most notably the e-commerce and the manufacturing of custom-made products. Customers look for personalized products and mass customization is pushing the industry to reduce time to market and enhance production flexibility, where the batch size tends to one. This fact is highly linked with warehouses management, where the exploitation costs increase with the value-added tasks, where Third Party Logistics Providers (TPLs) have to raise the service quality while maintaining the operation costs.

Automated warehouses are evolving into more intelligent storage systems without need of installing fixed structures that complicate and reduce the scalability and flexibility to future changes. The current trend is to make them as flexible as possible. For that purpose, the Automatic Guided Vehicles (AGVs) are the ideal automation technology [2].

The advantages of using AGVs continue to increase as new applications are researched and explored:

- **Flexibility.** One of the main advantages comes from the flexibility of the technology because it does not require conventional material-handling infrastructures.
- **Dynamic Design solution.** Vehicles can be quickly reprogrammed to change their tasks or the path of operation, eliminating the need for expensive physical equipment installation. New paths, nodes, tasks, and work cells can be created almost instantaneously without the need for expensive retrofitting.
- **Efficiency.** Using this technology we can optimize the transport work flows distributed dynamically between different AGVs. Also, with this solution we have the possibility of 24/7 operation without human intervention.
- **Modular systems.** AGVs can be added as required by the growth of the operation as demand increases, allowing for a gradual implementation depending on the workload. Besides, these systems can be easily integrated with robot/palletization robotic cells and other storage machinery.
- **Precision.** As technology improves a more precise space localization is available obtaining a good stock management precision. Time precision can also be obtained thanks to the optimization systems, allowing for a just-in-time delivery.
- **Economic.** An excellent price/quality ratio can be obtained decreasing the running and maintenance costs.
- **Safety.** AGVs offer a safe and predictable method of pallet management, while avoiding interference with human and building factors.

There are already many companies that provide different solutions depending on the kind of products to be handled, from small items such as the KIVA solution [3] to full pallets such as the AGVs provided by JBT Corporation, Savant Automation Inc, American in Motion, etc. The AGVs can be seen as autonomous mobile robots able to perform some transport operations and in this paper we present a hybrid approach that allows using the forklift in its original manual operation mode or in the new implemented autonomous operation mode, showing the advantages and drawbacks of using a commercial forklift adapted (in particular RIDE) to develop a warehouse control system based on AGVs. Furthermore, the proposed solution includes a vision system installed onboard the AGV to detect the pallets in the working environment. This approach reduces the number of sensors installed in the warehouse facilities, increases the flexibility and reduces both the installation costs and the deployment time.

The rest of this paper is organized as follows. Next section introduces the global description of the system. The control architecture is presented in section 3. After describing the AGV control architecture in section 4, the task programming details are shown in section 5. Finally, section 7 presents the results and concludes the paper.

2 Global Description of the System

The flow of the pallets in this system is represented in Fig. 1. Customers leave their pallets in the reception area and pick them up in the Shipping area. AGVs should be in charge of moving the pallets between the reception area, the permanent storage area, the temporal storage area, the working area and the shipping area.

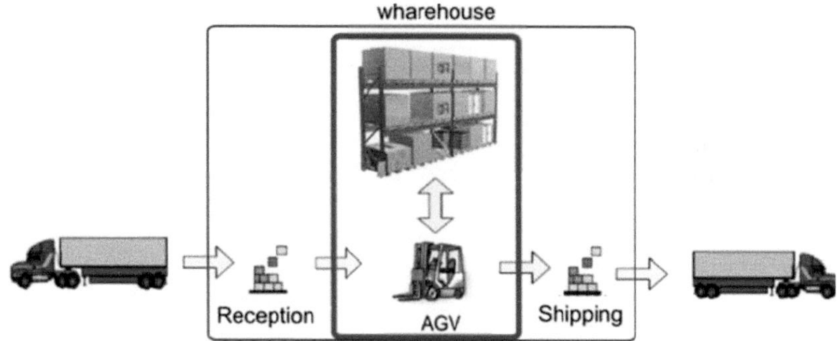

Fig. 1 Image of the system from the user point of view.

We carried out a detailed analysis of the flow of pallets in the logistics warehouse of KALEIDO SCM. In Fig. 2 the flow of pallets between the five zones that define the operation of the warehouse are represented. Then flow numbers represent the 12 different tasks that can be commanded to the AGVs. We will see later that one of the advantages of using RIDE is the task definition flexibility.

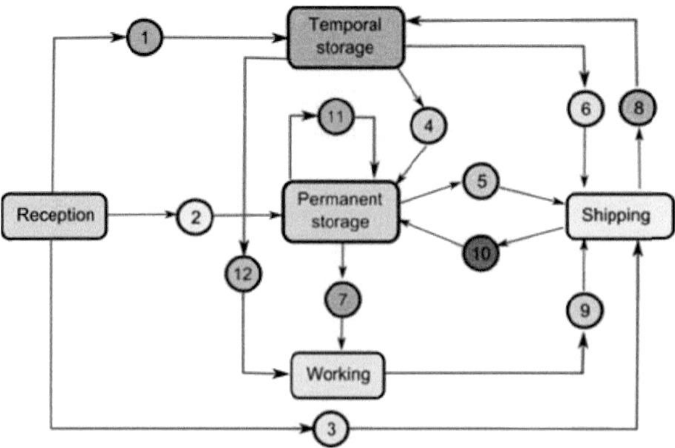

Fig. 2 General flow of the items in the warehouse.

According to the RIDE architecture [4], AGVs are connected to a central control server via Wi-Fi (Fig. 3). The management system requires robot services using an interface defined as a set of messages. For new task requests, the central server will decide which robot to send. Employees and maintenance staff can monitor the system from any terminal connected to the local network or Internet using a GUI. Since this user GUI is programmed in Java, it can be executed from a computer, Smartphone or any portable device.

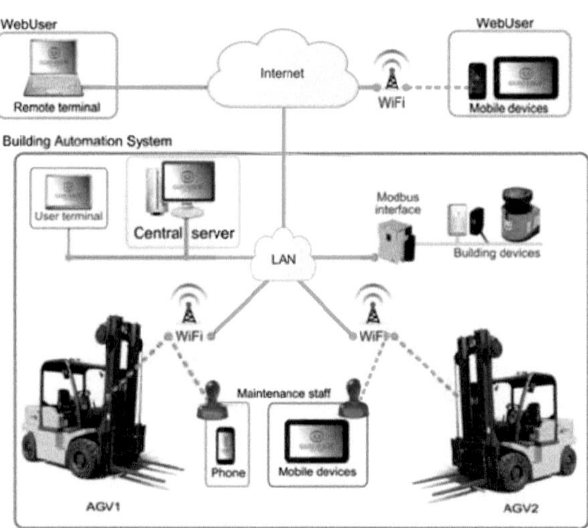

Fig. 3 General scheme of the warehouse automation system.

All the running GUIs and the Building Automation System (BAS) are also connected to the central server. Some of the elements controlled by the BAS that need to interact with the WCS system are:

- Occupancy detectors. Some sensors are settled to detect the presence of manual forklifts and people in some areas in order to control the traffic and avoid deadlocks. For example, some sensors are located to detect the presence of other forklifts in narrow corridors.
- Automatic doors. The path of some AGVs can include some doors. The AGVs request the opening of an automatic door to central server. This server interacts with the BAS to open the door.
- Charging sensors. Some sensors are located to detect the occupancy of some special spots such as the charging stations. This is necessary to be able to include manual and automatic vehicles.

AGVs can be connected and disconnected at any time. A central server computer keeps all the relevant information about the global state of the system. All the components or modules need to request and send data through the central server. For example, AGVs need to:

- Send information about their state (position in the map, task being executed, on-board sensor readings, etc.).
- Request other elements (modules) to execute some commands such as "open a door".
- Receive information about events such as "door opened" or "area occupied".
- Receive task execution requests.

3 System Control Architecture

The connections between different elements in the automation system are shown in Fig. 4. All the building devices in the BAS are connected to Modbus interfaces. These interfaces are connected to the LAN using Ethernet. A program in the central computer (central server) manages all the Modbus communications. The AGVs are also connected to the central server via Wi-Fi, even though other ways of communication such as GPRS or 3G could also be used.

Employees and maintenance staff can monitor the system using two different graphic user interfaces (GUIs). Robots that are not doing a task should be connected to the charging station. A typical task might start when the warehouse management System requests the Warehouse Control System to move a pallet.

The modules on the control architecture contained in RIDE can be divided into two distinct parts. The first corresponds to the set of modules that run on the host computer that is on board the mobile platform, while the second is the management system and centralized control that runs on a server.

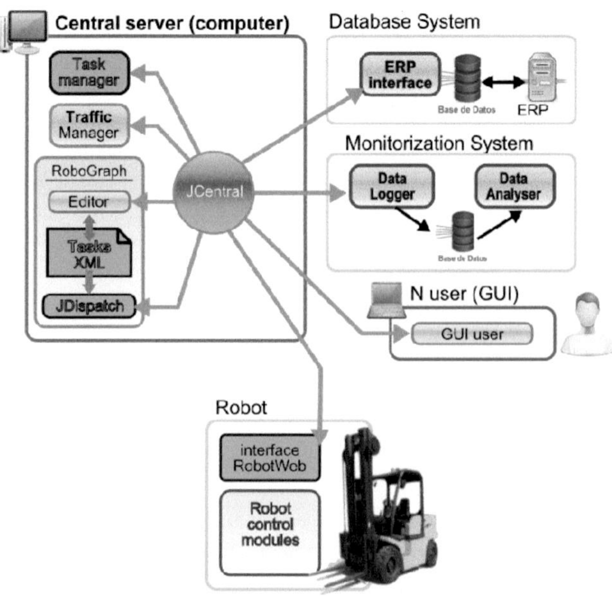

Fig. 4 Global control architecture.

This organizational structure facilitates the scalability of the system, both in terms of control modules, such as the deployment of more autonomous vehicles, to have a centralized management system multi-robot and multi-user.

This modular architecture has the same layers and shares most of the modules with other mobile robot applications developed with RIDE such as WatchBot [5] or GuideBot [6].

The global architecture shown in Fig. 4 is a modular centralized framework where modules are independent processes, most of them running in different CPUs. These modules exchange information using a message publish/subscribe mechanism named JIPC. JIPC provides a central communication process named JCentral (Fig. 4) and an interface (Java class) to be used by the different modules that want to communicate through JCentral.

The navigation control architecture in each robot, that is going to be explained in detail in the next section, includes a module to connect to the central unit using a wireless connection available in the building. So far we have been using a couple of alternatives: Wi-Fi because it is present in most modern buildings and a 3G modem for the rest of the cases.

The robotics development environment (RIDE [4]) includes a tool named RoboGraph [7] that implements the executive layer (Fig. 4). Besides RoboGraph and the robots, the main modules connected via JIPC are:

- **Task manager**. This module can start new tasks as a result of an event such as a request of an user. Also assigns tasks to robots and manages the queues of waiting tasks. An important issue in AGV systems is deciding what task should be assigned to a particular AGV [8][9].

- **Traffic manager**. This module does the path planning and manages the access of restricted areas where only a robot can be at the same time such as narrow corridors and elevators. The path obtained by this module is a sequence of topological nodes that can include doors. When a robot is following one of these paths, it should make a request to the traffic manager before entering a restricted area.
- **Building interface**. The communications with all the building devices including elevators are managed by this module.
- **ERP interface**. The Enterprise Resource Planning Software (ERP) handles end-to-end operational planning and is responsible with the Warehouse Management Systems (WMS) to send those tasks to the Warehouse Control System. This module is the interface between the Management and the control units of the warehouse.

4 AGV Control Architecture

The scheme of the on board control architecture is shown in 5. Each module running on-board the robot is a Linux process that exchanges information with other modules using IPC [10]. The framework implemented here uses IPC in centralized mode. IPC has been developed at Carnegie Mellon's Robotics Institute, and provides, among others, a publication–subscription model. An IPC based system consists of an application independent central server and a number of application-specific processes. Each process connects with the central server and specifies what types of messages it publishes and what types it listens for. Any message that is passed to the central server is immediately copied to all other processes subscribed.

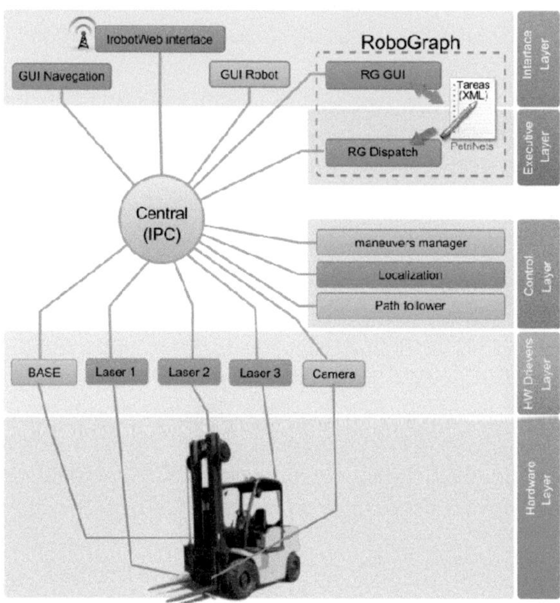

Fig. 5 Robot control architecture.

The architecture is divided into four levels software:

- **Hardware drivers layer.** This layer includes a series of programs aimed at controlling hardware devices. The main modules in this layer are:
  - BASE. All the movements of the base (wheels, fork, etc) are controlled by the PLC SmartController XL type R360. This module connects to the PLC using a USB-CAN interface to command the base and fork movements and integrates the information from the different encoders.
  - LASER 1 and LASER 3 are the drivers for the security lasers located in the front and rear of the AGV.
  - LASER 2 is the driver for the NAV 350 localization laser.
  - CAMERA module handles the two cameras installed in the AGV to detect and manipulate the pallets. The first camera is a RGB camera installed under the localization laser and the second camera is a NIR installed between the forks. The RGB camera is used to detect the pallets in the near environment while the NIR is used to perform precision positioning in the load and unload actions. The infrared camera is also used to read the pallet labels to ensure the traceability.
- **Control layer.** This layer includes the algorithms responsible for different functions such as determining the position of the robot (**Localization**) maneuvers collecting and delivering pallets manager that make use of the information coming from the cameras (**Maneuvers manager**) and the algorithm that is responsible for following the planned path (**Path follower**).
- **Executive layer.** It is responsible for the execution of the sequence of actions that are part of a task. It coordinates the actions of all the other modules to carry out each task. The modules in this layer are part of RoboGraph [7] and the tasks will be described later.
- **Interface layer.** There are two kind of modules in this layer. First, the graphic interfaces (GUI Robot and GUI Navigation) used primarily for debugging and monitoring the system. Second, there is a module (RobotWeb interface) used to connect with the central system.

5 Task Programming

We use RoboGraph [7] to define, execute, monitor and trace the execution of every task assigned to the WCS. RoboGraph uses PNs for Task Modeling. Even though PNs provide no means for modelling the connection between an algorithm and its environment, we use an extension of the Signal Interpreted Petri Nets (SIPNs) [11] to stress the fact that the influence of the environment on the system is based on messages sent between the different modules of the architecture [4].

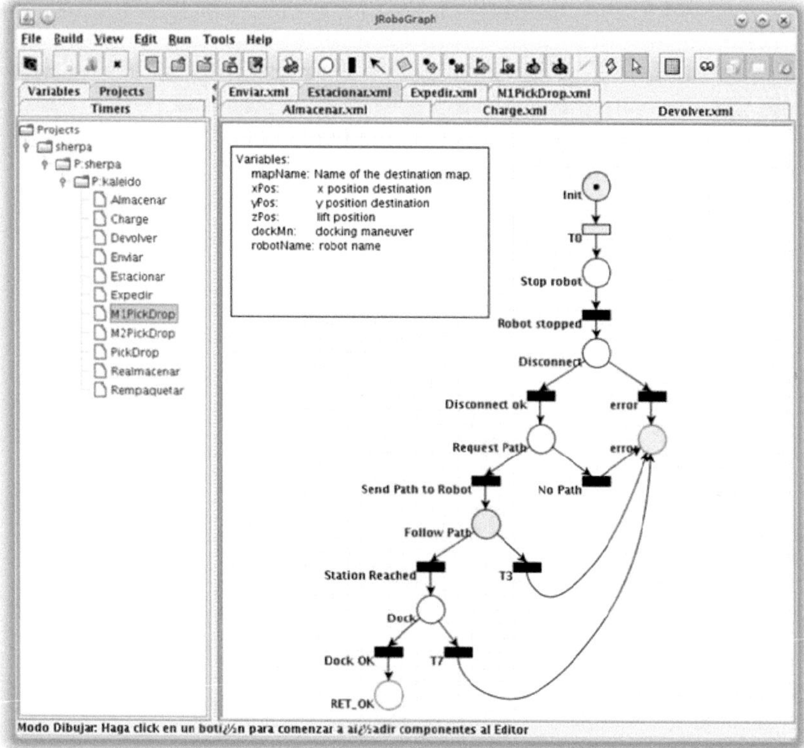

Fig. 6 RoboGraph GUI in editor mode.

The control architecture (Fig. 4 and Fig. 5) includes several independent modules that implement primitive actions and report events about their state. These modules are connected with two inter-process communication mechanisms. We use hierarchical interpreted binary Petri nets to coordinate the activity of these modules. Tasks are described using an interpreted Petri net editor and saved in a XML file. A dispatcher loads these files and executes the different Petri nets under user requests.

As shown in Fig. 4, RoboGraph includes two modules:

- **RoboGraph GUI** is the programming IDE for defining and debugging application tasks (Petri nets). Tasks are defined as Petri Nets and stored in an XML file. Fig. 6 shows this GUI in editor mode with the ShERPA project that includes 12 different tasks represented in Fig. 2 and other basic tasks such as "Charge" that sends the robot to the charging station and plugs it in.
- **Robograph dispatch** loads from an xml file and executes tasks defined as Petri nets. When executing a task this module schedules the different actions of the functional (basic actions), executive (other Petri nets) and interface layers (user and web interfaces) and receives information about the events produced. The interaction with other modules in the architecture is done by publishing and subscribing messages.

6 Results and Conclusions

A prototype of this Warehouse Control System is being tested in the storage facilities that the logistics company KALEIDO SCM manages in Valadares –Vigo (Spain). The forklift automation (Fig. 7) that includes the addition of a set of sensors and actuators connected to the PLC, has been carried out by the company GALMAN S.A.

Fig. 7 Automated forklift used in the project. The NAV350 is located on top of the fork and the control box that includes a computer is located on the roof.

The forks have fourth degrees of freedom (up-down, left-right, forward-backward, tilt) and the BASE driver can command the PLC with velocity commands for each one of them. At the same time, different sensors give the feedback about the fork position, reading the sensor data on input/output modules connected to the CAN bus.

The AGVs are connected via Wi-Fi to the central control system that is in charge of commanding new tasks and controlling the traffic. Therefore, the AGVs depends on the Wi-Fi to receive new orders but the integrity is still guaranteed without the Wi-Fi signal. The way the traffic is controlled is that to each robot a safe non-conflict portion of the route is provided at any time. When the robot is approaching the end of that portion of the route another safe portion of route is provided. Therefore, if there is any problem with the wiresless signal the robots will stop at safe positions waiting for the signal to come back again.

For the positioning system we use a commercial NAV350 from SICK with a set of TAGs made using reflective tape. At the starting step, the laser2 driver reads the map of tags from a HTML file and send it to the NAV350. Once the system is running, the laser2 driver is receiving periodically the position from the NAV350.

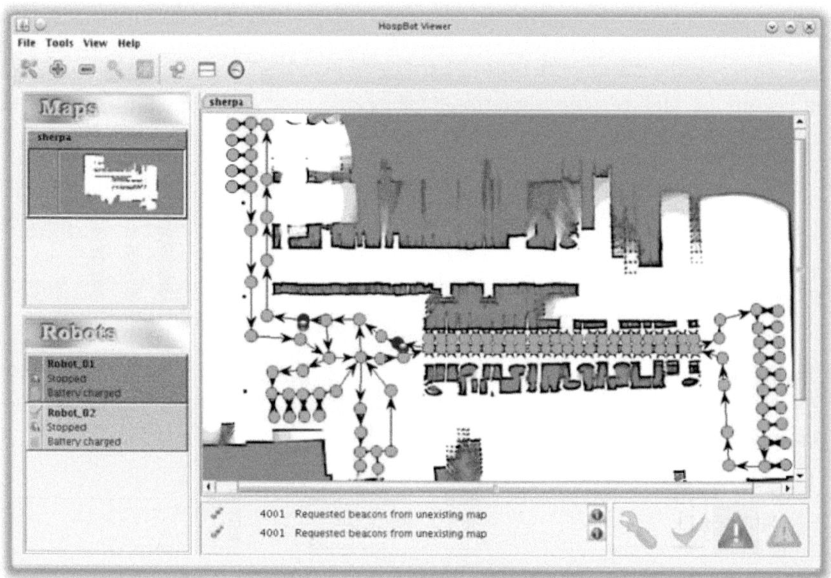

Fig. 8. User interface main window

Fig. 8 shows the User window with the nodes that constitute the possible paths and the robot positions. For the first tests we are using only one physical AGV and the other one is simulated.

As a conclusion, we have the following advantages on using RIDE to implement the warehouse control system:

- Hardware abstraction. Hardware server modules provide a hardware abstraction layer. Therefore, when a device is changed, the only part of the system that has to be updated is the driver if we keep the same interface (messages)
- Enhanced scalability. The framework is modular, flexible and easily extended.
- Maintainability. Modular systems are usually easy to maintain, update and scale. Tracing and debugging problems are easier when the system state can be seen by looking at the evolution of a Petri net rather than monitoring a set of variables.
- Module reusability. A key requirement to promote software reuse is to loosen the coupling between software modules. Each module in Fig. 3 is an independent Linux process. Basic modules, and even some tasks, will remain unchanged from application to application.
- Reduce development time. In similar applications, the modules remain without changes and only the edition of the Petri nets and configuration files will be necessary for a new application.
- Training. Almost everybody that has worked or learned to use IEC 61131-3 compliant programming environments (Siemens S7 Graph, Graphcet, etc.) will be able to program new tasks using RoboGraph.

A drawback when using the application every day is the lack of graphical tools for regular users to carry out simple operations. Since RIDE is oriented to developers, a few Graphical Interfaces should be programmed to deal with file configurations and regular operations for people that are not familiar with the implementation of the system. Our future work is oriented in two ways:

- Finding low-cost and reliable solutions to solve some problems such as localization that right now are solved with expensive and environment invasive solutions.
- Building a set of GUIs to help final user to execute regular operations without having to learn anything about the implementation of the system.

Acknowledgement This work has been partially supported by the FEDER-CONECTAPEME II Program under the Project "Sistema autónomo robotizado de transporte de pales (ShERPA)" IN852A 2014/36.

References

1. Gu, J., Goetschalckx, M., McGinnis, L.F.: Research on warehouse operations: A comprehensive review. European Journal of Operational Research **177**(1), 1–21 (2007)
2. Vis, I.F.: Survey of research in the design and control of automated guided vehicle systems. European Journal of Operational Research **170**(3), 677–709 (2006)
3. Guizzo, E.: Three engineers, hundreds of robots, one warehouse. In: Spectrum, IEEE, vol. 45(7), pp. 26–34 (2008)
4. Lopez, J., Perez, D., Zalama, E.: A framework for building mobile single and multi-robot applications. Robotics and Autonomous Systems **59**(3–4), 151–162 (2011)
5. López, J., Pérez, D., Paz, E., Santana, A.: WatchBot: A building maintenance and surveillance system based on autonomous robots. Robotics and Autonomous Systems **61**(12), 1559–1571 (2013)
6. López, J., Pérez, D., Santos, M., Cacho, M.: GuideBot. A tour guide system based on mobile robots. International Journal of Advanced Robotic System **10**, 381 (2013)
7. Fernández, J.L., Sanz, R., Paz, E., Alonso, C.: Using hierarchical binary Petri nets to build robust mobile robot applications: RoboGraph. In: IEEE International Conference on Robotics and Automation, ICRA 2008, pp. 1372–1377 (2008)
8. Binhardi, L., Reis, E., Pedrino, E.C., Morandin, O.: A multi-agent system using fuzzy logic to increase AGV fleet performance in warehouses. In: Proceedings of the III Brazilian Symposium on Computing Systems Engineering (SBESC 2013), Niteroy, December, 4–8, 2013, pp. 137–142 (2013)
9. Le-Anh, T., De Koster, M.: A review of design and control of automated guided vehicle systems. European Journal of Operational Research **171**(1), 1–23 (2006)
10. Simmons, R.: The interprocess communications system (IPC). http://www.cs.cmu.edu/afs/cs/project/TCA/www/ipc/ipc.html (accessed July 20, 2015)
11. Frey, G.: Design and Formal Analysis of Petri Net based Logic Controllers, Dissertation University of Kaiserslautern, Germany, Aachen, Shaker Verlag, April 2002

Author Index

Printed in the United States
By Bookmasters